# Wisconsin Flora

## FIELD GUIDE

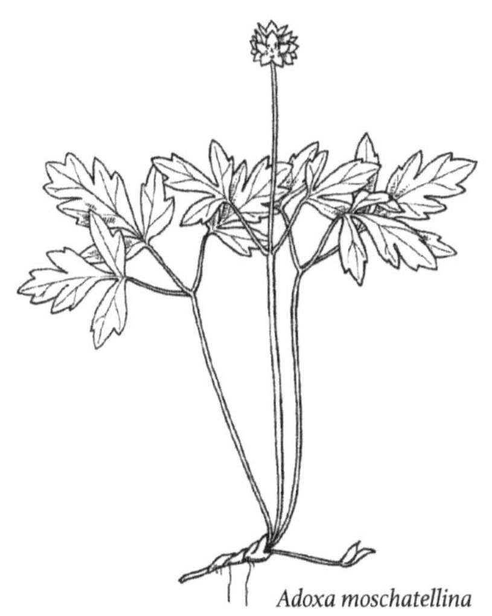

*Adoxa moschatellina*

STEVE W. CHADDE

WISCONSIN FLORA
*Field Guide*

Steve W. Chadde


Printed in the United States of America.

ISBN: 978-1951682651

Grateful acknowledgment is given to the Biota of North America Program (*www.bonap.org*) for permission to use their data to generate the distribution maps.

The author can be reached via email: *steve@chadde.net*

VERSION 2.0 (07/05/2022)

## CONTENTS

**MONOCOTS**

# INTRODUCTION

*Wisconsin Flora: Field Guide* is intended to be a stand-alone companion to the author's *Wisconsin Flora* (second edition, 2019). The present work, mirroring the plant species included in *Wisconsin Flora* (over 2,100 taxa), provides keys, distribution maps, habitat information, and illustrations to help you identify unknown plants. My goal has been to produce a more "field-friendly" version of the larger *Flora* – something easier to tuck into a glovebox or daypack – while allowing you to quickly and easily identify essentially any plant found growing in the wild in Wisconsin. The family key has been omitted from this work, as such keys are often unworkable in the field as they rely on characters which may not be present at the time of observation; however, a brief description of each of the 145 plant families treated in this *Flora* is provided beginning on page 7. Likewise, extensive descriptions of each species and references are not included, and a number of older synonyms have been omitted. If a family key or more information is needed, please refer to the 2019 *Wisconsin Flora*.

# ARRANGEMENT OF THE TAXA

All plants treated in the *Flora* belong to three informal groups, presented in order: (1) *ferns and fern relatives,* (2) *gymnosperms* (conifers), and, by far the largest group, (3) *angiosperms*. The angiosperms are subdivided into two classes, the *dicotyledons* or "dicots" (sometimes termed Magnoliopsida) and the *monocotyledons* or "monocots" (sometimes termed Liliopsida). These subdivision names derive from the observation that the dicots most often have two cotyledons, or embryonic leaves, within each seed. The monocots usually have only one, but the rule is not absolute either way. From a diagnostic point of view, the number of cotyledons is neither a particularly handy nor a reliable character, but provides a simple way to organize plant families into smaller groups. Dicots include many familiar trees, shrubs, and "wildflowers," such as those of the Aster Family; the monocots include the large grass and sedge families, and also smaller familes such as Juncaceae, Orchidaceae, and Typhaceae.

Within each of the divisions, families are listed alphabetically. Under each family, genera and species are also listed alphabetically. If there is more than one genus in a family, a key is provided to the genera. Likewise, if there is more than one species in a genus, a key to the species is provided. For each species treated in the text, the following information is provided: scientific name, common name, synonyms (a partial listing of other formerly accepted scientific names), whether of conservation concern (endangered or threatened), a description of the plant's habitat, and important characters of the plant itself. In addition, a map generated from the BONAP database (Biota of North America Program, www.bonap.org), shows verified county distribution patterns within the state.

Placement of angiosperms genera within families, with several exceptions, follows that of the Angiosperm Phylogeny Group III system (AGP III) of 2009. The APG was formed in the late 1990s, when researchers from major institutions around the world gathered with the goal of providing a modern, widely accepted classification of angiosperms. Their first attempt at a new system was published in 1998 (the AGP SYSTEM). To date, two revisions have been published, in 2003 (AGP II) and in 2009 (AGP III), each superseding the previous system.

The major exception to APG III in this work is the retention of the traditional Liliaceae (Lily Family) to facilitate field use of the Flora; however the APG III families segregated from the traditional Liliaceae are noted for each genera. Another exception is the retention of the Dipsacaceae (Teasel Family), and not including Wisconsin's two genera within Caprifoliaceae (Honeysuckle Family). Any other deviations from APG III regarding generic placement are noted in the text.

Similary, fern families have been updated to reflect recent realignment of some families and genera. In the past, ferns had been loosely grouped with other spore-bearing vascular plants, these often called "fern allies" or "lycophytes." However, recent studies suggest an important dichotomy within vascular plants, separating the fern relatives or lycophytes (less than 1% of all vascular plant species) from a group termed the euphyllophytes. Euphyllophytes comprise two major groups: the spermatophytes (seed plants), which number more than 260,000 species, and the monilophytes (ferns), with over 9,200 species, including horsetails, whisk ferns, and all "true ferns." Genetic studies also reveal surprises about the relationships among true ferns and fern allies. True ferns appear to be closely related to horsetails, and in fact these plants are now grouped within the true ferns. Also, plants commonly called fern relatives (clubmosses, spike-mosses and quillworts) are not closely related to the true ferns.

Nomenclature of genera and species is not based on a single source, but in general, conforms to that of the published volumes of The Flora of North America series (www.efloras.org), the BONAP database (Biota of North America Program, www.bonap.org), and The Plant List, a collaboration between the Royal Botanic Gardens (Kew), and the Missouri Botanical Garden (www.theplantlist.org). Common names largely reflect those of the BONAP database, or sometimes are names in popular use locally.

## CONSERVATION STATUS

Because of their rarity, primarily as a result of habitat loss, approximately 73 plant species are listed as endangered by the state of Wisconsin; a further 59 species are listed as threatened (as of 2019):

- *Endangered* – A species is considered endangered if the species is threatened with extinction throughout all or a significant portion of its range within Wisconsin.
- *Threatened* – A species is considered threatened if the species is likely to become endangered within the foreseeable future throughout all or a significant portion of its range within Wisconsin.

The website of the Wisconsin Dept. of Natural Resources (http://dnr.wi.gov) maintains a list of species listed as endangered, threatened, or of special concern (an additional category for plants but not afforded legal protection). Sightings of uncommon taxa should be reported to the state's Endangered Resources Program.

## FAMILY SYNOPSES

In lieu of a lengthy key to the families of vascular plants found in Wisconsin, below are brief descriptions of the plant families present in the state's flora and included in this guide. For a complete family key, please refer to the author's *Wisconsin Flora* (second edition, 2019).

### FERNS AND FERN RELATIVES (*17 families*)

**ASPLENIACEAE** (Spleenwort Family) Mostly small ferns, usually of rock crevices and where shaded and mossy, with short rootstocks covered by old petiole-bases, and a tuft of small to medium-sized leaves ("fronds"); the blades simple, pinnate, or 2-pinnate, and often evergreen. Sori (clusters of spore containers) elongate, occurring along the veinlets. Indusium (covering over sori) usually membranous, attached lengthwise along one side of the sorus.

**ATHYRIACEAE** (Lady Fern Family) Widely distributed, medium to large ferns.

**CYSTOPTERIDACEAE** (Bladder Fern Family) Small to medium ferns growing in soil or on rocks.

**DENNSTAEDTIACEAE** (Bracken Fern Family) Large, deciduous, colony-forming ferns, with leaves all alike, arising singly from creeping rhizomes.

**DRYOPTERIDACEAE** (Wood Fern Family) Medium to large ferns; rhizomes short, stout and scaly. Sori round, on underside veins of pinnae; indusia round to kidney-shaped.

**EQUISETACEAE** (Horsetail Family) Our single genus, *Equisetum,* are rushlike herbs with dark rhizomes. Stems annual or perennial, grooved, usually with large central cavity and smaller outer cavities, unbranched or with whorls of branches at nodes. Leaves reduced to scales, united into a sheath at each node.

**ISOETACEAE** (Quillwort Family) Perennial aquatic or emergent herbs. Leaves simple, entire, linear, from a 2–3 lobed rhizome (corm). Outermost and innermost leaves typically sterile. Outer fertile leaves have a pocketlike structure (sporangia) bearing whitish spores (megaspores; about 0.5 mm in diameter); inner fertile leaves have numerous small microspores.

**LYCOPODIACEAE** (Clubmoss Family) Low, trailing, evergreen herbs resembling large mosses. Leaves needlelike or scalelike. Spore-bearing leaves (sporophylls) similar to vegetative leaves or in conelike clusters at tips of upright stems.

**ONOCLEACEAE** (Sensitive Fern Family) Large coarse ferns of moist to wet places, with creeping hairy rhizomes or with stolons on ground surface; sterile and fertile fronds strongly different.

**OPHIOGLOSSACEAE** (Adder's-Tongue Family) Perennial herbs from short, erect rhizomes having several fleshy roots. Spores in numerous round sporangia.

**OSMUNDACEAE** (Royal Fern Family) Perennial ferns with large rootstocks and exposed crowns covered with old roots and stalks, sending up tufts of coarse leaves. Leaves 1–2-pinnate, differentiated into sterile and fertile segments. Sporangia in round clusters.

**POLYPODIACEAE** (Polypody Fern Family) Our single species (*Polypodium virginianum*) is an evergreen, colony-forming fern. Easily identified by the small evergreen blades and its colony-forming habit on rocky slopes, talus, boulders, and ledges.

**PTERIDACEAE** (Maidenhair Fern Family) Delicate to coarse ferns, deciduous, or evergreen. Sori marginal, or borne along the veins and lacking an indusium.

**SALVINIACEAE** (Water Fern Family) Our single species (*Azolla cristata*) is a small annual aquatic fern; plants free-floating or forming floating mats several cm thick, sometimes stranded on mud; roots few and unbranched. Sporocarps usually in pairs on underwater lobes of some leaves.

**SELAGINELLACEAE** (Selaginella Family) Trailing, evergreen herbs with branched, leafy stems, rooting at branching points. Leaves small and overlapping.

**THELYPTERIDACEAE** (Marsh Fern Family) Medium-sized deciduous ferns, spreading by rhizomes to form colonies; sterile and fertile fronds usually alike, with transparent needle-like hairs.

**WOODSIACEAE** (Cliff Fern Family) Small tufted ferns arising from compact rootstocks. Indusium of thread-like or plate-like segments, more or less arched over the round sori.

## CONIFERS (*3 families*)

**CUPRESSACEAE** (Cypress Family) Trees or shrubs; leaves opposite or whorled. Fruit a cone, or becoming fleshy and berry-like.

**PINACEAE** (Pine Family) Resinous trees with evergreen or deciduous, needlelike leaves. Male and female cones separate but borne on same tree.

**TAXACEAE** (Yew Family) Ours a straggling evergreen shrub. Leaves spirally arranged on the stem, linear, abruptly narrowed to a sharp point. Staminate flowers solitary in the axils; pistillate flowers in pairs. Fruit a fleshy red aril.

## DICOTS (*99 families*)

**ACANTHACEAE** (Acanthus Family) Our single species (*Ruellia humilis*, endangered in Wisc) is a perennial herb of prairies and dry upland woods. Leaves opposite. Flowers perfect, 5-merous; in sessile or subsessile, crowded, cymose clusters from the axils of several of the upper leaves.

**ADOXACEAE** (Muskroot Family) Ours a musk-scented herb, *Adoxa moschatellina*, from a scaly rhizome. Family now includes shrubby genera, *Sambucus* and *Viburnum*, previously included in Caprifoliaceae.

**AMARANTHACEAE** (Amaranth Family) Annual or perennial herbs. Leaves simple, alternate, or occasionally opposite. Flowers small, often aggregated into large spikes, panicles, or heads, in some species with conspicuous colored bracts. Flowers perfect or unisexual; sepals usually 5; petals absent. Fruit a 1-seeded utricle. Amaranthaceae now includes former members of the Chenopodiaceae.

**ANACARDIACEAE** (Sumac Family) Woody plants, juice often milky. Leaves alternate, chiefly compound. Flowers small, regular, perfect or unisexual, 5-merous. Fruit a 1-seeded, dry or fleshy drupe.

**APIACEAE** (Carrot Family) Biennial or perennial aromatic herbs with hollow stems, some very toxic. Leaves alternate and sometimes also from base of plant, mostly compound; petioles sheathing stems. Flowers small, perfect, regular, in flat-topped or rounded umbrella-like clusters (umbels); petals 5, white or greenish. Fruit 2-chambered.

**APOCYNACEAE** ( Dogbane Family) Herbs or twining woody vines; most species have milky juice. Leaves opposite, alternate, or sometimes whorled. Flowers 5-merous, regular, perfect. Fruit a capsule or follicle; seeds often bearing long hairs. Family now includes former members of Asclepiadaceae.

**AQUIFOLIACEAE** ( Holly Family) Shrubs. Leaves usually alternate, toothed or entire. Flowers from leaf axils, 4–8-parted, usually either staminate or pistillate, sometimes perfect, on same or different plants. Fruit a fleshy berrylike drupe.

**ARALIACEAE** ( Ginseng Family) Shrubs or herbs, rarely trees. Leaves usually alternate, compound or rarely simple. Flowers small, umbellate. Fruit a berry or a leathery drupe.

**ARISTOLOCHIACEAE** (Birthwort Family) Our single species (*Asarum canadense*) a perennial herb, producing annually a pair of cordate leaves, between which arises the solitary, short-peduncled flower.

**ASTERACEAE** (Aster Family) Annual, biennial or perennial herbs. Leaves simple or compound, opposite, alternate, or whorled. Flowers perfect or single-sexed (sometimes sterile) and of 2 types: ray (or ligulate) and disk (or tubular). Ray flowers joined at base and have a long, flat, segment above (the ray); disk flowers tube-shaped with 5 lobes or teeth at tip. Flowers are clustered in 1 of 3 types of heads resembling a single flower and attached to a common surface (receptacle): ray flowers only (as in dandelion, *Taraxacum*); disk flowers only (discoid, as in tansy, *Tanacetum*); and heads with both ray and disk flowers (radiate), the ray flowers surrounding the disk flowers (as in sunflower, *Helianthus*).

**BALSAMINACEAE** (Touch-Me-Not Family) Thin-leaved plants with watery juice and pendent, brightly colored irregular flowers. The seedpod, when ripe, pops open when touched to eject the seeds.

**BERBERIDACEAE** (Barberry Family) Herbs or shrubs. Leaves alternate or basal.

**BETULACEAE** (Birch Family) Medium to large trees, or shrubs. Leaves deciduous, alternate. Fruit a small, 1-seeded, winged nutlet.

**BIGNONIACEAE** (Trumpet-Creeper Family) Trees or woody vines; leaves opposite, simple or compound.

**BORAGINACEAE** (Borage Family) Annual or perennial herbs with usually bristly stems and alternate, bristly leaves. Flowers typically in a spirally coiled, spike-like head that uncurls as flowers mature.

**BRASSICACEAE** (Mustard Family) Annual, biennial or perennial herbs; leaves simple or compound, alternate on stems or basal; flowers with 4 sepals and 4 yellow, white, pink or purple petals. fruit a cylindrical (silique) or round (silicle) pod with 2 chambers.

**CABOMBACEAE** (Watershield Family) Our single species, *Brasenia schreberi*, is an aquatic plant with long slender stems that may attain a length of 2 m or more; all submerged parts coated with a gelatinous substance. Leaves oval, long-petioled, centrally peltate, floating, palmately veined.

CACTACEAE (Cactus Family) Thick fleshy plants of dry places; leaves reduced to spines. Blossoms large, cuplike, with many petals, numerous stamens and several stigmas.

CAMPANULACEAE (Bellflower Family) Perennial herbs. Stems usually with milky juice. Leaves simple, alternate. Flowers in racemes at ends of stems or single from upper leaf axils, perfect, 5-parted, regular and funnel-shaped (*Campanula*) or irregular (*Lobelia*); petals blue, white or scarlet.

CANNABACEAE (Hemp Family) Trees (*Celtis*), erect herbs (*Cannabis*), or twining herbs (*Humulus*). Leaves alternate or opposite, simple to palmately lobed or compound. Inflorescences axillary to the upper (often reduced) leaves, the staminate relatively loose, branched, and many-flowered; the pistillate more compact and few-flowered; petals absent.

CAPRIFOLIACEAE (Honeysuckle Family) Shrubs or vines, with opposite, mostly simple leaves. Flowers perfect, mostly 5-parted. Fruit a fleshy berry or dry capsule. Family now includes members of the former Valerianaceae (*Valeriana, Valerianella*); herbs with opposite, simple or divided leaves, and numerous small flowers in terminal, panicled or capitate cymes.

CARYOPHYLLACEAE (Pink Family) Annual or perennial herbs. Leaves simple, entire, mostly opposite but sometimes alternate or whorled. Stems often swollen at nodes. Flowers perfect or imperfect, in open or compact heads at ends of stems or from leaf axils. Fruit a few- to many-seeded capsule.

CELASTRACEAE (Bittersweet Family) Shrubs (*Euonymus*), vines (*Celastrus*), or glabrous perennial herbs (*Parnassia*) with simple, evergreen or deciduous, opposite or alternate leaves, and small, axillary or terminal, solitary or clustered flowers. Flowers perfect or unisexual, usually 4–5-merous. Celastraceae now includes members of genus *Parnassia.* In *Parnassia,* staminodes (infertile stamens) attached to base of petals and divided into threadlike segments tipped with glandular knobs.

CERATOPHYLLACEAE (Hornwort Family) Aquatic perennial herbs, often forming large patches; roots absent, but plants usually anchored to substrate by pale, modified leaves. Leaves in whorls, with more than 4 leaves per node, whorls crowded at ends of stems, dissected 2–3 times into narrow segments. Flowers small, inconspicuous in leaf axils, staminate and pistillate flowers separate on same plant. Our only aquatic vascular plant (*Ceratophyllum*) with whorled, forked leaves.

CISTACEAE (Rock-Rose Family) Herbs or shrubs. Leaves simple, alternate, opposite, or appearing whorled. Flowers cymose, perfect, regular except the calyx, 3–5-merous; petals small to large, soon deciduous, or lacking in some flowers. Fruit a capsule, usually separating completely to the base and enclosed by the persistent calyx.

CLEOMACEAE (Cleome Family) Annual herbs. Leaves alternate, compound. Flowers in terminal bracteate racemes.

CONVOLVULACEAE (Morning-Glory Family) Herbs (ours), often twining, with alternate simple leaves and small to large flowers. Flowers regular, perfect, mostly 5-merous. Corolla rotate, funnelform, salverform, or tubular, entire or deeply to shallowly lobed.

**CORNACEAE** (Dogwood Family) Trees (*Nyssa*) or shrubs (*Cornus*). Flowers 4- or 5-parted. Fruit a drupe.

**CRASSULACEAE** (Stonecrop Family) Plants usually succulent. Leaves simple. Flowers usually cymose, regular, 4–5-merous or occasionally more, usually perfect.

**CUCURBITACEAE** (Cucumber Family) Annual or perennial vines, trailing or climbing by tendrils, with mostly white or yellow or greenish flowers, and simple, alternate, often lobed leaves. Flowers monoecious or dioecious, regular. Fruit a dry or fleshy pepo, few to many-seeded.

**DIPSACACEAE** (Teasel Family) Herbs. Leaves opposite, simple or divided. Flowers in dense heads subtended by a many-leaved involucre. Flowers perfect or polygamo-monoecious. Included in Caprifoliaceae in the 2009 Angiosperm Phylogeny Group III system.

**DROSERACEAE** (Sundew Family) Low bog plants that supplement their diet by capturing insects, thereby absorbing nitrogen lacking in their habitat. Leaves stringy or spatulate, covered with sticky glands or hairs. Flowers 5-petaled, in slender cluster on separate stalk.

**ELAEAGNACEAE** (Oleaster Family) Shrubs or trees. Leaves opposite or alternate, covered with small scales (lepidote). Flowers small, solitary or clustered, perfect or unisexual. Petals none.

**ELATINACEAE** (Waterwort Family) Small, branched, annual herbs of shallow water, shores and mud flats. Leaves simple, opposite, entire or toothed, with small membranous stipules. Flowers small, 1 to several from leaf axils.

**ERICACEAE** (Heath Family) Ericaceae now includes former members of Monotropaceae and Pyrolaceae. The traditional Ericaceae are shrubs or scarcely woody shrubs. Leaves evergreen or deciduous, mostly alternate, simple. Flowers usually perfect, urn- or vase-shaped, mostly white, pink, or cream-colored. Fruit a berry or dry capsule. Former Monotropaceae (*Monotropa, Pterospora*) are mycotropic perennial herbs without chlorophyll, variously white to pink, red, purple, yellow or brown in color. Leaves much-reduced, scale-like, alternate. Flowers solitary or in a bracteate raceme; petals distinct or connate into a lobed tube, commonly about the same color as the stem. Fruit a capsule or berry; seeds numerous and tiny. Former Pyrolaceae (*Moneses, Orthilia, Pyrola*) are perennial herbs or half-shrubs, most dependent on wood-rotting fungi (mycotrophic). Leaves alternate to sometimes opposite or nearly whorled, often shiny, evergreen or deciduous. Flowers perfect, 5-parted, waxy and nodding. Fruit a capsule.

**EUPHORBIACEAE** (Spurge Family) Herbs (ours). Leaves usually alternate and simple. Flowers mostly tiny but in some species subtended by conspicuous bracts or involucral appendages. Plants monoecious or dioecious. Flowers commonly unisexual, very rarely perfect, regular. Fruit usually a dehiscent capsule.

**FABACEAE** (Pea Family) Perennial herbs, shrubs and trees. Leaves alternate, pinnately divided, the terminal leaflet sometimes modified as a tendril (*Lathyrus, Vicia*). Flowers in simple or branched racemes, perfect, irregular, 5-lobed (only 1 lobe in *Amorpha*), the upper lobe (banner) larger than the other lobes, with 2 outer, lateral petals (wings), and 2 inner petals which are partly joined (the keel); ovary 1-chambered, maturing into a pod.

**FAGACEAE** (Beech Family) Trees or shrubs. Leaves alternate, simple, entire to lobed. Plants monoecious. Staminate flowers in catkins or heads. Pistillate flowers solitary, or in small clusters or short spikes, more or less enclosed by an involucre of numerous bracts. Fruit a 1-seeded nut, wholly or partly surrounded by the expanded involucre.

**GENTIANACEAE** (Gentian Family) Annual, biennial or perennial herbs; plants usually glabrous. Leaves simple, entire, opposite or whorled, stem leaves without petioles. Flowers often showy, perfect, single at end of stems or in clusters; petals 4-5, blue, purple, white or green, joined for at least part of their length. Fruit a 2-chambered, many-seeded capsule enclosed by the withered, persistent petals.

**GERANIACEAE** (Geranium Family) Annual or perennial herbs. Leaves usually opposite, simple or compound, palmately toothed, lobed, or divided. Flowers 5-merous, regular or somewhat zygomorphic, all perfect or part of them sterile. Petals 5, pink to purple. Fruit a carpel prolonged at maturity into beaks.

**GROSSULARIACEAE** (Currant Family) Shrubs. Stems smooth, or with spines at nodes and sometimes also with bristles between nodes. Leaves alternate, palmately veined and palmately lobed, margins toothed. Flowers 1 to several in short clusters, or few to many in racemes; green to white or yellow, perfect, regular. Fruit a many-seeded berry, usually topped by persistent, dry flower parts.

**HALORAGACEAE** (Water-Milfoil Family) Perennial aquatic herbs. Leaves alternate or whorled, finely dissected. Flowers small, stalkless in axils of leaves or bracts, 3- or 4-parted, regular, perfect, or imperfect, petals small or absent. Fruit small and nutlike, dividing into 3 or 4 segments (mericarps).

**HAMAMELIDACEAE** (Witch-Hazel Family) Our single species (*Hamamelis virginiana*) is a tall shrub of moist woods. Leaves broadly obovate, with several to many rounded teeth. Flowers in short-pediceled axillary clusters; 4-merous; petals bright yellow or suffused with red. Seeds black, eventually discharged explosively from the capsule.

**HYDRANGEACEAE** (Hydrangea Family) Our single species (*Philadelphus coronarius*) is an escaped shrub. Leaves simple, opposite, short-petioled, ovate. Flowers in terminal racemes; perfect, regular, 4-merous or rarely 5-merous, white, fragrant.

**HYPERICACEAE** (St. John's-Wort Family) Glabrous annual or perennial herbs (shrubby in *Hypericum kalmianum* and *H. prolificum*). Leaves simple, opposite, dotted with dark or translucent glands (visible when held to light), especially on underside; margins entire; petioles absent. Flowers few to many in clusters at ends of stems or from upper leaf axils, perfect, regular, petals 5, yellow or pink to green or purple. Fruit a 3-chambered, many-seeded capsule.

**JUGLANDACEAE** (Walnut Family) Trees. Leaves alternate, odd-pinnate. Flowers monoecious. Staminate flowers in elongate catkins. Pistillate flowers terminating the young branches. Fruit large, consisting of a fleshy or woody exocarp enclosing a nut.

**LAMIACEAE** (Mint Family) Perennial, often aromatic, herbs. Stems usually 4-angled. Leaves simple, opposite, sharply toothed or deeply lobed. Flowers in leaf axils or in heads or spikes at ends of stems, perfect; petals white, pink, blue or purple, often 2-lipped.

**LENTIBULARIACEAE** (Bladderwort Family) Insectivorous herbs. Leaves in a basal rosette (*Pinguicula*), or floating, or in peat, muck, or wet soil (*Utricularia*). Flowers perfect, irregular, 2-lipped, sometimes with a spur, 1 to several on an erect stem. Fruit a capsule.

**LIMNANTHACEAE** (Meadowfoam Family) Our single species (*Floerkea proserpinacoides*) is an annual herb. Stems weak, diffuse or decumbent. Leaves deeply divided into 3–7 narrow lobes. Peduncles from the upper axils; petals white. Distinct among our dicots in its completely 3-merous flowers; distinguished from our monocots by the very deeply pinnately lobed leaves.

**LINACEAE** (Flax Family) Annual or perennial herbs. Leaves simple, alternate or opposite, narrow, petioles absent. Flowers regular, perfect, 5-parted. Petals yellow or blue. Fruit a 10-chambered capsule.

**LINDERNIACEAE** ( Lindernia Family) Our single species (*Lindernia dubia*) is an annual herb of wet places. Leaves opposite, ovate to obovate; margins entire or with small, widely spaced teeth; petioles absent. Flowers single, on slender stalks from leaf axils; corolla pale blue-purple, 2-lipped, the upper lip 2-lobed, the lower lip 3-lobed and wider than upper lip.

**LYTHRACEAE** ( Loosestrife Family) Annual or perennial herbs, sometimes woody at base (*Decodon*). Leaves simple, opposite, or both opposite and alternate, or whorled, margins entire. Flowers 1 or several in leaf axils or in spike-like heads at ends of stems; perfect, regular or irregular; petals 4 or 6, separate, pink or purple. Fruit a dry, many-seeded capsule.

**MALVACEAE** (Mallow Family) Annual or perennial herbs with upright stems; trees in *Tilia.* Leaves alternate, entire to lobed or dissected, often round or kidney-shaped, palmately veined. Flowers single or in small, narrow clusters from leaf axils, with 5 united sepals (separate in Tilia) and 5 petals. Fruit a capsule.

**MELASTOMATACEAE** (Melastome Family) Our single species, *Rhexia virginica,* is a perennial herb of wet places; roots often with tubers. Stems 4-angled and 4-winged, with bristly hairs at nodes. Leaves ovate, margins finely toothed. Flowers perfect, regular, 4-merous, in cymes from ends of stems and upper leaf axils; petals purple.

**MENISPERMACEAE** (Moonseed Family) Our single species, *Menispermum canadense,* is a dioecious woody twiner. Leaves simple, alternate, broadly ovate to nearly orbicular, palmately veined, shallowly 3–7-lobed to entire. Flowers small, unisexual, usually 3-merous, in racemes or panicles that arise just above the leaf-axils. Drupes grape-like, bluish-black.

**MENYANTHACEAE** (Buckbean Family) Our single species, *Menyanthes trifoliata,* is a perennial glabrous herb of wetlands. Leaves alternate along rhizomes, palmately divided into 3 leaflets, the base of petiole expanded and sheathing stem. Flowers in racemes on leafless stalks; petals white, often purple-tinged, bearded with white hairs on inner surface.

**MOLLUGINACEAE** (Carpetweed Family) Our single species, *Mollugo verticillata,* is an annual, mat-forming herb, weedy in moist soil. Leaves in whorls of 3–8. Flowers perfect, 2–5 from each node; petals 5, pale green to white.

**MONTIACEAE** (Montia Family) Now includes our 2 genera formerly within Portulacaceae. Plants from tubers (*Claytonia*) or a taproot (*Phemeranthus*). Petals 5, white to pink, often with pink veins in *Claytonia.*

**MORACEAE** (Mulberry Family) Trees (rarely herbs), juice milky or watery. Leaves alternate, simple or compound. Flowers small, crowded in dense clusters or heads; unisexual, the plants monoecious or dioecious. Corolla none. Fruits diverse.

**MYRICACEAE** (Bayberry Family) Monoecious or dioecious shrubs, with alternate simple leaves; leaves resinous-dotted and fragrant. Flowers unisexual, solitary in the axils of small bracts, aggregated into globose to cylindric catkins.

**NELUMBONACEAE** (Lotus-Lily Family) Our single species, *Nelumbo lutea,* is a perennial aquatic herb, in Wisc, mostly found near the Mississippi River. Leaves large, shield-shaped, floating on water surface or held above water; petioles attached at center of blade. Flowers pale yellow; receptacle flat-topped, to 1 dm wide; seeds acornlike.

**NYCTAGINACEAE** (Four-O'clock Family) Our sole genus, *Mirabilis,* are perennial herbs, sometimes woody at base. Leaves opposite. Flowers many in terminal panicles; petals rose to pink-purple, open in the morning, solitary or in clusters of 2–4. Fruit 5-ribbed, mucilaginous when wet.

**NYMPHAEACEAE** (Water-Lily Family) Aquatic, perennial herbs. Stems long and fleshy, from horizontal rhizomes rooted in bottom mud. Leaves large, leathery, mostly floating or emergent above water surface, heart-shaped to shield-shaped, notched at base. Flowers showy, single on long stalks and borne at or above water surface, perfect, white or yellow; petals numerous. Fruit a many-seeded, berrylike capsule, opening underwater when mature.

**OLEACEAE** (Olive Family) Trees or shrubs with opposite, simple or compound leaves. Flowers perfect or unisexual, regular. Calyx small or in some genera lacking. Corolla in our genera partially or wholly fused, or lacking (*Fraxinus*). Fruit a drupe, capsule, or samara.

**ONAGRACEAE** (Evening-Primrose Family) Annual or perennial herbs. Leaves opposite to alternate, simple to pinnately divided, stalkless or short-petioled. Flowers usually large and showy, perfect, regular, borne in leaf axils or in heads at ends of stems; petals 4, white, yellow, or pink to rose-purple. Fruit a 4-chambered capsule; seeds many, with or without a tuft of hairs (coma).

**OROBANCHACEAE** (Broom-Rape Family) Annual, biennial, or perennial herbs; some genera without green color, parasitic on the roots of other plants. Leaves opposite, alternate, or reduced to scales. Flowers mostly perfect, single or few from leaf axils, or numerous in clusters at ends of stems or leaf axils, usually with a distinct upper and lower lip; petals 4–5 (or sometimes absent). Fruit a several- to many-seeded 2-valved capsule. Now includes many former members of Scrophulariaceae.

**OXALIDACEAE** (Wood-Sorrel Family) Perennial herbs (our single genus Oxalis). Leaves basal or alternate on the stem, 3-foliolate and resembling leaves of a clover. Flowers solitary on axillary peduncles or in cymose or umbel-like clusters, 5-merous, white, yellow, pink, or purple.

**PAPAVERACEAE** (Poppy Family) Herbs or vines (*Adlumia*), with watery, milky, or colored juice. Leaves alternate or rarely opposite. Flowers regular, perfect. Sepals 2 or 3, early deciduous. Petals 4 or more (rarely absent), separate, conspicuous. Fruit a capsule, dehiscent by terminal valves or longitudinally (rarely otherwise). The Fumariaceae, now included as a subfamily of Papaveraceae, were previously recognized as a separate family, differing in bilateral symmetry of the flowers and watery juice. All of our members of Papaveraceae in the strict sense have colored juice (yellow to red-orange or milky).

**PENTHORACEAE** (Penthorum Family) Our single species, *Penthorum sedoides*, is a perennial herb of wet places, spreading by rhizomes; plants often red-tinged. Stems smooth and round in section below, upper stem often angled and with gland-tipped hairs. Leaves alternate, lance-shaped; margins with small, forward-pointing teeth. Flowers star-shaped, perfect, in branched racemes at ends of stems; petals usually absent.

**PHRYMACEAE** (Lopseed Family) Perennial herbs. Calyx tubular, 5-lobed. Fruit a dehiscent capsule. Previously, this family was monotypic with only genus *Phryma;* now includes Wisc genera *Mazus* and *Mimulus*.

**PHYTOLACCACEAE** (Pokeweed Family) Our single genus, *Phytolacca*, is a large, glabrous, perennial herb. Leaves alternate, oblong lance-shaped to ovate. Racemes peduncled, 1–2 dm long, nodding in fruit. Flowers perfect or plants dioecious, petals greenish white or suffused with pink. Berry dark purple.

**PLANTAGINACEAE** (Plantain Family) Annual or perennial herbs. Leaves simple, entire, all from base of plant. Flowers perfect in a narrow spike (*Plantago*), or single-sexed, the staminate and pistillate flowers on same plant (*Littorella*); flower parts mostly in 4s. Fruit a capsule opening at tip. Plantaginaceae is the accepted name for the family that encompasses not only the plantains with their reduced flowers, but also the related larger-flowered genera formerly placed in the Scrophulariaceae, as well as highly reduced aquatics, such as *Hippuris* (Hippuridaceae) and *Callitriche* (Callitrichaceae).

**PLATANACEAE** (Planetree Family) Our single species, *Platanus occidentalis* (Sycamore), is a large tree; bark red-brown when young, soon breaking into thin, flat sections which fall away to expose the distinctive white-green inner bark. Leaves alternate, divided into 3 or 5 shallow, sharp-pointed lobes, bright green and smooth on upper surface, underside paler. Staminate and pistillate flowers tiny, in dense clusters, separate but on same tree. Fruit a round, light brown head, on a long, drooping stalk.

**POLEMONIACEAE** (Phlox Family) Perennial herbs (ours). Leaves opposite (*Phlox*) or pinnately divided (*Polemonium*). Flowers perfect, single or in clusters at ends of stems and from leaf axils; sepals and petals 5-parted and joined for part of length. Fruit a 3-chambered capsule, with usually 1 seed per chamber.

**POLYGALACEAE** (Milkwort Family) Our single genus, *Polygala*, are annual, biennial, or perennial herbs. Leaves alternate or verticillate. Flowers perfect, in racemes; petals 3, all more or less united with each other and with the stamen-tube, the two upper ones similar, the lower one keel-shaped or boat-shaped with a fringe-like crest (in our species).

**POLYGONACEAE** (Buckwheat Family) Annual or perennial herbs, plants sometimes vining. Leaves alternate, simple, sometimes wavy-margined, otherwise entire; the nodes usually enlarged. Stipules joined to form a membranous or papery sheath (ocrea) around stem at each node. Flowers in spike-like racemes or small clusters from leaf axils (*Persicaria, Polygonum*), or in crowded panicles at ends of stems (*Rumex*). Flowers small, perfect, regular, petals absent. In *Rumex* the sepals herbaceous, green to brown, in inner and outer groups, each group with 3 sepals, the 3 inner enlarging after flowering, becoming broadly winged, persisting to enclose the achene; in other genera of family, sepals more or less petal-like, white to pink or yellow, mostly 5 (sometimes 4). Polygonaceae recognized by presence of a stipular sheath (ocrea), which surrounds the stem above the attachment of each leaf. The similar reduced structure in the inflorescence is called an ocreola.

**PORTULACACEAE** (Purslane Family) Our single genus, *Portulaca,* are succulent annual herbs. Leaves cauline, mostly alternate, the uppermost crowded and forming an involucre to the flowers. Flowers ephemeral, opening only in the sunshine, solitary or glomerate at ends of the stems and branches. Petals 4–6, commonly 5.

**PRIMULACEAE** (Primrose Family) Annual or perennial herbs. Leaves simple, opposite (sometimes whorled in *Lysimachia*), or leaves all basal. Flowers perfect, regular, single from leaf axils, or in clusters at ends of stems; petals mostly 5 (varying from 4–9), joined, tube-shaped below and flared above, deeply cleft to shallowly lobed at tip. Fruit a 5-chambered capsule.

**RANUNCULACEAE** (Buttercup Family) Annual or perennial, aquatic or terrestrial herbs (or vines in *Clematis*). Leaves usually alternate, sometimes opposite or whorled, or all at base of plant. Flowers mostly white or yellow, usually with 5 (occasionally more) separate petals and sepals, or petals absent and then with petal-like sepals; sepals leafy and green or petal-like and colored; flowers perfect, stamens usually numerous; pistils several to many, ripening into beaked achenes or dry capsules (follicles).

**RESEDACEAE** (Mignonette Family) Our single genus, *Reseda,* are introduced biennial herbs of waste places. Leaves alternate, oblong lance-shaped in outline, deeply and irregularly pinnatifid above the middle into a few narrow segments. Flowers perfect, in terminal racemes; petals usually 6, unequal, the upper the largest, greenish yellow.

**RHAMNACEAE** (Buckthorn Family) Shrubs, trees, or woody vines with simple, opposite or alternate leaves. Flowers perfect or unisexual, regular, 4–5-merous. Petals present or lacking, small, separate. Fruit a capsule or drupe.

**ROSACEAE** (Rose Family) Shrubs, and perennial, biennial, or annual herbs. Leaves evergreen or deciduous, mostly alternate and simple or compound. Flowers perfect, regular, with 5 sepals and petals; stamens numerous. Fruit an achene, capsule, or fleshy fruit with numerous embedded seeds (drupe), or a fleshy fruit with seeds within (pome).

**RUBIACEAE** (Madder Family) Shrubs (*Cephalanthus*) or herbs. Leaves simple, opposite or whorled. Flowers small, perfect, white to green, single or in loose or round clusters. Fruit a nutlet (*Cephalanthus, Diodia*), a capsule (*Galium, Houstonia*), or a berry (*Mitchella*).

**RUTACEAE** (Rue Family) Mostly trees or shrubs with alternate, simple or compound leaves and small flowers. Flowers perfect or unisexual. Fruit commonly separating into segments, in some genera a capsule, drupe, or berry. Most parts of the plant contain oil-glands; those of the leaves appear as translucent dots.

**SALICACEAE** (Willow Family) Deciduous trees or shrubs with alternate, stipulate leaves. Both staminate and pistillate flowers in aments (catkins). Fruit a capsule. Seeds having a dense coma of long, mostly white, silky hairs.

**SANTALACEAE** (Sandalwood Family) Herbs (ours); usually root-parasites. Leaves simple, alternate or opposite. Flowers perfect or unisexual, in terminal or axillary clusters, or solitary. Fruit a nut or drupe. Both our genera, *Comandra* and *Geocaulon,* though bearing green leaves, are hemiparasitic, and are apparently always attached (by means of modified roots, or haustoria) to some other plant. Both species also serve as alternate hosts for the canker-producing *Comandra* blister rust fungus (*Cronartium comandrae*), which in Wisc infects trees of jack pine.

**SAPINDACEAE** (Soapberry Family) Now includes former members of Aceraceae and Hippocastanaceae. Trees or shrubs. Leaves opposite, simple or compound. Staminate and pistillate flowers borne on same or separate plants. Flowers with 5 sepals and 5 petals (sometimes absent), clustered into a raceme or umbel. Fruit a samara with 2 winged achenes joined at base (*Acer*), or a prickly red-brown capsule (*Aesculus*).

**SARRACENIACEAE** (Pitcherplant Family) Our single species, *Sarracenia purpurea,* is a perennial insectivorous herb of bogs and fens. Flower stalks leafless. Leaves clumped, hollow and vaselike, green or veined with red-purple, winged, smooth on outside, upper portion of inside with downward-pointing hairs. Flowers large and nodding, single at ends of stalks.

**SAXIFRAGACEAE** (Saxifrage Family) Perennial herbs. Leaves alternate, opposite or basal. Flowers perfect, regular, single on stalks or in narrow heads. Sepals and petals 5 (4 in *Chrysosplenium*).

**SCROPHULARIACEAE** (Figwort Family) Annual, biennial, or perennial herbs. Leaves mostly opposite or alternate (*Verbascum*). Flowers single or few from leaf axils, or numerous in clusters at ends of stems or leaf axils, perfect, usually with a distinct upper and lower lip; sepals and petals 4–5 (petals sometimes absent). Fruit a several- to many-seeded capsule. Formerly a much larger family, many of our genera now segregated into other families, especially Orobanchaceae and Plantaginaceae.

**SIMAROUBACEAE** (Quassia-Wood Family) Our single species, *Ailanthus altissima,* is a rapidly growing, weedy tree. Leaves odd-pinnate; leaflets 11–41. Dioecious, with staminate and pistillate flowers borne on different individuals (sometimes flowers perfect). Flowers greenish or greenish yellow, in large terminal pyramidal panicles.

**SOLANACEAE** (Potato Family) Herbs or shrubs, rarely climbing. Leaves alternate or appearing opposite. Flowers perfect, almost always 5-merous, regular (in most of our genera) or irregular. Fruit a capsule or berry.

**STAPHYLEACEAE** (Bladdernut Family) Our single species, *Staphylea trifolia,* is an erect shrub with striped bark. Leaves opposite, 3-foliolate. Inflorescence a terminal drooping panicle. Flowers regular, perfect, 5-merous. Fruit a 3-lobed, inflated capsule.

**THYMELAEACEAE** (Mezereum Family) Our single species (apart from an uncommon introduced annual herb) is the woodland shrub *Dirca palustris.* Bark very tough and pliable, twigs jointed. Leaves alternate, entire. Flowers perfect, regular, pale yellow, subtended by hairy bud scales in early spring before the leaves appear. Fruit an ellipsoid drupe.

**ULMACEAE** (Elm Family) Our sole genus, *Ulmus,* are trees. Leaves alternate, simple, inequilateral; margins usually doubly serrate. Flowers perfect, in short racemes or, by abbreviation of the axis, in fascicles. Fruit a flat, 1-seeded samara. Hackberry (*Celtis*), a former member of this family, now included in Cannabaceae.

**URTICACEAE** (Nettle Family) Annual or perennial herbs with watery juice, sometimes with stinging hairs. Leaves alternate or opposite, simple, with petioles. Flowers small, green, in simple or branched clusters from leaf axils, staminate and pistillate flowers usually separate, on same or separate plants. Fruit an achene, often enclosed by the sepals which enlarge after flowering.

**VERBENACEAE** (Verbena Family) Perennial herbs with 4-angled stems. Leaves opposite, toothed. Flowers small, numerous, perfect, in branched or unbranched spikes or heads at ends of stems or from upper leaf axils, the spikes elongating as flowers open upward from the base. Fruit dry, enclosed by the sepals, splitting lengthwise into 2 or 4 nutlets when mature.

**VIOLACEAE** (Violet Family) Herbs. Leaves simple, alternate or rarely opposite. Flowers perfect, 5-merous, usually irregular, axillary or basal, usually nodding. Lower petal usually spurred or larger than the others. Fruit a capsule.

**VITACEAE** (Grape Family) Mostly woody vines, climbing by tendrils. Leaves alternate, simple or compound; tendrils and flower clusters produced opposite the leaves. Flowers regular, 4–5-merous, perfect or unisexual. Fruit a berry.

**ZYGOPHYLLACEAE** (Creosote-Bush Family) Our single species, *Tribulus terrestris,* is an uncommon introduced herb of waste places. Recognized by its prostrate habit, hairy stems and leaves, the leaves usually with 6–7 pairs of small leaflets, and the spiny nutlets.

## MONOCOTS (*20 families*)

**ACORACEAE** (Calamus Family) Our single genus, *Acorus,* are perennial wetland herbs; rhizomes and leaves pleasantly scented. Leaves sword-shaped, bright green, with 1–6 prominent veins parallel along length of leaf. Inflorescence a solitary spadix, borne from near midway of leaf; true spathe absent. Flowers bisexual; tepals 6, light brown. Fruit light brown to reddish berry with darker streaks. Seeds embedded in a mucilagenous jelly.

**ALISMATACEAE** (Water-Plantain Family) Perennial, aquatic or emergent herbs; plants swollen and tuberlike at base. Leaves all from base of plant and clasping an erect stem; underwater leaves often ribbonlike; emergent leaves broader. Flowers perfect or imperfect, in racemes or panicles at ends of stems, with 3 sepals and 3 petals. Fruit a compressed achene, usually tipped by the persistent style.

**BUTOMACEAE** (Flowering Rush Family) Our single species, *Butomus umbellatus,* is an introduced and potentially invasive perennial herb of wetlands, from creeping rhizomes. Leaves all from base of plant, erect when emersed, or floating when in deep water, linear, to 1 m long, parallel-veined. Flowers pink, perfect, in a many-flowered umbel, borne on a stalk 1–1.2 m tall; sepals 3, petal-like; petals 3. Fruit a dry capsule.

**CYPERACEAE** (Sedge Family) Mostly perennial, grasslike, rushlike or reedlike plants. Stems 3-angled, or more or less round in section, solid or pithy. Leaves 3-ranked or reduced to sheaths at base of stem; leaf blades, when present, grasslike, parallel-veined, often keeled; sheaths mostly closed around the stem. Flowers small, perfect, or single-sexed, each flower subtended by a bract (scale); perianth of 1 to many (often 6) small bristles, or a single perianth scale, or absent; stamens usually 3; ovary contained in a saclike covering (perigynium) in *Carex,* maturing into an achene, stigmas 3 or 2. Flowers arranged in spikelets (termed spikes in *Carex*), the spikelets single as a terminal or lateral spike, or several to many in various types of heads, the head often subtended by 1 to several bracts.

**DIOSCOREACEAE** (Yam Family) Our single species, *Dioscorea villosa,* is a perennial dioecious herb. Stems twining, to 5 m long. Leaves alternate, cordate-ovate. Flowers regular; unisexual; small, white to greenish yellow. Fruit a 3-winged capsule.

**ERIOCAULACEAE** (Pipewort Family) Our sole species, *Eriocaulon aquaticum,* is a perennial wetland plant, spongy at base, with fleshy roots. Stems usually single, leafless, slightly twisted, 5–7-ridged. Leaves grasslike, in a rosette at base of plant, thin and often translucent, with conspicuous cross-veins. Flowers either staminate or pistillate, grouped together in a single, more or less round head at end of stem, the heads white-woolly.

**HYDROCHARITACEAE** (Tape-Grass Family) Aquatic herbs. Stems leafy, the leaves opposite (*Najas*), whorled (*Elodea*), or plants stemless with clusters of long, linear, ribbonlike leaves (*Vallisneria*). Flowers usually either staminate or pistillate and borne on separate plants, small and stalkless, or in a spathe at end of a stalk. Fruit several-seeded, maturing underwater.

**HYPOXIDACEAE** (Liliid Monocot Family) Our single species, *Hypoxis hirsuta,* is a low perennial herb of moist to wet places, from a small corm. Leaves from base of plant, linear, hairy. Flowers 1–6 (usually 2), yellow, in racemes at ends of stems, tepals hairy on outside, persistent.

**IRIDACEAE** (Iris Family) Perennial herbs with rhizomes, bulbs, or fibrous roots. Leaves parallel-veined, narrow, 2-ranked, the margins joined to form an edge facing the stem (equitant). Flowers perfect, with 6 petal-like segments, single or in clusters at ends of stem. Fruit a 3-chambered capsule.

**JUNCACEAE** (Rush Family) Distinguished from grasses and sedges by the presence of a true perianth of 6 tepals and a 3–many-seeded capsule rather than a 1-seeded grain (grasses) or achene (sedges). No ligule (as in the grasses) is present at junction of leaf blade and sheath; however an auricle (an ear-like appendage) may occur at top of leaf sheath.

**JUNCAGINACEAE** (Arrow-Grass Family) Our single genus, *Triglochin,* are grasslike perennial herbs, often in brackish habitats. Stems slender, leafless. Leaves all from base of plant, slender, linear, round or somewhat flattened in section. Flowers perfect, regular, on short stalks in a spike-like raceme at end of stem.

**LILIACEAE** (Lily Family) Under the Angiosperm Phylogeny Group III system (APG III), genera in the Liliaceae have been placed into various new familes. However, the family designations are still in a state of flux, and may change in the future. As a convenience, our genera are retained within the former Liliaceae, as the plants are usually readily recognizable as belonging to the traditional Lily Family. Perennial herbs, from corms, bulbs or rhizomes. Stems leafy or leafless. Leaves linear to ovate, usually from base of plant, sometimes along stem, alternate to opposite or whorled. Flowers perfect, regular; sepals and petals of 6 petal-like tepals in 2 series of 3. Fruit a capsule or round berry.

**ORCHIDACEAE** (Orchid Family) Perennial herbs, from fleshy or tuberous roots, corms, or bulbs. Leaves simple, along the stem and alternate, or mostly at base of plant, stalkless and usually sheathing the stem, parallel-veined, often somewhat fleshy. Flowers perfect, irregular, showy in some species, in heads of 1 or 2 flowers at ends of stems, or with several to many flowers in a spike, raceme or panicle, each flower usually subtended by a bract. Fruit a many-seeded capsule, opening by 3 or sometimes 6 longitudinal slits, but remaining closed at tip and base; seeds very small.

**POACEAE** (Grass Family) Perennial or annual herbs, clumped or spreading by rhizomes. Stems (culms) usually hollow, with swollen, solid nodes. Leaves linear, parallel-veined, alternate in 2 ranks or rows, sheathing the stem, the sheaths usually split vertically, sometimes joined and tubular as in brome (*Bromus*) and mannagrass (*Glyceria*); with a membranous or hairy ring (ligule) at top of sheath between blade and stem, or the ligule sometimes absent; a pair of projecting lobes (auricles) sometimes present at base of blade. Flowers (florets) small, usually perfect, or sometimes either staminate or pistillate, the staminate and pistillate flowers separate on the same or different plants. Florets grouped into spikelets, each spikelet with 1 to many florets, with a pair of small bracts (glumes) at base of each spikelet (the glumes rarely absent); the glumes usually of different lengths, the lowermost (or first) glume usually smaller, the upper (or second) glume usually longer. Within the spikelet, each floret subtended by 2 bracts, the larger one (lemma) containing the flower, the smaller one (palea) covering the flower; the lemma and palea often enclosing the ripe fruit (grain or caryopsis); ovary superior, never enclosed in a sac (as in sedges); styles 2–3-parted, the stigmas often feathery. Spikelets grouped in a variety of heads, most commonly in branching heads (panicles), or stalked along an unbranched stem (rachis) in a raceme, or the spikelets stalkless along an unbranched stem in a spike.

**PONTEDERIACEAE** (Pickerelweed Family) Mostly perennial, aquatic or emergent herbs. Leaves alternate, stalkless and straplike, or with a petiole and broad blade. Flowers perfect, regular or irregular, single from leaf axils or in spikes or panicles, subtended by leaflike bracts (spathes), light yellow, white or blue-purple.

**POTAMOGETONACEAE** (Pondweed Family) This treatment includes two genera, *Ruppia* and *Zannichellia,* previously included in separate families. The most important and widespread genus, *Potamogeton,* are quatic perennial herbs, with only underwater leaves, or with both underwater and floating leaves, from rhizomes or tubers, sometimes reproducing and over-wintering by free-floating winter buds. Stems long, wavy, anchored to bottom by roots and rhizomes. Leaves alternate, or becoming opposite upward in some species, simple, with an open or closed sheath at base. Underwater leaves usually linear and threadlike, sometimes broader, margins often wavy, usually stalkless. Floating leaves, if present, oval or ovate, with a waxy upper surface. Flowers perfect, regular, green to red, in stalked spikes at ends of stems or from leaf axils, usually raised above water surface, the spikes with few to many small flowers.

**SCHEUCHZERIACEAE** (Scheuchzeria Family) Our single species, *Scheuchzeria palustris,* is a perennial rushlike herb, from creeping rhizomes, found in wet peatlands. Leaves alternate, several from base and 1–3 along stem. Flowers perfect, regular, green-white, in a several-flowered raceme.

**SMILACACEAE** (Greenbrier Family) Our single genus, *Smilax,* are perennial herbs (with annual stems), or vining shrubs, climbing by tendrils terminating the stipules, with wide, longitudinally nerved, net-veined, alternate leaves and axillary peduncled umbels of small yellow or greenish yellow flowers. Flowers dioecious, the staminate often the larger. Fruit a 1–6-seeded berry.

**TYPHACEAE** (Cat-Tail Family) Family now includes genus *Sparganium* from former family Sparganiaceae (discontinued under APG III). *Sparganium* are perennial sedgelike herbs, floating or emergent in shallow water, from rhizomes and forming colonies. Stems stout, usually erect, unbranched, round in section. Leaves long, broadly linear, sheathing stem at base. Flowers crowded in round heads, the heads with either staminate or pistillate flowers. Fruit a beaked, nutlet-like achene. *Typha* are large reedlike perennials, from fleshy rhizomes and forming colonies. Stems erect, unbranched, round in section, sheathed for most of length by overlapping leaf sheaths. Leaves mostly near base of plant, alternate in 2 ranks, erect, linear, spongy. Flowers tiny, either staminate or pistillate, separate on same plant. Heads with staminate flowers above pistillate in a single, dense, cylindric spike, the staminate and pistillate portions of the spike unalike, contiguous in broad-leaf cat-tail (*T. latifolia*) or separated in narrow-leaf cat-tail (*T. angustifolia*).

**XYRIDACEAE** (Yellow-Eyed-Grass Family) Our single genus, *Xyris,* are perennial rushlike herbs of shores and wet places. Stems erect, leafless, straight or sometimes ridged. Leaves all from base of plant, upright to spreading, linear, often twisted, usually dark green. Flowers small, perfect, yellow, in rounded or cylindric heads at ends of stems.

## ABBREVIATIONS

BOTANICAL

APG III Angiosperm Phylogeny Group III system of 2009

spp. species (plural)

subsp. subspecies

syn synonym

var. variety

× hybrid (or times)

* An asterisk following a species name in the key means that the species is listed under *Additional Species* and is not fully described (usually introduced or adventive species of limited occurrence in Wisconsin)

• Indicates an illustrated species (followed by the plate number)

GEOGRAPHICAL

c central

n northern

ne northeastern

nw northwestern

s southern

se southeastern

sw southwestern

w western

MEASUREMENT

cm centimeter

dm decimeter

m meter

mm millimeter

## ASPLENIACEAE *Spleenwort Family*

### ASPLENIUM *Spleenwort*

Mostly small ferns, with short rootstocks covered by old petiole-bases, and a tuft of small to medium-sized leaves ("fronds"); the blades firm, simple, pinnate, or 2-pinnate, and often evergreen. Sori (clusters of spore containers) elongate, occurring along the veinlets. Indusium (covering over sori) usually membranous, attached lengthwise along one side of the sorus. Species usually of rock crevices and where shaded and mossy.

*Key to species*

1 Blades simple or deeply pinnately lobed .................... 2
1 Blades once-pinnate .................... 3

2 Blades simple, rooting at tip to form new plants .. *A. rhizophyllum*
2 Blades pinnatifid .................... *A. pinnatifidum*

3 Leaves of two types, the fertile long and upright, the sterile shorter and spreading, pinnae with conspicuous basal lobes overlapping rachis .................... *A. platyneuron*
3 Leaves all alike; pinnae bases not overlapping rachis .......... 4

4 Rachis purple-brown throughout entire length .... *A. trichomanes*
4 Rachis green, dark only at base .................... *A. viride*

***Asplenium pinnatifidum*** Nutt. LOBED SPLEENWORT *Threatened.* Crevices in sandstone cliff faces.

***Asplenium platyneuron*** (L.) B.S.P. EBONY-SPLEENWORT Crevices of shady sandstone cliffs, in moss or in shallow soil over rocks, also in partial shade in open woods. Stiff, upright fertile blades are distinctive.

***Asplenium rhizophyllum*** L. WALKING FERN Shaded rocks, usually of limestone, less commonly on sandstone or rarely quartzite. The tips of the arching blades often root to form new plants, hence the name walking fern. *Camptosorus rhizophyllus* (L.) Link.

***Asplenium trichomanes*** L. MAIDENHAIR-SPLEENWORT Sheltered rock crevices on sandstone, limestone or quartzite, where moist or dry.

***Asplenium viride*** Huds. GREEN SPLEENWORT *Endangered.* Crevices of shaded, wet limestone cliffs and talus. A distinctive species less common than *Asplenium trichomanes,* with which it is sometimes associated. The green **rachis** (stem of the blade) distinguishes it from our other spleenworts. *Asplenium trichomanes-ramosum* L.

## ATHYRIACEAE *Lady Fern Family*

Medium to large ferns. *Deparia* and *Diplazium,* formerly considered as species of *Athyrium,* are now treated as separate species. In contrast to *Deparia, Athyrium* has a more deeply grooved rachis, which is continuous from rachis to costa (vs. discontinuous in *Deparia*).

*Key to genera*

1 Pinnae margins finely toothed, but otherwise undivided ......... .................................................. **DIPLAZIUM**

1 Pinnae deeply lobed or divided, margins sometimes finely toothed .................................................. 2

2 Blades 2-pinnate (with the pinnae again divided), the pinnules also sometimes deeply lobed .......................... **ATHYRIUM**

2 Blades with deeply lobed pinnae (1-pinnate-pinnatifid) **DEPARIA**

### ATHYRIUM *Lady Fern*

***Athyrium filix-femina*** L. [•1] LADY FERN Moist deciduous woods, thickets, streambanks, wetland margins, shaded rock outcrops. *Athyrium angustum* (Willd.) K. Presl.

### DEPARIA *Silvery Glade Fern*

***Deparia acrostichioides*** (Swartz) M. Kato [•1] SILVERY GLADE FERN Moist, rich deciduous or mixed woods, especially in swales, ravines, depressions; streambanks. *Athyrium thelypterioides* (Michx.) Desv.

### DIPLAZIUM *Glade Fern*

***Diplazium pycnocarpon*** (Spreng.) Broun GLADE FERN Rich moist woods and ravines. The sterile blades of glade fern somewhat resemble those of Christmas fern (*Polystichum acrostichoides*), but are deciduous rather than evergreen. *Athyrium pycnocarpon* (Spreng.) Tidestrom.

## CYSTOPTERIDACEAE *Bladder Fern Family*

Small to medium ferns; two genera in Wisc, *Cystopteris* and *Gymnocarpium.*

*Key to genera*

1 Blades ternate (divided into 3 more or less equal parts); indusium absent .................................... **GYMNOCARPIUM**

1 Blades 1-pinnate; indusium present ............. **CYSTOPTERIS**

### CYSTOPTERIS *Bladder Fern*

Delicate, medium-sized ferns, with 2-3-pinnate blades arising from short creeping rhizomes.

HYBRIDS ***Cystopteris × illinoensis*** R.C. Moran (Illinois bladder fern), hybrid between *C. bulbifera* and *C. tenuis;* known from s Wisc.

*Key to species*

1 Blades elliptic to lance-shaped, typically widest at or slightly below middle of blade; rachis and pinnule midribs without glandular hairs .................................................. 2

1 Blades elliptic to triangle-shaped, usually widest at base; rachis and pinnule midribs sparsely to densely covered with glandular hairs .................................................. 4

2 Stems covered with yellow hairs; leaves clustered 1–4 cm below protruding apex of stem .............................. *C. protrusa*

2 Stems without hairs; leaves clustered at apex of stem .......... 3

3 Pinnae usually at acute angle to rachis and often curving toward apex of blade; pinnae margins rounded-toothed......... *C. tenuis*

3 Pinnae usually perpendicular to rachis and not curving toward apex of blade; pinnae margins sharp-toothed ............... *C. fragilis*

4 Rachis often with bulblets, rachis and midribs usually densely glandular-hairy; blades narrowly to broadly triangle-shaped, apex of blade long-tapered .............................. *C. bulbifera*

4 Rachis occasionally with bulblets, rachis and midribs usually only sparsely glandular-hairy; blades narrowly triangle-shaped to ovate lance-shaped, apex of blade short-tapered..................... 5

5 Blades ovate to lance-shaped, widest above base ... *C. laurentiana*

5 Blades triangle-shaped, widest at or near base ... *C. tennesseensis*

***Cystopteris bulbifera*** (L.) Bernh. [•2] BULBLET FERN, BLADDER FERN Rocky streambanks, ravines, seepy slopes, cedar swamps, and moist, shaded, often calcium-rich rocks and cliffs. Distinguished from *Cystopteris fragilis,* a common fern of moist woods, by the blade broadest at base, most veins ending in a notch, and the small bulblets on underside of rachis. In fragile fern, blade broadest above its base, most veins end in a tooth, and bulblets are absent.

***Cystopteris fragilis*** (L.) Bernh. BRITTLE BLADDER FERN Sheltered crevices in cliffs, moist banks, and wooded talus slopes. Petioles translucent (when held to a light), with veins of the blade extending to the very tips of the teeth, and a smooth rachis.

***Cystopteris laurentiana*** (Weatherby) Blasdell ST. LAWRENCE BLADDER FERN Calcareous rock or slopes. This species combines the attributes of its presumed parents, *C. fragilis* var. *fragilis* and *C. bulbifera.* It is usually an upright, vigorous plant larger than typical *C. fragilis.* The veins extend both to the teeth-tips and to the sinuses. *Cystopteris fragilis* (L.) Bernh. var. *laurentiana* Weath.

***Cystopteris protrusa*** (Weatherby) Blasdell LOWLAND BLADDER FERN Under deciduous trees on riverbottom benches. Distinguished from *C. fragilis* by the long internodes on the rhizome, the greenish or straw-colored petioles, the softer, larger blades, and the lower pinnules, which taper to a stalk-like base. *Cystopteris fragilis* (L.) Bernh. var. *protrusa* Weatherby.

***Cystopteris tennesseensis*** Shaver TENNESSEE BLADDER FERN Ledges and crevices on rock cliffs. *Cystopteris fragilis* (L.) Bernh. var. *tennesseensis* (Shaver) McGregor.

***Cystopteris tenuis*** (Michx.) Desv. UPLAND BRITTLE BLADDER FERN Similar to *C. fragilis,* but the pinnules oblong to lance-shaped and evenly wedge-shaped at the base; the indusium about 0.5 mm long and shallowly toothed or entire at its tip. In habitats similar to *Cystopteris fragilis,* but more often on streambanks, rotted logs, and moist openings. *Cystopteris fragilis* (L.) Bernh. var. *mackayi* Lawson.

### GYMNOCARPIUM *Oak Fern*

Small ferns with 3-parted delicate blades, arising singly from slender rootstocks. Sori round. Indusium absent. Northern oak fern (*G. dryopteris*) is common; our other two species are much less frequent.

*Key to species*

1 The two lower divisions of the blade nearly as long as the terminal division; blades membranous and thin; rachis glabrous .......... **G. dryopteris**

1 The two lower divisions of the blade about half the length of terminal division; blades firm and somewhat stiff; rachis in part densely glandular .......... 2

2 Innermost pinnules of lowest pair of pinnae only slightly longer than opposite upper pinnules; upper blade surface glabrous; acidic or neutral rock .......... **G. jessoense**

2 Innermost pinnules of lowest pair of pinnae much longer than opposite upper pinnules; upper blade surface moderately glandular; limestone and calcareous rock .......... **G. robertianum**

***Gymnocarpium dryopteris*** (L.) Newman [•2] NORTHERN OAK FERN Cool, moist coniferous and mixed woods; base of talus slopes; swamp margins. The small, delicate, triangular blades oriented parallel to the ground and yellow-green in color are distinctive.

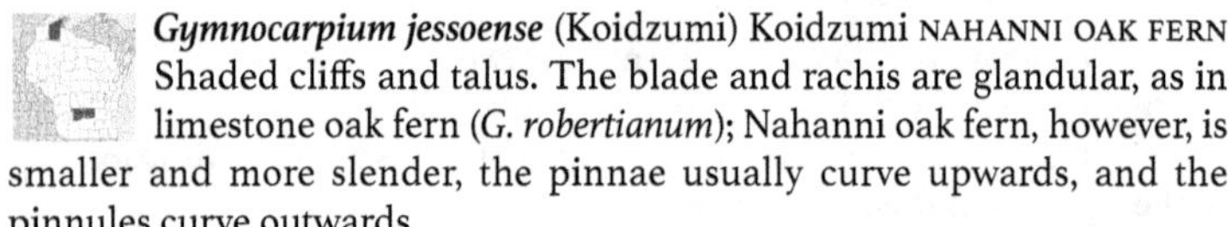

***Gymnocarpium jessoense*** (Koidzumi) Koidzumi NAHANNI OAK FERN Shaded cliffs and talus. The blade and rachis are glandular, as in limestone oak fern (*G. robertianum*); Nahanni oak fern, however, is smaller and more slender, the pinnae usually curve upwards, and the pinnules curve outwards.

***Gymnocarpium robertianum*** (Hoffmann) Newman LIMESTONE OAK FERN Limestone cliffs, outcrops and pavements (alvars). The long-triangular, glandular blades (including glands on the upper surface), with pinnules at right angles are distinctive.

## DENNSTAEDTIACEAE *Bracken Fern Family*

Ferns with leaves all alike, arising singly from creeping rhizomes.

*Key to genera*

1 Blades lance-shaped; sc Wisc only .......... DENNSTAEDTIA

1 Blades broadly triangular, 3-parted; abundant statewide .......... PTERIDIUM

### DENNSTAEDTIA *Hay-Scented Fern*

***Dennstaedtia punctilobula*** (Michx.) T. Moore EASTERN HAY-SCENTED FERN Well-drained sandy soils, open woods, pastures, and old fields. Sometimes confused with lady fern (*Athyrium filix-femina*) and bulblet fern (*Cystopteris bulbifera*); eastern hay-scented fern is identified by its scent when crushed, the satiny brown petioles, and the glandular hairs on the rachis (visible when held to a light); uncommon in Wisc, much more widespread in ne USA.

### PTERIDIUM *Bracken Fern*

***Pteridium aquilinum*** (L.) Kuhn BRACKEN FERN Ubiquitous in open drier woods, pine plantations, old fields, and sandy openings. Plants growing in shade tend to have more or less horizontal blades; blades of plants growing in sun tend to be upright and stiff.

## DRYOPTERIDACEAE *Wood Fern Family*

Medium to large ferns; rhizomes short, stout and scaly. Sori round, on underside veins of pinnae; indusia round to kidney-shaped.

*Key to genera*

1 Fronds 1-pinnate-pinnatifid to more divided, the pinnae pinnatifid or themselves fully divided, lacking a prominent basal lobe, light green to dark green, herbaceous to nearly leathery; indusia kidney-shaped .......... **DRYOPTERIS**

1 Fronds 1-pinnate, the pinnae toothed and each with a slight to prominent lobe near the base on the side towards the leaf tip, dark green, leathery or nearly so; indusia peltate (umbrella-like) .......... **POLYSTICHUM**

### DRYOPTERIS *Wood-Fern*

ADDITIONAL SPECIES ***Dryopteris clintoniana*** (D.C. Eat.) Dowell, historical records from se Wisc, likely no longer present in state.

HYBRIDS Hybrids may be recognized by an appearance intermediate between the parent species, and the presence of abortive spores. Three Dryopteris hybrids are fairly common in Wisc: ***Dryopteris* × *boottii*** (Tuckerman) Underwood: *D. cristata* × *D. intermedia;* leaves more dissected than *D. cristata.* ***Dryopteris* × *triploidea*** Wherry: *D. carthusiana* × *D. intermedia;* leaves similar to parents but often somewhat larger. ***Dryopteris* × *uliginosa*** (A. Braun ex Dowell) Druce: *D. carthusiana* × *D. cristata;* usually in swamps and wet woods.

*Key to species*

1 Blades small, very scaly on underside; old leaves forming conspicuous, persistent curled tufts at base of plant ... ***D. fragrans***

1 Blades large, scales few or absent .......... 2

2 Sori on margins of blade segments; blades leathery gray-green and paler on underside .......... ***D. marginalis***

2 Sori near middle of smallest blade segments; blades various .... 3

3 Lowest pinnules on lowest pinnae stalkless .......... 4

3 Lowest pinnules on lowest pinnae stalked .......... 7

4 Leaves of 2 types; the sterile leaves shorter than fertile; the pinnae of fertile leaves usually turned to a nearly horizontal position .......... ***D. cristata***

4 Sterile and fertile leaves similar; pinnae of fertile leaves in same plane as blade .......... 5

5 Blade widest near middle; petiole much shorter than length of blade .......... ***D. filix-mas***

5 Blade widest at or near base; petiole longer than blade ......... 6

6 Blades large, abruptly narrowed to a short, pointed tip; statewide ............................................... *D. goldiana*

6 Blades gradually reduced to long, tapering tips; historical records from se Wisc ...................................... *D. clintoniana*

7 Lowermost inner pinnule shorter than adjacent lower pinnule ... ............................................. *D. intermedia*

7 Lowermost inner pinnule longer than next outer one .......... 8

8 Lower basal pinnule on basal pinna closer to the second upper pinnule than to the inner or first upper pinnule ...... *D. expansa*

8 Lower basal pinnule on basal pinna closer to the inner upper pinnule than to the second upper pinnule ........ *D. carthusiana*

***Dryopteris carthusiana*** (Villars) H. P. Fuchs SPINULOSE WOOD-FERN Moist to wet woods, hummocks in swamps, thickets; also drier sand dunes and ridges. *Dryopteris spinulosa* (O.F. Müll.) Watt.

***Dryopteris cristata*** (L.) A. Gray CRESTED WOOD-FERN Swamps, thickets, open bogs, fens and seeps.

***Dryopteris expansa*** (K. Presl) Fraser-Jenkins & Jermy SPREADING WOOD-FERN Cool moist woods and thickets. *Dryopteris assimilis* S. Walker.

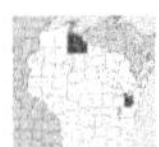

***Dryopteris filix-mas*** (L.) Schott MALE FERN Rich woods, rocky slopes, and at the base of shaded rock outcrops often on limestone. The taper of the blades at their base and tip and the presence of sori only on the upper half of fertile fronds are characteristic; male fern is somewhat similar to marginal wood fern (*D. marginalis*), but the blades are much less leathery.

***Dryopteris fragrans*** (L.) Schott FRAGRANT WOOD-FERN Cliffs and talus slopes (often somewhat calcareous). *D. fragrans* is somewhat similar to the smaller rusty cliff fern (*Woodsia ilvensis*) of similar habitats, but the curled, persistent old leaves drooping below fragrant wood fern are distinctive.

***Dryopteris goldiana*** (Hook.) A. Gray GOLDIE'S WOOD-FERN Moist hardwood forests, shaded streambanks, talus slopes; soils rich in humus and usually neutral.

***Dryopteris intermedia*** (Muhl.) A. Gray [•2] FANCY WOOD-FERN Moist hardwood and mixed hardwood-conifer forests, hummocks in swamps; soils rich in humus, slightly acid to neutral. *Dryopteris spinulosa* var. *intermedia* (Muhl. ex Willd.) Underw.

***Dryopteris marginalis*** (L.) A. Gray MARGINAL WOOD-FERN Rocky woods and ravines. The leathery or spongy character of the nearly evergreen blades and the nearly marginal sori are characteristic.

## POLYSTICHUM *Holly Fern*

Large, tufted ferns with mostly evergreen, leathery blades; the petioles usually scaly, arising from short, stout, chaffy rhizomes. Sori round; indusia round, attached at the center.

*Pinnae underside with indusia and sori*
***Athyrium filix-femina***
ATHYRIACEAE
***Deparia acrostichiodes***
ATHYRIACEAE
*Pinnae underside with indusia and sori*

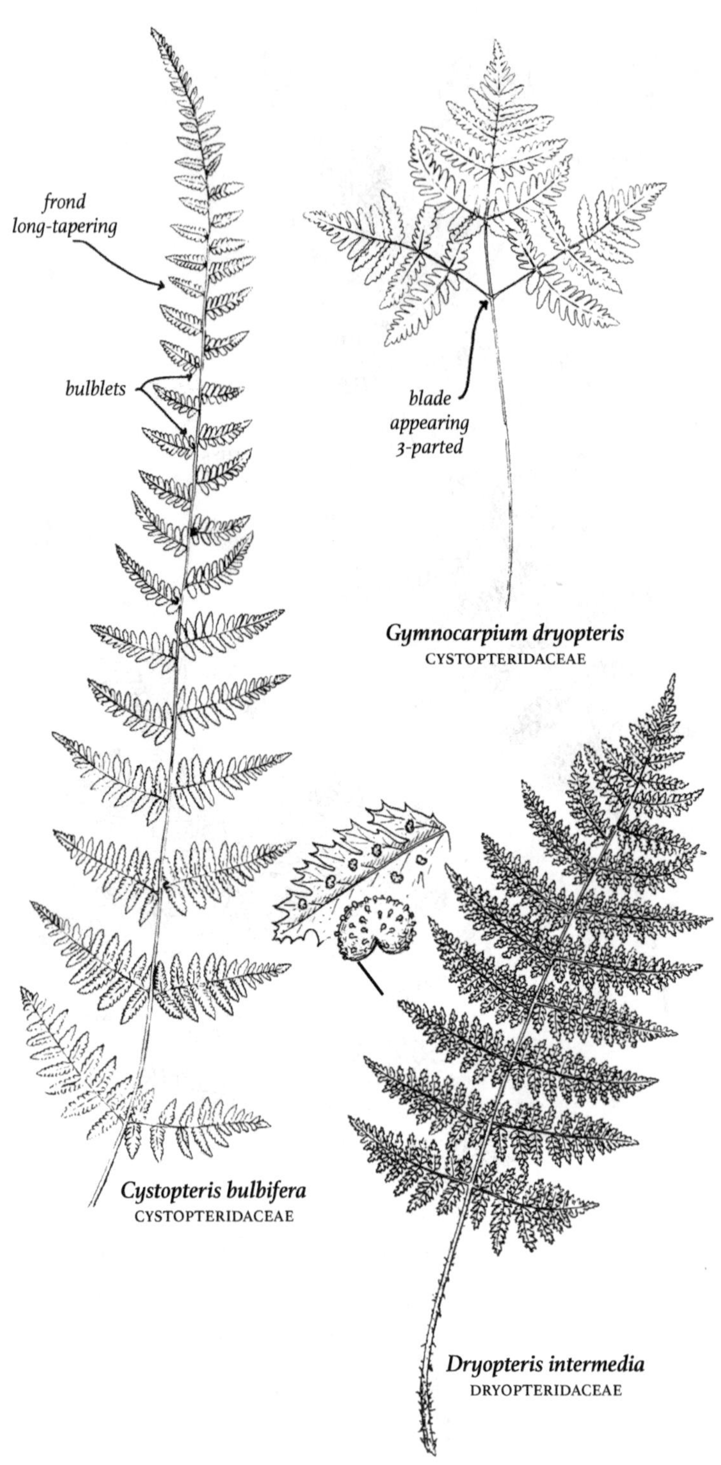


***Gymnocarpium dryopteris***
CYSTOPTERIDACEAE

***Cystopteris bulbifera***
CYSTOPTERIDACEAE

***Dryopteris intermedia***
DRYOPTERIDACEAE

*Key to species*

1 Sori located on reduced upper pinnae ........... ***P. acrostichoides***
1 Sori on underside of typical pinnae ........................... 2

2 Blades once-pinnate ................................ ***P. lonchitis***
2 Blades twice-pinnate ............................... ***P. braunii***

***Polystichum acrostichoides*** (Michx.) Schott CHRISTMAS FERN Rich woods and shaded rocky slopes.

***Polystichum braunii*** (Spenner) Fée BRAUN'S HOLLY FERN *Threatened.* Rocky woods, along rocky streams, and on shaded cliffs within moist northern forests.

***Polystichum lonchitis*** (L.) Roth HOLLY FERN Cliffs, moist rocky and talus slopes, occasionally in coniferous woods.

## EQUISETACEAE *Horsetail Family*

### EQUISETUM *Horsetail, Scouring-Rush*

Rushlike herbs with dark rhizomes. Stems annual or perennial, grooved, usually with large central cavity and smaller outer cavities, unbranched or with whorls of branches at nodes. Leaves reduced to scales, united into a sheath at each node; top of sheath divided into dark-colored teeth. Spores in cones at tips of green or brown fertile stems.

HYBRIDS ***Equisetum* × *ferrissii*** Clute: *E. hyemale* × *E. laevigatum,* common. ***Equisetum* × *nelsonii*** (A.A. Eat) Schaffn.: *E. laevigatum* × *E. variegatum,* mostly near Lake Michigan. ***Equisetum* × *litorale*** Kuehl ex Rupr.: *E. arvense* × *E. fluviatile,* sw Wisc.

*Key to species*

1 Stems evergreen (annual in *E. laevigatum*); unbranched or with a few scattered branches, branches not in regular whorls (*scouring rushes*) ................................................ 2
1 Stems annual; usually with regular whorls of branches, sometimes unbranched (*horsetails*) ................................. 5

2 Stems solid (central cavity absent); stems small, slender and sprawling ........................................ ***E. scirpoides***
2 Stems hollow (central cavity present); stems larger, usually upright. ............................................................ 3

3 Stems 1–3 dm tall, with 5–12 ridges, central cavity to 1/3 diameter of stem ........................................... ***E. variegatum***
3 Stems usually taller, with 16–50 ridges, central cavity more than half diameter of stem ........................................... 4

4 Cones with a distinct, small sharp tip; stem sheaths with a black band at tip and base ................................ ***E. hyemale***
4 Cones blunt-tipped, sheaths with black band at tip only ......... ........................................................ ***E. laevigatum***

5 Stems unbranched ........................................... 6
5 Stems with regular whorls of branches ......................... 10

6 Stems green .......... 7

6 Stems brown or flesh-colored .......... 8

7 Stems with 9–25 shallow ridges; central cavity more than half diameter of stem; sheath teeth entirely black or with narrow white margins .......... ***E. fluviatile***

7 Stems with 5–10 strongly angled ridges; central cavity less than 1/3 diameter of stem; sheath teeth with white margins and dark centers .......... ***E. palustre***

8 Sheath teeth papery and red-brown, teeth joined and forming several broad lobes .......... ***E. sylvaticum***

8 Sheath teeth black or brown, not papery, separate or joined in more than 4 small groups .......... 9

9 Stems withering after spores mature, remaining unbranched .......... ***E. arvense***

9 Stems persistent, becoming branched and green.......... ***E. pratense***

10 First internode of each branch shorter than the subtending sheath of the main stem.......... 11

10 First internode of each branch equal or longer than the subtending sheath of the main stem.......... 12

11 Stems with 9–25 shallow ridges; central cavity more than half diameter of stem; sheath teeth more than 12, entirely black or with narrow white margins .......... ***E. fluviatile***

11 Stems with 5–10 strongly angled ridges; central cavity about same size as outer cavities; sheath teeth 5–6, with white margins and dark centers .......... ***E. palustre***

12 Stem branches themselves branched; sheath teeth papery and red-brown, teeth joined and forming several broad lobes ***E. sylvaticum***

12 Stem branches unbranched; sheath teeth black or brown, not papery, separate or joined in more than 4 small groups.......... 13

13 Stem branches ascending; teeth of branch sheaths gradually tapering to a slender tip .......... ***E. arvense***

13 Stem branches spreading; teeth of branch sheaths broadly triangular .......... ***E. pratense***

***Equisetum arvense*** L. [•3] COMMON HORSETAIL, FIELD HORSETAIL Streambanks, meadows, moist woods, ditches, roadsides and along railroads; calcareous fens.

***Equisetum fluviatile*** L. WATER-HORSETAIL In standing water of marshes, ponds, peatlands, ditches and swales.

***Equisetum hyemale*** L. [•3] COMMON SCOURING-RUSH Often forming dense colonies in seeps, wet to moist meadows, shores and streambanks, ditches, roadsides and along railroads; usually where sandy or gravelly.

***Equisetum laevigatum*** A. Braun SMOOTH SCOURING-RUSH Wet meadows, low prairie, streambanks, floodplains, seeps, and ditches, often where sandy or gravelly.

***Equisetum palustre*** L. MARSH-HORSETAIL Wetland margins, streambanks, alder thickets, fens; often in shallow water.

***Equisetum pratense*** Ehrh. MEADOW-HORSETAIL Moist woods, streambanks and meadows.

***Equisetum scirpoides*** Michx. DWARF SCOURING-RUSH Mossy places and moist, shaded woods, the stems often partly buried in humus.

***Equisetum sylvaticum*** L. [•3] WOODLAND-HORSETAIL Wet or swampy woods, thickets, usually in partial shade.

***Equisetum variegatum*** Schleicher VARIEGATED SCOURING-RUSH Wet, calcareous open areas such as shores, low places in dunes, borrow pits and ditches. *Equisetum variegatum* commonly forms hybrids with *E. hyemale* and *E. laevigatum,* sometimes making identification of this species difficult.

## ISOETACEAE *Quillwort Family*

### ISOETES *Quillwort*

Perennial aquatic or emergent herbs. Leaves simple, entire, linear, from a 2–3 lobed rhizome (corm). Outermost and innermost leaves typically sterile. Outer fertile leaves have a pocketlike structure (sporangia) bearing whitish spores (megaspores; about 0.5 mm in diameter, magnification needed to see features); inner fertile leaves have numerous small microspores.

*Key to species*

1 Megaspores conspicuously covered with small spines .................................................... ***I. echinospora***

1 Megaspores not spiny .............................. ***I. lacustris***

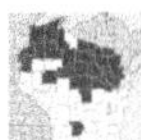

***Isoetes echinospora*** Durieu SPINY-SPORED QUILLWORT Shallow water (to 1 m deep) of lakes, ponds and slow-moving rivers; plants rooted in mud, sand, or gravel. *Isoetes muricata* Durieu.

***Isoetes lacustris*** L. LAKE QUILLWORT Underwater in shallow to deep water of cold lakes, ponds and streams. *Isoetes macrospora* Durieu.

ADDITIONAL SPECIES ***Isoetes hieroglyphica*** A.A. Eaton is sometimes treated as a separate species of northeastern USA and adjacent Canada, with a disjunct population reported from Wisc.

## LYCOPODIACEAE *Clubmoss Family*

Low, trailing, evergreen herbs resembling large mosses. Leaves needlelike or scalelike, alternate or opposite on stem. Spore-bearing leaves (sporophylls) similar to vegetative leaves or in conelike clusters at tips of upright stems.

*Key to genera*

1 Horizontal stems absent; sporangia in axils of unmodified leaves .................................................... **HUPERZIA**

1 Horizontal stems present; sporangia in axils of modified, reduced sporophylls, the sporophylls grouped into upright or nodding strobili .......... 2

2 Strobili upright on leafy peduncles, the peduncle leaves not reduced in size; wetland species .......... **LYCOPODIELLA**

2 Strobili sessile or on peduncles, the peduncles if present with scattered, small leaves; mostly upland species .......... 3

3 Ultimate shoots and their leaves 5–12 mm wide, rounded in cross-section; leaves not strongly overlapping .......... **LYCOPODIUM**

3 Ultimate shoots and their leaves to 6 mm wide, 4-angled or flattened in cross-section; leaves overlapping .......... **DIPHASIASTRUM**

## DIPHASIASTRUM *Ground-Pine*

Small plants of drier habitats resembling miniature trees; branches flattened or 4-angled in cross-section; leaves 4-ranked, neither spine- nor hair-tipped. Strobili (cones) stalked, the stalks branched into segments of equal length.

HYBRIDS ***Diphasiastrum* × *habereri*** (House) Holub, hybrid between *D. digitatum* × *D. tristachyum*. ***Diphasiastrum* × *zeilleri*** (Rouy) Holub, hybrid between *D. complanatum* × *D. tristachyum*.

*Key to species*

1 Stem branchlets cordlike, nearly square in cross-section, usually waxy blue-green color .......... ***D. tristachyum***

1 Stem branchlets flat in cross-section, usually green .......... 2

2 Branchlets regularly fan-shaped and arching, without conspicuous constrictions between seasonal growth; most strobili with sterile tips .......... ***D. digitatum***

2 Branchlets irregular, with conspicuous constrictions; most strobili without sterile tips .......... ***D. complanatum***

***Diphasiastrum complanatum*** (L.) Rothm. NORTHERN RUNNING-PINE Woodlands and clearings. Conspicuous annual constrictions present, giving plants a somewhat irregular appearance, in contrast to the regularity of fan ground-pine (*D. digitatum*). The strobili are also irregular in number per peduncle (varying from 1–4), and the naked peduncles are very slender. *Lycopodium complanatum* L.

***Diphasiastrum digitatum*** (Dill.) Holub FAN GROUND-PINE Dry woods and clearings. The branchlets are very regular and fan-like, annual constrictions are lacking, and the strobili are usually in groups of 4 on long, naked peduncles. *Lycopodium digitatum* Dill.

***Diphasiastrum tristachyum*** (Pursh) Holub [•4] DEEP-ROOT GROUND-PINE Dry, sometimes sandy woods and clearings. The branches are vase-shaped and crowded, bluish green and white waxy on their underside; annual constrictions are present along the branches; the peduncles often branch and then branch again, resulting in 4 strobili. *Lycopodium tristachyum* Pursh.

## HUPERZIA *Fir-Moss*

Low evergreen perennials with erect shoots; leaves spreading or appressed and upright. Spores borne at base of upper leaves.

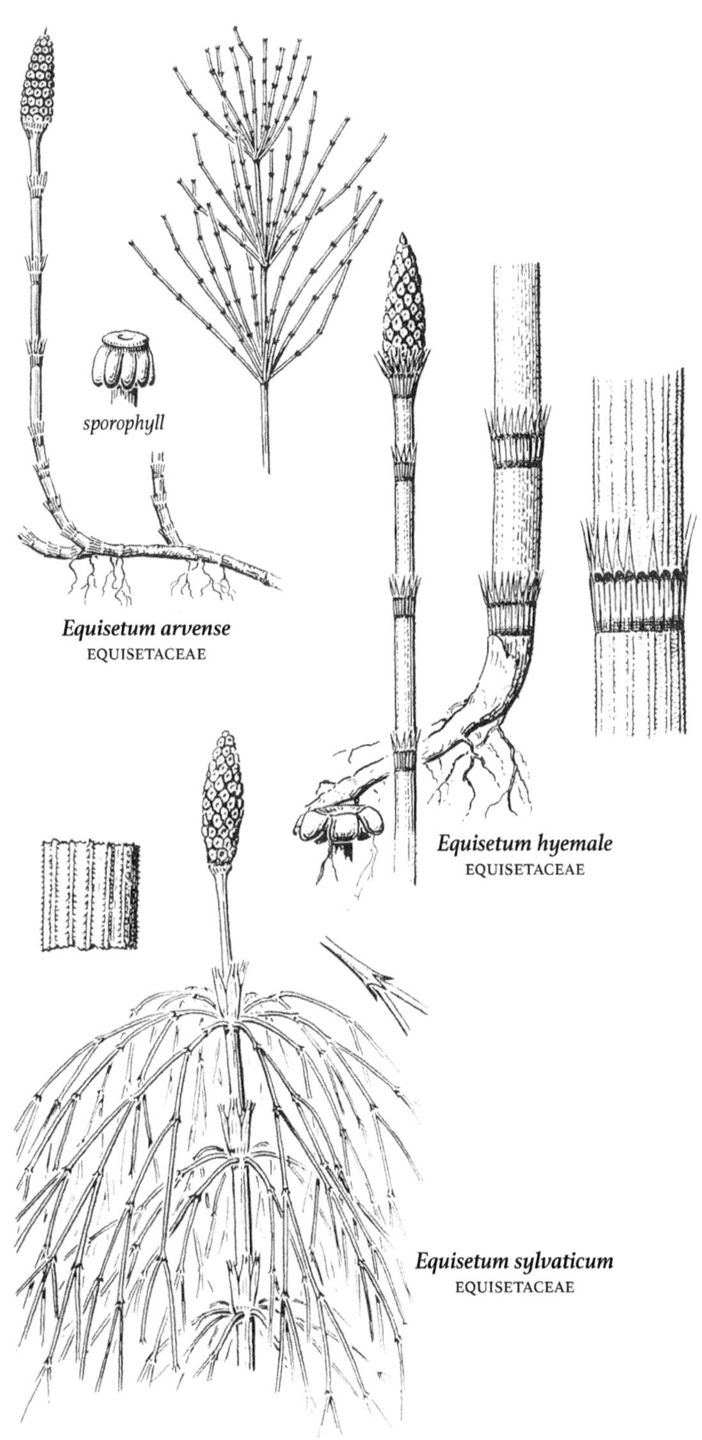
sporophyll
*Equisetum arvense*
EQUISETACEAE
*Equisetum hyemale*
EQUISETACEAE
*Equisetum sylvaticum*
EQUISETACEAE

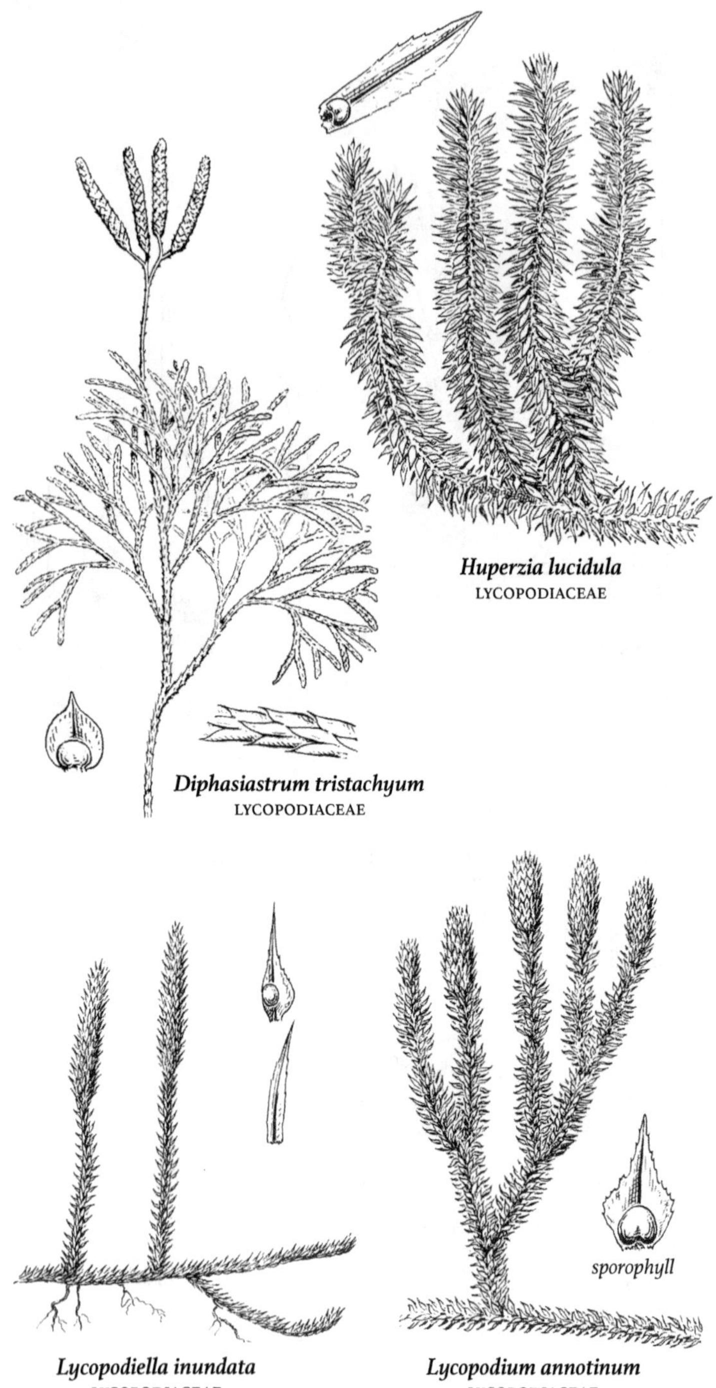


*Huperzia lucidula*
LYCOPODIACEAE

*Diphasiastrum tristachyum*
LYCOPODIACEAE

*Lycopodiella inundata*
LYCOPODIACEAE

*Lycopodium annotinum*
LYCOPODIACEAE

HYBRIDS ***Huperzia* × *bartleyi*** (Cusick) Kartesz & Gandhi, hybrid between *H. lucidula* × *H. porophila*. ***Huperzia* × *buttersii*** (Abbe) Kartesz & Gandhi, hybrid between *H. lucidula* × *H. selago;* nw Wisc; resembles a slender *H. lucidula,* but distinguished by presence of abortive spores and scattered stomata on upper leaf surface.

*Key to species*

1 Leaves obovate, widest above middle, spreading to ± reflexed, upper portion of at least the larger leaves with distinct teeth; shoots "shaggy" with conspicuous annual constrictions; usually growing on soil ............................................. ***H. lucidula***

1 Leaves lance-shaped, widest below the middle, leaves (at least those on the upper stem) often ascending, entire or with a few small teeth; annual constrictions absent or faint ......................... 2

2 Leaves lance-shaped with sides nearly parallel; stomates on upper surface of each leaf number 2-50 (view fresh leaves under 20x lens to see the light-colored, dot-like stomates) ........... ***H. porophila***

2 Leaves lance-shaped (as above) or ovate or triangular; if leaf shape is inconclusive, then number of stomates on upper leaf surface is greater than 60 ............................................ 3

3 Leaves near base of plant essentially same size as those on upper portion; gemmae formed in a single whorl at end of the annual growth ............................................. ***H. selago***

3 Leaves near base of plant conspicuously longer than those on upper portion; gemmae formed throughout upper portions of shoot ... ............................................. ***H. appressa***

***Huperzia appressa*** (Desv.) Á. & D. Löve MOUNTAIN FIR-MOSS Cliffs, talus slopes, where open and exposed, on moss or thin soil; local in n Wisc. *Huperzia appalachiana* Beitel & Mickel.

***Huperzia lucidula*** (Michaux) Trev. [•4] SHINING FIR-MOSS Moist to wet conifer and hardwood forests. Small two-lobed buds (gemmae) produced in some upper leaf-axils; these may sprout into new plants after falling onto moist humus. *Lycopodium lucidulum* Michx.

***Huperzia porophila*** (Lloyd & Underwood) Holub ROCK CLUBMOSS Shaded sandstone ledges. *Lycopodium porophilum* Lloyd & Underwood.

***Huperzia selago*** L. NORTHERN FIR-MOSS Cedar swamps, streambanks, sandy lake shores, usually where mossy; also in wet, sandy borrow pits. *Lycopodium selago* L.

## LYCOPODIELLA *Bog Clubmoss*

Small plants of wet places with horizontal stems creeping on ground surface, and fertile, upright stems. Strobili (cones) formed on upper part of upright stems, stalkless and covered with leaves.

*Key to species*

1 Horizontal stems less than 1 mm wide except at root nodes; horizontal stem leaves usually less than 6 mm long, teeth or bristles absent; erect shoots mostly less than 10 cm tall; mostly n and c Wisc ............................................. ***L. inundata***

1 Horizontal stems 2–3 mm wide; horizontal stem leaves 6–13 mm long, margins toothed or bristle-tipped; tallest erect shoots often more than 10 cm tall; uncommon in c Wisc ........ *L. marguerita*e

***Lycopodiella inundata*** (L.) Holub [•4] NORTHERN BOG CLUBMOSS Creeping perennial of wet habitats. Spores borne in terminal, leafy cones. Acidic sphagnum bogs, wet sandy shores and streambanks; disturbed wetlands, sandy borrow pits. *Lycopodium inundatum* L.

ADDITIONAL SPECIES ***Lycopodiella margueritae*** J.G. Bruce, W.H. Wagner & Bietel, newly defined and similar to *L. inundata,* is reported from Wood County.

### LYCOPODIUM *Ground-Pine*

Plants mainly trailing on ground. Horizontal stems on substrate surface or subterranean, long-creeping. Upright shoots scattered along horizontal stem, unbranched or with 1–4 lateral branchlets. Leaves not imbricate, linear to linear lance-shaped; leaves on horizontal stems scattered, appressed; leaves on lateral branchlets mostly 6-ranked or more. Gemmae absent. Strobili single and sessile, or multiple and pedunculate.

*Key to species*

1 Stroboli stalked; upright stems with 2–5 branches, not forming tree-like shapes; leaves tipped with hairs .......................... 2
1 Stroboli not stalked .......................................... 3

2 Stroboli mostly single on stalk (rarely in pairs and then nearly stalkless); leaves apppressed or ascending on stem ..... ***L. lagopus***
2 Stroboli 2–5; leaves spreading or somewhat ascending ***L. clavatum***

3 Stems creeping and horizontal; stroboli single at end of upright, mostly unbranched shoot ........................ ***L. annotinum***
3 Stems upright, much-branched and tree-like; stroboli 1–7 at end of shoot.................................................... 4

4 Branches flat in cross-section; leaves of unequal sizes ***L. obscurum***
4 Branches round in cross-section; leaves of equal sizes........... 5

5 Leaves on main stem below branches dark green and appressed to stem, soft to touch .................................. ***L. hickeyi***
5 Leaves on stem below branches pale green and spreading, prickly ............................................ ***L. dendroideum***

***Lycopodium annotinum*** L. [•4] STIFF GROUND-PINE Moist woods and clearings, subalpine forests, and exposed rocky and peaty habitats. *Spinulum annotinum* (L.) Haines.

***Lycopodium clavatum*** L. RUNNING GROUND-PINE Dry woods and clearings. Young or sterile plants sometimes confused with *L. annotinum.* The extended, soft, hair-like bristles on the leaf tips are useful for identification.

***Lycopodium dendroideum*** Michx. TREE GROUND-PINE Woods and clearings. Quickly identified by grasping the base of an aerial stem; this will feel distinctly prickly because of the stiff divergent leaves. *Dendrolycopodium dendroideum* (Michx.) Haines, *Lycopodium obscurum* L. var. *dendroideum* (Michx.) D.C. Eat.

***Lycopodium hickeyi*** W.H. Wagner, Beitel & Moran PENNSYLVANIA GROUND-PINE Woodlands. *Dendrolycopodium hickeyi* (W.H. Wagner, Beitel & R.C. Moran) A. Haines, *Lycopodium obscurum* L. var. *isophyllum* Hickey.

***Lycopodium lagopus*** (Laestad.) Zinserl. ONE-CONE GROUND-PINE Fields and openings in woods. *Lycopodium clavatum* L. var. *lagopus* Laestad.

***Lycopodium obscurum*** L. PRINCESS-PINE Woodlands. *Dendrolycopodium obscurum* (L.) Haines

## ONOCLEACEAE *Sensitive Fern Family*

Large coarse ferns with creeping hairy rhizomes (*Onoclea*) or with stolons on ground surface (*Matteuccia*); sterile and fertile fronds strongly different, the sterile fronds deciduous, pinnatifid to 1-pinnate-pinnatifid; fertile fronds persistent. Sori enclosed under recurved margin of pinna segment (outer false indusium) and a tiny true inner indusium (membranous or of hairs).

*Key to genera*

1 Sterile blades solitary from creeping rhizomes, deeply divided into lobes (or the lowermost divisions pinnae) ............ ONOCLEA

1 Sterile blades in a circle from a thick crown; pinnate with lobed pinnules ........................................ MATTEUCCIA

### MATTEUCCIA *Ostrich Fern*

***Matteuccia struthiopteris*** (L.) Todaro [•5] OSTRICH FERN Wet and swampy woods, streambanks, seeps, and ditches.

### ONOCLEA *Sensitive Fern*

***Onoclea sensibilis*** L. SENSITIVE FERN [•5] Swampy woods and low places in forests, wet meadows, calcareous fens, roadside ditches, wet or moist wheel ruts; sometimes weedy. Leaves of *Onoclea* susceptible to damage from even light frosts, hence the common name of sensitive fern.

## OPHIOGLOSSACEAE *Adder's-Tongue Family*

Perennial herbs from short, erect rhizomes having several fleshy roots. Plants produce one leaf each year on a single stalk (stipe), with bud for next year's leaf at base of stipe. Leaves divided into a fertile segment (sporophyll) and a sterile expanded blade. Sterile blades entire (*Ophioglossum*), or lobed or 1–3x pinnately divided (*Botrychium, Botrypus, Sceptridium*). Spores in numerous round sporangia.

*Key to genera*

1 Sterile blades simple, entire; veins netlike; sporangia embedded in rachis of spike ............................. OPHIOGLOSSUM

1 Sterile blades pinnately lobed or dissected; veins forked; sporangia exposed, often on a branched structure ...................... 2

2 Sterile blades somewhat leathery, persisting over winter; blades with distinct stalk, usually 5–25 cm long in fertile plants; fertile portion of frond joining sterile portion at or near ground level ... ............................................. **SCEPTRIDIUM**

2 Plants deciduous, withering in fall; blades usually ± unstalked, large (more than 5 cm long) or much smaller (in some species of Botrychium), herbaceous or sometimes fleshy; fertile portion of frond joining sterile portion well above ground level ........... 3

3 Sterile blades triangular, 3–4x pinnate, stalkless and mostly 5–25 cm wide; fertile portion erect, appearing to be a continuation of stipe ........................................ **BOTRYPUS**

3 Sterile blades short-triangular, oblong, or linear; lobed (simple) to 3-pinnate (usually 1-pinnate to 2-pinnate-pinnatifid), mostly 1-5 cm wide; fertile portion of blade upright or spreading **BOTRYCHIUM**

## BOTRYCHIUM *Grape-Fern; Moonwort*

Small plants with one leaf, the blade divided into sterile and fetile segments. Sterile portion of blade pinnately divided or lobed, fertile portion branched to form a panicle bearing the sporangia.

*Key to species*

1 Sterile blade (trophophore) simple to lobed, lobes rounded to square and angular, stalks usually 1/2 to 2/3 length of sterile blade; rare plant of shaded woods (*B. mormo*), or more common and often in open grassy fields (*B. simplex*) ............................ 2

1 Sterile blade pinnately lobed (either if actual pinnae or simply lobed), lobes of varying shapes, stalk usually less than 1/4 length of sterile blade; plants often in open sunny places; dunes, streambanks, roadsides, on trails and openings in forests, etc. ... 3

2 Segments of sterile leaves rounded, margins mostly entire; plants herbaceous in texture, green; habitats various, in forests or open ground ........................................... ***B. simplex***

2 Segments of sterile leaves angular, outer margins often coarsely dentate; plants ± succulent; shiny yellow-green; rare in shady forest understories, often concealed by leaf litter ............ ***B. mormo***

3 Basal pinnae or segments of sterile blade with venation like the ribs of a fan; midrib absent .................................... 4

3 Basal pinnae or segments of sterile blade with pinnate venation; midrib present ............................................ 8

4 Basal pinnae broadly fan-shaped (almost perfect half moons) with narrow stalks ....................................... ***B. lunaria***

4 Basal pinnae narrowly fan- or wedge-shaped to nearly linear ... 5

5 Sterile blade at least partially folded longitudinally when alive (conduplicate), usually not more than 4 cm long by 1 cm wide; pinnae up to 5 pairs; basal pinnae usually 2-parted............. 6

5 Sterile blade flat or folded only at base when alive, usually up to 10 long by 2.5 cm wide; pinnae up to 10 pairs; basal pinnae unlobed, or if lobed, not usually 2-parted.............................. 7

6 Sterile blade very fleshy; fertile portion of blade usually less than

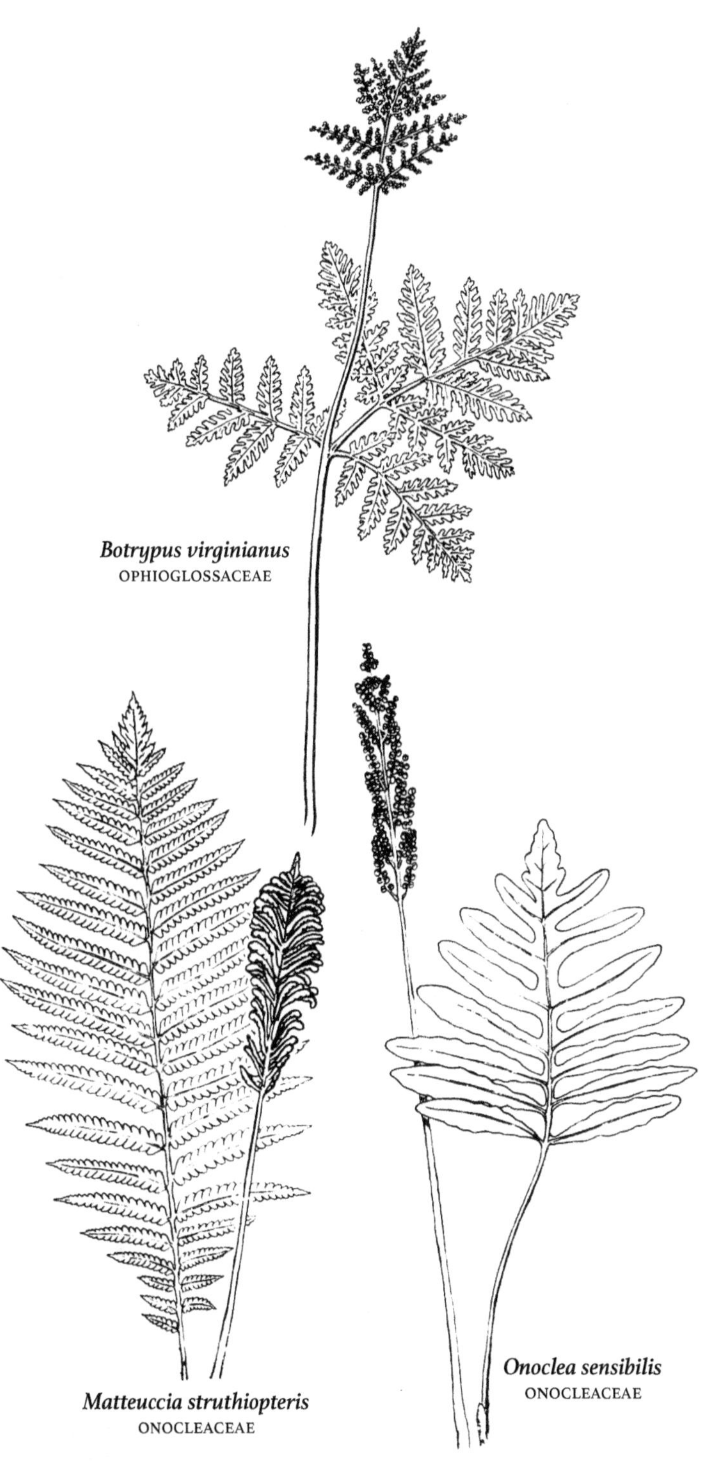
*Botrypus virginianus*
OPHIOGLOSSACEAE
*Matteuccia struthiopteris*
ONOCLEACEAE
*Onoclea sensibilis*
ONOCLEACEAE

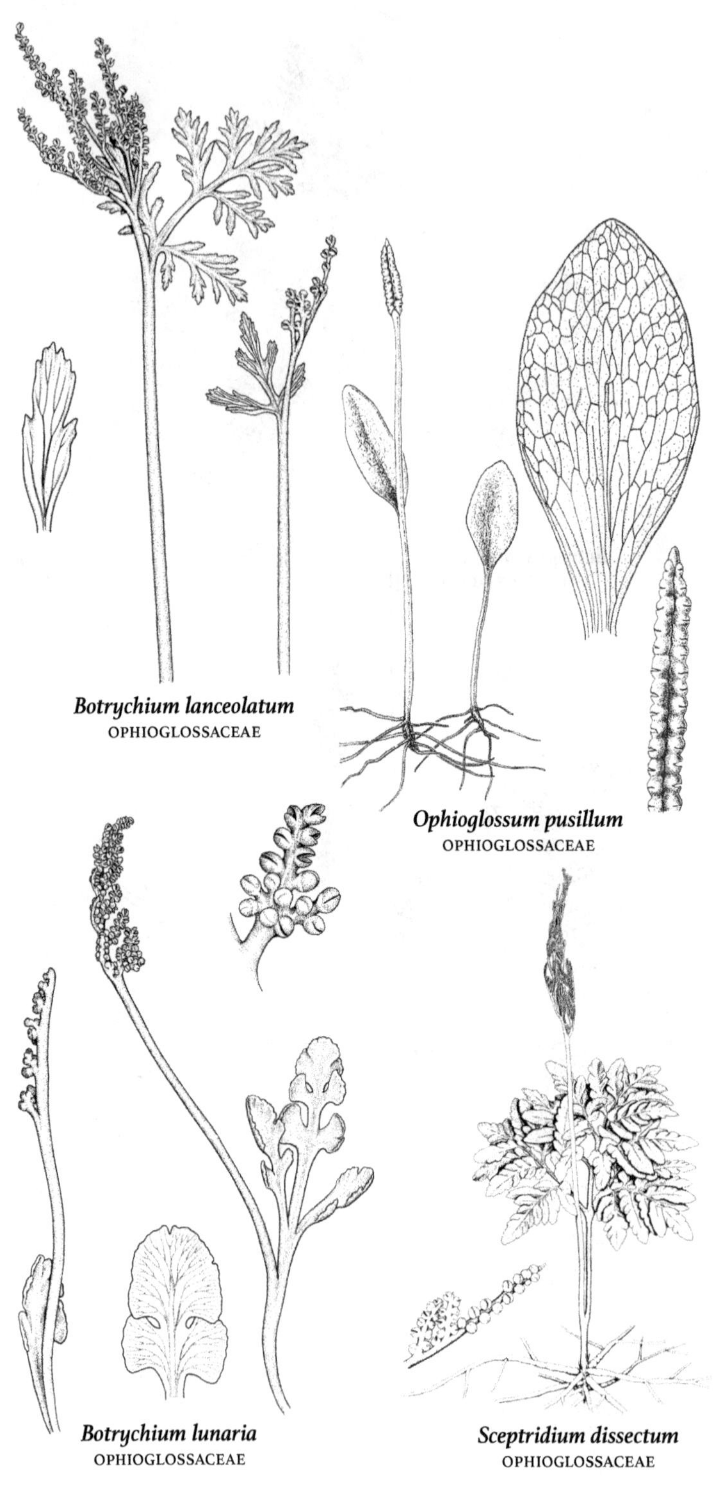
*Botrychium lanceolatum*
OPHIOGLOSSACEAE
*Ophioglossum pusillum*
OPHIOGLOSSACEAE
*Botrychium lunaria*
OPHIOGLOSSACEAE
*Sceptridium dissectum*
OPHIOGLOSSACEAE

1.5 times length of sterile blade; pinnae mostly linear; basal pinna lobes ± equal; plants appearing in late spring ........ ***B. campestre***

6 Sterile blade herbaceous; fertile portion usually 1.5–4 times the length of vegetative blades; pinnae asymmetrically fan-shaped; basal pinna lobes unequal; plants appearing in summer ..... ***B. pallidum***

7 Sterile blade narrowly oblong (sterile blade widest above base), firm to herbaceous; pinnae fan-shaped, margins shallowly crenate .... .............................................. ***B. minganense***

7 Sterile blade narrowly deltate (sterile blade widest at lowest pinna pair); pinnae spatulate to linear spatulate, margins entire to very coarsely and irregularly dentate; Door County .... ***B. spathulatum***

8 Fertile portion of blade 3-parted, with 3 major branches from near base of stalk at sterile blade ....................... ***B. lanceolatum***

8 Fertile portion of blade unbranched or with loosely pinnate branches smaller than the single main stem ... ***B. matricariifolium***

***Botrychium campestre*** W.H. Wagner & Farrar IOWA MOONWORT *Endangered.* Inconspicuous in prairies, dunes, grassy railroad sidings, and fields over limestone. Leaves appear in early spring and wither in late spring and early summer, long before those of other moonworts.

***Botrychium lanceolatum*** (Gmel.) Angstr. [•6] TRIANGLE MOONWORT Moist humus-rich woods, hummocks in swamps, streambanks. Ours var. *angustisegmentum,* found from Nfld and Ontario to Minn, becoming increasingly rare southward to NJ and Ohio.

***Botrychium lunaria*** (L.) Sw. [•6] COMMON MOONWORT *Endangered.* Cool, moist sandy soils in woods. Moonwort has a long and illustrious history in early herbals; the "seeds" could reputedly make one invisible or could be used to unlock doors. *Botrychium neolunaria* Stensvold & Farrar.

***Botrychium matricariifolium*** (A. Braun ex Dowell) A. Braun ex Koch DAISY-LEAF MOONWORT Acidic soil in old sandy and sterile fields, dry wooded slopes, rocky woods, moist cedar woods, and rich swamps. This species is somewhat larger than *B. simplex.* The shape of the blade is variable (deltoid to ovate) but it is stalked, and the toothed segments are distinctive (compare *B. lanceolatum*).

***Botrychium minganense*** Victorin MINGAN MOONWORT Moist hardwood forests, aspen-balsam fir woods, and old clearings; soils mostly circumneutral. *B. minganense* can be distinguished from *B. lunaria* by its yellowish green hue and by its trough-shaped sterile segments, which are ascending rather than at right angles to the stalk and which rarely overlap with each other.

***Botrychium mormo*** W.H. Wagner LITTLE GOBLIN MOONWORT *Endangered.* Extremely sporadic in mature deciduous forests, typically dominated by sugar maple or basswood, and sometimes with eastern hemlock or northern white cedar. Sites shaded and moist; soils are loams, with a rich litter layer. Our tiniest moonwort, uncommon; leaves appearing in late spring to fall or sometimes not appearing above the leaf litter.

***Botrychium pallidum*** W.H. Wagner PALE MOONWORT Open fields, dry sand and gravel ridges, roadsides, wet depressions, marshy lakeshores, tailings basins, second-growth forests; soils sandy.

***Botrychium simplex*** E. Hitchc. LEAST MOONWORT Pastures, meadows, lakeshores, and gravelly slopes; easily overlooked in the field due to its small size and grassy habitat.

***Botrychium spathulatum*** W.H. Wagner SPOON-LEAF MOONWORT Sand dunes, old fields, grassy railways, often where underlain by limestone. *B. spathulatum* has long been confused with the more common *B. minganense,* with which it often grows in the Lake Superior region; leaves appear later in *B. spathulatum* than in *B. minganense.*

## BOTRYPUS *Rattlesnake Fern*

***Botrypus virginianus*** (L.) Holub [•5] RATTLESNAKE FERN Plants appearing in spring, withering in autumn, not overwintering. Blade (trophophore) broadly triangular. Pinnae to 12 pairs, usually somewhat overlapping and slightly ascending. Occasional in swamps of cedar and black spruce; more common in moist to fairly dry deciduous woods. Rattlesnake fern is widespread in North America, occurring across Canada and most of the USA. *Botrychium virginianum* (L.) Sw.

## OPHIOGLOSSUM *Adder's-Tongue*

***Ophioglossum pusillum*** Raf. [•6] NORTHERN ADDER'S-TONGUE Plants erect, from slender rhizomes. Leaves 1, entire, on a stalk; blades upright, oval to ovate, conspicuously net-veined. Sporangia in 2 rows in a terminal, unbranched fertile segment. Wet sandy meadows and prairies, moist depressions, wetland margins, sandy beaches.

## SCEPTRIDIUM *Grape Fern*

Small to medium leathery ferns found in a variety of moist to dry, open to shaded habitats, often where sandy. Sterile blades dissected, winter-green; sporophore short-lived, withering by late summer. Sterile blade and sporophore joined near or below ground level.

*Key to species*

1 Sterile blade segments deeply cut more than half way to the midvein, the entire blade lacerate .................. ***S. dissectum***

1 Sterile blade segments finely to coarsely toothed .............. 2

2 Ultimate segments of blade ± uniform in size; sterile blade segments finely toothed to ± entire; dissection of blade into segments extending to within 1 cm of apex at tips of blades .............. 3

2 Ultimate segments of blades variable in size, the apical segments much longer than the laterals; sterile blade segments coarsely and ± irregularly toothed or cut; dissection of blade into segments stopping at ca. 1–2.5 cm from apex at tips of blades ............ 4

3 Segments of sterile blade rounded at base; symmetrically tapered to an often ± blunt or even rounded apex; larger segments mostly 9–17 mm long; mar- gins nearly entire or finely and inconspicuously toothed ........................................ ***S. multifidum***

3 Segments of sterile blade usually (obliquely) asymmetrical and angular, cuneate to the apex; larger segments mostly 4–9 mm long; margins clearly finely dentate, especially visible in immature leaves ............................................... *S. rugulosum*

4 Overwintering leaves green, not bronze; larger (terminal) segments of vegetative blades narrowly to broadly ovate, obtuse to rounded at apex, ± symmetrical at base; margins toothed but never lacerate ............................................... *S. oneidense*

4 Overwintering leaves bronze-colored (or green if covered by leaves); larger (terminal) segments of sterile blades lance-shaped, acute, and strongly asymmetric at base; margins toothed to irregularly cut .. ............................................... *S. dissectum*

***Sceptridium dissectum*** (Spreng.) Lyon [•6] CUT-LEAF GRAPE FERN Spores mature Sept–Nov. Sterile hilltops, dry pastures, dry woodlands, and grassy banks. Blades are often bronze or turn reddish in late fall. *Botrychium dissectum* Spreng.

***Sceptridium multifidum*** (Gmel.) Nishida ex Tagawa LEATHERY GRAPE FERN Spores mature in Aug and Sept. Grassy hillsides, sterile fields, exposed meadows, and sandy open places. *Botrychium multifidum* (Gmel.) Trev.

***Sceptridium oneidense*** (Gilbert) Holub BLUNT-LOBE GRAPE FERN Spores mature in Sept–Oct. Rich moist woodland. The broad, rounded divisions and the shaded habitat are characteristic. *Botrychium oneidense* (Gilbert) House.

***Sceptridium rugulosum*** (W.H. Wagner) Skoda & Holub TERNATE GRAPE FERN Spores mature Aug–Oct. Swampy woods, brushy fields, and wooded streambanks. *Botrychium rugulosum* W.H. Wagner.

## OSMUNDACEAE *Royal Fern Family*

### OSMUNDA *Royal Fern*

Perennial ferns with large rootstocks and exposed crowns covered with old roots and stalks, sending up tufts of coarse leaves. Leaves 1–2-pinnate, differentiated into sterile and fertile segments. Sporangia in round clusters. The fibrous roots (osmunda fibre) were formerly used as a medium for growing orchids and bromeliads.

*Key to species*

1 Leaves 2-pinnate, pinnae ± entire; sporangia on upper half of fertile leaves ............................................... *O. regalis*

1 Leaves 1-pinnate, sterile pinnae deeply cleft; sporangia only near middle of fertile leaves, or fertile and sterile leaves separate ..... 2

2 Fertile and sterile leaves separate, fertile leaves cinnamon-colored, sterile leaves with a tuft of wool in axil of pinnae .. *O. cinnamomea*

2 Fertile pinnae near middle of vegetative leaves, with sterile pinnae above and below fertile portion, fertile portion green-black, pinnae mostly without tuft of wool in axil ................. *O. claytoniana*

***Osmunda cinnamomea*** L. [•7] CINNAMON-FERN Swamps, bog-margins, wooded stream-banks, and low wet places; soils acid. *Osmundastrum cinnamomeum* (L.) K. Presl.

***Osmunda claytoniana*** L. [•7] INTERRUPTED FERN Moist or seasonally wet depressions in forests, hummocks in swamps, low prairie, wet roadsides; often in drier places than *O. cinnamomea* or *O. regalis.*

***Osmunda regalis*** L. [•7] ROYAL FERN Bogs, swamps, alder thickets and shallow pools; soils usually acidic.

## POLYPODIACEAE *Polypody Fern Family*

### POLYPODIUM *Polypody*

***Polypodium virginianum*** L. ROCK POLYPODY Evergreen, colony-forming fern. In shallow humus on rocks, in crevices, on woodland banks, and rarely on mossy stumps and in crotches of trees. Easily identified by the small evergreen blades and its colony-forming habit on rocky slopes, talus, boulders, and ledges. *Polypodium vulgare* auct. non L. p.p.

## PTERIDACEAE *Maidenhair Fern Family*

Delicate to coarse ferns, deciduous, or evergreen. Blades pinnate to decompound. Sori marginal, protected by the indusium, which opens toward the margin, or by the reflexed margins of the pinnae, or borne along the veins and lacking an indusium.

*Key to genera*

1 Blade segments separate from one another ........ ADIANTUM
1 Blade segments not separate and distinct ..................... 2

2 Leaves of two types, the fertile much longer than the sterile; petioles dark brown near base, green above ......... CRYPTOGRAMMA
2 Fertile and sterile leaves mostly similar; petioles dark brown to black ................................................ 3

3 Smallest blade segments more than 4 mm wide ...... PELLAEA
3 Smallest segments less than 4 mm wide ........ CHEILANTHES

### ADIANTUM *Maidenhair Fern*

***Adiantum pedatum*** L. [•8] NORTHERN MAIDENHAIR FERN Leaves in colonies arising from horizontal rhizomes. Petioles lustrous purple-brown, forking at the summit into two arching rachises, each of which is divided several times, thus forming a semicircular blade. Wooded, sometimes rocky slopes in humus-rich soil. The usually arching and palmately divided lustrous purple brown rachises and fan-shaped pinnules with the main vein along the lower margin are distinctive.

### CHEILANTHES *Lip Fern*

Small evergreen ferns of dry rocky places. Rhizomes with numerous slender, brown to blackish, hyaline scarious-margined scales. Sori marginal, covered

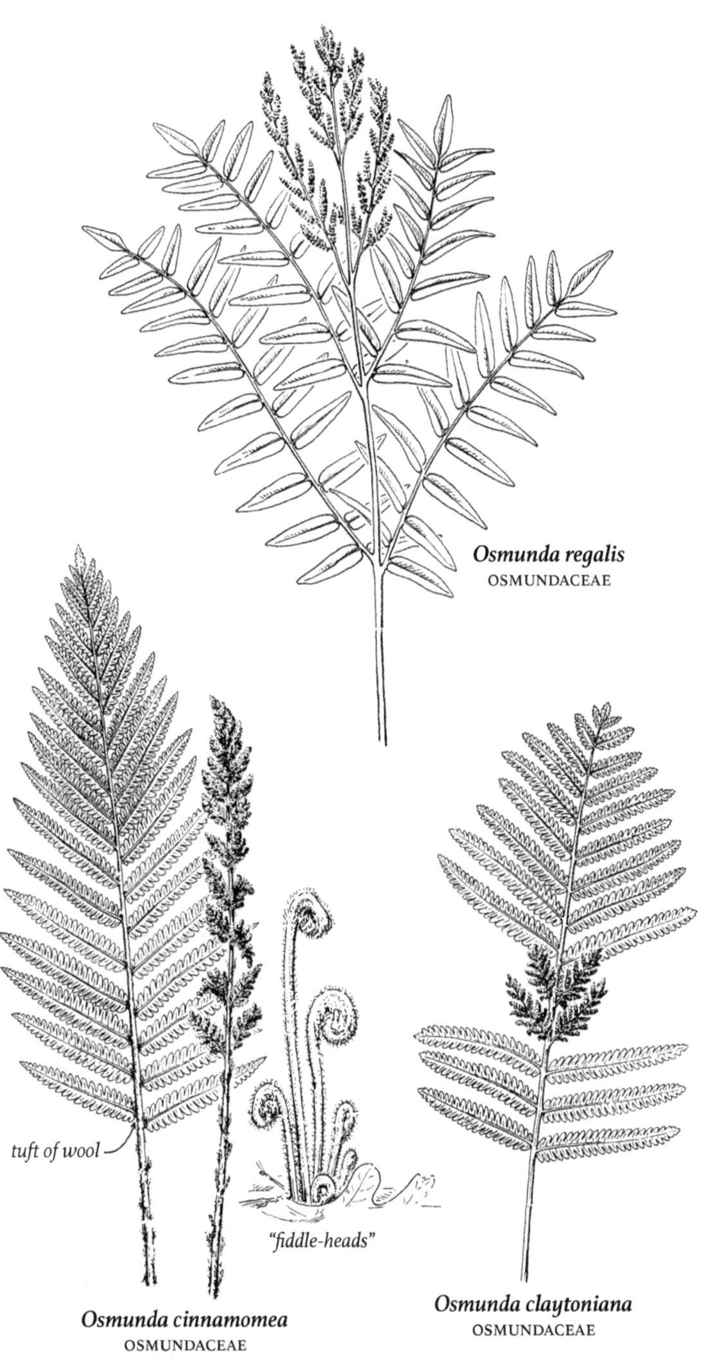
Osmunda regalis
OSMUNDACEAE
tuft of wool
"fiddle-heads"
Osmunda cinnamomea
OSMUNDACEAE
Osmunda claytoniana
OSMUNDACEAE

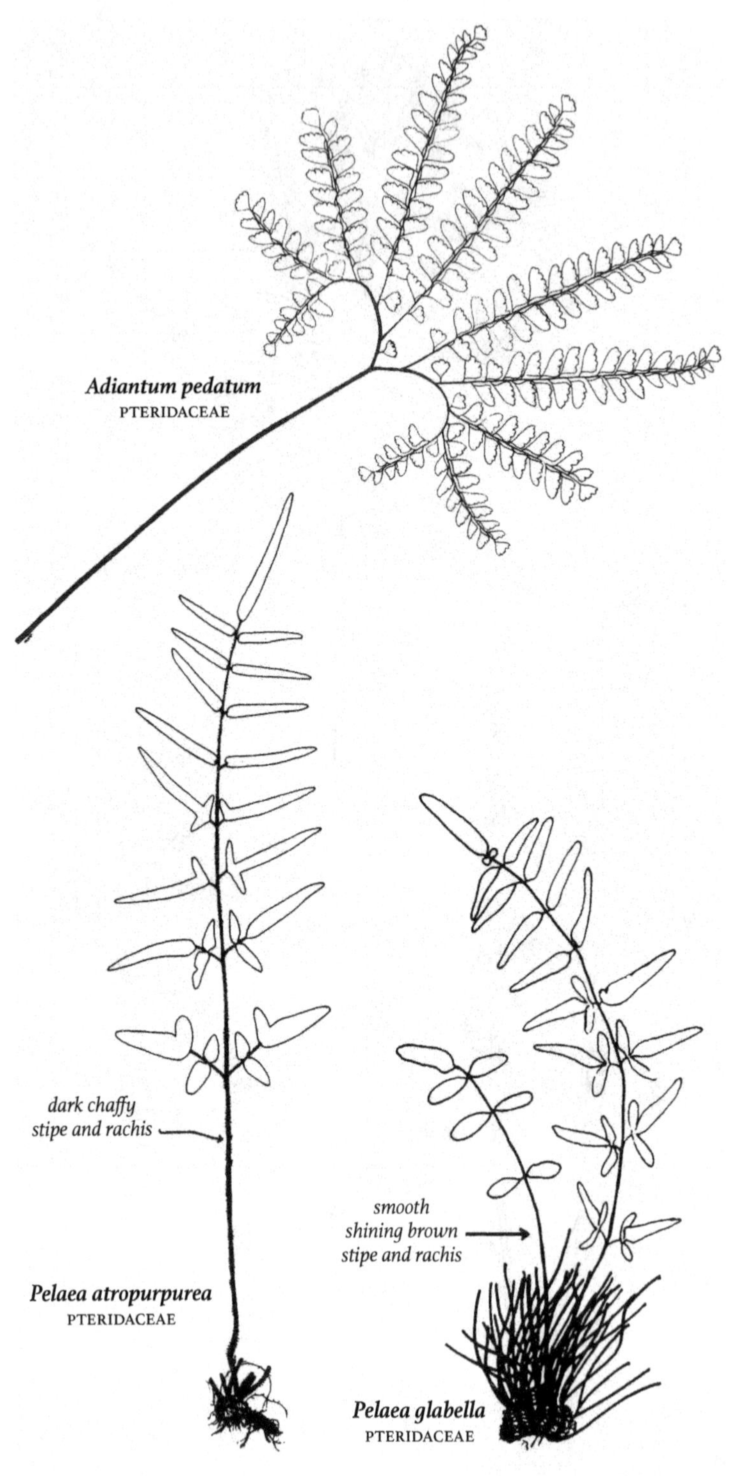
***Adiantum pedatum***
PTERIDACEAE
*dark chaffy*
*stipe and rachis*
*smooth*
*shining brown*
*stipe and rachis*
***Pelaea atropurpurea***
PTERIDACEAE
***Pelaea glabella***
PTERIDACEAE

by the inrolled margin of the pinnule. Distinguished from other ferns by the small bead-like segments of the pinnae.

*Key to species*

1 Smallest segments of fertile blades small, round, and beadlike, 1–3 mm wide .......... *C. feei*

1 Smallest segments of fertile blades elongate, not beadlike, mostly 3–5 mm wide .......... *C. lanosa*

***Cheilanthes feei*** T. Moore SLENDER LIP FERN Crevices of limestone or calcareous cliffs.

***Cheilanthes lanosa*** (Michx.) D.C. Eat. HAIRY LIP FERN Cliffs and shale outcrops, mostly in subacid soil.

## CRYPTOGRAMMA *Rockbrake*

***Cryptogramma stelleri*** (Gmel.) Prantl FRAGILE ROCKBRAKE Small rock fern with dimorphic leaves, from branched rhizomes. Leaves glabrous, deciduous, dimorphic, scattered along the horizontal rhizome. Fertile leaves stiffer than sterile blades. Sori marginal, covered by a continuous indusium formed by the reflexed margin. Moist, shaded, usually calcareous crevices and cliffs. Plants may be easily overlooked as they turn brown and wither later in the season.

## PELLAEA *Cliffbrake*

Small tufted plants from compact rootstocks. Blades firm; petioles and rachises wiry; pinnae gray green, blending well with the limestone rock crevices and ledges with which they are associated. Sori marginal and confluent under the inrolled margin of the fertile pinnules.

*Key to species*

1 Sterile and fertile leaves different; petiole and rachis scurfy and with appressed hairs .......... *P. atropurpurea*

1 Sterile and fertile leaves similar; petiole and rachis glabrous or with a few spreading hairs .......... *P. glabella*

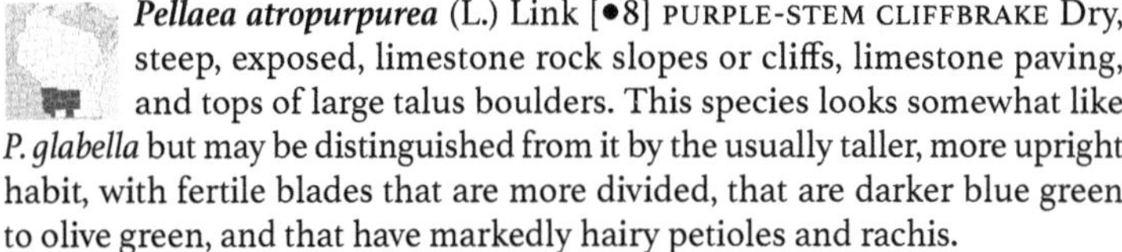

***Pellaea atropurpurea*** (L.) Link [•8] PURPLE-STEM CLIFFBRAKE Dry, steep, exposed, limestone rock slopes or cliffs, limestone paving, and tops of large talus boulders. This species looks somewhat like *P. glabella* but may be distinguished from it by the usually taller, more upright habit, with fertile blades that are more divided, that are darker blue green to olive green, and that have markedly hairy petioles and rachis.

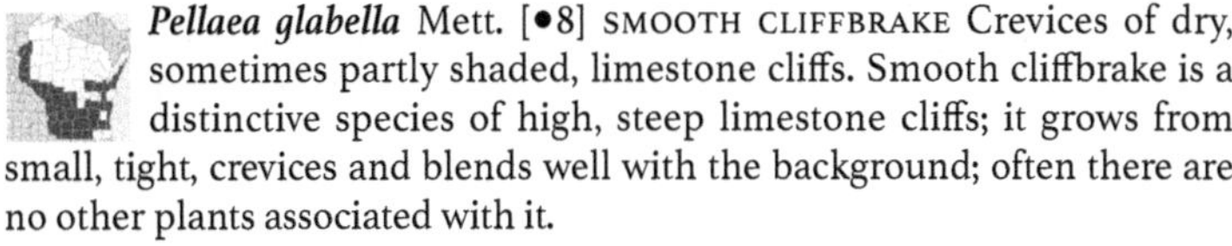

***Pellaea glabella*** Mett. [•8] SMOOTH CLIFFBRAKE Crevices of dry, sometimes partly shaded, limestone cliffs. Smooth cliffbrake is a distinctive species of high, steep limestone cliffs; it grows from small, tight, crevices and blends well with the background; often there are no other plants associated with it.

## SALVINIACEAE *Water Fern Family*

### AZOLLA *Mosquito-Fern*

***Azolla cristata*** Kaulfuss CRESTED MOSQUITO-FERN Small annual aquatic fern; plants free-floating or forming floating mats several cm thick, sometimes stranded on mud; roots few and unbranched. Sporocarps usually in pairs on underwater lobes of some leaves. Local in quiet water of river backwaters and ponds. *Azolla caroliniana* Willd., *Azolla mexicana* Schltdl. & Cham. ex K. Presl, *Azolla microphylla* Kaulfuss.

## SELAGINELLACEAE *Selaginella Family*

### SELAGINELLA *Spikemoss*

Trailing, evergreen herbs with branched, leafy stems, rooting at branching points. Leaves small and overlapping. Spore-bearing leaves similar to vegetative leaves and clustered in cones at ends of branches. Megaspores 4 in each sporangium, yellow or white; microspores numerous and very small, red or yellow, covered with small spines.

*Key to species*

1 Leaves and stems firm and evergreen, the plants forming small tufts 2.5–5 cm high; leaves crowded, very narrow, tipped by a bristle; cones four-angled .................................. ***S. rupestris***

1 Leaves and stems lax and subevergreen or deciduous; plants forming small mats; cones nearly round in cross-section ........ **2**

2 Leaves in four rows, of two kinds: large and spreading, and small and appressed to stem; cones 0.5–1 cm long; ne and se Wisc ...... ................................................. ***S. eclipes***

2 Leaves in many rows, all alike, with hairs on margins; fertile branches upright, cones 2–4 cm long; rare in Door County ...... ............................................. ***S. selaginoides***

***Selaginella eclipes*** W. R. Buck. HIDDEN SPIKEMOSS Open fens, wet meadows, sandy or marly lakeshores and riverbanks; especially where calcium-rich. *Selaginella apoda* subsp. *eclipes* (W.R.Buck) Skoda.

***Selaginella rupestris*** (L.) Spring [•9] LEDGE SPIKE-MOSS Sand dunes; dry, often igneous rocky bluffs. *Lycopodium rupestre* L.

***Selaginella selaginoides*** (L.) Link northern spikemoss *Endangered.* Calcium-rich fens and fen-like shores of Door County; mossy hummocks in cedar swamps.

## THELYPTERIDACEAE *Marsh Fern Family*

Medium-sized deciduous ferns, spreading by rhizomes to form colonies; sterile and fertile fronds usually alike, 1-pinnate to pinnate-pinnatifid, with transparent needle-like hairs; sori usually on veins (but not marginal) on pinna underside; indusia present and often soon withering, or absent (*Phegopteris*).

*Key to genera*

1 Leaf blades broadly triangular in outline, broadest at base, lowermost pinnae directed downward; indusia absent .......... PHEGOPTERIS
1 Blades lance-shaped in outline, broadest above base; indusia present .......... THELYPTERIS

## PHEGOPTERIS *Beech-Fern*

Deciduous ferns with creeping rhizomes. Leaf blades triangular in outline, 1-pinnate-pinnatifid (the pinnae deeply lobed); rachis winged. Sori round to oblong, indusia absent.

*Key to species*

1 Rachis winged for much of length; underside of blade sparsely hairy .......... *P. hexagonoptera*
1 Rachis not winged; underside of blade usually densely hairy and scaly .......... *P. connectilis*

***Phegopteris connectilis*** (Michaux) Watt NORTHERN BEECH-FERN Cool moist woods, thickets, streambanks, sphagnum moss hummocks, shaded rock crevices. *Dryopteris phegopteris* (L.) C.Chr., *Thelypteris phegopteris* (L.) Sloss.

***Phegopteris hexagonoptera*** (Michx.) Fée [•9] BROAD BEECH-FERN Rich, often rocky, woods and wooded slopes. *P. hexagonoptera* has all the divisions of the blade connected to the rachis, including the basal pair. In shape, the blade is more broadly triangular in broad beech fern than in northern beech fern. The shape of the basal segments is unlike that in *P. connectilis,* being widest in the middle and lobed again rather than entire. *Thelypteris hexagonoptera* (Michx.) Weatherby.

## THELYPTERIS *Marsh-Fern*

Small to medium ferns from slender rhizomes. Leaf blades 1-pinnate to pinnate-pinnatifid, pinnae entire to deeply lobed. Sori round, obong, or elongate along veins; indusia round kidney-shaped.

*Key to species*

1 Veins of pinnae mostly forked; indusia fringed with hairs; widespread .......... *T. palustris*
1 Veins not forked; indusia with glands on margin; w Wisc .......... *T. simulata*

***Thelypteris palustris*** Schott [•9] MARSH-FERN Swamps, low areas in forests, sedge meadows, open bogs, calcareous fens, marshes.

***Thelypteris simulata*** (Davenp.) Nieuwl. MASSACHUSETTS FERN Hummocks in white pine-red maple swamps, often in sphagnum moss; soils acid. In Wisc, disjunct from main range of Nova Scotia s to Virginia and Alabama. Similar to *Thelypteris palustris* but lower pinnae in *T. simulata* are narrowed at base next to rachis (only slightly narrowed at base in marsh-fern); veins in both sterile and fertile leaves of *T. simulata* are unbranched (veins in sterile leaves of *T. palustris* are mostly forked). *Parathelypteris simulata* (Davenport) Holttum.

## WOODSIACEAE *Cliff Fern Family*

### WOODSIA *Cliff Fern*

Small tufted ferns arising from compact rootstocks. Indusium of thread-like or plate-like segments, more or less arched over the round sori. To distinguish *Woodsia* from *Cystopteris,* check if the indusium is attached below the sorus (*Woodsia*) or is hooded (attached at one side and arched over the sorus, *Cystopteris*); petioles are opaque in *Woodsia* and translucent in *Cystopteris* (visible when held up to the light in the field); the veins in *Woodsia* are less distinct and appear to stop short of the margin; in *Cystopteris,* the veins clearly extend to the margin; in *Woodsia,* old petiole bases persist as either an even or uneven stubble (see key).

HYBRIDS ***Woodsia × abbeae*** Butters is a hybrid between *W. ilvensis* and *W. oregana,* known from nw Wisc.

*Key to species*

1 Petioles jointed at base, the persistent bases (stubble) all about same length ........ ***W. ilvensis***

1 Petioles not jointed at base; the persistent bases of the petioles of differing lengths........ **2**

2 Indusia of narrow, thread-like segments; rachis smooth or with only fine glandular hairs; fern of limestone cliffs and ledges ***W. oregana***

2 Indusia of several wide segments; rachis glandular-hairy; fern of various types of loose rock ........ ***W. obtusa***

***Woodsia ilvensis*** (L.) R. Br. [•10] RUSTY CLIFF FERN Dry, often exposed, rocks and crevices of cliff faces and talus, the rock usually acidic. Plants are both scaly and glandular.

***Woodsia obtusa*** (Spreng.) Torr. BLUNT-LOBED WOODSIA Usually on talus, occasionally on shaded ledges and rocky slopes. *Woodsia obtusa* is an erect, rather robust species. In aspect it resembles *Cystopteris fragilis,* with which it often grows. *W. obtusa* is stiffer, with glands and scales on the axes and veins.

***Woodsia oregana*** D.C. Eat. OREGON WOODSIA Crevices of calcareous ledges and cliffs. Somewhat similar to *Woodsia ilvensis* but usually without scales and on calcareous rather than acid rock; the stubble is uneven rather than even as in *W. ilvensis.*

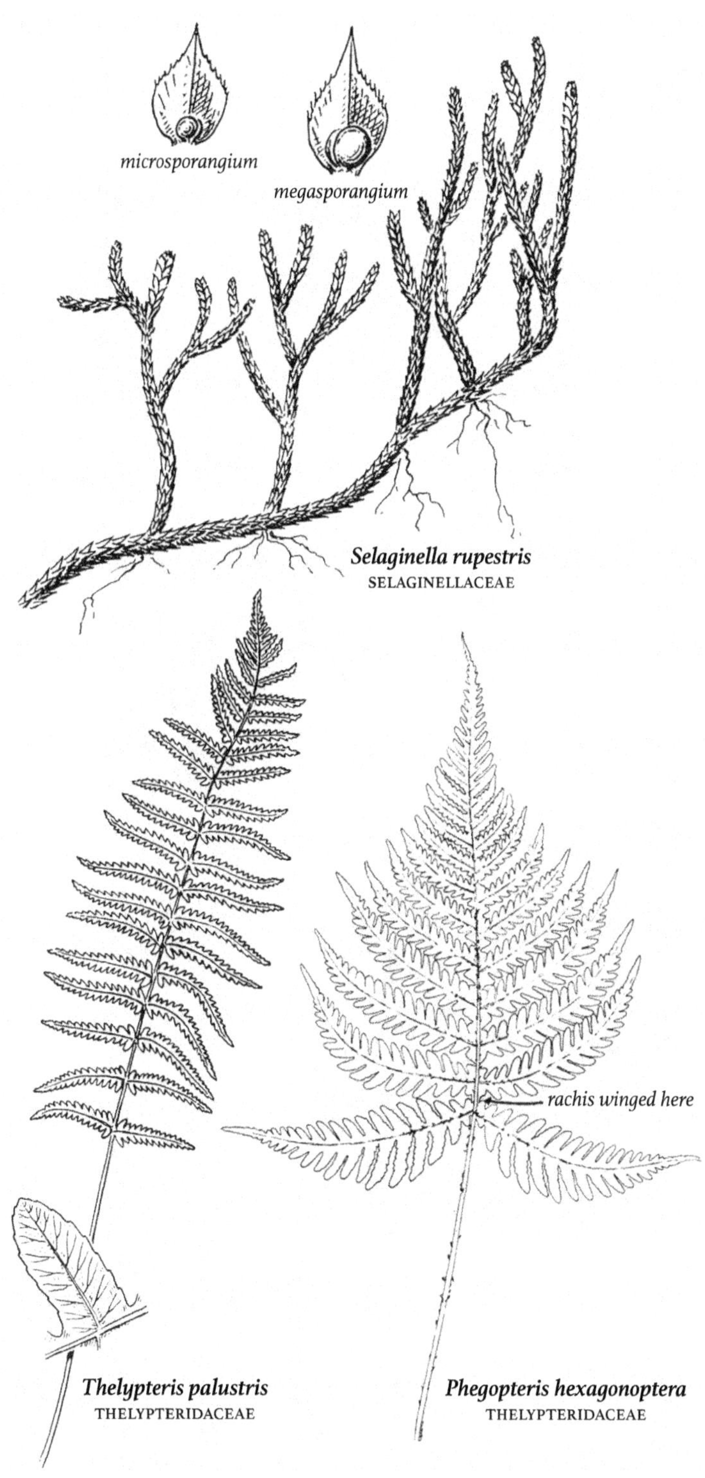
microsporangium
megasporangium
Selaginella rupestris
SELAGINELLACEAE
rachis winged here
Thelypteris palustris
THELYPTERIDACEAE
Phegopteris hexagonoptera
THELYPTERIDACEAE

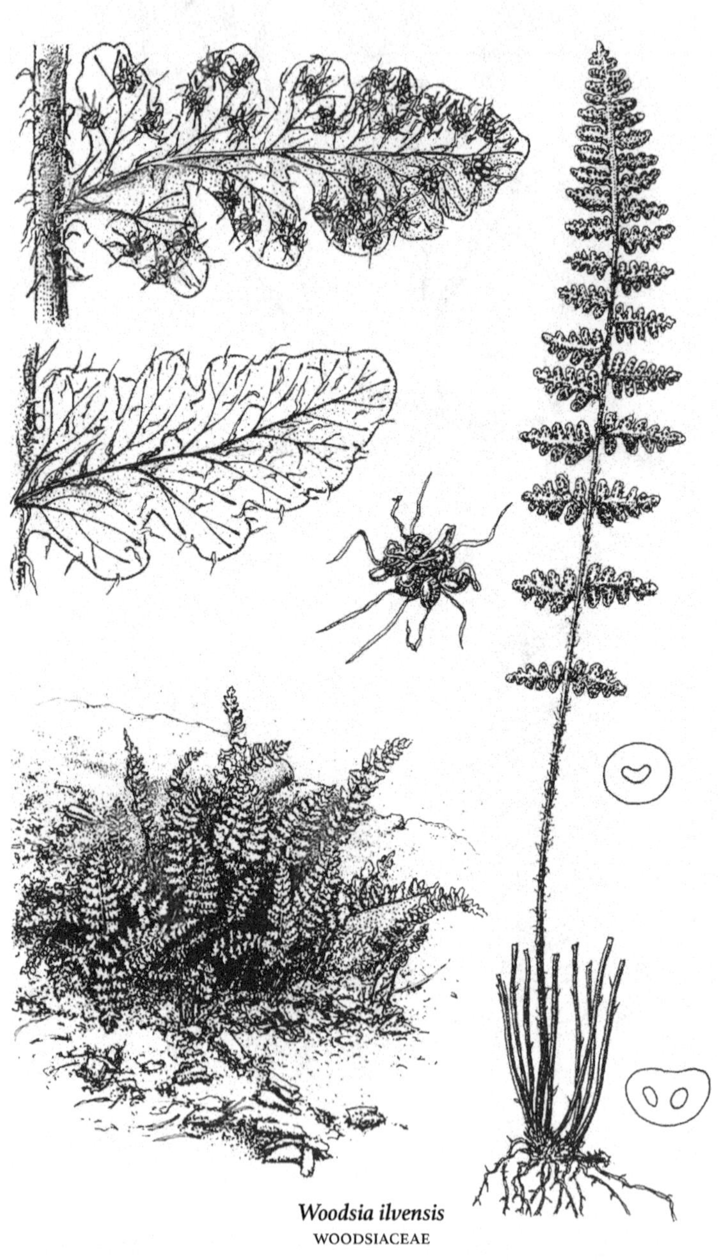

*Woodsia ilvensis*
WOODSIACEAE

## CUPRESSACEAE *Cypress Family*

Trees or shrubs; leaves opposite or whorled, sometimes dimorphic, separated by short internodes or overlapping. Flowers monoecious or dioecious, solitary, axillary or terminal. Fruit a cone, or becoming fleshy and berry-like.

*Key to genera*

1 Trees or shrubs; cones berry-like and fleshy ......... JUNIPERUS
1 Trees; cones woody or leathery ......................... THUJA

### JUNIPERUS *Juniper*

Trees or shrubs; leaves evergreen, scale-like or subulate, opposite or in whorls of 3. Cone-scales at maturity becoming fleshy and coalescent, forming a berry-like fruit with 1–10 seeds.

*Key to species*

1 Leaves in clusters of three, linear and sharp-pointed, jointed at base; cones in leaf axils .................................. *J. communis*
1 Leaves scale-like, not jointed; cones at end of branches ......... 2

2 Tree or upright shrub ............................. *J. virginiana*
2 Prostrate shrub .................................. *J. horizontalis*

***Juniperus communis*** L. COMMON JUNIPER Dry woods, old fields, dried bogs, rocky bluffs.

***Juniperus horizontalis*** Moench CREEPING JUNIPER On rocks, sandy openings, sandy or gravelly shores, sand dunes and beach ridges along Lake Michigan.

***Juniperus virginiana*** L. [•11] EASTERN RED CEDAR Old fields, open hillsides, dry woods, stabilized sand dunes.

### THUJA *Arbor-Vitae*

***Thuja occidentalis*** L. [•11] NORTHERN WHITE CEDAR; ARBOR-VITAE Cold, poorly drained swamps where *Thuja* may form dense stands; soils neutral or basic, usually highly organic, water not stagnant; also along streams, on gravelly and sandy shores of Great Lakes, and dry soils over limestone.

## PINACEAE *Pine Family*

Resinous trees with evergreen or deciduous, needlelike leaves. Male and female cones separate but borne on same tree. Male cones small and soft, falling after pollen is shed. Female cones larger, with woody scales arranged in a spiral. Seeds on upper surface of scales.

*Key to genera*

1 Leaves grouped into clusters.................................. 2
1 Leaves not in clusters, alternate on branches.................. 3

2 Leaves evergreen, in clusters of 2–5 needles .............. PINUS
2 Leaves deciduous, with many leaves in each cluster ...... LARIX

3 Cones upright; leaves attached directly to branch, not leaving a bump when shed ........................................ ABIES
3 Cones drooping; leaves attached to a persistent short stalk...... 4

4 Leaves flat in cross-section, soft ........................ TSUGA
4 Leaves four-sided in cross-section, stiff ................. PICEA

## ABIES *Fir*

***Abies balsamea*** (L.) Miller [•12] BALSAM FIR Shade-tolerant tree. Leaves evergreen, linear, blunt or with a small notch at tip, flat in cross-section, twisted at base. Cold boreal forests, swamps, and moist forests in n Wisc; in s, mostly restricted to fens.

## LARIX *Larch*

***Larix laricina*** (Duroi) K. Koch [•12] TAMARACK; EASTERN LARCH Shade-intolerant tree. Leaves deciduous, in clusters of 10–20, linear, blunt-tipped, bright green, turning yellow in fall. Seed cones ripening in fall and persisting on trees for 1 year. Cold, poorly drained swamps, bogs and wet lakeshores; s in Wisc, mostly confined to wet depressions.

## PICEA *Spruce*

Evergreen trees; resin blisters common in bark of white spruce (*Picea glauca*). Leaves linear, square in cross-section, stiff, spreading in all directions around twig, jointed at the base to a short projecting sterigma which persists on the leafless branches. Cones borne on last year's branches, drooping. Several species are commonly cultivated, especially Norway spruce (*Picea abies*).

*Key to species*

1 Leaves mostly 0.5–1.2 cm long, blunt-tipped; twigs with rust-colored hairs; cones ovoid, less than twice as long as wide when open ... ........................................ *P. mariana*
1 Leaves mostly 1–2 cm long, sharp-tipped; twigs glabrous; cones more than twice as long as wide ............................ 2

2 Cones 2.5–5 cm long; native species .................. *P. glauca*
2 Cones larger, 12–15 cm long; introduced species .......... *P. abies*

***Picea abies*** (L.) Karst. NORWAY SPRUCE Introduced from Europe, P. abies is the most widely cultivated spruce in North America, widely planted as a shade tree, sometimes used for reforestation, and now locally naturalized in woods.

***Picea glauca*** (Moench) Voss WHITE SPRUCE Moderately shade-tolerant tree. Leaves evergreen, linear, sharp-tipped. Moist to sometimes wet forests; absent from wetlands where water is stagnant.

***Picea mariana*** (Miller) BSP. BLACK SPRUCE Cold, acid, sphagnum bogs, swamps, and lakeshores; often where water is slow-moving and low in oxygen; less common in calcium-rich, well-aerated swamps dominated by northern white cedar (*Thuja occidentalis*). Black spruce can be distinguished from white spruce (*Picea glauca*) by its shorter needles, the branches with fine, white to red-brown hairs, the smaller, rounded seed cones with toothed scale margins, and its occurrence in generally wetter (and sometimes stagnant) habitats.

### PINUS *Pine*

Trees (ours) with leaves solitary or in clusters of 2–5. Staminate flowers catkin-like, in fascicles at the base of the current year's growth, each composed of numerous spirally imbricate stamens. Pistillate flowers forming a cone a hard woody cone, maturing at the end of the second or third season and often long persistent on the tree.

*Key to species*

1 Leaves in clusters of 5; cones 10–25 cm long ............ ***P. strobus***
1 Leaves in clusters of 2; cones less than 10 cm long .............. 2

2 Leaves 10–15 cm long ................................ ***P. resinosa***
2 Leaves 2.5–10 cm long ........................................ 3

3 Cones persistent on branches; bark dark gray ....... ***P. banksiana***
3 Cones not persistent on branches; bark of upper trunk becoming orange-brown ..................................... ***P. sylvestris***

***Pinus banksiana*** Lamb. JACK PINE Dry or sterile, sandy or rocky soil.

***Pinus resinosa*** Soland. RED PINE Dry sandy or rocky soil.

***Pinus strobus*** L. EASTERN WHITE PINE In many different habitats, but preferring fertile or well-drained, sandy soil.

***Pinus sylvestris*** L. SCOTCH PINE Native of Europe, where it is an important source of lumber; planted and persisting, and occasionally escaped in Wisc.

### TSUGA *Hemlock*

***Tsuga canadensis*** (L.) Carr. EASTERN HEMLOCK Tree; twigs pubescent. Leaves linear, blunt, marked beneath with two white strips of stomata. Cones at maturity pendulous on short peduncles. Moist soil, especially on rocky ridges and hillsides. Often cultivated in a number of horticultural forms.

## TAXACEAE *Yew Family*

### TAXUS *Yew*

***Taxus canadensis*** Marsh. CANADA YEW; GROUND HEMLOCK Straggling evergreen shrub. Leaves spirally arranged on the stem, linear, abruptly narrowed to a sharp point. Staminate flowers solitary in the axils; pistillate flowers in pairs. Fruit a fleshy red aril, open at the top. Coniferous and mixed woods. A favored winter browse for deer.

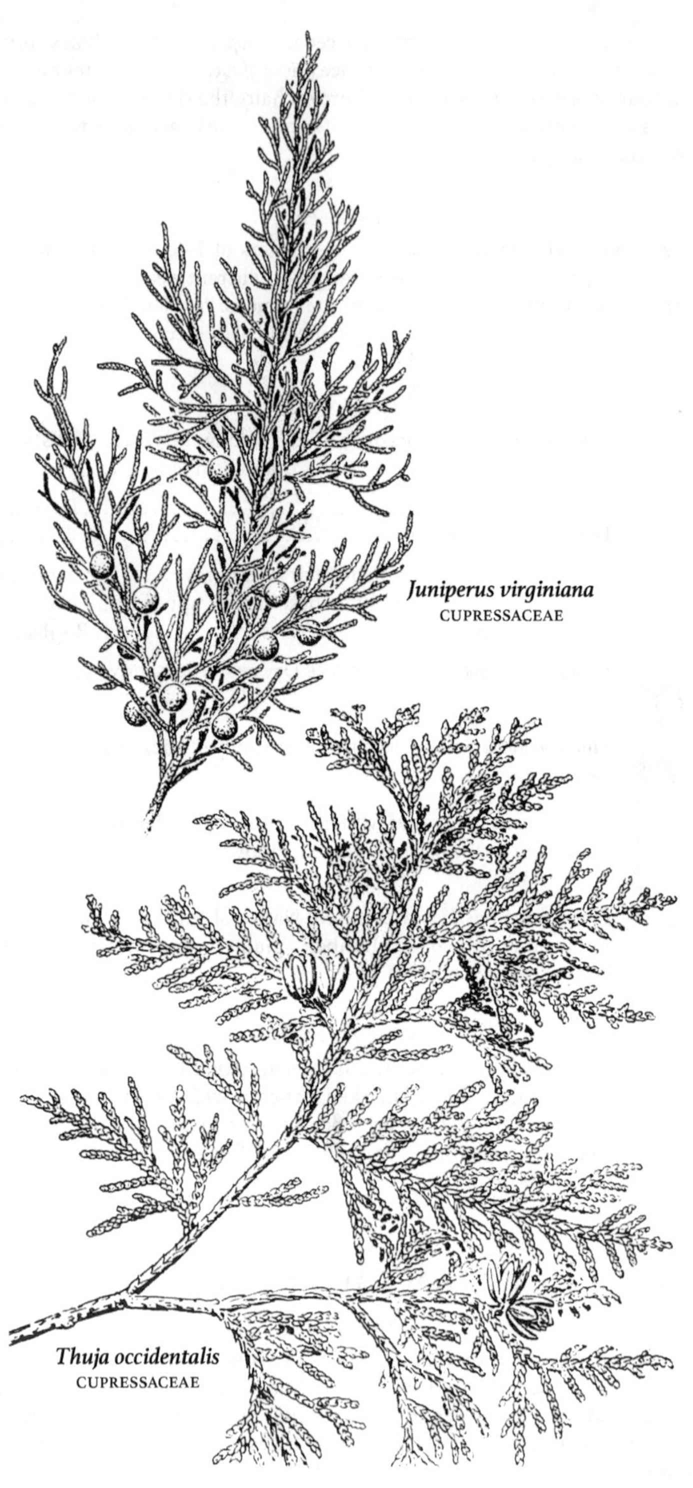
*Juniperus virginiana*
CUPRESSACEAE
*Thuja occidentalis*
CUPRESSACEAE

*Abies balsamea*
PINACEAE
*Larix laricina*
PINACEAE

# DICOTS

## ACANTHACEAE *Acanthus Family*

### RUELLIA *Wild Petunia*

***Ruellia humilis*** Nutt. FRINGE-LEAF WILD PETUNIA *Endangered.* Perennial herb. Leaves opposite. Flowers perfect, 5-merous; in sessile or subsessile, crowded, cymose clusters from the axils of several of the upper leaves. May–October. Prairies and dry upland woods. *Ruellia ciliosa* Pursh var. *longiflora* A. Gray.

## ADOXACEAE *Muskroot Family*

Our 2 shrubby genera, *Sambucus* and *Viburnum,* were previously included in Caprifoliaceae.

*Key to genera*

1 Plants small herbs .................................... **ADOXA**
1 Plants woody shrubs or small trees .......................... **2**

2 Leaves pinnately compound; fruit with 3 (or more) seed-like pits ........................................................ **SAMBUCUS**
2 Leaves simple; fruit with only 1 pit ................. **VIBURNUM**

### ADOXA *Moschatel*

***Adoxa moschatellina*** L. MUSKROOT *Threatened.* Delicate, musk-scented herb, from a scaly rhizome. Basal leaves long-petioled, 2–3 times divided; cauline leaves 1-pair, similar but smaller. Head with usually 5 greenish or yellowish flowers. Flowers perfect, dimorphic. Fruit a dry drupe, with 4 or 5 nutlets. May–July. Shaded, damp cliffs and slopes. An inconspicuous circumpolar species apart for the shiny quality of the pale green leaves.

### SAMBUCUS *Elder*

Shrubs or small trees. Stems pithy, the bark with wartlike lenticels. Leaves pinnately divided. Flowers in large, rounded terminal clusters; small, perfect. Fruit a red or dark purple, berrylike drupe with 3 nutlets.

*Key to species*

1 Flowers opening in summer after leaves developed, in broad, nearly flat clusters; fruit purple-black, edible; leaflets usually 7 ................................................................ ***S. canadensis***
1 Flowers opening in late spring with unfolding leaves, in pyramid-shaped or rounded clusters; fruit red, inedible; leaflets usually 5 ................................................................ ***S. racemosa***

***Sambucus canadensis*** L. [•13] COMMON ELDER Floodplain forests, swamps, wet forest depressions, thickets, shores, meadows, roadsides, fencerows. *Sambucus nigra* subsp. *canadensis* (L.) R. Bolli.

***Sambucus racemosa*** L. [•13] RED-BERRIED ELDER May–June (flowers opening with developing leaves). Occasional in swamps and thickets; more common in moist deciduous forests, roadsides and fencerows. *Sambucus pubens* Michx.

## VIBURNUM *Squashberry; Arrow-Wood*

Shrubs or small trees. Leaves simple, entire, toothed, often palmately lobed. Flowers white or pink, in rounded clusters at ends of stems, sometimes outer florets larger and sterile. Fruit a fleshy drupe with a single large seed; white, yellow, pink, or orange at first, maturing to orange, red, or blue-black.

ADDITIONAL SPECIES ***Viburnum recognitum*** Fern. (Smooth arrow-wood), native of e USA, considered adventive in Wisc; leaves coarsely toothed, with prominent veins on leaf underside.

*Key to species*

1 Leaves not lobed; pinnately veined.............................2
1 Leaves 3-lobed; palmately veined from base of leaf.............6

2 Leaves entire, wavy-margined or finely sharp-toothed; lateral veins not terminating in the teeth ..................................3
2 Leaves with large spreading teeth; lateral veins terminating in the teeth ......................................... ***V. rafinesquianum***

3 Leaf underside with branched hairs .................. ***V. lantana***
3 Leaf underside glabrous or scurfy, without branched hairs......4

4 Inflorescence stalked; leaf margins entire, wavy or with fine rounded teeth ...................................... ***V. nudum***
4 Inflorescence sessile or nearly so; leaf margins sharply toothed..5

5 Leaves tapered to sharp tips; branchlets slender and flexible ..... ........................................................ ***V. lentago***
5 Leaves obtuse, rounded or somewhat tapered to tip; branchlets stiff ................................................ ***V. prunifolium***

6 Outer flowers large and sterile, much larger than inner flowers .. ........................................................ ***V. opulus***
6 Flowers all similar ...........................................7

7 Leaf underside densely hairy with branched hairs, margins coarsely toothed; fruit blue-black ......................... ***V. acerifolium***
7 Leaf underside glabrous or hairy on veins only, margins with fine, sharp teeth; fruit yellow, orange or red .................. ***V. edule***

***Viburnum acerifolium*** L. [•14] MAPLE-LEAF ARROW-WOOD May–June. Moist or dry woods.

***Viburnum edule*** (Michx.) Raf. Squashberry *Endangered.* June–July; fruit ripening in late-summer. Moist, shaded talus slopes. *Viburnum pauciflorum* Bach. Pyl. ex Torr. & A. Gray

***Viburnum lantana*** L. WAYFARING-TREE June. Native of Eurasia, occasionally escaped.

***Viburnum lentago*** L. [•14] Nanny-Berry May–June. Woods, roadsides.

***Viburnum nudum*** L. Withe-Rod, Wild Raisin May–July. Cedar swamps, open bogs, fens, floodplain forests, wetland margins; occasional in drier woods. *Viburnum cassinoides* L.

***Viburnum opulus*** L. HIGH-BUSH CRANBERRY Native-introduced. June. Swamps, fens, streambanks, shores, ditches. Our native plants are var. *americanum:*

1 Larger petiolar glands less than 1 mm long (rarely absent), usually stalked and flat topped ........................ var. *americanum*

1 Larger petiolar glands (0.8–) 0.9–1.5 (–2) mm long, usually sessile and with the apex indented ......................... var. *opulus*

***Viburnum prunifolium*** L. SMOOTH BLACKHAW April–May. Woods, thickets, and roadsides, in moist or dry soil.

***Viburnum rafinesquianum*** J.A. Schultes [•14] DOWNY ARROW-WOOD May–June. Usually in dry sandy or rocky woods, less often in swamps and wet thickets.

## AMARANTHACEAE *Amaranth Family*

Our species annual or perennial herbs. Leaves simple, alternate, or occasionally opposite (*Salicornia*). Flowers small, often aggregated into large spikes, panicles, or heads, in some species with conspicuous colored bracts. Flowers perfect or unisexual; sepals usually 5; petals absent; ovary superior, 1-chambered. Fruit a 1-seeded utricle; seeds lenticular. Amaranthaceae now includes former members of the Chenopodiaceae.

*Key to genera*

1 Leaves opposite; either much reduced and scale like or with white silky-woolly hairs on both surfaces .......................... 2

1 Leaves alternate (or the lower sometimes opposite), well developed but without white silky-woolly pubescence ................... 3

2 Leaves linear-lanceolate; stem and leaves (both surfaces) with white silky-woolly hairs ............................. **FROELICHIA**

2 Leaves much reduced and scale-like, scarious, glabrous, connate; stem branches succulent, glabrous, appearing jointed, the flowers entirely sunk in the fleshy internodes ............ **SALICORNIA**

3 Leaf tips with a sharp spine over 0.5 mm (usually ca. 1 mm, even longer on bracts subtending flowers); leaves filiform, ± terete; fruit horizontal, 1–1.3 mm long, slightly broader, covered by the perianth; tepals with transverse keel or wing sometimes longer than body of tepal ............................................ **SALSOLA**

3 Leaf tips at most with mucro less than 0.5 mm long; leaves various in width, flat; fruit and perianth various ..................... 4

4 Flowers unisexual (plants monoecious or dioecious); tepals and bracts acute, scarious or fruit in most if not all flowers enveloped by a pair of bracteoles (perianth absent) ........................ 5

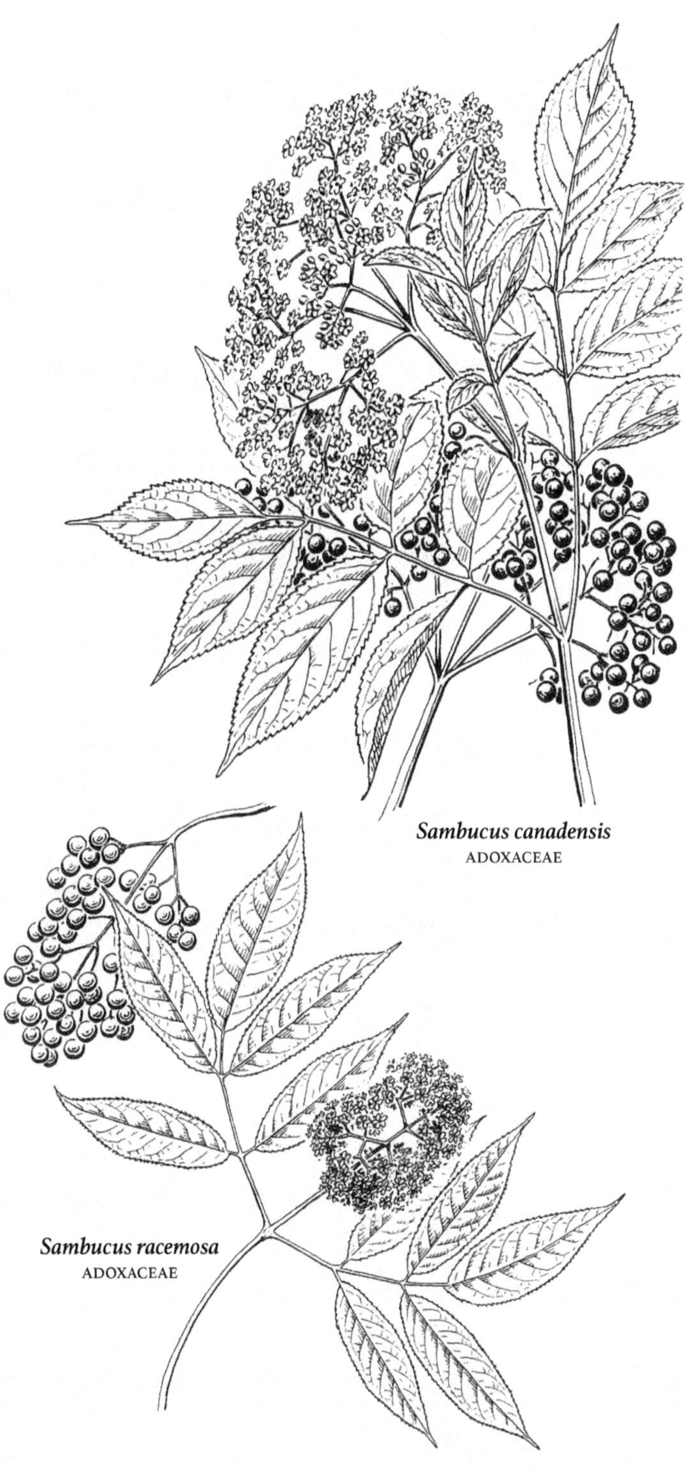

*Sambucus canadensis*
ADOXACEAE

*Sambucus racemosa*
ADOXACEAE

***Viburnum rafinesquianum***
ADOXACEAE

***Viburnum acerifolium***
ADOXACEAE

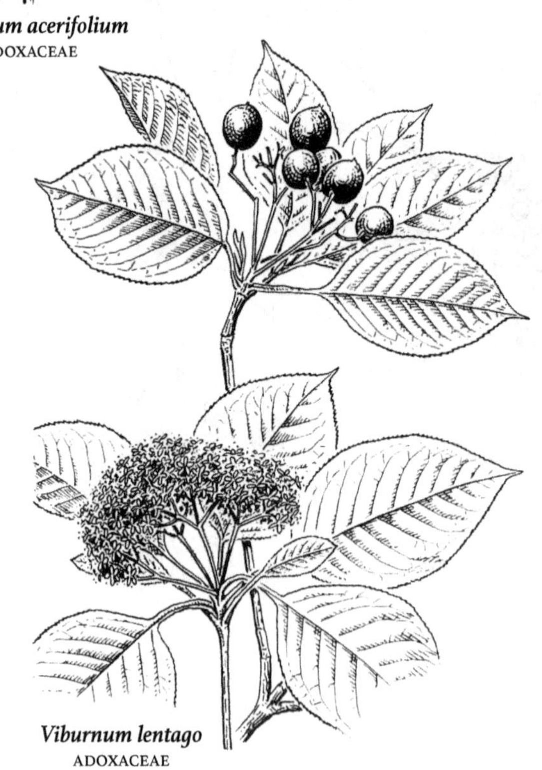

***Viburnum lentago***
ADOXACEAE

4 Flowers mostly bisexual; fruit not enveloped by bracts but perianth may cover it; bracts herbaceous or firm and hardened, not scarious ........................................................ 6

5 Bracts and tepals all acute, scarious ............ **AMARANTHUS**

5 Bracts beneath pistillate flowers broad and usually tuberculate and toothed with margins partly fused, obtuse to acute but herbaceous in texture (or even hardened in one species), tepals herbaceous .. ................................................... **ATRIPLEX**

6 Leaves linear to narrowly lanceolate, less than 4 (–6) mm broad, entire, 1 (–3)-nerved ........................................ 7

6 Leaves usually at least 4 mm broad, toothed to sinuate or crenulate on the margin (if entire, then pinnate- or 3-nerved and not linear) 10

7 Inflorescence and leaves beneath farinose; flowers crowded on short branches that exceed their subtending bracts .................. ................................ **CHENOPODIUM** (*C. pratericola*)

7 Inflorescence and leaves not farinose; flowers 1–3 in the axils of longer bracts................................................ 8

8 Leaves and bracts green to the tips, not mucronate; bracts long-ciliate, especially basally; fruit horizontal, round, less than 1 mm long, enclosed by the perianth (each tepal with a transverse wing) ........................................................ **BASSIA**

8 Leaves tipped with a non-green sharp mucro less than 0.5 mm long (no longer, or even absent, on bracts subtending flowers); bracts glabrous to pubescent but not long-ciliate; fruit various......... 9

9 Fruit vertical, flattened, usually narrowly wing-margined, 3–4.5 mm long, greatly exceeding the tiny scarious perianth; sandy (not saline) habitats ...................................... **CORISPERMUM**

9 Fruit horizontal (or mostly so), less than 1 mm long and 1.5 mm broad, enclosed by the perianth; saline habitats ........ **SUAEDA**

10 Fruit horizontal, completely encircled by the connate wing of the perianth; styles 3 ............................... **CYCLOLOMA**

10 Fruit horizontal or vertical, but the perianth without connate wing; styles usually 2 ............................................ 11

11 Tepals with transverse (but separate) wings; leaves entire, not over 5 mm wide; fruit horizontal; bracts long-ciliate, especially basally ........................................................ **BASSIA**

11 Tepals not transversely winged (may be keeled); leaves and fruit various; bracts not ciliate ....................................... 12

12 Leaves (and rest of plant) neither glandular nor pubescent, but farinose in some species ................. **CHENOPODIUM**

12 Leaves with yellow to orange resinous glands or gland-tipped hairs at least beneath, not farinose; bruised plant strongly aromatic ... ............................................... **DYSPHANIA**

## AMARANTHUS *Amaranth*

Annual herbs; stems usually much branched. Leaves alternate. Flowers in small clusters in the axils, or aggregated into axillary or terminal, simple or panicled spikes.

ADDITIONAL SPECIES In addition to the 6 most common species of *Amaranth* in the key below, the following introduced species are reported from one or several Wisc counties: ***Amaranthus arenicola*** I.M.Johnst., ***A. caudatus*** L., ***A. cruentus*** L., ***A. hypochondriacus*** L., and ***A. spinosus*** L.

*Key to species*

1 Plants dioecious (with staminate and pistillate flowers on separate plants) ........................................ ***A. tuberculatus***
1 Plants monoecious (with staminate and pistillate flowers separate but on same plants); the flowers intermixed or in separate inflorescences ........................................ 2

2 Flowers all or nearly all in small clusters from the leaf axils (a small terminal panicle may also be present) ........................ 3
2 Flowers mainly in elongate, spike-like, terminal clusters (small axillary clusters may be present) ............................ 4

3 Plants bushy tumbleweeds ............................ ***A. albus***
3 Plants prostrate .................................. ***A. blitoides***

4 Sepals obtuse, upper portion curved outward ...... ***A. retroflexus***
4 Sepals acute, straight or nearly so ............................ 5

5 Inflorescence lax, with many short, crowded branches ***A. hybridus***
5 Inflorescence stiff, unbranched or with several long branches ... ................................................ ***A. powellii***

***Amaranthus albus*** L. [•15] TUMBLEWEED Disturbed areas such as roadsides and railways; also sandy lakeshores and streambanks.

***Amaranthus blitoides*** S. Wats. MAT AMARANTH Disturbed areas such as yards and along roads and railways. Native of the w states, common as a weed throughout Wisc.

***Amaranthus hybridus*** L. SMOOTH AMARANTH Introduced and weedy, especially in corn and soybean fields.

***Amaranthus powellii*** S. Wats. GREEN AMARANTH Introduced and weedy in cultivated fields and on roadsides.

***Amaranthus retroflexus*** L. [•15] RED-ROOT AMARANTH Introduced and weedy along roadsides and in fields and gardens, rarely along sandy lakeshores. *A. powellii* resembles *A. retroflexus* in general habit but is nearly glabrous, with sharply acute sepals.

***Amaranthus tuberculatus*** (Moq.) Sauer ROUGH-FRUIT AMARANTH July–Sept. Exposed sandy or muddy shores, streambanks, wet meadows and ditches. *Acnida altissima* (Riddell) Moq. ex Standl.

## ATRIPLEX *Spearscale; Orache*

Annual or perennial herbs or shrubs, usually mealy or with bran-like scales; flowers minute, sessile or short-pediceled in glomerules at the nodes, in the upper axils, or in terminal spikes.

*Key to species*

1 Leaves and branches alternate ........................ ***A. rosea***

1 At least the lowest leaves and branches opposite . . . . . . . . . . . . . . . 2

2 Fruiting bracteoles nearly round, longer than 1 cm long when mature, net-veined . . . . . . . . . . . . . . . . . . . . . . . . . . . . . . . . *A. hortensis*

2 Fruiting bracteoles triangular, less than 1 cm long when mature, not strongly net-veined . . . . . . . . . . . . . . . . . . . . . . . . . . . . . . . . . . . . . . . . . . 3

3 Plants weedy; bracteoles not spongy-thickened near base . *A. patula*

3 Plants of saline habitats, not especially weedy; bracteoles spongy-thickened near base . . . . . . . . . . . . . . . . . . . . . . . . . . . . . . . *A. prostrata*

***Atriplex hortensis*** L. GARDEN ORACHE Native of Asia; cultivated as a potherb and an uncommon escape.

***Atriplex patula*** L. [•15] HALBERD-LEAF ORACHE Introduced (naturalized). Aug–Sept. Shores, streambanks and mud flats, usually where brackish; disturbed places.

***Atriplex prostrata*** Bouchér HASTATE ORACHE Introduced. Saline soil. Atriplex hastata L.

***Atriplex rosea*** L. TUMBLING ORACHE Native of the Old World; occasional in waste places.

## BASSIA *Smotherweed*

***Bassia scoparia*** (L.) A.J. Scott MEXICAN-FIREWEED Late summer. Native of Asia; occasionally escaped from cultivation, especially along railroads and highways where salt applied in winter; plants turn bright red in fall. *Kochia scoparia* (L.) Schrad.

ADDITIONAL SPECIES ***Bassia hyssopifolia*** (Pallas) Kuntz, an annual Eurasian introduction, recently found in a ditch along a railroad in Grant County.

## CHENOPODIUM *Goosefoot*

Taprooted annual herbs. Leaves alternate, mostly lance-shaped to broadly triangular, somewhat fleshy and often mealy on lower surface. Flowers perfect, small and numerous, green or red-tinged, in dense spike-like clusters from leaf axils or at ends of stems.

ADDITIONAL SPECIES ***Chenopodium foliosum*** (Moench) Aschers., uncommon introduction, Dane County. ***Chenopodium strictum*** Roth, native to n Great Plains, considered adventive in Wisc.

*Key to species*

1 Seeds erect; sepals mostly 3 . . . . . . . . . . . . . . . . . . . . . . . . . . . . . . . . . 2

1 Seeds horizontal; sepals 5 . . . . . . . . . . . . . . . . . . . . . . . . . . . . . . . . . . 4

2 Leaves white-mealy on underside, dull green above . . . *C. glaucum*

2 Leaves not white-mealy when mature, green on upper and lower sides . . . . . . . . . . . . . . . . . . . . . . . . . . . . . . . . . . . . . . . . . . . . . . . . . 3

3 Flower clusters few, 5–15 mm wide when mature, becoming fleshy . . . . . . . . . . . . . . . . . . . . . . . . . . . . . . . . . . . . . . . . . . . . . . *C. capitatum*

3 Flower clusters numerous, up to 5 mm wide, not fleshy *C. rubrum*

4 Mature sepals rounded to conform with fruit, the midvein not much raised . . . . . . . . . . . . . . . . . . . . . . . . . . . . . . . . . . . . . . . . . . . . . . . . 5

4 Mature sepals raised, folded, or hood-like, the calyx appearing somewhat star-shaped .......................................... 8

5 Leaves entire .................................. *C. polyspermum*
5 Leaves coarsely toothed........................................ 6

6 Fruit rounded on its margins ......................... *C. urbicum*
6 Fruit sharply angled on margins ............................ 7

7 Seeds shiny, 1.5–2.5 mm wide .......................... *C. simplex*
7 Seeds dull, to 1.5 mm wide ........................... *C. murale*

8 Leaves narrow, linear to lance-shaped, mostly entire *C. pratericola*
8 Leaves wider, lance-shaped to ovate or triangular .............. 9

9 Seed loosely enclosed by dry, brittle pericarp .... *C. standleyanum*
9 Seed tightly enclosed by thin, membranous pericarp .......... 10

10 Pericarp smooth (check with a hand lens) .............. *C. album*
10 Pericarp roughened (check with hand lens) ........ *C. berlandieri*

***Chenopodium album*** L. [•16] LAMB'S QUARTERS, PIGWEED Highly variable. Fields, gardens, roadsides, waste ground, dry woods, and barrens.

***Chenopodium berlandieri*** Moq. PITSEED GOOSEFOOT Much like *C. album,* but the pericarp evidently roughened and cellular-reticulate when viewed at 10–20x; a minute (0.1 mm) undivided style-base persistent on the fruit. Native but weedy.

***Chenopodium capitatum*** (L.) Ambrosi STRAWBERRY-BLITE Woodland clearings, often following a fire, roadsides, and waste places. *Blitum capitatum* L.

***Chenopodium glaucum*** L. [•16] OAK-LEAF GOOSEFOOT Aug–Oct. Shores, streambanks, and disturbed areas such as railroad ballast and barnyards, soils often brackish. Introduced from Eurasia.

***Chenopodium murale*** L. NETTLE-LEAF GOOSEFOOT Native of Europe; waste places.

***Chenopodium polyspermum*** L. MANY-SEED GOOSEFOOT Native of Eurasia; occasionally naturalized in waste ground.

***Chenopodium pratericola*** Rydb. DESERT GOOSEFOOT Occasional on lakeshores, prairies, barrens, and waste ground. Native of w USA, considered adventive in Wisc.

***Chenopodium rubrum*** L. RED GOOSEFOOT Aug–Oct. Lakeshores, streambanks, disturbed areas; probably introduced in our area from w USA.

***Chenopodium simplex*** (Torr.) Raf. MAPLE-LEAF GOOSEFOOT Disturbed ground and moist woods.

***Chenopodium standleyanum*** Aellen [•16] WOODLAND GOOSEFOOT Dry open woods.

***Chenopodium urbicum*** L. CITY GOOSEFOOT Native of Europe; waste places.

*Amaranthus retroflexus*
AMARANTHACEAE

*Amaranthus albus*
AMARANTHACEAE

*Atriplex patula*
AMARANTHACEAE

*Chenopodium album*
AMARANTHACEAE
*Chenopodium standleyanum*
AMARANTHACEAE
*Chenopodium glaucum*
AMARANTHACEAE

## CORISPERMUM *Bugseed*

***Corispermum americanum*** (Nutt.) Nutt. BUGSEED Sandy shores and soils, occasionally adventive along railways and waste places. *Corispermum orientale* Lam.

ADDITIONAL SPECIES ***Corispermum pallasii*** Steven, introduced, mostly along Lake Michigan shore.

## CYCLOLOMA *Winged-Pigweed*

***Cycloloma atriplicifolium*** (Spreng.) Coult. WINGED-PIGWEED Annual, branched herb. Dry or sandy ground, weedy.

## DYSPHANIA *Wormseed*

Previously included in genus *Chenopodium.*

*Key to species*

1 Leaf blades ± copiously covered with short spreading gland-tipped hairs; stem with abundant stalked glands; flowers in branched axillary cymes . . . . . . . . . . *D. botrys*

1 Leaf blades with mostly sessile glands on the underside; stem glabrous or pubescent, but not or only sparsely glandular; flowers in axillary and terminal spike-like inflorescences . . . . . . . . . . 2

2 Perianth with sessile glands; leaves with abundant spreading hairs beneath, especially on the veins . . . . . . . . . . *D. pumilio*

2 Perianth lacking glands; leaves with only sessile glands beneath (sometimes some appressed hairs on the main veins) . . . . . . . . . . *D. ambrosioides*

***Dysphania ambrosioides*** (L.) Mosyakin & Clemants [•17] MEXICAN TEA, WORMSEED Native of tropical America; naturalized in gardens, roadsides, and waste places. Highly variable. *Chenopodium ambrosioides* L.

***Dysphania botrys*** (L.) Mosyakin & Clemants [•17] JERUSALEM-OAK Native of Europe; a weed in waste places. *Chenopodium botrys* L.

***Dysphania pumilio*** (R. Br.) Mosyakin & Clemants CLAMMY GOOSEFOOT Native of Australia; becoming a weed in gardens and disturbed urban areas in s Great Lakes region. The larger leaves have 3–4 coarse teeth and the axillary inflorescences are bracteate with small toothed leaves. *Chenopodium pumilio* R. Br.

## FROELICHIA *Cottonweed*

Annual herbs (ours). Leaves narrow, opposite. Flowers perfect, each subtended by a scarious bract and 2 bractlets, in elongate, woolly, terminal spikes.

*Key to species*

1 Larger leaves mostly 1–2.5 cm wide; stems very slender, rarely more than 50 cm high . . . . . . . . . . *F. floridana*

1 Larger leaves mostly less than 1 cm wide; stems stout, usually more than 50 cm high . . . . . . . . . . *F. gracilis*

***Froelichia floridana*** (Nutt.) Moq. COMMON COTTONWEED Dry soil, especially where open and sandy.

***Froelichia gracilis*** (Hook.) Moq. SLENDER COTTONWEED Dry soil.

### SALSOLA *Russian-Thistle*

***Salsola tragus*** L. [•17] PRICKLY RUSSIAN-THISTLE Introduced (naturalized). Late summer. Along railroads and roads, on dunes and other sandy or cindery places. *Salsola kali* L. subsp. *tenuifolia* Moq.

ADDITIONAL SPECIES ***Salsola collina*** Pall., introduced, reported from Chippewa County; plants with long, narrow spikes of flowers with appressed bracts.

### SUAEDA *Sea-Blite*

***Suaeda calceoliformis*** (Hook.) Moq. PLAINS SEA-BLITE Introduced annual taprooted herb; leaves alternate, linear. July–Sept. Brackish wetlands and along salted highways.

---

## ANACARDIACEAE *Sumac Family*

Woody plants, juice often milky. Leaves alternate, chiefly compound. Flowers small, regular, perfect or unisexual, 5-merous. Fruit a 1-seeded, dry or fleshy drupe.

*Key to genera*

1 Flowers in dense inflorescences, these terminal or lateral on previous year's twigs; fruit red, glandular-hairy ........... RHUS
1 Flowers in loose clusters from leaf axils; fruit whitish, nearly smooth ........................................... TOXICODENDRON

### RHUS *Sumac*

Trees or shrubs. Leaves pinnately compound, of 3 to many leaflets. Flowers lateral or terminal, polygamo-dioecious. Petals 5, white or greenish. Fruit a drupe.

HYBRIDS ***Rhus × borealis*** Greene, *R. glabra* and *R. typhina,* statewide.

*Key to species*

1 Bushy shrubs with 3 sessile leaflets ................ *R. aromatica*
1 Sparsely branched shrubs or small trees; leaflets several to many 2

2 Rachis of the leaf winged ........................ *R. copallinum*
2 Rachis of the leaf not winged ................................ 3

3 Twigs and leaf petioles glabrous ..................... *R. glabra*
3 Twigs and petioles densely hairy .................... *R. typhina*

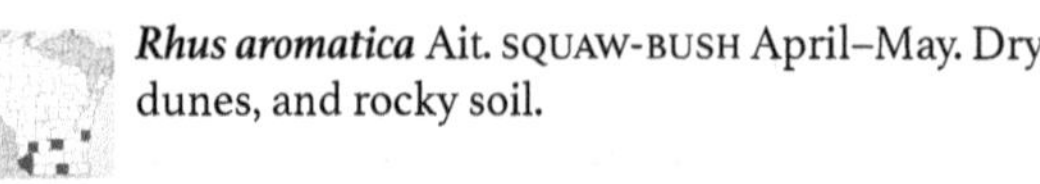

***Rhus aromatica*** Ait. SQUAW-BUSH April–May. Dry woods, hills, sand dunes, and rocky soil.

***Rhus copallinum*** L. SHINING SUMAC June–July. Dry soil.

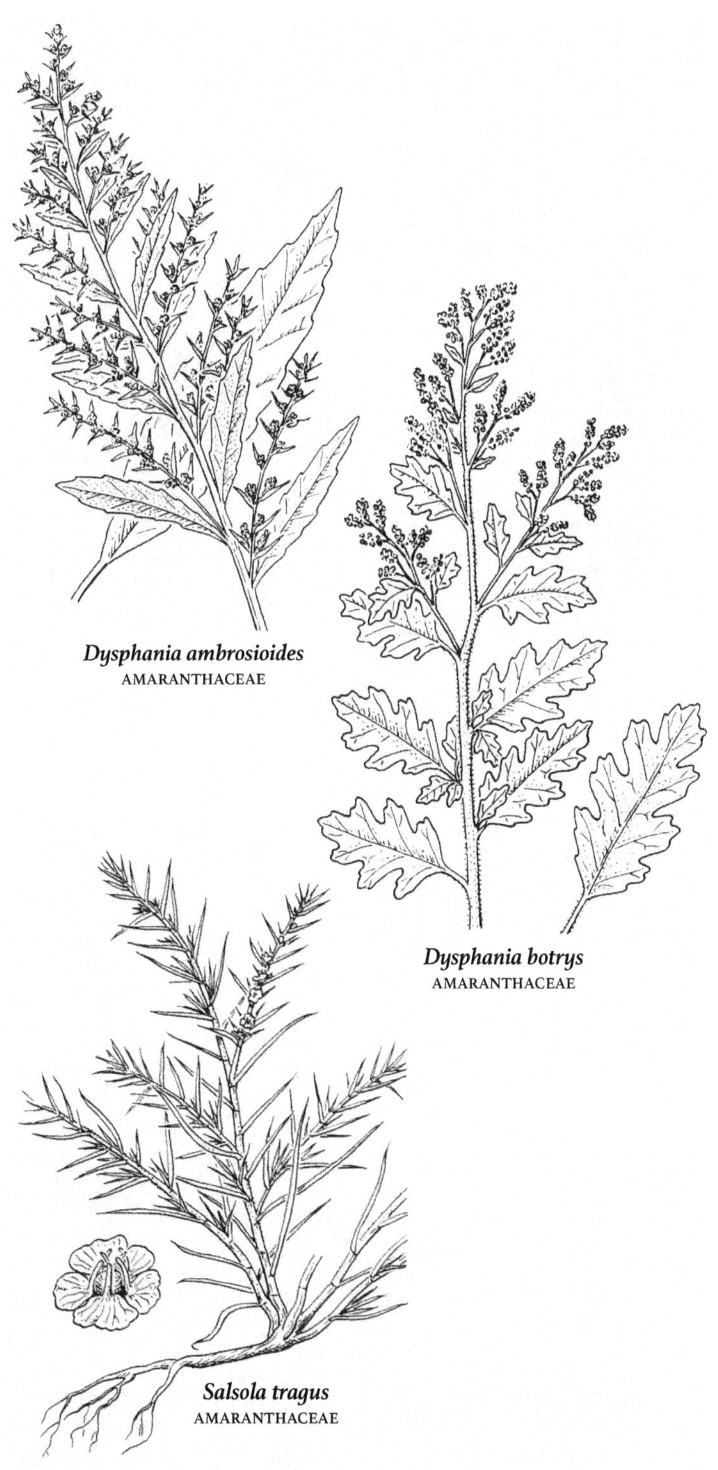

*Dysphania ambrosioides*
AMARANTHACEAE

*Dysphania botrys*
AMARANTHACEAE

*Salsola tragus*
AMARANTHACEAE

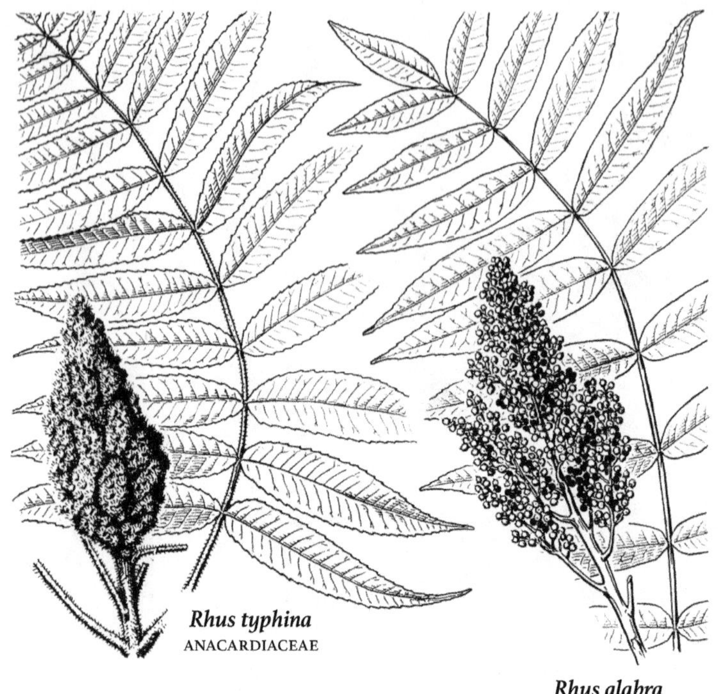

*Rhus typhina*
ANACARDIACEAE

*Rhus glabra*
ANACARDIACEAE

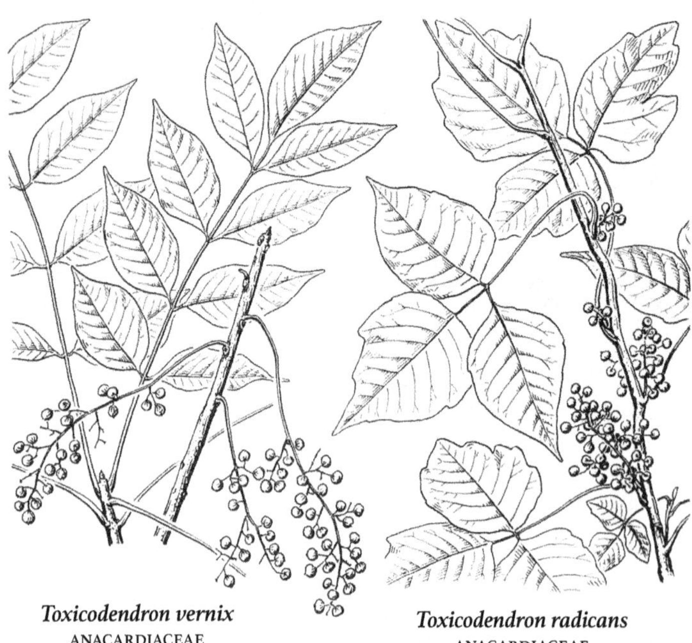

*Toxicodendron vernix*
ANACARDIACEAE

*Toxicodendron radicans*
ANACARDIACEAE

***Rhus glabra*** L. [•18] SMOOTH SUMAC June–July. Dry soil, old fields, roadsides, and margins of woods.

***Rhus typhina*** L. [•18] STAGHORN SUMAC June–July. Dry soil.

### TOXICODENDRON *Poison-Ivy*

Shrubs or vines, with axillary, rather loose inflorescences often drooping in fruit; otherwise much like *Rhus,* and sometimes included in that genus. Fruit a white or yellowish drupe, shining and glabrous or inconspicuously short-hairy. All parts of these plants may cause an allergic skin reaction.

*Key to species*

1 Leaflets 7–13, margins entire .......................... ***T. vernix***
1 Leaflets 3, margins entire, toothed or lobed .................... 2

2 Climbing or trailing vine, with aerial roots ........... ***T. radicans***
2 Somewhat erect shrubs without aerial roots, not climbing ....... ................................................ ***T. rydbergii***

***Toxicodendron radicans*** (L.) Kuntze [•18] COMMON POISON-IVY Moist to dry woods, thickets and open places. *Rhus radicans* L.

***Toxicodendron rydbergii*** (Small) Greene WESTERN POISON-IVY Dunes, shores, open sandy or rocky places. *Rhus radicans* L. var. *rydbergii* (Small ex Rydb.) Rehder.

***Toxicodendron vernix*** (L.) Kuntze [•18] POISON-SUMAC June–July. Tamarack swamps, thickets, floating bog mats and bog margins, often in partial shade. *Rhus vernix* L.

---

## APIACEAE *Carrot Family*

Biennial or perennial aromatic herbs with hollow stems, some very toxic. Leaves alternate and sometimes also from base of plant, mostly compound; petioles sheathing stems. Flowers small, perfect, regular, in flat-topped or rounded umbrella-like clusters (umbels); petals 5, white or greenish. Fruit 2-chambered, separating into 2, 1-seeded fruit when mature.

*Key to Apiaceae groups*

1 Inflorescence neither a true umbel nor a compound umbel ...... ................................................ GROUP A
1 Inflorescence a true umbel or a compound umbel.............. 2

2 Leaves all simple ................................. OXYPOLIS
2 Leaves, or many of them, compound, dissected, or deeply divided. 3

3 Ovary and fruit pubescent, tuberculate, bristly, or prickly ....... ................................................ GROUP B
3 Ovary and fruit glabrous........................................ 4

4 Leaves divided into distinct and separate leaflets of about uniform shape, these often more than 2 cm wide ............. GROUP C
4 Leaves much dissected or 2 or more times compound, the segments ovate, oblong, linear, or thread-like and less than 1 cm wide..... 5

5 Plants flowering ........................................ GROUP D
5 Plants fruiting ........................................ GROUP E

*Group A*

Inflorescence neither a true umbel nor a compound umbel.

1 Inflorescence of dense heads, each flower subtended by a small spine-tipped bract ................................ **ERYNGIUM**
1 Inflorescence otherwise; small bracts absent, or forming an a small involucre, not spine-tipped ................................ 2

2 Stem leaves deeply palmately divided .............. **SANICULA**
2 Stem leaves pinnately dissected ........................ **TORILIS**

*Group B*

Inflorescence a true umbel or compound umbel; most leaves compound, dissected, or deeply divided; fruit and ovary pubescent, covered with small bumps, or bristly or prickly.

1 Principal leaves palmately or once-pinnately compound or divided, the leaflets sometimes again divided .......................... 2
1 Principal leaves twice or more compound ...................... 4

2 Leaflets large, mostly 1 dm wide or more; fruit pubescent ........ ................................................ **HERACLEUM**
2 Leaflets less than 1 dm wide; fruit bristly or spiny .............. 3

3 Leaves palmately divided into 3–7 wide segments, the segments toothed or incised ................................ **SANICULA**
3 Leaves once-pinnate, some of the segments again divided **TORILIS**

4 Leaves with sharply toothed leaflets, the leaflets 1 cm wide or more ........................................................ 5
4 Leaves highly dissected into segments less than 1 cm wide ...... 6

5 Umbel branches 2–8; fruit not winged ............ **OSMORHIZA**
5 Umbel branches 18–35; fruit winged ................ **ANGELICA**

6 Branches of main umbel 20 or more; involucre of several conspicuous bracts ................................ **DAUCUS**
6 Umbel branches 1–3; involucral bracts absent or of only a few small bracts ................................ **CHAEROPHYLLUM**

*Group C*

Inflorescence a true umbel or a compound umbel; fruit and ovary glabrous; leaves divided into distinct leaflets of uniform shape, these often more than 2 cm wide.

1 Principal leaves once-compound (or sometimes simple in *Zizia* and *Thaspium*) ................................................ 2
1 Main leaves twice or three-times compound.................. 11

2 Upper leaf-sheaths expanded, 1 cm or more wide when flattened; flowers white; fruit flattened and wing-margined . **HERACLEUM**
2 Upper leaf sheaths not expanded, less than 1 cm wide; flowers and fruit various................................................ 3

3 Taprooted introduced weeds of waste places and disturbed areas 4

3 Native species with fibrous or tuberous-thickened roots, most common in woods or wetlands .............................. 5

4 Flowers yellow; fruit wing-margined ............... **PASTINACA**

4 Flowers white; fruit not winged .................. **PIMPINELLA**

5 Leaves with 3 leaflets (or the basal ones simple and toothed) .... 6

5 Leaves with 5 or more leaflets; flowers white .................. 9

6 Flowers white or greenish-white .............................. 7

6 Flowers yellow or cream-colored or purple .................... 8

7 Leaflets linear ................................. **OXYPOLIS**

7 Leaflets much wider ...................... **CRYPTOTAENIA**

8 Central flower of each umbellet sessile; fruit ribbed but not winged ........................................................ **ZIZIA**

8 Flowers all on pedicels; fruit winged .............. **THASPIUM**

9 Leaflets entire, or with 5 or fewer coarse teeth on each side; fruit winged ........................................................ **OXYPOLIS**

9 Leaflets with numerous teeth, or irregularly deeply toothed; fruit not winged.............................................. 10

10 Ribs of the fruit faint; leaflets of the upper leaves irregularly deeply toothed .......................................... **BERULA**

10 Ribs of the fruit prominent; leaflets of upper leaves usually regularly sharply toothed ..................................... **SIUM**

11 Leaflets entire; flowers yellow ...................... **TAENIDIA**

11 Leaflets toothed or lobed; flowers yellow or white ............ 12

12 Plants flowering ............................................ 13

12 Plants fruiting.............................................. 19

13 Flowers yellow or cream-colored............................. 14

13 Flowers white ............................................... 16

14 Central flower of each umbellet sessile .................. **ZIZIA**

14 Flowers all on pedicels ...................................... 15

15 Native perennial herb with fibrous or fleshy roots, not weedy; sepals present (check with a hand lens) ................. **THASPIUM**

15 Introduced biennial herb, taprooted weedy; sepals absent ....... ............................................... **PASTINACA**

16 Sheaths of the upper leaves expanded, 1 cm or more wide when flattened............................................ **ANGELICA**

16 Sheaths of the upper leaves less than 1 cm wide ............... 17

17 Rachis of the leaf conspicuously leafy-winged below each cluster of 3 leaflets ..................................... **FALCARIA**

17 Rachis of the leaf not leafy-winged ........................... 18

18 Main lateral veins of the leaflets oriented toward the sinuses between the teeth; some of the roots tuberous-thickened; base of stem thickened, hollow and cross-partitioned .......... **CICUTA**

18 Main lateral veins mostly oriented toward the teeth on the margin; stem and roots not modified as in *Cicuta* ........ **AEGOPODIUM**

19 Fruit evidently winged .................................. 20
19 Fruit not winged, or only slightly so .................................. 22

20 Dorsal and lateral ribs winged; fruit not much compressed .................................. **THASPIUM**
20 Dorsal ribs wingless, only the lateral ribs winged; fruit dorsally compressed .................................. 21

21 Dorsal ribs prominent; plants taprooted .................................. **ANGELICA**
21 Dorsal ribs very slender, only slightly raised above the fruit surface .................................. **PASTINACA**

22 Rachis of the leaf broadly winged below each group of 3 leaflets; fruit linear-oblong .................................. **FALCARIA**
22 Rachis of the leaf not winged; fruit various, often wider .................................. 23

23 Main lateral veins of the leaflets oriented toward the sinuses between the teeth; base of the stem thickened, hollow and cross-partitioned .................................. **CICUTA**
23 Main lateral veins mostly oriented toward the teeth on the margin, or more net-like and oriented toward neither the teeth or sinuses; base of stem not modified as in *Cicuta* .................................. 24

24 Stylopodium (disk-like swelling at base of style) absent; stem from fibrous or fleshy roots .................................. **ZIZIA**
24 Stylopodium well developed .................................. **AEGOPODIUM**

*Group D*

Inflorescence a true umbel or a compound umbel; fruit and ovary glabrous; leaves dissected or 2 or more times compound, the segments ovate, oblong, linear, or thread-like and less than 1 cm wide; plants flowering.

1 Flowers yellow .................................. 2
1 Flowers white (or rarely pink) .................................. 5

2 Introduced, often weedy plants; sepals absent .................................. 3
2 Native, non-weedy perennials; sepals evident (check with a hand lens) .................................. 4

3 Plants perennial .................................. **FOENICULUM**
3 Plants annual .................................. **ANETHUM**

4 Plants taprooted .................................. **POLYTAENIA**
4 Plants with a cluster of fibrous or fleshy roots .................................. **THASPIUM**

5 Stem purple-spotted; plants coarse, well-branched, biennial herbs to 3 m high .................................. **CONIUM**
5 Stem not purple-spotted .................................. 6

6 Plants annual or biennial .................................. 7
6 Plants perennial .................................. 10

7 Bractlets fringed with hairs .................................. **ANTHRISCUS**
7 Bractlets with entire margins, or bractlets absent .................................. 8

8 Involucel (secondary umbels which make up the larger umbel) scarcely developed .................................. **CARUM**
8 Involucel well developed .................................. 9

9 Bractlets arranged all around the umbellet . CHAEROPHYLLUM
9 Bractlets all on the outer side of the umbellet ........ AETHUSA

10 Plants small, less than 2 dm tall when in flower ...... ERIGENIA
10 Plants taller, well over 2 dm when in flower.................... 11

11 Plants with bulblets in axils of some of the upper leaves CICUTA
11 Plants not with bulblets in upper leaf axils.................... 12

12 Umbel branches few, mostly 3–10 in number; bractlets lance-ovate, fringed with hairs ............................ ANTHRISCUS
12 Umbel branches often more than 10; bractlets narrow and linear, not fringed with hairs ...................... CONIOSELINUM

*Group E*

Inflorescence a true umbel or a compound umbel; fruit and ovary glabrous; leaves dissected or 2 or more times compound, the leaf segments ovate, oblong, linear, or thread-like and less than 1 cm wide; plants fruiting.

1 Fruit dorsally flattened....................................... 2
1 Fruit nearly round in cross-section or somewhat compressed laterally ................................................ 5

2 Garden escape with thread-like leaf segments ...... ANETHUM
2 Native perennial herbs, not weedy; the leaf segments mostly wider ....................................................... 3

3 Plants taprooted; stylopodium absent; carpophore (prolonged part of receptacle, extending between the carpel segments) 2-parted to the base ................................... POLYTAENIA
3 Plants with a cluster of fibrous or fleshy-fibrous roots.......... 4

4 Carpophore and stylopodium (disk-like swelling at base of style) absent ........................................ THASPIUM
4 Carpophore present, 2-parted nearly to its base; stylopodium cone-shaped ................................. CONIOSELINUM

5 Stems purple-spotted; coarse, branched, biennial herb to 3 m high ..................................................... CONIUM
5 Stems not purple-spotted .................................. 6

6 Fruit lance-shaped or linear and with a beak 1–3 mm long, ribs absent; bractlets fringed with hairs .............. ANTHRISCUS
6 Fruit not beaked, the ribs evident; bractlets entire or absent .... 7

7 Plants perennial ............................................ 8
7 Plants annual or biennial ................................. 10

8 Fruit slightly wider than long; the main umbel usually only with 2–4 branches ........................................ ERIGENIA
8 Fruit as long as or longer than wide; umbel branches several or many ..................................................... 9

9 Plants with bulblets in axils of some of the upper leaves CICUTA
9 Plants not with bulblets in axils of the upper leaves.. THASPIUM

10 Smallest segments of the leaves linear or thread-like .... CARUM
10 Smallest segments of the leaves mostly wider ................ 11

II Fruit narrow, lance-shaped to elliptic, 5–10 mm long ........... ............................................ CHAEROPHYLLUM

II Fruit wider, broadly ovate to nearly round in outline, to 5 mm long ..................................................... AETHUSA

## AEGOPODIUM *Goutweed*

***Aegopodium podagraria*** L. BISHOP'S GOUTWEED Perennial herb from a creeping rhizome. Lower leaves mostly 1- or 2-times parted with 9 leaflets but often irregular; margins sharply serrate; upper leaves reduced. Flowers in dense umbels, these terminal and lateral; petals white. Native of Eurasia; cultivated in gardens and sometimes escaped, especially where moist and partially shaded.

## AETHUSA *Fool's-Parsley*

***Aethusa cynapium*** L. FOOL'S-PARSLEY Annual herb. Leaves shining, broadly triangular in outline, 2–3x pinnately dissected into narrow segments. Flowers in terminal and lateral compound umbels; petals white. June–Sept. Native of Eurasia; established as a weed of waste places; toxic.

## ANETHUM *Dill*

***Anethum graveolens*** L. DILL Strongly scented annual herb. Stems glabrous and more or less glaucous throughout. Leaves pinnately dissected into numerous filiform segments. Flowers in terminal and lateral compound umbels; petals yellow. July–Aug. Native of s Europe; cultivated commercially and in kitchen gardens and escaped into waste ground.

## ANGELICA *Angelica*

***Angelica atropurpurea*** L. [•19] PURPLE-STEM ANGELICA Perennial herb. Stems stout, often streaked with purple and green. Leaves alternate, lower leaves 3-parted; upper leaves smaller or reduced to bladeless sheaths. Flowers in rounded small clusters (umbelets), these grouped into large rounded umbels; petals white to green-white. May–July. Springs, seeps, calcareous fens, streambanks, shores, marshes, sedge meadows, wet depressions in forests; often where calcium-rich.

## ANTHRISCUS *Chervil*

***Anthriscus sylvestris*** (L.) Hoffmann WILD CHERVIL Annual or biennial herb. Leaves 2–3x compound; leaflets dentate to incised. Flowers in large compound umbels, terminal and from the upper axils; petals white. May–July. Native of Europe; rarely established in waste places.

## BERULA *Water-Parsnip*

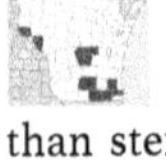

***Berula erecta*** (Huds.) Coville CUT-LEAF WATER-PARSNIP Perennial herb. Stems erect to trailing, often rooting along trailing portion. Leaves alternate, once-pinnate, basal leaves larger and less dissected than stem leaves. Flowers grouped into small clusters (umbelets), these grouped into umbels; flowers white. July–Sept. Shallow water, springs, spring-fed streams, marshes, swamps, often where calcium-rich. *Berula pusilla* (Nutt. ex Torr. & A. Gray) Fernald.

## CARUM *Caraway*

***Carum carvi*** L. CARAWAY Glabrous biennial herb, from a taproot. Leaflets pinnately dissected into linear segments. Flowers in terminal and lateral compound umbels; petals white or rarely pink. June–Aug. Native of Eurasia; sometimes cultivated and often weedy in waste places.

## CHAEROPHYLLUM *Chervil*

***Chaerophyllum procumbens*** (L.) Crantz SPREADING CHERVIL Annual or biennial herb. Stems spreading, often weak, usually branched from the base. Leaves 3-parted, pinnately compound. Flowers in lateral and terminal compound umbels; petals white. April–May. Moist woods and alluvial soil.

## CICUTA *Water-Hemlock*

Biennial or perennial toxic herbs. The tuberous roots, chambered stem base and young shoots of common water-hemlock (*C. maculata*) are especially toxic. Leaves alternate, 2–3-pinnate; leaflets narrow or lance-shaped, entire or toothed; leaf veins ending in the lobes (sinuses) and not at teeth as in other members of this family. Flowers white or green, in few to many umbels; umbels usually without bracts, umbellets bracted.

*Key to species*

1 Upper leaflet axils usually with bulblets; leaflets to 5 mm wide ........ ***C. bulbifera***

1 Bulblets absent; leaflets usually much more than 5 mm wide ........ ***C. maculata***

***Cicuta bulbifera*** L. [•19] BULBLET-BEARING WATER-HEMLOCK Aug–Sept. Streambanks, lake and pond shores, marshes, swamps, open bogs, thickets, springs, ditches.

***Cicuta maculata*** L. [•19] COMMON WATER-HEMLOCK June–Sept. Wet meadows, marshes, swamps, moist to wet forests, thickets, shores, streambanks, springs. Considered the most toxic plant in North America.

## CONIOSELINUM *Hemlock-Parsley*

***Conioselinum chinense*** (L.) B.S.P. CHINESE HEMLOCK-PARSLEY *Endangered.* Perennial herb. Leaves alternate, triangular in outline, 1–3x pinnate. Flowers white, in long-stalked umbels. Aug–Sept. Tamarack swamps, floodplain forests, streambanks, seeps, forested fens. Historical records from se Wisc, now likely no longer present in state.

## CONIUM *Poison-Hemlock*

***Conium maculatum*** L. Poison-Hemlock Introduced (invasive). Biennial herb with a strong, unpleasant odor. Stems stout, branched, purple-spotted. Leaves alternate, 2–4 dm long, 3–4x pinnately divided, the leaflets toothed or sharply lobed. Flowers white, in many umbelets, these grouped in umbels. June–July. Weed of shores, streambanks, waste ground and roadsides, especially on moist, fertile soil. Very toxic, fatal if eaten.

### CRYPTOTAENIA *Honewort*

***Cryptotaenia canadensis*** (L.) DC. CANADIAN HONEWORT Perennial glabrous herb. Leaves 3-foliate, lower leaves long-petioled, the upper on short petioles dilated as far as the leaflets. Flowers in numerous loose compound umbels; corolla white. June–July. Moist rich woods, swamps.

### DAUCUS *Carrot*

***Daucus carota*** L. QUEEN ANNE'S-LACE, WILD CARROT Biennial herb, with a stout taproot. Stems commonly rough-hairy. Leaves pinnately compound. Umbels compound, terminal and from the upper axils. Flowers white or rarely pinkish, the central one of each umbellet often purple. June–Sept. Native of Eurasia; established as a weed in fields, roadsides, waste ground, and open woods. The cultivated carrot is a race of this species.

### ERIGENIA *Harbinger-of-Spring*

***Erigenia bulbosa*** (Michx.) Nutt. HARBINGER-OF-SPRING *Endangered.* Delicate perennial herb from a globose tuber. Leaves 3-parted, the segments linear to spatulate. Flowers nearly sessile in a single compound terminal umbel; petals white. April–May. Rich moist woods.

### ERYNGIUM *Eryngo*

Biennial or perennial herbs of various aspect, often glaucous, often spinose on the margins or tips of the leaves, bracts, and bractlets. Inflorescence of dense heads instead of umbels, each subtended by bracts (involucre); each flower subtended by a separate bractlet; petals white to purple.

*Key to species*

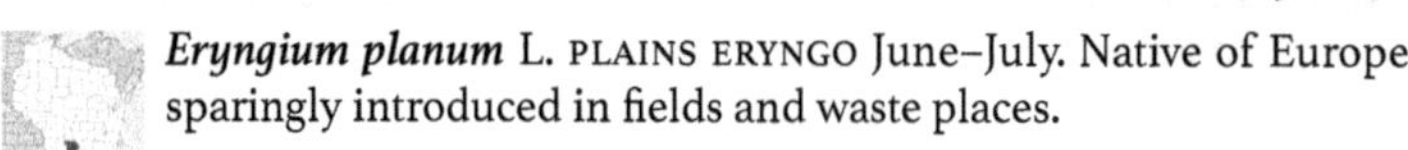

1 Leaves linear, parallel-veined ..................... ***E. yuccifolium***
1 Leaves net-veined ..................................... ***E. planum***

***Eryngium planum*** L. PLAINS ERYNGO June–July. Native of Europe, sparingly introduced in fields and waste places.

***Eryngium yuccifolium*** Michx. RATTLESNAKE MASTER July–Aug. Moist or dry sandy soil, open woods and prairies.

### FALCARIA *Sickleweed*

***Falcaria vulgaris*** Bernh. SICKLEWEED Glabrous herb. Lower leaves petioled, the upper sessile, each of the 3 principal divisions either simple or again divided into 2–5 linear segments. Umbels lateral and terminal, compound; petals white. July–Sept. Native of Europe and w Asia; weedy in waste places and fields.

## FOENICULUM *Fennel*

***Foeniculum vulgare*** P. Mill. SWEET FENNEL Stout perennial herb. Stems glabrous and glaucous. Leaves 3–4 times pinnately dissected into linear or threadlike segments. Umbels compound; petals yellow. Native of the Mediterranean region; widely introduced in the USA.

## HERACLEUM *Cow-Parsnip*

***Heracleum maximum*** Bartr. [•19] AMERICAN COW-PARSNIP Large perennial herb. Stems stout, hairy. Leaves alternate, nearly round in outline, divided into 3 leaflets. Flowers white, in large umbels. May–July. Streambanks, thickets, wet meadows, moist forest openings. *Heracleum lanatum* Michx., *Heracleum sphondylium* L. subsp. *montanum.*

## OSMORHIZA *Sweet-Cicely*

Erect perennial herbs from thickened roots; our 3 species similar in general appearance and foliage. Leaves ternate; leaflets several. Flowers in terminal and lateral umbels, these usually surpassing the leaves; petals white or greenish white. Fruit elongate, ribbed, the base prolonged into bristly tails.

*Key to species*

1 Umbels without bracts at base of umbel branches ..... ***O. chilensis***
1 Umbels with bracts at base of umbel branches ................ 2

2 Plants anise-scented; styles 2 mm long, becoming 3–4 mm long in fruit ........................................... ***O. longistylis***
2 Plants unscented; styles less than 1.5 mm long (even in fruit) ..... ................................................... ***O. claytonii***

***Osmorhiza chilensis*** Hook. & Arn. MOUNTAIN SWEET-CICELY June. Moist woods. *Osmorhiza berteroi* DC.

***Osmorhiza claytonii*** (Michx.) C.B. Clarke [•20] HAIRY SWEET-CICELY May–June. Moist woods.

***Osmorhiza longistylis*** (Torr.) DC. [•20] ANISEROOT May–June. Moist woods.

## OXYPOLIS *Cowbane*

***Oxypolis rigidior*** (L.) Raf. STIFF COWBANE Glabrous perennial herb from a cluster of tuberous roots. Stems with few branches and leaves. Leaves once-pinnate or reduced to bladeless phyllodes. Flowers in loose umbels; petals white. July–Sept. Swamps, thickets, marshes, moist or wet prairie, calcareous fens. Similar to water-parsnip (*Sium suave*) but differs in having entire to irregularly toothed leaves and a slightly grooved stem, while *Sium* has finely toothed leaf margins and a more deeply grooved stem.

## PASTINACA *Parsnip*

***Pastinaca sativa*** L. [•20] WILD PARSNIP Stout biennial herb. Lower leaves long-petioled, the upper on shorter, wholly sheathing petioles, all typically 1-pinnate. Umbels large, compound, the

terminal soon overtopped by the lateral ones; petals yellow. Native of Eurasia; long in cultivation and thoroughly established as a weed in waste places, fields, and roadsides. Skin irritant if handled.

### PIMPINELLA *Burnet Saxifrage*

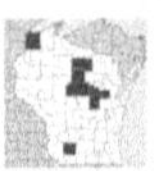

***Pimpinella saxifraga*** L. BURNET SAXIFRAGE Perennial herb. Stems filled with pith. Lower stem leaves 1-pinnate; upper leaves much reduced. Umbels terminal and lateral, compound; petals white. Native of Eurasia; escaped or adventive in waste places.

### POLYTAENIA

***Polytaenia nuttallii*** DC. NUTTALL'S PRAIRIE-PARSLEY *Threatened.* Stout perennial herb. Leaves pinnately dissected. Umbels terminal and axillary, compound; petals yellow. May–June. Mesic prairies and oak openings.

### SANICULA *Black-Snakeroot*

Biennial or perennial herbs. Leaves palmately divided into 3–5 segments. Umbels irregular, with spreading branches of unequal length; petals greenish white to greenish yellow. Fruit densely covered with hooked bristles.

*Key to species*

1 Styles longer than bristles of the fruit; staminate flowers 12–25 in each umbellet .......... 2
1 Styles shorter than bristles of the fruit; staminate flowers 2–7 ... 3

2 Staminate flowers longer than the fruit; sepals awl-shaped, 1–2 mm long .......... ***S. marilandica***
2 Staminate flowers shorter than the fruit; sepals lance-shaped or ovate, 0.5 mm long .......... ***S. odorata***

3 Calyx lobes of perfect flowers conspicuous, equal to or longer than bristles; pedicels of staminate flowers 2–4 times length of their sepals .......... ***S. trifoliata***
3 Calyx lobes of perfect flowers inconspicuous, shorter than the bristles; pedicels of staminate flowers 1–2 times length of their sepals .......... ***S. canadensis***

***Sanicula canadensis*** L. CANADA SANICLE June–Aug. Moist or dry woods.

***Sanicula marilandica*** L. MARYLAND BLACK-SNAKEROOT June–Aug. Moist or dry woods.

***Sanicula odorata*** (Raf.) Pryer & Phillippe CLUSTERED BLACK-SNAKEROOT June–Aug. Moist or dry woods.

***Sanicula trifoliata*** Bickn. LARGE-FRUIT BLACK-SNAKEROOT June–Aug. Moist or dry woods.

## SIUM *Water-Parsnip*

***Sium suave*** Walt. HEMLOCK WATER-PARSNIP Perennial emergent herb. Stems strongly ribbed upward; stem base thickened and hollow with cross-partitions. Leaves once-pinnate, on hollow stalks; leaflets 7–17 per leaf; finely dissected underwater leaves often present from spring to midsummer. Flowers white or green-white, in stalked umbels at ends of stems and from side branches. July–Sept. Wet forest depressions, marshes, swamps, streambanks, lakeshores, ditches; usually in shallow water.

## TAENIDIA *Pimpernel*

***Taenidia integerrima*** (L.) Drude YELLOW-PIMPERNEL Perennial herb. Stems branched, glabrous and somewhat glaucous. Lower leaves long-petioled, commonly 3x compound, the upper 1–2x compound. Umbels terminal and lateral; petals yellow. May–June. Dry woods and rocky hillsides.

## THASPIUM *Meadow-Parsnip*

Branched perennial herbs. Leaves variously compound or the lowest simple. Umbels terminal and lateral, compound; petals yellow or purple.

*Key to species*

1 Basal leaves simple or 3-parted; leaflet margins with 10–40 fine regular teeth on each side ........................ ***T. trifoliatum***

1 Basal leaves twice-compound; leaflet margins with 3–5 coarse irregular teeth on each side ........................ ***T. barbinode***

***Thaspium barbinode*** (Michx.) Nutt. HAIRY-JOINT MEADOW-PARSNIP *Endangered.* May–June. Moist or dry woods and woodland edges. Wisc at n edge of species' range.

***Thaspium trifoliatum*** (L.) Gray PURPLE MEADOW-PARSNIP June–July. Dry or moist woods.

## TORILIS *Hedge-Parsley*

***Torilis japonica*** (Houtt.) DC. ERECT HEDGE-PARSLEY Annual herb. Stems much branched, hispidulous with appressed hairs. Leaves ovate or triangular in outline, pinnately compound. Umbels lateral or also terminal; petals white. Fruit densely covered by rough hooked bristles. Native of Europe; established as a weed in fields and waste ground.

## ZIZIA *Alexanders*

Perennial herbs, glabrous or nearly so. Leaves mostly 1–3x compound. Umbels compound; petals bright yellow.

*Key to species*

1 Basal leaves simple, with heart-shaped blades; stem leaves with 3 leaflets ............................................ ***Z. aptera***

1 Basal leaves compound and similar to the stem leaves, with 5–11 leaflets ............................................ ***Z. aurea***

***Zizia aptera*** (Gray) Fern. HEART-LEAF ALEXANDERS May–June. Moist meadows and open woods.

***Zizia aurea*** (L.) W.D.J. Koch GOLDEN ALEXANDERS May–June. Moist fields and meadows.

## APOCYNACEAE *Dogbane Family*

Our species herbs or twining woody vines; most species have milky juice. Leaves opposite, alternate, or sometimes whorled. Flowers 5-merous, regular, perfect. Fruit a capsule or follicle; seeds often bearing long hairs. Family now includes former members of Asclepiadaceae; *Apocynum* differs by having corolla lobes overlapping and twisted in bud, and stamens without a crown.

*Key to genera*

1 Plants trailing, subwoody, evergreen; flowers solitary in leaf axils; corolla blue; seeds glabrous .......... **VINCA**

1 Plants erect or twining, herbaceous and not evergreen; flowers in terminal or axillary cymes or umbels; corolla various colors, not blue; seeds with tuft of silky hairs .......... 2

2 Plant a climbing vine, strongly twining at least apically; corolla lobes spreading or ascending, dark purple to nearly black .......... **VINCETOXICUM**

2 Plant erect or ascending, not twining; corolla lobes strongly reflexed at maturity, except in *Apocynum,* white, pink, purple, yellow, orange, or greenish .......... 3

3 Corolla lobes erect to spreading; flowers in small terminal (and sometimes axillary) cymes; mature fruits 3–5 mm wide .......... **APOCYNUM**

3 Corolla lobes strongly reflexed at maturity; flowers in umbels; mature fruits 6–35 mm in diameter .......... **ASCLEPIAS**

### APOCYNUM *Dogbane*

Perennial herbs with tough fibrous stems. Leaves opposite, mucronate. Flowers small, white or pink, in branched terminal cymes. Corolla white or pinkish, campanulate or short-cylindric, with 5 short lobes. Fruit a cylindric follicle, pendulous; seeds numerous, bearing long soft hairs (coma).

*Key to species*

1 Corolla pink, 5–8 mm long; leaves widely spreading or drooping .......... ***A. androsaemifolium***

1 Corolla white, 3–4 mm long; leaves ascending .... ***A. cannabinum***

***Apocynum androsaemifolium*** L. SPREADING DOGBANE May–Aug. Upland woods, occasionally in fields and roadsides.

***Apocynum cannabinum*** L. [•21] INDIAN-HEMP June–Sept. Dry or moist open places.

### ASCLEPIAS *Milkweed*

Perennial herbs from a thick root or deep rhizome and with milky juice

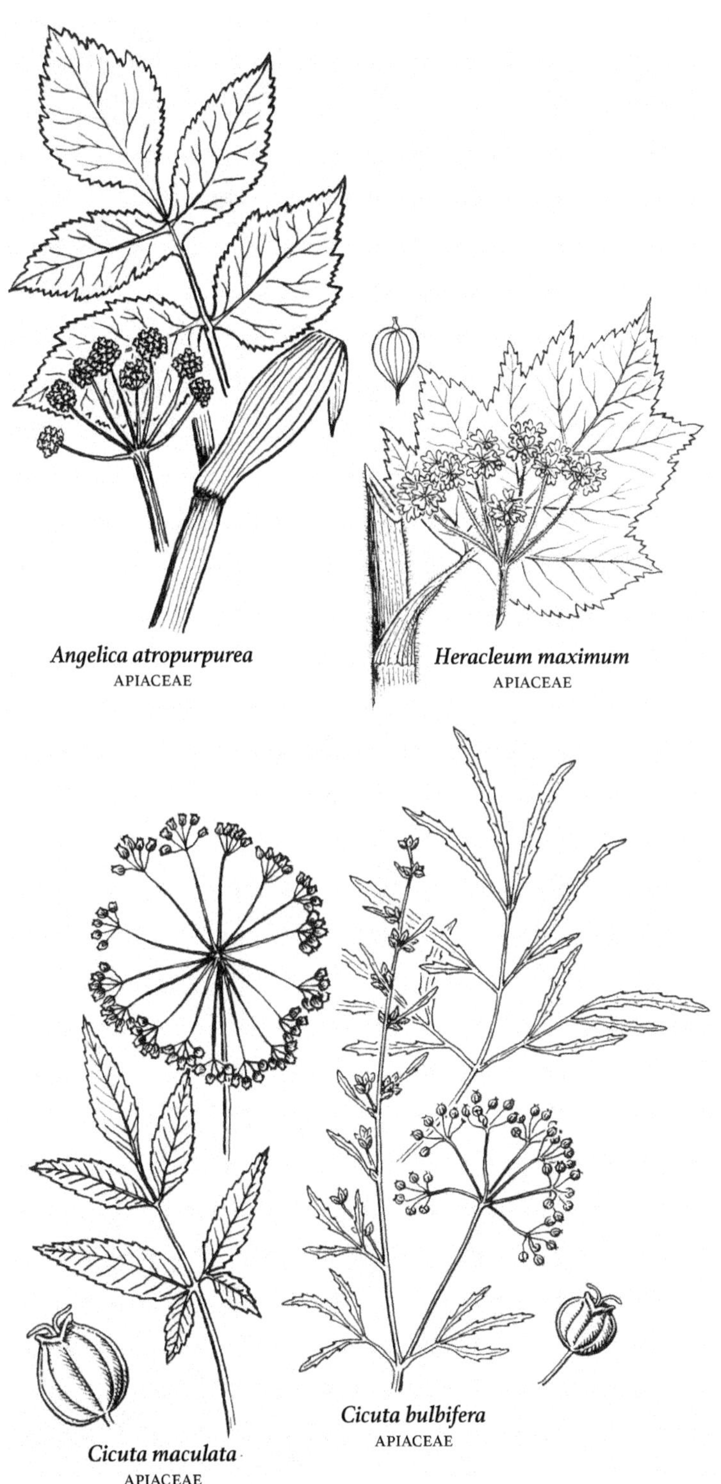
*Angelica atropurpurea*
APIACEAE

*Heracleum maximum*
APIACEAE

*Cicuta maculata*
APIACEAE

*Cicuta bulbifera*
APIACEAE

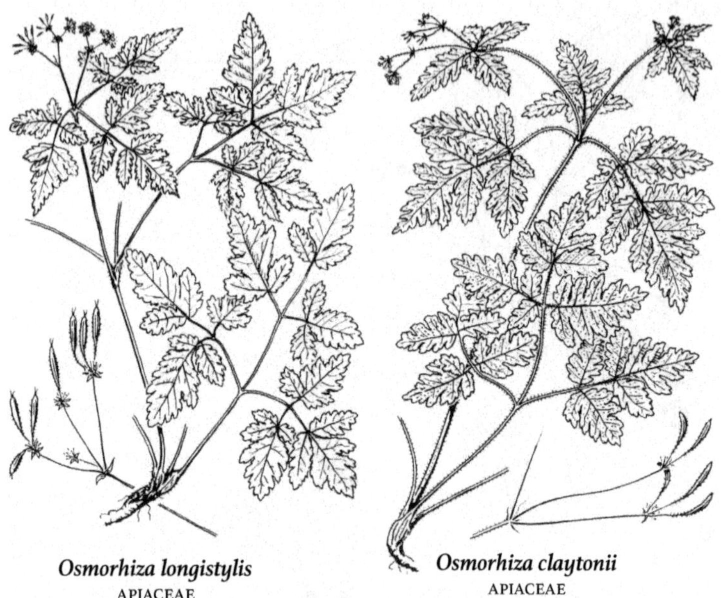

*Osmorhiza longistylis*
APIACEAE

*Osmorhiza claytonii*
APIACEAE

*Pastinaca sativa*
APIACEAE

(except in *A. tuberosa*). Stems usually simple. Leaves opposite (in some species whorled or rarely alternate), entire. Flowers in terminal or axillary umbels; anthers united with stigma forming an organ known as the gynostegium; pollen of each anther-sac united into a waxy mass known as a pollinium. Fruit a pod-like follicle, normally produced in pairs, commonly erect, lance-shaped; seeds with long silky hairs (coma).

ADDITIONAL SPECIES ***Asclepias meadii*** Torr. (Mead's milkweed), grasslands in Grant County, but probably no longer present in Wisc. Leaves at anthesis 3–5 pairs, ovate lance-shaped, margins scabrously ciliolate, otherwise glabrous. Umbel solitary, terminal, few-flowered; corolla greenish white; hoods purple. June.

*Key to species*

1 Leaves linear, less than 4 mm wide, mostly whorled ***A. verticillata***
1 Leaves more than 5 mm wide, alternate or opposite .......... 2

2 Leaves mostly alternate, linear to narrowly oblong .......... 3
2 Leaves opposite, oblong to ovate .......... 5

3 Flowers orange; juice not milky .......... ***A. tuberosa***
3 Flowers green, tinged with purple; juice milky .......... 4

4 Umbels single and terminal .......... ***A. lanuginosa***
4 Umbels 2 or more from leaf axils .......... ***A. hirtella***

5 Leaves heart-shaped or clasping at base, sessile or nearly so .......... 6
5 Leaves rounded or tapered at base, petioles present although often short .......... 9

6 Pedicels glabrous; horns (flattened or slender projections) shorter than hoods and mostly hidden .......... ***A. sullivantii***
6 Pedicels hairy; horns longer than hoods .......... 7

7 Umbels 2 or more from leaf axils; flowers red-purple .......... ***A. purpurascens***
7 Umbels 1 and terminal; flowers green-white or green-purple .......... 8

8 Peduncle strongly nodding at tip; leaves narrowed to an acute tip (likely extirpated from Wisc) .......... ***A. meadii****
8 Peduncle straight; leaves wider, obtuse or with a small sharp tip .......... ***A. amplexicaulis***

9 Reflexed corolla lobes red-purple, mostly less than 5 mm long .......... ***A. incarnata***
9 Reflexed corolla lobes various colors, 5 mm or more long .......... 10

10 Hoods without horns; corolla lobes pale green .......... 11
10 Hoods with slender horns; corolla lobes only rarely pale-green . 12

11 Plants small, stems up to 25 cm high; umbels single and terminal .......... ***A. lanuginosa***
11 Plants larger, stems more than 25 cm high; umbels 2 or more from leaf axils .......... ***A. viridiflora***

12 Umbels nodding; flowers white or greenish .......... ***A. exaltata***
12 Umbels mostly erect; flowers cream-colored or tinged with red or purple .......... 13

13 Upper stems, peduncles and pedicels white-woolly; hoods 1 cm long or more ........................................... *A. speciosa*

13 Pubescence varies but not white-woolly; hoods less than 1 cm long ........................................................ 14

14 Corolla lobes red-purple ....................... *A. purpurascens*

14 Corolla lobes greenish or purple-tinged ..................... 15

15 Plants small, slender, to 60 cm high; umbels usually single ...... ............................................... *A. ovalifolia*

15 Plants large and coarse, more than 60 cm high; umbels usually 2 or more ........................................... *A. syriaca*

***Asclepias amplexicaulis*** Sm. CLASPING MILKWEED June–Aug. Dry fields, prairies, and open woods, usually in sandy soil.

***Asclepias exaltata*** L. [•21] POKE MILKWEED June–July. Moist upland woods.

***Asclepias hirtella*** (Pennell) Woods. GREEN MILKWEED June–Aug. Dry sandy soil and prairies. *Asclepias longifolia* var. *hirtella* (Pennell) B.L. Turner.

***Asclepias incarnata*** L. SWAMP MILKWEED June–Aug. Openings in conifer swamps, marshes, streambanks, ditches, open bogs and fens; often in shallow water.

***Asclepias lanuginosa*** Nutt. SIDE-CLUSTER MILKWEED *Threatened.* May–June. Dry hillside prairies.

***Asclepias ovalifolia*** Dcne. DWARF MILKWEED June–July. Dry prairies.

***Asclepias purpurascens*** L. PURPLE MILKWEED *Endangered.* June–July. Open oak woods and roadsides in a range of soil moisture conditions. Similar to *A. syriaca* in general habit; usually identified by the purple corolla and the shape of the hood; the fruit lacks the conspicuous processes of *A. syriaca.*

***Asclepias speciosa*** Torr. SHOWY MILKWEED July–Aug. Moist prairies; common in w USA, considered introduced in Wisc.

***Asclepias sullivantii*** Engelm. SMOOTH MILKWEED *Threatened.* June–July. Moist prairies.

***Asclepias syriaca*** L. COMMON MILKWEED Fields, meadows, and roadsides; often weedy.

***Asclepias tuberosa*** L. [•21] BUTTERFLY WEED June–Aug. Dry or moist prairies and upland woods, especially in sandy soil. Our only *Asclepias* without milky juice. Plants variable in habit and shades of flower color.

***Asclepias verticillata*** L. WHORLED MILKWEED June–Aug. Dry or moist fields, roadsides, upland woods, and prairies.

***Asclepias viridiflora*** Raf. GREEN COMET MILKWEED July–Aug. Dry upland woods, prairies, and barrens, especially in sandy soil.

### VINCA *Periwinkle*

***Vinca minor*** L. LESSER PERIWINKLE Perennial herb. Stems trailing or scrambling, forming mats. Leaves leathery, opposite. Flowers blue or rarely white, solitary in 1 axil only of a pair of leaves. April–May. Native of s Europe, planted as a groundcover and escaped to roadsides and open woods.

### VINCETOXICUM *Swallow-Wort*

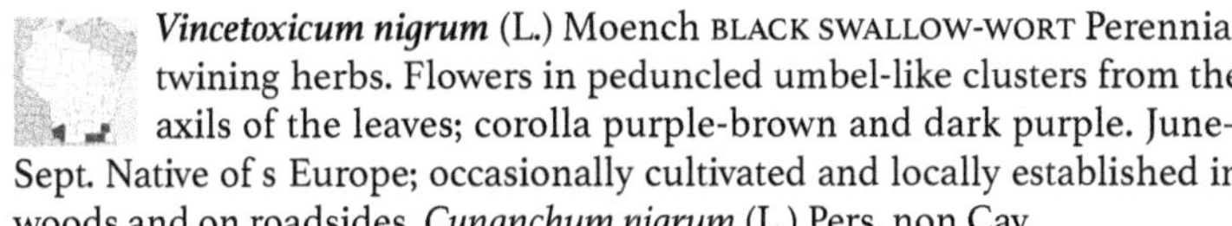

***Vincetoxicum nigrum*** (L.) Moench BLACK SWALLOW-WORT Perennial twining herbs. Flowers in peduncled umbel-like clusters from the axils of the leaves; corolla purple-brown and dark purple. June–Sept. Native of s Europe; occasionally cultivated and locally established in woods and on roadsides. *Cynanchum nigrum* (L.) Pers. non Cav.

## AQUIFOLIACEAE *Holly Family*

### ILEX *Holly*

Shrubs. Leaves usually alternate, toothed or entire, not lobed. Flowers from leaf axils, 4–8-parted, usually either staminate or pistillate, sometimes perfect, on same or different plants. Fruit a fleshy berrylike drupe with 4–9 stones.

*Key to species*

1 Leaves tipped with a short, sharp point, margins mostly entire or with a few scattered teeth; petals linear; sepals tiny or absent .... .................................................. ***I. mucronata***

1 Leaves not tipped with a short, sharp point, margins toothed; petals oblong; sepals evident ............................ ***I. verticillata***

***Ilex mucronata*** (L.) Powell, Savolainen & Andrews MOUNTAIN HOLLY, CATBERRY May–June. Open bogs (especially along outer moat), swamps, thickets, wet depressions in forests, lakeshores. *Nemopanthus mucronatus* (L.) Loes.

***Ilex verticillata*** (L.) Gray [•22] WINTERBERRY June. Swamps, open bogs, thickets, shores and streambanks.

## ARALIACEAE *Ginseng Family*

Shrubs or herbs, rarely trees. Leaves usually alternate, compound or rarely simple. Flowers small, umbellate. Fruit a berry or a leathery drupe.

*Key to genera*

1 Leaves simple, palmately lobed ............... **HYDROCOTYLE**

1 Leaves compound ................................................ **2**

2 Leaves alternate or basal, mostly 2–3 times compound; carpels 5 . ..................................................... **ARALIA**

2 Leaves in 1 whorl, once-palmately compound; carpels 2 or 3 ..... ........................................................ PANAX

## ARALIA *Sarsaparilla*

Herbs or shrubs (or rarely trees). Stems herbaceous or slightly woody at the base only; *A. hispida* bristly at the base. Leaves pinnately or 3-partedly compound. Flowers white or greenish, in 2–many umbels in each inflorescence. Fruit a berry, tipped by the persistent styles.

ADDITIONAL SPECIES ***Aralia elata*** (Miq.) Seem. (Angelica tree), introduced shrub or small tree with spiny stems and compound leaves; sometimes invasive; reported from Rock County in s Wisc.

*Key to species*

1 Plants with flowers on a leafless scape ............. ***A. nudicaulis***
1 Plants with leafy stems........................................ 2

2 Lower stems bristly; umbels several (3–13) in a loose cluster ..... ........................................................ ***A. hispida***
2 Stems smooth; umbels very many, in a large terminal panicle .... ........................................................ ***A. racemosa***

***Aralia hispida*** Vent. BRISTLY SARSAPARILLA June–July. Dry woods, especially in sandy or sterile soil.

***Aralia nudicaulis*** L. [•22] WILD SARSAPARILLA May–June. Moist or dry woods.

***Aralia racemosa*** L. SPIKENARD July. Rich woods.

## HYDROCOTYLE *Pennywort*

***Hydrocotyle americana*** L. AMERICAN MARSH-PENNYWORT Small perennial herb. Stems slender and creeping, often rooting at nodes. Leaves round to kidney-shaped; petioles long; margins with 7–12 shallow lobes. Flowers white, in nearly stalkless umbels from nodes. June–Sept. Conifer swamps, streambanks, shores, wet forest depressions. Formerly included in Apiaceae.

## PANAX *Ginseng*

Perennial herbs, the unbranched stems rising from a deep-seated, thickened or tuber-like root, bearing a single whorl of once palmately compound leaves, usually 3 in number. Flowers in usually a single long-peduncled terminal umbel; petals white or greenish. Fruit a small, 2–3-seeded berry.

*Key to species*

1 Leaflets long-stalked; uncommon .............. ***P. quinquefolius***
1 Leaflets sessile; common .......................... ***P. trifolius***

***Panax quinquefolius*** L. AMERICAN GINSENG June–July. Rich deciduous woods, rare due to heavy collecting of the roots.

***Panax trifolius*** L. [•22] DWARF GINSENG April–May. Rich woods.

*Apocynum cannabinum*
ASCLEPIADACEAE

*Asclepias tuberosa*
ASCLEPIADACEAE

*Asclepias exaltata*
ASCLEPIADACEAE

*Ilex verticillata*
AQUIFOLIACEAE

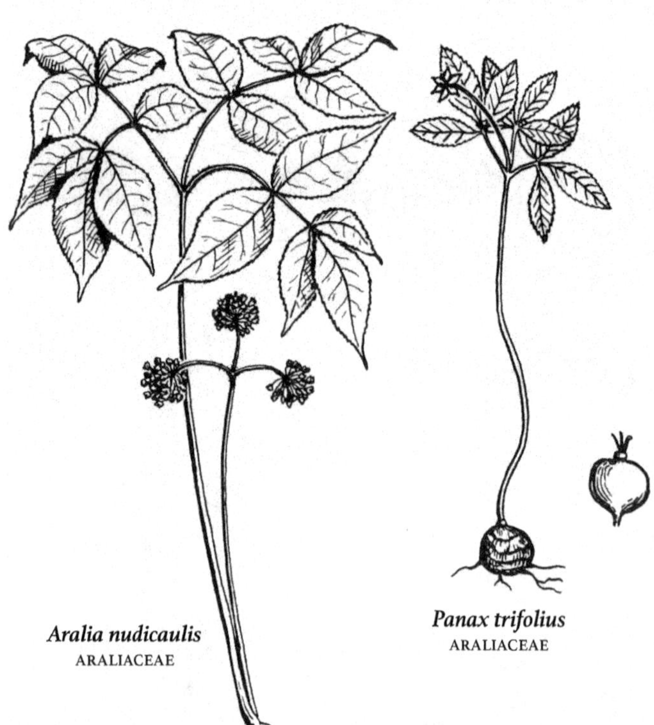

*Aralia nudicaulis*
ARALIACEAE

*Panax trifolius*
ARALIACEAE

## ARISTOLOCHIACEAE *Birthwort Family*

### ASARUM *Wild Ginger*

***Asarum canadense*** L. CANADIAN WILD GINGER Perennial herb; producing annually a pair of cordate leaves, between which arises the solitary, short-peduncled flower. April–May. Rich woods, usually in small colonies.

## ASTERACEAE *Aster Family*

Annual, biennial or perennial herbs. Leaves simple or compound, opposite, alternate, or whorled. Flowers perfect or single-sexed (sometimes sterile) and of 2 types: ray (or ligulate) and disk (or tubular). Ray flowers joined at base and have a long, flat, segment above (the ray); disk flowers tube-shaped with 5 lobes or teeth at tip.

Flowers are clustered in 1 of 3 types of heads resembling a single flower and attached to a common surface (receptacle): ray flowers only (as in dandelion, *Taraxacum*); disk flowers only (discoid, as in tansy, *Tanacetum*); and heads with both ray and disk flowers (radiate), the ray flowers surrounding the disk flowers (as in sunflower, *Helianthus*).

In addition to flowers, the receptacle may also have scales called chaff; if no scales present, the receptacle is termed naked. Each head is surrounded by involucral bracts (sometimes called phyllaries); collectively, the bracts are termed the involucre, comparable to the group of sepals (calyx) subtending an individual flower. Fertile flowers have 1 pistil tipped by a 2-cleft style (undivided in sterile flowers); stamens 5; ovary (and achene) often topped by several to many scales, awns or hairs (the pappus). Fruit a seedlike achene (sometimes termed cypsela in Asteraceae).

ADDITIONAL SPECIES A number of other members of the Aster Family occur in Wisc, and, if not included in further described genera, are listed below. Many are introduced garden escapes, and most are adventive and not truly established in our flora.

***Adenocaulon bicolor*** Hook., a species of the Pacific Northwest, known from Michigan's Upper Peninsula and reported for Wisc (but not verified); leaf underside conspicuously white-hairy.

***Calendula officinalis*** L. (Pot-marigold), reported for Kewaunee County but likely not persistent.

***Callistephus chinensis*** (L.) Nees (China-aster), introduced, doubtfully persistent, Dane County.

***Canadanthus modestus*** (Lindl.) Nesom (Canada-aster), native, Douglas County.

***Cosmos bipinnatus*** Cav. (Garden Cosmos), annual, leaves opposite, rays about 8, rose or lilac; cultivated and occasionally escaped.

***Dyssodia papposa*** (Vent.) A.S. Hitchc. (Fetid-marigold), collected once along Wisc River.

***Gamochaeta purpurea*** (L.) Cabrera (Spoon-leaf purple everlasting), annual or biennial native herb, Sheboygan County.

***Guizotia abyssinica*** (L. f.) Cass. (Nigerseed), reported for Wisc but likely not established in the flora.

***Leucanthemella serotina*** (L.) Tzvelev (Giant daisy), introduced in several Wisc counties.

***Madia glomerata*** Hook (Mountain tarplant), coarse glandular annual herb, leaves alternate, rays yellow; native in w USA, adventive in Wisc (Price and Sheboygan counties).

***Pluchea camphorata*** (L.) DC. (Camphorweed), native of s USA, adventive in Sheboygan County.

***Tagetes patula*** L. (French marigold), introduced, Dane and Marquette counties.

*Key to Asteraceae Tribes*

Because of its large size, the Asteraceae is often divided into smaller groups of related species called tribes. In Wisc, 11 tribes (and one subtribe) occur, and are identified in the following key (adapted from Kowal (2007).

1 Flowers wind-pollinated, not showy; rays absent; florets and most heads unisexual; anthers not united; involucral bracts typically connate, at least basally, where free (*Iva*), only 3-5 .............. *Tribe 8a.* **HELIANTHEAE** (*Subtribe* **AMBROSIINAE**, *Ragweed subtribe*)

1 Flowers insect-pollinated, usually showy; rays present or absent; heads and most florets bisexual; anthers united; involucral bracts free; if wind-pollinated, involucral bracts free, scarious and more than 5 (*Artemisia*) .............................................. 2

2 Plants with milky juice; heads ligulate (florets bisexual and with a 5-toothed ligule) ........... *Tribe 4.* **CICHORIEAE** (*Lettuce Tribe*)

2 Plants with watery juice; heads radiate (with disk florets surrounded by ray florets), or disciform but with ray florets without rays, or discoid (only disk florets)..................................... 3

3 Plants and/or heads usually prickly; heads discoid and corolla lobes of the disk florets at least 4 times longer than wide; receptacle densely bristly (naked in the prickly *Onopordum*); leaves alternate; style with a ring of hairs (sometimes merely with a thickened ring) below the branches ........... *Tribe 3.* **CARDUEAE** (*Thistle Tribe*)

3 Plants and heads not prickly; heads various, corolla lobes of the disk florets less than 4 times longer than wide; receptacle various, rarely bristly; leaves various; style otherwise ........................ 4

4 Heads discoid and corollas never yellow; style-branches long and slender (thread-like), conspicuously protruding from the corolla and often attractive; receptacle naked ......................... 5

4 Heads various, but if discoid, corollas yellow (or at least creamy); style- branches much shorter (relative to their widths); receptacle various .................................................. 6

5 Style-branches hispidulous, acute or acuminate at tip; corollas purple; inflorescence corymbose; leaves alternate................ ......................... *Tribe 11.* **VERNONIEAE** (*Ironweed Tribe*)

5 Style-branches merely papillate, blunt (to acutish) and sometimes thickened (clavate) towards the tip; corollas white, pink, rose or blue-violet; inflorescence various; leaves alternate, opposite or whorled ................. *Tribe 5.* **EUPATORIEAE** (*Boneset Tribe*)

6 Pappus absent or of awns, scales, or teeth; leaves alternate or opposite .......................................................... 10

6 Pappus of hairs or bristles; leaves alternate (in some Senecioneae directly from a rhizome) .................................... 7

7 Involucral bracts equal and in 1 row; rays yellow or absent; style-branches with a tuft of hairs at the end ........................ ...................... *Tribe 10.* **SENECIONEAE** (*Groundsel Tribe*)

7 Involucral bracts in 2-5 rows, equal or unequal, if (rarely) in 1 row, with conspicuous white, pink, purple, or blue rays ............ 8

8 Heads discoid or disciform; involucral bracts scarious, either virtually entirely or at least at the tip for a third of their lengths (styles and anthers as in Inuleae) .............................. ....................... *Tribe 6.* **GNAPHALIEAE** (*Pussytoes Tribe*)

8 Heads radiate (rays minute in *Conyza*); involucral bracts not scarious or scarious only on the margins ............................ 9

9 Giant perennial herb 1-3 m tall, with basal leaves often 1 m long, disk of head more than 2 cm wide, and rays more than 4 cm long, which are yellow, numerous, and narrowly linear; style branch slightly clavate and glabrous; anthers tailed at base; infrequent adventive .................. *Tribe 9.* **INULEAE** (*Elecampane Tribe*)

9 Plant smaller in all parts; rays of various colors and shapes; style branch with a lanceolate or elongate-deltoid hairy appendage; anthers rounded at base ......... *Tribe 2.* **ASTEREAE** (*Aster Tribe*)

10 Involucral bracts with scarious or hyaline margins; leaves alternate ................................................................ 11

10 Involucral bracts not scarious or hyaline, or if so, leaves opposite; leaves alternate or opposite ................................ 12

11 Leaves entire, not aromatic; receptacle naked; style branch with a lanceolate or elongate-deltoid hairy appendage (*Boltonia*) ........ ............................... *Tribe 2.* **ASTEREAE** (*Aster Tribe*)

11 Leaves toothed, lobed, or finely divided, often aromatic; receptacle chaffy or naked; style-branches mostly truncate, with a tuft of hairs at the end (like Senecioneae) ................................ ..................... *Tribe 1.* **ANTHEMIDEAE** (*Chamomile Tribe*)

12 Receptacle naked; rays present, widest at the prominently 3-lobed apex; leaves alternate (opposite in *Arnica*), lanceolate to ovate .... ......................... *Tribe 7.* **HELENIEAE** (*Sneezeweed Tribe*)

12 Receptacle chaffy (absent in *Dyssodia,* with unremarkable rays and opposite pinnatisect leaves); rays present or absent, but when present usually not as above, but if so, then the leaves opposite and either lobed or pinnatifid .................................... ...................... *Tribe 8.* **HELIANTHEAE** (*Sunflower Tribe*)

### *Tribe 1.* ANTHEMIDEAE *(Chamomile Tribe)*

*Achillea, Anthemis, Artemisia, Leucanthemum, Matricaria, Tanacetum*

1 Receptacle chaffy; heads radiate ............................. 2

1 Receptacle naked or villous; heads radiate, disciform or discoid . 4

2 Heads small, 5 mm or less in diameter, densely corymbose; receptacle flat; achenes compressed ................. **ACHILLEA**

2 Heads rather large, 1-4 cm in diameter, solitary and terminal on long peduncles; receptacle conic at maturity; achenes terete or angled ........................................................ 3

3 Ray florets white; disk 0.5-1.2 cm in diameter ....... **ANTHEMIS**
3 Ray florets yellow; disk 1-2 cm in diameter ..... ***Anthemis tinctoria***

4 Inflorescence paniculate, racemose or spike-like with inconspicuous discoid heads; florets green .............. **ARTEMISIA**
4 Inflorescence corymbose or heads terminal on long peduncles; ray florets showy, yellow or white (sometimes obsolete) ........... 5

5 Receptacle conic at maturity; leaves pinnatisect .. **MATRICARIA**
5 Receptacle flat or low-convex ................................ 6

6 Heads several or many, in corymbs, disk 4-9 mm wide, with or without rays; leaves often highly lobed ........... **TANACETUM**
6 Heads solitary at tips of stem or long branches, large, disk 1-2.5 cm wide, with conspicuous white rays; leaves toothed to lobed ...... ........................................ **LEUCANTHEMUM**

### *Tribe 2.* ASTEREAE *(Aster Tribe)*

*Bellis, Boltonia, Conyza, Doellingeria, Eurybia, Erigeron, Euthamia, Grindelia, Heterotheca, Ionactis, Solidago, Symphyotrichum*

1 Ray corollas yellow, conspicuous; disk corollas yellow .......... 2
1 Ray corollas white, pink, violet, bluish or purple; disk corollas various ........................................................ 5

2 Pappus of 2-8 caducous awns; involucre more or less glutinous ... ................................................ **GRINDELIA**
2 Pappus of numerous capillary bristles or hairs; involucre not glutinous ...................................................... 3

3 Pappus double with long capillary bristles surrounded by short, somewhat chaffy bristles; heads wider than 5 mm and rays longer than 4 mm ................................ **HETEROTHECA**
3 Pappus simple; heads small with disks no wider than 5 mm and with rays no longer than 4 mm ............................... 4

4 Inflorescence corymbiform; leaves glandular punctate, linear to narrowly oblong, only slightly reduced upwards on stem; ray florets more numerous than the disk florets .............. **EUTHAMIA**
4 Plants not with both inflorescence corymbiform and ray florets more numerous than the disk florets; leaves usually broader, not glandular punctate .............................. **SOLIDAGO**

5 Pappus absent or inconspicuous (2-4 awns up to 2 mm long and several minute bristles); receptacle conic, low-conical or hemispherical ................................................ 6
5 Pappus of long capillary bristles or hairs; receptacles flat........ 7

6 Plant 3-15 dm tall, with many heads; receptacle low-conical or hemispherical .................................... **BOLTONIA**
6 Plant 1.5 dm or less tall, with one head on a scape; receptacle conic ........................................................ **BELLIS**

7 Rays minute, shorter than the corolla tube and barely longer than

the pappus; heads small with involucres less than 4 mm long, disks no more than 4 mm wide, and disk florets numbering no more than 21 .............................................. CONYZA

7 Rays conspicuous, larger; heads larger ........................ 8

8 Involucral bracts approximately in one series, neither chartaceous at base nor with herbaceous green tip; style appendages roundish or obtuse, no longer than 0.3 mm; rays very numerous and narrowly linear (mostly 1.3 mm or less wide); plants blooming chiefly in spring and early summer (when later, plants also with heads past fruiting) ........................................... ERIGERON

8 Involucral bracts clearly imbricated or with a foliaceous outer series; style appendages longer and more acute; rays in one or two series and relatively broader; plants blooming in late summer and fall ...................................................... 9

9 Middle and lower stem leaves distinctly petioled, most of the petioles more than 1 cm long, wingless, or winged but less than 1/4 as wide as the blades; blades (except the uppermost) more than 12 mm wide, abruptly narrowed to a truncate or cordate base ..... 10

9 Middle and lower stem leaves not distinctly petioled; sessile or subsessile on petioles less than 0.5 cm long, or apparently on broad-winged petioles more than ¼ as wide as the blades; or with long narrow tapering petiole-like bases, but the blades not more than 12 mm wide ................................................. 11

10 Involucral bracts narrowly to broadly ovate-lanceolate, outer ones 1.0-2.5 mm wide, less than 2.5x as longer than wide; inflorescence corymbiform ....................................... EURYBIA

10 Involucral bracts linear-deltoid to lanceolate, outer ones 0.2-1.0 mm wide, more than 2.5x longer than wide; inflorescence elongate (paniculate or racemose) ................ SYMPHYOTRICHUM

11 Pappus double, the inner of long capillary bristles, the outer of short bristles, 1 mm long or less (very obscure); middle and upper involucral bracts with scarious margins extending to tip, central green line not or only slightly expanded towards tip; inflorescence corymbose (heads sometimes few or solitary in *Ionactis linariifolia*) ........................................................ 12

11 Pappus not double; involucral bracts various, but in most species with the central green line conspicuously dilated at tip; inflorescence various, but in most species not corymbose (*Symphyotrichum*) ........................................... 13

12 Leaves rigid, linear, veinless except for the strong midrib; rays pink or violet; scarious margins and tips of involucral bracts often purplish; pappus obscurely double, bristles slender to tip Ionactis

12 Leaves herbaceous, veiny and broad; rays white; scarious margins and tips of involucral bracts whitish; pappus distinctly double ... ............................................. DOELLINGERIA

13 Rays much reduced or absent (heads "disciform"); pappus conspicuous at anthesis; plants annual, with taproots .......... .................................... *Symphyotrichum ciliatum*

13 Rays present, heads conspicuously radiate; pappus inconspicuous at anthesis or at most barely overtopping disk corollas; plants perennial, forming clumps or with rhizomes ......... **SYMPHYOTRICHUM**

### *Tribe 3.* CARDUEAE *(Thistle Tribe)*

*Arctium, Carduus, Centaurea, Cirsium, Echinops, Onopordum*

1 Neither plants nor heads prickly or involucral bracts spine-tipped and corollas yellow; achenes obliquely attached to the receptacle; marginal disk florets often enlarged and showy; involucral bracts often with margins scarious and deeply cleft at tip (laciniate); pappus hairs mostly less than 3 mm long or lacking ....... **CENTAUREA**

1 Plants and/or heads prickly; corollas not yellow; achenes attached by the base to the receptacle; florets all alike; involucral bracts not laciniate at tip; pappus hairs usually more than 5 mm long ..... 2

2 Leaves unarmed, broadly rounded at base; tip of phyllary a hook Arctium

2 Leaves prickly, lanceolate to ovate; tip of phyllary a straight spine or merely mucronate ......................................... 3

3 Heads 1-flowered, aggregated into globose secondary heads ..... ................................................... **ECHINOPS**

3 Heads many-flowered, only rarely sessile .................... 4

4 Pappus plumose; involucral bracts with needle-like spiny tips or merely mucronate, often with a glutinous ridge on back ......... ................................................... **CIRSIUM**

4 Pappus barbellate to capillary; involucral bracts not glutinous .. 5

5 Receptacle bristly; leaves and stem wings glabrous or nearly so; pappus capillary ................................. **CARDUUS**

5 Receptacle alveolate (pitted), not bristly; leaves and stem wings densely cottony-velutinous; pappus barbellate; extremely rare adventive.................................... **ONOPORDUM**

### *Tribe 4.* CICHORIEAE *(Lettuce Tribe)*

*Cichorium, Crepis, Hieracium, Hypochaeris, Krigia, Lactuca, Lapsana, Leontodon, Nothocalais, Prenanthes, Sonchus, Taraxacum, Tragopogon*

1 Pappus absent...................................... **LAPSANA**

1 Pappus present ............................................. 2

2 Pappus of numerous simple hairlike (capillary) bristles only .... 3

2 Pappus otherwise (plumose bristles, scales, scales mixed with bristles, or a ring of numerous minute bristles)................. 9

3 Achenes flattened or compressed ........................... 4

3 Achenes cylindrical, fusiform or terete, not flattened ........... 5

4 Achenes not beaked, not enlarged at the tip; heads yellow with many florets (80 or more) ......................... **SONCHUS**

4 Achenes beaked or unbeaked, but constricted below enlarged tip; heads yellow or blue, with relatively few florets (5-56) . **LACTUCA**

5 Plants scapose; achenes beaked, or tapered and the beak lacking; pappus white; involucral bracts in more than one series ........ 6

5 Stems branched or unbranched and leafy or subscapose; achenes truncate or tapered, rarely short-beaked; pappus pale yellow, red-brown, tannish or white; involucral bracts in 1 or 2 series........ 7

6 Achenes tuberculate-muricate above with a long filiform beak; scapes hollow; leaves variously runcinate-pinnatifid ..... **TARAXACUM**

6 Achenes not tuberculate-muricate above, slightly tapered, but not beaked; scapes solid; leaves grasslike, the margins pubescent .... ............................................ **NOTHOCALAIS**

7 Annuals or biennials with well developed, usually pinnatifid basal leaves; inflorescences open corymbs or panicles of yellow campanulate heads; pappus white; main involucral bracts uniseriate ............................................ **CREPIS**

7 Perennials; cauline leaves lanceolate to palmately lobed, or unlobed and dentate to entire; inflorescences branched racemes, panicles of cylindrical drooping heads, or corymbs with erect campanulate heads; pappus tawny to brown, not pure snowy white; main involucral bracts biseriate.................................... **8**

8 Leaves lanceolate to palmately lobed; heads cylindrical, nodding; corolla pink, purplish to yellow or white; pappus pale yellow to red-brown; plants sometimes tomentose, not glandular .. Prenanthes

8 Leaves spatulate to oblanceolate, not lobed; heads campanulate, erect; corolla yellow to red-orange; pappus tannish; plants usually glandular-pubescent .......................... **HIERACIUM**

9 Pappus of plumose (feathery) bristles only .................. **10**

9 Pappus of scales and/or bristles (sometimes minute) ........... **12**

10 Plants leafy stemmed, branched, not scaly-bracted above; leaves cauline, grasslike ........................... **TRAGOPOGON**

10 Plants scapose, scaly bracted above; leaves basal, coarsely dentate ........................................................ **11**

11 Inner and outer achenes uniform, not slender-beaked; receptacle ............................................ **LEONTODON**

11 Inner achenes with long, slender beaks; receptacle chaffy ........ ............................................ **HYPOCHAERIS**

12 Pappus of 5 to numerous outer scales alternating with 5 to numerous scabrous hairs; plants scapose or sub-scapose, branched or not branched; corolla yellow ....................... **KRIGIA**

12 Pappus a ring of numerous minute (0.2 mm) scales or bristles; plants profusely branched; corolla blue, rarely pink or white **CICHORIUM**

### *Tribe 5.* EUPATORIEAE *(Boneset Tribe)*

*Ageratina, Brickellia, Eupatorium, Eutrochium, Liatris*

1 Leaves alternate; plants from a stout taproot or enlarged corm; achenes 10-ribbed; pappus of plumose or barbellate bristles; involucral bracts weakly or strongly ribbed .................... **2**

1 Leaves opposite or whorled; roots fibrous; achenes 5-angled; pappus of capillary bristles; involucral bracts not ribbed ............. **3**

2 Plants from stout taproots; pappus plumose; involucral bracts strongly ribbed; inflorescence corymbiform, the heads creamy-white .......................................... **BRICKELLIA**

2 Plants from enlarged corms; pappus plumose or barbellate; involucral bracts weakly ribbed; inflorescence spicate or racemose, the heads purple and often very showy ............... **LIATRIS**

3 Leaves in whorls of 3, 4 or 5; heads purple or dull rose; involucral bracts in 5-6 series ............................ **EUTROCHIUM**

3 Leaves opposite (rarely in 3s in E. perfoliatum); heads white (rarely purple in *E. perfoliatum*); involucral bracts in 2-3 series ........ 4

4 Leaves long-petioled, ovate; involucral bracts nearly uniseriate, narrowly linear, any basal ones usually much less than half the length of the longest; heads with 15-30 florets; amber resin glands absent ........................................ **AGERATINA**

4 Leaves sessile (except *E. serotinum*), narrowly ovate or lanceolate; involucral bracts in 2-3 series, not narrowly linear, many roughly half the length of the longest; heads with 15 or fewer florets; tiny amber resin glands on leaf undersides, involucral bracts, corollas, and achenes ................................. **EUPATORIUM**

### *Tribe 6.* GNAPHALIEAE *(Pussytoes Tribe)*

*Anaphalis, Antennaria, Gnaphalium, Pseudognaphalium*

1 Stem leaves few, much smaller than those of the persistent basal rosette, strongly ascending; stolons present; plants either staminate or pistillate, populations dioecious ............. **ANTENNARIA**

1 Stem leaves many, about the same size as the basal leaves, which soon wither; stolons absent ................................ 2

2 Involucral bracts pure white, with conspicuous, longitudinal creases creating the appearance of wrinkled tissue paper; populations dioecious, although pistillate plants often with heads having a few staminate florets in the center; dried plants without a strong odor ............................................ **ANAPHALIS**

2 Involucral bracts grayish white, yellow or brown, scarious, with very small longitudinal ridges but no conspicuous creases; heads bisexual, with pistillate florets marginally and staminate heads in center; dried plants with strong tobacco-like odor ............. 3

3 Perennial with narrow, spiciform or subcapitate inflorescence; achenes sparsely strigose; boreal species known from Apostle Islands ............................... ***Gnaphalium sylvaticum***

3 Annual or biennial; inflorescence various; achenes smooth or papillate ................................................ 4

4 Heads 2-3 mm long, in leafy-bracted clusters; upper stems very densely white-floccose-tomentose, obvious to the naked eye; stems usually much branched, 1-2 dm tall ...... ***Gnaphalium uliginosum***

4 Heads 4-6 mm long, capitate or corymbose; upper stems with appressed or nearly microscopic loose-spreading tomentum; stems erect, seldom branching except within a corymbose inflorescence, 1-10 dm tall ......................... **PSEUDOGNAPHALIUM**

*Tribe 7.* **HELENIEAE** *(Sneezeweed Tribe)*
*Gaillardia, Helenium*

1 Receptacle naked; style branches truncate, without an appendage . ........................................................ **HELENIUM**
1 Receptacle bristly; style branches with a subulate appendage ...... ........................................................ **GAILLARDIA**

*Tribe 8.* **HELIANTHEAE** *(Sunflower Tribe)*
*Bidens, Coreopsis, Echinacea, Eclipta, Galinsoga, Helianthus, Heliopsis, Parthenium, Polymnia, Ratibida, Rudbeckia, Silphium, Verbesina*

1 Involucral bracts (some or all) highly modified, either folded around outer achenes or united into a cup or tube; strongly scented annuals; rare adventives or escapes from cultivation ........... 2
1 Involucral bracts not highly modified, free and not infolding outer achenes; annuals and perennials, not strongly scented.......... 4

2 Involucral bracts free, outer (or larger) laterally compressed and folded around the laterally compressed achene; stem leaves mostly alternate and unlobed; stems viscid and glandular pubescent ... ........................................................ **MADIA***
2 Involucral bracts (at least innermost) united into a cup or tube, none inclosing the opposite flower or achene; cauline leaves mostly opposite and pinnately lobed; stems glabrous or glabrate ....... 3

3 Involucre a cup, with loose involucral bracts at base of the united series; receptacle bearing slender chaff; pappus of chaffy scales dissected into numerous, long bristles .............. **DYSSODIA***
3 Involucre a tube, naked at base; receptacle honeycombed; scales of pappus entire ........................................ **TAGETES***

4 Involucre distinctly double, the outer larger (or minute, 2 mm or less long), foliaceous, somewhat spreading, the inner broader and appressed, nearly membranous .............................. 5
4 Involucre not double, involucral bracts all about equal in length, the inner and outer similar in texture ......................... 7

5 Pappus absent or of a few teeth .................... **COREOPSIS**
5 Pappus of 2 to 4 barbed awns ................................ 6

6 Achenes beakless, flattened or slender and 4-sided (rarely subterete) ........................................................ **BIDENS**
6 Achenes long-beaked, slenderly fusiform, 5-angled and subterete ........................................................ **COSMOS***

7 Rays white or absent, if present, 1-10 mm long; disk small, 3-10 mm wide ........................................................ 8
7 Rays yellow, orange or purple, generally 1-6 cm long; disk generally large, (4-) 10-40 mm wide ................................. 11

8 Leaves alternate; heads whitish; leaves large, rough **PARTHENIUM**
8 Leaves opposite ............................................... 9

9 Lower leaves deeply lobed, with connate-perfoliate expanded blade tissue at the nodes ................................ **POLYMNIA**
9 Leaves not lobed, toothed, without such a foliaceous expansion at the nodes ................................................ 10

10 Leaves, except the uppermost, petioled; blades less than three times as long as wide .................................... GALINSOGA

10 Leaves tapered to the base, not distinctly petioled, the blades more than three times as long as wide; Mississippi River, rare in Wisc .. ....................................................... ECLIPTA

11 Rays purple, the receptacular bracts spiny-pointed . ECHINACEA

11 Rays yellow or orange ........................................ 12

12 Disk florets staminate; ray florets pistillate, their large achenes broadly ovate, winged, strongly flattened parallel with the adjoining involucral bracts; plants large, usually resinous ...... SILPHIUM

12 Disk florets bisexual; ray florets neuter or pistillate; achenes wingless, sub-terete or angled ................................ 13

13 At least some of the leaves opposite or all basal ............... 14

13 Leaves all alternate ............................................ 15

14 Outer involucral bracts shorter than the inner; ray florets neuter, their rays thin and easily wilting, deciduous ..... HELIANTHUS

14 Outer involucral bracts longer than the inner; ray florets pistillate, their rays marcescent (thickish and persistent after flowering) ... .................................................... HELIOPSIS

15 Disk flat or convex; leaves neither lobed nor divided .......... 16

15 Disk conical, hemispheric or columnar; leaves simple in *Rudbeckia* otherwise lobed, cleft, laciniate or pinnately parted ........... 17

16 Leaves not decurrent; achenes 3- or 4-angled, wingless, forming a flat head ........................................ HELIANTHUS

16 Leaves decurrent down the stem; achenes flat, usually winged, forming a globose head ........................... VERBESINA

17 Leaves simple, 3-lobed, or -cleft, or laciniate; rays not subtended by receptacular bracts; achenes 4-sided .............. RUDBECKIA

17 Leaves pinnately divided; rays subtended by receptacular bracts; achenes laterally flattened .......................... RATIBIDA

### *Tribe 8a.* HELIANTHEAE *(Ambrosiinae, Ragweed Subtribe)*
*Ambrosia, Iva, Xanthium*

1 Staminate and pistillate florets in the same head; ray florets pistillate, disk florets staminate ............................ IVA

1 Staminate and pistillate florets borne in separate heads ........ 2

2 Pistillate heads 2-flowered, with many, sharp-hooked spines; staminate heads lacking involucral bracts ......... XANTHIUM

2 Pistillate heads 1 (-2) -flowered with a few vestigial spines or none; staminate heads with involucres of connate involucral bracts .... .................................................. AMBROSIA

### *Tribe 9.* INULEAE *(Elecampane Tribe)*
*one genus: Inula*

### *Tribe 10.* SENECIONEAE *(Groundsel Tribe)*

*Arnoglossum, Erechtites, Hasteola, Packera, Petasites, Senecio, Tussilago*

1 Perennials with green leaves arising individually from the ground from an underground rhizome; aerial stems consisting of scaly bracted flowering scapes arising before or as the leaves develop in early spring ........................................ 2

1 Habit various but with well developed cauline leaves (though these may differ from the basal leaves) ........................... 3

2 Heads solitary, yellow, radiate with ray florets pistillate and disk florets staminate; radical leaves rounded-cordate, dentate and very shallowly lobed; rare adventive .................. **TUSSILAGO**

2 Heads numerous, creamy-white, radiate but with disk florets largely staminate or largely pistillate on different plants (imperfectly dioecious); radical leaves either merely dentate or deeply lobed . ................................................ **PETASITES**

3 Corollas yellow to orange; heads usually with rays ............. 4

3 Corollas whitish or creamy; heads without rays ................ 6

4 Stem leaves progressively reduced upward and lobed (unlike the basal leaves); perennials, usually with obvious vegetative reproduction ....................................... **PACKERA**

4 Leaves more or less equal in size up the stem; annuals (perhaps rarely biennials) ........................................... 5

5 Rays conspicuous; leaves entire to weakly toothed; pubescence often copious; historical records only, now likely absent from Wisc ........................................... **TEPHROSERIS**

5 Rays inconspicuous or absent; leaves, or some of them, lobed to pinnatifid; pubescence short and often sparse; introduced weeds ................................................... **SENECIO**

6 Annuals; heads disciform, with 2 to several marginal rows of pistillate florets with filiform corollas; leaves roughly the same size up the stem .................................. **ERECHTITES**

6 Perennials; heads discoid, containing only bisexual florets with 5-lobed corollas ........................................... 7

7 Heads with ca. 13 involucral bracts and 20-40 florets; receptacle flat; larger leaves hastate; leaves roughly the same size up the stem ... ............................................... **HASTEOLA**

7 Heads with ca. 5 involucral bracts and ca. 5 florets; receptacle with a short conic projection in the center; leaves not hastate; leaves largest at base of the stem and becoming smaller upwards ...... ........................................... **ARNOGLOSSUM**

### *Tribe 11.* VERNONIEAE *(Ironweed Tribe)*

*one genus: Vernonia*

## ACHILLEA *Yarrow*

Perennial herbs.

*Key to species*

1 Leaves finely dissected; plants tomentose; ubiquitous ........... ................................................ ***A. millefolium***

1 Leaves nearly entire; plants nearly hairless; uncommon adventive species ........................................... ***A. ptarmica***

***Achillea millefolium*** L. [•23] COMMON YARROW June–Oct. Common in fields, prairies, lawns, beaches, and waste places.

***Achillea ptarmica*** L. SNEEZEWEED July–Sept. Beaches, roadsides, and waste places; native to n Europe and Asia. Forms escaped from cultivation are often "double," with more than the usual number of ray flowers.

## AGERATINA *Snakeroot*

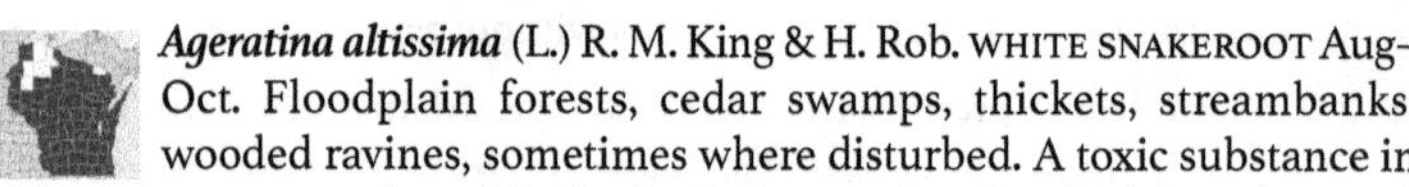

***Ageratina altissima*** (L.) R. M. King & H. Rob. WHITE SNAKEROOT Aug–Oct. Floodplain forests, cedar swamps, thickets, streambanks, wooded ravines, sometimes where disturbed. A toxic substance in this plant can cause "trembles," a fatal disease of cattle which have browsed on it and transmittable to humans by their milk, in whom the consequent "milk sickness" caused many deaths in the 19th century. *Eupatorium rugosum* Houtt.

## AMBROSIA *Ragweed*

The pollen of Ambrosia is wind-borne, and some species are among the most important causes of hay-fever in the USA.

*Key to species*

1 Pistillate involucre with 2 flowers and 2 sharp beaks; uncommon ................................................ ***A. tomentosa***

1 Pistillate involucre with 1 flower and 1 beak; widespread ........ 2

2 Leaves palmately 3–5 lobed or unlobed; large annual plant to 2 m or more tall ............................................ ***A. trifida***

2 Leaves 1–2 times pinnately lobed or divided; plants usually less than 1 m tall ............................................... 3

3 Plants perennial, forming colonies from creeping underground roots; leaves usually coarsely lobed ............... ***A. psilostachya***

3 Plants taprooted annuals; leaves finely divided ... ***A. artemisiifolia***

***Ambrosia artemisiifolia*** L. COMMON RAGWEED Aug–Oct. Waste places.

***Ambrosia psilostachya*** DC. PERENNIAL RAGWEED Similar to *A. artemisiifolia.* July–Oct. Waste places, usually in dry or sandy soil.

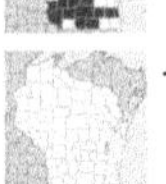

***Ambrosia tomentosa*** Nutt. SKELETON-LEAF BUR-RAGWEED May–Aug. Plains species, casually introduced as far east as se Wisc and Illinois.

***Ambrosia trifida*** L. [•23] GIANT RAGWEED July–Oct. Invasive in moist soil and waste places.

## ANAPHALIS *Pearly-Everlasting*

***Anaphalis margaritacea*** (L.) Benth. [•23] PEARLY-EVERLASTING July–Aug. Chiefly in dry woods and clearings.

## ANTENNARIA *Pussytoes*

Most of our species are partly or wholly apomictic, producing seeds without fertilization.

*Key to species*

1 Rosette leaves small, 1-nerved or obscurely 3-nerved (best viewed on old basal leaves from the previous year)..................... 2

1 Rosette leaves larger, 3- or 5-nerved ............................ 3

2 New basal leaves of the season essentially glabrous above or very soon becoming so (may appear hairy along the margin from tomentum of underside) ............................ ***A. howellii***

2 New basal leaves pubescent above when young (becoming glabrous only in age) ........................................ ***A. neglecta***

3 Basal leaves glabrous or tomentose on upper surface, underside green-glabrous; pistillate involucres (7–)8–13 mm; staminate corollas 3.5–5 mm; pistillate corollas 4–7 mm; young stolons mostly decumbent ........................................ ***A. parlinii***

3 Underside of basal leaves tomentose; pistillate involucres 5–7 mm; staminate corollas 2–3.5 mm; pistillate corollas 3–4 mm; young stolons mostly ascending....................... ***A. plantaginifolia***

***Antennaria howellii*** Greene SMALL PUSSYTOES Many types of dry, open places, including rock ledges and outcrops, openings in sandy or rocky woods; sometimes on moist shores, roadsides, and in fields and lawns. *Antennaria neglecta* var. *howellii* (Greene) Cronquist.

***Antennaria neglecta*** Greene FIELD PUSSYTOES April–June. Dry woods and open places. *Antennaria neodioica* Greene.

***Antennaria parlinii*** Fernald PARLIN'S PUSSYTOES Our most common Antennaria, found in dry open places, including rock outcrops, banks, grassy roadsides, hillsides, and open woods; sometimes in shaded forests. Similar to *A. plantaginifolia* and sometimes included in it as *A. plantaginifolia* var. *parlinii* Cronquist.

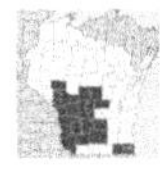

***Antennaria plantaginifolia*** (L.) Richards. [•23] PLANTAIN-PUSSYTOES Spring–early summer. Dry, open, woodlands, banks, bluff tops. Similar to *A. parlinii* except for smaller heads, underside of basal leaves gray-pubescent.

## ANTHEMIS *Chamomile*

*Anthemis* and *Matricaria* are very similar, separated by the receptacle chaffy in *Anthemis* and naked in *Matricaria.*

*Key to species*

1 Ray flowers yellow .................................. *A. tinctoria*
1 Ray flowers white .................................. 2
2 Ray flowers pistillate and fertile; receptacle chaffy throughout ... .................................. *A. arvensis*
2 Ray flowers usually neutral, sterile; receptacle chaffy only near middle .................................. *A. cotula*

***Anthemis arvensis*** L. CORN CHAMOMILE May–Aug. Fields and waste places; native of Europe.

***Anthemis cotula*** L. STINKING CHAMOMILE, DOGFENNEL May–Oct. Native of Europe; fields and waste places.

***Anthemis tinctoria*** L. GOLDEN CHAMOMILE June–July. Fields and waste places; native of Europe.

## ARCTIUM *Burdock*

Coarse biennial herbs. Leaves large, alternate, heart-shaped, entire or toothed. Flowers all tubular and perfect, the corolla pink or purplish; pappus of numerous short bristles. Any of our introduced species will hybridize with any of the others.

1 Heads 1–1.5 cm long, sessile or on short peduncles; common weed *A. minus*
1 Heads 1.5–2.5 cm long; on long peduncles; uncommmon........ 2
2 Heads about 1.5 cm long; bracts densely cottony-hairy .......... .................................. *A. tomentosum*
2 Heads about 2.5 cm long; bracts glabrous ............... *A. lappa*

***Arctium lappa*** L. GREATER BURDOCK Aug–Oct. Native of Eurasia, sparingly established as a weed along along roadsides and in waste places.

***Arctium minus*** Bernh. LESSER BURDOCK Introduced (invasive). Roadsides, railroads, fields, fencerows, farmyards, around old buildings, disturbed places.

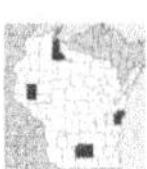

***Arctium tomentosum*** P. Mill. WOOLLY BURDOCK Similar to *A. lappa,* but smaller. June–Oct. Native of Eurasia, sparingly established on roadsides and in waste places.

## ARNOGLOSSUM *Indian Plantain*

Large perennial herbs with basal or alternate leaves. Flower heads with white disk flowers only, the ray flowers absent. Fruit an achene, tipped by a pappus of numerous, slender bristles.

*Key to species*

1 Main leaves clearly longer than wide, margins nearly entire to toothed .................................. *A. plantagineum*
1 Main leaves broadly ovate to kidney-shaped, about as wide as long, margins coarsely toothed or undulate .......................... 2

2 Leaf undersides strongly glaucous-whitened; lower leaves broadly ovate ........................................ *A atriplicifolium*

2 Leaves not glaucous; lower leaves kidney-shaped .... *A. reniforme*

***Arnoglossum atriplicifolium*** (L.) H.E. Robins. PALE INDIAN-PLANTAIN Similar to *A. reniforme.* July–Sept. Woodlands and moist or rather dry open places. *Cacalia atriplicifolia* L.

***Arnoglossum plantagineum*** Raf. GROOVE-STEM INDIAN-PLANTAIN June–July. Wet to sometimes dry prairies; marshy or boggy places. *Cacalia plantaginea* (Raf.) Shinners.

***Arnoglossum reniforme*** (Hook.) H.Rob. [•24] GREAT INDIAN-PLANTAIN June–Sept. Woodlands. *Arnoglossum muhlenbergii* (Schultz-Bip.) H.E. Robins., *Cacalia muhlenbergii* (Schultz-Bip.) Fern.

## ARTEMISIA *Wormwood; Sage*

Annual, biennial, or perennial herbs, or shrubs, usually aromatic, with alternate, entire to dissected leaves and few to numerous small heads in a spiciform, raceme-like, or panicle-like inflorescence.

ADDITIONAL SPECIES ***Artemisia abrotanum*** L. (Southern wormwood), introduced, reported from Racine and Sheboygan counties.

*Key to species*

1 Plants perennial and somewhat woody at base; leaves covered with silky hairs; receptacle hairy.................................. 2

1 Plants annual, biennial, or perennial; leaves hairy to glabrous; receptacle naked ........................................... 3

2 Leaves to 2 cm long, the segments thread-like, to 1 mm wide; flowering stems ascending, to 5 dm tall ............... *A. frigida*

2 Leaves 5–15 cm long, the segments 2–3 mm wide; flowering stems erect, to 9 dm tall ................................. *A. absinthium*

3 Disk flowers sterile; mature plants usually glabrous ............ 4

3 Disk flowers fertile .............................................. 5

4 First-year lower leaves in a basal rosette; leaves usually densely hairy; sandy habitats .......................... *A. campestris*

4 Lower leaves not in a rosette; leaves glabrous; uncommon ....... ................................................ *A. dracunculus*

5 Leaves glabrous or nearly so, pinnately divided or dissected..... 6

5 Leaves densely hairy at least on one surface, simple or dissected 7

6 Inflorescence a dense raceme-like panicle, the heads erect; common ................................................... *A. biennis*

6 Inflorescence a wide panicle, the heads nodding; uncommon ... ..................................................... *A. annua*

7 Leaves linear lance-shaped, unlobed, margins regularly toothed to entire, densely white hairy on underside .............. *A. serrata*

7 Leaves deeply lobed or cut, or entire with irregular teeth on margin ........................................................ 8

8 Leaves finely divided into thread-like segments; uncommon garden escape ............................................ *A. pontica*

8 Leaves entire or the leaf segments broader ....................9

9 Leaves green and nearly glabrous above, white hairy below; uncommon weed ....................................*A. vulgaris*
9 Leaves hairy on upper and lower sides .......................10

10 Leaves entire or irregularly toothed; common .....*A. ludoviciana*
10 Leaves obtusely lobed; uncommon garden escape near shores of Lake Michigan.....................................*A. stelleriana*

***Artemisia absinthium*** L. COMMON WORMWOOD July–Sept. Native of Europe; fields and waste places.

***Artemisia annua*** L. ANNUAL WORMWOOD Aug–Nov. Fields and waste places; native to Asia and eastern Europe, now naturalized.

***Artemisia biennis*** Willd. [•24] BIENNIAL WORMWOOD Aug–Sept. Sandy lakeshores, streambanks, ditches, mud flats, disturbed areas; often where seasonally flooded. Native to nw USA, throughout Wisc as a weed.

***Artemisia campestris*** L. FIELD SAGEWORT July–Sept. Open places, often in sandy soil.

***Artemisia dracunculus*** L. DRAGON WORMWOOD July–Sept. Dry open places.

***Artemisia frigida*** Willd. PRAIRIE SAGEWORT July–Sept. Prairies and dry open places.

***Artemisia ludoviciana*** Nutt. WHITE SAGE July–Oct. Prairies, dry ground, and waste places.

***Artemisia pontica*** L. ROMAN WORMWOOD Aug–Sept. Native of Europe, escaped from cultivation and sparingly established in dry open places.

***Artemisia serrata*** Nutt. TOOTHED SAGE Aug–Oct. Prairies and low ground.

***Artemisia stelleriana*** Bess. DUSTY MILLER May–Sept. Sandy beaches; native of Asia, escaped from cultivation.

***Artemisia vulgaris*** L. [•24] MUGWORT July–Oct. Fields, roadsides, and waste places; Old World native, now established throughout most of e North America.

### BELLIS *English Daisy*

***Bellis perennis*** L. LAWN DAISY, ENGLISH DAISY Introduced perennial herb. Leaves basal. Heads solitary atop a scape. Rays many, pistillate, white to pink or purple; disk flowers yellow. April–Nov. Weedy in lawns or waste places.

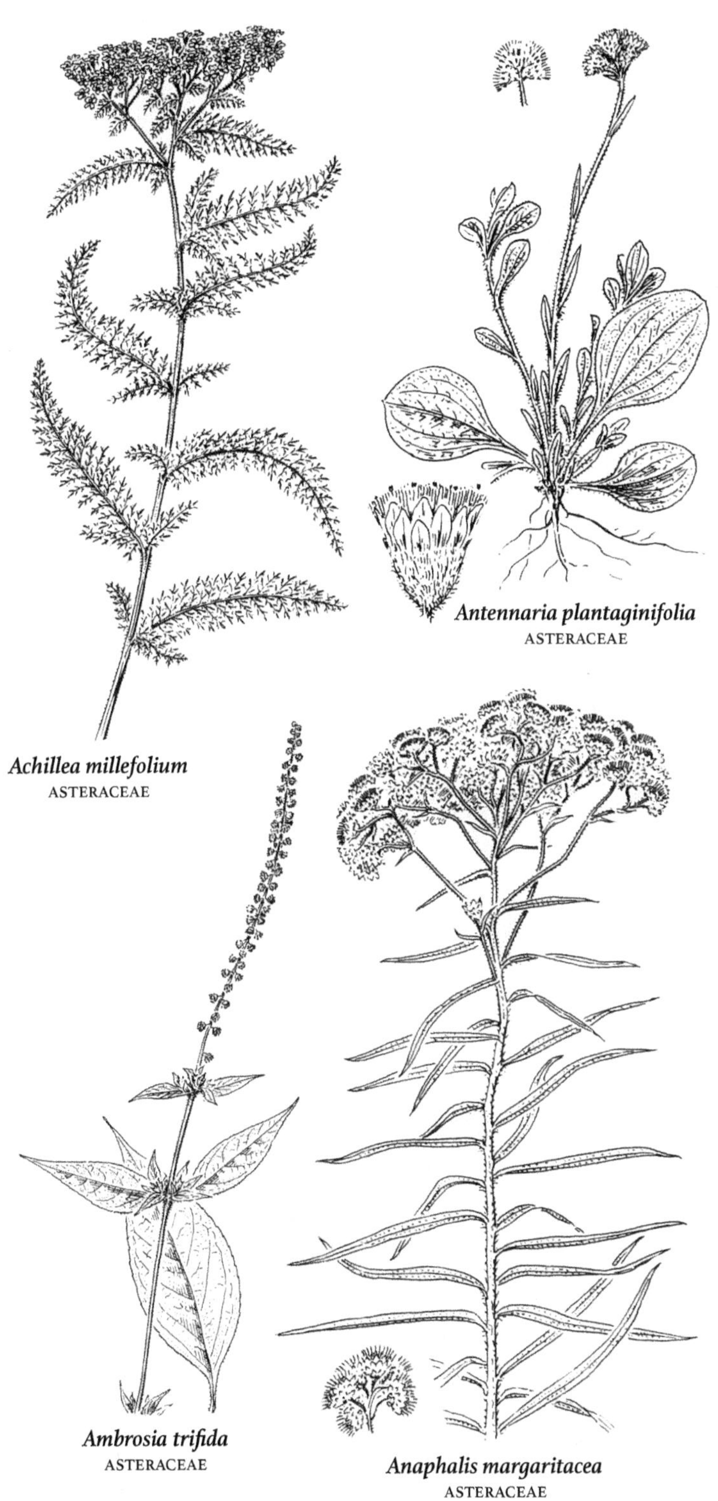
*Antennaria plantaginifolia*
ASTERACEAE
*Achillea millefolium*
ASTERACEAE
*Ambrosia trifida*
ASTERACEAE
*Anaphalis margaritacea*
ASTERACEAE

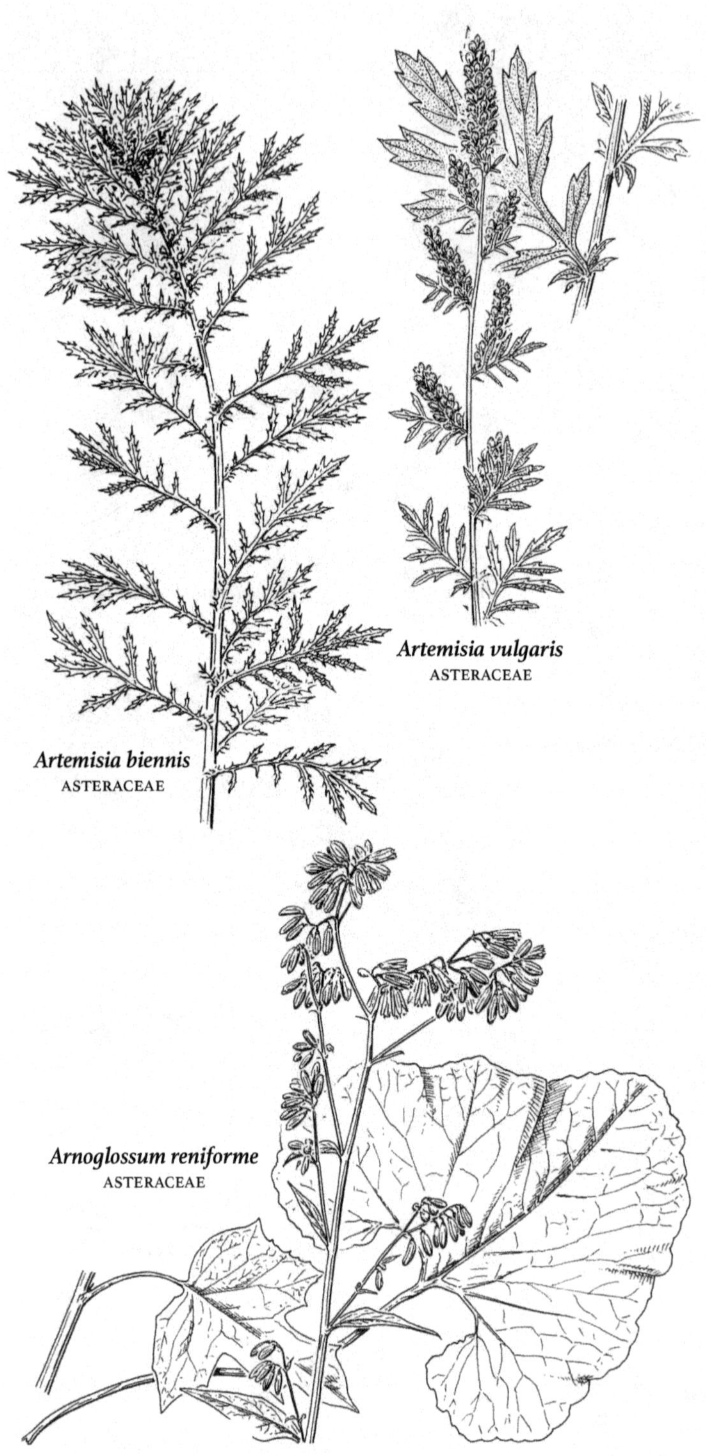
*Artemisia vulgaris*
ASTERACEAE
*Artemisia biennis*
ASTERACEAE
*Arnoglossum reniforme*
ASTERACEAE

## BIDENS *Beggarticks*

Weedy annual herbs; perennial in the aquatic *B. beckii.* Leaves opposite (or whorled in *B. beckii*), simple, lobed, or pinnately divided. Flower heads with both disk and ray flowers, or with disk flowers only. Fruit a flattened achene; pappus of 2–5 barbed awns which persist atop the achene; the body of achene barbed or with stiff hairs (at least on the angles), the "stick-tights" facilitating dispersal of seed by animals.

ADDITIONAL SPECIES ***Bidens pilosa*** L. (Spanish-needles); reported from Rock County in s Wisc where adventive.

*Key to species*

1 Plants aquatic; underwater leaves whorled, dissected into narrow segments .................... ***B. beckii***

1 Plants not aquatic (sometimes emergent); leaves not as above ... 2

2 Leaves simple and toothed, or sometimes lobed; achenes 3–4-awned.................... 3

2 Leaves all (or mostly) pinnately divided or compound; achenes 2-awned.................... 6

3 Leaves mostly sessile .................... 4

3 Leaves with a petiole 1–4 cm long.................... 5

4 Heads nodding when mature; outer involucral bracts widely spreading .................... ***B. cernua***

4 Heads mostly upright; outer involucral bracts erect or nearly so .................... ***B. tripartita***

5 Disk flowers 4-lobed, pale yellow; stamens shorter than the lobes; achenes mostly 3-awned .................... ***B. tripartita***

5 Disk flowers 5-lobed, yellow-orange; stamens longer than lobes; achenes mostly 4-awned .................... ***B. connata***

6 Heads with disk flowers only, or with short rays less than 5 mm long .................... 7

6 Heads with both disk and ray flowers, the rays over 1 cm long ... 9

7 Outer involucral bracts 2–5 (usually 4), not fringed with hairs .................... ***B. discoidea***

7 Outer involucral bracts 6 or more, fringed with hairs (at least near base) .................... 8

8 Disk flowers orange; outer involucral bracts mostly 6–8 ***B. frondosa***

8 Disk flowers yellow; outer involucral bracts 10 or more . ***B. vulgata***

9 Achenes broad and ovate, mostly more than 3 mm wide ***B. aristosa***

9 Achenes narrow and nearly straight-sided, less than 3 mm wide .................... ***B. trichosperma***

***Bidens aristosa*** (Michx.) Britt. BEARDED BEGGARTICKS Aug–Sept. Marshy areas, ditches, disturbed wetlands.

***Bidens beckii*** Torr. ex Spreng. BECK'S WATER-MARIGOLD June–Sept. Quiet, shallow to deep water of lakes, ponds, rivers and streams. *Megalodonta beckii* (Torr.) Greene.

***Bidens cernua*** L. NODDING BUR-MARIGOLD July–Oct. Exposed, sandy or muddy shores, streambanks, marshes, forest depressions, wet meadows, ditches and other wet places.

***Bidens connata*** Muhl. PURPLE-STEM BEGGARTICKS Aug–Oct. Exposed muddy shores, streambanks, marshes, pond, forest depressions, wet meadows, ditches and other wet places. Sometimes included in *B. tripartita.*

***Bidens discoidea*** (Torr. & Gray) Britt. [•25] SMALL BEGGARTICKS Aug–Sept. Hummocks or logs in swamps, exposed muddy shores; usually where shaded.

***Bidens frondosa*** L. DEVIL'S-PITCHFORK July–Oct. Wet, sandy or gravelly shores, forest depressions, streambanks, pond margins; weedy in wet disturbed areas.

***Bidens trichosperma*** (Michx.) Britton [•25] CROWNED BEGGARTICKS July–Oct. Open bogs, fens, tamarack swamps, shores, streambanks, marshes, sand bars. *Bidens coronata* (L.) Britt.

***Bidens tripartita*** L. THREE-LOBE BEGGARTICKS Aug–Oct. Exposed shores, streambanks, mudflats, forest depressions, pond, wet meadows, ditches and other wet places. *Bidens comosa* (Gray) Wieg.

***Bidens vulgata*** Greene TALL BEGGARTICKS Aug–Oct. Streambanks, wet meadows, wet forests; weedy in moist disturbed areas. Similar to *B. frondosa,* but usually larger. *Bidens puberula* Wieg.

### BOLTONIA *Doll's Daisy*

***Boltonia asteroides*** (L.) L'Hér. WHITE BOLTONIA Perennial herb. Aug–Sept. Seasonally flooded muddy shores, wet meadows, marshes, low prairie.

### BRICKELLIA *Brickellbush*

***Brickellia eupatorioides*** (L.) Shinners FALSE BONESET Perennial herbs from a stout taproot. Leaves mostly alternate. Heads discoid, the flowers all tubular and perfect, mostly in small clusters terminating the branches. Flowers creamy-white. Aug–Oct. Dry open places, especially in sandy soils. *Kuhnia eupatorioides* L.

### CARDUUS *Thistle*

Annual, biennial, or perennial spiny herbs. Stems generally winged by the decurrent leafbases. Closely related to *Cirsium,* from which it is distinguished primarily by the non-plumose pappus.

*Key to species*

1 Heads nodding; involucral bracts 2 mm wide or more .. ***C. nutans***
1 Heads not nodding; involucral bracts narrow, less than 2 mm wide .......................................... ***C. acanthoides***

***Carduus acanthoides*** L. SPINY PLUMELESS-THISTLE July–Oct. Roadsides, pastures, and waste places; native of Europe.

***Carduus nutans*** L. MUSK THISTLE June–Oct. Native of Europe and w Asia, established on roadsides and waste places.

## CENTAUREA *Knapweed; Star-Thistle*

Annual, biennial, or perennial herbs, often weedy.

ADDITIONAL SPECIES ***Centaurea americana*** Nutt. (American basket-flower), native of south-central USA, Waukesha County. ***Centaurea benedicta*** (L.) L. (Blessed-thistle), introduced annual, known from Dane County. ***Centaurea macrocephala*** Puschk. ex Willd. (Globe knapweed), introduced, reported from Bayfield County. ***Centaurea montana*** L. (Mountain cornflower), introduced and known from several Wisc counties. ***Centaurea scabiosa*** L. (Greater knapweed), introduced, Shawano County.

*Key to species*

1 Involucral bracts tipped by short to long spines ................ 2
1 Involucral bracts not tipped by long spines; leaf bases not decurrent ................ 4

2 Leaf bases not decurrent; stem angled, not winged ..... *C. diffusa*
2 Leaf bases decurrent (extending downward) on winged stem .... 3

3 Central spines of bracts stout, to 2 cm long .......... *C. solstitialis*
3 Central spines very slender, less than 1 cm long ...... *C. melitensis*

4 Leaves pinnately divided into linear-elliptic lobes; common weed ................ *C. stoebe*
4 Leaves entire or toothed, sometimes few-lobed................ 5

5 Margins of involucral bracts entire or nearly so ......... *C. repens*
5 Margins of involucral bracts toothed, jagged or fringed ......... 6

6 Plants annual (or winter-annuals); leaves linear, less than 1 cm wide ................ *C. cyanus*
6 Plant perennial; leaves wider, at least some lower leaves more than 1 cm wide ................ 7

7 Chaffy tips of involural bracts tan to dark brown ......... *C. jacea*
7 Chaffy tips of involural bracts black ........................ 8

8 Chaffy tips small, 1–3 mm long ........................ *C. dubia*
8 Chaffy tips larger, 4–6 mm long ........................ *C. nigra*

***Centaurea cyanus*** L. GARDEN CORNFLOWER, BACHELOR'S BUTTON May–Oct. Native to the Mediterranean region, cultivated as an ornamental, and now weedy in fields, roadsides, and waste places.

***Centaurea diffusa*** Lam. WHITE KNAPWEED July–Sept. Native of Europe, adventive weed in waste places.

***Centaurea dubia*** Suter SHORT-FRINGED KNAPWEED July–Oct. Native of Europe; widely established in fields, roadsides, and waste places in ne USA and se Canada. *Centaurea nigrescens* Willd.

***Centaurea jacea*** L. BROWN-RAY KNAPWEED June–Sept. Fields, roadsides, and waste places; native of Europe.

***Centaurea melitensis*** L. MALTESE STAR-THISTLE European, weedy in waste places. Reported, but distribution unknown (*no map*).

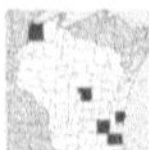

***Centaurea nigra*** L. LESSER KNAPWEED July–Oct. Fields, roadsides, and waste places; native of Europe, now widely established in ne USA and se Canada.

***Centaurea repens*** L. RUSSIAN KNAPWEED June–Sept. Native of Asia; fields, roadsides, and waste places. *Rhaponticum repens* (L.) Hidalgo.

***Centaurea solstitialis*** L. YELLOW STAR-THISTLE Native to the Mediterranean region; weedy in fields and waste places.

***Centaurea stoebe*** L. [•25] SPOTTED KNAPWEED June–Oct. Native of Europe; aggressive weed of fields, roadsides, waste places. *Centaurea maculosa* Lam.

## CICHORIUM *Chicory*

***Cichorium intybus*** L. [•26] CHICORY Perennial with milky juice, from a long deep taproot. July–Oct. Roadsides, fields, and waste places; native of Eurasia, now a cosmopolitan weed. The root is used as an adulterant or substitute for coffee.

## CIRSIUM *Thistle*

Biennial or perennial herbs. Stems and leaves often spiny.

*Key to species*

1 Involucral bracts tipped by spines mostly more than 2 mm long . **2**
1 Involucral bracts tipped by short spines to only 1 mm long ..... **8**

2 Leaves coarsely hairy, with cobwebby hairs on underside ....... **3**
2 Leaves densely white-hairy on both sides, especially on underside ........ **6**

3 Stem leaves decurrent at base, forming conspicuous spiny wings on stem ........ ***C. vulgare***
3 Stem leaves not decurrent; stems not appearing winged ........ **4**

4 Leaves green, with coarse hairs; stems 3–5 dm tall from persistent basal rosettes; rare in dry or moist prairies ............... ***C. hillii***
4 Leaves densely white-hairy on underside; stems 6 dm tall or more, the basal rosettes not persistent.............................. **5**

5 All leaves deeply lobed, the lobes tipped with stout spines 3–7 mm long; open habitats ................................. ***C. discolor***
5 Leaves mostly only shallowly lobed, tipped with small weak spines; wooded habitats ................................ ***C. altissimum***

6 Middle stem leaves conspicuously decurrent; rare native species along Lake Michigan dunes ......................... ***C. pitcheri***
6 Leaves not decurrent or only slightly so; uncommon introduced weeds ........ **7**

7 Leaves narrowed to their base, rarely clasping ....... ***C. flodmanii***
7 Leaves broadest near base, partially clasping ....... ***C. undulatum***

8 Colony-forming perennial herb from deep, creeping rhizomes; common weed of dry to moist places ................. ***C. arvense***
8 Biennial herbs; moist to wet habitats ........................ **9**

9 Leaf bases not decurrent; stem not winged; involucral bracts usually with cobwebby hairs; flowers deep rose-purple; native and not weedy ........................................ *C. muticum*

9 Leaf bases decurrent, forming spiny wings on stem; involucral bracts usually without cobwebby hairs; flowers pale pink-purple; introduced and weedy ................................ *C. palustre*

***Cirsium altissimum*** (L.) Spreng. TALL THISTLE July–Oct. Fields, waste places, river bottoms, and open woods.

***Cirsium arvense*** (L.) Scop. CANADIAN THISTLE July–Aug. A noxious weed of fields and waste places; native of Eurasia, now statewide.

***Cirsium discolor*** (Muhl.) Spreng. FIELD THISTLE July–Oct. Fields, open woods, river bottoms, and waste places. Similar to *C. altissimum* and intergrading with it; perhaps not specifically distinct.

***Cirsium flodmanii*** (Rydb.) Arthur PRAIRIE THISTLE June–Sept. Prairies, dry meadows; common in Great Plains, considered adventive in Wisc. Similar to *C. undulatum,* but smaller and more delicate.

***Cirsium hillii*** (Canby) Fern. HILL'S THISTLE *Threatened.* June–Aug. Prairies and other open places.

***Cirsium muticum*** Michx. SWAMP THISTLE Aug–Oct. Swamps, thickets, calcareous fens, sedge meadows, streambanks, shores.

***Cirsium palustre*** (L.) Scop. EUROPEAN SWAMP THISTLE June–Aug. Roadside ditches and adjacent wetlands, including swamps, thickets and fens; resembling the native *C. muticum* in these habitats. Introduced and spreading into wetlands, especially where disturbed.

***Cirsium pitcheri*** (Torr.) Torr. & Gray DUNE THISTLE *Threatened.* June–Aug. Uncommon plants of sand dunes along Lakes Michigan, Huron, and Superior. In their first year plants form a rosette of leaves flattened against the dune sands; the following year or after several years, plants bloom and then die.

***Cirsium undulatum*** (Nutt.) Spreng. WAVY-LEAF THISTLE Introduced. June–Sept. Prairies, railroad tracks, and other open places.

***Cirsium vulgare*** (Savi) Ten. [•26] BULL THISTLE June–Oct. Pastures, fields, roadsides, and waste places; native of Eurasia, now widely established as a weed in North America.

## CONYZA *Horseweed*

Annual or perennial herbs, often weedy. Closely related to *Erigeron.*

*Key to species*

1 Plants with spreading branches from base of plant; plants less than 3 dm tall ................................ *C. ramosissima*

1 Plants unbranched to the inflorescence; plants more than 3 dm tall ................................................ *C. canadensis*

*Conyza canadensis* (L.) Cronq. [•26] HORSEWEED Introduced. Late summer and autumn. Weed of waste places. *Erigeron canadensis* L.

*Conyza ramosissima* Cronq. DWARF FLEABANE Summer–fall. A weed in waste places, particularly in sandy soil or along streams. *Erigeron divaricatus* Michx.

## COREOPSIS *Tickseed*

Annual or perennial herbs or subshrubs.

*Key to species*

1 Stems with less than 5 pairs of leaves, the leaves clustered on the lower two-thirds of the stem, simple or sometimes with 1 or 2 lobes ........ ***C. lanceolata***

1 Stems with 5 or more pairs of leaves, the main leaves with 3 or more lobes or divisions ........ 2

2 Leaves stiffly 3-pronged, the blade tissue decurrent along the leaf midrib ........ ***C. palmata***

2 Leaves not stiffly 3-pronged, divided to the midvein into 3 or more narrow leaflets ........ 3

3 Plants more than 1 m high; leaves more than 5 mm wide ***C. tripteris***

3 Plants less than 1 m high; leaves less than 5 mm wide ........ 4

4 Disk flowers yellow; ray flowers yellow and more than 1.5 cm long ........ ***C. grandiflora***

4 Disk flowers red-brown; ray flowers yellow and often red-spotted at base, less than 1.5 cm long ........ ***C. tinctoria***

***Coreopsis grandiflora*** Hogg BIGFLOWER TICKSEED Introduced. May–June. Rather dry, often sandy places. The typical form is essentially glabrous.

***Coreopsis lanceolata*** L. LANCE-LEAF TICKSEED May–July. Dry, often sandy places.

***Coreopsis palmata*** Nutt. STIFF TICKSEED June–July. Prairies and open woods.

***Coreopsis tinctoria*** Nutt. PLAINS TICKSEED Introduced. June–Aug. Dry ground and waste places.

***Coreopsis tripteris*** L. TALL TICKSEED. July–Sept. Various habitats, sometimes used in prairie restorations and grown in gardens; reported from several locations in Wisc.

## CREPIS *Hawk's-Beard*

***Crepis tectorum*** L. NARROW-LEAF HAWK'S-BEARD Annual taprooted herb with milky juice. June–July. Native of Eurasia; naturalized in waste places.

ADDITIONAL SPECIES ***Crepis capillaris*** (L.) Wallr. (Smooth hawk's-beard), introduced in several Wisc locations. ***Crepis pulchra*** L. (Small hawk's-beard), an Eurasian introduction, known in Wisc from along a railroad in Grant

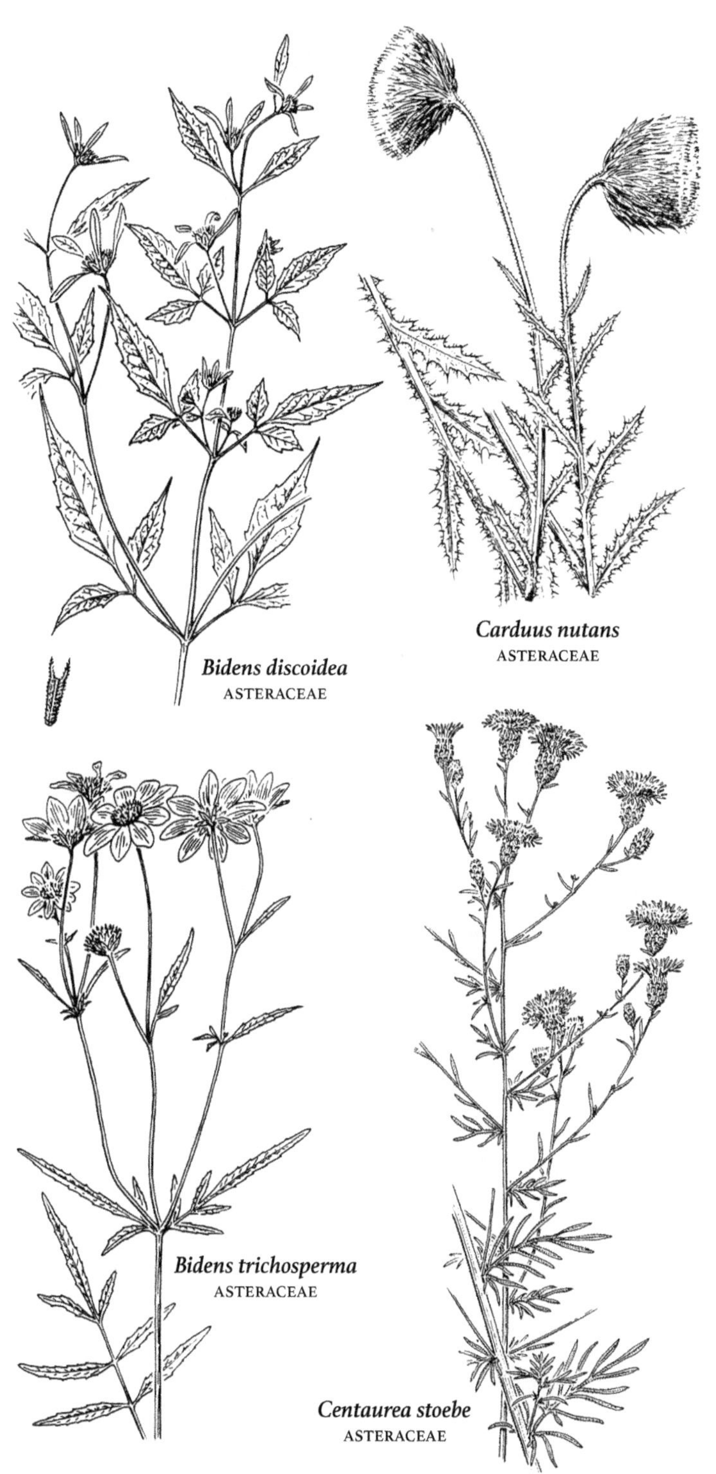
Bidens discoidea
ASTERACEAE
Carduus nutans
ASTERACEAE
Bidens trichosperma
ASTERACEAE
Centaurea stoebe
ASTERACEAE

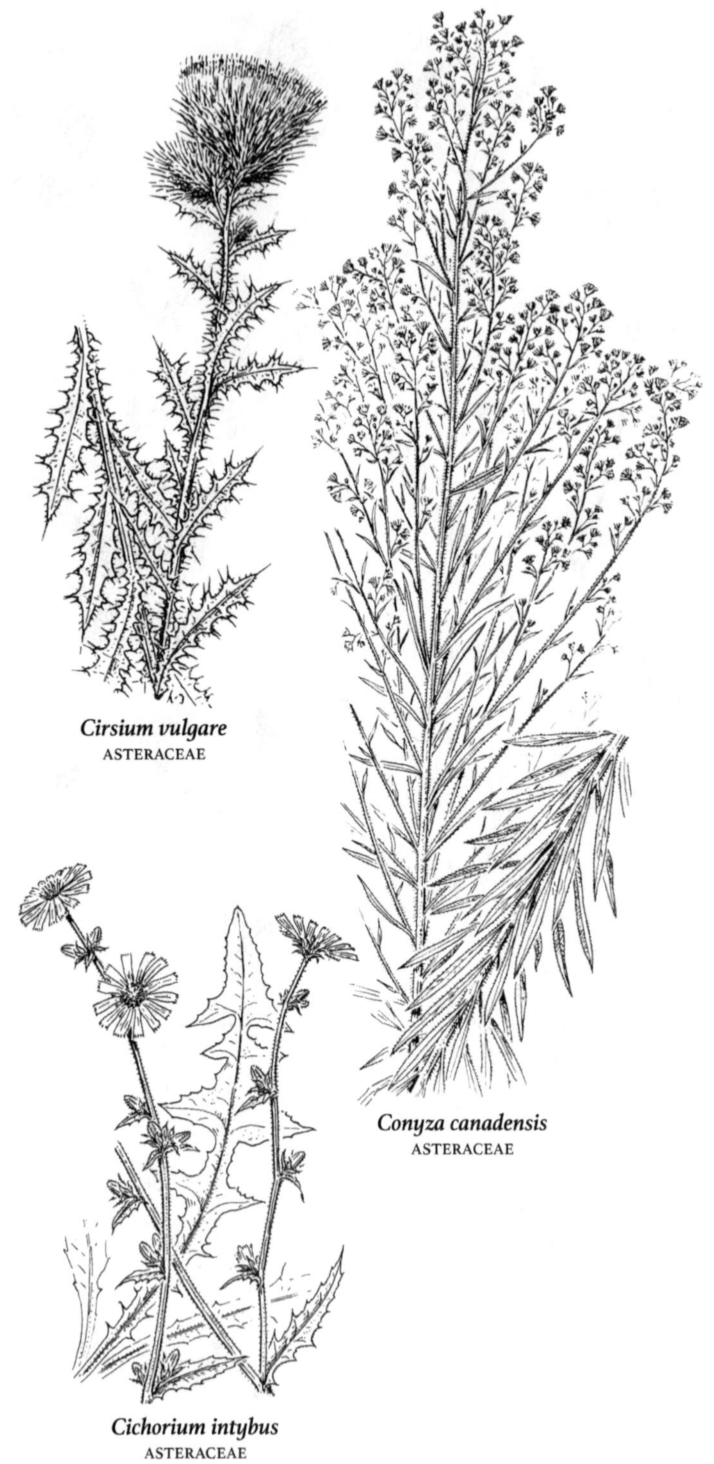

*Cirsium vulgare*
ASTERACEAE

*Conyza canadensis*
ASTERACEAE

*Cichorium intybus*
ASTERACEAE

County. ***Crepis setosa*** Haller f. (Bristly hawk's-beard), introduced, reported from Fond du Lac County.

### DOELLINGERIA *Flat-Topped White Aster*

***Doellingeria umbellata*** (P. Mill.) Nees TALL FLAT-TOPPED WHITE ASTER Perennial herb, from thick rhizomes. July–Sept. Openings in swamps and moist forests, thickets, streambanks, sedge meadows, calcareous fens, roadside ditches. *Aster umbellatus* P. Mill.

### ECHINACEA *Coneflower*

Perennial herbs. Leaves simple, alternate.

*Key to species*

1 Leaf margins entire, the blades linear to narrowly lance-shaped, 5 times or more longer than wide ...................... ***E. pallida***

1 Leaf margins mostly toothed, the blades ovate, less than 5 times longer than wide ................................ ***E. purpurea***

***Echinacea pallida*** (Nutt.) Nutt. PRAIRIE CONEFLOWER *Threatened.* May–Aug. Dry or dry-mesic prairies, roadsides, and along railroad rights-of-way.

***Echinacea purpurea*** (L.) Moench PURPLE CONEFLOWER Introduced. June–Oct. Woodlands and prairies.

### ECHINOPS *Globe-Thistle*

***Echinops sphaerocephalus*** L. GREAT GLOBE-THISTLE Coarse perennial, to 2.5 m tall. July–Sept. Native of Eurasia, several species are in cultivation; casually established in waste places in e Wisc.

### ECLIPTA *False Daisy*

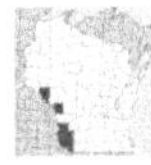

***Eclipta prostrata*** (L.) L. FALSE DAISY, YERBA-DE-TAJO Annual herb; leaves opposite. July–Oct. Mud flats, muddy streambanks and ditches (where somewhat weedy). *Verbesina alba* L.

### ERECHTITES *Fireweed*

***Erechtites hieraciifolius*** (L.) Raf. [•27] FIREWEED Fibrous-rooted annual herb. Aug–Sept. Various habitats: dry woods, marshes, waste places, often abundant after fires. *Senecio hieraciifolius* L.

### ERIGERON *Daisy; Fleabane*

Biennial to perennial herbs with simple, alternate leaves.

ADDITIONAL SPECIES ***Erigeron acris*** L. (Bitter fleabane) Ashland and Bayfield counties.

*Key to species*

1 Pappus of the ray flowers short, less than 1 mm long; weedy annual herbs...................................................... 2

1 Pappus of long bristles; biennial or perennial herbs ............ 3

2 Plants 6 dm or more tall; stems leafy; pubescence on middle of stem long and spreading ................................ ***E. annuus***

2 Plants to 7 dm tall; stem leaves few; pubescence mostly short and appressed ........................................ *E. strigosus*

3 Disk flowers less than 4 mm long; rays very narrow, less than 1 mm wide ........................................ *E. philadelphicus*

3 Disk flowers 4–6 mm long; rays about 1 mm wide .............. 4

4 Plants with shallow rhizomes or stolons; rays 50–100 . *E. pulchellus*

4 Plants without rhizomes or stolons; rays 125 or more .. *E. glabellus*

***Erigeron annuus*** (L.) Pers. EASTERN DAISY FLEABANE Early and middle summer. A weed over most of n USA and s Canada.

***Erigeron glabellus*** Nutt. STREAMSIDE FLEABANE Meadows, prairies.

***Erigeron philadelphicus*** L. [•27] PHILADELPHIA DAISY May–Aug. Wet meadows, shores, wet woods, floodplains, springs; also weedy in open disturbed areas and lawns.

***Erigeron pulchellus*** Michx. ROBIN'S PLANTAIN Spring. Woodlands and streambanks.

***Erigeron strigosus*** Muhl. [•27] ROUGH FLEABANE Early and mid-summer. A weed in much of the USA and s Canada.

## EUPATORIUM *Joe-Pye-Weed; Boneset*

Perennial herbs from a thick rhizome; leaves opposite, and sometimes joined at base, the stem passing through the joined leaves.

*Key to species*

1 Leaf bases (except sometimes the uppermost) joined around the stem ........................................ *E. perfoliatum*

1 Leaf bases entirely free and separate from the stem ............ 2

2 Leaves with distinct petioles; florets 8–50 (or more) per head ..... ........................................ *E. serotinum*

2 Leaves sessile or nearly so (at most tapering to a winged base); florets fewer than 8 (usually 5) per head ...................... 3

3 Stems densely finely hairy on middle internodes; leaves tapered to narrow base (or short-winged petiole), with 3 prominent longitudinal veins (at least on lower portion of blade) *E. altissimum*

3 Stems glabrous on middle internodes (or sparsely puberulent); leaves truncate to broadly rounded at base, with only 1 long vein (the midrib) prominent beneath ................. *E. sessilifolium*

***Eupatorium altissimum*** L. THROUGHWORT Aug–Oct. Mesic prairies, wooded bluffs, mine tailings.

***Eupatorium perfoliatum*** L. BONESET July–Sept. Marshes, wet meadows, low prairie, shores, streambanks, ditches, cedar swamps, thickets, calcareous fens. Often growing with spotted joe-pye-weed (*Eutrochium maculatum*).

***Eupatorium serotinum*** Michx. LATE-FLOWERING THOROUGHWORT Introduced. Aug–Oct. Mostly in bottomlands and moist woods, sometimes where drier.

***Eupatorium sessilifolium*** L. UPLAND BONESET Aug–Sept. Woodlands.

### EURYBIA *Wood-Aster*

Perennial herbs, spreading via rhizomes. Stems typically simple. Leaves basal and alternate on the stem. Flower heads radiate and corymbiform. Ray florets pistillate and fertile; white to purple; disk florets bisexual and fertile, yellow, becoming purple at maturity. Pappus persistent and bristly.

*Key to species*

1 Inflorescence glandular; rays light-purple ......... ***E. macrophylla***
1 Inflorescence not glandular; rays white ....................... 2

2 Plants with well developed clusters of basal leaves, the clusters on separate short shoots .............................. ***E. schreberi***
2 Plants not with well developed clusters of basal leaves .. ***E. furcata***

***Eurybia furcata*** (Burgess) Nesom FORKED WOOD-ASTER *Threatened.* Aug–Oct. Oak woods, moist woods and woodland edges, often near streams, and sometimes associated with disturbance. *Aster furcatus* Burgess.

***Eurybia macrophylla*** (L.) Cass. LARGE-LEAF WOOD-ASTER July–Oct. Woodlands. *Aster macrophyllus* L.

***Eurybia schreberi*** (Nees) Nees SCHREBER'S WOOD-ASTER July–Oct. Woodlands. *Aster schreberi* Nees.

### EUTHAMIA *Flat-Topped Goldenrod*

Perennial herbs, spreading by rhizomes; leaves alternate, covered with resinous dots.

*Key to species*

1 Largest stem leaves 4 mm or more wide, with 3 conspicuous longitudinal veins; leaves and upper stem short-hairy; upper leaves dull, glandular dots usually indistinct ............. ***E. graminifolia***
1 Largest stem leaves less than 3 mm wide, with single longitudinal vain (midrib) and sometimes with faint pair of longitudinal veins; leaves and upper stem smooth; upper leaves shiny, with conspicuous glandular dots ................. ***E. gymnospermoides***

***Euthamia graminifolia*** (L.) Greene [•28] COMMON FLAT-TOPPED GOLDENROD Aug–Sept. Shores, wet meadows, low prairie, springs, fens, swamps, interdunal wetlands, often where sandy or gravelly; also weedy in abandoned fields. *Solidago graminifolia* (L.) Salisb.

***Euthamia gymnospermoides*** Greene TEXAS GOLDENTOP Aug.–Oct. Open, often sandy places. *Solidago gymnospermoides* (Greene) Fern.

### EUTROCHIUM *Joe-Pye-Weed*

Perennial herbs from a thick rhizome; leaves whorled.

*Key to species*

1 Stems green, purple only at nodes, not purple-spotted; heads with usually 3–6 flowers; dry woods .................... ***E. purpureum***

1 Stems purple throughout or purple-spotted; heads with more than 8 flowers; common and widespread in wet habitats . ***E. maculatum***

***Eutrochium maculatum*** (L.) E. Lamont [•27] SPOTTED JOE-PYE-WEED July–Sept. Wet meadows, marshes, shores, streambanks, ditches, cedar swamps, bogs, calcareous fens. *Eupatorium maculatum* L.

***Eutrochium purpureum*** (L.) E. Lamont [•28] PURPLE-NODE JOE-PYE-WEED July–Sept. Thickets and open woods, often in drier habitats than *E. maculatum. Eupatorium purpureum* L.

## GAILLARDIA *Blanket-Flower*

***Gaillardia aristata*** Pursh COMMON BLANKET-FLOWER Introduced. Commonly perennial. May–Sept. Plains, meadows, and other open places; in Wisc, escaped from cultivation.

## GALINSOGA *Quickweed*

Annual herbs; leaves opposite.

*Key to species*

1 Leaves nearly entire to shallowly toothed; pappus of ray flowers absent ............................................ ***G. parviflora***

1 Leaves all sharply toothed; ray flowers with well developed pappus ........................................... ***G. quadriradiata***

***Galinsoga parviflora*** Cav. GALLANT-SOLDIER Introduced. June–Nov. Waste places.

***Galinsoga quadriradiata*** Cav. [•28] SHAGGY-SOLDIER Native of South and Central America; now found as a weed. Similar to and more common than *G. parviflora.*

## GNAPHALIUM *Cudweed*

Annual or perennial woolly herbs; leaves alternate, entire.

*Key to species*

1 Plants perennial; inflorescence narrow; achenes sparsely hairy ... ................................................... ***G. sylvaticum***

1 Plants annual; inflorescence various; achenes smooth or covered with small bumps ............................... ***G. uliginosum***

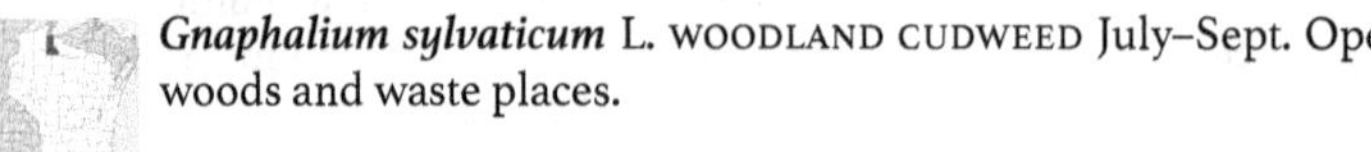

***Gnaphalium sylvaticum*** L. WOODLAND CUDWEED July–Sept. Open woods and waste places.

***Gnaphalium uliginosum*** L. MARSH CUDWEED Introduced (naturalized). July–Oct. Streambanks and waste places, wet or dry.

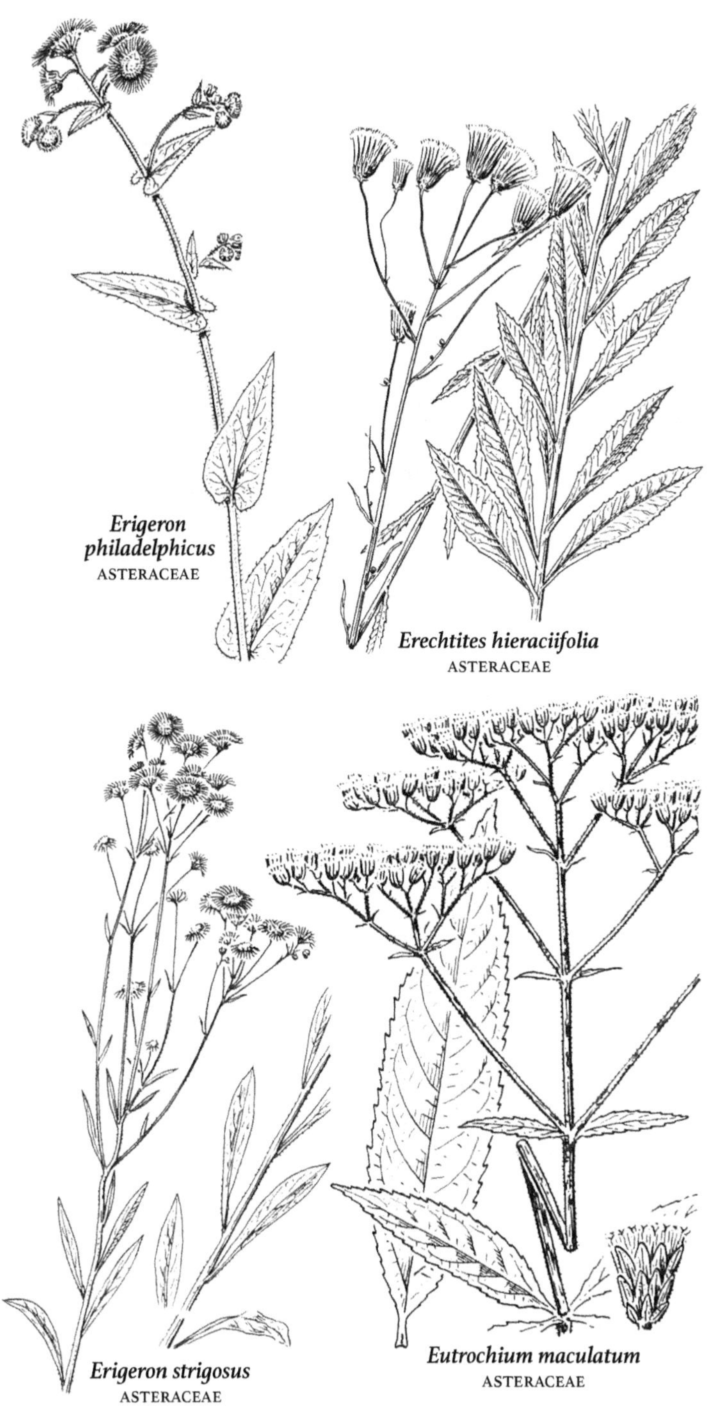
*Erigeron philadelphicus*
ASTERACEAE
*Erechtites hieraciifolia*
ASTERACEAE
*Erigeron strigosus*
ASTERACEAE
*Eutrochium maculatum*
ASTERACEAE

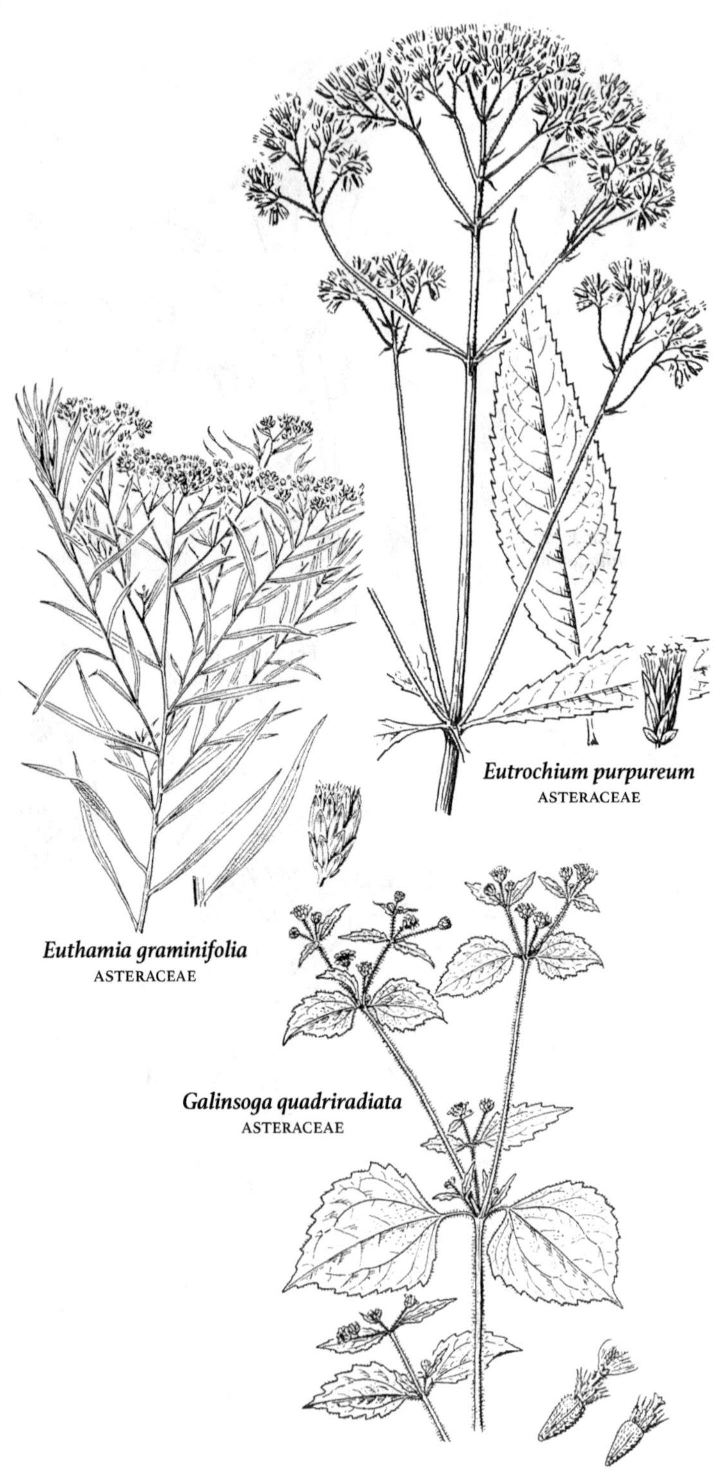

*Eutrochium purpureum*
ASTERACEAE

*Euthamia graminifolia*
ASTERACEAE

*Galinsoga quadriradiata*
ASTERACEAE

## GRINDELIA *Gumweed*

***Grindelia squarrosa*** (Pursh) Dunal CURLY-TOP GUMWEED Introduced. Biennial or sometimes perennial herb; leaves alternate, punctate and resinous. July–Sept. Open or waste places.

ADDITIONAL SPECIES ***Grindelia lanceolata*** Nutt., native of south-central USA, reported from Dane County.

## HASTEOLA *False Indian Plantain*

***Hasteola suaveolens*** (L.) Pojark. FALSE INDIAN PLANTAIN Perennial herb, from fleshy roots; leaves alternateJuly–Sept. Riverbanks, shores, calcareous fens, wet low areas. *Cacalia suaveolens* L.

## HELENIUM *Sneezeweed*

Annual or perennial herbs; leaves alternate, glandular-dotted, usually decurrent on stem.

*Key to species*

1 Leaves thread-like, less than 2 mm wide; stems not winged; uncommon introduced species ..................... ***H. amarum***

1 Leaves lance-shaped; stems winged by decurrent leaf bases ..... 2

2 Disk flowers yellow, 5-lobed at tip; stem leaves more than 1 cm wide; common ........................................ ***H. autumnale***

2 Disk flowers dark brown, 4-lobed; stem leaves to 1 cm wide; uncommon introduced species ..................... ***H. flexuosum***

***Helenium amarum*** (Raf.) H. Rock NARROW-LEAVED SNEEZEWEED Introduced (invasive). June–Oct. Prairies, open woods, fields, and waste places, especially in sandy soil.

***Helenium autumnale*** L. [•29] COMMON SNEEZEWEED July–Sept. Wet meadows, shores, streambanks, marshes, fens, tamarack swamps.

***Helenium flexuosum*** Raf. PURPLE-HEAD SNEEZEWEED Introduced (naturalized). June–Oct. Moist ground and waste places.

## HELIANTHUS *Sunflower*

Large perennial herbs (annual in several species); leaves usually opposite on lower part of stem and alternate above.

HYBRIDS Several *Helianthus* hybrids are reported for Wisc: ***H. × intermedius*** R.W. Long, ***H. × kellermanii*** Britt. (pro sp.), ***H. × laetiflorus*** Pers. (pro sp.), and ***H. × luxurians*** E.E. Wats. (pro sp.).

*Key to species*

1 Disk flowers red-purple to brown; receptacle flat or nearly so; leaves mostly alternate; plants annual .............................. 2

1 Disk flowers yellow, or rarely red-brown or purple; receptacle convex to conical; leaves opposite or alternate; plants perennial . 3

2 Involucral bracts lance-shaped, gradually tapered to a tip; chaff of receptacle bearded at tip with white hairs ........... ***H. petiolaris***

2 Involucral bracts ovate, abruptly narrowed to a slender tip; chaff not with white hairs ........................................ ***H. annuus***

3 Leaves linear, less than 5 mm wide, crowded on stems *H. salicifolius*
3 Leaves lance-shaped to ovate, more than 5 mm wide . . . . . . . . . . . 4

4 Disk flowers reddish-brown or yellow; stem leaves few to several, reduced in size upward on the stem . . . . . . . . . . . . . . . . . . . . . . . . . . 5
4 Disk flowers yellow; stem leaves numerous, well developed . . . . . 6

5 Disk flowers reddish brown; stems with more than 6 pairs of leaves, the leaves only gradually reduced in size upward on the stem . . . . . . . . . . . . . . . . . . . . . . . . . . . . . . . . . . . . . . . . . . . . . . . . . . . . . . *H. pauciflorus*
5 Disk flowers yellow; stems with less than 6 pairs of leaves, greatly reduced in size upward on stem . . . . . . . . . . . . . . . . . . *H. occidentalis*

6 Stems glabrous or nearly so, sometimes glaucous; fine hairs may be present within the inflorescence . . . . . . . . . . . . . . . . . . . . . . . . . . . . . . 7
6 Stems pubescent . . . . . . . . . . . . . . . . . . . . . . . . . . . . . . . . . . . . . . . . . 10

7 Stem leaves alternate, firm-textured, undersides pale and densely hairy . . . . . . . . . . . . . . . . . . . . . . . . . . . . . . . . . . . . . . . . *H. grosseserratus*
7 Leaves opposite, the undersides glabrous to finely hairy . . . . . . . . 8

8 Leaves sessile or with short petioles less than 5 mm long; the blades less than 5 cm wide; heads solitary or few, the disk less than 1.5 cm wide . . . . . . . . . . . . . . . . . . . . . . . . . . . . . . . . . . . . . . . . . . . . . *H. divaricatus*
8 Leaves abruptly narrowed at base to a petiole 5 mm or more long; the blades often more than 5 cm wide; heads few to many; disks more than 1.5 cm wide. . . . . . . . . . . . . . . . . . . . . . . . . . . . . . . . . . . . . . . . . . . . 9

9 Leaves lance-shaped, thick and coarse, the underside pale and densely hairy; petioles less than 3 cm long . . . . . . . . . . *H. strumosus*
9 Leaves ovate, thin and membranous, the underside glabrous or only finely hairy; petioles often more than 3 cm long . . . *H. decapetalus*

10 Leaves ovate, sessile, heart-shaped to clasping at base, the blades densely gray-hairy . . . . . . . . . . . . . . . . . . . . . . . . . . . . . . . . . . *H. mollis*
10 Leaves lance-shaped, contracted or tapered to a short petiole, the blades pubescent but not densely covered with gray hairs . . . . . . 11

11 Leaves lance-shaped, less than 3.5 cm wide, mostly alternate . . . 12
11 Leaves ovate, often more than 3.5 cm wide, the upper leaves opposite or alternate . . . . . . . . . . . . . . . . . . . . . . . . . . . . . . . . . . . . . . 13

12 Leaves often somewhat folded along their midrib; stems pubescent, the hairs fine, white, and appressed . . . . . . . . . . . . . . *H. maximiliani*
12 Leaves not folded; stems pubescent, the hairs coarse and spreading . . . . . . . . . . . . . . . . . . . . . . . . . . . . . . . . . . . . . . . . . . . . . . . . . . . . *H. giganteus*

13 Upper leaves alternate, tapered at base to a winged petiole more than 1.5 cm long; involucral bracts becoming dark with age, especially near their base . . . . . . . . . . . . . . . . . . . . . . . . . *H. tuberosus*
13 Leaves opposite, tapered at base to an unwinged petiole less than 1.5 cm long; involucral bracts remaining green . . . . . . . . *H. hirsutus*

***Helianthus annuus*** L. COMMON SUNFLOWER Introduced (naturalized). July–Sept. Prairies and dry places.

***Helianthus decapetalus*** L. THIN-LEAF SUNFLOWER Aug–Oct. Woodlands and along streams.

***Helianthus divaricatus*** L. WOODLAND SUNFLOWER July–Sept. Dry woodlands.

***Helianthus giganteus*** L. GIANT SUNFLOWER July–Sept. Wet meadows, low prairie, sedge meadows, fens, floodplain forests, streambanks.

***Helianthus grosseserratus*** Martens SAWTOOTH SUNFLOWER July–Oct. Wet meadows, low prairie, streambanks, swamps, ditches, roadsides.

***Helianthus hirsutus*** Raf. [•29] HAIRY SUNFLOWER July–Oct. Dry wooded or open places. Similar to *H. divaricatus.*

***Helianthus maximiliani*** Schrad. [•29] MAXIMILIAN SUNFLOWER Introduced (naturalized). June–Sept. Prairies and waste ground, often in sandy soil.

***Helianthus mollis*** Lam. ASHY SUNFLOWER July–Sept. Mostly in dry, often sandy places.

***Helianthus occidentalis*** Riddell NAKED-STEMMED SUNFLOWER Aug–Oct. Dry soil.

***Helianthus pauciflorus*** Nutt. STIFF SUNFLOWER Aug–Sept. Dry prairies and plains.

***Helianthus petiolaris*** Nutt. PLAINS SUNFLOWER June–Sept. Prairies, plains, and waste places. Similar to *H. annuus* but smaller, seldom over 1 m tall.

***Helianthus salicifolius*** A. Dietr. WILLOW-LEAF SUNFLOWER Introduced. Aug–Oct. Prairies and dry places, reported as adventive from Dane Co.

***Helianthus strumosus*** L. PALE-LEAF WOODLAND Sunflower July–Sept. Chiefly in woodlands.

***Helianthus tuberosus*** L. JERUSALEM-ARTICHOKE Aug–Oct. Moist soil and waste places; escaped from cultivation and also native. Cultivated since pre-Columbian times for its edible tubers.

### HELIOPSIS *Sunflower-Everlasting*

***Heliopsis helianthoides*** (L.) Sweet [•29] SUNFLOWER-EVERLASTING Short-lived perennial; leaves opposite. June–Oct. Dry woodlands, prairies, and waste places.

### HETEROTHECA *Golden Aster*

***Heterotheca villosa*** (Pursh) Shinners HAIRY GOLDEN ASTER Perennial herb, pubescent throughout. July–Oct. Dry, open, often sandy places. *Chrysopsis villosa* (Pursh) Nutt.

### HIERACIUM *Hawkweed*

Fibrous-rooted perennial herbs with milky juice; leaves alternate or all basal.

ADDITIONAL SPECIES Several other introduced *Hieracium* are reported for Wisc: ***Hieracium lachenalii*** K.C. Gmel., Walworth and Wood counties. ***Hieracium murorum*** L., Door County. ***Hieracium pilosella*** L., Ashland County.

*Key to species*

1 Plants with flowers on a naked stalk (scape); leaves clustered at base .......... 2
1 Flowers not with on a scape; leaves not clustered at base.......... 4

2 Flowers bright orange-red .......... *H. aurantiacum*
2 Flowers yellow .......... 3

3 Leaves glabrous, narrowly oblong lance-shaped; stolons absent .......... *H. piloselloides*
3 Leaves with tan or white hairs, oblong lance-shaped; stolons present, arching .......... *H. caespitosum*

4 Leaves mostly basal, strongly reduced in size upwards, stems and leaves long-hairy; peduncles with yellow gland-tipped hairs .......... *H. longipilum*
4 Leaves mostly on stem; plants glabrous or hairy; peduncles various .......... 5

5 Leaves broadly elliptic, coarsely toothed, tapered to long, hairy petioles; stems glabrous .......... *H. lachenalii**
5 Leaves various, petioles short or absent; stems glabrous or hairy. 6

6 Leaves spatula-shaped, lower leaves with petioles, upper leaves sessile; involucres and peduncles with black glands ... *H. scabrum*
6 Leaves lance-shaped to oblong lance-shaped, sessile, toothed; involucres and peduncles without glands .......... *H. umbellatum*

***Hieracium aurantiacum*** L. [•30] ORANGE KING-DEVIL, DEVIL'S-PAINTBRUSH June–Sept. Native of Europe; fields, roadsides.

***Hieracium caespitosum*** Dumort. YELLOW KING-DEVIL, MEADOW HAWKWEED May–Sept. A weed in fields, pastures, and along roadsides, occasionally in dry woods; native of Europe. *Hieracium pratense* Tausch.

***Hieracium longipilum*** Torr. HAIRY HAWKWEED July–Aug. Dry prairies, open woodlands, and fields, especially in sandy soil.

***Hieracium piloselloides*** Vill. TALL HAWKWEED June–Sept. Mostly in fields, meadows, pastures, roadsides, and waste places; native of Europe.

***Hieracium scabrum*** Michx. ROUGH HAWKWEED July–Sept. Open ground and dry woods, especially in sandy soil.

***Hieracium umbellatum*** L. NARROW-LEAF HAWKWEED July–Sept. Woodlands, beaches, and fields, especially in sandy soil. *Hieracium canadense* Michx.

## HYPOCHAERIS *Cat's-Ear*

***Hypochaeris radicata*** L. HAIRY CAT'S-EAR Perennial from a caudex.

May–Sept. Roadsides, pastures, fields, and waste places; native of Eurasia; now widely established. Similar to *Leontodon,* from which it is distinguished primarily by its chaffy-bracted receptacle.

### INULA *Elecampane*

***Inula helenium*** L. ELECAMPANE Coarse perennial herb; leaves alternate. July–Aug. Introduced from Europe; cultivated and escaped to fields and waste places where sometimes forming large colonies.

### IONACTIS *Ankle-Aster*

***Ionactis linariifolia*** (L.) Greene FLAX-LEAF ANKLE-ASTER Perennial; leaves linear or nearly so, entire. July–Oct. Dry ground and open woods, especially in sandy soil. *Aster linariifolius* L.

### IVA *Marsh-Elder*

Annual herbs; leaves opposite (or the upper alternate).

ADDITIONAL SPECIES ***Iva axillaris*** Pursh, introduced and adventive in Dane County.

*Key to species*

1 Leaves ovate, coarsely toothed; heads subtended by conspicuous bracts .......... ***I. annua***

1 Leaves nearly heart-shaped to ovate, usually coarsely lobed and toothed; heads without subtending bracts .......... ***I. xanthifolia***

***Iva annua*** L. ROUGH MARSH-ELDER Introduced. Sept–Oct. Waste ground, especially in moist soil.

***Iva xanthifolia*** Nutt. CARELESSWEED Aug–Oct. Bottomlands and moist waste places. *Cyclachaena xanthiifolia* (Nutt.) Fresen.

### KRIGIA *Dwarf-Dandelion*

Annual or perennial herbs with milky juice.

*Key to species*

1 Plants annual; scapes leafless; local in s Wisc .......... ***K. virginica***

1 Plants perennial; scapes with 1–2 small, sessile leaves; common statewide .......... ***K. biflora***

***Krigia biflora*** (Walt.) Blake [•30] ORANGE DWARF-DANDELION May–Oct. Woodlands, roadsides, and fields.

***Krigia virginica*** (L.) Willd. VIRGINIA DWARF-DANDELION April–July. Sandy places.

### LACTUCA *Lettuce*

Annual, biennial, or perennial herbs with milky juice; leaves alternate.

ADDITIONAL SPECIES ***Lactuca sativa*** L., the common cultivated lettuce, may rarely persist after cultivation, distinguished from our other species by its very broad, merely toothed leaves.

*Key to species*

1 Achenes with one central nerve on each side .................... 2
1 Achenes with several prominent nerves on each face ........... 3

2 Leaves with small spines on margins and underside midrib; involucre relatively large, 15–22 mm wide; achene 7–10 mm long . ............................................ *L. ludoviciana*
2 Leaves without spines; involucre smaller, to 15 mm wide; achenes to 6 mm long ......................................... *L. canadensis*

3 Plants perennial, heads large, the involucre 15–20 mm wide; flowers blue or purple; leaves thickened, entire or with backward-pointing clefts, not sagittate ................................. *L. pulchella*
3 Plants annual or biennial; heads smaller; leaves various ........ 4

4 Achene tipped by a slender beak as long as or longer than achene body ............................................... *L. serriola*
4 Achene with a short beak, shorter than body, or the beak absent. 5

5 Leaves with petioles, the margins lobed; flowers blue; pappus white; s Wisc ............................................ *L. floridana*
5 Leaves sessile, lower leaves lobed, the upper entire; flowers pale blue to creamy white; pappus brown; common statewide .... *L. biennis*

***Lactuca biennis*** (Moench) Fern. TALL BLUE LETTUCE July–Sept. Moist places.

***Lactuca canadensis*** L. [•30] TALL LETTUCE July–Sept. Fields, waste places, woods.

***Lactuca floridana*** (L.) Gaertn. WOODLAND LETTUCE June–Sept. Thickets, woodlands, and moist open places.

***Lactuca ludoviciana*** (Nutt.) Riddell PRAIRIE LETTUCE July–Sept. Prairies and other open places.

***Lactuca pulchella*** (Pursh) DC. BLUE LETTUCE Introduced. June–Sept. Mostly in meadows, thickets, and other moist low places. *Lactuca tatarica* (L.) C.A. Mey. subsp. *pulchella.*

***Lactuca serriola*** L. PRICKLY LETTUCE July–Sept. A weed in fields and waste places; native of Europe, now naturalized throughout most of USA. *Lactuca scariola* L.

## LAPSANA *Nipplewort*

***Lapsana communis*** L. COMMON NIPPLEWORT Annual herb with milky juice; leaves alternate. June–Sept. Native of Eurasia; now established in woods, fields, and waste ground.

## LEONTODON *Hawkbit*

***Leontodon saxatilis*** Lam. LITTLE HAWKBIT Perennial herb with milky juice; leaves basal. June–Sept. Native of Europe; a weed in lawns and waste places. *Leontodon taraxacoides* L. (Vill.) Mérat.

Helenium autumnale
ASTERACEAE
Heliopsis helianthoides
ASTERACEAE
Helianthus hirsutus
ASTERACEAE
Helianthus maximiliani
ASTERACEAE

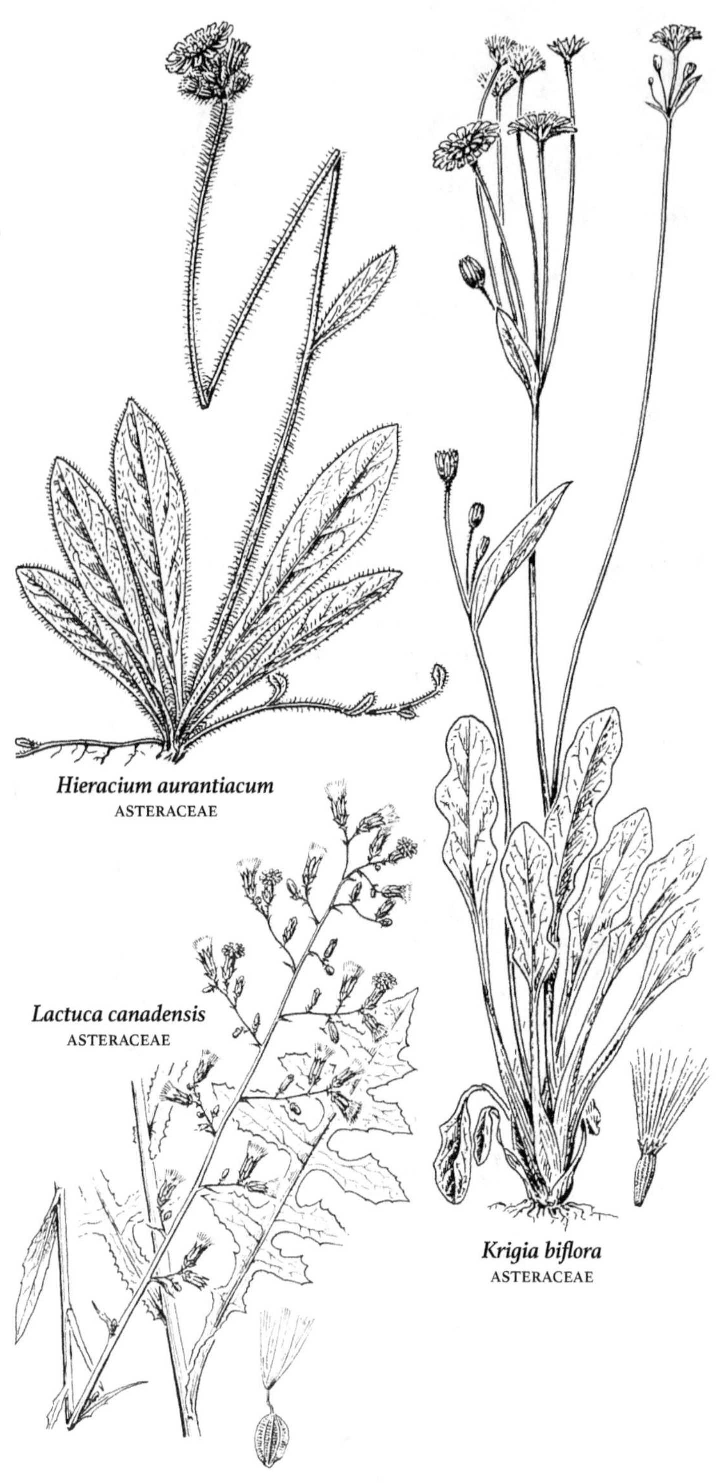
*Hieracium aurantiacum*
ASTERACEAE
*Lactuca canadensis*
ASTERACEAE
*Krigia biflora*
ASTERACEAE

ADDITIONAL SPECIES ***Leontodon autumnalis*** L., introduced, Douglas and Sheboygan counties.

## LEUCANTHEMUM *Ox-Eye Daisy*

***Leucanthemum vulgare*** Lam. OX-EYE DAISY Rhizomatous perennial; leaves alternate. May–Oct. Fields, roadsides, and waste places; native of Europe and Asia, naturalized throughout most of temperate North America. *Chrysanthemum leucanthemum* L.

## LIATRIS *Blazing Star; Gay Feather*

Perennial herbs, mostly with an evident corm; leaves alternate.

*Key to species*

1 Pappus covered with short, stiff hairs less than 0.5 mm long ..... 2
1 Pappus with softer feather-like hairs 0.5–1 mm long, the hairs with lateral branches ........ 6

2 Inflorescence a dense spike, heads with mostly 5–10 flowers; leaves mostly less than 1 cm wide ........ 3
2 Inflorescence an open spike or raceme; heads with 14 or more flowers; larger leaves 1–4 cm wide........ 4

3 Rachis of inflorescence glabrous; involucral bracts obtuse, not reflexed at tip; se Wisc only ........ ***L. spicata***
3 Rachis with coarse hairs; involucral bracts acute, reflexed at tip; widespread in Wisc ........ ***L. pycnostachya***

4 Middle involucral bracts green and herbaceous throughout, or with entire or slightly fringed chaffy margins ........ ***L. scariosa***
4 Middle involucral bracts with distinctly chaffy, torn margins .... 5

5 Heads with 30 or more flowers; terminal head distinctly larger than the others ........ ***L. ligulistylis***
5 Heads with 14–35 flowers; heads similar in size ......... ***L. aspera***

6 Flowers 10–35 per head; corolla lobes coarsely hairy on inner surface; widespread in s Wisc ........ ***L. cylindracea***
6 Flowers 4–6 per head; corolla lobes glabrous; rare in wc Wisc .... ........ ***L. punctata***

***Liatris aspera*** Michx. [•31] TALL GAYFEATHER Aug–Oct. Dry open places and thin woods, especially in sandy soil.

***Liatris cylindracea*** Michx. FEW-HEADED BLAZING STAR July–Sept. Dry open places.

***Liatris ligulistylis*** (A. Nels.) K. Schum. NORTHERN PLAINS BLAZING STAR Aug–Sept. Mostly in damp low places, occasionally in drier soil.

***Liatris punctata*** Hook. DOTTED GAYFEATHER *Endangered.* July–Sept. Dry open places, Sandy and gravelly prairies, roadsides; Wisc at eastern edge of species' range.

***Liatris pycnostachya*** Michx. THICK-SPIKE BLAZING STAR July–Sept. Moist or dry prairies and open woods.

***Liatris scariosa*** (L.) Willd. DEVIL'S-BITE Aug–Sept. Dry open places; Rock and Walworth counties.

***Liatris spicata*** (L.) Willd. SESSILE BLAZING STAR July–Sept. Wet meadows and other moist open places.

## MATRICARIA *Mayweed*

Small annual or perennial herbs; leaves alternate.

*Key to species*

1 Heads with greenish disk flowers only; common ..... ***M. discoidea***
1 Heads with white rays; disk flowers yellow..................... **2**

2 Receptacle hemispheric; achenes with wing-like ribs; uncommon along Lake Superior .............................. ***M. maritima***
2 Receptacle cone-shaped at maturity; achenes ribbed but these not enlarged and wing-like; infrequent statewide ...... ***M. chamomilla***

***Matricaria chamomilla*** L. WILD CHAMOMILE May–Sept. Native of Europe and Asia, now on roadsides and in waste places. Similar to *Anthemis cotula. Matricaria recutita* L.

***Matricaria discoidea*** DC. [•31] PINEAPPLE-WEED Introduced (naturalized); pineapple-scented. May–Sept. Roadsides and waste places. *Matricaria matricarioides* (Less.) Porter.

***Matricaria maritima*** L. SCENTLESS CHAMOMILE July–Sept. Native of Europe; waste places. *Chamomilla inodora* (L.) Gilib.

## NOTHOCALAIS *False Dandelion*

***Nothocalais cuspidata*** (Pursh) Greene FALSE DANDELION Scapose, taprooted perennial with milky juice; leaves crowded at base of plant. May–June. Prairies, bluff tops, hillsides and other dry open places, often in gravelly soil. The combination of leaves all basal and grasslike, and the yellow, blunt-tipped or fringed, ray flowers distinguish this species. *Agoseris cuspidata* (Pursh) Raf., *Microseris cuspidata* (Pursh) Schultz-Bip.

## ONOPORDUM *Scotch-Thistle*

***Onopordum acanthium*** L. SCOTCH-THISTLE Coarse, spiny biennial; stems broadly winged; leaves alternate. July–Oct. Native of Europe and e Asia; waste places.

## PACKERA *Groundsel*

Erect perennial, biennial, or annual herbs; leaves alternate or from base of plant.

*Key to species*

1 Plants with persistent covering of dense woolly hairs, especially on stem and leaf undersides .......................... ***P. plattensis***
1 Plants not persistently woolly hairy, except sometimes in leaf axils ........................................................... **2**

2 Heads with disk flowers only; Apostle Islands ......... ***P. indecora***
2 Heads with both ray and disk flowers......................... 3

3 Basal leaves mostly oblong lance-shaped or elliptic, tapered at base to petiole ....................................... ***P. paupercula***
3 Basal leaves ovate or heart-shaped ........................... 4

4 Basal leaves heart-shaped at base; widespread ........... ***P. aurea***
4 Basal leaves ovate; uncommon in se Wisc along Lake Michigan .... ..................................................... ***P. pseudaurea***

***Packera aurea*** (L.) Á. & D. Löve [•31] HEART-LEAVED GROUNDSEL May–July. Floodplain forests, wet forest depressions, swamp openings and hummocks, sedge meadows, thickets, fens, ditches. *Senecio aureus* L.

***Packera indecora*** (Greene) Á. & D. Löve RAYLESS MOUNTAIN GROUNDSEL *Threatened.* July–Aug. Moist woodlands, streambanks, swales, and bogs. *Senecio indecorus* Greene.

***Packera paupercula*** (Pursh) Á. & D. Löve RAYLESS ALPINE GROUNDSEL May–July. Meadows, prairies, streambanks, beaches, and cliffs. *Senecio pauperculus* Michx.

***Packera plattensis*** (Nutt.) W.A. Weber & Á. Löve PRAIRIE GROUNDSEL May–July. Mostly in dry open places. *Senecio plattensis* Nutt.

***Packera pseudaurea*** (Rydb.) W.A. Weber & Á. & D. Löve WESTERN HEART-LEAVED GROUNDSEL May–July. Wet meadows, low prairie, fens. Ours var. *semicordata* (Mack. & Bush) Trock & T.M.Barkley. *Senecio pseudaureus* Rydb.

## PARTHENIUM *Feverfew*

***Parthenium integrifolium*** L. EASTERN PARTHENIUM, WILD QUININE Perennial from a tuberous-thickened, usually short root; leaves alternate. June–Sept. Prairies, dry woods, roadsides, and along railroads.

## PETASITES *Sweet Colt's-Foot*

Perennial herbs, spreading by rhizomes. Leaves mostly from base of plant on long petioles, arrowhead-shaped or palmately lobed, with white woolly hairs on underside.

*Key to species*

1 Leaf blades palmately lobed ......................... ***P. frigidus***
1 Leaf blades arrowhead-shaped, toothed and not lobed . ***P. sagittatus***

***Petasites frigidus*** (L.) Fries NORTHERN SWEET COLT'S-FOOT May–June. Wet conifer forests and swamps, wet trails and clearings, aspen woods. *Petasites palmatus* (Ait.) Gray.

***Petasites sagittatus*** (Banks) Gray ARROW-LEAF SWEET COLT'S-FOOT *Threatened.* May–June. Wet meadows, marshes, sedge meadows, open swamps. *Petasites frigidus* var. *sagittatus* (Banks ex Pursh) Cherniawsky.

## POLYMNIA *Leafcup*

***Polymnia canadensis*** L. WHITE-FLOWER LEAFCUP Coarse perennial herb; leaves opposite (or the upper alternate). June–Oct. Moist woodlands. *Polymnia radiata* (Gray) Small.

## PRENANTHES *Rattlesnake-Root*

Perennial herbs with milky juice and tuberous-thickened roots; leaves alternate. Our species sometimes placed in genus *Nabalus.*

*Key to species*

1 Inflorescence an open panicle; lower leaves on long petioles, broadly ovate to triangular .................................... 2

1 Inflorescence a narrow raceme-like panicle; lower leaves spatula-shaped, gradually narrowed to petiole; uncommon prairie species 3

2 Basal leaves deeply palmately lobed; involucral bracts purple-tinged; pappus cinnamon-brown; common ............... ***P. alba***

2 Basal leaves coarsely toothed; involucral bracts green; pappus tan or pale yellow; rare in sc Wisc ..................... ***P. crepidinea***

3 Flowers purplish to white; stem and leaves glabrous .. ***P. racemosa***

3 Flowers creamy-yellow; stem and leaf undersides rough-hairy ... ................................................ ***P. aspera***

***Prenanthes alba*** L. [•32] WHITE RATTLESNAKE-ROOT Aug–Sept. Woodlands.

***Prenanthes aspera*** Michx. ROUGH RATTLESNAKE-ROOT *Endangered.* Aug–Oct. Dry prairies, generally on lower slopes.

***Prenanthes crepidinea*** Michx. NODDING RATTLESNAKE-ROOT *Endangered.* Aug–Sept. Moist prairies, seeps in mesic forests.

***Prenanthes racemosa*** Michx. GLAUCOUS WHITE LETTUCE Aug–Sept. Sandy or gravelly shores, streambanks, wet meadows, low prairie, fens.

## PSEUDOGNAPHALIUM *Rabbit-Tobacco*

Biennial glandular herbs (ours), sometimes aromatic; leaves basal and along the stem or mostly cauline, alternate.

ADDITIONAL SPECIES ***Pseudognaphalium saxicola*** (Fassett) H.E. Ballard & Feller (state threatened), occurs in sc Wisc along cliffs and in ravines. Plants less than 25 cm tall, with broader leaves less hairy beneath than in *P. obtusifolium*, and with the involucre commonly less imbricate; syn: *Gnaphalium obtusifolium* var. *saxicola* (Fassett) Cronquist.

*Key to species*

1 Leaves decurrent at base; stems glandular-hairy, sometimes also woolly hairy ........................................ ***P. macounii***

1 Leaves not decurrent at base.................................. 2

2 Stem woolly hairy, not glandular except sometimes near base ..... ............................................. ***P. obtusifolium***

2 Stem glandular-hairy, not woolly except in inflorescence *P. helleri*

***Pseudognaphalium helleri*** (Britt.) A. Anderb. HELLER'S RABBIT-TOBACCO Aug–Oct. Dry, commonly sandy soil, often in woods. Similar to *P. obtusifolium,* less common, averaging a little smaller. *Gnaphalium helleri* Britt.

***Pseudognaphalium macounii*** (Greene) Kartesz CLAMMY RABBIT-TOBACCO July–Sept. Open places. Similar to *P. obtusifolium. Gnaphalium macounii* Greene.

***Pseudognaphalium obtusifolium*** (L.) Hilliard & Burtt [•32] FRAGRANT RABBIT-TOBACCO July–Oct. Open, often sandy places. When crushed, plants have a characteristic maple syrup scent. *Gnaphalium obtusifolium* L.

## RATIBIDA *Coneflower*

Perennial herbs (ours); leaves alternate, pinnatifid.

*Key to species*

1 Disk columnar, 2–4 times longer than wide; plants taprooted ...... *R. columnifera*

1 Disk less than 2 times longer than wide; plants fibrous-rooted ... *R. pinnata*

***Ratibida columnifera*** (Nutt.) Woot. & Standl. COLUMNAR CONEFLOWER Introduced. June–Aug. Prairies and waste ground; also along railroads.

***Ratibida pinnata*** (Vent.) Barnh. GLOBULAR CONEFLOWER June–Aug. Prairies and dry woods.

## RUDBECKIA *Coneflower*

Perennial herbs; stems and leaves rough-hairy; leaves alternate. The genus includes the well-known black-eyed Susan (*Rudbeckia hirta*), widespread on dry sites.

*Key to species*

1 Main leaves deeply lobed .... 2

1 Leaves unlobed .... 4

2 Disk yellow; stems glabrous or nearly so; largest leaves 5–7 lobed . .... *R. laciniata*

2 Disk dark purple-red; stems pubescent; largest leaves 3-lobed ... 3

3 Rays usually orange at base; chaff of receptacle tapered to a sharp point, glabrous .... *R. triloba*

3 Rays yellow; chaff rounded at tip and with fine glandular hairs at apex .... *R. subtomentosa*

4 Leaves ovate, margins toothed; chaff of receptacle glabrous, tapered to a sharp prolonged point .... *R. triloba*

4 Leaves and margins various; chaff of receptacle not tapered to a sharp point .... 5

5 Stems and leaves with scattered soft hairs; mid-stem leaves with ovate blades and distinct petioles .... *R. fulgida*

5 Stems and leaves pubescent with coarse stiff hairs; mid-stem leaves lance-shaped or oblong lance-shaped, sessile or tapered to a winged petiole .......... *R. hirta*

***Rudbeckia fulgida*** Ait. EASTERN CONEFLOWER Aug–Sept. Sedge meadows, calcareous fens, wet streambanks, low prairies.

***Rudbeckia hirta*** L. BLACK-EYED SUSAN June–Oct. Various habitats, chiefly in disturbed or waste places, meadows, and roadsides.

***Rudbeckia laciniata*** L. [•32] CUTLEAF CONEFLOWER July–Sept. Floodplain forests, swamps, streambanks, thickets, ditches; usually in partial or full shade.

***Rudbeckia subtomentosa*** Pursh SWEET CONEFLOWER July–Sept. Prairies and low ground.

***Rudbeckia triloba*** L. THREE-LOBED CONEFLOWER July–Oct. Woodlands and moist soil. Our form is var. *triloba,* with the larger leaves merely 3-lobed.

## SENECIO *Groundsel; Ragwort*

Erect annual herbs (ours); leaves alternate or from base of plant. Several former members of this genus now placed in *Packera.*

ADDITIONAL SPECIES ***Senecio congestus*** (R. Br.) DC. [syn: Tephroseris palustris (L.) Reichenb., Clustered marsh ragwort]; historical records from mostly n Wisc; presumed extirpated from the state; annual or biennial of muddy shores, plants with stout stems to 1 m tall. ***Senecio viscosus*** L. (Sticky ragwort), introduced, Douglas County.

*Key to species*

1 Rays absent; bracts below involucral bracts with distinct black tips . .......... *S. vulgaris*

1 Rays present but tiny; bracts below involucral bracts not black-tipped .......... *S. sylvaticus*

***Senecio sylvaticus*** L. WOODLAND RAGWORT July–Sept. Native of Europe, established in dry soil and waste places.

***Senecio vulgaris*** L. OLD-MAN-IN-THE-SPRING May–Oct. Native of the Old World; now in waste places.

## SILPHIUM *Rosinweed*

Tall perennial herbs, with resinous juice; leaves opposite or all from base of plant.

*Key to species*

1 Main leaves all opposite .......... 2

1 Main leaves alternate or basal .......... 3

2 Leaves joined at base and perforated by stem; stem 4-angled ..... .......... *S. perfoliatum*

2 Leaf bases sessile but not joined; stem not 4-angled *S. integrifolium*

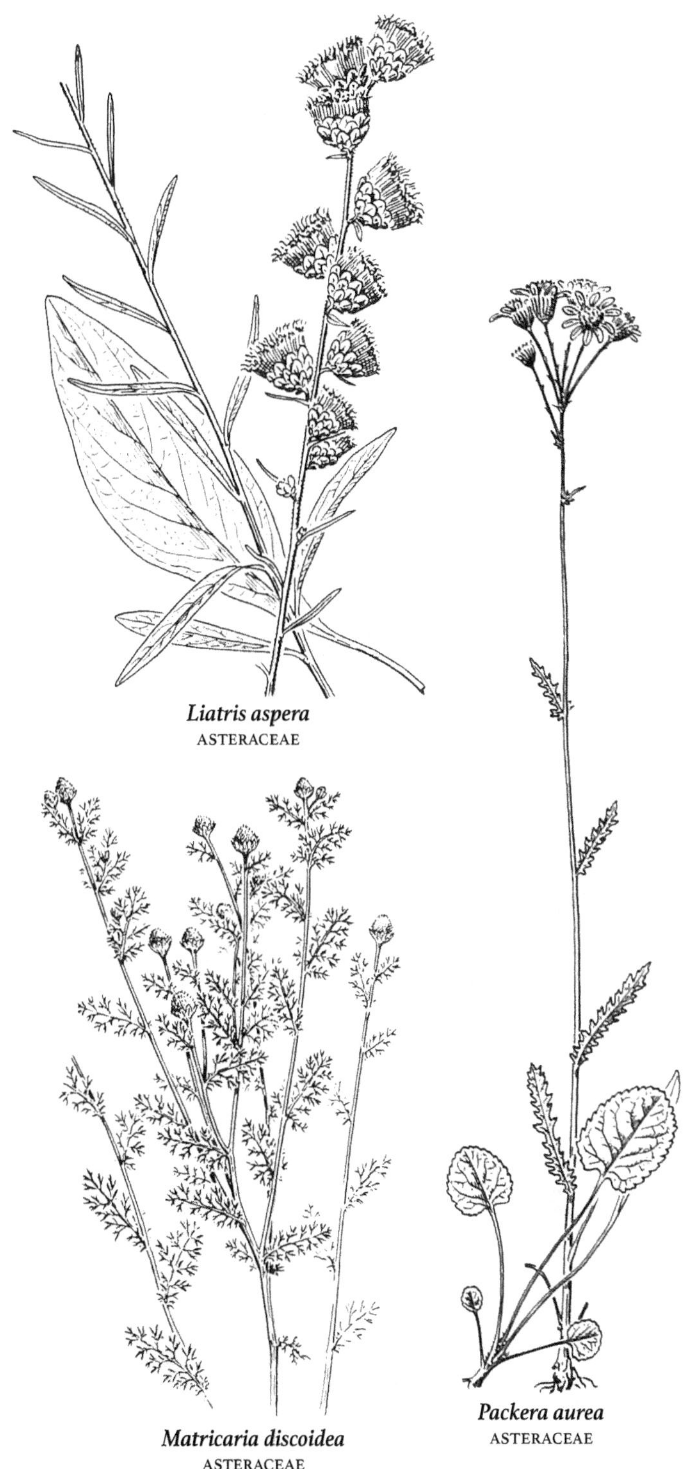

*Liatris aspera*
ASTERACEAE

*Matricaria discoidea*
ASTERACEAE

*Packera aurea*
ASTERACEAE

*Prenanthes alba*
ASTERACEAE

*Pseudognaphalium obtusifolium*
ASTERACEAE

*Rudbeckia laciniata*
ASTERACEAE

3 Leaves alternate, deeply lobed .................... *S. laciniatum*
3 Leaves basal, unlobed ...................... *S. terebinthinaceum*

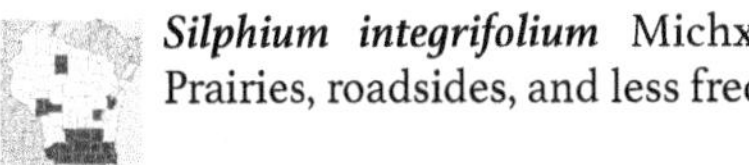

***Silphium integrifolium*** Michx. PRAIRIE ROSINWEED July–Sept. Prairies, roadsides, and less frequently woodlands.

***Silphium laciniatum*** L. COMPASS-PLANT July–Sept. Prairies. The large basal leaves tend to align themselves facing in an east-west direction.

***Silphium perfoliatum*** L. CUP-PLANT July–Sept. Floodplain forests, streambanks, springs.

***Silphium terebinthinaceum*** Jacq. BASAL-LEAVED ROSINWEED July–Sept. Low prairie, fens; especially where calcium-rich.

## SOLIDAGO *Goldenrod*

Erect perennials, spreading by rhizomes or from a crown; leaves alternate.

ADDITIONAL SPECIES ***Solidago mollis*** Bartl. (Velvet goldenrod), Great Plains native, adventive in Washington County.

*Key to species*

1 Heads in a more or less flat-topped cluster at end of stem ....... 2
1 Heads in an elongate or pyramid-shaped cluster .............. 5

2 Leaf blades of middle and upper stem ovate to elliptic; stems and leaves densely hairy; common species of dry to mesic habitats, mostly south of tension zone ......................... *S. rigida*
2 Leaf blades linear to lance-shaped or oblong lance-shaped, glabrous apart from rough leaf margins; stems glabrous or nearly so, or slightly hairy below inflorescence ..................... 3

3 Rays white, 12 or more; upper stem leaves broadest above their middle.......................................... *S. ptarmicoides*
3 Rays yellow, 10 or less; upper stem leaves broadest at or below middle ............................................... 4

4 Pedicels rough-hairy; leaves folded inward along midrib, with 3 or more veins from base ............................. *S. riddellii*
4 Pedicels smooth or nearly so; leaves flat, not 3-veined from base .. ..................................................... *S. ohioensis*

5 Inflorescence terminal, usually more or less pyramid-shaped and slightly nodding at top; inflorescence branches curving; the heads mostly on upper side of the branches ....................... 6
5 Flower heads spiraled around branches of inflorescence and not all on one side of branch ....................................... 17

6 Stem leaves with 3 prominent veins (midrib plus 2 distinct lateral veins) ..................................................... 7
6 Stem leaves with prominent midrib and weaker lateral veins ... 12

7 Lower leaves linear lance-shaped; inflorescence branches and pedicels glabrous; prairies south of tension zone . *S. missouriensis*

7 Lower leaves elliptic, much larger than leaves of middle stem; inflorescence branches and pedicels at least sparsely hairy, the hairs short .......................................... 8

8 Stem pubescent for all or most of its length .................... 9
8 Stem glabrous (or nearly so) below inflorescence .............. 11

9 Involucres 2-3 mm long .......................... *S. canadensis*
9 Involucres 3-6 mm long.......................................... 10

10 Mid to upper leaves serrate, glabrous or scabrous above, pubescent on the veins beneath; stem pilose chiefly above the middle ...... ................................................ *S. canadensis*
10 Mid to upper leaves minutely serrate to entire, scabrous above, densely pubescent beneath; stem grayish with close puberulence throughout, except sometimes near the base .......... *S. altissima*

11 Basal leaves absent; stem leaves elliptic, withering by flowering time, numerous, not reduced in size; distinctly 3-nerved; inflorescence branches densely hairy; flowering in Aug–Sept ....... *S. gigantea*
11 Basal leaves present; basal and lower stem leaves oblong lance-shaped to elliptic, with long petioles, persistent, middle and upper stem leaves few, smaller than basal leaves; leaves obscurely 3-nerved; inflorescence branches glabrous or nearly so; plants begin flowering in July ....................................... *S. juncea*

12 Stems pubescent, at least on upper half....................... 13
12 Stems glabrous (or sometimes with fine hairs in inflorescence) . 14

13 Margins of stem leaves entire or with rounded teeth; stems and leaves finely and densely short-hairy; lower leaves oblong lance-shaped, tapered to a winged petioles; upper stem leaves reduced in size; plants of dry habitats ......................... *S. nemoralis*
13 Margins sharply toothed; lower leaves elliptic to lance-shaped, same size as upper stem leaves, but withering during season; plants of moist or shaded places ............................... *S. rugosa*

14 Bases of lowest stem leaves clasping stem; wet habitats *S. uliginosa*
14 Lower leaf bases not clasping stem .......................... 15

15 Stems strongly ridged; leaf upperside very rough-to-touch; wet habitats ............................................ *S. patula*
15 Stems round with only small ridges; leaf upperside smooth or only slightly roughened; dry to mesic meadows and woods ......... 16

16 Basal and lower stem leaves much larger than leaves of middle stem, persistent; flowering begins early (July) ................ *S. juncea*
16 Basal and lower stem leaves not much larger than leaves of middle stem, withered by flowering time; flowering late (August–September) ........................................ *S. ulmifolia*

17 Basal and lower stem leaves smaller than mid-stem leaves; middle and upper stem leaves with sharp teeth, longer than the axillary inflorescences............................................ 18
17 Basal and lower stem leaves larger than mid-stem leaves; middle and upper stem leaves entire or with rounded teeth, not longer than inflorescences............................................ 19

18 Stems ribbed, zigzagged from node to node; leaves ovate, narrowed at base to a winged petiole; common ................ *S. flexicaulis*

18 Stems not ribbed or zigzagged; leaves narrowly elliptical, sessile; rare in se Wisc .......................................... *S. caesia*

19 Stem and leaves pubescent.................................. **20**

19 Stems glabrous or nearly so, sparse fine hairs may be present in inflorescence........................................... **22**

20 Rays white ........................................ *S. bicolor*

20 Rays yellow ............................................... **21**

21 Achenes glabrous .................................. *S. hispida*

21 Achenes densely covered with upward-pointing appressed hairs . ................................................ *S. simplex*

22 Lower stem leaves (including petioles) mostly 7–16 times longer than wide; petioles clasping stem; wet habitats ....... *S. uliginosa*

22 Lower stem leaves usually 3–8 times longer than wide; petioles not clasping stem; dry, often sandy habitats ...................... **23**

23 Basal and lower stem leaves broadly spatula-shaped to obovate; achenes short-hairy; plants mostly on cliffs in driftless area ..... ................................................ *S. sciaphila*

23 Basal and lower stem leaves ovate to oblong lance-shaped; achenes glabrous; widespread .............................. *S. speciosa*

***Solidago altissima*** L. TALL GOLDENROD Many types of wet to dry habitats. The short hairs on the leaves give fresh plants a gray-green color not seen in *S. canadensis.* Subject to insect galls on the stems. *Solidago canadensis* var. *scabra* Torr. & Gray.

***Solidago bicolor*** L. WHITE GOLDENROD Aug–Oct. Dry woods and open, often rocky places.

***Solidago caesia*** L. WREATH GOLDENROD *Endangered.* Sept–Oct. Deciduous forests near Lake Michigan.

***Solidago canadensis*** L. [•33] COMMON GOLDENROD July–Sept. Open, moist or dry places.

***Solidago flexicaulis*** L. ZIGZAG GOLDENROD Aug–Oct. Woodlands.

***Solidago gigantea*** Ait. [•33] SMOOTH GOLDENROD July–Sept. Wet meadows, streambanks, swamps, floodplain forests, thickets, marshes, calcareous fens, ditches; also in moist to dry open woods and roadsides. Canada goldenrod (*S. canadensis*) similar but generally smaller and densely short-hairy on leaf undersides and upper stem. *Solidago serotina* Ait. non Retz.

***Solidago hispida*** Muhl. HAIRY GOLDENROD July–Oct. Dry woodlands and rocky shores.

***Solidago juncea*** Ait. EARLY GOLDENROD June–Oct. Dry open places and open woods, especially in sandy soil. One of the earliest goldenrods to flower.

***Solidago missouriensis*** Nutt. MISSOURI GOLDENROD July–Oct. Prairies and other dry, open or sparsely wooded places.

***Solidago nemoralis*** Ait. GRAY GOLDENROD Aug–Oct. Dry woods and open places, especially in sandy soil.

***Solidago ohioensis*** Frank OHIO FLAT-TOPPED GOLDENROD July–Sept. Wet, sandy or gravelly shores, streambanks, sedge meadows, calcareous fens, low prairie; soils often calcium-rich. *Oligoneuron ohioense* (Frank) G.N. Jones

***Solidago patula*** Muhl. ROUGH-LEAVED GOLDENROD Aug–Sept. Swamps, thickets, calcareous fens, sedge meadows.

***Solidago ptarmicoides*** (Nees) Boivin PRAIRIE FLAT-TOPPED GOLDENROD July–Sept. Prairies and other open, usually dry places. *Oligoneuron album* (Nutt.) Nesom.

***Solidago riddellii*** Frank RIDDELL'S FLAT-TOPPED GOLDENROD Aug–Oct. Wet meadows, calcareous fens, low prairie, lakeshores, streambanks. *Oligoneuron riddellii* (Frank) Rydb.

***Solidago rigida*** L. STIFF GOLDENROD July–Oct. Prairies and other dry open places, especially in sandy soil. *Oligoneuron rigidum* (L.) Small.

***Solidago rugosa*** P. Mill. WRINKLE-LEAVED GOLDENROD Introduced. Aug–Oct. Various habitats; common in e USA, considered adventive in Wisc.

***Solidago sciaphila*** Steele DRIFTLESS AREA GOLDENROD Aug–Sept. Calcareous or sandy cliffs, sw Wisc and adjacent Ill, Minn, and Iowa.

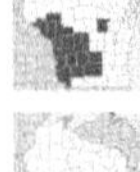

***Solidago simplex*** Kunth MOUNT ALBERT GOLDENROD Aug–Oct. Sand dunes and beaches. Ours mostly subsp. *randii* (Porter). Plants of var. *gillmanii* (state threatened, known from Door and Sheboygan counties) are robust, mostly 3–9 dm tall, often more hairy than the other varieties, with large heads (involucre 6–9 mm high) in a long, often branched inflorescence, and mostly found on sand dunes along Lake Michigan. *Solidago spathulata* DC.

***Solidago speciosa*** Nutt. SHOWY GOLDENROD Aug–Oct. Open woods, fields, prairies, and plains.

***Solidago uliginosa*** Nutt. NORTHERN BOG-GOLDENROD Aug–Sept. Conifer swamps, fens, open bogs, low prairie, wet meadows, interdunal wetlands, Lake Superior rocky shore.

***Solidago ulmifolia*** Muhl. [•33] ELM-LEAVED GOLDENROD Aug–Oct. Woodlands.

### SONCHUS *Sow-Thistle*

Annual or perennial herbs with milky juice; leaves alternate or all basal. The perennial species are troublesome farm weeds.

*Key to species*

1 Perennial with creeping rhizomes; leaf bases auriculate and clasping stem, the auricles small ..................... *S. arvensis*

1 Taprooted annuals; leaf bases auriculate and clasping stem, the auricles large and conspicuous ............................... 2

2 Leaf margins sparsely prickly ....................... *S. oleraceus*

2 Leaf margins with numerous spine-tipped teeth ......... *S. asper*

***Sonchus arvensis*** L. PERENNIAL SOW-THISTLE July–Oct. A cosmopolitan weed of European origin.

***Sonchus asper*** (L.) Hill [•33] SPINY-LEAF SOW-THISTLE July–Oct. A cosmopolitan weed; native of Europe. Similar to *S. oleraceus,* but usually more prickly.

***Sonchus oleraceus*** L. COMMON SOW-THISTLE July–Oct. A cosmopolitan weed; native of Europe.

## SYMPHYOTRICHUM *Wild Aster*

Mostly perennial herbs (annual in *S. ciliatum*); leaves alternate, simple. The traditional genus Aster has been split into several segregate genera to reflect differences with European species. Most species native to e USA are now placed within the genus *Symphyotrichum,* with the following exceptions for Wisconsin species: *A. umbellatus* in the genus *Doellingeria; A. furcatus, A. macrophyllus,* and *A. schreberi* in the genus *Eurybia;* and *A. linariifolius* in the genus *Ionactis.* Initially controversial, this classification is now widely accepted and is followed here.

ADDITIONAL SPECIES ***Symphyotrichum racemosum*** (Elliott) G.L.Nesom, native, known from several central Wisc locations.

*Key to species*

1 Leaves, at least the lower ones, heart-shaped at base and with petioles.............................................. 2

1 Leaves not both heart-shaped and petioled ................... 8

2 Leaves entire or nearly so; involucral bracts with a short, diamond-shaped green tip ........................................... 3

2 Leaves toothed; involucral bracts various ..................... 4

3 Nearly all leaves below the inflorescence heart-shaped or nearly so ...................................................... *S. shortii*

3 Only the lower leaves heart-shaped ........... *S. oolentangiense*

4 Inflorescence with relatively few heads, often less than 50; peduncles and inflorescence branches with only a few bracts .... ............................................. *S. ciliolatum*

4 Inflorescence with many heads, often over 100; peduncles and inflorescence branches with many bracts ...................... 5

5 Plants glabrous or nearly so (sometimes slightly finely hairy in the inflorescence) .................................. *S. urophyllum*

5 Plants hairy to rough-hairy, at least in part..................... 6

6 Leaf margins with large, sharp teeth; petiole not winged or only slightly so; involucral bracts often strongly purple-tinged, rounded at tip .................................................. *S. cordifolium*

6 Leaf margins with only shallow, usually rounded teeth; petioles of stem leaves often strongly winged; involucral bracts green, tapered to a sharp tip .................................................. 7

7 Rays bright blue; plants often densely hairy ....... *S. drummondii*

7 Rays pale-blue, pale-purple, or white; plants thinly hairy .................................................. *S. urophyllum*

8 Involucre strongly glandular .................................. 9

8 Plants not glandular .......................................... 10

9 Base of leaf strongly clasping stem .............. *S. novae-angliae*

9 Base of leaf not clasping stem or only slightly so . *S. oblongifolium*

10 Base of leaf strongly clasping stem............................ 11

10 Base of leaf not clasping stem (or only slightly so) ............. 13

11 Involucral bracts (at least the inner), long-tapered to a slender tip .. .................................................. *S. puniceum*

11 Involucral bracts rounded or short-tapered to the tip........... 12

12 Perennial herbs from an enlarged base or short rhizome; involucral bracts appressed and strongly overlapping; plants glabrous or with lines of hairs in the inflorescence, often glaucous ........ *S. laeve*

12 Perennial herbs from long creeping rhizomes; involucral bracts equal lengths or overlapping, not appressed, often loose; plants glabrous or hairy, not glaucous .................. *S. prenanthoides*

13 Leaves densely silvery-hairy; margins entire .......... *S. sericeum*

13 Leaves glabrous or pubescent, not silvery-hairy; margins entire to toothed.................................................. 14

14 Taprooted annual herb; rays absent ................. *S. ciliatum*

14 Perennial herbs with fibrous roots and from rhizomes or crowns; rays well developed .......................................... 15

15 Most involucral bracts with a slender green tip, the margins inrolled .................................................. *S. pilosum*

15 Involucral bracts flat, the margins not inrolled ................ 16

16 Involucre 7–12 mm long; rays violet or blue .......... *S. praealtum*

16 Involucre 3–6 mm long; rays usually white, sometimes pink or blue .................................................. 17

17 Tips of outer involucral bracts loose or recurved, tapered to a very small spine-tip; leaves entire................................. 18

17 Involucral bracts appressed or only slightly loose, the tips not spine-tipped; leaves various ........................................ 19

18 Heads many, often arranged on 1 side of the branches; rays 8–20 . .................................................. *S. ericoides*

18 Heads fewer, solitary or clustered at branch ends; rays 20–35 ..... .................................................. *S. falcatum*

19 Leaves hairy on underside, at least along the midvein ......... 20

19 Leaf underside glabrous (except sometimes hairy in *S. praealtum*). 21

20 Leaf underside hairy; plants with creeping rhizomes *S. ontarionis*

20 Leaf underside glabrous except for hairs on the midvein; plants without creeping rhizomes ........................ *S. lateriflorum*

21 Heads very small and numerous, the rays 3–6 mm long; heads often arranged on 1-side of the inflorescence branches .. *S. lanceolatum*

21 Heads larger or few in number, not arranged on 1 side of the branches .................................................. **22**

22 Underside leaf venation forming a distinct net-like pattern, the spaces enclosed by the secondary veins of about equal length and width ............................................... *S. praealtum*

22 Underside leaf venation not clearly net-veined, or if so, the spaces enclosed by the secondary veins longer than wide............. **23**

23 Peduncles long, with many large bracts along its length, the bracts often more than 2 cm long .......................... *S. dumosum*

23 Peduncles either short or with only a few bracts along its length **24**

24 Rays bright blue-violet; stem leaves ± clasping (most bases, especially lower on the stem, circling more than half the circumference of the stem); involucral bracts nearly equal in length (or the outer ones over half as long as the inner ones) .......... ............................................ *S. robynsianum*

24 Rays usually white; stem leaves not clasping (bases circling half or less the stem circumference); involucral bracts of different lengths, imbricate ................................................ **25**

25 Slender plants of bogs and other wetlands; inflorescence short-stalked and wide in outline .......................... *S. boreale*

25 Plants stouter, not in bogs; inflorescence elongate .. *S. lanceolatum*

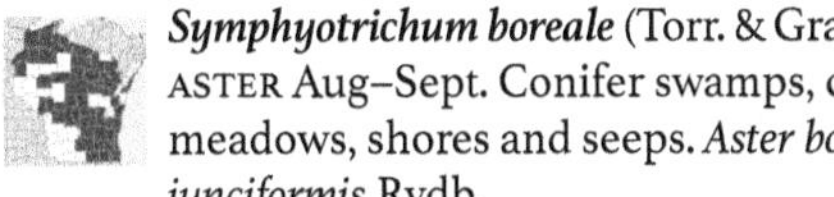

***Symphyotrichum boreale*** (Torr. & Gray) A. & D. Löve NORTHERN BOG-ASTER Aug–Sept. Conifer swamps, calcareous fens, open bogs, wet meadows, shores and seeps. *Aster borealis* (Torr. & Gray) Prov., *Aster junciformis* Rydb.

***Symphyotrichum ciliatum*** (Ledeb.) Nesom WESTERN ANNUAL ASTER Aug–Sept. Shores (including along Great Lakes), streambanks, wet meadows, roadside ditches, usually where brackish. Native of w North America; considered adventive in Wisc. *Aster brachyactis* Blake.

***Symphyotrichum ciliolatum*** (Lindl.) A. & D. Löve NORTHERN HEART-LEAVED ASTER July–Oct. Woods and clearings. *Aster ciliolatus* Lindl.

***Symphyotrichum cordifolium*** (L.) Nesom [•34] COMMON BLUE HEART-LEAVED ASTER Aug–Oct. Woodlands. *Aster cordifolius* L.

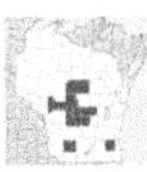

***Symphyotrichum drummondii*** (Lindl.) Nesom HAIRY HEART-LEAVED ASTER Sept–Oct. Clearings and open woodland. *Aster drummondii* Lindl.

***Symphyotrichum dumosum*** (L.) Nesom BUSHY ASTER Aug–Oct. Moist to wet sandy or mucky shores, interdunal swales, sedge meadows, sometimes where calcium-rich; also in drier oak and jack pine woods. *Aster dumosus* L.

***Symphyotrichum ericoides*** (L.) Nesom WHITE HEATH ASTER Dry, open places. *Aster ericoides* L.

***Symphyotrichum falcatum*** (Lindl.) Nesom WHITE PRAIRIE ASTER Prairies and other open places. Similar to *S. ericoides* but smaller, seldom over 6 dm tall, with fewer and larger heads. *Aster falcatus* Lindl.

***Symphyotrichum laeve*** (L.) A. & D. Löve SMOOTH BLUE ASTER Aug–Oct. Open, usually dry places. Aster laevis L.

***Symphyotrichum lanceolatum*** (Willd.) Nesom EASTERN LINED ASTER Aug–Oct. Marshes, wet meadows, fens, swamp openings, low prairie, streambanks and shores. One of our most common asters. *Aster lanceolatus* Willd., *Symphyotrichum simplex* (Willd.) A.& D. Löve.

***Symphyotrichum lateriflorum*** (L.) A. & D. Löve GOBLET-ASTER, FAREWELL-SUMMER Aug–Oct. Various habitats, most commonly in open woodlands, dry open places, and on beaches. *Aster lateriflorus* (L.) Britt.

***Symphyotrichum novae-angliae*** (L.) Nesom [•34] NEW ENGLAND ASTER Aug–Oct. Wet meadows, low prairie, shores, thickets, calcareous fens, roadsides; usually in moist or wet open areas. *Aster novae-angliae* L.

***Symphyotrichum oblongifolium*** (Nutt.) Nesom [•34] AROMATIC ASTER Aug–Oct. Dry, usually open places. *Aster oblongifolius* Nutt.

***Symphyotrichum ontarionis*** (Wieg.) Nesom ONTARIO ASTER Sept–Oct. Floodplain forests, river terraces, thickets. Similar to *S. lateriflorum,* but with long rhizomes rather than a crown or short rhizomes. *Aster ontarionis* Wieg.

***Symphyotrichum oolentangiense*** (Riddell) Nesom PRAIRIE HEART-LEAVED ASTER Aug–Oct. Prairies and dry open woods. *Aster oolentangiensis* Riddell.

***Symphyotrichum pilosum*** (Willd.) Nesom WHITE OLDFIELD ASTER Sandy and gravelly shores, interdunal swales, wet meadows; often where calcium-rich; sometimes weedy in disturbed fields and roadsides. *Aster pilosus* Willd.

***Symphyotrichum praealtum*** (Poir.) Nesom WILLOW-LEAF ASTER Sept–Oct. Wet meadows, low prairie, moist fields, thickets. *Aster praealtus* Poir.

***Symphyotrichum prenanthoides*** (Muhl.) Nesom [•34] ZIGZAG ASTER Aug–Oct. Streambanks, meadows, and moist woodlands. *Aster prenanthoides* Muhl.

***Symphyotrichum puniceum*** (L.) A. & D. Löve PURPLE-STEM ASTER Aug–Sept. Swamps, sedge meadows, thickets, calcareous fens, streambanks, shores, springs, roadside ditches. *Aster puniceus* L.

***Symphyotrichum robynsianum*** (J.Rousseau) Brouillet & Labrecque LONG-LEAVED ASTER Moist open sandy, gravelly, or rocky places, including shores, limestone alvars, seasonally wet swales, rocky shores of Lake Superior. The long leaves usually overtop most or all of the inflorescence. *Aster longifolius* sensu Semple & Heard, non Lam.

***Symphyotrichum sericeum*** (Vent.) Nesom WESTERN SILVERY ASTER Aug–Oct. Dry prairies and other open places. *Aster sericeus* Vent.

***Symphyotrichum shortii*** (Lindl.) Nesom SHORT'S ASTER Aug–Oct. Woodlands. *Aster shortii* Lindl.

***Symphyotrichum urophyllum*** (Lindl.) Nesom ARROW-LEAVED ASTER Aug–Oct. Streambanks, woodlands, and less often in open places. *Aster sagittifolius* Willd.

## TANACETUM *Tansy*

Annual or perennial herbs, sometimes somewhat woody at the base; leaves alternate, pinnately dissected.

*Key to species*

1 Leaves undivided (though regularly toothed) ........ ***T. balsamita***
1 Leaves pinnatifid or bipinnatifid .............................. 2

2 Rays present, white .............................. ***T. parthenium***
2 Rays absent or yellow........................................ 3

3 Heads 13–20 mm wide; leaves ± hairy; rare along Lake Michigan beaches ........................................ ***T. bipinnatum***
3 Heads 5–10 mm wide; leaves glabrous or nearly so; common .... ................................................... ***T. vulgare***

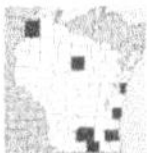

***Tanacetum balsamita*** L. COSTMARY Aug–Oct. Native of s Europe and the Orient, escaped from cultivation to roadsides and other waste places. *Chrysanthemum balsamita* L.

***Tanacetum bipinnatum*** (L.) Sch.Bip. EASTERN TANSY *Endangered.* July–Aug. Sandy beaches, dunes, and cracks in limestone pavement (alvars); rare in Door County. *Tanacetum huronense* Nutt.

***Tanacetum parthenium*** (L.) Schultz-Bip. FEVERFEW June–Sept. Native of Europe, sometimes escaping from cultivation to waste places. *Chrysanthemum parthenium* (L.) Bernh.

***Tanacetum vulgare*** L. COMMON TANSY Aug–Oct. Native of the Old World, escaped from cultivation to roadsides, fields and waste places.

## TARAXACUM *Dandelion*

Perennial, scapose, taprooted herbs with milky juice; leaves all basal, forming a rosette, entire to pinnatifid.

*Key to species*

1 Leaves generally deeply lobed or cut to midrib; mature achenes red-brown ...................................... ***T. erythrospermum***
1 Leaves various, deeply lobed to entire; mature achenes tan or olive-green ............................................... ***T. officinale***

***Taraxacum erythrospermum*** Andrz. ex Besser RED-SEED DANDELION April–June. Native of Eurasia, now established throughout Wisc in fields, pastures, lawns, and other disturbed places. Similar to *T. officinale,* but less common and plants often more slender. *Taraxacum laevigatum* (Willd.) DC.

***Taraxacum officinale*** G.H. Weber COMMON DANDELION March–Dec. Native of Europe and adjacent Asia, now a cosmopolitan weed of lawns and disturbed sites.

## TRAGOPOGON *Goat's-Beard*

Biennial or perennial lactiferous herbs with a taproot; leaves alternate, linear, entire, and clasping. Our species may hybridize with the others where they grow together.

*Key to species*

1 Flowers purple; uncommon garden escape .......... ***T. porrifolius***
1 Flowers yellow; common weeds .............................. **2**

2 Peduncle enlarged or inflated below the head; leaf tips not recurved .................................................. ***T. dubius***
2 Peduncle not enlarged; leaf tips recurved ............ ***T. pratensis***

***Tragopogon dubius*** Scop. MEADOW GOAT'S-BEARD May–July. Native of Europe; roadsides and waste places. Similar to *T. porrifolius* but with yellow flowers and often smaller and less robust.

***Tragopogon porrifolius*** L. SALSIFY April–Aug. Native of Europe; escaped from cultivation to roadsides and waste places, mostly in moist soil.

***Tragopogon pratensis*** L. JACK-GO-TO-BED-AT-NOON May–Aug. Native of Europe; roadsides, fields, and waste places.

## TUSSILAGO *Colt's-Foot*

***Tussilago farfara*** L. COLT'S-FOOT Perennial herb from a creeping rhizome; leaves basal, long-petioled, cordate to suborbicular. April–June. Native of the Old World, naturalized in disturbed and waste places.

## VERBESINA *Wingstem*

***Verbesina alternifolia*** (L.) Britt. WINGSTEM Introduced. Large perennial herb; stems usually winged; leaves alternate. Aug–Oct. Floodplain forests, wet thickets, streambanks. *Actinomeris alternifolia* (L.) DC.

## VERNONIA *Ironweed*

Perennial herbs (ours); leaves alternate.

ADDITIONAL SPECIES ***Vernonia arkansana*** DC. (Arkansas ironweed), reported as adventive in Douglas and Manitowoc counties. Stems 1–3 m tall, commonly glabrous and somewhat glaucous; leaves linear or linear lance-shaped; low woods and streambanks. ***Vernonia baldwinii*** Torr. (Western ironweed), adventive in Calumet County.

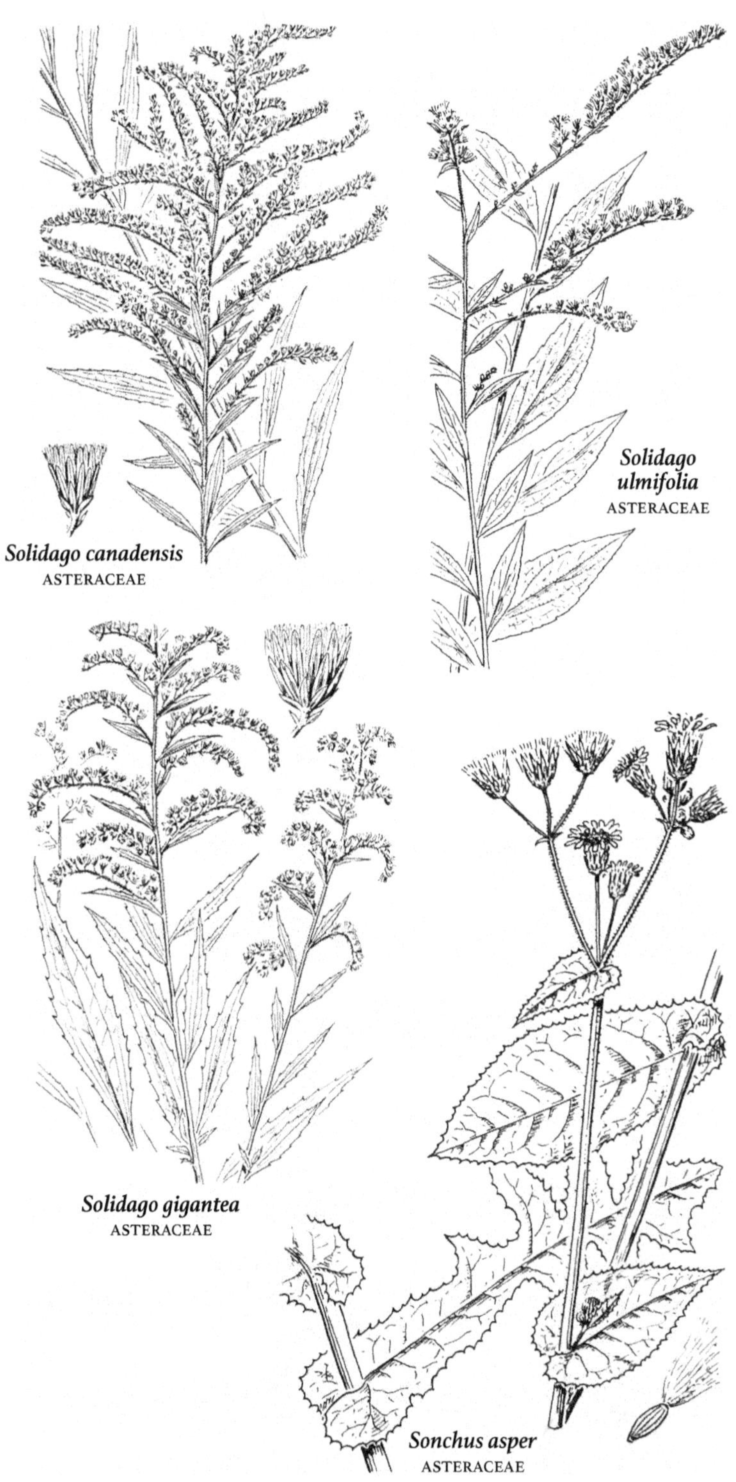

*Solidago canadensis*
ASTERACEAE

*Solidago ulmifolia*
ASTERACEAE

*Solidago gigantea*
ASTERACEAE

*Sonchus asper*
ASTERACEAE

*Symphyotrichum cordifolium*
ASTERACEAE

*Symphyotrichum oblongifolium*
ASTERACEAE

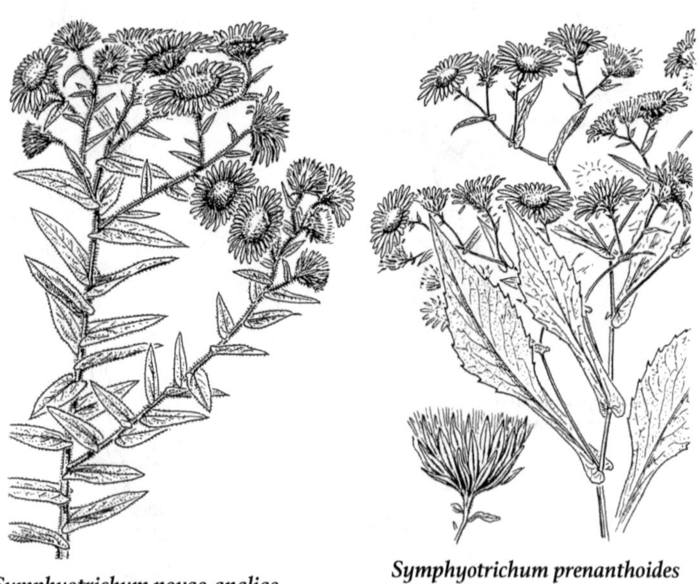

*Symphyotrichum novae-angliae*
ASTERACEAE

*Symphyotrichum prenanthoides*
ASTERACEAE

*Key to species*

1 Involucral bracts prolonged into a thread-like tip ... ***V. arkansana****

1 Involucral bracts blunt-tipped or tapered to a sharp tip .................................................................. ***V. fasciculata***

***Vernonia fasciculata*** Michx. SMOOTH IRONWEED Stout perennial herb, from a thick rootstock. Stems red or purple, smooth but short-hairy on branches of the head. Leaves alternate. Flower heads usually many, crowded in flat-topped clusters, with purple disk flowers only. July–Sept. Marshes, low prairie, streambanks.

### XANTHIUM *Cocklebur*

Coarse annual herbs; leaves alternate, often shallowly lobed.

ADDITIONAL SPECIES ***Xanthium spinosum*** L. (Spiny cocklebur), introduced, reported from Sheboygan County.

*Key to species*

1 Leaves pinnately lobed; stems bearing long, golden, three-rayed, axillary spines ....................................... ***X. spinosum****

1 Leaves coarsely palmately lobed; stems unarmed .. ***X. strumarium***

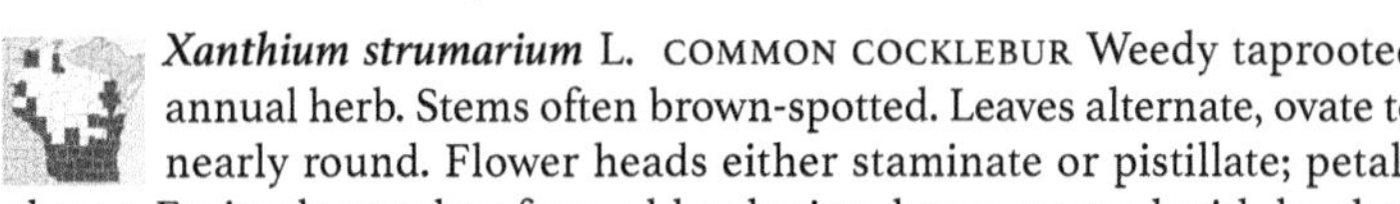

***Xanthium strumarium*** L. COMMON COCKLEBUR Weedy taprooted annual herb. Stems often brown-spotted. Leaves alternate, ovate to nearly round. Flower heads either staminate or pistillate; petals absent. Fruit a brown bur formed by the involucre, covered with hooked prickles. Aug–Sept. Shores, streambanks, wet meadows, sand bars, dried depressions, often where disturbed; also in cultivated and abandoned fields, roadsides and waste places.

## BALSAMINACEAE *Touch-Me-Not Family*

### IMPATIENS *Touch-Me-Not*

Smooth annual herbs with hollow, succulent stems; leaves alternate. Small, cleistogamous (self-fertile) flowers lacking petals are sometimes produced in summer and are often the only flowers on plants growing in shaded situations.

ADDITIONAL SPECIES ***Impatiens balfourii*** Hook. f. (Balfour's touch-me-not), introduced, reported from Dane County.

*Key to species*

1 Flowers single or paired in leaf axils; fruit pubescent ***I. balsamina***

1 Flowers in racemes from leaf axils; fruit glabrous............... 2

2 Flowers orange-yellow, usually with red-brown spots; common .. ..................................................... ***I. capensis***

2 Flowers pale yellow, spots faint or absent; uncommon ... ***I. pallida***

***Impatiens balsamina*** L. GARDEN-BALSAM Native of s Asia; planted for ornament and occasionally escaped.

***Impatiens capensis*** Meerb. SPOTTED TOUCH-ME-NOT, JEWELWEED July–Sept. Swamps, low areas in woods, floodplain forests, thickets,

streambanks, shores, marshes, fens, springs; often where disturbed. *Impatiens biflora* Walt.

***Impatiens pallida*** Nutt. PALE TOUCH-ME-NOT July–Sept. Floodplain forests, low spots in woods, swamps, streambanks, shores; often where somewhat disturbed. Similar to *I. capensis* but less common and plants typically larger.

---

## BERBERIDACEAE *Barberry Family*

Herbs or shrubs. Leaves alternate or basal.

*Key to genera*

1 Plants spiny shrubs ................................ BERBERIS
1 Plants smooth perennial herbs ................................ 2

2 Flowers in a small panicle-like cyme ......... CAULOPHYLLUM
2 Flowers single................................................ 3

3 Leaves all basal, the flowering stem naked ....... JEFFERSONIA
3 Flowering stem with a pair of opposite leaves below inflorescence ........................................ PODOPHYLLUM

### BERBERIS *Barberry*

Spiny shrubs; leaves of the shoots reduced to alternate, simple or 3-branched spines, with clusters of small foliage leaves in their axils. Fruit a red, one to few-seeded berry.

*Key to species*

1 Leaves entire; flowers single or in clusters of 2–4 ..... ***B. thunbergii***
1 Leaves tipped by a small spine; flowers in racemes of 10–20 flowers. ........................................ ***B. vulgaris***

***Berberis thunbergii*** DC. JAPANESE BARBERRY May. Native of Japan; commonly planted for low hedges and frequently escaped along roadsides and in thickets.

***Berberis vulgaris*** L. EUROPEAN BARBERRY Native of Europe; formerly widely planted and frequently escaped along roadsides and fences and in open woods; now largely purposefully exterminated as the alternate host of black rust of wheat.

### CAULOPHYLLUM *Blue Cohosh*

***Caulophyllum thalictroides*** (L.) Michx. [•35] BLUE COHOSH Smooth perennial herb. Stems bearing above the middle a single large, sessile, 3-parted leaf, and another smaller leaf just below the panicle. April–May. Rich moist woods.

### JEFFERSONIA *Twinleaf*

***Jeffersonia diphylla*** (L.) Pers. [•35] TWINLEAF Smooth perennial herb. Leaves basal, deeply divided into 2 half-ovate segmentsApril–May. Rich woods, preferring calcareous soil.

### PODOPHYLLUM *May-Apple*

***Podophyllum peltatum*** L. [•35] MAY-APPLE Herb, from a perennial rhizome, usually colony-forming. Flowering stem bearing a pair of leaves and a solitary flower. Sterile plants bearing a single, large, peltate, deeply lobed leaf; fertile plants bearing 2 half-round, similarly lobed leaves. May; fruit ripe in Aug. Moist, preferably open woods. The ripened fruit is edible in small amounts, toxic if consumed in large quantities; rhizome, leaves and seeds toxic.

## BETULACEAE *Birch Family*

Medium to large trees, or shrubs. Leaves deciduous, alternate. Fruit a small, 1-seeded, winged nutlet.

*Key to genera*

1 Plants in flower. . . . . . . . . . 2
1 Plants in fruit . . . . . . . . . . 6

2 Pistillate flowers 1 or several in a cluster . . . . . . . . . . **CORYLUS**
2 Pistillate flowers in catkins . . . . . . . . . . 3

3 Each bract of staminate catkin with 1 flower, this without sepals. 4
3 Each bract of staminate catkin with 3–6 flowers, each with sepals. 5

4 Staminate catkins in groups of 1 . . . . . . . . . . **CARPINUS**
4 Staminate catkins usually in clusters of 3 . . . . . . . . . . **OSTRYA**

5 Pistillate bracts 3-lobed; stamens 2 . . . . . . . . . . **BETULA**
5 Pistillate bracts 5-lobed; stamens 3–5 . . . . . . . . . . **ALNUS**

6 Each fruit (nut) subtended by leaf-like bracts . . . . . . . . . . 7
6 Fruit without leafy bracts, in the axil of a small scaly bract . . . . . . 9

7 Shrubs; nut 1 cm long or more . . . . . . . . . . **CORYLUS**
7 Trees; nut to 6 mm long . . . . . . . . . . 8

8 Bark furrowed and shredding, gray-brown, bracts saclike, enclosing the nut . . . . . . . . . . **OSTRYA**
8 Bark smooth and gray; bracts not enclosing the nut . **CARPINUS**

9 Bracts woody, widely spreading from rachis of cone . . . . . **ALNUS**
9 Bracts papery, ascending . . . . . . . . . . **BETULA**

### ALNUS *Alder*

Thicket-forming shrubs, or an introduced tree; leaves deciduous, ovate; fruit a flattened achene with winged or thin margins.

*Key to species*

1 Introduced tree; leaves broadly rounded, tip rounded to blunt or notched . . . . . . . . . . ***A. glutinosa***
1 Shrubs; leaves ovate to oval, tapered to a sharp tip. . . . . . . . . . 2

2 Twigs and young leaves sticky, leaves with small, sharp teeth; catkins on long stalks; fruit broadly winged . . . . . . . . . . ***A. viridis***
2 Twigs and young leaves not sticky, leaves unevenly double-toothed; catkins stalkless or on short stalks; fruit narrowly winged ***A. incana***

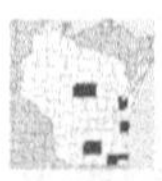

***Alnus glutinosa*** (L.) Gaertn. EUROPEAN ALDER April–May. Floodplain forests, riverbanks; also in drier places. Introduced from Eurasia and planted as an ornamental; occasionally escaping and naturalizing.

***Alnus incana*** (L.) Moench [•36] SPECKLED ALDER, TAG ALDER April–June. Swamps, thickets, bog margins, shores and streambanks. *Alnus rugosa* (Du Roi) Spreng.

***Alnus viridis*** (Vill.) Lam. & DC. GREEN ALDER Lakeshores, wet depressions in woods, rock outcrops, beaches along Lake Superior. *Alnus crispa* (Ait.) Pursh

## BETULA *Birch*

Trees or shrubs, often with multiple stems from base; bark sometimes peeling in thin layers; leaves deciduous, alternate. Fruit a wing-margined achene (samara).

ADDITIONAL SPECIES AND HYBRIDS ***Betula pendula*** Roth (European white birch), introduced, known from c and s Wisc. In contrast to native *B. papyrifera, B. pendula* has glabrous, broadly triangular and long-acuminate leaf blades. ***Betula × purpusii*** Schneid., hybrid between *B. alleghaniensis* and *B. pumila,* s Wisc. ***Betula × sandbergii*** Britt., hybrid between *B. papyrifera* and *B. pumila,* statewide.

*Key to species*

1 Shrub to 2 m tall; bark not shredding; leaves to 5 cm long . ***B. pumila***
1 Small to large trees; bark shredding with age. . . . . . . . . . . . . . . . . . 2

2 Bark white; samara wings as wide or wider than body ***B. papyrifera***
2 Bark red-brown or yellow-gray; samara wings narrower than body . . . . . . . . . . . . . . . . . . . . . . . . . . . . . . . . . . . . . . . . . . . . . . . . . 3

3 Bark red-brown; leaves wedge-shaped at base, margins wavy-toothed . . . . . . . . . . . . . . . . . . . . . . . . . . . . . . . . . . . . . . . . . . . . . ***B. nigra***
3 Bark yellow-gray; leaves rounded at base, margins not wavy-toothed . . . . . . . . . . . . . . . . . . . . . . . . . . . . . . . . . . . . . . . . . ***B. alleghaniensis***

***Betula alleghaniensis*** Britt. [•36] YELLOW BIRCH April–May. Moist forests with sugar maple; also occasional in swamps, thickets, and forest depressions with red maple, black ash, black spruce, eastern hemlock and *Alnus incana.*

***Betula nigra*** L. RIVER BIRCH May. Floodplain forests, riverbanks, swamps.

***Betula papyrifera*** Marsh. WHITE BIRCH, PAPER BIRCH Late spring. Moist, open, upland forest, especially where rocky; also on sand dunes swamps and sometimes in swampy woods; especially characteristic after fire or timber harvests, when seedlings are often abundant. Includes *B. papyrifera* Marsh. var. *cordifolia* (Regel) Fern., sometimes considered a separate species (*B. cordifolia* Regel). The bark, which has a high oil content making it waterproof, was used for a wide variety of building and clothing purposes by Native Americans.

***Betula pumila*** L. BOG BIRCH May. Swamps, bogs, fens, seeps; often where calcium-rich. *Betula glandulosa* var. *glandulifera* (Regel) Gleason.

### CARPINUS *Hornbeam*

***Carpinus caroliniana*** Walt. HORNBEAM, IRONWOOD Tall shrub or small tree, with fluted trunk and smooth, blue-gray or ashy gray bark; leaf margins sharply and often doubly serrate. Moist woods.

### CORYLUS *Hazelnut*

Shrubs or small trees; leaf margins doubly serrate.

*Key to species*

1 Twigs and leaf petioles with glandular bristles ...... *C. americana*
1 Twigs and petioles not glandular-bristly .............. *C. cornuta*

***Corylus americana*** Walt. AMERICAN HAZELNUT Dry or moist woods and thickets.

***Corylus cornuta*** Marsh. [•36] BEAKED HAZELNUT Moist woods and thickets.

### OSTRYA *Hop-Hornbeam*

***Ostrya virginiana*** (P. Mill.) K. Koch HOP-HORNBEAM Tree or tall shrubs, with light brown scaly bark; leaves alternate, margins sharply and often doubly serrate. Moist or dry woods, streambanks.

## BIGNONIACEAE *Trumpet-Creeper Family*

Trees or woody vines; leaves opposite, simple or compound.

*Key to genera*

1 Vines; leaves compound ............................ CAMPSIS
1 Trees; leaves simple .............................. CATALPA

### CAMPSIS *Trumpet-Creeper*

***Campsis radicans*** (L.) Seem. TRUMPET-CREEPER Introduced. Woody vine, trailing or climbing by rootlets along the stem; leaves pinnately compound; flowers large, orange to red, in crowded terminal clusters. July–Aug. Moist woods, fencerows, and roadsides; escaped from cultivation. *Bignonia radicans* L.

### CATALPA *Catalpa*

***Catalpa speciosa*** Warder NORTHERN CATALPA Tree; leaves simple, opposite or whorled. May–June. Alluvial forests; native to south-central USA, adventive in Wisc.

ADDITIONAL SPECIES ***Catalpa ovata*** G. Don (Chinese catalpa), introduced, reported from sw Wisc.

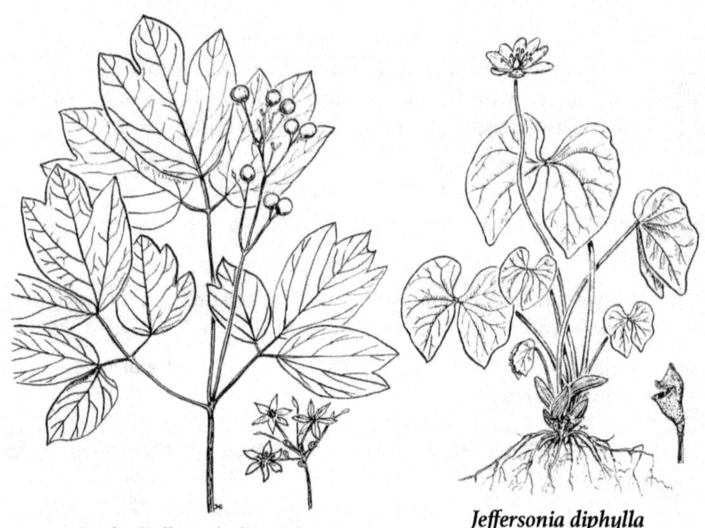

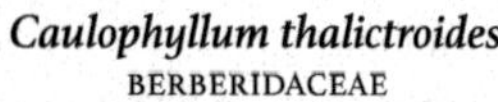

***Caulophyllum thalictroides***
BERBERIDACEAE

***Jeffersonia diphylla***
BERBERIDACEAE

***Podophyllum peltatum***
BERBERIDACEAE

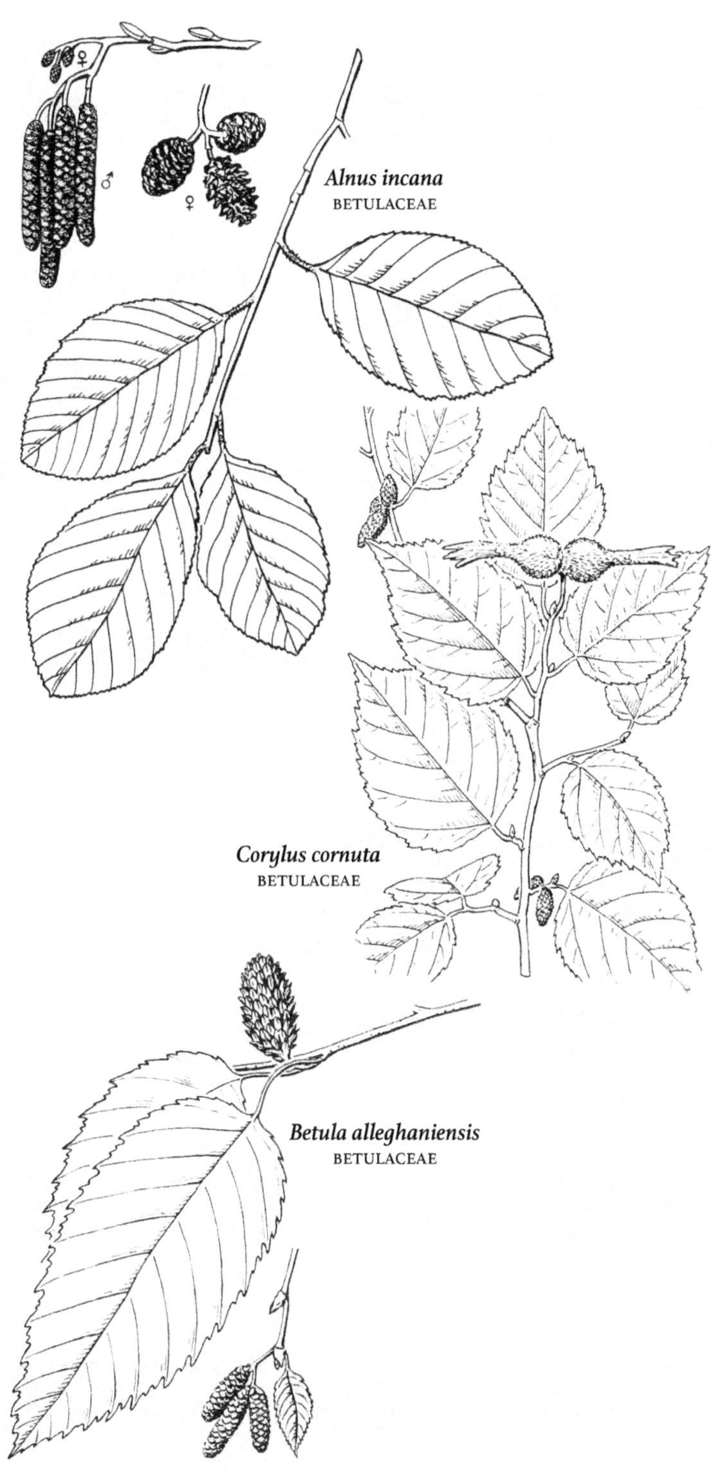
♀
♂
♀
*Alnus incana*
BETULACEAE
*Corylus cornuta*
BETULACEAE
*Betula alleghaniensis*
BETULACEAE

## BORAGINACEAE *Borage Family*

Annual or perennial herbs with usually bristly stems and alternate, bristly leaves (plants glabrous in Mertensia virginica); flowers typically in a spirally coiled, spike-like head that uncurls as flowers mature. *Ellisia* and *Hydrophyllum* previously included in Hydrophyllaceae and lack the deeply 4-lobed ovary of other Boraginaceae.

ADDITIONAL SPECIES ***Borago officinalis*** L. (Borage), annual herb, 2–6 dm tall, with hirsute stem and bright blue flowers in large terminal cymes; sometimes cultivated as a salad herb; escaped in waste places and roadsides; se Wisc. ***Buglossoides arvensis*** (L.) I.M. Johnston (Corn-gromwell), annual from a slender taproot; flowers solitary in axils of the crowded upper leaves, white or bluish white; native of Eurasia, introduced as a weed in waste places; e Wisc. ***Phacelia distans*** Benth. (Distant scorpion-weed), introduced and adventive in nw Wisc.

*Key to genera*

1 Leaves shallowly palmately lobed to deeply pinnately divided .. 2
1 Leaves simple, entire .......... 3

2 Flowers on solitary pedicels opposite alternate leaves; stem leaves at lowest nodes opposite .......... ELLISIA
2 Flowers in terminal inflorescences; stem leaves all alternate .......... HYDROPHYLLUM

3 Plants glabrous .......... MERTENSIA
3 Plants hairy.......... 4

4 Plants in flower.......... 5
4 Plants in fruit .......... 19

5 Corolla rotate, the upper portion of the corolla lobes reflexed and disk-like .......... ***Borago officinalis****
5 Corolla not rotate .......... 6

6 Flowers irregular; stamens conspicuously exserted .... ECHIUM
6 Flowers regular; stamens not longer than corolla .......... 7

7 Corolla blue or purple, or leaf base extending downward along stem (decurrent) .......... 8
7 Corolla not blue or purple; the leaf bases not decurrent along the stem.......... 14

8 Flowers more than 1 cm long .......... 9
8 Flowers less than 1 cm long; leaf bases not decurrent .......... 10

9 Leaf bases decurrent along stem .......... SYMPHYTUM
9 Leaf bases not decurrent along stem .......... MERTENSIA

10 Leaves 2 cm or more wide, or calyx lobes 5 mm long or more ... 11
10 Leaves less than 2 cm wide and calyx lobes less than 5 mm long 12

11 Calyx lobes downy-hairy .......... CYNOGLOSSUM
11 Calyx lobes coarsely hairy .......... ANCHUSA

12 Flowers all subtended by bracts .......... LAPPULA
12 Only lowest flowers with bracts .......... 13

13 Calyx much longer than pedicel and with coarse, appressed hairs .............................................. PLAGIOBOTHRYS
13 Calyx shorter or longer than pedicel, with spreading or glandular hairs ........................................... MYOSOTIS

14 Style well-exserted beyond corolla .... *Lithospermum onosmodium*
14 Style included or only slightly exserted from corolla ........... 15

15 Style 2-lobed, stigmas 2 ..................... LITHOSPERMUM
15 Style not lobed, stigma 1 .................................... 16

16 Corolla yellow .................................. AMSINCKIA
16 Corolla white.................................................. 17

17 Leaves 2 cm or more wide ......................... HACKELIA
17 Leaves less than 2 cm wide .................................. 18

18 Flowers with subtending bracts ..................... LAPPULA
18 Most flowers without subtending bracts ............ MYOSOTIS

19 Nutlets covered with bristly hairs, the hairs hooked at tip ...... 20
19 Nutlets not covered with bristly hairs.......................... 22

20 Leaves less than 1 cm wide ......................... LAPPULA
20 Leaves more than 1 cm wide ................................. 21

21 Sepals when mature more than 5 mm long .... CYNOGLOSSUM
21 Sepals when mature less than 5 mm long ........... HACKELIA

22 All flowers subtended by bracts.............................. 23
22 All, or at least upper, flowers without subtending bracts ....... 27

23 Lateral veins on leaves conspicuous.......................... 24
23 Lateral veins absent or very faint............................. 25

24 Stems covered with stiff, spreading hairs *Lithospermum onosmodium*
24 Stems with short, stiff, appressed hairs ...... LITHOSPERMUM

25 Plants covered with coarse, stiff hairs, the hairs 2–3 mm long .... ....................................................... ECHIUM
25 Pubescence various, the hairs less than 2 mm long ............ 26

26 Larger leaves more than 8 cm long ................. ANCHUSA
26 Larger leaves less than 8 cm long ........... LITHOSPERMUM

27 Calyx lobes to 6 mm long .................................... 28
27 Calyx lobes more than 6 mm long ........................... 30

28 Nutlets smooth and shiny ......................... MYOSOTIS
28 Nutlets wrinkled and dull.................................... 29

29 Perennial; stems erect, to 1 m; leaves ovate, 5 cm or more long .... .................................................. MERTENSIA
29 Plants annual; stems prostrate to ascending, to 20 cm long; leaves linear, to 6 cm long ........................ PLAGIOBOTHRYS

30 Leaves less than 1 cm wide, the petioles not winged AMSINCKIA
30 Leaves mostly more than 1 cm wide, the petioles usually winged 31

31 Plants usually more than 60 cm high; flowers in scorpioid (coiled) cymes ........................................ SYMPHYTUM
31 Plants shorter, less than 60 cm high; flowers not in scorpioid cymes ............................................ *Borago officinalis**

## AMSINCKIA *Fiddleneck*

***Amsinckia lycopsoides*** Lehm. BUGLOSS FIDDLENECK Introduced. Hispid, annual herb, the hairs irritating to touch. Adventive in waste places. *Amsinckia barbata* Greene.

## ANCHUSA *Bugloss*

Annual or perennial herbs; stems and leaves hirsute or hispid.

*Key to species*

1 Plants annual; corolla irregular ........................ ***A. arvensis***
1 Plants perennial; corolla regular ..................... ***A. officinalis***

***Anchusa arvensis*** (L.) Bieb. SMALL BUGLOSS June–Sept. Introduced as a weed in waste places. *Lycopsis arvensis* L.

***Anchusa officinalis*** L. COMMON BUGLOSS May–July. Native of Europe; escaped from cultivation in fields and roadsides.

## CYNOGLOSSUM *Hound's-Tongue*

Biennial or perennial herbs; leaves large, usually pubescent; nutlets with conspicuous, stout, hooked bristles.

*Key to species*

1 Flowers red-purple; leaves many, continuing upward on stem into inflorescence, not clasping stem ..................... ***C. officinale***
1 Flowers blue; leaves few, not in inflorescence, the upper leaves clasping at base ....................................... ***C. boreale***

***Cynoglossum boreale*** Fern. WILD COMFREY May–June. Upland woods. *Cynoglossum virginianum* subsp. *boreale* (Fern). A. Haines.

***Cynoglossum officinale*** L. HOUND'S-TONGUE May–Aug. Native of Eurasia; established in fields, meadows, and open woods.

## ECHIUM *Viper's-Bugloss*

***Echium vulgare*** L. [•37] COMMON VIPER'S-BUGLOSS Biennial, very hispid herb. Native of s Europe, weedy in waste places, roadsides, and meadows, usually where sandy or gravelly.

## ELLISIA *Aunt Lucy*

***Ellisia nyctelea*** (L.) L. WATER-POD Annual herb; leaves deeply pinnatifid. May. Moist alluvial woods.

## HACKELIA *Stickseed; Beggar's-Lice*

Perennial herbs with numerous, usually paired racemes terminating the axillary branches.

*Key to species*

1 Corolla blue; widest stem leaves to 2.5 cm broad ....... ***H. deflexa***
1 Corolla white; widest stem leaves to 3–5 cm broad ... ***H. virginiana***

***Hackelia deflexa*** (Wahlenb.) Opiz NODDING STICKSEED May–Aug. Moist woods, thickets, and hillsides.

***Hackelia virginiana*** (L.) I. M. Johnston [•37] BEGGAR'S-LICE July–Sept. Dry or moist upland woods.

### HYDROPHYLLUM *Waterleaf*

Perennial herbs; leaves large, lobed or divided.

*Key to species*

1 Leaves pinnately compound, divided into 5–7 leaflets or lobes...... ........................................... ***H. virginianum***

1 Leaves palmately lobed ..................... ***H. appendiculatum***

***Hydrophyllum appendiculatum*** Michx. GREAT WATERLEAF May–June. Rich moist woods.

***Hydrophyllum virginianum*** L. EASTERN WATERLEAF May–June. Moist or wet woods, or open wet places.

### LAPPULA *Stickseed; Beggar's-Lice*

Roughly pubescent annual herbs; stems branched, each branch terminating in an elongate bracteate raceme of small blue or occasionally white flowers. Our species can be distinguished accurately only when in fruit; fruit is also necessary for the separation of *Lappula* from *Hackelia.* A single nutlet viewed from the end will show whether the bristles are in a single or double row on each margin.

*Key to species*

1 Nutlets with a single row of bristles, these usually joined at their base ............................................ ***L. redowskii***

1 Nutlets with a double row of bristles, the bristles distinct ........ ............................................ ***L. squarrosa***

***Lappula redowskii*** (Hornem.) Greene WESTERN STICKSEED Native of the western states; introduced in Wisc in waste places and along railways. Very similar to *L. squarrosa* in habit, size, leaf-shape, and inflorescence.

***Lappula squarrosa*** (Retz.) Dumort. TWO-ROW STICKSEED May–Sept. Native of Asia and the Mediterranean region; established as a weed in waste places.

### LITHOSPERMUM *Gromwell; Puccoon; Stoneseed*

Perennial herbs (ours) with pubescent stem and foliage; nutlets bony, ovoid to nearly globose, smooth or pitted.

*Key to species*

1 Corolla cylindrical, the lobes acute, barely if at all spreading ..... ........................................... ***L. onosmodium***

1 Corolla funnelform or salverform, the lobes (especially if corolla over 10 mm long) usually spreading, not pointed ............... **2**

2 Flowers light to deep yellow or yellow-orange in a terminal inflorescence.................................................3

2 Flowers white to pale yellow from leaf axils....................5

3 Leaves linear, mostly less than 4 mm wide; corolla lobes fringed.. .................................................. *L. incisum*

3 Leaves wider; corolla lobes entire.............................4

4 Plants soft-hairy; calyx lobes less than 5 mm long .... *L. canescens*

4 Plants stiffly hairy; corolla lobes more than 5 mm long ........... ............................................. *L. caroliniense*

5 Largest leaves 2–4 cm wide; stem leaves below inflorescence number 20 or fewer ............................... *L. latifolium*

5 Largest leaves less than 2 cm wide; stem leaves 25 or more ....... ................................................ *L. officinale*

***Lithospermum canescens*** (Michx.) Lehm. [•37] HOARY PUCCOON April–May. Moist or dry prairies and dry open woods.

***Lithospermum caroliniense*** (Walt.) MacM. PLAINS PUCCOON May–July. In dry, moist or preferably sandy soil, upland woods, shores, and prairies.

***Lithospermum incisum*** Lehm. NARROW-LEAVED PUCCOON April–May. Dry prairies and barrens.

***Lithospermum latifolium*** Michx. AMERICAN GROMWELL May–June. Dry woods and thickets.

***Lithospermum officinale*** L. EUROPEAN GROMWELL May–Aug. Native of Eurasia; introduced as a weed of waste places.

***Lithospermum onosmodium*** J. Cohen FALSE GROMWELL May–June. Dry limestone hills and barrens. Lithospermum molle (Michx.) Muhl., *Onosmodium molle* Michx.

## MERTENSIA *Bluebells*

Perennial herbs; plants smooth or hairy; leaves alternate.

*Key to species*

1 Leaves and sepals hairy ........................... *M. paniculata*

1 Leaves and sepals without hairs ..................... *M. virginica*

***Mertensia paniculata*** (Ait.) G. Don NORTHERN Bluebells June–July. Conifer swamps, streambanks, seeps.

***Mertensia virginica*** (L.) Pers. [•38] VIRGINIA BLUEBELLS April–May. Floodplain forests, moist deciduous forests, streambanks; sometimes escaping from gardens where grown as an ornamental.

## MYOSOTIS *Forget-Me-Not; Scorpion Grass*

Perennial (sometimes annual) herbs; plants with short, appressed hairsl; leaves alternate.

*Key to species*

1 Calyx hairs all straight-tipped, appressed . . . . . . . . . . . . . . . . . . . . . . 2
1 Calyx hairs mostly hooked at tip, spreading . . . . . . . . . . . . . . . . . . . 3

2 Plants without stolons; lobes of sepals as long or longer than corolla tube; flowers up to 6 mm wide; nutlets longer than style . . *M. laxa*
2 Plants creeping and spreading by stolons; lobes of sepals shorter than corolla tube; flowers mostly 6 mm or more wide; nutlets shorter than style . . . . . . . . . . . . . . . . . . . . . . . . . . . . . . *M. scorpioides*

3 Pedicels equal to length of calyx (when in full-flower or fruit); corolla 3 mm or more wide . . . . . . . . . . . . . . . . . . . . . . . . . . . . . . . . . . 4
3 Pedicels shorter than calyx; corolla 1–2 mm wide . . . . . . . . . . . . . . . 5

4 Expanded part of petal cupped, less than 4 mm wide . *M. arvensis*
4 Expanded part of petal flat, 5–10 mm wide . . . . . . . . . . *M. sylvatica*

5 Corolla white; calyx appearing 2-lipped, the 3 upper lobes shorter than the 2 lower . . . . . . . . . . . . . . . . . . . . . . . . . . . . . . . . . . . . *M. verna*
5 Corolla blue, or yellowish becoming blue; calyx lobes nearly equal . . . . . . . . . . . . . . . . . . . . . . . . . . . . . . . . . . . . . . . . . . . . . . . . . . . . . . 6

6 Corolla blue; flowers on nearly entire length of stem . . . *M. stricta*
6 Corolla yellow when young, turning blue; flowers only on upper stem . . . . . . . . . . . . . . . . . . . . . . . . . . . . . . . . . . . . . . . . . . . . . . *M. discolor*

***Myosotis arvensis*** (L.) Hill ROUGH FORGET-ME-NOT Summer. Native of Eurasia; established in fields and roadsides.

***Myosotis discolor*** Pers. YELLOW AND BLUE SCORPION-GRASS May–July. Native of Europe and w Asia; locally introduced in fields and roadsides.

***Myosotis laxa*** Lehm. SMALLER FORGET-ME-NOT June–Sept. Cedar swamps, wet shores and streambanks.

***Myosotis scorpioides*** L. [•38] TRUE FORGET-ME-NOT Introduced (invasive). May–Sept. Streambanks, shores, ditches, swamps, wet depressions in forests. *Myosotis palustris* (L.) Hill.

***Myosotis stricta*** Link BLUE SCORPION-GRASS Native of Eurasia; locally introduced in dry waste places.

***Myosotis sylvatica*** Ehrh. GARDEN FORGET-ME-NOT April–Sept. Native of Eurasia; commonly cultivated for ornament and sometimes escaped near gardens.

***Myosotis verna*** Nutt. SPRING FORGET-ME-NOT April–July. Dry soil of upland woods and fields. *Myosotis virginica* auct. non (L.) B.S.P.

## PLAGIOBOTHRYS *Popcorn-Flower*

***Plagiobothrys hispidulus*** (Greene) I.M. Johnston POPCORN-FLOWER Annual herb; stems prostrate or ascending, terminating in an elongate, loosely flowered false raceme or spike; leaves essentially all cauline. May–Aug. Native to w USA; adventive in Wisc. *Plagiobothrys scouleri* (Hook. & Arn.) I.M. Johnston

*Lithospermum canescens*
BORAGINACEAE
*Hackelia virginiana*
BORAGINACEAE
*Echium vulgare*
BORAGINACEAE

*Symphytum officinale*
BORAGINACEAE
*Myosotis scorpioides*
BORAGINACEAE
*Mertensia virginica*
BORAGINACEAE

### SYMPHYTUM *Comfrey*

Perennial pubescent herbs; leaves large; flowers numerous, nodding, dull blue or dull yellow.

*Key to species*

1 Leaves distinctly decurrent on the stem (leaf petiole extending downward on stem) .................................. ***S. officinale***
1 Leaves not distinctly decurrent on the stem ........... ***S. asperum***

***Symphytum asperum*** Lepechin PRICKLY COMFREY June–Aug. Native of Europe; waste places. Very similar to *S. officinale* in size, habit, foliage, inflorescence, and nutlets; flowers averaging larger than those of *S. officinale.*

***Symphytum officinale*** L. [•38] COMMON COMFREY June–Aug. Native of Europe; sometimes escaping to fields, roadsides, and waste places.

## BRASSICACEAE *Mustard Family*

Annual, biennial or perennial herbs; leaves simple or compound, alternate on stems or basal; flowers with 4 sepals and 4 yellow, white, pink or purple petals. fruit a cylindrical (silique) or round (silicle) pod with 2 chambers.

ADDITIONAL SPECIES ***Bunias orientalis*** L. (Turkish warty-cabbage); introduced and adventive in Green County.

*Key to genera*

1 Petals yellow, yellow-tinged, or orange ........................ 2
1 Petals white, greenish, pink, purple, or absent................. 23

2 Leaves simple, not deeply lobed ............................. 3
2 At least the lower leaves lobed, pinnately lobed, or pinnately compound ................................................. 12

3 Plants glabrous throughout .................................. 4
3 Plants pubescent, at least near the base ...................... 7

4 Stem leaves not clasping at their base ................ RORIPPA
4 Stem leaves lobed at base and clasping the stem................ 5

5 Leaves entire .................................... CONRINGIA
5 At least some leaves wavy-margined or finely toothed .......... 6

6 Uppermost leaves entire or nearly so, narrow (5–10 times as long as wide); fruit glabrous; petals 7–10 mm long ........... BRASSICA
6 Uppermost leaves (e.g., at base of main branches of inflorescence) coarsely toothed, 2–4 times as long as wide; fruit glabrous or densely hairy; petals mostly 10–15 mm long ............ SINAPIS

7 At least some of the stem leaves clasping the stem ............. 8
7 None of the leaves clasping stem ............................. 9

8 Upper stem glabrous or nearly so; fruit with many seeds and more than 3 mm wide ................................. CAMELINA
8 Upper stem densely hairy; fruit with only 1 or 2 seeds, up to 3 mm wide ............................................. NESLIA

9 Ovary and fruit soon becoming much longer than wide **ERYSIMUM**
9 Ovary and fruit rounded in outline and usually to 1–2 times longer than wide.......................................................10

10 Petals 6 mm or more long .................... **LESQUERELLA**
10 Petals less than 3 mm long...................................11

11 Fruit about as long as wide; pubescence of well-branched hairs .. ............................................... **ALYSSUM**
11 Fruit much longer than wide; leaves loosely pubescent, the hairs with few branches ................................ **DRABA**

12 Pedicels in the lower portion of the raceme subtended by leafy bracts ........................................ **ERUCASTRUM**
12 Pedicels not subtended by bracts.............................13

13 Petals up to 5 mm long (including the slender or tapered claw), or if 6 mm long then plant a creeping or rhizomatous perennial.....14
13 Petals more than 5 mm long; plants never creeping perennials..17

14 Lobes of leaves rounded, nearly round in outline to broadly oval or obovate, the terminal lobe much larger than the lateral lobes; plants glabrous throughout ............................. **BARBAREA**
14 Lobes of leaves pointed, mostly distinctly longer than wide, the terminal lobe similar in size to the lateral lobes; plants glabrous or pubescent.....................................................15

15 Leaves 2–3 times pinnately dissected; plants with tiny stalked glands, or pubescent with branched hairs, or both **DESCURAINIA**
15 Leaves 1–2 times pinnately dissected; plants without glands, glabrous or pubescent, the hairs unbranched .................16

16 Fruits linear, more than 5x longer than wide; style very short; plants taprooted annuals or winter annuals ........... **SISYMBRIUM**
16 Fruits spherical to oblong, less than 5x longer than wide (if longer, then the plant a rhizomatous perennial); style stout to elongate; plants annuals or perennials ....................... **RORIPPA**

17 Leaves all on lower half of stem ................. **DIPLOTAXIS**
17 Leaves present above the middle of stem......................18

18 Basal rosette leaves often present at flowering time; plants glabrous throughout or nearly so; uppermost leaves usually with at least one pair of lobes ................................... **BARBAREA**
18 Basal rosette leaves usually absent or mostly withered at flowering time; plants glabrous or pubescent; uppermost leaves variously toothed or wavy-margined, but not distinctly lobed...........19

19 Fruit short, with the style forming a beak about as long as the nearly round lower portion, the whole fruit usually less than 1 cm long ............................................ **RAPISTRUM**
19 Fruit elongate, the lower portion much longer than wide, the whole fruit more than 1 cm long ..................................20

20 Fruit bristly hairy, or the leaves clasping; plants glabrous throughout ........................................ **BRASSICA**
20 Fruit not bristly hairy; leaves not clasping; plants not glabrous throughout.................................................21

21 Petals pale yellow with dark veins; fruit strongly twisted and long-beaked .......... RAPHANUS

21 Petals pale to deep yellow but without dark veins; fruit not twisted (or only slightly so), beak present or absent .......... 22

22 Fruit widely spreading, beakless or nearly so, up to 1 mm wide; sepals to 4 mm long .......... SISYMBRIUM

22 Plants not as above .......... BRASSICA

23 Leaves 3-parted or deeply palmately divided ..... CARDAMINE

23 Leaves neither 3-parted nor palmately divided .......... 24

24 Petals pink or purple .......... 25

24 Petals white, greenish, or absent .......... 30

25 At least the lower and middle stem leaves pinnately lobed or dissected .......... 26

25 None of the stem leaves pinnately lobed or dissected .......... 27

26 Plants glabrous or nearly so; middle and lower leaves clasping the stem at their base; mature pedicels to 1 cm long; fruit slender, about 1 mm wide .......... IODANTHUS

26 Plants at least sparsely hairy, the hairs stiff; none of the leaves clasping at base; pedicels soon longer than 1 cm; fruit much wider than 1 mm .......... RAPHANUS

27 Plants glabrous or nearly so .......... 28

27 Plants pubescent, at least near the base or on the leaves .......... 29

28 Plants succulent, the leaves thick and leathery, oblong lance-shaped .......... CAKILE

28 Plants not succulent, the leaves not leathery ...... IODANTHUS

29 Stem leaves numbering less than 10, ovate or lance-ovate, entire to wavy-margined or wavy-toothed; petals less than 1.5 cm long .......... CARDAMINE

29 Stem leaves more than 10, lance-shaped to ovate or oblong lance-shaped, entire to finely toothed; petals mostly 1.5 cm or more long .......... HESPERIS

30 At least the upper leaves sessile and clasping at their base ...... 31

30 Leaves not clasping the stem .......... 41

31 Stem leaves pinnately compound; plants aquatic or subaquatic .......... NASTURTIUM

31 Leaves not all pinnately compound; plants never aquatic ...... 32

32 Ovaries and fruit becoming linear, more than 5 times longer than wide .......... 33

32 Ovaries and fruit up to 2 times longer than wide .......... 36

33 Fruiting pedicels ± spreading, divaricate, or reflexed; fruit straight or somewhat curved and clearly spreading from the axis or even pendent .......... BOECHERA

33 Fruiting pedicels strongly ascending to appressed; fruit straight, erect and closely appressed to the stem .......... 34

34 Stem and leaves entirely glabrous or with a very few scattered hairs at the very base of the plant (especially on leaf margins and

petioles); sepals ca. half as long as the petals; mature fruit 1.4–2.5 (–3.3) mm wide, with seeds in 2 rows in each locule. . *Boechera stricta*

34 Stem and leaves pubescent, at least at the base, with spreading simple or stellate hairs; sepals ca. 2/3 as long as the petals; mature fruit less than 1.3 mm wide, with seeds crowded into 1 row in each locule .................................................. 35

35 Fruit rather strongly flattened; style-beak clearly narrower than mature fruit; stem pubescent with simple and/or forked (or stellate) hairs on at least the lower half or third, and leaves on the same portion ± pubescent (often stellate) ........... *Arabis pycnocarpa*

35 Fruit ± terete or 4-angled, slightly if at all flattened at maturity; style-beak nearly or quite as wide as the fruit; stem pubescence only on the lowermost 1–3 full-grown internodes, and only the lowermost leaves pubescent .................................. TURRITIS

36 Plants glabrous throughout................................ 37

36 Plants pubescent, at least below the middle.................. 39

37 Upper leaves nearly round in outline, less than 1.5 times longer than wide, clasping, the stem appearing to perforate the leaf LEPIDIUM

37 Upper leaves lance-shaped to narrowly ovate, more than 2 times longer than wide, clasping the stem.......................... 38

38 Inflorescence of corymb-like racemes; fruit widest at their base, less than 5 mm long; style present ...................... LEPIDIUM

38 Inflorescence of unbranched racemes, or branched 1–2 times but not corymb-like; fruit oval or obovoid, more than 5 mm long; style absent ............................................ THLASPI

39 Pedicels densely hairy; stem leaves unlobed ......... LEPIDIUM

39 Pedicels glabrous or nearly so; leaves lobed or unlobed........ 40

40 Basal leaves deeply incised to pinnate, present at flowering time; stem leaves strongly reduced in size, pinnate (or if entire then less than 1 cm wide); fruit triangle-shaped, truncate at the tip .... CAPSELLA

40 Basal leaves absent or shriveled at flowering time; stem leaves well developed, entire to finely toothed, mostly more than 1 cm wide; fruit nearly round in outline ...................... LEPIDIUM

41 Plants aquatic, the submersed leaves dissected into numerous linear, thread-like segments; rare in northern and east-central Wisc ............................................ *Rorippa aquatica*

41 Plants not aquatic, or if aquatic then the leaves not as above.... 42

42 Nearly all stem leaves deeply lobed, or pinnately divided or compound, or stem leaves absent .......................... 43

42 All or nearly all of the stem leaves simple, not lobed or pinnately divided.................................................. 48

43 Lowest leaves 2–3 times pinnately divided or pinnately compound .................................................... 44

43 Lowest leaves only once-pinnate (the segments sometimes with a few teeth)................................................. 45

44 Upper leaves simple to lobed or once-pinnate; fruit nearly round in outline ........................................... LEPIDIUM

44 All leaves 2–3 times pinnately divided or pinnately compound; fruit linear ........................................ **DESCURAINIA**

45 Petals more than 6 mm long.................................... **46**
45 Petals to 5 mm long.............................................. **47**

46 Leaves pinnately compound, the lobes entire and all similar, the lobes of the upper leaves linear, the lobes of the lower leaves round in outline or nearly so ......................... **CARDAMINE**
46 Leaves pinnately lobed, the terminal lobe much larger than the smaller lateral lobes; upper leaves similar to the lower leaves, though usually smaller ........................... **RAPHANUS**

47 Plants aquatic, rooting at the nodes; stems succulent; petals about 5 mm long ............................... **NASTURTIUM**
47 Plants not aquatic; stems slender and firm; petals to 4 mm long ... ................................................ **CARDAMINE**

48 Basal and lower leaves with distinct petioles, rounded to heart-shaped at their base ......................................... **49**
48 Leaves narrowed to the sessile or nearly sessile base ........... **52**

49 Stem leaves deeply toothed to pinnately divided; fruit to 6 mm long, but usually soon falling ......................... **ARMORACIA**
49 Stem leaves shallowly toothed or wavy-margined, or nearly entire; fruit more than 6 mm long, persistent........................ **50**

50 Stem leaves truncate to heart-shaped at base, petioles 5 mm or more long; petals less than 7 mm long; fruiting pedicels stout, about 5 mm long .................................................. **ALLIARIA**
50 Stem leaves not both heart-shaped at base and on petioles as long as 5 mm; petals more thann 7mm long; fruiting pedicels slender, or if stout then much longer than 5 mm .......................... **51**

51 Stem leaves fewer than 10, ovate, the margins entire to wavy or wavy-toothed; petals less than 1.5 cm long ............ **CARDAMINE**
51 Stem leaves more than 10, lance-shaped to ovate or oblong lance-shaped, the margins finely toothed; petals usually 1.5 cm or more long ............................................... **HESPERIS**

52 Ovaries and fruit less than 2 times longer than wide ........... **53**
52 Ovaries and fruit more than 2 times longer than wide.......... **55**

53 Fruit ascending or erect; the fruit and stems densely woolly hairy .. ................................................ **BERTEROA**
53 Fruit widely spreading; plants glabrous to pubescent .......... **54**

54 Plants glabrous or with tiny hairs, the hairs unbranched ......... ................................................ **LEPIDIUM**
54 Plants pubescent, the hairs branched ................ **ALYSSUM**

55 Fruit linear, more or less round in cross-section, to 1 mm wide; plants blooming and then withering by late spring ...... **ARABIDOPSIS**
55 Fruit linear, flat, 1 mm or more wide; plants withering or persisting ...................................................... **56**

56 Fruit to 18 mm long; plants usually not leafy above the middle; early blooming and withering by late spring ................. **DRABA**

56 Fruit longer than 18 mm; plants usually leafy throughout; plants persisting to late summer or early fall .............. **BOECHERA**

## ALLIARIA *Garlic-Mustard*

***Alliaria petiolata*** (Bieb.) Cavara & Grande [•39] GARLIC-MUSTARD Biennial, garlic-scented herb, glabrous or nearly so. basal leaves in more or less evergreen rosettes; stem leaves deltoid, coarsely toothed. petals white. May–June. Native of Europe; invasive in rich, moist, shaded soil; also on roadsides or rarely in swamps. *Alliaria officinalis* Andrz.

## ALYSSUM *Madwort*

***Alyssum alyssoides*** (L.) L. PALE MADWORT Annual herb. Stem, leaves, inflorescence, and fruits stellate-pubescent. May–June. Native of Europe; weedy in waste places.

## ARABIDOPSIS *Thalecress*

Annual or biennial herbs, more or less pubescent with branched hairs; leaves mostly in a basal rosette. Sepals oblong, obtuse. Petals white, spatulate. Fruit linear, nearly terete, many-seeded.

*Key to species*

1 Petals mostly 5–8 mm long, seeds ca. 1 mm long ......... ***A. lyrata***
1 Petals mostly 2–4 mm long, seeds to 0.5 mm long ...... ***A. thaliana***

***Arabidopsis lyrata*** (L.) O'Kane & Al-Shehbaz LYRE-LEAF ROCKCRESS April–June. Dry woods and fields, especially in sandy soil; sand dunes. *Arabis lyrata* L.

***Arabidopsis thaliana*** (L.) Heynh. MOUSE-EAR CRESS April–June. Widespread in e USA; introduced in s Wisc on disturbed, usually sandy sites, including cultivated land, fields, and oak forests.

## ARABIS *Rockcress*

***Arabis pycnocarpa*** M. Hopkins HAIRY ROCK CRESS Biennial herb. Stems pubescent at least at base with simple or branched hairs. Petals white. Fruit erect, flat, linear. May–June. Woods, often where calcareous. *Arabis hirsuta* (L.) Scop. Most of our native species formerly in genus *Arabis* are now placed in genus *Boechera,* with either broader fruit (if tightly appressed) or the fruit spreading, reflexed or pendent; *Arabis glabra* (L.) Bernh. now placed in genus *Turritis.*

ADDITIONAL SPECIES ***Arabis caucasica*** Willd. (Gray rockcress), garden escape, reported from Door County.

## ARMORACIA *Horse-Radish*

***Armoracia rusticana*** P.G. Gaertn. B. Mey. & Scherb. HORSE-RADISH Glabrous perennial herb from thick roots. May–July. Native of se Europe and w Asia; commonly cultivated and escaped into moist soil of ditches, shores, roadsides, and disturbed places.

## BARBAREA *Yellow-Rocket*

Biennial herbs, smooth or with a few simple hairs; basal leaves pinnatifid with a large terminal lobe and 2 to several small lateral lobes; stem leaves smaller, entire to pinnatifid; petals yellow.

*Key to species*

1 Petals 6–8 mm long; beak of fruit 2–3 mm long ........ ***B. vulgaris***
1 Petals to 5 mm long; beak of fruit less than 2 mm long ***B. orthoceras***

***Barbarea orthoceras*** Ledeb. AMERICAN YELLOW-ROCKET June–July. Rocky shores, swamps and wet woods.

***Barbarea vulgaris*** Ait. f. [•39] GARDEN YELLOW-ROCKET April–June. Native of Europe; naturalized as a weed in damp soil of fields, roadsides, and gardens.

## BERTEROA *Hoary Alyssum*

***Berteroa incana*** (L.) DC. HOARY ALYSSUM Annual herb; stem, foliage, and inflorescence finely canescent, the hairs stellate with radiating branches. May–Sept. Native of Europe; now established as a weed.

## BOECHERA *Rockcress*

Biennial or perennial herbs; basal leaves petioled, the stem leaves smaller and usually sessile. Includes former native members of genus *Arabis.*

*Key to species*

1 Leaves not clasping at base ......................... ***B. canadensis***
1 Upper leaves clasping at their base............................ 2

2 Pedicels becoming distinctly reflexed before the petals wither, the fruit pendent; sepals ca. half as long as mature petals or a little shorter ........................................... ***B. grahamii***
2 Pedicels spreading or ascending to strongly appressed, even after anthesis, the fruit spreading to erect; sepals various ............ 3

3 Fruiting pedicels strongly ascending to appressed, the fruit straight, erect and closely appressed to the stem ................ ***B. stricta***
3 Fruiting pedicels ± spreading, the fruit straight or somewhat curved and clearly spreading from the axis.......................... 4

4 Upper stem leaves (i.e., below the lowermost pedicels or branches) ± dentate and pubescent on both surfaces ............. ***B. dentata***
4 Upper stem leaves entire or nearly so, glabrous ................ 5

5 Basal leaves lyrate-pinnatifid, with at least a few simple hairs at tips of the teeth or lobes; stem leaves below the inflorescence many (ca. 30–40); sepals ca. half as long as the petals........ ***B. missouriensis***
5 Basal leaves entire or merely serrate (or absent at anthesis), completely glabrous or stellate-pubescent on both surfaces; stem leaves various; sepals various ................................. 6

6 Stem leaves ca. 25–35 or more below the inflorescence, the longest 2.5–5 cm; stem at the base and both surfaces of basal leaves ± stellate-pubescent; petals pink or pale purple; sepals at most barely more than half as long as the petals .................. ***B. grahamii***

6 Stem leaves ca. 10–15 (–20) below the inflorescence, the longest 9–15 cm; stem and leaves completely glabrous at base of plant (and elsewhere); petals white; sepals much more than half as long as petals ........................................ ***B. laevigata***

***Boechera canadensis*** (L.) Al-Shehbaz SICKLEPOD May–July. Moist or dry woods. *Arabis canadensis* L.

***Boechera dentata*** (Raf.) Al-Shehbaz & Zarucchi SHORT'S ROCKCRESS April–May. Rich moist woods. *Arabis shortii* (Fern.) Gleason

***Boechera grahamii*** (Lehmann) Windham & Al-Shehbaz SPREADING ROCKCRESS June–July. Sandy or rocky soil. *Arabis divaricarpa* A. Nels.

***Boechera laevigata*** (Muhl. ex Willd.) Al-Shehbaz SMOOTH BANK CRESS May–June. Moist or dry woods. *Arabis laevigata* (Muhl.) Poir.

***Boechera missouriensis*** (Greene) Al-Shehbaz GREEN ROCKCRESS May–June. Dry sandy woods. *Arabis missouriensis* Greene

***Boechera stricta*** (Graham) Al-Shehbaz DRUMMOND'S ROCKCRESS May–Aug. Moist or dry places, often where calcareous. *Arabis drummondii* A. Gray

## BRASSICA *Mustard*

Coarse annual or biennial herbs. Leaves (at least the lower) pinnatifid. Petals yellow (ours), varying to nearly white in some cultivated species. Many species have been long-cultivated, and may persist in gardens over winter, blooming the second year. The oilseeds known as canola are sometimes varieties of *Brassica rapa* but are mostly *B. napus* and *B. juncea.*

ADDITIONAL SPECIES ***Brassica napus*** L. (Turnip), cultivated and sometimes escaping.

*Key to species*

1 Middle and upper leaves clasping stem .................. ***B. rapa***
1 Leaves not clasping stem ........................................ 2

2 Fruit to 2 cm long, strongly appressed .................. ***B. nigra***
2 Fruit becoming more than 2 cm long, not strongly appressed .... .................................................... ***B. juncea***

***Brassica juncea*** (L.) Czern. BROWN MUSTARD, CHINESE MUSTARD Introduced (naturalized). June–Oct. Established as a weed in waste places and fields.

***Brassica nigra*** (L.) W.D.J. Koch BLACK MUSTARD Introduced. Summer and fall. Naturalized in fields and waste places.

***Brassica rapa*** L. FIELD MUSTARD, TURNIP Introduced. May–Oct. Naturalized as a weed of fields and waste ground. *Brassica campestris* L.

## CAKILE *Sea-Rocket*

***Cakile edentula*** (Bigelow) Hook. SEA-ROCKET Succulent annual. Stems much branched and bushy. Petals pale purple. Atlantic and Great Lakes coastal species, in Wisc mostly along sandy beaches and low dunes near Lake Michigan, often spreading by pieces of the floating fruit. *Cakile lacustris* (Fern.) Pobed.

## CAMELINA *False Flax*

Annual or winter-annual herbs, bearing both simple and branched hairs. Basal leaves narrowly spatulate; stem leaves linear to lance-shaped, clasping by a sagittate-auriculate base. Petals yellow.

*Key to species*

1 Fruit to 7 mm long and 5 mm wide; lower stem pubescent with both spreading and appressed hairs .................. ***C. microcarpa***

1 Fruit more than 7 mm long and more than 5 mm wide; stem glabrous or with tiny hairs ............................ ***C. sativa***

***Camelina microcarpa*** DC. LITTLE-POD FALSE FLAX Introduced. April–June. Fields and waste places, usually in sandy soil.

***Camelina sativa*** (L.) Crantz LARGE-SEED FALSE FLAX Introduced. May–June. Fields and waste places, usually where sandy.

## CAPSELLA *Shepherd's-Purse*

***Capsella bursa-pastoris*** (L.) Medik. [•39] SHEPHERD'S-PURSE Introduced (naturalized). Annual herb, pubescent with stellate hairs. Petals white. Spring. Lawns, gardens, waste places. Where sheltered, this is one of the first plants to bloom in spring.

## CARDAMINE *Bittercress; Toothwort*

Annual, biennial or perennial herbs, smooth or with short hairs near base of stem. Leaves simple to pinnately divided, the basal leaves often different in shape than stem leaves. Flowers in racemes or umbel-like clusters; petals usually white.

*Key to species*

1 Leaves simple to pinnately compound......................... 2

1 Leaves palmately 3–5 parted or compound .................... 6

2 Leaves simple; plants from a shallow tuber-like rhizome........ 3

2 Stem leaves pinnately dissected, with 2 or more deep lobes; plants without a tuber-like base ................................. 4

3 Petals pink to purple; sepals purple, turning brown with age ..... ............................................... ***C. douglassii***

3 Petals white; sepals green, turning yellow with age ..... ***C. bulbosa***

4 Petals 8 mm or more long ......................... ***C. pratensis***

4 Petals to 4 mm long...................................... 5

5 Leaflets of stem leaves linear ...................... ***C. parviflora***

5 Leaflets ovate ............................. ***C. pensylvanica***

6 Leaves divided into 4–7 linear segments ........... *C. concatenata*
6 Leaves divided into 3 ovate segments ......................... 7

7 Leaves usually 2 ................................... *C. diphylla*
7 Leaves usually 3 ................................... *C. maxima*

***Cardamine bulbosa*** (Schreb.) B.S.P. BULBOUS BITTERCRESS May–June. Wet forest depressions, floodplain forests, streambanks, wet meadows, swamps, calcareous fens. *Cardamine rhomboidea* (Pers.) DC.

***Cardamine concatenata*** (Michx.) Sw. CUT-LEAF TOOTHWORT April–May. Moist rich woods. *Dentaria laciniata* Muhl.

***Cardamine diphylla*** (Michx.) Wood [•40] BROAD-LEAF TOOTHWORT April–May. Rich woods. *Dentaria diphylla* Michx.

***Cardamine douglassii*** Britt. LIMESTONE BITTERCRESS April–May. Floodplain forests and low deciduous woods, often in shade.

***Cardamine maxima*** (Nutt.) Wood THREE-LEAF TOOTHWORT April–May. Rich deciduous woods, often along streams. *Dentaria maxima* Nutt.

***Cardamine parviflora*** L. SMALL-FLOWERED BITTERCRESS Usually in dry soil. Pedicels and fruit as in *C. pensylvanica. Cardamine arenicola* Britt.

***Cardamine pensylvanica*** Muhl. [•40] PENNSYLVANIA BITTERCRESS May–Sept. Streambanks, swamps, and wet forests (often where seasonally flooded); wet, disturbed areas.

***Cardamine pratensis*** L. CUCKOO-FLOWER May–June. Peatlands, tamarack and cedar swamps, wet depressions in forests.

## CONRINGIA *Hare's-Ear-Mustard*

***Conringia orientalis*** (L.) Dumort. HARE'S-EAR-MUSTARD Annual glabrous herb, often glaucous. Petals yellowish white. May–Aug. Native of Eurasia; naturalized or adventive in waste places.

## DESCURAINIA *Tansy-Mustard*

Annual or biennial herbs, more or less pubescent or canescent with wholly or partly branched hairs; leaves 1–3-pinnate with very numerous small segments. Petals yellow.

*Key to species*

1 Plants green, with glandular hairs; fruit less than 13 mm long .... ................................................ *D. pinnata*
1 Plants gray-green, with stellate hairs, the hairs not glandular; fruit 13 mm or more long ................................ *D. sophia*

***Descurainia pinnata*** (Walt.) Britt. [•40] TANSY-MUSTARD Usually in disturbed places; roadsides, railroads, fields, gravel pits, shores.

***Descurainia sophia*** (L.) Webb HERB-SOPHIA Introduced (naturalized). Disturbed places, roadsides, railroads.

## DIPLOTAXIS *Wallrocket*

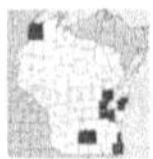

***Diplotaxis muralis*** (L.) DC. ANNUAL WALLROCKET Annual herb. Leaves chiefly basal or near the base, shallowly or deeply toothed or pinnatifid. Petals yellow. May–Sept. Native of Europe; naturalized in waste places, especially in sandy soil.

## DRABA *Whitlow-Grass*

Annual, biennial, or perennial herbs, in some species woody at base. Leaves entire or dentate, more or less pubescent with simple, branched, or stellate hairs, or with 2 types of hairs together. Petals yellow or white.

*Key to species*

1 Plants annuals or winter-annuals .............................. 2
1 Plants perennial................................................ 4

2 Petals cleft to near middle; leaves all basal .............. *D. verna*
2 Petals not cleft; at least some leaves along stem................. 3

3 Petals white; pedicels about half length of fruit ........ *D. reptans*
3 Petals yellow; pedicels equal to or longer than fruit .. *D. nemorosa*

4 Fruit densely covered with stellate hairs; rare in Door County .... ............................................................ *D. cana*
4 Fruit glabrous or only sparsely hairy .......................... 5

5 Fruit usually twisted .............................. *D. arabisans*
5 Fruit straight ...................................... *D. glabella*

***Draba arabisans*** Michx. ROCK WHITLOW-GRASS May–June. Rocks and cliffs.

***Draba cana*** Rydb. hoary whitlow-grass *Endangered.* May–July. Rocky limestone ledges, cliffs, and gravelly or rocky soil; known from Door County; more common in Rocky Mtns.

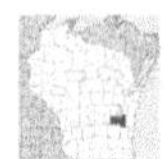

***Draba glabella*** Pursh SMOOTH WHITLOW-GRASS May–July. Reported from Fond du Lac County.

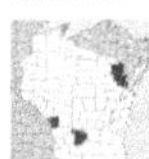

***Draba nemorosa*** L. WOODLAND WHITLOW-GRASS Introduced. Dry soil, prairies and hillsides.

***Draba reptans*** (Lam.) Fern. CAROLINA WHITLOW-GRASS April–May. Dry, sterile or sandy soil.

***Draba verna*** L. SPRING WHITLOW-GRASS April–May. Native of Europe; naturalized in fields and roadsides.

## ERUCASTRUM *Dog-Mustard*

***Erucastrum gallicum*** (Willd.) O.E. Schulz COMMON DOG-MUSTARD Introduced (naturalized). Annual or biennial; pubescence of simple hairs or none. Mature racemes greatly elongate; petals yellow, spatulate, about 7 mm long. May–Sept. Waste places.

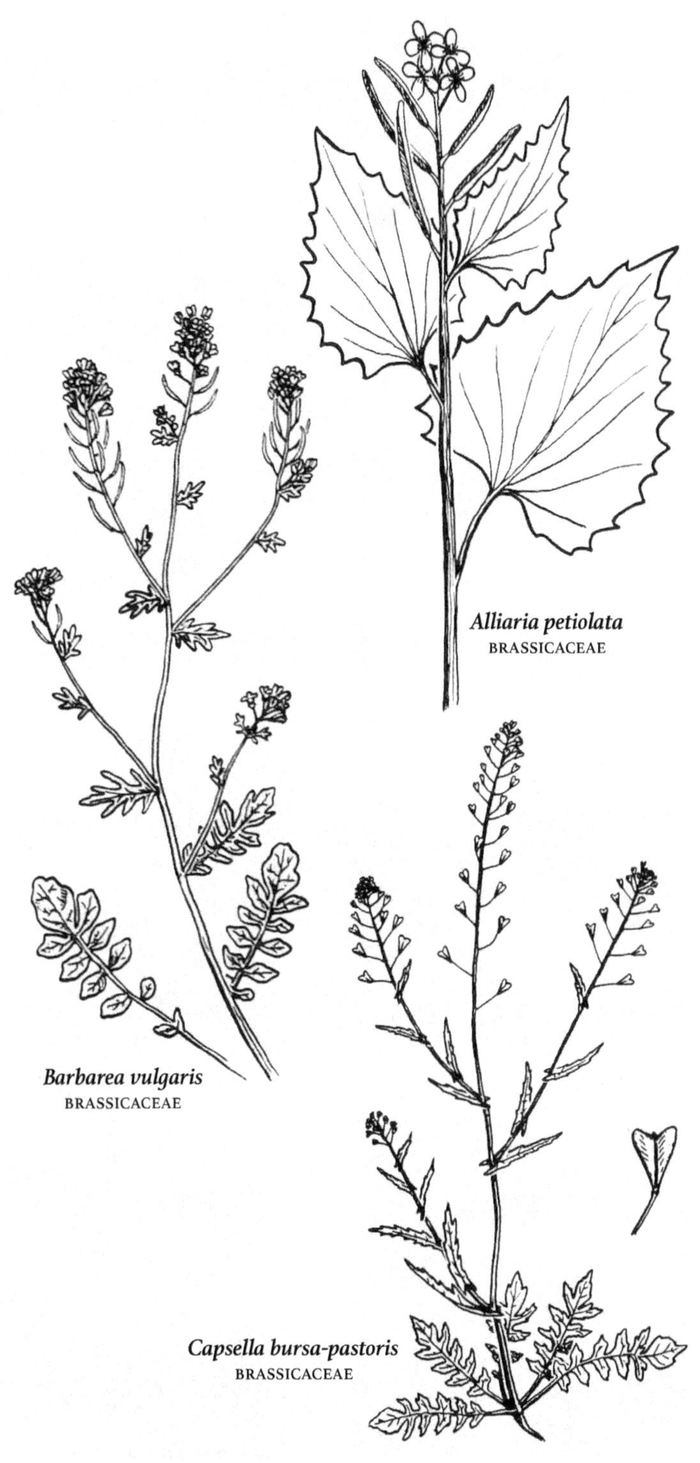
*Alliaria petiolata*
BRASSICACEAE
*Barbarea vulgaris*
BRASSICACEAE
*Capsella bursa-pastoris*
BRASSICACEAE

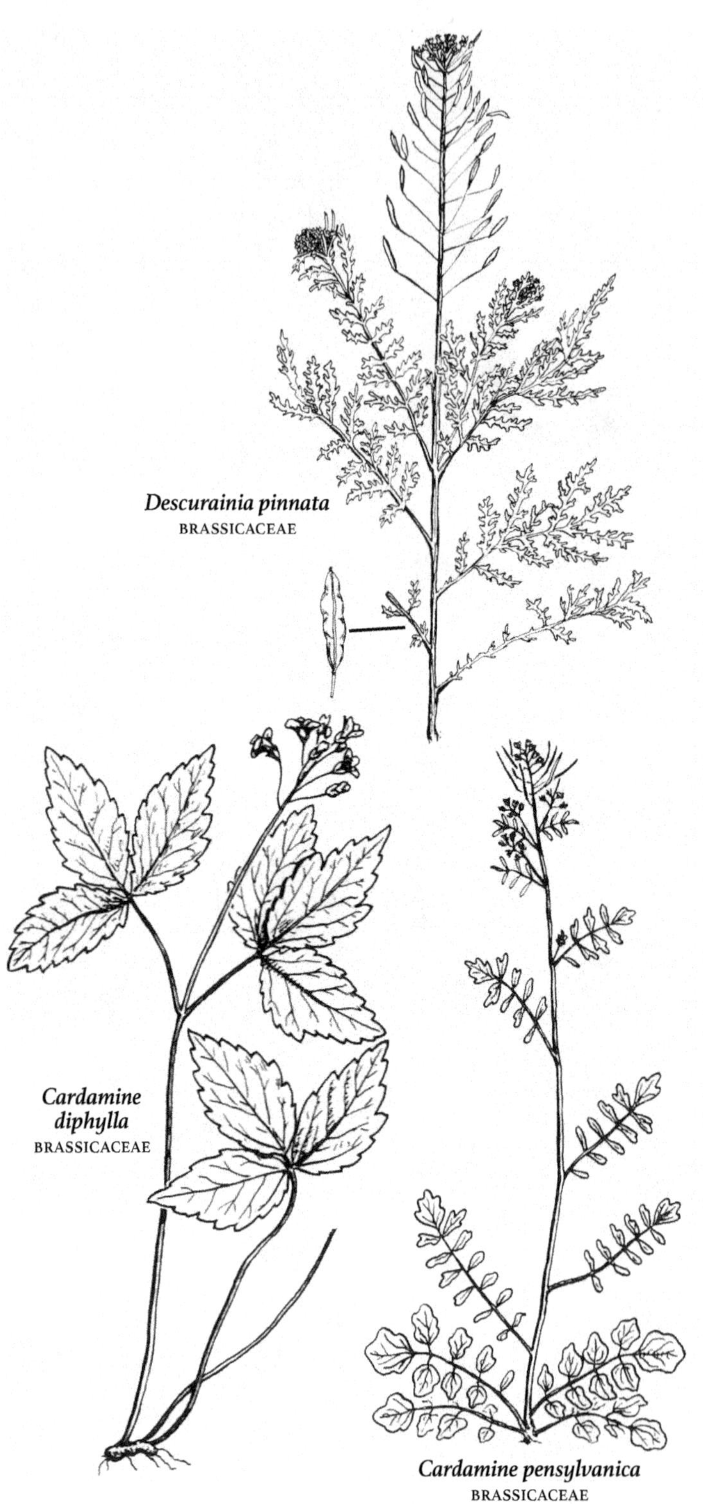
*Descurainia pinnata*
BRASSICACEAE
*Cardamine diphylla*
BRASSICACEAE
*Cardamine pensylvanica*
BRASSICACEAE

## ERYSIMUM *Wallflower*

Annual to perennial herbs, with narrow, entire, dentate, or pinnatifid leaves. Petals yellow to orange (in our species). All our species are more or less densely pubescent on the stem, leaves, sepals, and fruit, and more or less so on the back of the petals, especially at the base of the blade.

*Key to species*

1 Petals 15–25 mm long; fruit 5–10 cm long ........... ***E. capitatum***
1 Petals 10 mm long or less........................................ 2

2 Petals less than 6 mm long; fruit to 3 cm long ... ***E. cheiranthoides***
2 Petals 6 mm or more long ........................................ 3

3 Leaves very finely toothed ...................... ***E. hieraciifolium***
3 Leaves entire .................................... ***E. inconspicuum***

***Erysimum capitatum*** (Dougl.) Greene WESTERN WALLFLOWER Introduced. May–June. Prairies, sand hills, open woods. *Erysimum asperum* (Nutt.) DC.

***Erysimum cheiranthoides*** L. WORM-SEED WALLFLOWER Introduced (naturalized). June–Aug. Usually in wet soil, but also appearing as a weed in fields and roadsides.

***Erysimum hieraciifolium*** L. EUROPEAN WALLFLOWER June–July. European weed. Much like *E. inconspicuum,* usually (but not always) with somewhat broader leaves.

***Erysimum inconspicuum*** (S. Wats.) MacM. SHY WALLFLOWER Introduced (naturalized). May–Aug. Dry soil of prairies, plains, and upland woods.

## HESPERIS *Dame's Rocket*

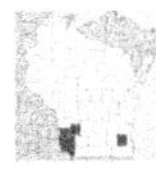

***Hesperis matronalis*** L. DAME'S ROCKET Introduced (invasive), perennial herb. Leaves lance-shaped, remotely and sharply denticulate, pubescent above with simple hairs, below chiefly with branched hairs. Flowers fragrant; petals purple, varying to pink or white. May–June. Formerly cultivated for ornament; frequently escaped along roads and fencerows and in open woods.

## IODANTHUS *Purple-Rocket*

***Iodanthus pinnatifidus*** (Michx.) Steud. Purple-Rocket Perennial herb; plants smooth. Leaves lance-shaped to oval or oblong, leaf base often with lobes which clasp stem. Petals pale violet to white. June–July. Wet or moist floodplain forests.

## LEPIDIUM *Pepperwort*

Annual, biennial, or perennial herbs. Leaves linear to elliptic, entire, toothed, or pinnatifid. Petals small, white (rarely yellowish). Fruit a flattened silicle, ovate to circular or obovate, often winged, commonly notched at tip and tipped by the persistent style or stigma.

ADDITIONAL SPECIES ***Lepidium appelianum*** Al-Shehbaz [syn: *Cardaria pubescens* (C.A. Mey.) Jarmolenko] introduced, reported from Walworth County.

*Key to species*

1 At least upper stem leaves sessile and auriculate, sagittate, or clasping at base . . . . . . . . . . 2
1 Stem leaves petiolate or subsessile, never auriculate, sagittate, or clasping at base . . . . . . . . . . 5

2 Flowers yellow; seeds winged; uppermost leaves cordate-clasping; plants glabrous above . . . . . . . . . . ***L. perfoliatum***
2 Flowers white; seeds wingless; uppermost leaves auriculate to sagittate; plants pubescent above . . . . . . . . . . 3

3 Plants annual, not rhizomatous; fruit broadly winged at apex; racemes elongated in fruit . . . . . . . . . . ***L. campestre***
3 Plants perennial, rhizomatous; fruit wingless at apex; racemes not elongated in fruit . . . . . . . . . . 4

4 Fruit pubescent, inflated, spherical, valves not veined ***L. appelianum****
4 Fruit glabrous, flattened, cordate, valves veined . . . . . . . . . . ***L. draba***

5 Fruit elliptic; fruiting pedicels puberulent all around; basal leaves (often withered in fruit) 1-or 2-pinnatisect . . . . . . . . . . ***L. ruderale***
5 Fruit orbicular or obovate; fruiting pedicels glabrous at least below; basal leaves dentate or pinnatifid . . . . . . . . . . 6

6 Fruit obovate, widest above middle; petals absent or often rudimentary; rachis of raceme puberulent with cylindrical or clavate hairs . . . . . . . . . . ***L. densiflorum***
6 Fruit orbicular, widest at middle; petals present or rarely rudimentary; rachis of raceme puberulent with curved hairs, rarely glabrous . . . . . . . . . . ***L. virginicum***

***Lepidium campestre*** (L.) Ait. f. FIELD-CRESS May–June. Native of Europe; a weed of sandy waste ground, fields, and roadsides.

***Lepidium densiflorum*** Schrad. PRAIRIE PEPPERWORT May–June. Dry, sandy or gravelly disturbed places; perhaps native to w USA, considered adentive in Wisc and e USA.

***Lepidium draba*** L. HEART-POD HOARYCRESS Introduced. Cardaria draba (L.) Desv. May–July. Fields, roadsides, and waste places.

***Lepidium perfoliatum*** L. CLASPING PEPPERWORT May–June. Native of Europe; occasional in Wisc in dry disturbed places.

***Lepidium ruderale*** L. ROADSIDE PEPPERWORT May–June. Native of Europe; introduced in waste places.

***Lepidium virginicum*** L. [•41] POOR-MAN'S PEPPER May–June. Dry fields, gardens, roadsides, and waste places.

## LESQUERELLA *Bladderpod*

***Lesquerella ludoviciana*** (Nutt.) S. Wats. LOUISIANA BLADDERPOD *Threatened.* May–July. Rare on dry dolomitic bluff edges along the Miss River.

## NASTURTIUM *Watercress*

***Nasturtium officinale*** R.Br. WATERCRESS Introduced (invasive). Perennial herb; plants smooth, often forming large, tangled colonies. Stems underwater, floating, or trailing on mud; rooting from lower nodes. Leaves pinnately divided, the terminal segment largest. Flowers in 1 to several racemes per stem; petals white, sometimes purple-tinged. Fruit a linear, often curved pod (silique). May–Sept. Seeps, slow-moving streams, ditches, cedar swamps, especially in cold spring-fed waters. Naturalized throughout most of USA and s Canada. *Rorippa nasturtium-aquaticum* (L.) Hayek.

ADDITIONAL SPECIES ***Nasturtium microphyllum*** (Boenn. ex Rchb.) Rchb., introduced perennial known from Dane and Rock counties; very similar to *N. officinale* (sometimes treated as a variety) and found in same aquatic habitats; distinguished by narrower mature fruit and 2 rows of seed under each valve rather than 1 row.

## NESLIA *Ball-Mustard*

***Neslia paniculata*** (L.) Desv. YELLOW BALL-MUSTARD Introduced annual herb. Stems pubescent with branched hairs. Flowers small, petals yellow. June–July. Introduced and weedy, sometimes in cultivated fields.

## RAPHANUS *Radish*

***Raphanus raphanistrum*** L. WILD RADISH Coarse annual herb from a stout taproot, pubescence of simple hairs. Lower leaves obovate in outline, pinnatifid. Petals yellow, becoming white in age. June–Aug. Native of Eurasia; weedy on roadsides, fields, waste places.

ADDITIONAL SPECIES ***Raphanus sativus*** L. (Radish), cultivated and rarely escaping from gardens.

## RAPISTRUM *Turnipweed*

***Rapistrum rugosum*** (L.) All. TURNIPWEED Annual herb. Leaves mostly in a basal rosette; stem leaves few. Petals yellow. Summer. Native of Eurasia; rarely adventive in waste places, Milwaukee County.

## RORIPPA *Yellowcress*

Annual, biennial or perennial herbs; plants smooth or with unbranched hairs. Leaves sometimes in a basal rosette in young plants, toothed to pinnately divided, petioles short or absent. Flowers small, in racemes at ends of stems or from lateral branches; petals yellow or white.

ADDITIONAL SPECIES ***Rorippa austriaca*** (Crantz) Bess. (Austrian yellowcress), perennial from a rhizome; native of Europe; weedy in fields where the plants spread by fragments of the rhizome.

*Key to species*

1 Plant truly aquatic, the submersed leaves dissected in a bipinnate pattern into filiform segments (midvein present, the lateral segments again dissected), frequently detaching readily from the stem; petals white .................................. ***R. aquatica***

1 Plant terrestrial or aquatic but even if in water the leaves with definite flat lobes (not bipinnately dissected) and not falling from the stem; petals yellow .......... 2

2 Plants annual or biennial, taprooted; petals shorter than or equal to sepals .......... 3

2 Plants perennial, roots creeping; petals longer than sepals .......... 4

3 Pedicels of fruit 3 mm or more long; fruit to 1.5 times longer than its pedicel .......... ***R. palustris***

3 Pedicels of fruit 1–2 mm long; fruit more than 1.5 times longer than its pedicel .......... ***R. sessiliflora***

4 Stems sprawling or spreading; lateral leaf segments entire or with a few shallow teeth; beak of fruit 1–2 mm long .......... ***R. sinuata***

4 Stems erect or nearly so; lateral leaf segments with sharp teeth; beak of fruit to 1 mm long .......... ***R. sylvestris***

***Rorippa aquatica*** (Eaton) E.J. Palmer & Steyerm. LAKECRESS *Endangered.* June–Aug. Lake Superior estuaries, quiet water in lakes, rivers and streams; muddy shores. Armoracia aquatica (Eaton) Wieg., *Neobeckia aquatica* (Eaton) Greene.

***Rorippa palustris*** (L.) Bess. COMMON YELLOWCRESS June–Sept. Marshes, wet meadows, streambanks, ditches and other wet places.

***Rorippa sessiliflora*** (Nutt.) A.S. Hitchc. STALKLESS YELLOWCRESS June–July. Muddy shores and streambanks.

***Rorippa sinuata*** (Nutt.) A.S. Hitchc. SPREADING YELLOWCRESS June–Aug. Stream and riverbanks, ditches, and other low places, especially where sandy.

***Rorippa sylvestris*** (L.) Bess. [•41] CREEPING YELLOWCRESS June–Aug. Introduced from Europe; sometimes weedy in wet forests, lakeshores, muddy streambanks and ditches.

## SINAPIS *White-Mustard*

Annual introduced herbs. Similar to *Brassica* (and previously included in that genus) but the flowers tend to be larger.

*Key to species*

1 Fruit glabrous, 2 mm wide .......... ***S. arvensis***

1 Fruit bristly, 4 mm wide .......... ***S. alba***

***Sinapis alba*** L. WHITE-MUSTARD Introduced. May–July. Occasionally cultivated for its seeds which are used to make the condiment mustard; established as a weed in fields and waste places but seldom abundant. *Brassica alba* Rabenh. non L.

***Sinapis arvensis*** L. CORN-MUSTARD Introduced (naturalized). May–July. Common weed of fields, gardens, and waste ground. *Brassica arvensis* Rabenh. non L.

## SISYMBRIUM *Hedge-Mustard*

Ours annual or winter-annual herbs, with simple hairs. Leaves (at least the lower) deeply pinnatifid. Petals yellow.

*Key to species*

1 Fruit erect, appressed; pedicels erect, 2–3 mm long ... ***S. officinale***
1 Fruit widely spreading; pedicels spreading, 5 mm long or more.. 2

2 Fruit 5 cm long or more; pedicels about as thick as fruit .......... ...................................................... ***S. altissimum***
2 Fruit to 5 cm long; pedicels thinner than fruit ........... ***S. loeselii***

***Sisymbrium altissimum*** L. [•41] TUMBLING MUSTARD June–July. Native of Eurasia; established as a weed in fields and waste ground.

***Sisymbrium loeselii*** L. TALL HEDGE-MUSTARD June–July. Native of se Europe and w Asia; occasionally adventive.

***Sisymbrium officinale*** (L.) Scop. HEDGE-MUSTARD May–Oct. Native of Europe; established as a weed in gardens, roadsides, and waste ground.

### THLASPI *Pennycress*

***Thlaspi arvense*** L. [•42] FIELD PENNYCRESS April–June. Native of Europe; roadsides and waste ground.

### TURRITIS *Tower-Mustard*

***Turritis glabra*** L. TOWER-MUSTARD Biennial herb. Lower leaves more or less pubescent, usually with Y-shaped hairs; stem leaves overlapping in the lower part of the stem, more remote above. Fruit erect, nearly terete, overlapping. May–June. Dry sandy fields, gravel pits, roadsides, gravelly shores. *Arabis glabra* (L.) Bernh.

## CABOMBACEAE *Watershield Family*

### BRASENIA *Watershield*

***Brasenia schreberi*** J.F. Gmel. [•42] WATERSHIELD Perennial aquatic herb; underwater portions of plant with a slippery, jellylike coating. Leaf blades floating, oval; petiole attached to center of blade underside. Flowers perfect, dull-purple, on emergent stalks. July. Quiet ponds and lakes; water usually acid.

## CACTACEAE *Cactus Family*

### OPUNTIA *Prickly Pear*

Branched and jointed perennial plants, the joints ("pads") varying from cylindric to greatly flattened, with conspicuous yellow flowers (ours), and with short prickles from the areoles (glochids). Fruit a dry, pulpy, or juicy berry.

*Key to species*

1 Spines 1–2, borne at only a few areoles .............. ***O. humifusa***
1 Spines several per cluster, borne at most areoles ............... 2

2 Pads flattened, 5–10 cm long, not easily detached from one another ............................................... *O. macrorhiza*

2 Pads only slightly flattened, 2–5 cm long, easily detached ........ ............................................... *O. fragilis*

***Opuntia fragilis*** (Nutt.) Haw. LITTLE PRICKLY PEAR *Threatened.* May–July. Dry sand prairies and shallow, dry soil over rock outcrops.

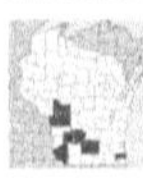

***Opuntia humifusa*** (Raf.) Raf. EASTERN PRICKLY PEAR June–July. On rocks, shores, sand dunes, or sandy prairies.

***Opuntia macrorhiza*** Engelm. PLAINS PRICKLY PEAR Sand dunes, sandy prairies.

## CAMPANULACEAE *Bellflower Family*

Perennial herbs. Stems usually with milky juice. Leaves simple, alternate. Flowers in racemes at ends of stems or single from upper leaf axils, perfect, 5-parted, regular and funnel-shaped (*Campanula*) or irregular (*Lobelia*); petals blue, white or scarlet. *Lobelia* is sometimes placed in the Lobeliaceae, but that family discontinued under APG III.

*Key to genera*

1 Flowers irregular; stamens joined to form a tube around the style ............................................... **LOBELIA**

1 Flowers regular; stamens separate ............................ 2

2 Plants annual; flowers sessile in leaf axils; leaves clasping stem ... ............................................... **TRIODANIS**

2 Plants biennial or perennial; flowers on slender pedicels; leaves not clasping stem ............................................... 3

3 Flowers rotate (flattened and disk-like) ... **CAMPANULASTRUM**

3 Flowers bell-shaped or funnel-shaped .......... **CAMPANULA**

### CAMPANULA *Bellflower; Harebell*

Annual, biennial, or perennial herbs. Leaves alternate. Flowers conspicuous, solitary or in various types of inflorescence. Corolla rotate, campanulate, or funnelform, in our species blue or violet to white. A number of introduced species are cultivated for their handsome flowers and are occasionally reported as escaped (e.g., *C. cericaria* L., *C. latifolia* L., *C. persicifolia* L., and *C. trachelium* L.).

*Key to species*

1 Stems weak, reclining on other plants ............ *C. aparinoides*

1 Stems upright............................................... 2

2 Flowers on short pedicles in an erect, 1-sided raceme ........... ............................................... *C. rapunculoides*

2 Flowers solitary or in loose, open clusters on slender pedicels ... 3

3 Plants glabrous or nearly so; stem leaves linear .... *C. rotundifolia*

3 Stems bristly; stem leaves lance-shaped ........... *C. trachelium*

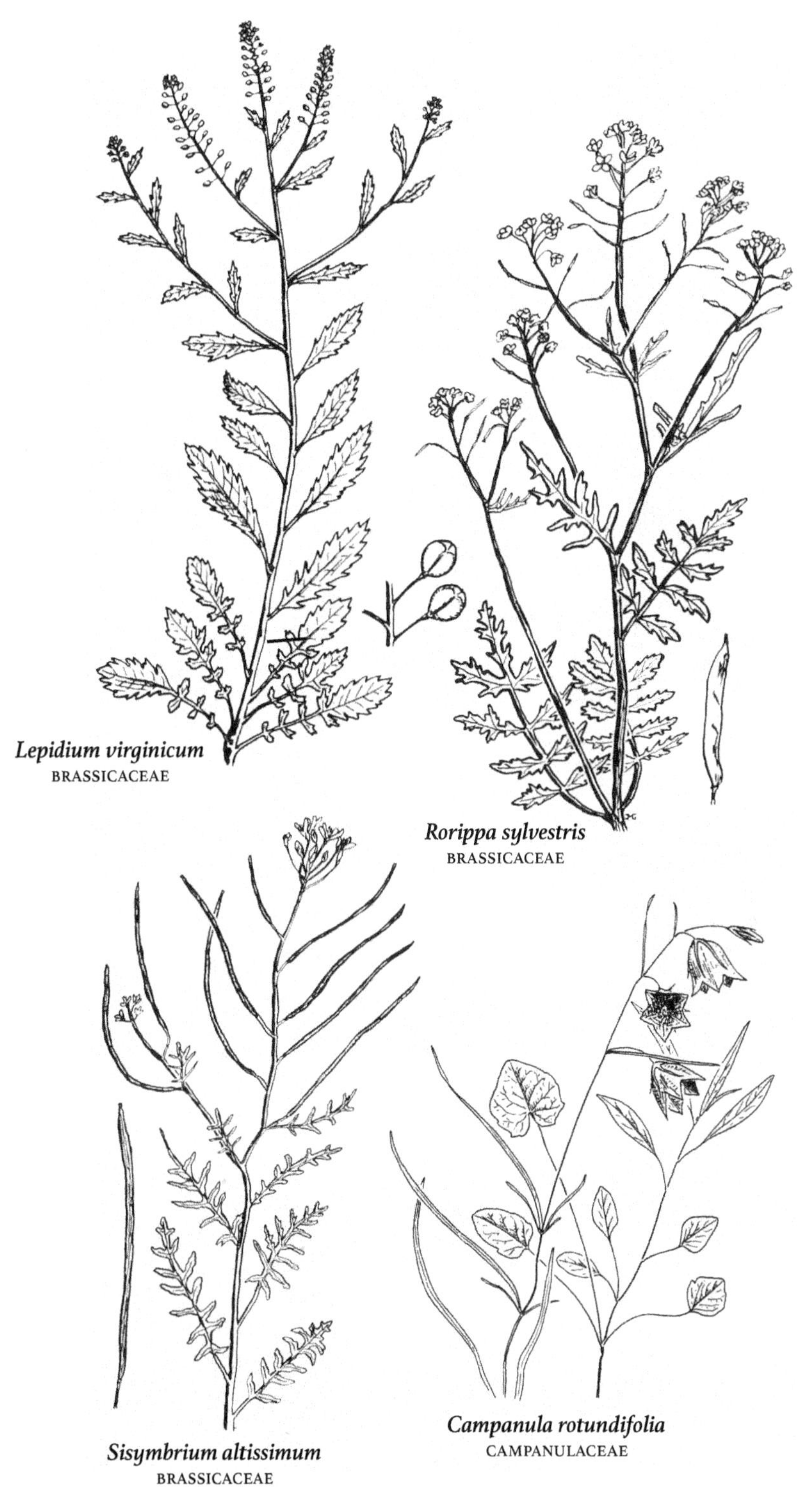
*Lepidium virginicum*
BRASSICACEAE
*Rorippa sylvestris*
BRASSICACEAE
*Sisymbrium altissimum*
BRASSICACEAE
*Campanula rotundifolia*
CAMPANULACEAE

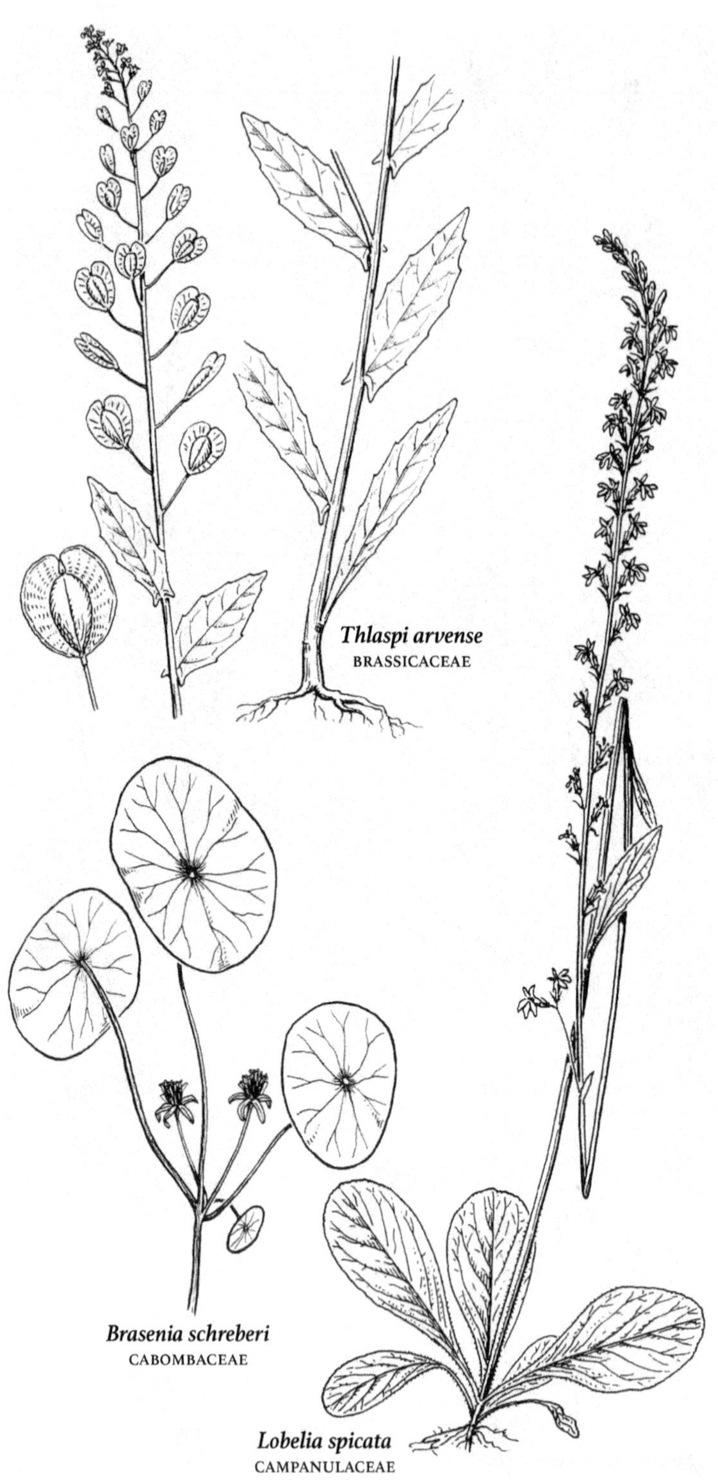

*Thlaspi arvense*
BRASSICACEAE

*Brasenia schreberi*
CABOMBACEAE

*Lobelia spicata*
CAMPANULACEAE

***Campanula aparinoides*** Pursh MARSH BELLFLOWER July–Sept. Sedge meadows, marshes, calcareous fens, conifer swamps (cedar, tamarack), thickets, open bogs; soils often calcium-rich.

***Campanula rapunculoides*** L. CREEPING BELLFLOWER July–Aug. Native of Eurasia, forming persistent weedy colonies on roadsides, railroads, and in disturbed places.

***Campanula rotundifolia*** L. [•41] BLUEBELL-OF-SCOTLAND June–Sept. Dry woods, meadows, cliffs and beaches. Variable in habit, stature, and number and size of flowers.

***Campanula trachelium*** L. THROATWORT Aug. Native of Eurasia and n Africa; roadsides and waste places.

## CAMPANULASTRUM *American-Bellflower*

***Campanulastrum americanum*** (L.) Small AMERICAN-BELLFLOWER Biennial herb. Leaves thin, lance-shaped to ovate-oblong, serrate. Flowers blue, solitary or in small clusters. July–Sept. Moist woods and streambanks, especially in openings, and in disturbed areas such as trails, field edges and railroads. *Campanula americana* L.

## LOBELIA *Lobelia*

Mostly perennial herbs. Stems usually with milky juice. Leaves alternate. Flowers irregular, in racemes at ends of stems; white, bright red, or pale to dark blue, often with white or yellow markings; 2-lipped, the 3 lobes of lower lip spreading, the 2 lobes of upper lip erect or pointing forward, divided to base. Most species toxic if eaten.

*Key to species*

1 Stem leaves narrow, to 4 mm wide, margins entire or with a few small teeth; or leaves all from base of plant .................... 2
1 Stem leaves broader, 1–5 cm wide, margins toothed............ 3

2 Leaves all from base of plant, hollow, round in cross-section; plants usually underwater ............................. *L. dortmanna*
2 Leaves all from stem, flat and linear; wetland habitats ... *L. kalmii*

3 Flowers small, to 1.5 cm long ................................ 4
3 Flowers larger, 2–4 cm long.................................... 5

4 Inflorescence usually branched; hypanthium equalling the corolla or nearly so, inflated in fruit .......................... *L. inflata*
4 Inflorescence unbranched; hypanthium shorter than corolla, not much inflated in fruit ................................ *L. spicata*

5 Flowers bright red (rarely white), 3 cm or more long .. *L. cardinalis*
5 Flowers blue with white stripes on lower lip, less than 2.5 cm long ............................................... *L. siphilitica*

***Lobelia cardinalis*** L. CARDINAL-FLOWER July–Sept. Floodplain forests, swamps, thickets, streambanks, shores and ditches; sometimes in shallow water.

***Lobelia dortmanna*** L. WATER LOBELIA July–Sept. Shallow water of acid lakes and ponds; wet, sandy shores.

***Lobelia inflata*** L. INDIAN-TOBACCO July–Oct. Open woods in moist or dry soil, disturbed places such as roadsides, ditches, borrow pits, trails, utility line clearings. The long, irregular hairs at base of stem are distinctive. Long used in herbal medicine.

***Lobelia kalmii*** L. BROOK LOBELIA July–Oct. Wet, sandy or gravelly shores, wet meadows, interdunal wetlands, conifer swamps (cedar, tamarack), rock ledges and crevices; usually where calcium-rich.

***Lobelia siphilitica*** L. GREAT BLUE LOBELIA Aug–Sept. Swamps, streambanks, calcareous fens, wet meadows.

***Lobelia spicata*** Lam. [•42] SPIKED LOBELIA May–Aug. Moist to wet prairies (sometimes where disturbed), swamp margins.

### TRIODANIS *Venus'-Looking-Glass*

***Triodanis perfoliata*** (L.) Nieuwl. CLASPING-LEAF VENUS'-LOOKING-GLASS Annual herb. Stems angular, glabrous to finely rough-hairy. Leaves orbicular or broadly ovate, cordate-clasping at the base, palmately veined. Flowers in sessile axillary cymes. May–June. Open woods, old fields, roadsides.

---

## CANNABACEAE *Hemp Family*

Trees (*Celtis*), erect herbs (*Cannabis*), or twining herbs (*Humulus*). Leaves alternate or opposite, simple to palmately lobed or compound. Inflorescences axillary to the upper (often reduced) leaves, the staminate relatively loose, branched, and many-flowered; the pistillate more compact and few-flowered. Flowers unisexual, small and inconspicuous; petals absent.

*Key to genera*

1 Trees or shrubs; leaves simple; fruits pediceled drupes ... CELTIS
1 Herbs or herbaceous vines; larger leaves compound or lobed; fruits sessile achenes subtended by bracts .......................... 2

2 Plants erect; leaves compound ...................... CANNABIS
2 Plants twining; leaves not compound ............... HUMULUS

### CANNABIS *Hemp*

***Cannabis sativa*** L. HEMP, MARIJUANA Annual herb. Leaves parted to the base into 5–9 serrate leaflets, the lower commonly opposite, the upper smaller, usually alternate. Flowers green, ordinarily dioecious; staminate flowers numerous in small clusters from the upper axils; pistillate flowers in small clusters on short lateral leafy branches from the upper axils. Native of Asia; cultivated for its valuable fiber in many parts of the world and frequently escaped. The pistillate inflorescence is the source of the narcotic marijuana.

### CELTIS *Hackberry*

***Celtis occidentalis*** L. COMMON HACKBERRY Tree; bark cork-like with warty protuberances. Leaves alternate, simple, ovate lance-shaped to broadly ovate,

distinctly asymmetrical at base, 3-nerved. Fruit an ellipsoid drupe, dark red to nearly black at maturity. May (appearing after the leaves). Usually in rich moist forests, on riverbanks, and in ravines. Trees sometimes with "witch's brooms" when disease causes a proliferation of branch tips.

### HUMULUS *Hops*

Twining, annual or perennial vines with rough stems, broad, opposite, usually lobed leaves, and axillary clusters of small flowers. Dioecious. Pistillate flowers in short spikes, in pairs, each pair subtended by a foliaceous bract. Fruit an achene enclosed within the persistent calyx and covered by the expanded bracts.

*Key to species*

1 Main leaves with 3 lobes; plants perennial ............ ***H. lupulus***
1 Main leaves with 5–7 lobes; plants annual ........... ***H. japonicus***

***Humulus japonicus*** Sieb. & Zucc. JAPANESE HOPS Native of eastern Asia; occasional escaped from cultivation to waste places.

***Humulus lupulus*** L. COMMON HOPS Low lying thickets and floodplains but most common around old homesteads where once cultivated. Yellow glands secreting a bitter substance, lupulin, occur on many parts of the plant but are most numerous on the fruit, which is important in beer-making.

## CAPRIFOLIACEAE *Honeysuckle Family*

Shrubs or vines, with opposite, mostly simple leaves. Flowers perfect, mostly 5-parted. Fruit a fleshy berry or dry capsule. Family now includes members of the former Valerianaceae (*Valeriana, Valerianella*); herbs with opposite, simple or divided leaves, and numerous small flowers in terminal, panicled or capitate cymes.

*Key to genera*

1 Flowers numerous, in rather dense terminal inflorescences (at ends of stem and branches) ........................................ 2
1 Flowers axillary or on paired pedicels on a peduncle ........... 3

2 Annual herb, the leaves entire ................ VALERIANELLA
2 Perennial herbs, leaves mostly pinnately divided or compound .... ............................................. VALERIANA

3 Plants small, creeping, evergreen; flowers paired and nodding at tips of slender stalks ................................... LINNAEA
3 Plants larger shrubs or coarse herbs, upright, deciduous ........ 4

4 Plants herbaceous; flowers from leaf axils ......... TRIOSTEUM
4 Plants woody vines or shrubs; flowers various.................. 5

5 Leaf margins toothed or lobed .................... DIERVILLA
5 Leaf margins entire.......................................... 6

6 Corolla bell-shaped, less than 1 cm long ... SYMPHORICARPOS
6 Corolla tube-shaped, mostly more than 1 cm long ... LONICERA

## DIERVILLA *Bush-Honeysuckle*

***Diervilla lonicera*** P. Mill. [•43] NORTHERN BUSH-HONEYSUCKLE Shrub. Leaves opposite. Corolla funnelform, 5-lobed, at first yellow, becoming reddish in age. June–July. Dry or rocky soil.

## LINNAEA *Twinflower*

***Linnaea borealis*** L. [•43] TWINFLOWER Low trailing vine. Stems slightly woody, with numerous short, erect, leafy branches. Leaves opposite, simple, evergreen, oval to round, upper surface and margins with short, straight hairs; margins rolled under, with a few rounded teeth near tip. Flowers small, pink to white, bell-shaped, in nodding pairs atop a Y-shaped stalk. June–Aug. Hummocks in cedar swamps and thickets, moist conifer woods, on rotten logs and mossy boulders.

## LONICERA *Honeysuckle*

Shrubs or woody vines. Leaves opposite, simple, entire. Flowers long and tubular or funnel-shaped, in pairs from leaf axils. Fruit a few-seeded, blue or red berry.

HYBRIDS ***Lonicera* × *bella*** Zabel, a hybrid between *L. morrowii* and *L. tatarica*, is common (and invasive) statewide. Other reported honeysuckle hybrids include: ***Lonicera* × *minutiflora*** Zabel (*L. morrowii* × *L. xylosteoides*), ***L.* × *notha*** Zabel (*L. ruprechtiana* × *L. tatarica*), and ***L.* × *xylosteoides*** Tausch (*L. tatarica* × *L. xylosteum*).

*Key to species*

1 Flowers in opposite, sessile, 3-flowered clusters, producing a whorl of 6 flowers; plants woody, climbing vines .................... 2
1 Flowers paired at ends of peduncles from leaf axils; shrubs or vines .................... 4

2 Leaves hairy on upper surface .................... ***L. hirsuta***
2 Leaves glabrous on upper surface .................... 3

3 Uppermost joined leaves waxy on upper surface, rounded or notched at tip; flowers yellowish .................... ***L. reticulata***
3 Uppermost joined leaves green on upper surface, pointed at tip; flowers reddish .................... ***L. dioica***

4 Trailing or climbing vine; corolla 3–5 cm long .................... ***L. japonica***
4 Upright shrubs; corolla to 2.5 cm long .................... 5

5 Style glabrous; fruit black, red, or blue; native species .................... 6
5 Style with coarse, stiff hairs; fruit red; introduced species (except *L. oblongifolia*) .................... 8

6 Fruit black; bracts below flowers oval-shaped .................... ***L. involucrata***
6 Fruit red or blue; bracts below flowers awl-shaped to spatula-shaped .................... 7

7 Fruit red; ovaries appearing separate and divergent .................... ***L. canadensis***
7 Fruit blue; ovaries appearing united .................... ***L. caerulea***

8 Flower peduncles less than 5 mm long, shorter than the petioles .................... ***L. maackii***
8 Flower peduncles 5 mm or more long .................... 9

9 Corolla strongly 2-lipped; the upper lip lobed to half its length . 10

9 Corolla only weakly 2-lipped; the upper lip lobed to base or nearly so . . . . . . . . . . . . . . . . . . . . . . . . . . . . . . . . . . . . . . . . . . . . . . . . . . . . . 11

10 Ovaries separate, gland-covered . . . . . . . . . . . . . . . . . . . *L. xylosteum*

10 Ovaries partly or entirely joined, glabrous . . . . . . . . . *L. oblongifolia*

11 Leaves glabrous on underside; flower peduncles 1.5 cm long or more . . . . . . . . . . . . . . . . . . . . . . . . . . . . . . . . . . . . . . . . . . . . . . . . . . *L. tatarica*

11 Leaves hairy on underside; flower peduncles to 1.5 cm long . . . . . . . . . . . . . . . . . . . . . . . . . . . . . . . . . . . . . . . . . . . . . . . . . . . . . *L. morrowii*

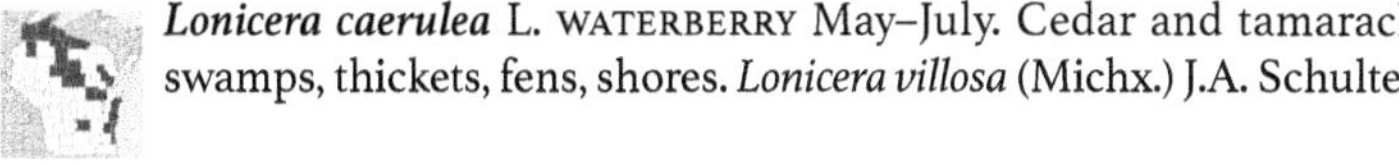

***Lonicera caerulea*** L. WATERBERRY May–July. Cedar and tamarack swamps, thickets, fens, shores. *Lonicera villosa* (Michx.) J.A. Schultes

***Lonicera canadensis*** Bartr. [•43] FLY-HONEYSUCKLE May–June. Dry or moist woods, occasionally swamps.

***Lonicera dioica*** L. LIMBER HONEYSUCKLE May–June. Moist woods and thickets, occasionally on dunes or in swamps.

***Lonicera hirsuta*** Eat. HAIRY HONEYSUCKLE June–July. Moist woods, particularly on margins and in clearings, often where sandy or rocky; occasionally in white cedar swamps. The pubescent upper surface of the leaves distinguish this species from *L. dioica* and *L. reticulata* which also have connate terminal leaves.

***Lonicera involucrata*** Banks BLACK TWINBERRY *Endangered.* Cold moist conifer woods and rocky openings.

***Lonicera japonica*** Thunb. JAPANESE HONEYSUCKLE Introduced (invasive). May–Sept. Woods and fields. Its densely tangled stems are capable of smothering shrubs and small trees. Milwaukee County; common south and east of Wisc.

***Lonicera maackii*** (Rupr.) Herder AMUR HONEYSUCKLE Native of Asia, escaped and established in wet to dry forests, fencerows, and thickets.

***Lonicera morrowii*** Gray [•43] MORROW'S HONEYSUCKLE May–June. Native of Asia; escaping and invasive along roads, ditches, and forest margins.

***Lonicera oblongifolia*** (Goldie) Hook. SWAMP FLY-HONEYSUCKLE May–June. Cedar and tamarack swamps, fens, open bogs, wet streambanks and shores; often over limestone.

***Lonicera reticulata*** Raf. GRAPE HONEYSUCKLE May–June. Moist woods and thickets.

***Lonicera tatarica*** L. [•43] TARTARIAN HONEYSUCKLE May–June. Native of e Europe and Asia, an old favorite in cultivation and frequently escaped and invasive in forests, fields, roadsides, and railways.

***Lonicera xylosteum*** L. EUROPEAN FLY-HONEYSUCKLE May–June. Native of Eurasia, escaping to forests, thickets, and swamps.

## SYMPHORICARPOS *Snowberry*

Low bushy shrubs. Leaves ovate-oblong to rotund, short-petioled. Flowers small, white or pink, terminating the stem or also in the upper axils. Corolla funnelform or campanulate, 4- or 5-lobed, usually bearded within. Fruit a 2-seeded white or red berry.

*Key to species*

1 Fruit red; corolla to 4 mm long .................... ***S. orbiculatus***
1 Fruit white; corolla 5 mm or more long ........................ 2

2 Style exserted, 4 mm or more long ................ ***S. occidentalis***
2 Style included, to 3 mm long ............................ ***S. albus***

***Symphoricarpos albus*** (L.) Blake [•44] SNOWBERRY Dry or rocky soil.

***Symphoricarpos occidentalis*** Hook. WESTERN SNOWBERRY June–Aug. Dry or rocky soil.

***Symphoricarpos orbiculatus*** Moench CORALBERRY Introduced. June–Aug. Dry or rocky soil and margin of woods.

## TRIOSTEUM *Feverwort; Horse-Gentian*

Coarse, pubescent, perennial herbs. Leaves large, connate or united by a ridge around the stem. Flowers greenish yellow to dull red, solitary or in small clusters in their axils. Fruit a yellow, red, or greenish dry berry, crowned by the persistent sepals.

*Key to species*

1 Main leaves joined at base and perforated by stem; hairs on stem short, less than 0.5 mm long ...................... ***T. perfoliatum***
1 Leaves not joined at base or perforated by stem, mostly tapered to a narrow base; hairs on stem longer, more than 0.5 mm long ..... ............................................. ***T. aurantiacum***

***Triosteum aurantiacum*** Bickn. [•44] HORSE-GENTIAN May–July, flowering a little after *T. perfoliatum.* Rich woods and thickets. Much like *T. perfoliatum,* and not always clearly separable, but the leaves distinct, tapering to a narrow base.

***Triosteum perfoliatum*** L. [•44] FEVERWORT May–July. Woods and thickets, often in shallow or rocky soils.

## VALERIANA *Valerian*

Perennial, strongly scented herbs. Leaves from base of plant and opposite along stem, simple to pinnately divided. Flowers somewhat irregular, in branched heads at ends of stems; petals joined into a tube-shaped, 5-lobed corolla.

*Key to species*

1 Basal and stem leaves mostly with 6 or more pairs of leaflets ..... ............................................... ***V. officinalis***
1 Basal leaves entire or with a single pair of lobes at base; stem leaves sparse, with only 1–5 pairs of lobes........................... 2

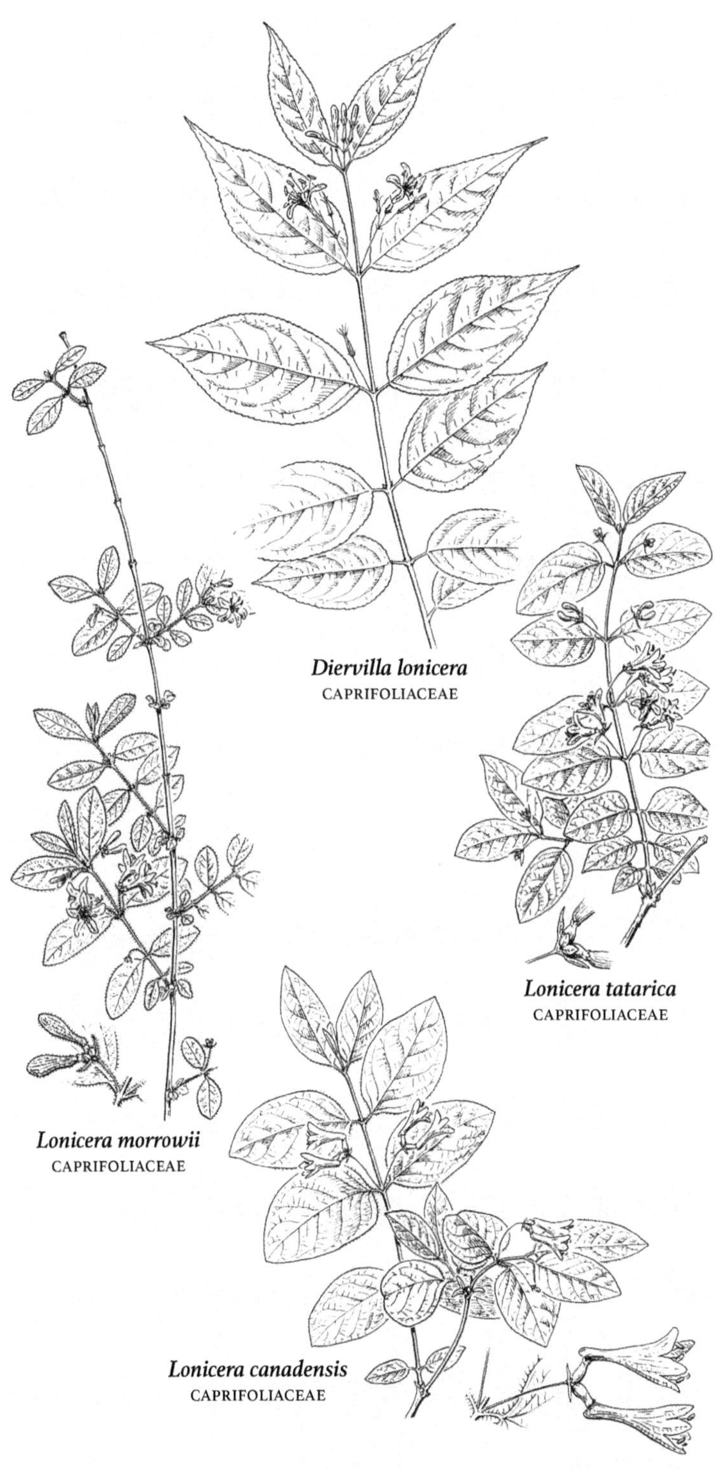
*Diervilla lonicera*
CAPRIFOLIACEAE
*Lonicera tatarica*
CAPRIFOLIACEAE
*Lonicera morrowii*
CAPRIFOLIACEAE
*Lonicera canadensis*
CAPRIFOLIACEAE

*Linnaea borealis*
CAPRIFOLIACEAE

*Triosteum perfoliatum*
CAPRIFOLIACEAE

*Triosteum aurantiacum*
CAPRIFOLIACEAE

*Symphoricarpos albus*
CAPRIFOLIACEAE

2 Leaves thick, parallel-veined, margins densely fringed with hairs; plants from a stout, carrotlike taproot .................... *V. edulis*

2 Leaves thin, net-veined, the margins smooth or with sparse hairs; plants from a creeping or ascending rhizome ......... *V. uliginosa*

***Valeriana edulis*** Nutt. ex Torr. & Gray COMMON VALERIAN, TOBACCO-ROOT May–June. Wet meadows, calcareous fens, low prairie. Ours var. *ciliata* (Torr. & Gray) Cronq. *Valeriana ciliata* Torr. & Gray

***Valeriana officinalis*** L. ALLHEAL May–Aug. Native of Europe and Asia; escaped from gardens, where it is commonly cultivated, to roadsides, ditches, fields, shores, and forest margins.

***Valeriana uliginosa*** (Torr. & Gray) Rydb. BOG VALERIAN *Threatened.* May–July. Openings in conifer swamps (especially white cedar and tamarack), marshes, calcareous fens, wet meadows; soils often alkaline. *Valeriana sitchensis* subsp. *uliginosa* (Torr. & Gray) F.G. Mey.

### VALERIANELLA *Cornsalad*

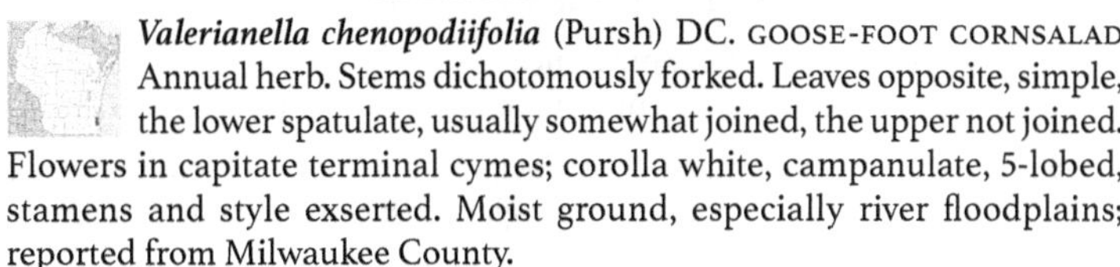

***Valerianella chenopodiifolia*** (Pursh) DC. GOOSE-FOOT CORNSALAD Annual herb. Stems dichotomously forked. Leaves opposite, simple, the lower spatulate, usually somewhat joined, the upper not joined. Flowers in capitate terminal cymes; corolla white, campanulate, 5-lobed, stamens and style exserted. Moist ground, especially river floodplains; reported from Milwaukee County.

## CARYOPHYLLACEAE *Pink Family*

Annual or perennial herbs. Leaves simple, entire, mostly opposite but sometimes alternate or whorled. Stems often swollen at nodes. Flowers perfect or imperfect, in open or compact heads at ends of stems or from leaf axils. Fruit a few- to many-seeded capsule.

ADDITIONAL SPECIES ***Herniaria hirsuta*** L. (Hairy rupturewort), introduced in Door County.

*Key to genera*

1 Leaves with chaffy or membranous stipules.................... 2

1 Leaves without stipules ........................................ 4

2 Flowers with petals; fruit a several-seeded capsule ............. 3

2 Flowers without petals; fruit a 1-seeded utricle ... PARONYCHIA

3 Leaves whorled; styles 5 ........................... SPERGULA

3 Leaves opposite; styles 3 ...................... SPERGULARIA

4 Sepals joined to form a lobed tube .......................... 5

4 Sepals not joined, distinct from one another ................. 12

5 Styles 2 ..................................................... 6

5 Styles 3–5 .................................................. 10

6 Calyx subtended by 1–3 pairs of bracts ...................... 7

6 Calyx not subtended by bracts ............................... 8

7 Calyx with 20 or more nerves ..................... DIANTHUS

7 Calyx 5-nerved ............................... PETRORHAGIA

8 Flowers to 1 cm long .... GYPSOPHILA
8 Flowers 2 cm long or more .... 9

9 Calyx tube-shaped, nerved .... SAPONARIA
9 Calyx ovoid, angled and wing .... VACCARIA

10 Styles mostly 3 .... SILENE
10 Styles mostly 5.... 11

11 Lobes of calyx much longer than tube formed by sepals .... AGROSTEMMA
11 Lobes of calyx much shorter than tube .... SILENE

12 Petals absent, or entire or somewhat fringed at tip .... 13
12 Petals deeply cleft at tip into 2 segments .... 18

13 Petals absent; fruit a 1-seeded utricle .... SCLERANTHUS
13 Petals usually present; fruit a several- to many-seeded capsule.. 14

14 Styles 4 or 5, equaling the number of sepals .... SAGINA
14 Styles (or style-branches) 3, fewer than the sepals.... 15

15 Petals entire at tip; inflorescence of single flowers or raceme-like 16
15 Petals ragged-fringed at tip; inflorescence umbel-like HOLOSTEUM

16 Leaves linear-subulate, the principal stem leaves subtending dense axillary clusters; plant entirely glabrous; capsule dehiscing into 3 valves .... MINUARTIA
16 Leaves ovate to elliptic or lanceolate, mostly without axillary tufts; plants puberulent at least on the stem; capsule with the 3 valves again split, resulting in a total of 6 teeth .... 17

17 Ripe seeds minutely and regularly roughened (tuberculate) and unappendaged; leaves ovate-elliptic, acute to acuminate, but not over 7 (–9) mm long; petals shorter than sepals; annuals .. ARENARIA
17 Ripe seeds smooth and shiny, with a pale appendage at the point of attachment; leaves lance-shaped to elliptic, mostly over 10 mm long; petals exceeding sepals; perennials .... MOEHRINGIA

18 Fruit cylindric; styles usually 5 (3 in *Cerastium nutans*) CERASTIUM
18 Fruit ovoid or oblong; styles usually 3 (5 in *Stellaria aquatica*) .... STELLARIA

## AGROSTEMMA *Corncockle*

***Agrostemma githago*** L. COMMON CORNCOCKLE Annual herb. Leaves opposite, entire, linear or lance-shaped. Flowers reddish. solitary at the ends of the branches. July–Sept. Native of Europe, originally a weed in grainfields and waste places, but our collections are old and the plant may no longer be present in the state.

## ARENARIA *Sandwort*

***Arenaria serpyllifolia*** L. [•45] THYME-LEAF SANDWORT Finely hairy annual herb. Leaves usually 8–10 pairs, ovate, sparsely rough hairy, often pustulate. Native of Eurasia; sandy or rocky places.

## CERASTIUM *Mouse-Ear Chickweed*

Low annual or perennial herbs. Leaves opposite. Flowers solitary or more commonly in terminal cymes. Petals 5, notched at tip to bifid, seldom entire.

ADDITIONAL SPECIES ***Cerastium tomentosum*** L. (Snow-in-summer), uncommon ornamental escape in ne and se Wisc.

*Key to species*

1 Petals large and showy, longer than the sepals.................. 2
1 Petals about equal to the sepals.............................. 3

2 Plants covered with long, soft, straight, glandular hairs . *C. **arvense***
2 Plants covered with tangled hairs, these not glandular ........... .............................................. *C. **tomentosum****

3 Bracts of inflorescence green and herbaceous .......... *C. **nutans***
3 Bracts of upper inflorescence with papery, non-green margins and tips ..................................................... 4

4 Matted perennial; stamens 10 ...................... *C. **fontanum***
4 Small annual; stamens 4–5 ................... *C. **semidecandrum***

***Cerastium arvense*** L. FIELD CHICKWEED April–Aug. Rocky, gravelly, or sandy areas, chiefly in calcium- or magnesium-rich soils, weedy in abandoned fields and meadows.

***Cerastium fontanum*** Baumg. [•45] MOUSE-EAR CHICKWEED April–Oct. Native of Eurasia; widely naturalized in fields, woods, and waste places and frequently a troublesome weed, especially in lawns. *Cerastium vulgatum* L.

***Cerastium nutans*** Raf. NODDING MOUSE-EAR CHICKWEED April–June. Moist or dry woodlands or open places. *Cerastium brachypodum* (Engelm.) B.L. Robins.

***Cerastium semidecandrum*** L. FIVE-STAMEN MOUSE-EAR CHICKWEED Native of Eurasia; dry, sandy disturbed places.

## DIANTHUS *Pink*

Biennial or perennial, usually glaucous herbs. Leaves narrow. Flowers solitary or in paniculate or capitate cymes. Petals 5. Many species are well known in cultivation; the carnation is *D. caryophyltus* L.

ADDITIONAL SPECIES ***Dianthus carthusianorum*** L. (Clusterhead), introduced; Bayfield County.

*Key to species*

1 Leaves not linear, more than 9 mm wide ............ *D. **barbatus***
1 Leaves linear, less than 9 mm wide............................ 2

2 Annual; calyx and bracts hairy ...................... *D. **armeria***
2 Perennial; calyx and bracts glabrous or only very finely hairy.... 3

3 Leaves hard and stiff; flowers clove-scented ......... *D. **plumarius***
3 Leaves soft and lax; flowers not clove-scented ........ *D. **deltoides***

***Dianthus armeria*** L. DEPTFORD PINK Summer. Native of Europe, established as a weed.

***Dianthus barbatus*** L. SWEETWILLIAM Summer. Native of the Old World; escaped from cultivation and locally established.

***Dianthus deltoides*** L. MAIDEN PINK Summer. Native of Europe; often cultivated and locally escaped into waste places.

***Dianthus plumarius*** L. FEATHERED PINK Summer. Native of Europe; escaped from cultivation and locally established.

## GYPSOPHILA *Baby's-Breath*

Annual or perennial Eurasian herbs. Petals white to pinkish.

ADDITIONAL SPECIES ***Gypsophila acutifolia*** Stev. (Sharp-leaf baby's-breath), reported for Wisc. ***Gypsophila elegans*** Bieb. (Showy baby's-breath), reported from Oneida County.

*Key to species*

1 Plants annual . . . . . . . . . . . . . . . . . . . . . . . . . . . . . . . . . . . . . . . . ***G. muralis***
1 Plants perennial . . . . . . . . . . . . . . . . . . . . . . . . . . . . . . . . . . . . . . . . . **2**

2 Larger leaves to 1 cm wide, narrowed and not clasping at base . . . . . . . . . . . . . . . . . . . . . . . . . . . . . . . . . . . . . . . . . . . . . . . . ***G. paniculata***
2 Leaves 1–2 cm wide, broad and clasping at base . ***G. scorzonerifolia***

***Gypsophila muralis*** L. LOW BABY'S-BREATH June–Sept. Native of Eurasia; established locally as a weed.

***Gypsophila paniculata*** L. TALL BABY'S-BREATH Introduced (invasive). Sandy roadsides, fields, ditches and railroad embankments, a troublesome weed of sand dunes.

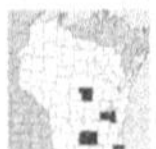

***Gypsophila scorzonerifolia*** Ser. GARDEN BABY'S-BREATH Introduced. Roadsides, shores, dunes, quarries, sometimes where very calcareous.

## HOLOSTEUM *Jagged-Chickweed*

***Holosteum umbellatum*** L. JAGGED-CHICKWEED Annual herb. Stems glabrous or nearly so below, glandular-viscid above. Flowers in terminal clusters, much surpassing the leaves; petals 5, white, erose toward the tip. April–May; plants often shrivel after fruiting in May. Native of Eurasia; roadsides, lawns, and cultivated ground.

## MINUARTIA *Stitchwort*

Previously included in genus *Arenaria.*

*Key to species*

1 Leaves in 2s opposite each other on the stem, mostly without additional leaves in axillary tufts . . . . . . . . . . . . . . . . . . . . . . ***M. patula***
1 Leaves with more than 2 clustered together, and with tufts of smaller leaves in the main axils . . . . . . . . . . . . . . . . . . . . . . . . . . . . . . . . . . . . **2**

2 Petals clearly longer than the sepals; seeds 0.7–0.8 mm wide . . . . . . . . . . . . . . . . . . . . . . . . . . . . . . . . . . . . . . . . . . . . . . . . . . . ***M. michauxii***
2 Petals shorter than to equaling the sepals; seeds 0.6–0.7 mm wide . . . . . . . . . . . . . . . . . . . . . . . . . . . . . . . . . . . . . . . . . . . . . . . ***M. dawsonensis***

***Minuartia dawsonensis*** (Britt.) House ROCK STITCHWORT Late spring–summer. Moist to dry gravelly, rocky, or sometimes calcareous places.

***Minuartia michauxii*** (Fenzl) Farw. MICHAUX'S STITCHWORT July–Sept. Dry woods, sand ridges and dunes. *Arenaria michauxii* (Fenzl) Hook. f.

***Minuartia patula*** (Michx.) Mattf. PITCHER'S STITCHWORT Reported for Wisc; more common in sc USA. *Arenaria patula* Michx. (*no map*)

## MOEHRINGIA *Grove-Sandwort*

Perennial herbs. Leaves not congested at or near base of flowering stem. Inflorescence an open cyme. Seeds have strophioles, spongy seed appendages that attract ants; foraging ants gather the seeds, eat only the strophiole, and "plant" the seeds in their nests.

*Key to species*

1 Leaves acute; sepals often longer than the petals .. ***M. macrophylla***
1 Leaves mostly blunt-tipped; sepals much shorter than the petals .. ................................................ ***M. lateriflora***

***Moehringia lateriflora*** (L.) Fenzl BLUNT-LEAF GROVE-SANDWORT May–July. Woodlands or sometimes open areas. *Arenaria lateriflora* L.

***Moehringia macrophylla*** (Hook.) Fenzl LARGE-LEAF GROVE-SANDWORT *Endangered.* May–Aug. Dry to partly shaded rocks and cliffs in the Penokee Range of n Wisc. Similar in habit and pubescence to *M. lateriflora. Arenaria macrophylla* Hook.

## PARONYCHIA *Nailwort*

Annual or perennial herbs. Leaves small, opposite, entire, with conspicuous hyaline stipules. Flowers perfect, often remaining closed.

*Key to species*

1 Stems glabrous; sepals oval ........................ ***P. canadensis***
1 Stems finely hairy; sepals ovate .................... ***P. fastigiata***

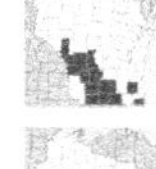

***Paronychia canadensis*** (L.) Wood [•45] SMOOTH FORKED NAILWORT June–Sept. Sandy soil and dry, open places.

***Paronychia fastigiata*** (Raf.) Fern. FORKED CHICKWEED July–Sept. Dry places.

## PETRORHAGIA *Saxifrage-Pink*

***Petrorhagia saxifraga*** (L.) Link SAXIFRAGE-PINK Cespitose perennial. Leaves linear-subulate. Flowers solitary or in cymes; petals purple to pink, the blades broadly notched. Summer. Native of Europe; rarely escaped from cultivation and established mostly as a roadside weed.

## SAGINA *Pearlwort*

***Sagina procumbens*** L. BIRD-EYE PEARLWORT Introduced. Perennial or perhaps sometimes annual herb. Leaves opposite, linear. Flowers

lateral or in terminal cymes, on filiform pedicels, often nodding after anthesis, at length becoming erect; petals white. May–Sept. Moist soil and rocky places, weedy in paths and pavements.

### SAPONARIA *Soapwort; Bouncing-Bet*

***Saponaria officinalis*** L. [•46] BOUNCING-BET Perennial herb from a horizontal rhizome and forming colonies. Leaves elliptic to elliptic-ovate, glabrous, rarely puberulent. Flowers fragrant, frequently double; petals white or pinkish; stamens exsert. Summer. Native of the Old World; formerly in cultivation and now commonly weedy on roadsides, railways and waste places. Name from the Latin, *sapo,* soap, alluding to the mucilaginous juice which forms a lather with water.

### SCLERANTHUS *Knawel*

Herbs with forked stems. Leaves opposite, subulate, joined at the base. Flowers perfect.

*Key to species*

1 Perennial; sepals blunt-tipped, with a broad white margin ....... ........ ***S. perennis***
1 Annual; sepals pointed, with only a tiny white margin .. ***S. annuus***

***Scleranthus annuus*** L. ANNUAL KNAWEL Summer. Native of Eurasia; a weed of dry, usually sandy fields, roadsides, and waste places.

***Scleranthus perennis*** L. PERENNIAL KNAWEL Introduced species of dry, open places in sc Wisc. Differs from *S. annuus* by its perennial habit, its obtuse or rounded sepals with a conspicuous white-scarious border to 0.5 mm wide near the tip.

### SILENE *Catchfly; Campion*

Annual or perennial herbs. Leaves opposite, entire. Inflorescence simple or branched, sometimes reduced to a few-flowered or 1-flowered cyme. Petals 5, the claw narrow, expanded distally into more or less prominent auricles.

*Key to species*

1 Calyx not glandular or inflated ........ 2
1 Calyx glandular, often inflated ........ 4

2 Plants densely covered with white woolly hairs ...... ***S. coronaria***
2 Plants hairy but not white-woolly ........ 3

3 Stem leaves many, 2 cm wide or more ........ ***S. chalcedonica***
3 Stem leaves few, less than 1.5 cm wide ........ ***S. viscaria***

4 Styles 5 ........ ***S. latifolia***
4 Styles 3 ........ 5

5 Main leaves whorled; petals fringed ........ ***S. stellata***
5 Leaves opposite; petals not fringed........ 6

6 Calyx glabrous ........ 7
6 Calyx hairy or glandular ........ 11

7 Corolla pink ........ ***S. armeria***

7 Corolla white . . . . . . . . . . . . . . . . . . . . . . . . . . . . . . . . . . . . . . . . . . . . . . . . 8

8 Annual; calyx tight around capsule . . . . . . . . . . . . . . . . *S. antirrhina*
8 Perennial; calyx somewhat inflated . . . . . . . . . . . . . . . . . . . . . . . . . . . 9

9 Flowers 5 or fewer, single in leaf axils . . . . . . . . . . . . . . . . . . *S. nivea*
9 Flowers in clusters of more than 5 . . . . . . . . . . . . . . . . . . . . . . . . . . 10

10 Perennial; calyx very inflated . . . . . . . . . . . . . . . . . . . . . . . . . *S. vulgaris*
10 Biennial; calyx only slightly inflated . . . . . . . . . . . . . . . . . . . . *S. csereii*

11 Petals crimson . . . . . . . . . . . . . . . . . . . . . . . . . . . . . . . . . . . . *S. virginica*
11 Petals white . . . . . . . . . . . . . . . . . . . . . . . . . . . . . . . . . . . . . . . . . . . . . . . 12

12 Flowers in long, upright, often one-sided racemes . . . *S. dichotoma*
12 Flowers in an open cyme, often nodding . . . . . . . . . . . . . *S. noctiflora*

***Silene antirrhina*** L. SLEEPY CATCHFLY Summer. Waste places or sandy soil.

***Silene armeria*** L. NONE-SO-PRETTY June–July. Native of Europe; once popular in cultivation; escaped as a weed in waste places.

***Silene chalcedonica*** (L.) E.H.L. Krause MALTESE CROSS June–Sept. Native of Asia; escaped from cultivation and occasionally spontaneous. *Lychnis chalcedonica* L.

***Silene coronaria*** (Desr.) Clairv. ex Rchb. ROSE CAMPION June–Aug. Native of Europe; escaped from cultivation and established in many places. *Lychnis coronaria* (L.) Desr.

***Silene csereii*** Baumg. BALKAN CATCHFLY Summer. Native of Europe; disturbed places. Often confused with *S. vulgaris.*

***Silene dichotoma*** Ehrh. FORKED CATCHFLY Summer. Native of Eurasia; disturbed places such as fields, roadsides, and railways.

***Silene latifolia*** Poir. WHITE CAMPION Summer. Native of Europe; a common weed. *Lychnis alba* Mill.

***Silene nivea*** (Nutt.) Muhl. SNOWY CATCHFLY June–July. Streambanks, wooded ravines, calcareous fens.

***Silene noctiflora*** L. [•46] NIGHT-FLOWERING CATCHFLY July–Sept. Native of Europe; disturbed places such as roadsides, railways, fields. Plants superficially resemble *S. latifolia* which normally has 5 styles.

***Silene stellata*** (L.) Ait. f. STARRY CAMPION Dry oak woods.

***Silene virginica*** L. FIRE-PINK *Endangered.* May–Sept. Rich woods, open woodlands and rocky slopes.

***Silene viscaria*** (L.) Jess. CLAMMY CAMPION Native of Europe, reported from Bayfield County. *Lychnis viscaria* L.

***Silene vulgaris*** (Moench) Garcke BLADDER-CAMPION Summer. Native of Europe, weedy in waste places.

### SPERGULA *Spurry*

***Spergula arvensis*** L. CORN SPURRY Fleshy annual herb. Leaves whorled, narrowly linear, clustered at the nodes in two opposite sets of 6–8; petals 5, white. May–Aug. Native of Europe; a weed of cultivated ground and waste places.

### SPERGULARIA *Sandspurry*

Low-growing succulent herbs, often found where salted in winter. Leaves opposite, linear or reduced to bristles. Flowers in branched terminal cymes. Sepals and petals each 5.

*Key to species*

1 Stamens 2–5 .............................................. ***S. salina***
1 Stamens usually 10 .................................... ***S. rubra***

***Spergularia rubra*** (L.) J.& K. Presl ROADSIDE SANDSPURRY May–Sept. Native of Europe; sandy or gravelly soil.

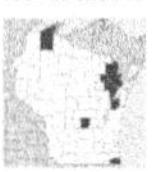

***Spergularia salina*** J.& K. Presl SALTMARSH SANDSPURRY Introduced. Summer. Highway ditches where salted. *Spergularia marina* (L.) Griseb.

### STELLARIA *Chickweed*

Low, spreading or erect perennials (ours), mostly without hairs. Stems 4-angled. Flowers single in forks of stems or in few-flowered clusters at ends of stems; sepals green with translucent margins; petals white, lobed or deeply cleft (sometimes absent in *S. borealis*).

*Key to species*

1 Plants large, the stems to 8 dm long; styles 5 ........... ***S. aquatica***
1 Plants smaller; styles 3–4 .......................................... 2

2 Leaves wider, not linear, mostly more than 1 cm wide ... ***S. media***
2 Leaves narrow, linear or lance-shaped, less than 1 cm wide...... 3

3 Flowers in branched cymes .......................................... 4
3 Flowers single in forks of stems .................................. 9

4 Petals usually absent, or shorter than the sepals ....... ***S. borealis***
4 Petals much longer than sepals ................................... 5

5 Bracts herbaceous, not papery; seeds 2 mm long ....... ***S. holostea***
5 Bracts membranous and papery; seeds about 1.5 mm long ...... 6

6 Inflorescence open and branched; pedicels spreading .......... 7
6 Inflorescence less open; pedicels erect or ascending ............ 8

7 Flowers numerous; sepals 4.5–5.5 mm long, with 3 prominent nerves; seeds bumpy ............................... ***S. graminea***
7 Flowers few; sepals to 4.5 mm long, only weakly 3-nerved; seeds smooth ............................................ ***S. longifolia***

8 Sepals 5–8 mm long .................................. *S. palustris*
8 Sepals 4–4.5 mm long .................................. *S. longipes*

9 Stems 25 cm or more long; seeds smooth .............. *S. borealis*
9 Stems to 20 cm long; seeds rough .................. *S. crassifolia*

***Stellaria aquatica*** (L.) Scop. [•46] GIANT CHICKWEED Introduced (naturalized). June–Oct. Streambanks, ponds, wet or moist disturbed areas, often in partial shade. *Myosoton aquaticum* (L.) Moench.

***Stellaria borealis*** Bigelow NORTHERN STITCHWORT June–Aug. Openings and hollows in conifer forests, margins of ponds and marshes.

***Stellaria crassifolia*** Ehrh. FLESHY STITCHWORT June–July. Streambanks and wet shores.

***Stellaria graminea*** L. GRASS-LEAF STITCHWORT May–July. Native of Europe; introduced in grassy places, fields, roadsides, and waste land.

***Stellaria holostea*** L. EASTER-BELL April–June. Native of Eurasia; reported from Milwaukee County.

***Stellaria longifolia*** Muhl. [•46] LONG-LEAVED STITCHWORT May–July. Wet meadows and marshes, shrub thickets, swamps, streambanks, pond margins.

***Stellaria longipes*** Goldie LONG-STALK STARWORT May–July. Wet meadows, ditches and thickets.

***Stellaria media*** (L.) Vill. COMMON CHICKWEED Introduced from the Old World but often appearing to be native; now a cosmopolitan weed of waste places, cultivated areas, meadows, and woodlands.

***Stellaria palustris*** (Murr.) Retz. MEADOW STARWORT Spring. Native of Eurasia; adventive in grassy places. Similar to *S. holostea* but less robust and entirely glabrous.

### VACCARIA *Cowcockle*

***Vaccaria hispanica*** (P. Mill.) Rauschert COWCOCKLE Introduced annual herb from a slender taproot. Stems glabrous and glaucous. Inflorescence an open, paniculate cyme; petals pink, without auricles or appendages. Native of Europe; often a weed in grain-fields.

## CELASTRACEAE *Bittersweet Family*

Shrubs (*Euonymus*), vines (*Celastrus*), or glabrous perennial herbs (*Parnassia*) with simple, evergreen or deciduous, opposite or alternate leaves, and small, axillary or terminal, solitary or clustered flowers. Flowers perfect or unisexual, usually 4–5-merous. In *Parnassia,* staminodes (infertile stamens) attached to base of petals and divided into threadlike segments tipped with glandular knobs. Celastraceae now includes members of genus *Parnassia.*

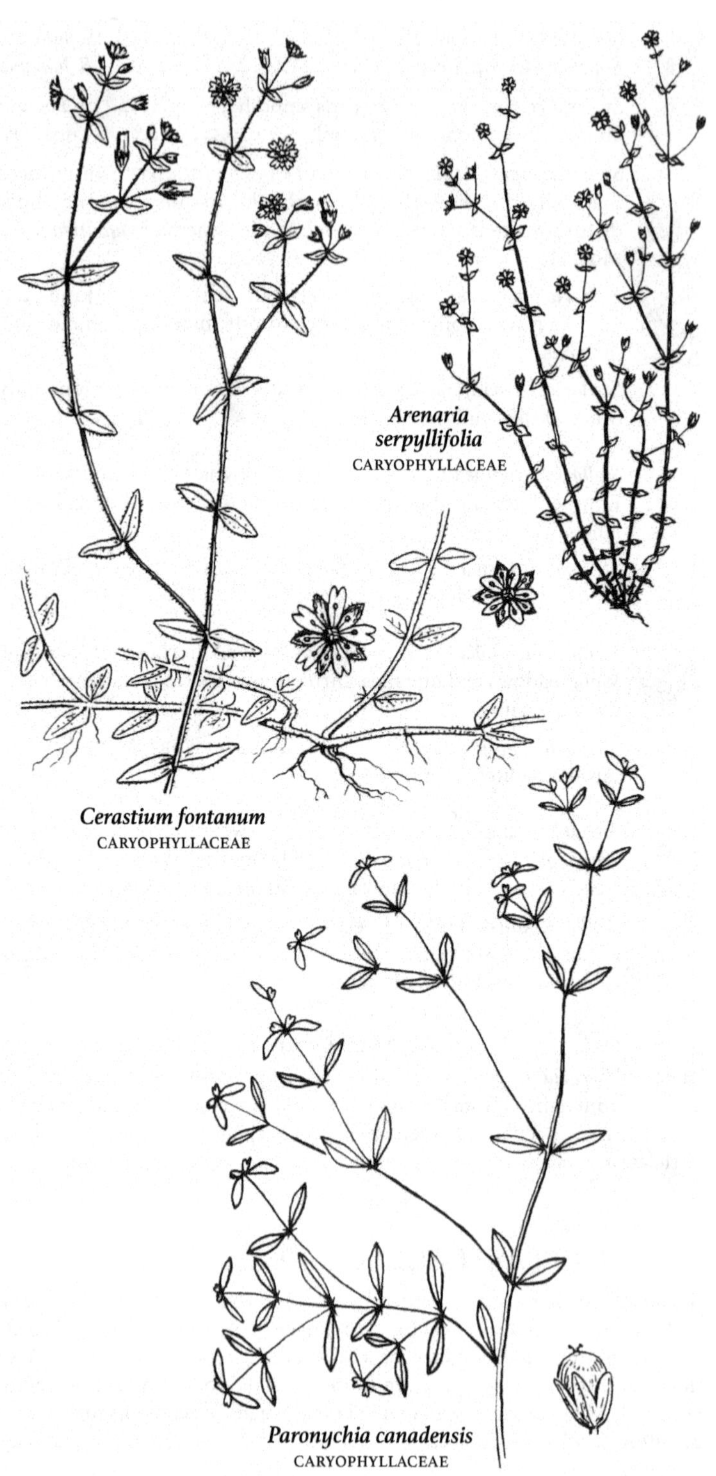
*Arenaria serpyllifolia*
CARYOPHYLLACEAE
*Cerastium fontanum*
CARYOPHYLLACEAE
*Paronychia canadensis*
CARYOPHYLLACEAE

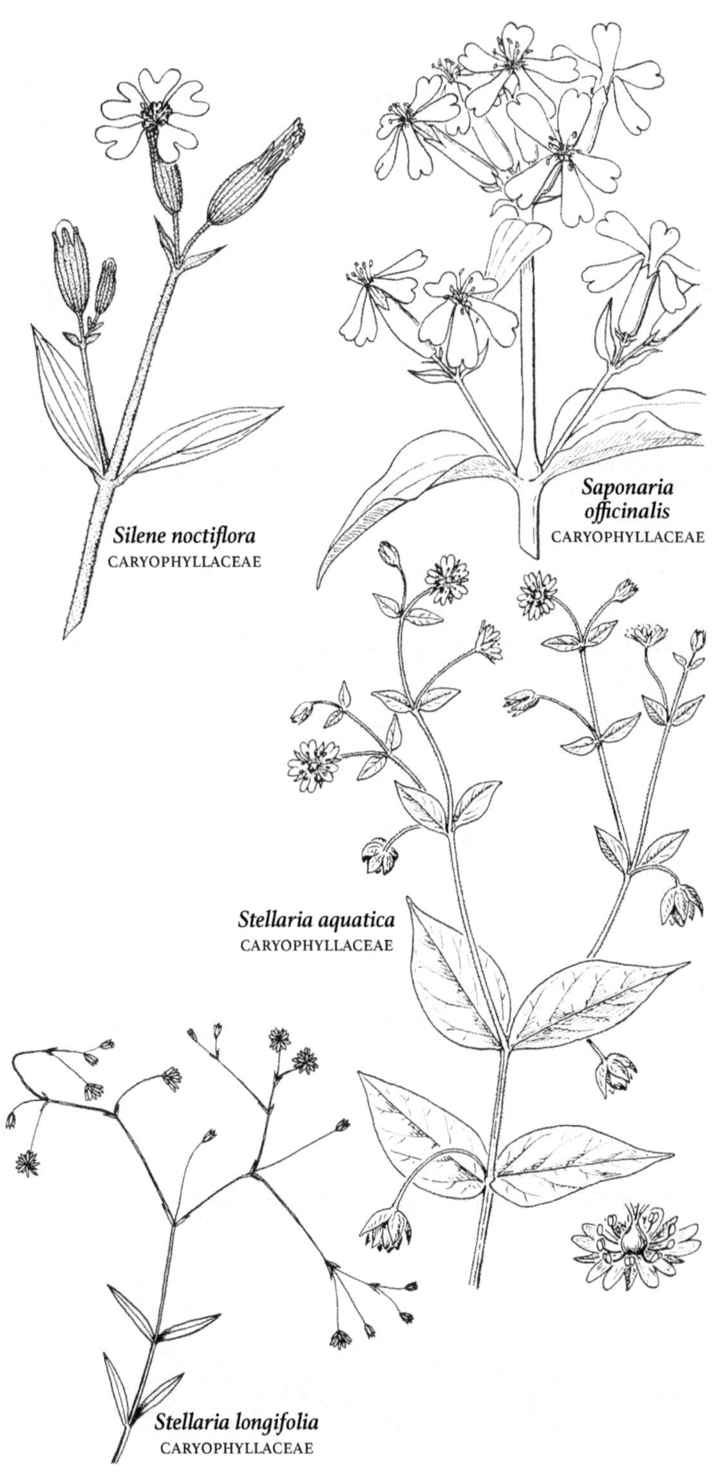
*Silene noctiflora*
CARYOPHYLLACEAE
*Saponaria officinalis*
CARYOPHYLLACEAE
*Stellaria aquatica*
CARYOPHYLLACEAE
*Stellaria longifolia*
CARYOPHYLLACEAE

*Key to genera*

1 Perennial herbs; leaves basal with a single stem leaf . **PARNASSIA**
1 Shrubs or twining vines; leaves cauline, alternate or opposite . . . 2

2 Twining vines; leaves alternate . . . . . . . . . . . . . . . . . . . **CELASTRUS**
2 Shrubs; leaves opposite . . . . . . . . . . . . . . . . . . . . . . . . . . **EUONYMUS**

## CELASTRUS *Bittersweet*

Woody twiners. Leaves deciduous, alternate, serrulate. Fruit 3-valved, each valve covering 1 or 2 seeds enclosed in a fleshy bright red-orange aril.

*Key to species*

1 Leaves obovate to nearly round; inflorescences from leaf axil . . . . . . . . . . . . . . . . . . . . . . . . . . . . . . . . . . . . . . . . . . . . . . . . . . . . ***C. orbiculatus***
1 Leaves ovate; inflorescences at ends of new twigs . . . . . ***C. scandens***

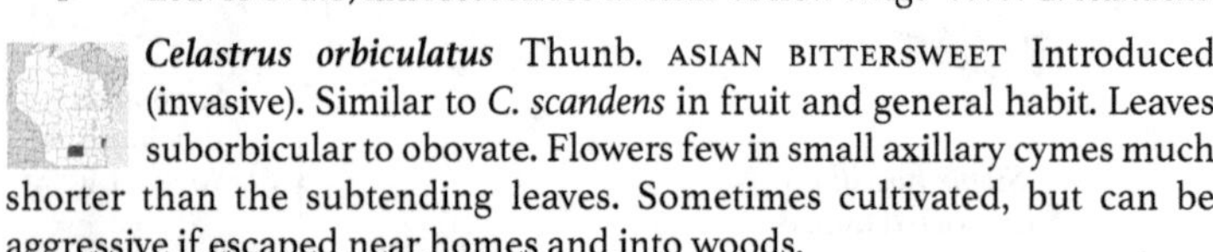

***Celastrus orbiculatus*** Thunb. ASIAN BITTERSWEET Introduced (invasive). Similar to *C. scandens* in fruit and general habit. Leaves suborbicular to obovate. Flowers few in small axillary cymes much shorter than the subtending leaves. Sometimes cultivated, but can be aggressive if escaped near homes and into woods.

***Celastrus scandens*** L. [•47] AMERICAN BITTERSWEET May–June. Roadsides and thickets, usually in rich soil; occasionally cultivated and now established in open woods and thickets.

## EUONYMUS *Spindletree*

Shrubs or small trees, with opposite, deciduous or evergreen, finely serrate leaves and small flowers solitary or cymose in the leaf-axils. Flowers perfect, 4–5-merous. Fruit 3–5-lobed, colored; aril completely covering the seed.

*Key to species*

1 Branches with corky wings; petals yellowish; fruit completely divided into 1–4 lobes . . . . . . . . . . . . . . . . . . . . . . . . . . . . . . . . ***E. alatus***
1 Branches without corky wings; petals green to purple; fruit only partially divided into 4 lobes . . . . . . . . . . . . . . . . . . . . . . . . . . . . . . . . 2

2 Leaves hairy on underside; petals purple . . . . . . . . ***E. atropurpureus***
2 Leaves glabrous on underside or hairy on veins; petals green-white . . . . . . . . . . . . . . . . . . . . . . . . . . . . . . . . . . . . . . . . . . . . . . . . . . . . ***E. europaeus***

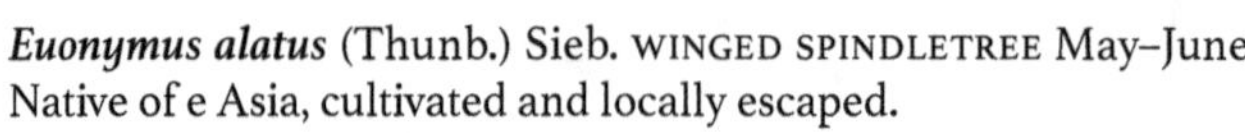

***Euonymus alatus*** (Thunb.) Sieb. WINGED SPINDLETREE May–June. Native of e Asia, cultivated and locally escaped.

***Euonymus atropurpureus*** Jacq. BURNING BUSH June. Moist woods.

***Euonymus europaeus*** L. EUROPEAN SPINDLETREE May–June. Native of Europe; planted for ornament and occasional escape.

## PARNASSIA *Grass-of-Parnassus*

Glabrous perennial herbs. Leaves all from base of plant but often with 1 stalkless leaf near middle of stalk. Flowers large, white, single at ends of stalks; petals white, veined.

*Key to species*

1 Sepals with narrow translucent margins; staminodes (sterile stamens) 3-parted, not widened at base; petals 12–16 mm long; leaves leathery and somewhat succulent ............... ***P. glauca***

1 Sepal margins green; staminodes 5 to many-parted; petals 5–13 mm long; leaves thin and membranous............................ **2**

2 Leaves broadly rounded or heart-shaped at base; petals 8–13 mm long ............................................. ***P. palustris***

2 Leaves narrowed to base; petals 5–9 mm long ........ ***P. parviflora***

***Parnassia glauca*** Raf. FEN GRASS-OF-PARNASSUS Aug–Sept. Calcareous fens and wet meadows.

***Parnassia palustris*** L. ARCTIC GRASS-OF-PARNASSUS *Threatened.* July–Sept. Calcareous fens, shores, streambanks and wet, seepy meadows.

***Parnassia parviflora*** DC. SMALL-FLOWERED GRASS-OF-PARNASSUS *Endangered.* July–Aug. Lake Michigan shoreline in cracks in wet limestone pavement or on open, moist, sandy beaches and dunes; Door Peninsula.

---

# CERATOPHYLLACEAE *Hornwort Family*

## CERATOPHYLLUM *Coon's-Tail; Hornwort*

Aquatic perennial herbs, often forming large patches; roots absent, but plants usually anchored to substrate by pale, modified leaves. Leaves in whorls, with more than 4 leaves per node, whorls crowded at ends of stems, dissected 2–3 times into narrow segments. Flowers small, inconspicuous in leaf axils, staminate and pistillate flowers separate on same plant. Our only genus of aquatic vascular plants with whorled, forked leaves.

*Key to species*

1 Leaves usually stiff, forked 1–2 times, margins coarsely toothed; achenes with 2 spines near base ................... ***C. demersum***

1 Leaves limp, some larger leaves forked 3–4 times, margins not toothed; achenes with 2 spines near base and several spines on margin ........................................... ***C. muricatum***

***Ceratophyllum demersum*** L. [•47] COON'S-TAIL Shallow to deep water of lakes, ponds, backwater areas, ditches; water typically neutral or alkaline.

***Ceratophyllum muricatum*** Cham. SPINELESS HORNWORT Lakes, ponds and quiet water of rivers and streams; water typically acid. Similar to *C. demersum,* but leaves usually limp, larger leaves usually 3- or sometimes 4-forked, the segments narrower and mostly without teeth. *Ceratophyllum echinatum* Gray.

# CISTACEAE *Rock-Rose Family*

Herbs or shrubs. Leaves simple, alternate, opposite, or appearing whorled. Flowers cymose, perfect, regular except the calyx, 3–5-merous; petals small to large, soon deciduous, or lacking in some flowers. Fruit a capsule, usually separating completely to the base and enclosed by the persistent calyx.

*Key to genera*

1 Plants shrubby; leaves small and scale-like ......... HUDSONIA
1 Plants herbaceous; leaves linear to ovate........................ 2

2 Leaves densely white-hairy, the hairs branched; petals 5, yellow, conspicuous .............................. CROCANTHEMUM
2 Leaves nearly glabrous to densely hairy, the hairs unbranched; petals 3, dark red, tiny ................................ LECHEA

## CROCANTHEMUM *Frostweed*

Perennial herbaceous or suffrutescent plants from rhizomes. Stems and leaves with stellate-pubescence. Leaves narrow. Petals yellow, 5 in the first flowers of the season, soon deciduous. Later, plants produce numerous smaller apetalous flowers. Both types are normally followed by mature capsules, those from apetalous flowers usually much smaller, containing similar seeds. Our species formerly included in genus *Helianthemum*.

*Key to species*

1 Petal-bearing flowers 1 .............................. *C. canadense*
1 Petal-bearing flowers 3 or more ..................... *C. bicknellii*

***Crocanthemum bicknellii*** (Fernald) Janch. HOARY FROSTWEED June–July; flowering 2–3 weeks later than *C. canadense.* Dry, usually sandy soil. The terminal flowers are not much surpassed by the lateral branches, and their capsules, if persistent in late summer, are near the top of the plant. *Helianthemum bicknellii* Fern.

***Crocanthemum canadense*** (L.) Britton [•47] LONG-BRANCH FROSTWEED Late May–June. Dry sandy soil, open upland woods. The terminal flowers are soon surpassed by the lateral branches and their capsules, usually long persistent, are then far below the top of the plant. Resembling *C. bicknellii* in habit and foliage. *Helianthemum canadense* (L.) Michx.

## HUDSONIA *Golden-Heather*

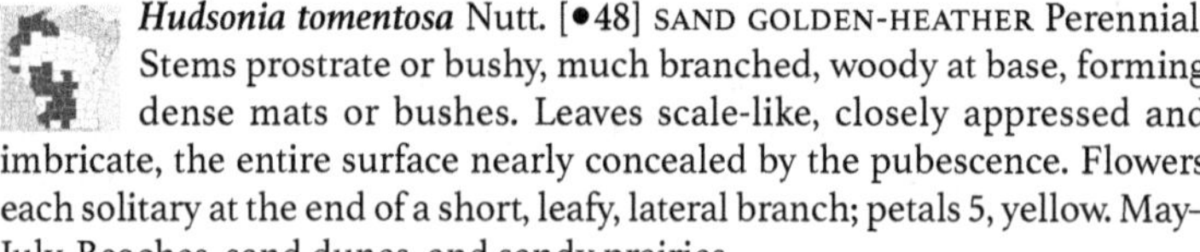

***Hudsonia tomentosa*** Nutt. [•48] SAND GOLDEN-HEATHER Perennial. Stems prostrate or bushy, much branched, woody at base, forming dense mats or bushes. Leaves scale-like, closely appressed and imbricate, the entire surface nearly concealed by the pubescence. Flowers each solitary at the end of a short, leafy, lateral branch; petals 5, yellow. May–July. Beaches, sand dunes, and sandy prairies.

## LECHEA *Pinweed*

Perennial herbs. Leaves small, alternate (occasionally appearing opposite or whorled), entire, sessile or short-petioled. Flowers many, tiny, red, in leafy panicles. Late in the season basal shoots with numerous crowded leaves are produced.

*Key to species*

1 Plants with long, spreading hairs .................. ***L. mucronata***
1 Plants with mostly appressed hairs............................. 2

2 Outer sepals narrow, equal to or longer than the wider inner sepals ................................................. ***L. tenuifolia***
2 Outer sepals narrow, shorter than the wider inner sepals ....... 3

3 Branches stiffly ascending to nearly erect; leaves densely gray-hairy ..................................................... ***L. stricta***
3 Branches mostly widely spreading from main stem; leaves sparsely hairy on underside ................................ ***L. intermedia***

***Lechea intermedia*** Leggett SAVANNA PINWEED Dry sterile or sandy soil.

***Lechea mucronata*** Raf. HAIRY PINWEED Fields and open woods in dry or sandy soil.

***Lechea stricta*** Leggett PRAIRIE PINWEED Dry sandy woods, prairies, and shores.

***Lechea tenuifolia*** Michx. [•48] NARROW-LEAF PINWEED Dry soil; upland woods and barrens.

## CLEOMACEAE *Cleome Family*

Annual herbs. Leaves alternate, compound. Flowers in terminal bracteate racemes.

*Key to genera*

1 Petals entire; stamens 6 ............................... CLEOME
1 Petals notched at tip; stamens more than 6 .......... POLANISIA

### CLEOME *Beeplant*

Annual herbs. Leaves palmately compound. Stamens normally 6.

*Key to species*

1 Plants glabrous, not sticky; leaflets 3 ................. ***C. serrulata***
1 Plants sticky to touch; leaflets usually 5 ............. ***C. houtteana***

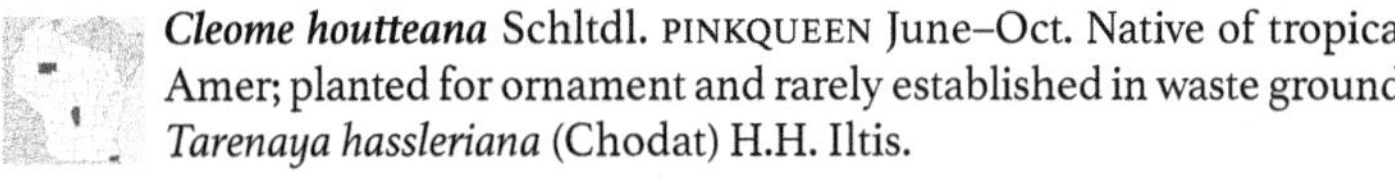

***Cleome houtteana*** Schltdl. PINKQUEEN June–Oct. Native of tropical Amer; planted for ornament and rarely established in waste ground. *Tarenaya hassleriana* (Chodat) H.H. Iltis.

***Cleome serrulata*** Pursh ROCKY MOUNTAIN BEEPLANT July–Aug. Native of w USA, adventive in Wisc. *Peritoma serrulata* (Pursh) DC.

### POLANISIA *Clammyweed*

Viscid-pubescent annual herbs. Leaves trifoliolate. Flowers small, white or pinkish, in terminal racemes. Stamens 6–many.

*Key to species*

1 Leaflets to 4 mm wide; bracts 3-parted .................. ***P. jamesii***
1 Leaflets 5 mm or more wide; bracts simple .......... ***P. dodecandra***

***Polanisia dodecandra*** (L.) DC. LARGE CLAMMYWEED July–Sept. Dry sandy or gravelly soil, especially along streams, also in waste places and along railroads.

***Polanisia jamesii*** (Torr. & Gray) Iltis JAMES' CLAMMYWEED June–July. Sand hills.

## CONVOLVULACEAE *Morning-Glory Family*

Herbs (ours), often twining, with alternate simple leaves and small to large flowers. Flowers regular, perfect, mostly 5-merous. Corolla rotate, funnelform, salverform, or tubular, entire or deeply to shallowly lobed.

*Key to genera*

1 Plants leafless, non-green, annual parasitic vines ..... CUSCUTA
1 Plants leafy, green, not parasitic ................................ 2

2 Stigma 1, expanded into a head-like tip .............. IPOMOEA
2 Stigmas 2, linear and not expanded at tip........................ 3

3 Bracts leaf-like, attached just below the calyx and nearly concealing it ............................................ CALYSTEGIA
3 Bracts small, attached much below the calyx ... CONVOLVULUS

### CALYSTEGIA *Bindweed*

Much like *Convolvulus*, but the bracts usually large, inserted just beneath the calyx, and more or less concealing it; flowers usually solitary; ours rhizomatous perennials.

*Key to species*

1 Petioles of leaves subtending flowers with petiole more than half length of blade midvein .............................. ***C. sepium***
1 Petioles of leaves subtending flowers with petiole much less than half length of midvein ........................... ***C. spithamaea***

***Calystegia sepium*** (L.) R. Br. HEDGE-BINDWEED Along streams and rivers and on wetland margins, sometimes forming large masses of tangled stems.

***Calystegia spithamaea*** (L.) Pursh LOW BINDWEED May–July. Dry rocky or sandy soil, fields and open woods.

### CONVOLVULUS *Bindweed*

***Convolvulus arvensis*** L. [•48] FIELD BINDWEED Perennial, deeply rooted herb. Stems trailing or climbing, often forming dense tangled mats. Leaves variable, triangular to oblong in outline. Flowers borne mostly 1–2 together on axillary peduncles; corolla funnelform, usually white, sometimes pink. May–Sept. Native of Europe; naturalized in fields, roadsides, and waste places; often a troublesome weed.

*Celastrus scandens*
CELASTRACEAE
*Ceratophyllum demersum*
CERATOPHYLLACEAE
*Crocanthemum canadense*
CISTACEAE

*Convolvulus arvensis*
CONVOLVULACEAE
*Lechea tenuifolia*
CISTACEAE
*Hudsonia tomentosa*
CISTACEAE

## CUSCUTA *Dodder*

Ours annual, yellow or brown, parasitic, twining vines. Leaves reduced to minute scales. Flowers small, yellow or whitish, in cymose clusters. Only a few species of the genus cause serious damage to crop plants, but all members of the genus are considered noxious weeds. Our species bloom in late summer and most of them live on a wide variety of host-plants. Identification of dodders is easiest using flowering rather than fruiting plants. Distribution of our species is not well known.

*Key to species*

1 Each flower subtended by 1 or several bracts; sepals free to their bases . . . . . . . . 2
1 Individual flowers without bracts; sepals joined at base . . . . . . . . 3

2 Flowers without pedicels, in dense, rope-like clusters *C. glomerata*
2 Flowers on pedicels in loose panicles . . . . . . . . *C. cuspidata*

3 Flowers 5-parted . . . . . . . . 4
3 Flowers 4-parted . . . . . . . . 5

4 Corolla lobes acute . . . . . . . . *C. pentagona*
4 Corolla lobes obtuse . . . . . . . . *C. gronovii*

5 Corolla lobes obtuse or rounded . . . . . . . . *C. cephalanthi*
5 Corolla lobes acute . . . . . . . . 6

6 Tips of corolla lobes erect . . . . . . . . *C. polygonorum*
6 Tips of corolla lobes bent inward . . . . . . . . *C. coryli*

***Cuscuta cephalanthi*** Engelm. BUTTONBUSH DODDER Parasitic on a number of hosts, including species of *Cephalanthus, Sambucus, Amphicarpaea, Spiraea, Salix, Equisetum, Boehmeria, Populus, Lycopus, Lythrum, Stachys,* and a number of Asteraceae.

***Cuscuta coryli*** Engelm. HAZEL DODDER Parasitic on numerous hosts, including species of *Mentha, Euthamia, Symphyotrichum, Stachys, Ceanothus, Amphicarpaea, Solidago, Bidens, Monarda, Symphoricarpos,* and *Corylus.*

***Cuscuta cuspidata*** Engelm. CUSP DODDER Parasitic on a variety hosts.

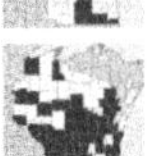

***Cuscuta glomerata*** Choisy ROPE DODDER Parasitic on members of Asteraceae.

***Cuscuta gronovii*** Willd. COMMON DODDER Usually in marshy or swampy places. Hosts include species of *Impatiens, Salix, Cephalanthus, Decodon,* and *Eupatorium.*

***Cuscuta pentagona*** Engelm. FIELD DODDER On a wide variety of hosts.

***Cuscuta polygonorum*** Engelm. SMARTWEED DODDER Often on *Persicaria,* but also on *Cephalanthus* and likely other genera. Distinctive among our species by its very short style.

### IPOMOEA *Morning-Glory*

Twining vines with broad, cordate or lobed leaves and showy flowers produced singly or few together at the summit of axillary peduncles. Corolla funnelform or campanulate.

ADDITIONAL SPECIES ***Ipomoea lacunosa*** L. (Small white morning glory), reported from Grant County. It has heart-shaped leaves, and, as its common name suggests, small white flowers.

*Key to species*

1 Leaves usually 3-lobed; sepals 15 mm or more long, recurved at tip .................................................. ***I. hederacea***

1 Leaves entire; sepals less than 15 mm long, not recurved at tip .... .................................................. ***I. purpurea***

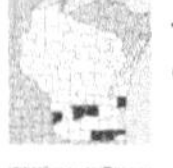

***Ipomoea hederacea*** Jacq. Ivy-Leaf Morning-Glory July–Sept. Native of tropical America; in waste places, along roads, and in cornfields.

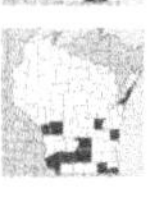

***Ipomoea purpurea*** (L.) Roth Common Morning-Glory July–Sept. Native of tropical America; formerly commonly cultivated; now escaped into fields, roadsides, and waste places; often a pernicious weed.

---

## CORNACEAE *Dogwood Family*

Trees (*Nyssa*) or shrubs (*Cornus*). Flowers 4- or 5-parted. Fruit a drupe.

*Key to genera*

1 Tree, uncommon in se Wisc ............................ **NYSSA**

1 Shrubs, mostly widespread ........................ **CORNUS**

### CORNUS *Dogwood*

Shrubs, or herbaceous shoots from a woody rhizome in bunchberry (Cornus canadensis). Leaves mostly opposite, sometimes alternate or whorled, simple, entire. Flowers in a rounded or flat-topped cluster, 4-parted, sepals and petals small. Fruit a berrylike drupe with 1–2 hard seeds.

*Key to species*

1 Plants herbaceous from a woody rhizome, less than 3 dm tall; leaves whorled ............................................ ***C. canadensis***

1 Taller shrubs, 5 dm or more tall; leaves opposite or alternate .... 2

2 Leaves alternate on stems ........................ ***C. alternifolia***

2 Leaves opposite ............................................ 3

3 Twigs yellow or yellow-green with purple spots; leaves round in outline or nearly so ................................. ***C. rugosa***

3 Twigs not yellow, or if yellow not spotted; leaves longer than wide .................................................. 4

4 Leaves with stiff, rough hairs on upper surface; s Wisc ........... .............................................. ***C. drummondii***

4 Leaves smooth, not rough-hairy above; widespread............ 5

5 Fruit white; young twigs densely short-hairy .......... *C. obliqua*
5 Fruit blue; young twigs smooth or nearly so .................... 6

6 Twigs gray; leaves with fewer than 5 pairs of lateral veins ........ *C. racemosa*
6 Twigs red; leaves with 5 or more pairs of lateral veins ..... *C. alba*

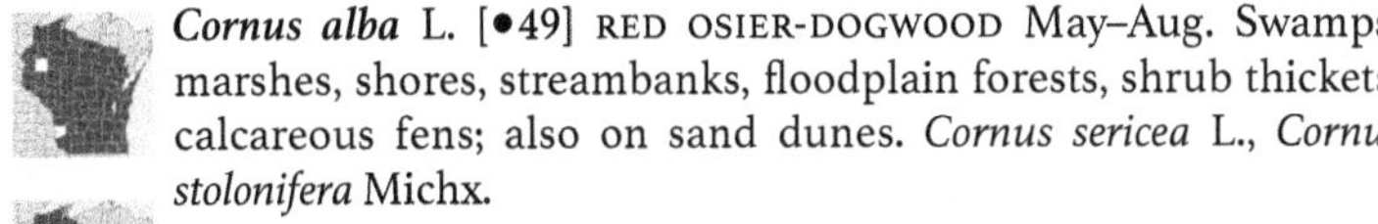

***Cornus alba*** L. [•49] RED OSIER-DOGWOOD May–Aug. Swamps, marshes, shores, streambanks, floodplain forests, shrub thickets, calcareous fens; also on sand dunes. *Cornus sericea* L., *Cornus stolonifera* Michx.

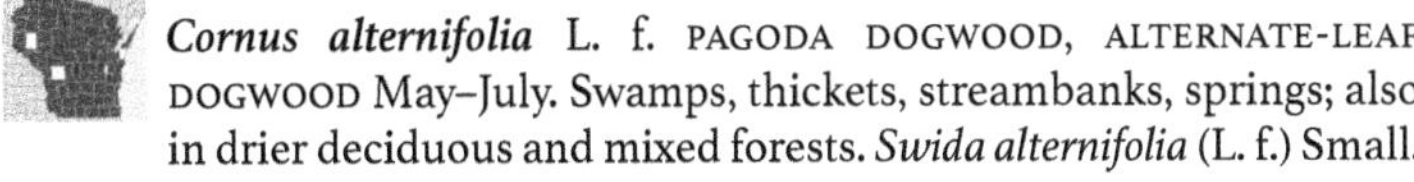

***Cornus alternifolia*** L. f. PAGODA DOGWOOD, ALTERNATE-LEAF DOGWOOD May–July. Swamps, thickets, streambanks, springs; also in drier deciduous and mixed forests. *Swida alternifolia* (L. f.) Small.

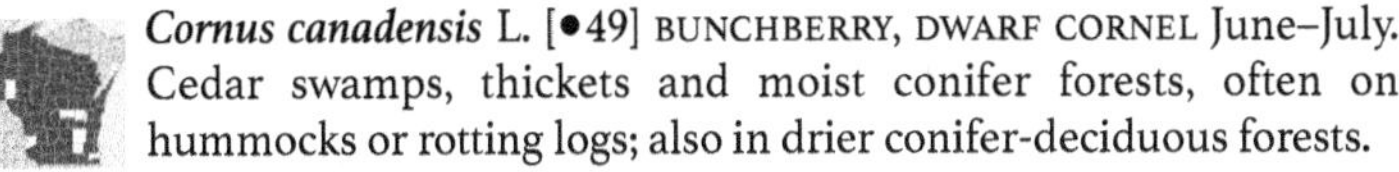

***Cornus canadensis*** L. [•49] BUNCHBERRY, DWARF CORNEL June–July. Cedar swamps, thickets and moist conifer forests, often on hummocks or rotting logs; also in drier conifer-deciduous forests.

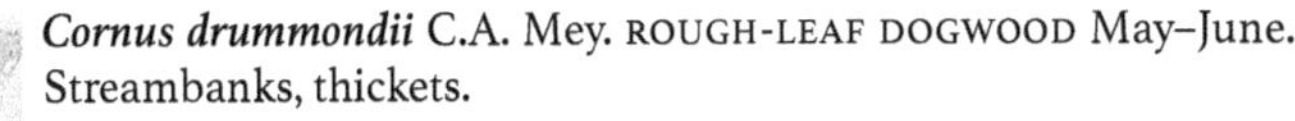

***Cornus drummondii*** C.A. Mey. ROUGH-LEAF DOGWOOD May–June. Streambanks, thickets.

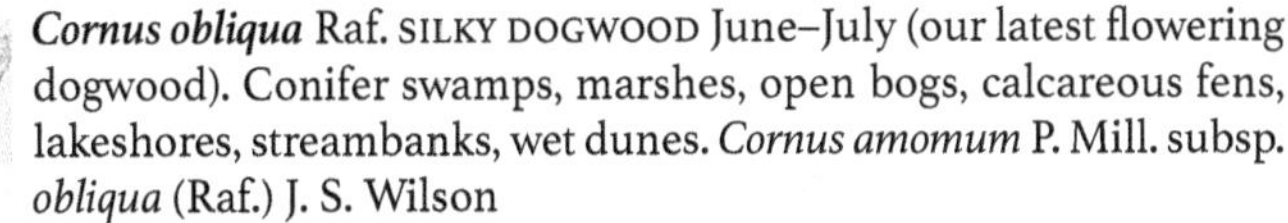

***Cornus obliqua*** Raf. SILKY DOGWOOD June–July (our latest flowering dogwood). Conifer swamps, marshes, open bogs, calcareous fens, lakeshores, streambanks, wet dunes. *Cornus amomum* P. Mill. subsp. *obliqua* (Raf.) J. S. Wilson

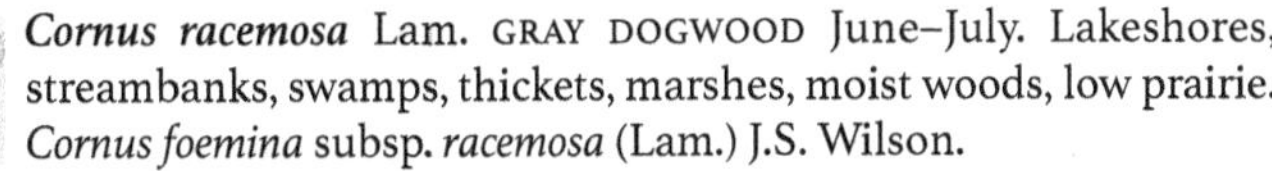

***Cornus racemosa*** Lam. GRAY DOGWOOD June–July. Lakeshores, streambanks, swamps, thickets, marshes, moist woods, low prairie. *Cornus foemina* subsp. *racemosa* (Lam.) J.S. Wilson.

***Cornus rugosa*** Lam. ROUND-LEAF DOGWOOD May–July. Moist or dry, sandy or rocky soil.

### NYSSA *Sour Gum*

***Nyssa sylvatica*** Marsh. BLACK GUM, BLACK TUPELO Tree. Leaves crowded toward the end of the fertile branches or widely spaced on rapidly growing shoots, elliptic to obovate, usually abruptly short-acuminate to an obtuse or acute tip, entire or rarely with a few coarse teeth; leaves turn brilliant red in autumn. Flowers polygamo-dioecious (having unisexual flowers with staminate and pistillate flowers borne on different trees, but also with some perfect flowers on each tree). Drupes dark blue or black. May–June. Dry or moist woods and swamps. In Wisc, known only from Kenosha County; Wisc at w edge of species' range, more common s and e of the state.

## CRASSULACEAE *Stonecrop Family*

Usually succulent plants of diverse habit and aspect. Leaves simple. Flowers usually cymose, regular, 4–5-merous or occasionally more, usually perfect.

ADDITIONAL SPECIES ***Hylotelephium erythrostrictum*** (Garden stonecrop), garden escape in se Wisc. ***Sempervivum heuffelii*** Schott (Hen-and-chickens), garden escape, Ashland County.

### SEDUM *Stonecrop*

Succulent perennial herbs. Leaves thick or terete, alternate, opposite, or whorled. Flowers yellow or red-purple, 4–5-merous.

*Key to species*

1 Petals deep pink .................................. *S. purpureum*
1 Petals yellow .................................................. 2

2 Leaves small, less than 6 mm long, thick and nearly round in cross-section .................................................. *S. acre*
2 Leaves flat, thickened, 2–3 cm long .............. *S. sarmentosum*

***Sedum acre*** L. [•49] MOSSY STONECROP Native of Eurasia; cultivated and escaped in dry, sandy soil.

***Sedum purpureum*** (L.) J.A. Schultes LIVE FOREVER Late summer. A highly variable Eurasian species, long cultivated for ornament and occasionally escaped. *Hylotelephium telephium* (L.) H. Ohba.

***Sedum sarmentosum*** Bunge STRINGY STONECROP Summer. Native of Asia; cultivated for ornament and escaped into dry or rocky soil near gardens.

---

## CUCURBITACEAE *Cucumber Family*

Annual or perennial vines, trailing or climbing by tendrils, with mostly white or yellow or greenish flowers, and simple, alternate, often lobed leaves. Flowers monoecious or dioecious, regular. Fruit a dry or fleshy pepo, few–many-seeded.

ADDITIONAL SPECIES Several species of the Old World have been cultivated since antiquity, of which *Citrullus lanatus* (Thunb.) Matsum. & Nakai (watermelon) and *Cucumis melo* L. (muskmelon or canteloupe) occur in Wisc. Native of America and of aboriginal cultivation here are *Cucurbita pepo* L., (pumpkin), and *Cucurbita maxima* Duchesne (squashes). These species may grow in waste ground, especially where their seeds have been discarded, but none are long-persistent.

*Key to genera*

1 Leaves triangular-ovate or round; corolla yellow; fruit smooth, not prickly, longer than 5 cm .......................... CUCURBITA
1 Leaves palmately 3–7 lobed; corolla white or greenish; fruit prickly, less than 5 cm long .................................................. 2

2 Leaves divided about halfway to petiole; corolla 6-lobed; fruit 3–5 cm long, inflated, with weak prickles, 4-seeded . ECHINOCYSTIS
2 Leaves divided less than half distance to petiole; corolla 5-lobed; fruit to 1.5 cm long, bur-like, spiny, 1-seeded ............. SICYOS

### CUCURBITA *Gourd*

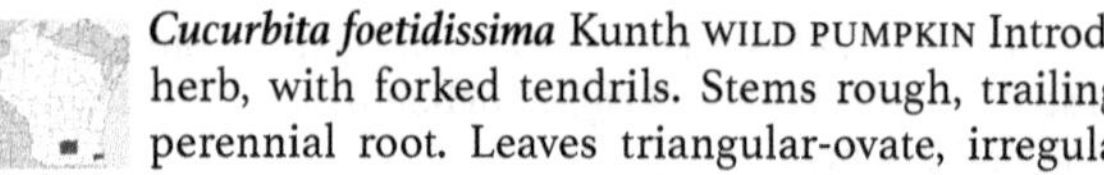

***Cucurbita foetidissima*** Kunth WILD PUMPKIN Introduced. Perennial herb, with forked tendrils. Stems rough, trailing, from a thick perennial root. Leaves triangular-ovate, irregularly and finely toothed, rough on both sides. Flowers monoecious, solitary in the axils; corolla yellow. Fruit a fleshy, nearly globose pepo, 5–10 cm long, greenish orange, smooth. Summer. Dry soil and also along railways.

### ECHINOCYSTIS *Wild Cucumber*

***Echinocystis lobata*** (Michx.) Torr. & Gray WILD CUCUMBER Annual vining herb. Leaves round in outline, with 3–7 (usually 5) sharp, triangular lobes. Flowers white; staminate flowers in long, upright racemes; pistillate flowers 1 to several on short stalks from leaf axils. Fruit green, ovate, inflated, with soft prickles. Aug–Sept. Floodplain forests, wet deciduous forests, streambanks, thickets, and waste ground.

### SICYOS *Bur-Cucumber*

***Sicyos angulatus*** L. [•50] ONE-SEED BUR-CUCUMBER Annual vining herb, to 2 m long. Stems angled, sticky-hairy, with branched tendrils. Leaves round in outline, with 3–5 shallow, toothed lobes, rough on both sides. Flowers monoecious, green or white; staminate flowers in corymbiform racemes; pistillate flowers in small capitate clusters. Fruit yellow, ovate, hairy and covered by prickly bristles. Aug–Sept. Floodplain forests, wet deciduous forests, streambanks, thickets and waste ground.

## DIPSACACEAE *Teasel Family*

Herbs. Leaves opposite, simple or divided. Flowers in dense heads subtended by a many-leaved involucre. Flowers perfect or polygamo-monoecious. Corolla tubular to narrowly campanulate. Included in Caprifoliaceae in the 2009 Angiosperm Phylogeny Group III system.

*Key to genera*

1 Stems prickly ........................................ DIPSACUS
1 Stems not prickly .................................. KNAUTIA

### DIPSACUS *Teasel*

Coarse, tall, biennial or perennial herbs, little branched, with prickly stems. Leaves large, sessile or connate. Flowers small, in dense ovoid to cylindric heads.

*Key to species*

1 Flowers white; leaves deeply pinnately lobed ....... *D. laciniatus*
1 Flowers purple; leaves not divided .................. *D. fullonum*

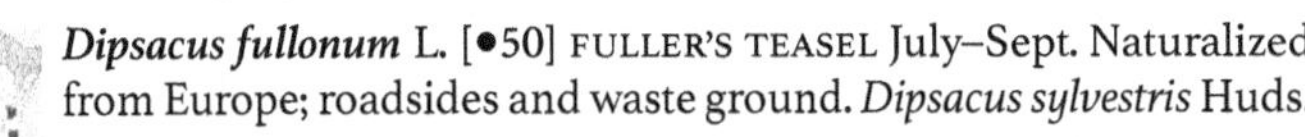

***Dipsacus fullonum*** L. [•50] FULLER'S TEASEL July–Sept. Naturalized from Europe; roadsides and waste ground. *Dipsacus sylvestris* Huds.

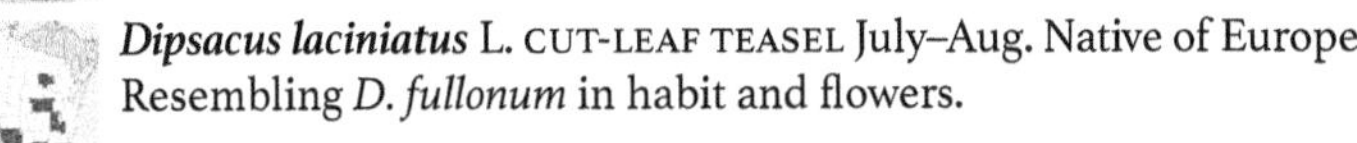

***Dipsacus laciniatus*** L. CUT-LEAF TEASEL July–Aug. Native of Europe. Resembling *D. fullonum* in habit and flowers.

### KNAUTIA *Bluebuttons*

***Knautia arvensis*** (L.) Coult. BLUEBUTTONS Perennial herb. Leaves pinnately divided into 5–15 narrowly lance-shaped segments.

Flowers in dense hemispheric heads terminating elongate peduncles; corolla lilac-purple. June–Sept. Native of Europe; fields, roadsides, waste places.

## DROSERACEAE *Sundew Family*

### DROSERA *Sundew*

Perennial herbs. Leaves all from base of plant, covered with stalked, sticky glands that trap and digest insects. Flowers white, several, on 1 side of erect, leafless stalks, the stalks nodding at tip.

*Key to species*

1 Leaves widely spreading, the blades round, wider than long ..... ........ ***D. rotundifolia***
1 Leaves upright, blades linear or broad at tip and tapered to base, longer than wide ........ 2

2 Leaf blades linear, 10–20 times longer than wide; young petals pink ........ ***D. linearis***
2 Leaf blades broad near tip and narrowed to base, 2–7 times longer than wide; young petals white ........ 3

3 Blades 2–3 times longer than wide, petioles without hairs; flower stalks from side of plant base and curving upward .. ***D. intermedia***
3 Blades 5–7 times longer than wide, petioles with some hairs; flower stalks erect from center of plant base ........ ***D. anglica***

***Drosera anglica*** Huds. ENGLISH SUNDEW *Threatened.* June–Aug. Floating sphagnum mats, calcareous fens, wet areas between dunes. Similar to *D. intermedia* but rarely occurring together; plants of *D. anglica* are generally larger, with shorter petioles (1–3x as long as blades vs. 2.5–3.5x as long in *D. intermedia*).

***Drosera intermedia*** Hayne SPOON-LEAF SUNDEW July–Sept. Low spots in open bogs, sandy shores, often in shallow water.

***Drosera linearis*** Goldie [•50] SLENDER-LEAF SUNDEW *Threatened.* June–Aug. Calcareous fens, wet areas between dunes near Great Lakes; rarely in sphagnum moss.

***Drosera rotundifolia*** L. [•50] ROUND-LEAF SUNDEW July–Aug. Swamps and open bogs, usually in sphagnum; wet sandy shores and openings.

## ELAEAGNACEAE *Oleaster Family*

Shrubs or trees. Leaves opposite or alternate, covered with small scales (lepidote). Flowers small, solitary or clustered, perfect or unisexual. Petals none.

*Key to genera*

1 Leaves alternate; stamens 4 ........ ELAEAGNUS
1 Leaves opposite; stamens 8 ........ SHEPHERDIA

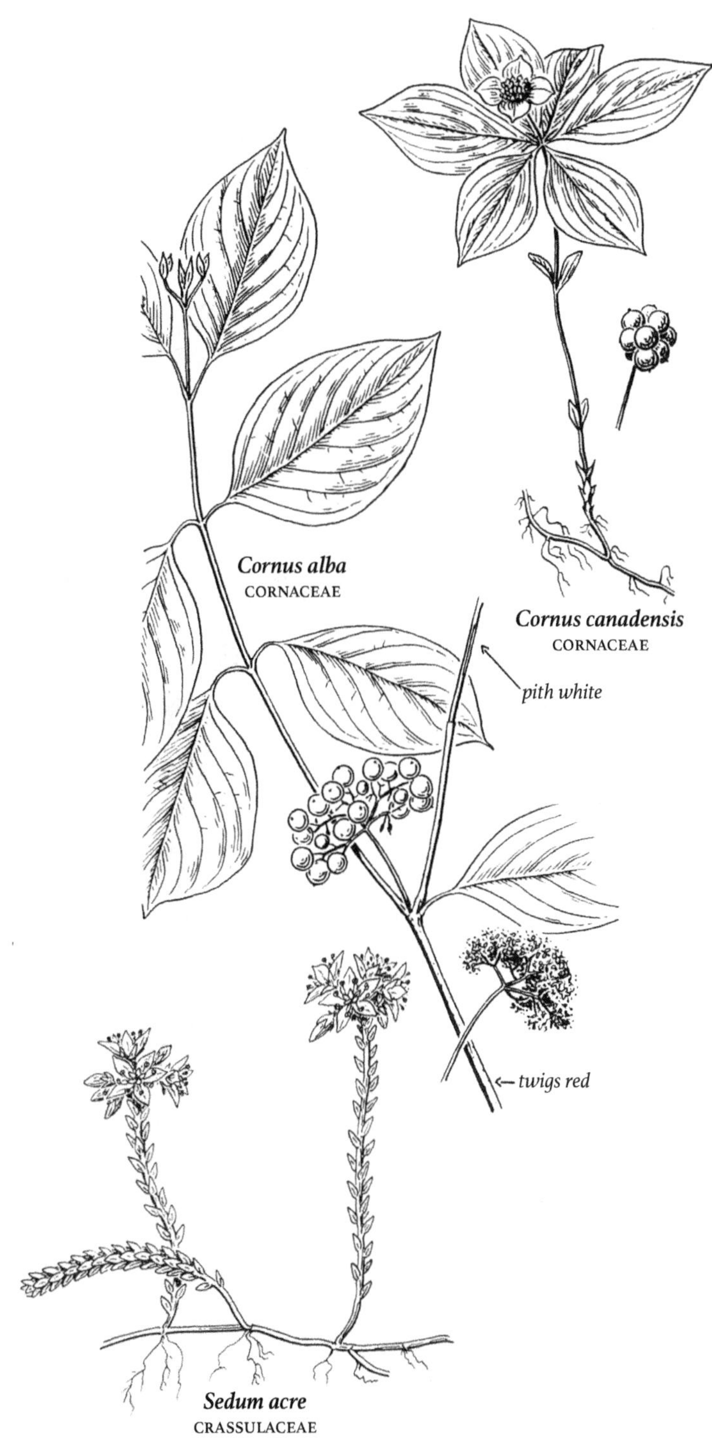
Cornus alba
CORNACEAE
Cornus canadensis
CORNACEAE
pith white
twigs red
Sedum acre
CRASSULACEAE

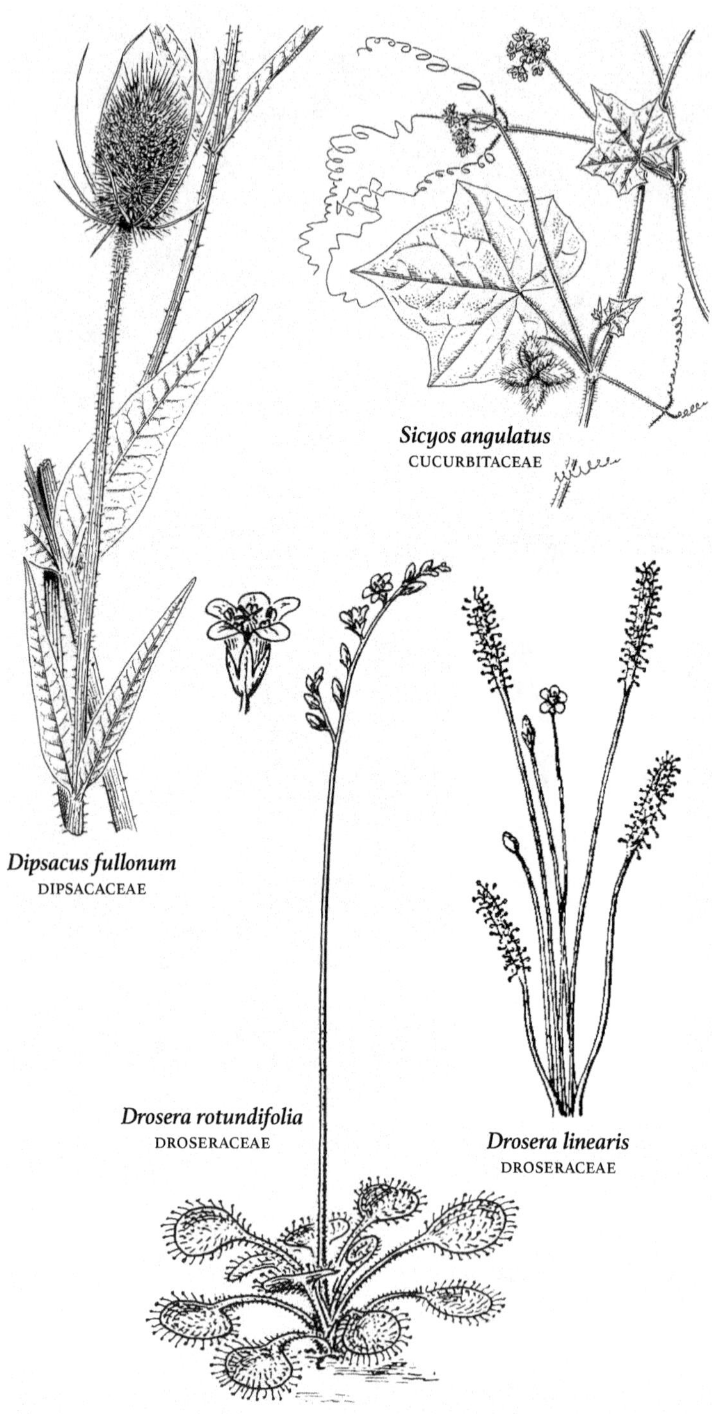
Sicyos angulatus
CUCURBITACEAE
Dipsacus fullonum
DIPSACACEAE
Drosera rotundifolia
DROSERACEAE
Drosera linearis
DROSERACEAE

### ELAEAGNUS *Russian-Olive; Silver-Berry*

Shrubs or small trees. Leaves alternate. Flowers perfect or unisexual in small lateral clusters on twigs of the current year.

*Key to species*

1 Small tree to 7 m high; upper and lower leaf surfaces covered with silvery scales .......................................... ***E. angustifolia***

1 Shrub to 4 m high; upper leaf surface green and nearly glabrous ........................................................ ***E. umbellata***

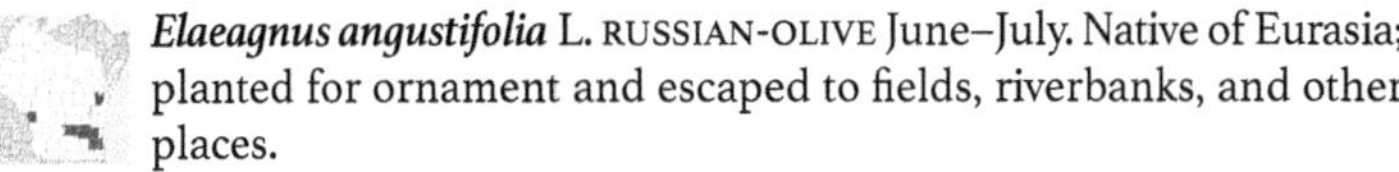

***Elaeagnus angustifolia*** L. RUSSIAN-OLIVE June–July. Native of Eurasia; planted for ornament and escaped to fields, riverbanks, and other places.

***Elaeagnus umbellata*** Thunb. AUTUMN OLIVE May–June. Native of e Asia, originally introduced as an ornamental shrub and for wildlife habitat, now spreading to many dry and wet habitats.

### SHEPHERDIA *Buffalo-Berry*

Shrubs or small trees. Leaves opposite, usually silvery and scaly. Flowers dioecious, in small clusters on twigs of the previous season. Fruit a red or yellow-red berry.

*Key to species*

1 Upper surface of leaf green and nearly glabrous .... ***S. canadensis***

1 Upper and lower leaf surfaces covered with silvery scales ........ ........................................................ ***S. argentea***

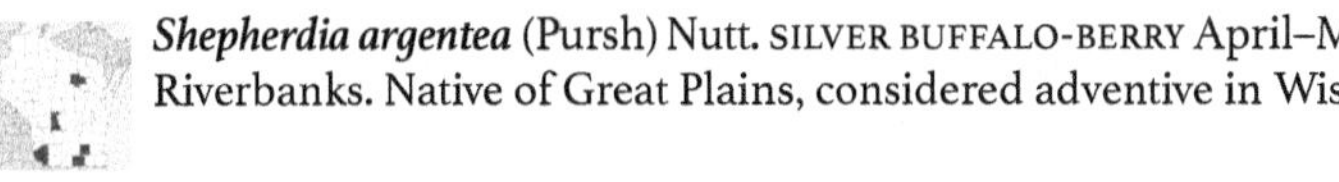

***Shepherdia argentea*** (Pursh) Nutt. SILVER BUFFALO-BERRY April–May. Riverbanks. Native of Great Plains, considered adventive in Wisc.

***Shepherdia canadensis*** (L.) Nutt. RUSSET BUFFALO-BERRY April–May. Dry, sandy or stony, calcareous soil.

## ELATINACEAE *Waterwort Family*

### ELATINE *Waterwort*

Small, branched, annual herbs of shallow water, shores and mud flats. Leaves simple, opposite, entire or toothed, with small membranous stipules. Flowers small, 1 to several from leaf axils.

*Key to species*

1 Flowers with 2 sepals and 2 petals, seeds all at base of fruit ....... ........................................................ ***E. minima***

1 Flowers with 3 sepals and 3 petals; seeds at differing levels in fruit ........................................................ ***E. rubella***

***Elatine minima*** (Nutt.) Fisch. & C.A. Mey. SMALL WATERWORT Shallow water and wet shores along lakes and ponds, usually where sandy or mucky.

***Elatine rubella*** Rydb. RED-STEM WATERWORT July–Sept. Mud flats or in shallow water of lakes and ponds. *Elatine triandra* auct. non Schkuhr p.p.

# ERICACEAE *Heath Family*

Ericaceae now includes former members of Monotropaceae and Pyrolaceae. The traditional Ericaceae are shrubs or scarcely woody shrubs. Leaves evergreen or deciduous, mostly alternate, simple. Flowers usually perfect, urn- or vase-shaped, mostly white, pink, or cream-colored. Fruit a berry or dry capsule.

Former Monotropaceae (*Monotropa, Pterospora*) are mycotropic perennial herbs without chlorophyll, variously white to pink, red, purple, yellow or brown in color. Leaves much-reduced, scale-like, alternate. Flowers solitary or in a bracteate raceme; petals distinct or connate into a lobed tube, commonly about the same color as the stem. Fruit a capsule or berry; seeds numerous and tiny.

Former Pyrolaceae (*Moneses, Orthilia, Pyrola*) are perennial herbs or half-shrubs, most dependent on wood-rotting fungi (mycotrophic). Leaves alternate to sometimes opposite or nearly whorled, often shiny, evergreen or deciduous. Flowers perfect, 5-parted, waxy and nodding. Fruit a capsule.

ADDITIONAL SPECIES ***Calluna vulgaris*** (L.) Hull (Heather), introduced shrub, known from several n and w Wisc locations.

*Key to genera*

1 Leaves reduced to non-green scales; plants entirely white, yellow, reddish, orange, or maroon .................................. 2
1 Leaves green (these sometimes small and needle- or scale-like); plants normal green color .................................. 3

2 Stems to 2 dm tall; petals free; fruit erect ......... **MONOTROPA**
2 Stems 3–10 dm tall; petals joined and urn-like; fruit nodding on curved pedicels; rare .......................... **PTEROSPORA**

3 Leaves scale-like, less than 1.5 mm wide; fruit a capsule; introduced small shrub .................................. ***Calluna vulgaris****
3 Leaves with expanded flat blades, more than 1.5 mm wide; plants herbaceous or woody .................................. 4

4 Leaves in a basal rosette; plants herbaceous.................... 5
4 Leaves opposite, alternate, or whorled; plants woody (sometimes small subshrubs woody only at the base or prostrate creepers) .. 8

5 Style ± strongly bent downward, at least 4 mm long; inflorescence a ± symmetrical raceme ................................ **PYROLA**
5 Style straight, short or long; inflorescence various (usually 1-flowered or a 1-sided raceme) ................................ 6

6 Inflorescence 1-flowered, the corolla 15–20 (–22) mm broad, flat (petals widely spreading); anthers prolonged into a short cylindrical tube below the pore; valves of capsule glabrous; style (not including prominent stigma lobes) 3–5 mm long .............. **MONESES**
6 Inflorescence racemose, the corolla 3–7 mm broad, ± bell-shaped (petals close about reproductive parts); anthers not prolonged into tubes; valves of capsule with cobwebby fibers on the margins when dehiscing; style various (but stigma only very shallowly lobed) .. 7

7 Raceme 1-sided; style 2.5–6 (–6.5) mm long, protruding at maturity from the corolla; sepal margins finely toothed or erose **ORTHILIA**

7 Raceme symmetrical; style 1.5 mm or less long, scarcely if at all protruding beyond the corolla; sepal margins entire . *Pyrola minor*

8 Leaves opposite or whorled; flowers 8–20 mm broad ........... 9
8 Leaves all alternate; flowers in most species less than 8 mm wide 10

9 Leaves coarsely few-toothed; woody only at base, forming colonies by rhizomes .................................. CHIMAPHILA
9 Leaves entire; true woody, clump-forming shrub ....... KALMIA

10 Leaves narrow, linear to linear lance-shaped, more thann 7 times longer than wide; margins revolute .............. ANDROMEDA
10 Leaves ovate to oblong; margins various ..................... 11

11 Leaves less than 7 times longer than wide, dark green and leathery on upper surface, densely covered with woolly rust-red hairs on underside; margins revolute ........ *Rhododendron groenlandicum*
11 Leaves never both revolute and with rust colored hairs on underside ...................................................... 12

12 Leaves scurfy, densely scale-covered (especially on upper surface) .................................... CHAMAEDAPHNE
12 Leaves various but not scurfy ................................ 13

13 Fruit fleshy; leaves evergreen or deciduous.................... 14
13 Fruit dry or mealy; leaves evergreen ........................... 19

14 Plants trailing; leaves evergreen ................................ 15
14 Upright shrubs; leaves deciduous ............................. 18

15 Flowers violet; uncommon in Wisc Dells area .................. .................................... *Rhododendron lapponicum*
15 Flowers white; widespread species ............................ 16

16 Leaves with small bristles; fruit white ........... GAULTHERIA
16 Leaves glabrous; fruit red when ripe.......................... 17

17 Wet habitats; leaves small, to 15 mm long ...................... .................................. VACCINIUM (*Cranberries*)
17 Drier habitats; leaves 2 cm long or more ... *Gaultheria procumbens*

18 Leaves with shiny, orange-yellow resinous glands (especially on underside) .................................. GAYLUSSACIA
18 Leaves without glands ................ VACCINIUM (*Blueberries*)

19 Plants with stiff hairs; leaves ovate to broadly elliptic .. EPIGAEA
19 Plants nearly glabrous or only finely hairy; leaves spatula-shaped, mostly widest above middle ............. ARCTOSTAPHYLOS

### ANDROMEDA *Bog-Rosemary*

***Andromeda glaucophylla*** Link [•51] BOG-ROSEMARY Low upright or trailing shrub. Leaves evergreen and leathery, often blue-green, linear or narrowly oval, tipped with a small spine, dark green above and whitened below by short stiff hair,; margins distinctly rolled under. Flowers in drooping clusters at ends of branches, white or often pink, urn-shaped. Fruit a rounded capsule, the style persistent from indented top of capsule; fruit drooping at first, but erect when mature. May–June. Sphagnum bogs, black spruce and tamarack swamps. *Andromeda polifolia* L. var. *glaucophylla* (Link) DC.

## ARCTOSTAPHYLOS *Bearberry*

***Arctostaphylos uva-ursi*** (L.) Spreng. RED BEARBERRY Shrub, forming low mats. Leaves leathery, evergreen, alternate, entire, obovate, 1–3 cm long, obtuse or rounded at tip. Flowers in short terminal racemes; corolla white or tinged with pink, the 5 rounded lobes spreading or recurved. Fruit a bright red drupe, dry or mealy. May–June. Sandy or rocky soil. Bearberry has a long history of medicinal uses, especially amongst American Indians; it is also the main component of a smoking mix known as kinnikinnick.

## CHAMAEDAPHNE *Leatherleaf*

***Chamaedaphne calyculata*** (L.) Moench LEATHERLEAF Upright shrub. Leaves evergreen and leathery, becoming smaller toward ends of flowering branches, oval to ovate, brown-green and smooth above, pale brown with a covering of small, round scales below. Flowers white, urn-shaped or cylindric, in 1-sided, leafy racemes, hanging from axils of reduced leaves near ends of branches. Fruit a brown, rounded capsule, capsules persisting on branches for several years. May–June. Open bogs, lakeshores and streambanks, often forming low, dense thickets.

## CHIMAPHILA *Prince's Pine*

***Chimaphila umbellata*** (L.) W. Bart. PRINCE'S PINE, PIPSISSEWA Low, perennial, evergreen half-shrubs, from a creeping rhizome. Leaves thick, oblong lance-shaped, sharply dentate especially toward the tip. Flowers white or pink, corymbose on long peduncles; petals widely spreading. Capsule erect, globose, opening from the top downward. June–Aug. Dry woods, especially in sandy soil.

## EPIGAEA *Trailing Arbutus*

***Epigaea repens*** L. [•51] TRAILING ARBUTUS Prostrate, creeping, evergreen shrub, often dioecious. Stems hirsute. Leaves leathery, alternate, entire, ovate or oblong, more or less pilose, especially when young. Flowers pink to white, fragrant, perfect or unisexual, in short, crowded spikes. Capsule white-pulpy within. April–May. Sandy or rocky acid soil.

## GAULTHERIA *Teaberry*

Shrubs, with alternate persistent leaves and usually white flowers in racemes or particles or (in our species) solitary in or just above the axils. Capsule completely enclosed in the fleshy, white or colored, expanded calyx, forming a dry or mealy berry.

*Key to species*

1 Leafy stems upright; flowers 5-parted; berries red . . ***G. procumbens***
1 Leafy stems prostrate; flowers 4-parted; berries white . ***G. hispidula***

***Gaultheria hispidula*** (L.) Muhl. [•51] CREEPING SNOWBERRY May–June. Open bogs, swamps, wet conifer woods, often in moss on hummocks or downed logs.

***Gaultheria procumbens*** L. [•51] WINTERGREEN July–Aug. Dry or moist woods in acid soil.

## GAYLUSSACIA *Huckleberry*

***Gaylussacia baccata*** (Wangenh.) K. Koch BLACK HUCKLEBERRY Medium shrub. Leaves alternate, deciduous, leathery, oval, both sides with shiny, orange-yellow resinous dots. Flowers yellow-orange or red-tinged, cylindric, 5-lobed, in more or less 1-sided racemes from lateral branches. Fruit a red-purple to black, berrylike drupe; edible but seedy. May–June. Open bogs, usually with tamarack and leatherleaf (*Chamaedaphne calyculata*); more common in dry, acid, sandy or rocky habitats.

## KALMIA *Laurel*

***Kalmia polifolia*** Wangenh. BOG-LAUREL Low evergreen shrub. Twigs swollen at nodes, flattened and 2-edged in section. Leaves opposite, evergreen and leathery, linear to narrowly oval, dark green and smooth above, white below with a covering of short, white hairs, midrib on underside with large purple, stalked glands. Flowers showy, pale to rose-pink, in terminal clusters at ends of current year's branches. Fruit a rounded capsule, tipped by the persistent style, the capsules in upright clusters. May–June. Sphagnum peatlands, black spruce and tamarack swamps.

## MONESES *Single-Delight*

***Moneses uniflora*** (L.) Gray ONE-FLOWERED SHINLEAF Low perennial herb from a slender creeping rhizome. Leaves deciduous, mostly at base of plant, opposite or in whorls of 3, nearly round. Flowers white, single at end of long stalk, nodding. July–Aug. Cedar swamps, wet conifer or mixed conifer and deciduous forests. *Pyrola uniflora* L.

## MONOTROPA *Indian-Pipe*

White, yellow, pink, or red plants, turning black in drying, parasitic on soil-fungi. Leaves small, scale-like. Flowers nodding, of the same color as the stem. Corolla urn-shaped or broadly tubular.

*Key to species*

1 Flowers single .................................... ***M. uniflora***
1 Flowers few to many in a raceme ................ ***M. hypopithys***

***Monotropa hypopithys*** L. [•52] PINESAP Moist or dry woods, usually in acid soil. *Hypopitys americana* (DC.) Small.

***Monotropa uniflora*** L. [•52] ONE-FLOWER INDIAN-PIPE June–Aug. Rich woods in leaf-mold.

## ORTHILIA *Sidebells*

***Orthilia secunda*** (L.) House ONE-SIDED SHINLEAF Perennial herb. Leaves elliptic to subrotund, often separated by conspicuous internodes. Scape bearing a crowded secund raceme; petals white or greenish; style elongate, exsert at anthesis. June–July. Moist woods and mossy bogs. Separated from other members of family by its ovary subtended by a 10-lobed hypogynous disk, and petals with 2 rounded projections at base. *Pyrola secunda* L.

## PTEROSPORA *Pinedrops*

***Pterospora andromedea*** Nutt. WOODLAND PINEDROPS *Endangered.* Parasitic on soil fungi. Stems erect, simple, brown or purplish, glandular-pubescent, bearing numerous scale-like leaves especially toward the base. Raceme many-flowered; corolla nodding, urn-shaped, white. Capsule depressed-globose. June–Aug. Under conifer trees (especially white pine as saprophytic on pines), usually on clayey soils.

## PYROLA *Wintergreen; Shinleaf*

Perennial herbs from creeping rhizomes. Leaves nearly basal. Flowers in an erect, terminal raceme.

*Key to species*

1 Style straight, the stamens closely surrounding style; rare in n Wisc .......... ***P. minor***
1 Style curved downward; anthers of stamen not surrounding style **2**

2 Sepal lobes longer than wide .......... **3**
2 Sepal lobes shorter than wide .......... **4**

3 Leaves to 3 cm long; sepals ovate .......... ***P. chlorantha***
3 Leaves 3–7 cm long; sepals triangular .......... ***P. elliptica***

4 Petals white; sepals oblong .......... ***P. americana***
4 Petals pink; sepals triangular .......... ***P. asarifolia***

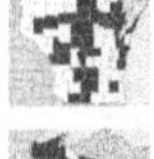

***Pyrola americana*** Sweet [•52] AMERICAN WINTERGREEN July–Aug. Dry or moist woods, rarely bogs. *Pyrola rotundifolia* L.

***Pyrola asarifolia*** Michx. PINK SHINLEAF June–Aug. Cedar swamps, peatlands, marly wetlands, and interdunal wetlands.

***Pyrola chlorantha*** Sw. GREEN-FLOWER WINTERGREEN June–Aug. Dry woods.

***Pyrola elliptica*** Nutt. ELLIPTIC SHINLEAF June–Aug. Dry upland woods.

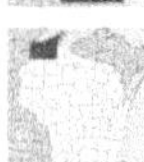

***Pyrola minor*** L. LITTLE SHINLEAF *Endangered.* June–Aug. Moist, northern boreal forests near Lake Superior; Wisc at s edge of range.

## RHODODENDRON *Rhododendron*

Shrubs (ours), with alternate, evergreen leaves and conspicuous flowers in terminal racemes or umbel-like clusters from scaly buds.

*Key to species*

1 Corollas purple; rare in central Wisc .......... ***R. lapponicum***
1 Corollas white to cream; common wetland species of north and central Wisc .......... ***R. groenlandicum***

***Rhododendron groenlandicum*** (Oeder) Kron & Judd RUSTY LABRADOR-TEA May–June. Sphagnum bogs, swamps and wet conifer forests. *Ledum groenlandicum* Oeder.

***Rhododendron lapponicum*** (L.) Wahlenb. LAPPLAND ROSE-BAY *Endangered.* Dwarf shrub. Corolla bright purple. April–May. Exposed sunny crevices of sandstone cliffs. Widely distributed through the arctic regions of North America and Eurasia, s to the mtns of New England and s NY; c Wisc.

## VACCINIUM *Blueberry*

Deciduous or evergreen shrubs. Leaves alternate, simple. Flowers 4- or 5-parted, single in leaf axils or in clusters in axils or at ends of branches; ovary inferior. Fruit a many-seeded, red, blue, or black berry. The genus may be divided into 3 subgroups: *Blueberries (V. angustifolium, V. corymbosum, V. myrtilloides, V. pallidum), cranberries (V. macrocarpon, V. oxycoccos, V. vitis-idaea),* and *bilberries (V. caespitosum).*

*Key to species*

1 Leaves deciduous; berries blue to blue-black.................. 2
1 Leaves evergreen; berries red ................................ 7

2 Leaves less than 2.5 cm long; flowers solitary in leaf axils or 2–3 from axillary buds .................................. ***V. caespitosum***
2 Leaves 2.5 cm long or more; flowers many in terminal or lateral clusters .................................................. 3

3 Tall shrubs (usually 1–2 m tall) .................. ***V. corymbosum***
3 Low shrubs (usually less than 0.5 m tall)...................... 4

4 Leaf margins with small bristle-tipped teeth .................. 5
4 Leaves entire; margins sometimes finely hairy ................ 6

5 Margin teeth many, closely spaced; branches spreading ......... ........................................ ***V. angustifolium***
5 Margin teeth few, margins finely hairy; branches stiffly erect .... ................................................ ***V. pallidum***

6 Stems and leaves velvety-hairy .................... ***V. myrtilloides***
6 Stems and leaves glabrous or with sparse hairs ....... ***V. pallidum***

7 Leaf underside with black bristly glands; rare in nw Wisc ........ ............................................ ***V. vitis-idaea***
7 Leaf underside without black glands; widespread, especially in n Wisc .................................................. 8

8 Leaves blunt or rounded at tip (and sometimes notched), pale below; bracts on flower stalk green and leaflike (more than 1 mm wide) ........................................ ***V. macrocarpon***
8 Leaves tapered to pointed tip, white below; bracts on flower stalk red and narrow (less than 1 mm wide) ............... ***V. oxycoccos***

***Vaccinium angustifolium*** Ait. LOWBUSH BLUEBERRY Flowering April–June, fruit ripening July–Aug. Sphagnum peatlands and wetland margins; also in dry, sandy openings and forests.

***Vaccinium caespitosum*** Michx. DWARF BILBERRY *Endangered.* May–June, fruit ripens in Aug. Openings in pine barrens, often with bracken fern.

***Vaccinium corymbosum*** L. HIGHBUSH BLUEBERRY Flowering May–June, fruit ripening July–Aug. Moist, low forests and swamps, shrubby peatlands and wetland margins.

***Vaccinium macrocarpon*** Ait. LARGE CRANBERRY Flowering June–July, fruit ripening Aug–Sept. Sphagnum bogs, swamps and peaty pond margins. *V. macrocarpon* is the cultivated cranberry.

***Vaccinium myrtilloides*** Michx. [•52] VELVET-LEAF BLUEBERRY Flowering May–July, fruit ripening July–Sept. Sphagnum bogs and swamps; also in dry to moist woods and clearings.

***Vaccinium oxycoccos*** L. SMALL CRANBERRY Flowering June–July, fruit ripening Aug–Sept. Wet, acid, sphagnum bogs.

***Vaccinium pallidum*** Ait. EARLY LOWBUSH BLUEBERRY Dry upland woods; known from Green and Monroe counties.

***Vaccinium vitis-idaea*** L. LINGONBERRY, MOUNTAIN-CRANBERRY *Endangered.* June–July. Sphagnum bogs; also in drier, sandy or rocky places; Wisc at s edge of range. Mountain cranberry can be distinguished from the more common cranberries (*V. macrocarpon* and *V. oxycoccos*) by the black, bristly, glandular dots on leaf underside. Gathered in Europe (where known as lingen or red whortleberry) and North America (where available) and cooked and eaten like commercial cranberries.

---

## EUPHORBIACEAE *Spurge Family*

Herbs (ours). Leaves usually alternate and simple. Flowers mostly tiny but in some species subtended by conspicuous bracts or involucral appendages. Plants monoecious or dioecious. Flowers commonly unisexual, very rarely perfect, regular. Fruit usually a dehiscent capsule. A large, chiefly tropical family, with about 6,500 species, some of which are of economic importance.

*Key to genera*

1 Stem and leaves with some (or all) hairs forked or stellate **CROTON**
1 Stem and leaves glabrous or with only simple hairs ............ 2

2 Plant with watery juice; stem pubescent with incurved hairs ..... ............................................... **ACALYPHA**
2 Plant with milky juice; stems glabrous or variously pubescent .... ............................................... **EUPHORBIA**

### ACALYPHA *Copperleaf*

Ours annual herbs, blooming from midsummer to fall. Leaves alternate. Flowers minute in axillary or terminal spikes or racemes or panicles. Plants monoecious. Petals none.

*Key to species*

1 Petioles to 5 mm long ................................ ***A. gracilens***
1 Petioles more than 5 mm long .................... ***A. rhomboidea***

***Acalypha gracilens*** Gray SHORT-STALK COPPERLEAF Moist or dry sandy soil, open woods, fields, and meadows. Native to e and sc USA; adventive in several Wisc locations.

*Andromeda glaucophylla*
ERICACEAE

*Epigaea repens*
ERICACEAE

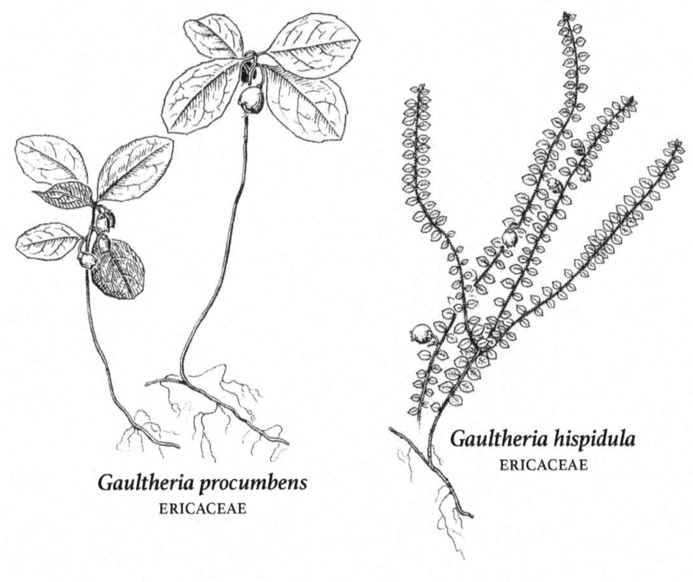

*Gaultheria hispidula*
ERICACEAE

*Gaultheria procumbens*
ERICACEAE

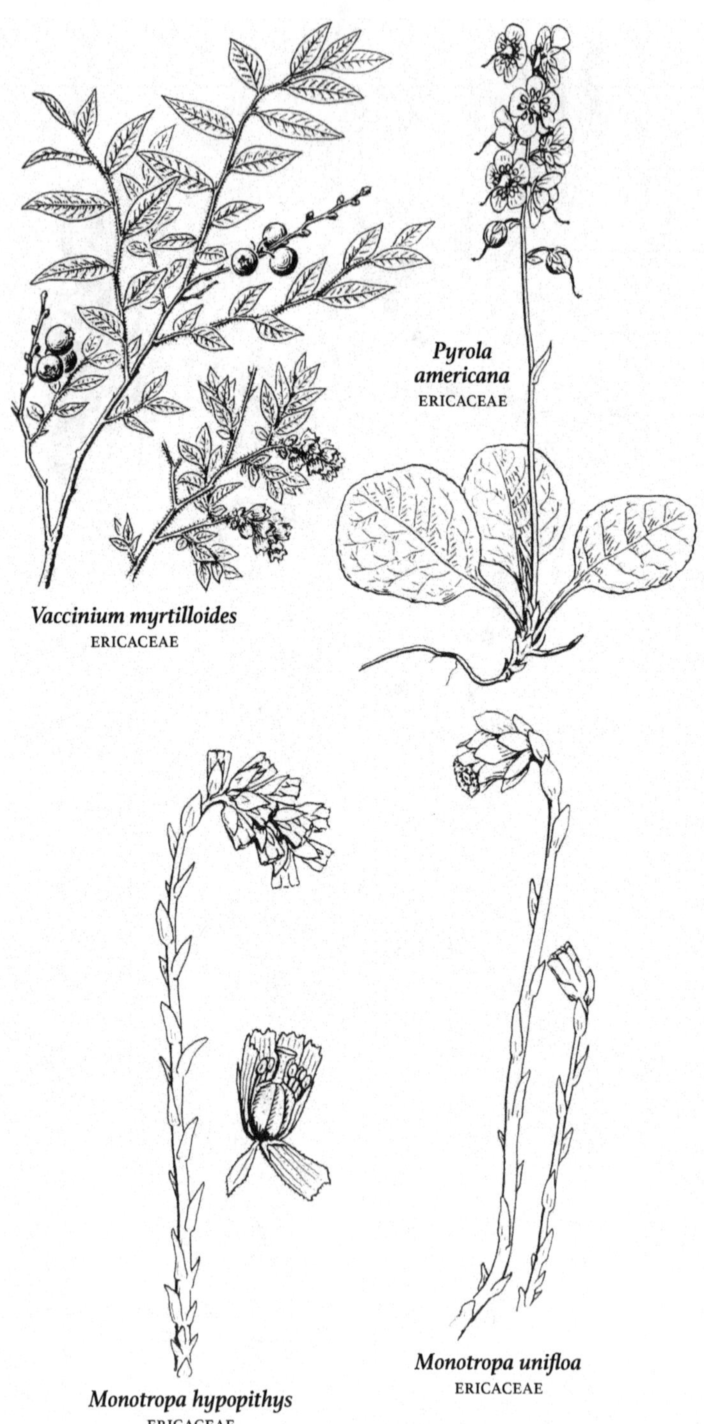

*Vaccinium myrtilloides*
ERICACEAE

*Pyrola americana*
ERICACEAE

*Monotropa hypopithys*
ERICACEAE

*Monotropa unifloa*
ERICACEAE

***Acalypha rhomboidea*** Raf. RHOMBIC COPPERLEAF Dry or moist soil of open woods, roadsides, waste places.

## CROTON *Croton*

Our species, blooming from July to October, are erect, strong-scented, freely and often dichotomously branched annuals, densely beset with stellate hairs and often distinctly gray or silvery in color. Lower leaves and branches alternate, but the upper tend to be opposite or whorled; terminal inflorescences soon surpassed by the branches and appear lateral.

ADDITIONAL SPECIES ***Croton texensis*** (Klotzsch) Muell.-Arg., introduced in southeastern Wisc.

*Key to species*

1 Leaf margins toothed; with 1 or 2 glands on tip of petiole . . . . . . . . . *C. glandulosus*

1 Leaves entire, without glands . . . . . . . . . . . . . . . . *C. monanthogynus*

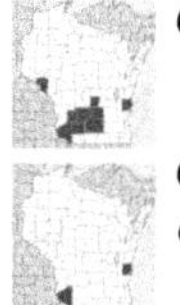

***Croton glandulosus*** L. TOOTH-LEAVED CROTON Dry or sandy soil.

***Croton monanthogynus*** Michx. PRAIRIE-TEA Dry or sterile soil. Native of sc USA, considered adventive in Wisc.

## EUPHORBIA *Spurge*

Annual or perennial herbs of diverse form; with milky, often highly acrid juice. Flowers greatly reduced, the staminate flowers consisting of a single stamen only, the pistillate flowers of a single pistil only. Several staminate flowers surround one pistillate flower inserted at the base of a cup-shaped involucre to form an inflorescence termed a cyathium. Around the margin of the cyathium are 4–5 glands, and in some species these have petal-like appendages, so that the whole cyathium mimics a single flower. The milky juice is toxic and in some people produces a dermatitis similar to that caused by poison-ivy.

ADDITIONAL SPECIES ***Euphorbia hexagona*** Nutt ex Spreng. (Six-angle spurge), native of central Great Plains; Pierce County. ***Euphorbia hirta*** Lam. (Pill-pod sandmat), introduced, Waushara County. ***Euphorbia myrsinites*** L. (Myrtle spurge), introduced, Winnebago County.

*Key to species*

1 Annual herbs; leaf bases symmetrical; glands of cyathium 1, without appendages; stems erect, with ascending branches . . . . . . . . . . . 2

1 Annual or perennial herbs; leaf bases various; glands of cyathium 4 or 5, not appendaged or with petal-like appendages; stems erect to prostrate . . . . . . . . . . . . . . . . . . . . . . . . . . . . . . . . . 4

2 Plants glabrous or upper stems softly hairy; leaves all alternate or nearly so, linear to narrowly lance-shaped, the margins entire . . . . . . . . . . . . . . *E. cyathophora*

2 Plants usually with hairs, especially on upper stems and leaf undersides; leaves all opposite or nearly so, ovate to linear, the margins toothed. . . . . . . . . . . . . . . . . . . . . . . . . . . . . . 3

3 Underside of leaves with relatively stiff hairs that are conspicuously widened at the base . . . . . . . . . . . . . . . . . . . . . . . *E. davidii*

3 Undersides of leaves with soft hairs that are ± the same width most of their length (only slightly widened at base) ......... *E. dentata*

4 Annual herbs, the stems prostrate or ascending; leaves all opposite; leaf bases typically unequal ................................. 5

4 Annual or perennial herbs, the stems erect, leaves mostly alternate; leaf bases equal or nearly so ................................. 11

5 Stems with long soft hairs; capsules with stiff appressed hairs .... ................................................ *E. maculata*

5 Stems glabrous or pubescent; capsules glabrous ............... 6

6 Leaves entire; seeds round, smooth, with a white seed coat ..... 7

6 Leaves finely toothed, at least along upper portion of blade; seeds angular, smooth, bumpy, or ridged, the seed coat brown or blackish ................................................ 8

7 Plants of Lake Michigan shoreline; capsule 3–4 mm long ........ ................................................ *E. polygonifolia*

7 Plants of w Wisc; capsule less than 2 mm long .......... *E. geyeri*

8 Stems pubescent, at least on the upper portions, leaves mostly more than 1 cm long, the margins toothed ....................... 9

8 Stems glabrous; leaves mostly less than 1 cm long, the margins toothed only near leaf tip and along one side near leaf base .... 10

9 Stems erect or ascending, nearly glabrous or with lines of hairs; mature leaves more than 15 mm long ................. *E. nutans*

9 Stems prostrate or nearly so, sparsely hairy; mature leaves less than 15 mm long .................................... *E. vermiculata*

10 Leaves linear, finely sharp-toothed, rounded at tip; seeds with small ridges........................................... *E. glyptosperma*

10 Leaves oblong to ovate, the lower two-thirds of the blade entire, finely toothed at the blunt tip; seeds smooth, or faintly wrinkled or pitted .................................... *E. serpyllifolia*

11 Inflorescence glands with conspicuous white, petal-like appendages ................................................ 12

11 Inflorescence glands without white, petal-like appendages ..... 13

12 Upper leaves and bracts green with conspicuous white patches; capsules pubescent; plants annual ................. *E. marginata*

12 Leaves and bracts green; capsules glabrous; plants perennial .... ................................................ *E. corollata*

13 Leaves finely sharp-toothed ...................... *E. helioscopia*

13 Leaves entire............................................... 14

14 Plants annual or short-lived perennials; stem leaves broadly ovate to obovate; inflorescence usually with 3 main branches; seeds pitted ................................................ *E. peplus*

14 Plants perennial; stem leaves linear to lance-shaped; inflorescence with 5 or more branches; seeds smooth ..................... 15

15 Main stem leaves slender and linear, 1–3 cm long and 1–3 mm wide, crowded ........................................ *E. cyparissias*

15 Main stem leaves 3–7 cm long, mostly 3–10 mm wide, less crowded ................................................ *E. virgata*

***Euphorbia corollata*** L. [•53] FLOWERING SPURGE Perennial from a deep root. June–Sept. Dry woods and old fields. Highly variable.

***Euphorbia cyathophora*** Murr. FIRE-ON-THE-MOUNTAIN Erect annual. Summer. Moist soil, often in shade; cultivated and sometimes escaped. *Poinsettia cyathophora* Murr.

***Euphorbia cyparissias*** L. CYPRESS SPURGE Perennial by horizontal rhizomes. April–July. Native of Eurasia; established on roadsides and waste ground. Fruit is seldom produced.

***Euphorbia davidii*** Subils DAVID'S SPURGE Introduced. Erect annual. July–Sept. Dry soil; established as a weed on roadsides and waste places, especially in cindery soil.

***Euphorbia dentata*** Michx. TOOTHED SPURGE Annual. A weed in disturbed, dry open ground; common south of Wisc but much less common in the state than the similar *E. davidii. Poinsettia dentata* Michx.

***Euphorbia geyeri*** Engelm. DUNE SPURGE Annual. Sandy prairies and dunes. *Chamaesyce geyeri* (Engelm.) Small

***Euphorbia glyptosperma*** Engelm. RIB-SEED SANDMAT Annual. Dry sandy soil. The teeth of the leaf are often visible only under a lens. *Chamaesyce glyptosperma* (Engelm.) Small

***Euphorbia helioscopia*** L. WARTWEED, MAD-WOMAN'S-MILK Annual. Native of Europe; fields and waste places.

***Euphorbia maculata*** L. [•53] SPOTTED SANDMAT Annual. Weedy in lawns, gardens, and waste places, also in meadows and open woods. *Chamaesyce maculata* (L.) Small

***Euphorbia marginata*** Pursh SNOW-ON-THE-MOUNTAIN Annual. Summer. Native of Great Plains; cultivated for ornament and often escaped farther east; considered adventive in Wisc. The milky juice is extremely acrid.

***Euphorbia nutans*** Lag. EYEBANE Annual. June–Oct. Dry or moist soil; weedy in lawns and gardens. *Chamaesyce nutans* (Lag.) Small

***Euphorbia peplus*** L. PETTY SPURGE Annual. Summer. Native of Eurasia; yards, gardens.

***Euphorbia polygonifolia*** L. SEASIDE SANDMAT Annual. Sand dunes and sandy beaches along Lake Michigan. *Chamaesyce polygonifolia* (L.) Small.

***Euphorbia serpyllifolia*** Pers. THYME-LEAVED SPURGE Introduced annual. Dry rocky soil. *Chamaesyce serpyllifolia* (Pers.) Small.

***Euphorbia vermiculata*** Raf. WORM-SEED SANDMAT Annual. Fields, roadsides, and waste ground. Native in ne USA; considered adventive in Wisc. *Chamaesyce vermiculata* (Raf.) House.

***Euphorbia virgata*** Waldst. & Kit. [•53] LEAFY SPURGE Perennial from a deep root. Summer. Native of Eurasia; widely established in North America; a troublesome noxious weed, sometimes infesting large areas. *Euphorbia esula* L.

## FABACEAE *Pea Family*

Perennial herbs, shrubs and trees. Leaves alternate, pinnately divided, the terminal leaflet sometimes modified as a tendril (*Lathyrus, Vicia*). Flowers in simple or branched racemes, perfect, irregular, 5-lobed (only 1 lobe in *Amorpha*), the upper lobe (banner) larger than the other lobes, with 2 outer, lateral petals (wings), and 2 inner petals which are partly joined (the keel); ovary 1-chambered, maturing into a pod.

ADDITIONAL SPECIES ***Anthyllis vulneraria*** L. (Common kidney-vetch), introduced perennial herb, reported from Dane County. Leaflets mostly 5–11, the terminal leaflet often much larger than the lateral ones; flowers yellow. ***Caragana arborescens*** (Siberian peashrub), introduced and sometimes escaped from hedgerow plantings. ***Genista tinctoria*** L. (Dyer's greenweed), introduced, Door County. ***Glycine max*** (L.) Merr. (Soybean), escape from cultivation but not persisting in Wisc flora. ***Thermopsis rhombifolia*** (Nutt. ex Pursh) Nutt. ex Richards., adventive in Wisc; native to Great Plains.

*Key to genera*

1 Trees or shrubs .......... 2
1 Herbs .......... 7

2 Trees .......... 3
2 Shrubs .......... 6

3 Leaves simple; petals pink .......... CERCIS
3 Leaves all or mostly compound .......... 4

4 Leaves odd-pinnate, with 5-17 leaflets; flowers white, 1-2.5 cm long .......... ROBINIA
4 Leaves once- or twice-pinnate .......... 5

5 Leaflets ovate, tapered to a pointed tip, the margins entire; flowers pinkish-white, 1.5 cm long, in many-flowered racemes; trees without spines .......... GYMNOCLADUS
5 Leaflets oval or lance-shaped, rounded at tip, margins with scattered small teeth; flowers small, green-yellow, in axillary spikes; trees usually with spines on trunk and branches .......... GLEDITSIA

6 Twigs and petioles covered with stiff hairs; petals 5; pods linear, stiffly hairy, with several seeds .......... ROBINIA
6 Twigs and petioles not covered with stiff hairs; corolla of a single purple petal; pods short, with 1 or 2 seeds .......... AMORPHA

7 Leaves simple; petals yellow .......... CROTALARIA
7 Leaves compound .......... 8

8 Leaves even-pinnate or leaflets 2 .......... 9
8 Leaves not even-pinnate .......... 14

9 Leaves ending in a tendril; flowers pea-like (that is, like those of garden peas) .......... 10

9 Leaves not ending in a tendril; leaflets many, small; flowers not at all pea-like, in rounded heads .................................. 11

10 Style round in cross-section, pubescent near the tip ...... VICIA
10 Style flattened, pubescent along the inner side ...... LATHYRUS

11 Leaves bipinnate; flowers white, pink, or rose-colored.......... 12
11 Leaves pinnate; flowers yellow .................................. 13

12 Plants glabrous or nearly so; flowers green-white; pods flat and smooth .................................. DESMANTHUS
12 Plants with recurved prickles; flowers rose-pink; pods 4-angled to nearly round, prickly .............................. MIMOSA

13 Stamens all with normal anthers; pods opening when mature .... ........................................ CHAMAECRISTA
13 Upper 3 stamens sterile; pods remaining closed when mature .... ................................................ SENNA

14 Leaves divided into 3 or rarely 5 leaflets (except *Onobrychis*)..... 15
14 Leaves with 5 or more leaflets.................................. 25

15 Leaves (and other parts of the plant) more or less covered with glandular dots; leaflets 3-5, the margins entire ... PEDIOMELUM
15 Leaves not dotted with glands .................................. 16

16 Leaflets toothed ................................................ 17
16 Leaflets entire.................................................. 19

17 Flowers reflexed in long slender racemes, white or yellow; pods small, straight, reflexed .......................... MELILOTUS
17 Flowers in rounded clusters .................................... 18

18 Pods straight; stamens joined to the corolla ........ TRIFOLIUM
18 Pods curved or coiled; stamens free from the corolla MEDICAGO

19 Fruit a loment, breaking into 1-seeded segments, or consisting of a single segment only ...................................... 20
19 Fruit a legume (like a pea pod), not breaking into 1-seeded segments ...................................................... 22

20 Leaflets 3 ..................................................... 21
20 Leaflets 11 or more .......................... ONOBRYCHIS

21 Pods 1–several-jointed and -seeded; leaflets usually petioled; flowers purple or white ............................ DESMODIUM
21 Pods of a single 1-seeded joint (the lower joint when present empty and stalk-like); leaflets without stipules, usually prominently veined ............................................ LESPEDEZA

22 Leaflets without stipule-like appendages at their base; plants not twining or vining ........................................ 23
22 Leaflets with stipule-like appendages at their base; plants twining or vining ................................................... 24

23 Flowers in heads; pods small, often included in the calyx ........ ............................................ TRIFOLIUM
23 Flowers in racemes, or solitary ..................... BAPTISIA

24 Style glabrous ............................ AMPHICARPAEA
24 Style pubescent on upper surface .......... STROPHOSTYLES

25 Leaves dotted with small pits or depressions .................. 26
25 Leaves not dotted with small pits ............................ 28

26 Pods covered with hooked prickles; flowers whitish; stamens 10 .. ............................................. GLYCYRRHIZA
26 Pods without hooked prickles .................................. 27

27 Corolla of 1 petal; stamens 10 ........................ AMORPHA
27 Corolla of 5 petals ........................................ DALEA

28 Leaflets 5–11 ................................................. 29
28 Leaflets 11–31 ................................................ 32

29 Flowers yellow, borne in umbels............................... 30
29 Flowers not yellow, borne in axillary or terminal racemes ...... 31

30 Bracts subtending flower heads entire .................. LOTUS
30 Bracts subtending flower heads 3-parted .... *Anthyllis vulneraria**

31 Stems twining or climbing; leaflets usually 5–7; flowers in axillary racemes ............................................... APIOS
31 Stems erect; leaflets 7–11; flowers blue (or pink or white), in terminal racemes ........................................... LUPINUS

32 Flowers in terminal racemes ..................... TEPHROSIA
32 Flowers in axillary racemes or head-like umbels .............. 33

33 Flowers in umbel-like clusters; pods linear, 4-angled, jointed ..... ........................................... SECURIGERA
33 Flowers in racemes; pods neither 4-angled nor jointed......... 34

34 Keel of flower tipped with a sharp point; rare in n half of Wisc ... ........................................... OXYTROPIS
34 Keel of flower not tipped with a sharp point ...... ASTRAGALUS

## AMORPHA *Indigo-Bush*

Shrubs. Leaves pinnately compound. Flowers purple, rarely white, terminal and from the upper axils in dense spike-like racemes. The foliage, pods, and often the calyx are glandular-punctate; the glands on the pod are especially conspicuous.

*Key to species*

1 Shrub, woody throughout, mostly more than 1.5 m tall; leaflets 2–5 cm long; pods usually 2-seeded, 6–8 mm long ........ *A. fruticosa*
1 Plants woody only near base, herbaceous upwards, less than 1 m tall; leaflets less than 2 cm long; pods 1-seeded, 3–4 mm long ..... ............................................... *A. canescens*

***Amorpha canescens*** Pursh LEADPLANT Sandy open woods and dry prairies.

***Amorpha fruticosa*** L. [•54] FALSE INDIGO-BUSH June–July. Wet meadows, shores, streambanks, ditches, mostly near St. Croix, Wisconsin, and Mississippi Rivers.

## AMPHICARPAEA *Hog-Peanut*

***Amphicarpaea bracteata*** (L.) Fern. AMERICAN HOG-PEANUT Twining, annual

to short-lived perennial herb. Leaves pinnately compound into 3 petioled leaflets. Flowers in racemes or panicles peduncled from many of the axils, bearing several to many pale purple to whitish flowers. Besides the pod-producing petaliferous flowers, the plants bear nearly or completely apetalous flowers near the base of the stem, producing, often under the ground, small 1-seeded pods. Aug–Oct. Woods and thickets.

## APIOS *Groundnut*

*Apios americana* Medik. [•54] GROUNDNUT Perennial herbaceous vine, rhizomes with a necklace-like series of 2 or more edible tubers; plants with milky juice. Stems climbing over other plants. Leaves pinnately divided; main leaves with 5–7 leaflets. Flowers brown-purple, single or paired, in crowded racemes from leaf axils. July–Aug. Floodplain forests, thickets, shores, wet meadows, low prairie.

## ASTRAGALUS *Milk-Vetch*

Ours perennial herbs from a stout taproot, caudex, or rhizome. Leaflets numerous. Flowers white, yellowish white, or purple, in long or short axillary racemes.

ADDITIONAL SPECIES *Astragalus cicer* L. (Chickpea milk-vetch), native of Europe, reported from Taylor County.

*Key to species*

1 Flowers purple; rare in n Wisc ........................ *A. alpinus*
1 Flowers white or yellow-white; rare or widespread species ...... 2

2 Pod one-chambered; se Wisc ........................ *A. neglectus*
2 Pod divided to form 2 chambers ........................ 3

3 Flowers in racemes shorter than the subtending leaf; introduced species ........................ *A. cicer**
3 Racemes as long as or longer than the subtending leaf; native species ........................ 4

4 Plants erect, to 1 m tall or more; pods elliptic, held erect or nearly so, opening at tip when mature; widespread in Wisc *A. canadensis*
4 Plants sprawling, the stems to 6 dm long; pods globose, spreading, not opening when mature; rare in w Wisc ......... *A. crassicarpus*

*Astragalus alpinus* L. ALPINE MILK-VETCH *Endangered.* May–Aug. Sandy or gravelly, sparsely wooded lakeshores and riverbanks.

*Astragalus canadensis* L. [•54] CANADIAN MILK-VETCH May–Aug. Open woodlands, river banks and shores, usually in moist soil.

*Astragalus crassicarpus* Nutt. GROUND-PLUM *Endangered.* Pods subglobose, about 2 cm wide, abruptly pointed. April–May. Dry prairies and bluffs.

*Astragalus neglectus* (Torr. & Gray) Sheldon COOPER'S MILK-VETCH *Endangered.* June. Riverbanks and lakeshores, especially on limestone; disturbed forests and fields.

*Euphorbia maculata*
EUPHORBIACEAE
*Euphorbia corollata*
EUPHORBIACEAE
*Euphorbia virgata*
EUPHORBIACEAE

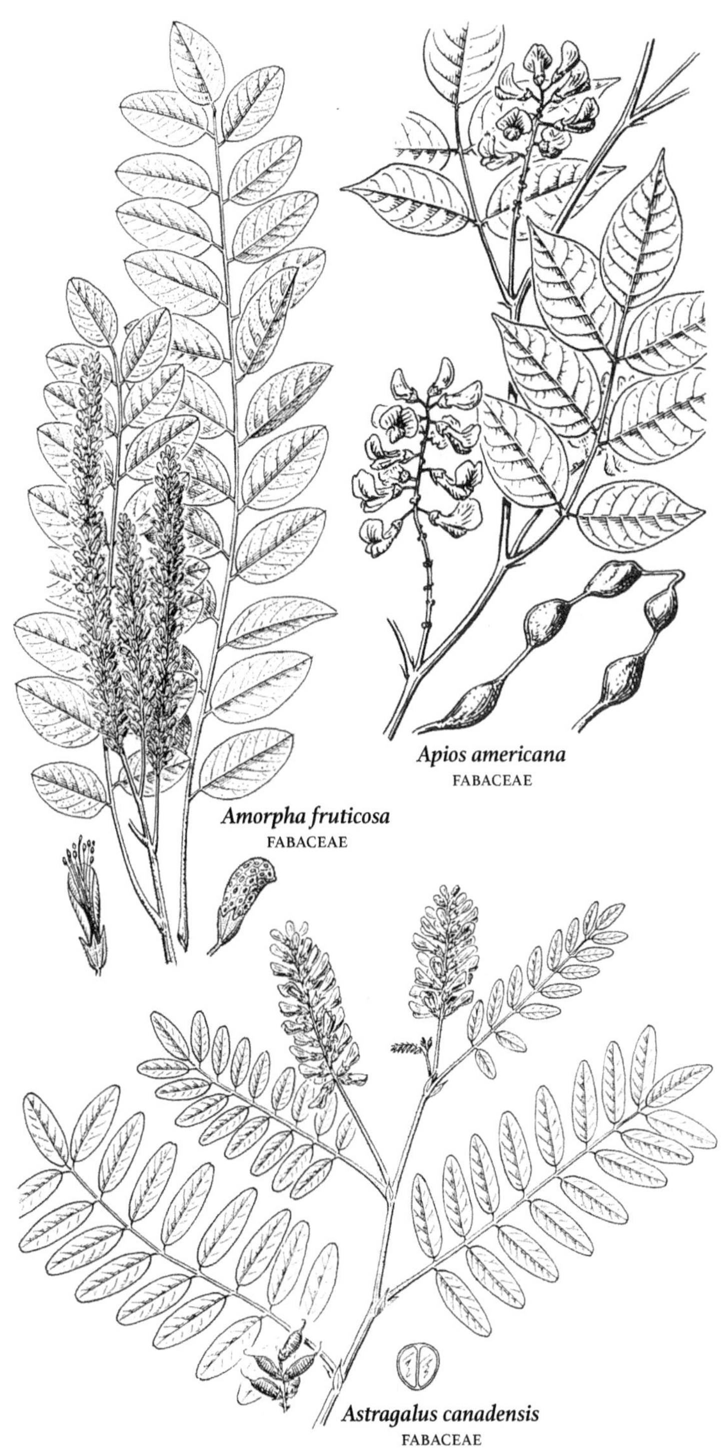
*Apios americana*
FABACEAE
*Amorpha fruticosa*
FABACEAE
*Astragalus canadensis*
FABACEAE

## BAPTISIA *Wild Indigo*

Perennial herbs from thick rhizomes. Stems usually much branched, the lateral branches bearing white, yellow, or violet flowers in long or short racemes. Pod papery to woody in texture, globose to cylindric, terminating in a curved beak, elevated on a distinct petiole.

ADDITIONAL SPECIES ***Baptisia australis*** (L.) R. Br. adventive in southern Wisc.

*Key to species*

1 Bract at base of each pedicel 1–3 cm long, persistent and leaf-like, distincly net-veined ................................ ***B. bracteata***
1 Bracts smaller, less than 1 cm long, usually soon deciduous, not net-veined.................................................... 2

2 Flowers white ........................................... ***B. alba***
2 Flowers yellow ........................................ ***B. tinctoria***

***Baptisia alba*** (L.) Vent. WHITE WILD INDIGO Dry sandy woods.

***Baptisia bracteata*** Muhl. PLAINS WILD INDIGO Prairies, open dry woods.

***Baptisia tinctoria*** (L.) R. Br. YELLOW WILD INDIGO June–July. Dry sterile or sandy soil.

## CERCIS *Redbud*

***Cercis canadensis*** L. eastern REDBUD Deciduous shrub or small tree; twigs slender and zigzagged, nearly black, with lighter lenticels. Leaves alternate, simple, heart-shaped; margins entire. Flowers showy, light to dark magenta pink in color, on bare stems before the leaves. Fruit flattened, brown, pea-like pods. April–May. Moist woods. Common in se USA, considered adventive in s Wisc. Formerly included in the Caesalpiniaceae.

## CHAMAECRISTA *Sensitive-Pea*

***Chamaecrista fasciculata*** (Michx.) Greene PARTRIDGE-PEA Annual. Leaflets 6–18 pairs, oblong, mucronate. Flowers 1–6 in short, axillary racemes; petals bright yellow. July–Sept. Variable in size and pubescence. Moist or dry, especially sandy soil; prairies, open woods, old fields, and roadsides.

## CROTALARIA *Rattlebox*

***Crotalaria sagittalis*** L. ARROW-HEAD RATTLEBOX Annual herb. Stems pubescent with spreading hairs. Leaves pubescent, sessile or nearly so, lance-shaped to linear. Racemes 2–4-flowered, terminating the stem and branches. Pods oblong, much inflated. June–Sept. Dry open soil and waste places.

## DALEA *Prairie-Clover*

Perennial herbs (ours). Leaves small, odd-pinnate, glandular-punctate. Flowers numerous, small, in dense terminal spikes.

*Key to species*

1 Flowers white; calyx tube glabrous .................. ***D. candida***
1 Flowers purple or violet; calyx tube densely hairy .............. 2

2 Main leaves with mostly 5 leaflets .................. ***D. purpurea***
2 Main leaves with mostly 11–17 leaflets ................ ***D. villosa***

***Dalea candida*** Michx. WHITE PRAIRIE-CLOVER Foliage and stems glabrous. Leaflets 5–9, commonly 7. Spikes 1 to few; petals white. Dry prairies or dry upland woods. *Petalostemum candidum* (Willd.) Michx.

***Dalea purpurea*** Vent. PURPLE PRAIRIE-CLOVER Foliage and stem usually glabrous. Leaflets 3–7, usually 5, linear, glandular-punctate beneath. Petals rose-purple, rarely varying to white. Dry prairies. *Petalostemum purpureum* (Vent.) Rydb.

***Dalea villosa*** (Nutt.) Spreng. DOWNY PRAIRIE-CLOVER Stems densely villous. Leaflets 11–17, oblong, glandular-punctate beneath, softly villous. Petals rose-purple, varying to white. Aug. Sandy prairies. *Petalostemum villosum* Nutt.

## DESMANTHUS *Bundle-Flower*

***Desmanthus illinoensis*** (Michx.) MacM. PRAIRIE BUNDLE-FLOWER Perennial herb. Stems strongly angled, glabrous to finely hairy. Leaves 2-pinnate, pinnae 6–12 pairs; leaflets 20–30 pairs, often ciliate. Flowers whitish or greenish in long-peduncled axillary heads. Pods strongly curved or somewhat twisted together in a dense subglobose head. Summer. Moist or dry soil, riverbanks, prairies, and pastures. Considered introduced and adventive in Wisc; native s and w of the state.

## DESMODIUM *Tick-Trefoil*

Ours perennial herbs. Leaves 3-parted. Flowers small, white to purple or violet, sometimes marked with yellow, in elongate, simple or panicled racemes. Fruit an indehiscent pod, more or less beset with hooked hairs.

*Key to species*

1 Pods conspicuously long-stalked, the stalk 2–3 times the length of the calyx.................................................. 2
1 Pods sessile or short-stalked ..................................... 3

2 Flowers in a panicle on a leafy stem .............. ***D. glutinosum***
2 Panicle on a long leafless stalk ................... ***D. nudiflorum***

3 Stipules large, persistent, lance-shaped to ovate, persistent...... 4
3 Stipules small, awl-shaped, soon deciduous ................... 6

4 Joints of the pod diamond-shaped, longer than wide ........... 5
4 Joints of the pod oval-shaped ...................... ***D. illinoense***

5 Leaflets rounded at tip, about same length as petiole ***D. canescens***
5 Leaflets tapered to a tip, longer than petiole ....... ***D. cuspidatum***

6 Pods distinctly stalked, the stalk longer than the calyx ..........
........................................ ***D. perplexum***

6 Pods sessile or short-stalked, the stalk shorter than the lobes of the calyx ........................................ *D. canadense*

***Desmodium canadense*** (L.) DC. [•55] SHOWY TICK-TREFOIL July–Aug. Moist soil, thickets, and riverbanks.

***Desmodium canescens*** (L.) DC. HOARY TICK-TREFOIL July–Aug. Moist or dry soil.

***Desmodium cuspidatum*** (Muhl.) DC. BIG TICK-TREFOIL July–Aug. Mesic woods of oak, sugar maple and basswood.

***Desmodium glutinosum*** (Muhl.) Wood POINTED-LEAF TICK-TREFOIL July. Rich woods.

***Desmodium illinoense*** Gray ILLINOIS TICK-TREFOIL July–Aug. Rich prairie soil.

***Desmodium nudiflorum*** (L.) DC. NAKED-FLOWER TICK-TREFOIL July–Sept. Rich woods. *Hylodesmum nudiflorum* (L.) H. Ohashi & R.R. Mill.

***Desmodium perplexum*** Schub. PERPLEXED TICK-TREFOIL July–Aug. Dry woods.

## GLEDITSIA *Honey-Locust*

***Gleditsia triacanthos*** L. [•55] HONEY-LOCUST Tree, stems and branches usually thorny. Leaves pinnate or 2-pinnate; leaflets oblong lance-shaped. Flowers unisexual or rarely perfect, greenish yellow, in spike-like racemes. Pods curved and usually twisted, dark brown, rather papery in texture. Rich moist woods.

## GLYCYRRHIZA *Licorice*

***Glycyrrhiza lepidota*** Pursh AMERICAN LICORICE Glandular-dotted perennial herb. Leaves odd-pinnate; leaflets 11–19, oblong or lance-shaped. Flowers in dense axillary racemes shorter than the subtending leaves; pale yellow. Pods brown, densely beset with hooked prickles. May–June. Moist prairies, along railroads and in waste places.

## GYMNOCLADUS *Coffeetree*

***Gymnocladus dioicus*** (L.) K. Koch KENTUCKY COFFEETREE Tree, polygamo-dioecious. Single leaves sometimes nearly 1 m long, with 3–7 pairs of pinnae, each with several pairs of ovate, abruptly acuminate leaflets, or the lowest pinnae replaced by single leaflets. Flowers regular, perfect or unisexual, greenish-white, softly pubescent, in terminal panicles. Pods shed in early spring; seeds very hard, nearly black. May. Rich moist woods.

## LATHYRUS *Vetchling; Wild Pea*

Perennial herbs (ours). Leaves terminated by a tendril. Flowers few to many in a raceme; corolla red-purple to white or yellow. Most of our species superficially resemble *Vicia*.

ADDITIONAL SPECIES ***Lathyrus sylvestris*** L. (Narrow-leaf vetchling), introduced and occasional across much of Wisc.

*Key to species*

1 Leaflets 1-pair; introduced species .......................... 2
1 Leaflets 2 or more pairs; native species ...................... 4

2 Stem winged .................................. ***L. latifolius***
2 Stem not winged ............................................ 3

3 Flowers yellow ................................ ***L. pratensis***
3 Flowers red-purple ............................ ***L. tuberosus***

4 Flowers yellow-white ......................... ***L. ochroleucus***
4 Flowers purple, rarely white.................................. 5

5 Stipules leafy, nearly as large as the adjacent leaflets .. ***L. japonicus***
5 Stipules much smaller than leaflets ........................... 6

6 Stems usually winged; racemes with usually 2–6 flowers; moist habitats ................................. ***L. palustris***
6 Stems not winged; racemes with 10–20 flowers; drier woods ..... ................................................ ***L. venosus***

***Lathyrus japonicus*** Willd. BEACH-PEA June–Aug. Beaches and lakeshores. *Lathyrus maritimus* Bigelow.

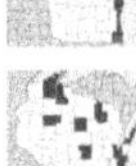

***Lathyrus latifolius*** L. EVERLASTING-PEA June–Aug. Native of s Europe; cultivated and escaping to roadsides and vacant land.

***Lathyrus ochroleucus*** Hook. CREAM VETCHLING May–July. Dry upland woods and thickets.

***Lathyrus palustris*** L. [•55] MARSH VETCHLING June–Aug. Conifer swamps, wet meadows, marshes, streambanks, fens, low prairie.

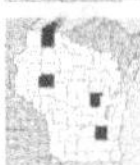

***Lathyrus pratensis*** L. MEADOW VETCHLING Native of Eurasia; introduced as a weed in waste places.

***Lathyrus tuberosus*** L. EARTH-NUT VETCHLING Introduced. June–Aug. Native of Europe and w Asia.

***Lathyrus venosus*** Muhl. VEINY VETCHLING June–July. Moist woods and thickets.

## LESPEDEZA *Bush-Clover*

Annual or perennial herbs, or some exotic species shrubs. Leaves small, trifoliolate. Flowers purple to yellowish white, in sessile or peduncled axillary clusters. The genus differs from *Desmodium* by its 1-seeded fruits and in the absence of leaflet petioles.

ADDITIONAL SPECIES ***Lespedeza × longifolia*** DC. (pro sp.), hybrid between *L. capitata* and *L. hirta,* reported from several Wisc locations. Plants with stems about 1 m tall, pubescent with spreading hairs; flowers yellowish white. ***Lespedeza thunbergii*** (DC.), introduced for ornament and known from sw Wisc; plants shrubby with racemes of attractive purple flowers. ***Lespedeza*** *violacea* (L.) Pers., native species usually on sandstone bluffs in sw Wisc, more common south and east of the state; leaflets glabrous; flowers purple, in loose racemes.

*Key to species*

1 Plants woody, at least at base, 1–3 m tall ............... *L. cuneata*
1 Plants entirely herbaceous .................................... 2
2 Flowers purple, but some flowers without petals ............... 3
2 Flowers white or yellow-white, all alike......................... 5

3 Flower clusters sessile or nearly so ................. *L. virginica*
3 Flower clusters on slender stalks distincly longer than the subtending leaves ............................................. 4

4 Stems glabrous ...................................... *L. repens*
4 Stems softly hairy ............................... *L. procumbens*

5 Leaflets linear; spikes slender, loosely flowered .... *L. leptostachya*
5 Leaflets elliptic; flowers in densely flowered rounded heads ...... ................................................. *L. capitata*

***Lespedeza capitata*** Michx. [•56] ROUND-HEAD BUSH-CLOVER July–Sept. Open dry woods, sand dunes, and prairies. Plants on sand dunes are occasionally prostrate.

***Lespedeza cuneata*** (Dum.-Cours.) G. Don CHINESE BUSH-CLOVER Sept–Oct. Native of e Asia; introduced into cultivation and reported from Green County.

***Lespedeza leptostachya*** Engelm. PRAIRIE BUSH-CLOVER *Endangered.* July–Aug. Dry, gravelly or sandy hillside prairies. Federally threatened.

***Lespedeza procumbens*** Michx. TRAILING BUSH-CLOVER Aug–Sept. Dry upland woods; Grant County. Closely resembling *L. repens* in habit, foliage, and flowers.

***Lespedeza repens*** (L.) W. Bart. CREEPING BUSH-CLOVER June–Sept. Dry woods and fields.

***Lespedeza virginica*** (L.) Britt. SLENDER BUSH-CLOVER *Threatened.* Aug–Sept. Dry prairies, open upland woods, red cedar glades. Similar to *L. capitata* except the leaves are more crowded, and the leaflets are longer and more hairy on upper surface.

## LOTUS *Trefoil*

Herbaceous or suffrutescent (woody at base) herbs. Leaves pinnately compound. Flowers solitary or umbellate.

*Key to species*

1 Leaflets 5 ........................................ *L. corniculatus*
1 Leaflets 3 ........................................ *L. unifoliolatus*

***Lotus corniculatus*** L. BIRD'S-FOOT-TREFOIL June–Aug. Native of Europe; established in fields, meadows, and roadsides.

***Lotus unifoliolatus*** (Hook.) Benth. SPANISH CLOVER Summer. Dry prairies. Lotus purshianus F.E. & E.G. Clem.

## LUPINUS *Lupine*

Perennial herbs. Leaves palmately compound. Flowers white, yellow, pink, or blue, in terminal racemes or spikes. Several species cultivated for ornament.

*Key to species*

1 Plants large, to 1 m or more tall; leaflets 11–17; inflorescence 2–4 dm long ........................................ ***L. polyphyllus***

1 Plants smaller, 3–6 dm tall; leaflets 7–11; inflorescence 1–2 dm long . ........................................ ***L. perennis***

***Lupinus perennis*** L. SUNDIAL-LUPINE May–June. In dry or moist sandy soil of prairies, clearings, and savannas.

***Lupinus polyphyllus*** Lindl. BLUE-POD LUPINE Escape from cultivation, especially along roadsides. Plants taller and coarser than *L. perennis,* to 1 m or more tall.

## MEDICAGO *Alfalfa; Medick*

Herbs with 3-foliolate leaves, the terminal leaflet stalked. Flowers in axillary heads or short head-like racemes of small yellow or blue flowers. Pod straight or coiled, glabrous or spiny.

*Key to species*

1 Plants perennial from a long taproot; flowers blue-violet or sometimes yellow ........................................ ***M. sativa***

1 Plants annual or biennial; flowers yellow ............ ***M. lupulina***

***Medicago lupulina*** L. BLACK MEDICK May–Sept. Native of Europe and w Asia; common as a troublesome weed of roadsides, lawns, fields, railroads, and disturbed places.

***Medicago sativa*** L. ALFALFA June–Sept. Native probably of c and w Asia; long in cultivation and valued for hay and forage; commonly escaped or introduced on roadsides, fields, and disturbed places.

## MELILOTUS *Sweet-Clover*

Annual or biennial herbs. Leaves 3-foliolate, the terminal leaflet stalked. Flowers white or yellow, in elongate peduncled racemes from the upper axils.

*Key to species*

1 Flowers white ........................................ ***M. albus***

1 Flowers yellow ........................................ 2

2 Pods finely hairy (under a hand lens) ............... ***M. altissimus***

2 Pods glabrous ........................................ 3

3 Flowers 4–7 mm long; pods 3–5 mm long; common ............. ........................................ ***M. officinalis***

3 Flowers to 3 mm long; pods to 3 mm long; local in s Wisc ........ ........................................ ***M. indicus***

***Melilotus albus*** Medik. WHITE SWEET-CLOVER Summer and fall. Native of Europe and w Asia; throughout Wisc in waste places and along roads, especially in calcareous soil (*no map*).

***Melilotus altissimus*** Thuill. TALL YELLOW SWEET-CLOVER Native of Eurasia; probably often mistaken for *M. officinalis* as similar in size, foliage, inflorescence, and flowers, but with pubescent, more conspicuously reticulate pods.

***Melilotus indicus*** (L.) All. INDIAN SWEET-CLOVER Summer. Native of the Mediterranean region, reported for Wisc (*no map*).

***Melilotus officinalis*** (L.) Lam. YELLOW SWEET-CLOVER Summer. Native of Eurasia; established as a weed of waste places.

## MIMOSA *Mimosa*

***Mimosa nuttallii*** (DC.) B.L. Turner NUTTALL'S MIMOSA Perennial herb. Stems branched, arched or decumbent, strongly ribbed, beset with stout hooked thorns, otherwise glabrous. Leaves sensitive to touch; leaflets 8–15 pairs. Flowers perfect or unisexual; numerous in globose, long-peduncled, axillary heads. Pods linear, strongly ribbed, thorny on the ribs, tipped with a slender beak. May–Sept. Native of sc USA; adventive in Wisc on dry sterile or sandy soil. *Mimosa quadrivalvis* L. var. *nuttallii* (DC.) Barneby.

## ONOBRYCHIS *Sainfoin*

***Onobrychis viciifolia*** Scop. SAINFOIN Perennial herb. Leaves odd-pinnate; leaflets 11–25. Peduncles from the upper axils, 1–3 dm long. Raceme dense. Flowers rose-colored. Pods broadly oval, compressed, strongly nerved, the nerves with short stout spines. June–July. Native of s Europe; occasionally cultivated for forage; escaped or introduced.

## OXYTROPIS *Locoweed*

***Oxytropis campestris*** (L.) DC. FASSETT'S LOCOWEED *Endangered.* Perennial herb. Leaflets lance-shaped, densely to sparsely villous or nearly glabrous. Spikes dense; corolla purple. Pods ovoid, pubescent. May–June. Sandy lakeshores where the water level fluctuates. Wisc plants differ from the typical variety by being more pubescent and with a smaller pod, the body only about 1 cm long; it has been differentiated as var. *chartacea* (Fassett) Barneby. *Federally threatened.*

## PEDIOMELUM *Indian-Breadroot*

Perennial herbs from rhizomes or thickened roots, more or less glandular-punctate. Leaves compound with mostly 3 or 5 leaflets. Flowers in peduncled spikes or racemes from the upper axils, usually blue. Fruit short, flattened or turgid, 1-seeded, sometimes terminated by the persistent style.

*Key to species*

1 Plants from a tuberous globose root; flowers 11–15 mm long ...... **P. esculentum**

1 Plants from woody rhizomes; flowers to 10 mm long **P. argophyllum**

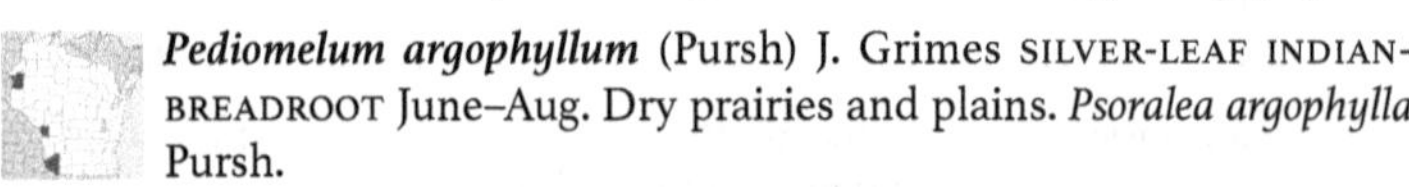

***Pediomelum argophyllum*** (Pursh) J. Grimes SILVER-LEAF INDIAN-BREADROOT June–Aug. Dry prairies and plains. *Psoralea argophylla* Pursh.

***Pediomelum esculentum*** (Pursh) Rydb. LARGE INDIAN-BREADROOT May–July. Dry prairies and plains. *Psoralea esculenta* Pursh.

## ROBINIA *Locust*

Trees or shrubs. Leaves odd-pinnate. Flowers white, pink, or purple in axillary racemes. Pods elongate, flat, many-seeded.

*Key to species*

1 Tree to 25 m tall; twigs and petioles glabrous; flowers white. . . . . . . . . . . . . . . . . . . . . . . . . . . . . . . . . . . . . . . . . . . . . . . . . . ***R. pseudoacacia***

1 Shrubs to 3 m tall; stems gummy or bristly; flowers pink or rose-colored . . . . . . . . . . . . . . . . . . . . . . . . . . . . . . . . . . . . . . . . . . . . . . . . . **2**

2 Stems gummy or sticky . . . . . . . . . . . . . . . . . . . . . . . . . . . . . . ***R. viscosa***

2 Stems bristly with short stiff hairs . . . . . . . . . . . . . . . . . . . . ***R. hispida***

***Robinia hispida*** L. BRISTLY LOCUST Introduced. June–July. Sometimes planted and occasionally escaped to roadsides and open woods.

***Robinia pseudoacacia*** L. BLACK LOCUST Introduced (invasive). June. Commonly planted and escaped to roadsides, open woods, and waste land.

***Robinia viscosa*** Vent. CLAMMY LOCUST Introduced. June. Reported for Washington County, but not verified.

## SECURIGERA *Crown-Vetch*

***Securigera varia*** (L.) Lassen PURPLE CROWN-VETCH Perennial herb. Leaves sessile; leaflets 11–21, oblong to obovate. Flowers in long-peduncled axillary umbels of 10–15 flowers, pink, the keel tipped with purple. Pods linear, 4-angled, with 3–7 joints. May–Sept. Native of Eurasia and n Africa; introduced or escaped along roadsides. *Coronilla varia* L.

## SENNA *Wild Sensitive-Plant*

Pernnial herbs (ours). Leaves evenly 1-pinnate, usually with one or more glands on the petiole. Flowers yellow (ours), in the axils or in terminal or axillary racemes.

ADDITIONAL SPECIES ***Senna obtusifolia*** (L.) Irwin & Barneby (Coffeeweed), introduced and adventive in Pierce County; also known in the Chicago region of Illinois so may eventually be found in se Wisc.

*Key to species*

1 Leaflets obovate and rounded at tip; leaf with gland between lowest pair of leaflets; pod only slightly flattened . . . . . . . . . . ***S. obtusifolia****

1 Leaflets oblong, elliptic or ovate; leaf with large gland near base of petiole; pod flattened . . . . . . . . . . . . . . . . . . . . . . . . . . . . . . . . . . . . . . . . **2**

2 Gland club-shaped on a short stalk; joints of the pod about as long as wide . . . . . . . . . . . . . . . . . . . . . . . . . . . . . . . . . . . . . . . . . . ***S. hebecarpa***

2 Gland rounded or short-cylindric; joints of the pod about 2-times longer than wide . . . . . . . . . . . . . . . . . . . . . . . . . . . . . . . . ***S. marilandica***

***Senna hebecarpa*** (Fern.) Irwin & Barneby AMERICAN WILD SENSITIVE-PLANT Moist open woods, roadsides, and streambanks. *Cassia hebecarpa* Fern.

***Senna marilandica*** (L.) Link MARYLAND WILD SENSITIVE-PLANT Moist open woods and streambanks. *Cassia marilandica* L.

## STROPHOSTYLES *Fuzzy-Bean*

Annual herbs. Stems twining or trailing. Leaves pinnately compound, 3-foliolate. Flowers pink-purple to white, in a short, head-like, long-peduncled raceme. Pods coiled after dehiscence, with several woolly seeds.

*Key to species*

1 Leaflets all entire; pods to 5 cm long ................ ***S. leiosperma***
1 At least some leaflets with shallow lobes; pods 5–9 cm long ...... ........................................................ ***S. helvula***

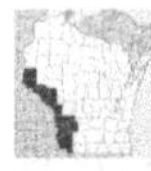

***Strophostyles helvula*** (L.) Ell. TRAILING FUZZY-BEAN Late summer. Dry or sandy uplands.

***Strophostyles leiosperma*** (Torr. & Gray) Piper SLICK-SEED FUZZY-BEAN Late summer. Dry or moist sandy soil, upland woods, dunes, and shores.

## TEPHROSIA *Goat's-Rue*

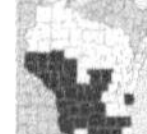

***Tephrosia virginiana*** (L.) Pers. GOAT'S-RUE Perennial herb. Stems and pods villous. Leaflets 9–27, commonly about 21. Flowers in single (or rarely compound) terminal racemes; standard yellow; wings pink or pale purple. June–July. Plants vary in the pubescence of the leaves and stems. Old fields, open woods, dunes, often in sandy soil.

## TRIFOLIUM *Clover*

Annual, biennial, or perennial herbs. Leaves 3-foliolate, serrulate. Flowers in heads, spikes, or head-like racemes or umbels. Petals all separate or more or less united into a tube. Valuable for forage and several species extensively cultivated. In the absence of fruit, our yellow-flowered species may be distinguished from *Medicago* by their strongly bilabiate calyx.

ADDITIONAL SPECIES Several clovers are introduced and known from Wisc locations: ***Trifolium dubium*** Sibthorp (Suckling clover) and ***T. incarnatum*** L. (Crimson clover).

*Key to species*

1 Flowers white, pink, or purple ................................ 2
1 Flowers yellow, turning brown with age ....................... 7

2 Flowers on short pedicels, these becoming reflexed with age .... 3
2 Flowers sessile or nearly so .................................. 4

3 Flowers white; very common .......................... ***T. repens***
3 Flowers pink or purple-tinged ...................... ***T. hybridum***

4 Heads cylindric-shaped ........................................ 5
4 Heads nearly globose to ovoid ................................. 6

5 Flowers white, shorter than the calyx ................ *T. arvense*
5 Flowers crimson, longer than the calyx .......... *T. incarnatum**

6 Flower heads 2–3 cm wide; flowers magenta or rarely white; very common .......................................... *T. pratense*
6 Flower heads smaller; flowers rose-colored ....... *T. resupinatum*

7 Leaflets sessile; stipules linear ........................ *T. aureum*
7 Terminal leaflet on a short petiole; stipules ovate to lance-shaped **8**

8 Heads with 20–40 flowers; the banner (upper, larger petal) distinctly finely striped ................................... *T. campestre*
8 Heads with 3–15 flowers; banner only faintly striped .. *T. dubium**

***Trifolium arvense*** L. RABBIT-FOOT CLOVER May–Sept. Native of Eurasia and n Africa; a weed of sterile soil, roadsides, old fields, and waste places.

***Trifolium aureum*** Pollich GREATER HOP CLOVER May–Sept. Native of Eurasia, weedy on roadsides and in waste places.

***Trifolium campestre*** Schreb. LESSER HOP CLOVER May–Sept. Native of Eurasia and n Africa; weedy on roadsides and in waste places.

***Trifolium hybridum*** L. ALSIKE CLOVER Summer. Native of Eurasia, commonly escaped.

***Trifolium pratense*** L. RED CLOVER May–Aug. Native of Europe, commonly planted for forage and escaped to fields and on roadsides.

***Trifolium repens*** L. WHITE CLOVER Summer. Native of Eurasia, commonly planted and escaped to lawns and roadsides.

***Trifolium resupinatum*** L. REVERSED CLOVER Early summer. Native of Europe, introduced in grass seed.

## VICIA *Vetch*

Annual or perennial herbs. Leaves 1-pinnate, with small stipules, the terminal leaflets in most species metamorphosed into tendrils. Flowers in racemes from the axils, or in sessile or subsessile, few-flowered, axillary clusters.

*Key to species*

1 Flowers single or in pairs, sessile from leaf axils ......... *V. sativa*
1 Flowers in racemes on stalks from leaf axils.................... 2

2 Flowers white or white tinged with purple, 3–7 mm long........ 3
2 Flowers blue or white, 8 mm long or more......................... 4

3 Lobes of calyx nearly equal in length; pods hairy, mostly with 2 seeds ............................................ *V. hirsuta*
3 Calyx lobes unequal; pods glabrous, mostly with 4 seeds ........ ............................................ *V. tetrasperma*

4 Calyx with a large swollen bump on one side of base .... *V. villosa*
4 Calyx only slightly swollen on one side of base ................ 5

5 Margins of stipules sharply toothed; flowers 15–30 mm long ..... .................................................. *V. americana*
5 Margins of stipules entire; flowers to 13 mm long............... 6

6 Flowers white, in loose racemes .................... *V. caroliniana*
6 Flowers blue, in dense racemes ........................ *V. cracca*

***Vicia americana*** Muhl. AMERICAN VETCH [•56] May–July. Moist woods.

***Vicia caroliniana*** Walt. CAROLINA VETCH May–June. Moist woods and thickets.

***Vicia cracca*** L. BIRD-VETCH Introduced (naturalized). June–Aug. Fields, roadsides, meadows.

***Vicia hirsuta*** (L.) S.F. Gray TINY VETCH May–Aug. Native of Europe; introduced in fields, roadsides, and waste places.

***Vicia sativa*** L. COMMON VETCH Native of Europe; in cultivation since antiquity; may persist after cultivation or escape into fields and roadsides.

***Vicia tetrasperma*** (L.) Schreb. LENTIL VETCH May–Aug. Eurasian introduction in Door County.

***Vicia villosa*** Roth [•56] HAIRY VETCH June–Aug. Native of Europe; introduced in fields, roadsides, and waste places.

---

## FAGACEAE *Beech Family*

Trees or shrubs. Leaves alternate, simple, entire to lobed. Plants monoecious. Staminate flowers in catkins or heads. Pistillate flowers solitary, or in small clusters or short spikes, more or less enclosed by an involucre of numerous bracts. Fruit a 1-seeded nut, wholly or partly surrounded by the expanded involucre.

*Key to genera*

1 Leaves entire, or toothed or lobed with fewer than 9 pairs of lateral veins ............................................ QUERCUS
1 Leaf margins various; lateral veins more than 9 pairs ........... 2

2 Teeth mostly tipped with an incurved awn 1 mm or more long; fruit prickly, the spines more than 1 cm long ............. CASTANEA
2 Teeth not awn-tipped; fruit not with spines more than 1 cm long 3

3 Bark smooth, silvery-gray; terminal buds single, narrow and more than 1 cm long; fruit bristly, the nut 3-angled ........... FAGUS
3 Bark becoming furrowed or scaly; terminal buds 2 or more, less than 1 cm long; fruit not bristly, the nuts rounded ......... QUERCUS

### CASTANEA *Chestnut*

***Castanea dentata*** (Marsh.) Borkh. AMERICAN CHESTNUT Tree. Leaves simple, alternate, straight-veined, lance-shaped, coarsely and sharply serrate with

*Gleditsia triacanthos*
FABACEAE
*Desmodium canadense*
FABACEAE
*Lathyrus palustris*
FABACEAE

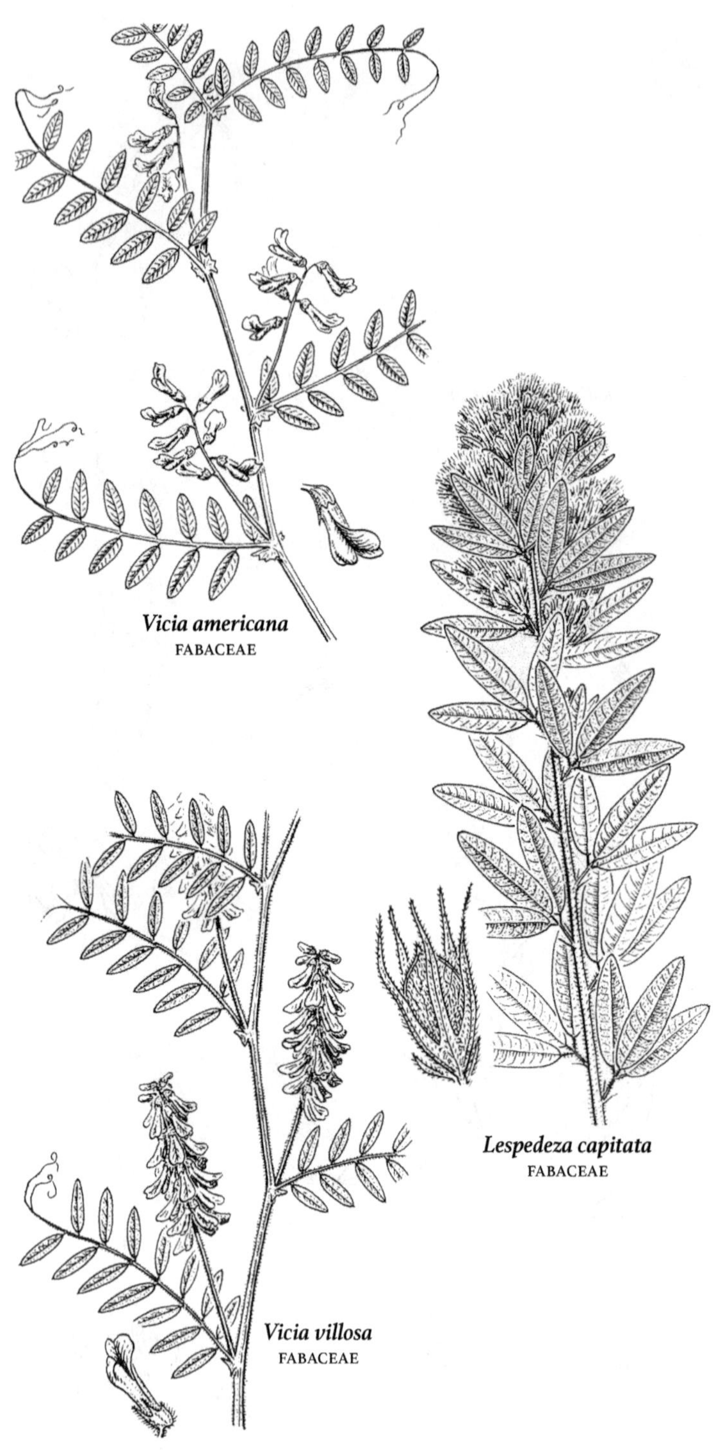

*Vicia americana*
FABACEAE
*Lespedeza capitata*
FABACEAE
*Vicia villosa*
FABACEAE

ascending or incurved teeth, glabrous or nearly so on both sides. Flowers appearing after the leaves; staminate catkins to 20 cm long; pistillate flowers borne on the base of the staminate catkins or from separate axils, within an ovoid prickly involucre. Fruit a 1-seeded nut, flattened on 1 or 2 sides, within the long-spined, 2–4-valved involucre. Usually in acid upland soils. Native to e USA, considered adventive in Wisc.

This formerly important forest tree has been nearly exterminated throughout its natural range by the chestnut blight, a parasitic fungus introduced from abroad and first noted in the midwest about 1930. Stump sprouts, not growing to maturity, are known, and where planted beyond its original native range, some chestnuts have escaped the blight and may grow to a large size.

## FAGUS *Beech*

***Fagus grandifolia*** Ehrh. AMERICAN BEECH Tree, with smooth gray bark. Leaves simple, alternate, straight-veined, a vein running to each tooth, ovate to obovate, serrate to denticulate. Flowers appearing with the leaves, the staminate flowers from the lower axils in small heads on drooping peduncles; pistillate flowers from the upper axils, usually in pairs at the end of a short peduncle. Fruit a sharply 3-angled, 1-seeded nut, borne in pairs enclosed within the expanded 4-valved involucre. Beech-maple forests mostly near Lake Michigan; Wisc at western edge of species' range.

## QUERCUS *Oak*

Deciduous trees (ours). Leaves alternate, simple, lobed, pinnately veined. Plants monoecious (staminate and pistillate flowers separate but on same tree). Staminate flowers in slender naked catkins, catkins appearing with the leaves. Pistillate flowers solitary or in small spikes, each subtended by a bract and surrounded by an involucre of many scales. Fruit a nut (acorn) partially enclosed by a cuplike structure (cupule).

HYBRIDS ***Quercus × bebbiana*** C.K.Schneid., *Q. alba* and *Q. macrocarpa,* sc and w Wisc. ***Quercus × deamii*** Trel., *Q. macrocarpa* and *Q. muehlenbergii,* Grant County. ***Quercus × hawkinsiae*** Sudw., *Q. rubra* and *Q. velutina,* s Wisc. ***Quercus × palaeolithicola*** Trel., *Q. ellipsoidalis* and *Q. velutina,* s half of Wisc. ***Quercus × schuettei*** Trel., *Q. bicolor* and *Q. macrocarpa,* mostly s Wisc.

*Key to species*

1 Leaves deeply lobed .......... 2
1 Leaves not deeply lobed .......... 8

2 Lobes rounded, not tipped by bristles .......... 3
2 Lobes acute, bristle-tipped .......... 4

3 Acorn 1.5–3 cm long within a deep, fringed cup; branches often with corky ridges .......... ***Q. macrocarpa***
3 Acorn 1.3–2 cm long, only about one-quarter covered by warty cup; branches not corky-ridged .......... ***Q. alba***

4 Leaf underside hairy .......... ***Q. velutina***
4 Leaf underside glabrous or nearly so .......... 5

5 Acorn cup covering lower one-third to lower one-half of acorn .. 6

5 Acorn cup saucer-shaped; covering only base of acorn ......... 7

6 Buds glabrous and satiny-shiny; acorn kernel yellow; trunk often with branch stubs near ground .................. *Q. ellipsoidalis*

6 Buds hairy above middle; acorn kernel whitish; trunk without branch stubs near ground .......................... *Q. coccinea*

7 Upper leaf surface satiny-shiny; acorn to 1 cm long ... *Q. palustris*

7 Upper leaf surface not shiny; acorn 2–3 cm long ......... *Q. rubra*

8 Leaf margins with wavy rounded teeth, not sharp-toothed; acorns on stalks 2.5 cm or more long ....................... *Q. bicolor*

8 Leaf margins with coarse teeth; acorns sessile or on stalks less than 1 cm long ................................... *Q. muehlenbergii*

***Quercus alba*** L. [•57] WHITE OAK Woods of oak-hickory and beech-maple, sometimes on sandy plains with other oaks and jack pine.

***Quercus bicolor*** Willd. [•57] SWAMP WHITE OAK May. Floodplain forests, low woods and swamps.

***Quercus coccinea*** Muenchh. SCARLET OAK Dry upland soils.

***Quercus ellipsoidalis*** E.J. Hill NORTHERN PIN OAK Dry upland soil.

***Quercus macrocarpa*** Michx. BUR-OAK Moist woods and alluvial floodplains.

***Quercus muehlenbergii*** Engelm. CHINKAPIN OAK Maple forests, forests on stabilized dunes, often where calcareous.

***Quercus palustris*** Muenchh. PIN OAK May. Floodplain forests, low wet woods, swamps; tolerant of periodic flooding.

***Quercus rubra*** L. [•57] NORTHERN RED OAK Rich mesic forests, ridges, sandy plains with jack pine.

***Quercus velutina*** Lam. [•57] BLACK OAK Usually in dry or sterile upland soil and on dunes.

## GENTIANACEAE *Gentian Family*

Annual, biennial or perennial herbs; plants usually glabrous. Leaves simple, entire, opposite or whorled, stem leaves without petioles. Flowers often showy, perfect, single at end of stems or in clusters; petals 4-5, blue, purple, white or green, joined for at least part of their length. Fruit a 2-chambered, many-seeded capsule enclosed by the withered, persistent petals.

*Key to genera*

1 Leaves reduced to small, narrow scales less than 3 mm long ..... ........................................................ BARTONIA

1 Leaves not scale-like, well developed ........................... 2

2 Flowers pink ................................. CENTAURIUM
2 Flowers blue, green tinged with purple, or white ............... 3

3 Petals 4, spurred at base; flowers green, tinged with purple ...... ........................................................ HALENIA
3 Petals 4, with fringed lobes; or petals 5 and not spurred; blue, purple or white ................................................ 4

4 Petals 4, fringed; flowers on stalks longer than the flowers; seeds covered with small bumps .................... GENTIANOPSIS

4 Petals 5, not fringed; flower stalks short or absent; seeds smooth . 5
5 Flowers 2.5–4 cm long, on short stalks; seeds flattened and winged ........................................................ GENTIANA
5 Flowers 1–2 cm long, stalkless; seeds round ..... GENTIANELLA

## BARTONIA *Screwstem*

Slender annual or biennial herbs. Stems pale green to yellow or purple. Leaves reduced to small opposite or alternate scales. Flowers small, 4-parted, green-white to green-yellow, bell-shaped, in slender panicles or racemes at ends of stems.

*Key to species*

1 Mid-stem leaves alternate; anthers 0.3–0.5 mm long . ***B. paniculata***
1 Mid-stem leaves opposite or subopposite; anthers 0.5–0.9 (–1.1) mm long ................................................ ***B. virginica***

***Bartonia paniculata*** (Michx.) Muhl. TWINING SCREWSTEM Aug–Sept. Tamarack swamps, fens, sphagnum bogs, open wetlands; more common in se USA.

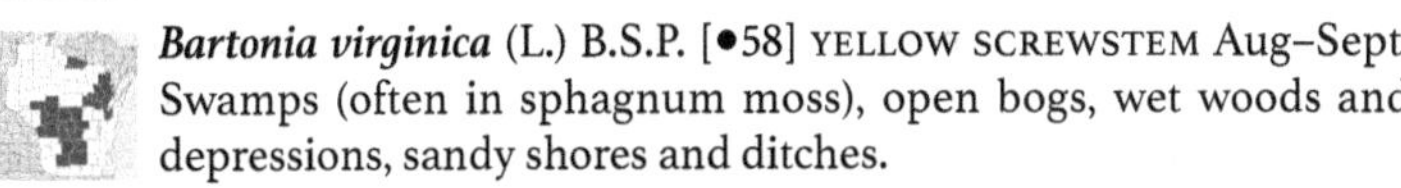

***Bartonia virginica*** (L.) B.S.P. [•58] YELLOW SCREWSTEM Aug–Sept. Swamps (often in sphagnum moss), open bogs, wet woods and depressions, sandy shores and ditches.

## CENTAURIUM *Centaury*

***Centaurium pulchellum*** (Sw.) Druce BRANCHED CENTAURY Annual herb. Stems much branched, often from the base. Leaves sessile, lance-shaped or ovate lance-shaped. Inflorescence a many-flowered terminal cyme. Flowers 4-merous; corolla lobes pink. June–Sept. Native of Europe; local in fields and waste places, often where salted in winter.

## GENTIANA *Gentian*

Perennial herbs, with thick, fibrous roots. Leaves opposite or whorled, simple, margins entire, petioles absent. Flowers large, blue, green-white or yellow, 5-parted, in clusters near ends of stems; petals forming a tubelike, shallowly lobed flower, the lobes alternating with a folded membrane as long or longer than petal lobes. Fruit a 2-chambered capsule.

ADDITIONAL SPECIES ***Gentiana saponaria*** L., native, known from Adams County.

*Key to species*

1 Flowers blue (rarely white), remaining closed, the corolla lobes absent or reduced to small points .................. ***G. andrewsii***

1 Flowers blue, white or yellowish, opening, the corolla lobes prominent .......................................................2

2 Corolla lobes spreading; flowers deep blue .......*G. puberulenta*

2 Corolla lobes erect; flowers vary from white or yellow to blue or purple; stems glabrous.........................................3

3 Flowers yellowish or white, with greenish veins; leaves 5-veined; mostly s Wisc .............................................*G. alba*

3 Flowers blue to purple; leaves 1–5-veined; mostly n Wisc ........ ..............................................*G. rubricaulis*

***Gentiana alba*** Muhl. [•58] YELLOW GENTIAN Aug–Sept. Clay soils in wooded ravines, open woodlands and woodland edges, bluffs, wet sandy prairies, and along railroads and in roadside ditches.

***Gentiana andrewsii*** Griseb. [•58] BOTTLE-GENTIAN Aug–Sept. Wet meadows, swamps and wet woods, thickets, low prairie, shores, ditches.

***Gentiana puberulenta*** J. Pringle DOWNY GENTIAN Aug–Oct. Dry upland woods and prairies.

***Gentiana rubricaulis*** Schwein. GREAT LAKES GENTIAN Wet meadows, peatlands, streambanks, thickets, conifer swamps, Lake Superior rocky shores; soils usually calcium-rich. *Gentiana linearis* var. *lanceolata* A. Gray.

## GENTIANELLA *Gentian*

***Gentianella quinquefolia*** (L.) Small [•58] STIFF GENTIAN Annual or biennial herb. Stems 4-angled. Leaves opposite, without petioles; lower leaves spatula-shaped, upper leaves lance-ovate; margins entire. Flowers 4–5-parted, blue (rarely white), in clusters of 1–7 flowers at ends of stems or from upper leaf axils; petals withering and persistent around capsule. Aug–Sept. Wet meadows, streambanks, moist woods; often where calcium-rich.

## GENTIANOPSIS *Fringed-Gentian*

Smooth, taprooted, annual or biennial herbs. Leaves opposite, stalkless. Flowers 1 to several, showy, blue, sometimes tinged with white on outside, long-stalked at ends of stems and branches, 4-parted; petals deeply lobed, forming a tubular or bell-shaped flower, the lobes ragged or fringed at tips, without a folded membrane between the lobes (present in *Gentiana*). Fruit a capsule.

*Key to species*

1 Upper leaves lance-shaped to ovate; petal lobes long-fringed across tip and sides, the fringes 2–5 mm long ................*G. crinita*

1 Upper leaves linear; tips of petal lobes ragged with short, fine teeth, and often fringed on sides ..........................*G. virgata*

***Gentianopsis crinita*** (Froel.) Ma GREATER FRINGED-GENTIAN Aug–Oct. Wet meadows, streambanks, ditches, wet woods; soils usually calcium-rich and sandy or gravelly.

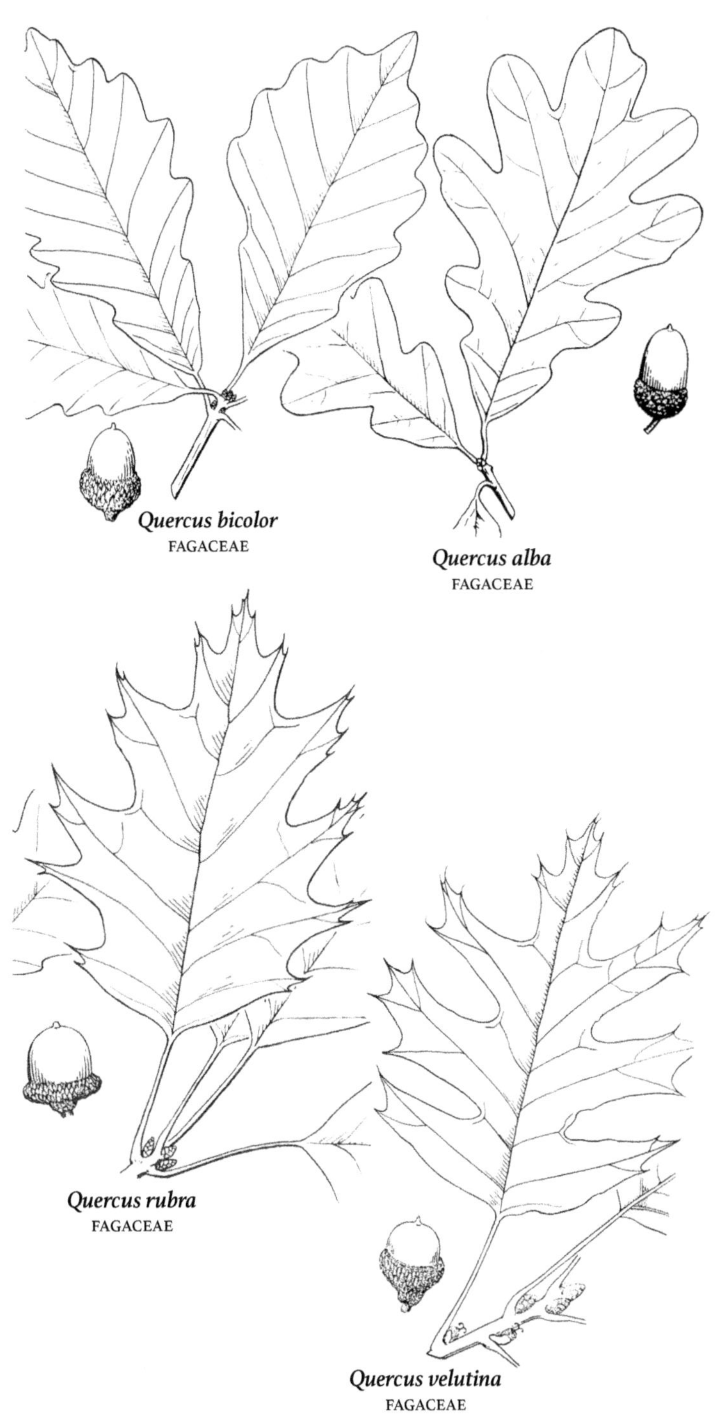

*Quercus bicolor*
FAGACEAE

*Quercus alba*
FAGACEAE

*Quercus rubra*
FAGACEAE

*Quercus velutina*
FAGACEAE

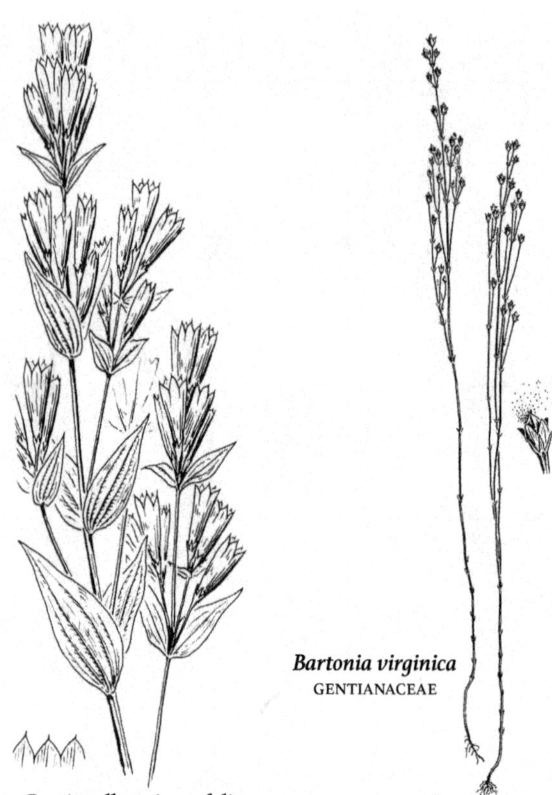

*Bartonia virginica*
GENTIANACEAE

*Gentianella quinquefolia*
GENTIANACEAE

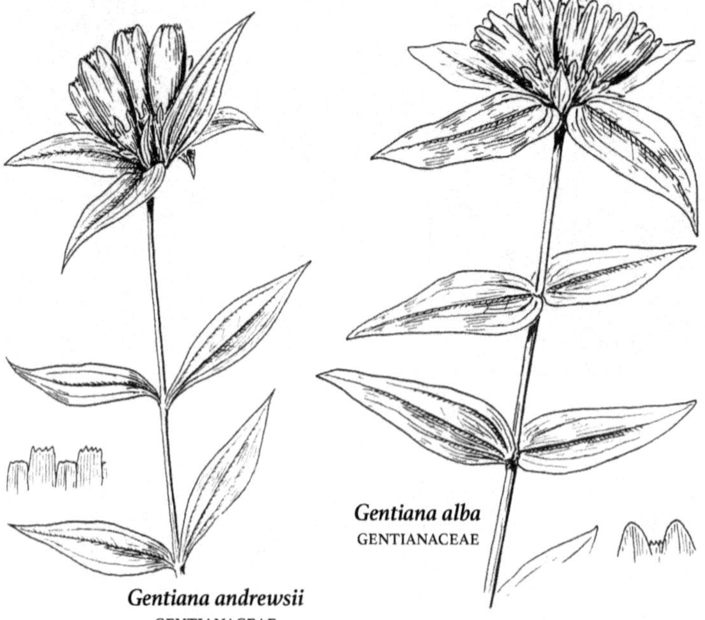

*Gentiana alba*
GENTIANACEAE

*Gentiana andrewsii*
GENTIANACEAE

***Gentianopsis virgata*** (Raf.) Holub LESSER FRINGED-GENTIAN Sept–Oct. Sandy and gravelly shores, wet meadows, fens, intradunal wetlands near Lake Michigan; soils usually calcium-rich. Similar to *G. crinita* but smaller. *Gentianopsis procera* (Holm) Ma.

### HALENIA *Spurred Gentian*

***Halenia deflexa*** (Sm.) Griseb. SPURRED GENTIAN Annual herb. Stems rounded 4-angled. Leaves opposite, lower leaves spatula-shaped; stem leaves lance-shaped to ovate; margins entire. Flowers green, tinged with purple, 4-parted, in loose clusters of 5–9 flowers at ends of stems; petals usually with downward-pointing spurs at base. July–Aug. Cedar swamps, moist conifer woods (especially along shores), old logging roads.

## GERANIACEAE *Geranium Family*

Annual or perennial herbs. Leaves usually opposite, simple or compound, palmately toothed, lobed, or divided. Flowers 5-merous, regular or somewhat zygomorphic, all perfect or part of them sterile. Petals 5, pink to purple. Fruit a carpel prolonged at maturity into beaks.

*Key to genera*

1 Leaves simple, palmately lobed, or 3-parted; anthers usually 10 .... GERANIUM
1 Leaves pinnately compound; anthers 5 .............. ERODIUM

### ERODIUM *Stork's Bill; Filaree*

***Erodium cicutarium*** (L.) L'Hér. REDSTEM-FILAREE Winter-annual or biennial herb. April–Sept. Native of the Mediterranean region; weedy, especially in fallow fields.

### GERANIUM *Wild Geranium; Crane's-Bill*

Annual or perennial herbs. Leaves palmately lobed, cleft, or divided, the stem leaves chiefly opposite. Flowers usually pink to purple, usually in pairs at the ends of axillary peduncles. Axis of the ovary prolonged at maturity into a long beak.

ADDITIONAL SPECIES ***Geranium molle*** L., introduced in Sheboygan County. ***Geranium pratense*** L.; introduced, Douglas County.

*Key to species*

1 Perennial, spreading by rhizomes; petals more than 11 mm long; leaves few, large .................. ***G. maculatum***
1 Annuals or short-lived perennials; petals less than 11 mm long; stem leaves several to many .................. 2

2 Sepals rounded or acute, not awn-tipped ............ ***G. pusillum***
2 Sepals narrowed to long, awn-like tips .................. 3

3 Leaves divided to their base .................. ***G. robertianum***
3 Leaves deeply divided but not to their base .................. 4

4 Flower pedicels covered with glandular, spreading hairs .................. ***G. bicknellii***

| | | |
|---|---|---|
| 4 | Pedicel hairs without glands | 5 |
| 5 | Pedicels less than twice length of calyx | *G. carolinianum* |
| 5 | Pedicels mostly more than twice length of calyx | 6 |
| 6 | Pedicel hairs spreading; beak of style short, 1 mm long | *G. sibiricum* |
| 6 | Pedicel hairs appressed; beak of style 3–5 mm long | *G. columbinum* |

***Geranium bicknellii*** Britt. NORTHERN CRANE'S-BILL May–Sept. Open woods and fields, usually where sandy or gravelly.

***Geranium carolinianum*** L. CAROLINA CRANE'S-BILL May–Aug. Dry, barren, rocky or sandy soil and disturbed places.

***Geranium columbinum*** L. LONG-STALK CRANE'S-BILL May–Aug. Native of Europe; Racine County.

***Geranium maculatum*** L. [•59] SPOTTED CRANE'S-BILL April–June. Dry or moist woods.

***Geranium pusillum*** L. SMALL-FLOWER CRANE'S-BILL Summer. Native of Europe; established as a weed in fields and waste land.

***Geranium robertianum*** L. HERB-ROBERT May–Sept. Damp rich woods.

***Geranium sibiricum*** L. SIBERIAN CRANE'S-BILL July–Sept. Introduced; native of Asia.

---

## GROSSULARIACEAE *Currant Family*

### RIBES *Currant; Gooseberry*

Shrubs. Stems smooth, or with spines at nodes and sometimes also with bristles between nodes. Leaves alternate, palmately veined and palmately 3–5-lobed, margins toothed. Flowers 1 to several in short clusters, or few to many in racemes; green to white or yellow, perfect, regular. Fruit a many-seeded berry, usually topped by persistent, dry flower parts. *Ribes* are of two types: currants and gooseberries. Currants lack spines and bristles (except in *R. lacustre*) and the stalk of berry is jointed at its tip so that berries detach from stalks. Gooseberries have spines and bristles and the berry stalk is not jointed so that stalks remain attached to berries when picked.

*Key to species*

| | | |
|---|---|---|
| 1 | Stems with spines or bristles, at least at the nodes; flowers single or in corymb-like clusters of 2–3 (gooseberries) | 2 |
| 1 | Flowers in racemes of 5 or more; stems without spines or bristles (except in *R. lacustre;* currants) | 5 |
| 2 | Ovary and fruit usually bristly; calyx lobes shorter than hypanthium | *R. cynosbati* |
| 2 | Ovaries and fruit smooth (or bristly in the rare *R. oxyacanthoides*); calyx lobes longer than corolla tube | 3 |

3 Flowers whitish; stamens exserted, about 2x longer than calyx lobes; spines at nodes stout ..................... *R. missouriense*

3 Flowers greenish or purplish; stamens not exserted; spines at nodes absent, or if present, weak and slender ........................ 4

4 Leaves with glands, at least on underside veins; fruit bristly or with gland-tipped hairs to sometimes smooth; rare in n Wisc ......... ........................................... *R. oxyacanthoides*

4 Leaves without glands; fruit smooth ................ *R. hirtellum*

5 Ovary and fruit bristly with gland-tipped hairs................ 6

5 Ovary and fruit neither bristly nor with gland-tipped hairs ..... 7

6 Stems densely bristly ............................... *R. lacustre*

6 Stems unarmed ........................... *R. glandulosum*

7 Leaf underside dotted with shiny resinous glands; fruit black ... 8

7 Leaf underside without resinous glands ..................... 10

8 Flowers yellow to greenish; calyx glabrous or sparsely hairy; inflorescence bracts longer than pedicels ........ *R. americanum*

8 Flowers white to greenish-white; calyx hairy; inflorescence bracts much shorter than pedicels ................................ 9

9 Flowers and fruit in upright racemes; native species ............ ........................................... *R. hudsonianum*

9 Flowers and fruit in drooping racemes; occasional garden escape ....................................................... *R. nigrum*

10 Flowers golden-yellow; fruit black .................. *R. odoratum*

10 Flowers yellow-green; fruit red.............................. 11

11 Pedicels with scattered hairs and short-stalked glands .... *R. triste*

11 Pedicels glabrous .................................. *R. rubrum*

***Ribes americanum*** P. Mill. WILD BLACK CURRANT April–June. Moist to wet forests, swamps, marsh and lake borders, streambanks.

***Ribes cynosbati*** L. EASTERN PRICKLY GOOSEBERRY May–June. Occasional in wet woods, swamps, thickets and streambanks; more typical in moist hardwood forests (where our most common gooseberry).

***Ribes glandulosum*** Grauer SKUNK CURRANT June. Cedar and tamarack swamps, cool wet woods, thickets and streambanks.

***Ribes hirtellum*** Michx. [•59] HAIRY-STEM GOOSEBERRY June. Cedar and tamarack swamps, thickets, shores, rocky openings.

***Ribes hudsonianum*** Richards. HUDSON BAY CURRANT June. Cedar swamps, wet conifer woods and streambanks.

***Ribes lacustre*** (Pers.) Poir. BRISTLY BLACK GOOSEBERRY May–June. Moist conifer woods, swamps, thickets, and rock outcrops.

***Ribes missouriense*** Nutt. [•59] MISSOURI GOOSEBERRY May. Moist or dry upland woods.

***Ribes nigrum*** L. GARDEN BLACK CURRANT Native of Eurasia; occasionally planted and rarely escaped.

***Ribes odoratum*** H. Wendl. BUFFALO-CURRANT Introduced. April–June. Cliffs and rocky hillsides; widely cultivated and sometimes escaped. *Ribes aureum* Pursh var. *villosum* DC.

***Ribes oxyacanthoides*** L. NORTHERN GOOSEBERRY *Threatened.* June. Rocky and sandy shores, rocky openings, cold moist woods.

***Ribes rubrum*** L. GARDEN RED CURRANT Native of the Old World; long in cultivation and occasionally escaped. *Ribes sativum* Syme.

***Ribes triste*** Pallas SWAMP RED CURRANT May–June. Wet woods swamps, alder thickets, seeps.

## HALORAGACEAE *Water-Milfoil Family*

Perennial aquatic herbs. Leaves alternate or whorled, finely dissected. Flowers small, stalkless in axils of leaves or bracts, 3- or 4-parted, regular, perfect, or imperfect, petals small or absent. Fruit small and nutlike, dividing into 3 or 4 segments (mericarps).

*Key to genera*

1 Flowers 4-parted; leaves mostly whorled, emersed leaves reduced to small bracts ............................ **MYRIOPHYLLUM**

1 Flowers 3-parted; leaves alternate, emersed leaves not bract-like .. .......................................... **PROSERPINACA**

### MYRIOPHYLLUM *Water-Milfoil*

Perennial aquatic herbs. Stems submerged, sparsely branched, freely rooting at lower nodes. Leaves mostly whorled (alternate in *M. farwellii*), pinnately divided into threadlike segments, upper leaves often reduced to bracts. Flowers small, stalkless in axils of upper emersed leaves (the floral bracts) or axils of underwater leaves; staminate flowers above pistillate flowers. Fruit nutlike, 4-lobed.

*Key to species*

1 Leaves simple, reduced to small, blunt-tipped scales; stems erect and crowded from creeping rhizomes ............... ***M. tenellum***

1 Leaves dissected into narrow segments ........................ 2

2 Leaves alternate, opposite, or scattered on stem ................ 3

2 Foliage leaves all whorled (or appearing so in *M. heterophyllum*). 4

3 Mature mericarps (fruit sections) with distinct bumps on back; uncommon ......................................... ***M. farwellii***

3 Mericarps rounded on back, smooth or nearly so; reported from Chippewa County .................................. ***M. humile***

4 Flowers and bracts below flowers alternate on stem ............ ............................................ ***M. alterniflorum***

4 Flowers and bracts below flowers whorled..................... 5

5 Bracts surrounding staminate flowers deeply cleft *M. verticillatum*
5 Bracts surrounding staminate flowers sharply toothed or entire . 6

6 Bracts sharply toothed and much longer than flowers .......... ........................................... *M. heterophyllum*
6 Bracts surrounding staminate flowers entire and not longer than flowers ............................................... 7

7 Leaf segments mostly 5–12 on each side of midrib; small bulbs (turions) produced at ends of stems and in upper leaf axils ...... ............................................... *M. sibiricum*
7 Leaf segments many, 12–20 on each side of midrib; turions absent ............................................... *M. spicatum*

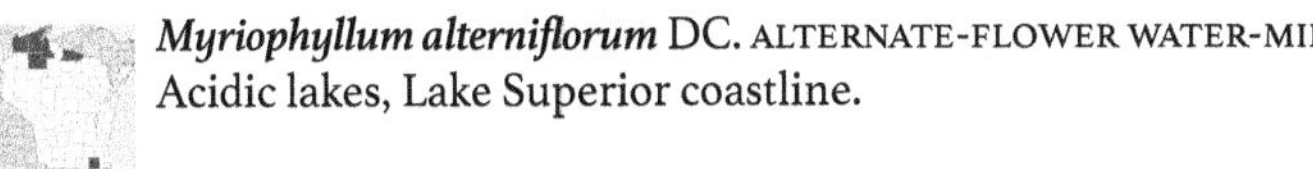

***Myriophyllum alterniflorum*** DC. ALTERNATE-FLOWER WATER-MILFOIL Acidic lakes, Lake Superior coastline.

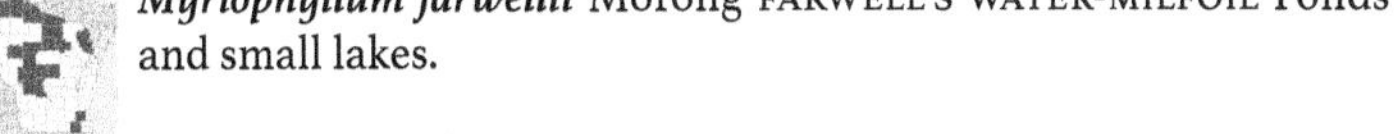

***Myriophyllum farwellii*** Morong FARWELL'S WATER-MILFOIL Ponds and small lakes.

***Myriophyllum heterophyllum*** Michx. [•60] TWO-LEAF WATER-MILFOIL June–Aug. Lakes, ponds and pools in streams; sometimes where calcium-rich.

***Myriophyllum humile*** (Raf.) Morong LOW WATER-MILFOIL Reported from Chippewa County; more common in ne USA.

***Myriophyllum sibiricum*** Komarov COMMON WATER-MILFOIL June–Sept. Shallow to deep water of lakes, ponds, marshes, ditches and slow-moving streams; sometimes where calcium-rich. When flowering, the numerous red spikes of this species are conspicuous on water surface. *M. spicatum,* introduced from Eurasia, is similar but has more finely divided leaves (12–24 threadlike segments on each side of midrib) and larger floral bracts. *Myriophyllum exalbescens* Fern.

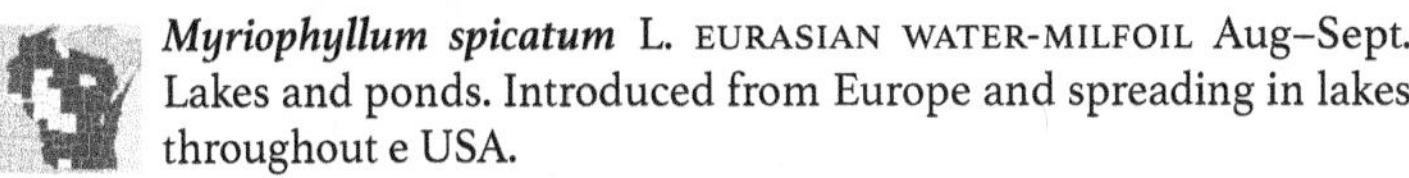

***Myriophyllum spicatum*** L. EURASIAN WATER-MILFOIL Aug–Sept. Lakes and ponds. Introduced from Europe and spreading in lakes throughout e USA.

***Myriophyllum tenellum*** Bigelow [•60] SLENDER WATER-MILFOIL Acidic lakes; often forming large colonies, especially in deep water.

***Myriophyllum verticillatum*** L. WHORLED WATER-MILFOIL July–Sept. Lakes, ponds, quiet rivers.

## PROSERPINACA *Mermaid-Weed*

***Proserpinaca palustris*** L. [•60] COMMON MERMAID-WEED Perennial aquatic herb, often forming large colonies. Stems horizontal at base and often rooting; the flower-bearing branches erect. Leaves alternate; underwater leaves deeply divided into linear segments; emersed leaves narrowly lance-shaped, margins with sharp, forward-pointing teeth. Flowers small, perfect, green or purple-tinged, 1–3 in axils of emersed leaves. Fruit nutlike, 3-angled. June–Aug. Shallow water of ponds, streambanks and ditches, muddy shores, sedge meadows; usually where seasonally flooded.

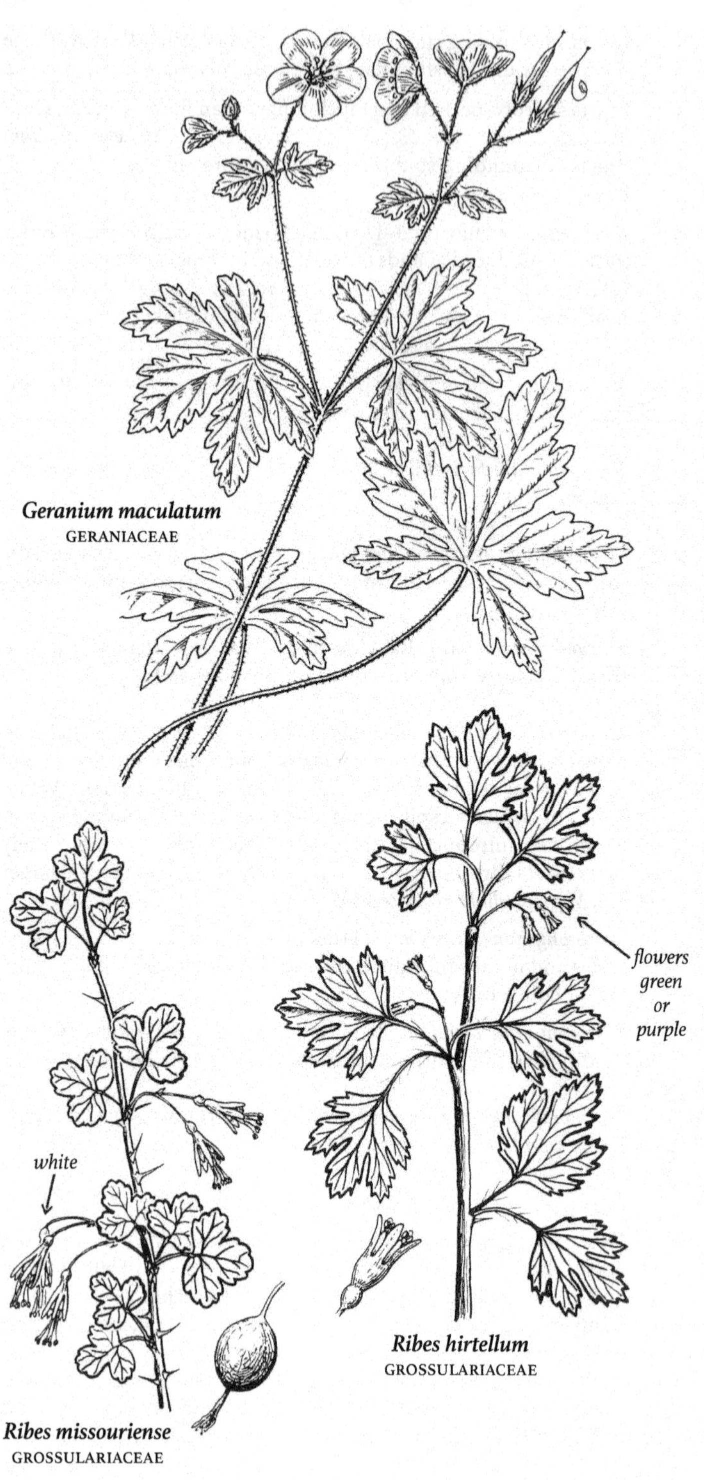
Geranium maculatum
GERANIACEAE
flowers green or purple
white
Ribes hirtellum
GROSSULARIACEAE
Ribes missouriense
GROSSULARIACEAE

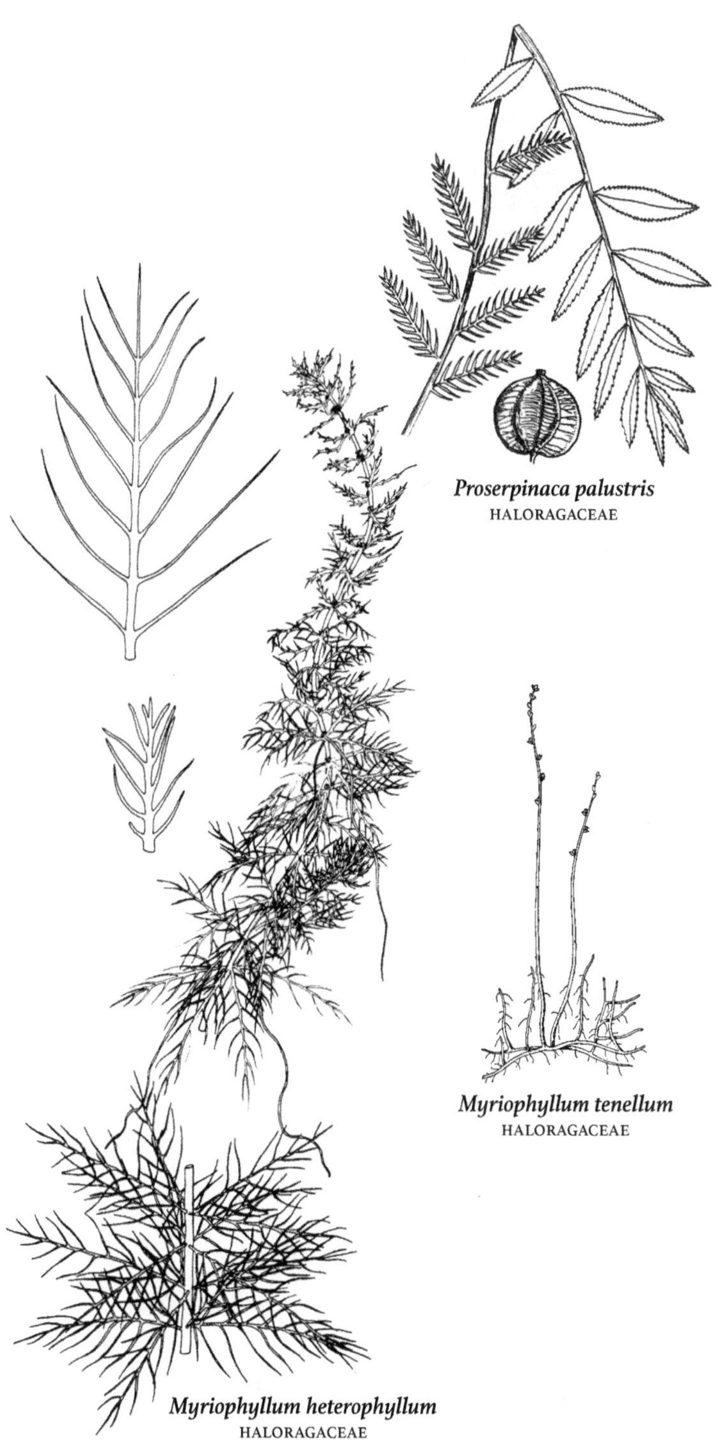

*Proserpinaca palustris*
HALORAGACEAE

*Myriophyllum tenellum*
HALORAGACEAE

*Myriophyllum heterophyllum*
HALORAGACEAE

## HAMAMELIDACEAE *Witch-Hazel Family*

### HAMAMELIS *Witch-Hazel*

***Hamamelis virginiana*** L. AMERICAN WITCH-HAZEL Tall shrub, with scurfy or glabrous twigs. Leaves broadly obovate, with several to many rounded teeth, glabrous or stellate-pubescent beneath. Flowers in short-pediceled axillary clusters; 4-merous; petals bright yellow or suffused with red. Seeds black, eventually discharged explosively from the capsule. Oct–Nov; fruit ripe a year later. Moist woods. Witch hazel extract is derived from the bark.

## HYDRANGEACEAE *Hydrangea Family*

### PHILADELPHUS *Mock-Orange*

***Philadelphus coronarius*** L. Sweet MOCK-ORANGE Shrub. Leaves simple, opposite, short-petioled, ovate or ovate-oblong. Flowers in terminal racemes of 5–7; perfect, regular, 4-merous or rarely 5-merous, white, fragrant. Fruit an obovoid capsule. June. Native of se Europe; long a favorite ornamental shrub, sometimes escaped, usually near gardens; reported from Walworth County.

## HYPERICACEAE *St. John's-Wort Family*

Glabrous annual or perennial herbs (shrubby in *Hypericum kalmianum* and *H. prolificum*). Leaves simple, opposite, dotted with dark or translucent glands (visible when held to light), especially on underside; margins entire; petioles absent. Flowers few to many in clusters at ends of stems or from upper leaf axils, perfect, regular, petals 5, yellow or pink to green or purple. Fruit a 3-chambered, many-seeded capsule.

*Key to genera*

1 Petals yellow; stamens 15–many .................. **HYPERICUM**
1 Petals pink or purple; stamens 9 ................. **TRIADENUM**

### HYPERICUM *St. John's-Wort*

Shrubs or herbs. Leaves opposite, sometimes dotted with black and/or small transparent glands. Flowers in clusters at ends of stems and upper leaf axils; petals yellow.

*Key to species*

1 Styles joined at base, persisting on capsule as a straight beak; stamens many, distinct.......................................... 2
1 Styles free to base, the capsules not beaked; stamens few to many, joined at base into 3 or 5 bundles ............................. 5

2 Small shrubs to 1 m tall ........................................ 3
2 Perennial herbs, slightly woody at base........................ 4

3 Leaves sessile; flowers in mostly terminal cymes; styles 5 .........

. . . . . . . . . . . . . . . . . . . . . . . . . . . . . . . . . . . . . . . . . . . . . . . *H. kalmianum*

3 Leaves with short petioles; flowers in terminal and axillary cymes; styles 3 . . . . . . . . . . . . . . . . . . . . . . . . . . . . . . . . . . . . . . . . . . . . *H. prolificum*

4 Plants 30–60 cm tall; leaves linear; rare in moist prairies in sc Wisc . . . . . . . . . . . . . . . . . . . . . . . . . . . . . . . . . . . . . . . . . . . . *H. sphaerocarpum*

4 Plants smaller, mostly less than 30 cm tall; leaves elliptic to ovate; shores and streambanks in n and c Wisc . . . . . . . . . . . . *H. ellipticum*

5 Plants 1–2 m tall; leaves 5 cm long or more; flowers 4 cm or more wide; styles 5 . . . . . . . . . . . . . . . . . . . . . . . . . . . . . . . . . *H. pyramidatum*

5 Plants usually less than 1 m tall; leaves less than 5 cm long; flowers to 3 cm wide; styles 3 . . . . . . . . . . . . . . . . . . . . . . . . . . . . . . . . . . . . 6

6 Petals spotted with black dots; stamens in 3 weak groups. . . . . . . . 7

6 Petals not spotted with black dots; stamens in 5 weak groups . . . 8

7 Flowers 15 mm or more wide, petals black-dotted only on margins; capsules oblong cone-shaped; common introduced weed . . . . . . . . . . . . . . . . . . . . . . . . . . . . . . . . . . . . . . . . . . . . . . . . . . . . *H. perforatum*

7 Flowers 6–10 mm wide, petals and sepals with black dots and lines; capsules nearly round to ovate; native species . . . . . . *H. punctatum*

8 Leaves reduced to tiny scales to 3 mm long; dry, sandy habitats in Driftless Area . . . . . . . . . . . . . . . . . . . . . . . . . . . . . . . . *H. gentianoides*

8 Leaves larger, linear to elliptic-ovate . . . . . . . . . . . . . . . . . . . . . . . . . 9

9 Sepals broadest near or above middle; capsule rounded at tip . . 10

9 Sepals lance-shaped, broadest below middle; capsule tapered to tip . . . . . . . . . . . . . . . . . . . . . . . . . . . . . . . . . . . . . . . . . . . . . . . . . . . . . . . . . 11

10 Bracts leafy and oval, uppermost 0.5–2 mm wide; sepals much shorter than fruit . . . . . . . . . . . . . . . . . . . . . . . . . . . . . . . . . . . *H. boreale*

10 Bracts narrow and awl-shaped, uppermost to 0.2 mm wide; sepals same length as fruit . . . . . . . . . . . . . . . . . . . . . . . . . . . . . . . . *H. mutilum*

11 Leaves 1-nerved (sometimes 3-nerved), tapered to base; sepals 2–4 mm long . . . . . . . . . . . . . . . . . . . . . . . . . . . . . . . . . . . . . . . . *H. canadense*

11 Leaves 5–7-nerved, rounded at base and broadest below middle; sepals 5–6 mm long . . . . . . . . . . . . . . . . . . . . . . . . . . . . . . . . . . *H. majus*

***Hypericum boreale*** (Britt.) Bickn. NORTHERN ST. JOHN'S-WORT July–Sept. Pond and marsh margins, low areas between dunes, open bogs.

***Hypericum canadense*** L. LESSER CANADIAN ST. JOHN'S-WORT July–Sept. Sandy shores, wetland margins, ditches.

***Hypericum ellipticum*** Hook. [•61] PALE ST. JOHN'S-WORT July–Aug. Streambanks, sandy shores and flats, thickets, bogs.

***Hypericum gentianoides*** (L.) B.S.P. ORANGE-GRASS June–Sept. Sterile, especially sandy soil.

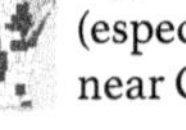

***Hypericum kalmianum*** L. KALM'S ST. JOHN'S-WORT June–Sept. Dunes (especially wet areas between dunes) and rocky lakeshores, mostly near Great Lakes, often on limestone or where calcium-rich.

***Hypericum majus*** (Gray) Britt. GREATER CANADIAN ST. JOHN'S-WORT July–Sept. Streambanks, sandy, mucky or calcareous shores, low areas between dunes, marshes, wetland margins.

***Hypericum mutilum*** L. DWARF ST. JOHN'S-WORT July–Sept. Streambanks, wet meadows, marshes, ditches; usually where sandy.

***Hypericum perforatum*** L. COMMON ST. JOHN'S-WORT June–Sept. Native of Europe; common as a weed in fields, meadows, and roadsides.

***Hypericum prolificum*** L. SHRUBBY ST. JOHN'S-WORT July–Sept. In a variety of habitats from margins of swamps to rocky woods or cliffs.

***Hypericum punctatum*** Lam. [•61] SPOTTED ST. JOHN'S-WORT June–Aug. Moist or dry soil, fields and open woods.

***Hypericum pyramidatum*** Ait. GREAT ST. JOHN'S-WORT July–Aug. Streambanks, ditches, fen and marsh margins.

***Hypericum sphaerocarpum*** Michx. ROUND-SEED ST. JOHN'S-WORT *Threatened.* June–Aug. Wet prairies, sedge meadows, and moist sites often where periodically disturbed.

### TRIADENUM *Marsh St. John's-Wort*

Glabrous perennial herbs. Leaves opposite, entire, oval-shaped, ours dotted with small dark and transparent glands. Flowers pink to green-purple, in clusters at ends of stems and from leaf axils. Fruit a cylindric capsule.

1 Sepals 3–4 mm long, oval and rounded at tip; styles mostly less than 1 mm long ........................................ ***T. fraseri***

1 Sepals 5–8 mm long, lance-shaped and tapered to a tip; styles 2–3 mm long ........................................ ***T. virginicum***

***Triadenum fraseri*** (Spach) Gleason Fraser's MARSH-ST. JOHN'S-WORT July–Aug. Marshes, sedge meadows, open bogs, fens, sandy and calcium-rich shores. *Hypericum virginicum* L. var. *fraseri* (Spach) Fern.

***Triadenum virginicum*** (L.) Raf. [•61] VIRGINIA MARSH-ST. JOHN'S-WORT July–Aug. Sphagnum bogs, wet meadows, shores; disjunct from Atlantic coast. *Hypericum virginicum* L.

## JUGLANDACEAE *Walnut Family*

Trees. Leaves alternate, odd-pinnate. Flowers monoecious. Staminate flowers in elongate catkins. Pistillate flowers terminating the young branches. Fruit large, consisting of a fleshy or woody exocarp enclosing a nut.

*Key to genera*

1 Leaflets mostly 5–9, the terminal leaflet largest .......... **CARYA**

1 Leaflets 11–23, the lateral leaflets largest .............. **JUGLANS**

### CARYA *Hickory*

Trees with hard, heavy wood. All species more or less stellate-pubescent, at least when young, and leaves, buds, and fruit also copiously covered with resin when young. Leaves odd-pinnate, the 3 terminal leaflets the largest. Flowers appear in spring as the leaves open. Staminate catkins elongate, borne in peduncled groups of 3 at the summit of the previous year's growth or the base of that of the current year. Pistillate flowers solitary or in spikes of 2–10, terminating the branches.

*Key to species*

1 Leafletss 7–9; bud scales sulfur-yellow; bark smooth or with shallow ridges .......................................... *C. cordiformis*

1 Leaflets 5; bud scales not yellow; bark shaggy ........... *C. ovata*

***Carya cordiformis*** (Wangenh.) K. Koch [•62] BITTERNUT HICKORY Dry or moist forests.

***Carya ovata*** (P. Mill.) K. Koch SHAGBARK HICKORY Rich moist soil. Variable in the size and shape of the nuts and in the pubescence of the leaves.

### JUGLANS *Walnut*

Trees. Leaves glandular-pubescent, odd-pinnate, the median lateral leaflets the largest. Staminate catkins protruding from the buds in autumn, elongating in spring, pendulous. Pistillate flowers in short spikes terminating the branches. Fruit clammy-glandular.

*Key to species*

1 Leaflets 11–17, terminal leaflet usually present; pith of twigs brown; fruit sticky-downy .................................... *J. cinerea*

1 Leaflets 13–23, terminal leaflet often absent; pith of twigs cream colored; fruit globose, not sticky-downy .................. *J. nigra*

***Juglans cinerea*** L. [•62] BUTTERNUT, WHITE WALNUT Rich moist soil.

***Juglans nigra*** L. BLACK WALNUT Rich moist soil. Its handsome durable wood is highly prized.

## LAMIACEAE *Mint Family*

Perennial, often aromatic, herbs. Stems usually 4-angled. Leaves simple, opposite, sharply toothed or deeply lobed. Flowers in leaf axils or in heads or spikes at ends of stems, perfect; petals white, pink, blue or purple, often 2-lipped.

ADDITIONAL SPECIES ***Collinsonia canadensis*** L. (Northern horse-balm), perennial herb of rich woods; lower lobe of the corolla fringed; historical records from s Wisc, now presumed extirpated from state. ***Satureja hortensis*** L. (Summer savory), corolla pale pink-purple to white, 5–7 mm long; native of Mediterranean region and sw Asia, and long cultivated as a culinary herb, reported from s Wisc. ***Thymus pulegioides*** L. (Lemon thyme), introduced and known from several Wisc locations.

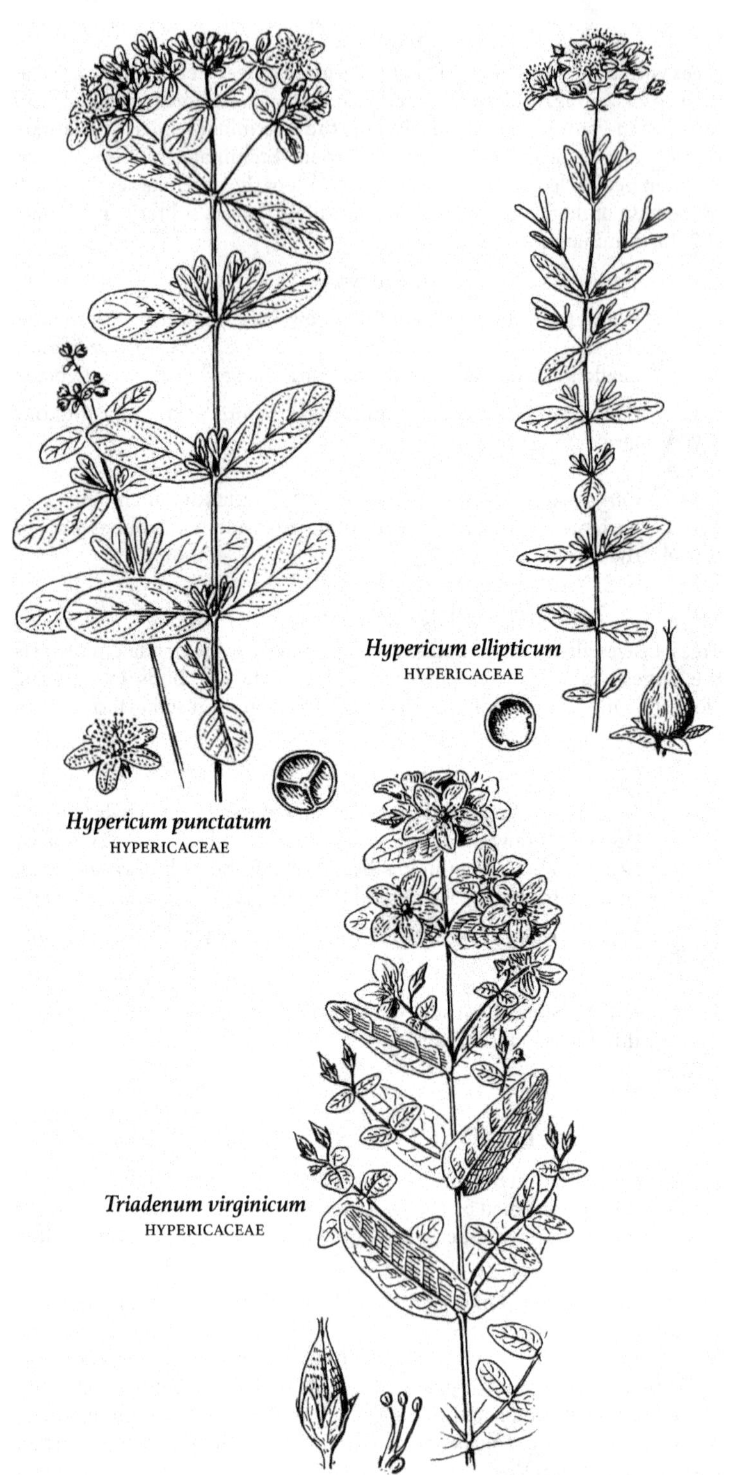
Hypericum ellipticum
HYPERICACEAE
Hypericum punctatum
HYPERICACEAE
Triadenum virginicum
HYPERICACEAE

*Carya cordiformis*
JUGLANDACEAE
*Juglans cinerea*
JUGLANDACEAE

*Key to genera*

1 Calyx with a distinct cap or protuberance on the upper side of the tube .......... **SCUTELLARIA**
1 Calyx without a cap or protuberance on the tube.......... 2

2 Upper lip of the corolla very short, or its lobes adjacent to the margins of the lower lip, the corolla thus appearing to be 1-lipped .......... 3
2 Upper lip of the corolla well developed, entire or 2-lobed, or the corolla regular or nearly so.......... 4

3 Lower lip 5-lobed, the 2 lobes nearest its base representing the upper lip .......... **TEUCRIUM**
3 Lower lip 3-lobed, or appearing 4-lobed if the center lip is notched .......... **AJUGA**

4 Stamens included and hidden within the corolla tube .......... 5
4 Stamens exserted beyond the throat of the corolla.......... 6

5 Calyx lobes 10, awl-shaped and hooked at the tip **MARRUBIUM**
5 Calyx lobes 5, broader and not hooked at tip ...... **GLECHOMA**

6 Stamens 2 .......... **GROUP A**
6 Stamens 4 .......... 7

7 Inflorescence appearing axillary, the verticils (whorls of flowers around the stem) several to many, subtended by normal leaves and separated from one another by normal interodes, or the uppermost subtending leaves smaller and internodes shorter (not including plants with axillary spikes or racemes) .......... **GROUP B**
7 Inflorescence appearing terminal, the verticils 1 to many, all or mostly subtended by bract-like leaves different from the main leaves, or separated by much shorter internodes (and including plants with lateral or axillary spikes) .......... 8

8 Flowers single in the axils of each bract-like leaf, the verticils with 1 or 2 flowers .......... **GROUP C**
8 Flowers 2–many in the axil of each bract-like leaf, the verticils with 4 or more flowers .......... **GROUP D**

*Group A*

1 Calyx distinctly 2-lipped .......... 2
1 Calyx regular or nearly so, the lobes alike in size and shape ..... 6

2 Flowers in loose, few-flowered verticils in the axils of foliage leaves, blue, 3–4 mm long .......... **HEDEOMA**
2 Flowers in terminal inflorescences .......... 3

3 Flowers in terminal panicles, yellow, 12–15 mm long; lower lip fringed; historically known from s Wisc ... ***Collinsonia canadensis****
3 Flowers in terminal racemes or spikes, or terminal heads ....... 4

4 Flowers single or paired at each node of the slender raceme ..... .......... **SALVIA**
4 Flowers 3 or more at each flower-bearing node (but not all blooming at the same time) .......... 5

5 Verticils usually numerous, the flowers at each node usually less than 12, not tightly clustered .......................... **SALVIA**
5 Verticils 1–5, dense and head-like, of numerous crowded flowers . ............................................. **BLEPHILIA**

6 Flowers in loose panicles or panicle-like cymes, each flower distinctly pediceled; historically known from s Wisc ............ .................................... *Collinsonia canadensis**
6 Flowers in dense, head-like clusters or verticils, the pedicels absent or very short ............................................ 7

7 Corolla very irregular, 15–50 mm long .............. **MONARDA**
7 Corolla regular or nearly so, to 5 mm long ............ **LYCOPUS**

*Group B*

1 Calyx regular or nearly so, the lobes of the upper and lower lips similar in shape and size .................................... 2
1 Calyx distinctly 2-lipped, the lobes of the upper and lower lips of different size and shape .................................. 8

2 Corolla about equally 4- or 5-lobed ......................... 3
2 Corolla strongly 2-lipped, the upper lip concave and arched over the stamens.............................................. 4

3 Flowers 1–3 in each axil, and 2–6 in each verticil **TRICHOSTEMA**
3 Flowers many in each axil ......................... **MENTHA**

4 Flowers distinctly pediceled, forming loosely flowered cymules (the clusters making up a cyme)................................. 5
4 Flowers sessile in the cymules .............................. 6

5 Stems creeping; flowers usually 3 in each axil ...... **GLECHOMA**
5 Stems erect; flowers usually 4–8 in each axil .......... **BALLOTA**

6 Calyx lobes tapered to a slender tip but not spiny ..... **LAMIUM**
6 Calyx lobes prolonged into short but stiff spines ............... 7

7 Lower corolla lip with 2 yellow or white protuberances at its base .. ............................................ **GALEOPSIS**
7 Lower corolla lip without protuberances ........... **LEONURUS**

8 Stamens projecting beyond the corolla ............... **MENTHA**
8 Stamens ascending under the upper lip of the corolla but not longer than the lip ............................... **CLINOPODIUM**

*Group C*

1 Main leaves linear to narrowly oblong, sessile or nearly so; flowers usually pink-purple ......................... **PHYSOSTEGIA**
1 Main leaves broadly ovate to oblong-ovate, with long petioles ... 2
2 Lower corolla lip yellow, fringed on the margin; racemes all terminal; historically known from s Wisc *Collinsonia canadensis**
2 Lower corolla lip blue to white, not fringed; racemes both terminal and lateral ........................................ **PERILLA**

*Group D*

1 Stamens ascending under the upper corolla lip but not longer than the lip ................................................. 2
1 At least some of the stamens protruding from the corolla ....... 8

2 Calyx distinctly 2-lipped and irregular ........................ 3
2 Calyx regular or nearly so, the lobes all alike or differing in size only ........................................................ 5

3 One calyx lobe (the upper center lobe) longer and wider than the other 4 .................................. DRACOCEPHALUM
3 Three calyx lobes (which form the upper lip) differing from the other 2 lobes ............................................. 4

4 Bracts broadly rounded, abruptly tapered at the tip to a short sharp point ........................................... PRUNELLA
4 Bracts awl-shaped, coarsely hairy ............. CLINOPODIUM

5 Leaves linear, entire, sessile ................................. 6
5 Leaves wider than linear, or the margins toothed, or petioled ... 7

6 Stems finely pubescent ....................... CLINOPODIUM
6 Stems glabrous, or with small hairs on the angles only . STACHYS

7 Calyx 15-nerved; lower verticils often with distinct peduncles ... ................................................... NEPETA
7 Calyx 5-10-nerved; lower verticils sessile............... STACHYS

8 Inflorescence a dense or loose raceme in which the component verticils are plainly visible; flowers on distinctly short pedicels .. 9
8 Inflorescence otherwise ..................................... 11

9 Main foliage leaves entire ......................... HYSSOPUS
9 Main foliage leaves toothed .................................. 10

10 Flowers blue to lavender, or sometimes white ........ MENTHA
10 Flowers yellow; historically known from s Wisc ................ ........................................ *Collinsonia canadensis**

11 Inflorescence a group of terminal heads or crowded cymes, often with secondary heads or cymes in some of the upper axils, never a spike or raceme ......................... PYCNANTHEMUM
11 Inflorescence a dense spike, or with one or 2 lower verticils sometimes separate from the others; flowers sessile or nearly so 12

12 Spike with flowers grouped all around the axis of the stem ...... ............................................. AGASTACHE
12 Spike with flowers grouped mainly on 1 side of the axis of the stem ............................................. ELSHOLTZIA

## AGASTACHE *Giant-Hyssop*

Perennial herbs. Flowers numerous in dense verticils, forming terminal, continuous or interrupted spikes.

*Key to species*

1 Flowers blue; leaf underside densely covered with white felt-like hairs ........................................... *A. foeniculum*
1 Flowers rose, purple or yellow; leaf underside smooth to hairy, the pubescence if present not dense and felt-like................... 2

2 Flowers rose or purple; stems often red-tinged *A. scrophulariifolia*
2 Flowers yellow; stems green ...................... *A. nepetoides*

***Agastache foeniculum*** (Pursh) Kuntze BLUE GIANT-HYSSOP July–Aug. Dry upland woods and prairies.

***Agastache nepetoides*** (L.) Kuntze YELLOW GIANT-HYSSOP July–Oct. Open woods and woodland edges.

***Agastache scrophulariifolia*** (Willd.) Kuntze [•63] PURPLE GIANT-HYSSOP Aug–Sept. Upland woods. Very similar to *A. nepetoides* in habit, size, and foliage.

## AJUGA *Bugle*

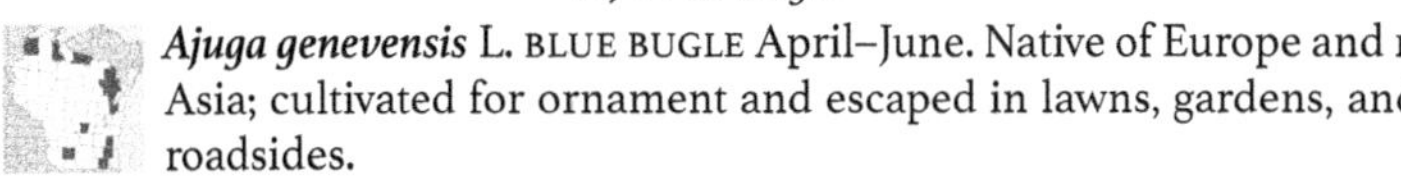

***Ajuga genevensis*** L. BLUE BUGLE April–June. Native of Europe and n Asia; cultivated for ornament and escaped in lawns, gardens, and roadsides.

ADDITIONAL SPECIES ***Ajuga reptans*** L., introduced in several Wisc locations.

## BALLOTA *Black Horehound*

***Ballota nigra*** L. BLACK HOREHOUND Perennial herb. Principal leaves short-petioled, ovate, coarsely crenate. Flowers in loose, short-peduncled cymes from the axils of the upper leaves; corolla strongly 2-lipped, pink or purple June–Sept. Native of the Mediterranean region and w Asia; introduced in waste places. June–Sept.

## BLEPHILIA *Pogoda-Plant*

Perennial aromatic herbs. Flowers pale purple, crowded in dense glomerules in axils of upper leaves.

*Key to species*

1 Leaves ovate, sharply toothed, with petioles 1–2 cm long; bracts green; stems unbranched ............................ ***B. hirsuta***

1 Leaves lance-shaped to ovate, entire or very finely toothed, the upper leaves sessile or nearly so; bracts often purplish; stems usually branched .................................... ***B. ciliata***

***Blephilia ciliata*** (L.) Benth. DOWNY PAGODA-PLANT May–July. Moist or dry open woods, meadows, barrens, borders of fens, thin soil over limestone.

***Blephilia hirsuta*** (Pursh) Benth. HAIRY PAGODA-PLANT May–Aug. Rich forests, swamps, floodplains, usually in moist shaded places.

## CLINOPODIUM *Wild Basil*

Annual or perennial herbs. Calyx tubular to campanulate, conspicuously 10–13-nerved, often hairy in the throat. Corolla tube widened toward the summit.

*Key to species*

1 Stem leaves broadly elliptic-ovate, mostly 1 cm or more wide; flowers subtended by narrow, awl-shaped bracts, their margins fringed with long hairs .............................. ***C. vulgare***

1 Stem leaves linear to lance-shaped, usually less than 1 cm wide; flowers not with narrow bracts................................ **2**

2 Plants glabrous .................................. *C. arkansanum*
2 Plants pubescent .................................. *C. acinos*

***Clinopodium acinos*** (L.) Kuntze BASIL-THYME June–Sept. Native of Europe; weedy on roadsides and waste places. *Satureja acinos* (L.) Scheele.

***Clinopodium arkansanum*** (Nutt.) House LIMESTONE WILD BASIL May–Aug. Sandy beaches, calcareous soil. *Satureja arkansana* (Nutt.) Briq.

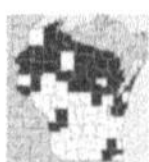

***Clinopodium vulgare*** L. [•63] WILD BASIL Dry or moist upland woods. *Satureja vulgaris* (L.) Fritsch

## DRACOCEPHALUM *Dragonhead*

Erect perennial herbs. Leaves serrate. Flowers small, blue, in loose or dense verticils. Corolla weakly bilabiate, the tube elongate, gradually widened upwards.

*Key to species*

1 Inflorescence compact and densely flowered; corolla barely longer than calyx .................................. *D. parviflorum*
1 Inflorescence elongate and open; corolla much longer than calyx .................................. *D. moldavica**

***Dracocephalum parviflorum*** Nutt. AMERICAN DRAGONHEAD Native of w USA, considered adventive in Wisc where occurs in disturbed places.

ADDITIONAL SPECIES ***Dracocephalum moldavica*** L., introduced, adventive in several Wisc locations.

## ELSHOLTZIA *Late-Summer-Mint*

***Elsholtzia ciliata*** (Thunb.) Hyl. CRESTED LATE-SUMMER-MINT July–Sept. Native of Asia; reported from Sauk County.

## GALEOPSIS *Hemp-Nettle*

Annual herbs, mostly pubescent. Flowers small, crowded in dense verticils in the axils of the upper foliage leaves. Calyx tube broadly tubular to campanulate, with 10 conspicuous ribs and usually 10 intermediate ones. Corolla strongly 2-lipped.

*Key to species*

1 Stems stiffly hairy, the nodes swollen; leaves ovate, the margins coarsely toothed .................................. *G. tetrahit*
1 Stems with fine recurved hairs, nodes not swollen; leaves linear to lance-shaped, the margins shallowly toothed to entire .................................. *G. angustifolia*

***Galeopsis angustifolia*** Ehrh. NARROW-LEAF HEMP-NETTLE June–Sept. Native of Eurasia; introduced in waste places, Calumet and Door counties.

***Galeopsis tetrahit*** L. [•63] BRITTLE-STEM HEMP-NETTLE June–Sept. Native of Eurasia; introduced as a weed of gardens, roadsides, waste

places, and forests. Variable in density of pubescence, presence of glandular hairs, shape of leaf, and size of calyx and corolla.

## GLECHOMA *Ground-Ivy*

***Glechoma hederacea*** L. GROUND-IVY Perennial herb. Stems creeping. Leaves rotund to kidney-shaped, conspicuously crenate, long-petioled. Flowers blue, usually 3 in each axil. April–June. Native of Eurasia; widely naturalized in yards, roadsides, cemeteries, and moist woods.

## HEDEOMA *False Pennyroyal*

Small, strongly scented, annual herbs (ours). Flowers blue, in axillary few-flowered verticils. Corolla tubular, weakly 2-lipped.

*Key to species*

1 Leaves lance-shaped, the main leaves with petioles and toothed margins .......................................... ***H. pulegioides***

1 Leaves linear, sessile, entire ........................... ***H. hispida***

***Hedeoma hispida*** Pursh ROUGH FALSE PENNYROYAL May–Aug. Dry soil, sand dunes and barrens.

***Hedeoma pulegioides*** (L.) Pers. AMERICAN FALSE PENNYROYAL July–Sept. Moist or dry woods.

## HYSSOPUS *Hyssop*

***Hyssopus officinalis*** L. HYSSOP Perennial from a stout woody rhizome. Leaves nearly sessile, often with smaller ones in their axils, lance-shaped, entire. Flowers 3–7 in the upper axils, the clusters sessile or nearly so, forming a terminal, spike-like, continuous or interrupted inflorescence; corolla blue, about 1 cm long, strongly 2-lipped. July–Oct. Native of Eurasia; long cultivated for its reputed medicinal properties and occasionally escaped along roadsides and in waste places.

## LAMIUM *Dead Nettle*

Annual or perennial herbs, commonly spreading or decumbent. Leaves broad, crenate. Flowers white to red or purple, in verticils of 6–12, subtended by scarcely reduced leaves, forming a short, crowded or somewhat interrupted, terminal spike.

ADDITIONAL SPECIES ***Lamium purpureum*** L. (Red henbit), Annual introduced herb, known from fields and waste places in s Wisc, resembling *L. amplexicaule* in habit and size.

*Key to species*

1 Upper leaves sessile and clasping stem, lower leaves on long petioles ................................................ ***L. amplexicaule***

1 All leaves with petioles ......................... ***L. maculatum***

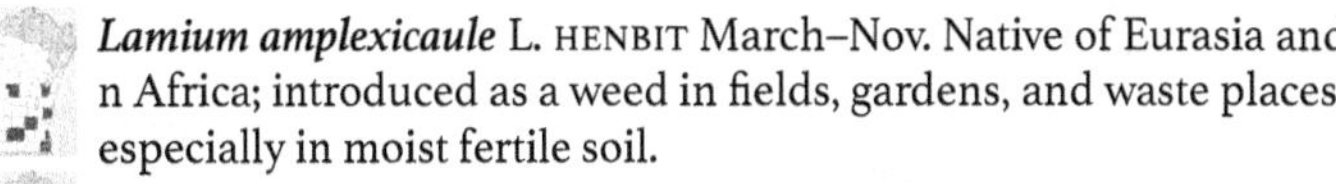

***Lamium amplexicaule*** L. HENBIT March–Nov. Native of Eurasia and n Africa; introduced as a weed in fields, gardens, and waste places, especially in moist fertile soil.

***Lamium maculatum*** L. SPOTTED DEAD NETTLE April–Sept. Native of Eurasia, escaped from cultivation on roadsides and waste places.

## LEONURUS *Motherwort*

Biennial or perennial, strongly scented herbs. Leaves dentate to laciniate. Flowers white to pink, crowded in dense verticils subtended by bracteal leaves and by linear bracts, forming long, interrupted, terminal spikes. Corolla strongly bilabiate.

*Key to species*

1 Main leaves palmately divided and coarsely toothed; calyx glabrous or nearly so, with 2 calyx teeth usually strongly deflexed ........ *L. cardiaca*

1 Main leaves entire to coarsely toothed; calyx pubescent, none of the teeth strongly deflexed ........................ ***L. marrubiastrum***

***Leonurus cardiaca*** L. MOTHERWORT June–Aug. Native of c Asia; formerly cultivated as a home remedy and now established in waste places, roadsides, and gardens.

***Leonurus marrubiastrum*** L. LION'S-TAIL June–Sept. Native of Europe and n Asia; introduced in waste places. *Chaiturus marrubiastrum* (L.) Reichenb.

## LYCOPUS *Water-Horehound*

Perennial, unscented herbs. Stems erect, 4-angled. Leaves opposite, coarsely toothed or deeply lobed, smaller on upper stems; petioles short or absent. Flowers small, in clusters in middle and upper leaf axils, often appearing whorled; white to pink, the sepals and petals often dotted on outer surface.

HYBRIDS ***Lycopus × sherardii*** Steele, hybrid between *L. uniflorus* and *L. virginicus,* found across much of Wisc.

*Key to species*

1 Sepal lobes broad, triangular to ovate, to 1 mm long, shorter than to about as long as nutlets, the midvein not prominent ......... 2

1 Sepal lobes slender, 1–3 mm long, longer than nutlets, the midvein prominent ................................................ 3

2 Leaves mostly less than 3 cm wide; stamens and styles visible, longer than petals; outer rim of nutlets taller than the inner rim ........ ***L. uniflorus***

2 Larger leaves 3 cm or more wide; stamens and styles hidden by petals; inner and outer rim of nutlets same height, the 4 nutlets appearing flat-topped across tops .................. ***L. virginicus***

3 Main leaves sessile ..................................... ***L. asper***

3 Leaves with petioles ........................................... 4

4 Upper surface of leaves with appressed hairs ........ ***L. europaeus***

4 Upper surface of leaves more or less smooth ....... ***L. americanus***

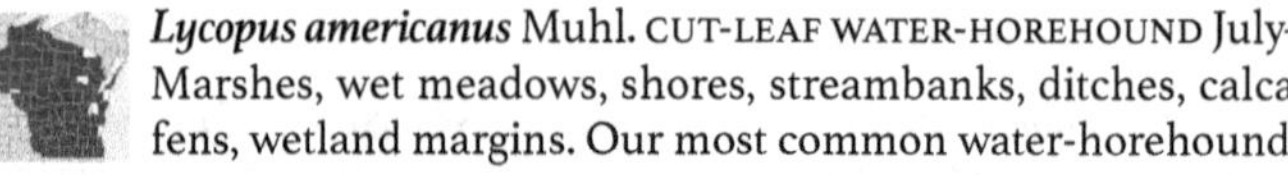

***Lycopus americanus*** Muhl. CUT-LEAF WATER-HOREHOUND July–Sept. Marshes, wet meadows, shores, streambanks, ditches, calcareous fens, wetland margins. Our most common water-horehound.

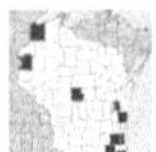

***Lycopus asper*** Greene ROUGH WATER-HOREHOUND Introduced. July–Sept. Shores and ditches, especially where disturbed, often with *L. americanus.*

***Lycopus europaeus*** L. EUROPEAN WATER-HOREHOUND Introduced. July–Aug. Moist to wet areas, often where disturbed. Similar to *L. americanus,* but often with slender stolons as well as rhizomes.

***Lycopus uniflorus*** Michx. NORTHERN WATER-HOREHOUND Aug–Sept. Swamps, streambanks, thickets, wet meadows, open bogs, calcareous fens, ditches; often with *L. americanus.* Hybrids common with *L. virginicus* where ranges overlap producing a hybrid swarm known as *L.* × *sherardii* Steele.

***Lycopus virginicus*** L. VIRGINIA WATER-HOREHOUND July–Sept. Floodplain forests.

## MARRUBIUM *Horehound*

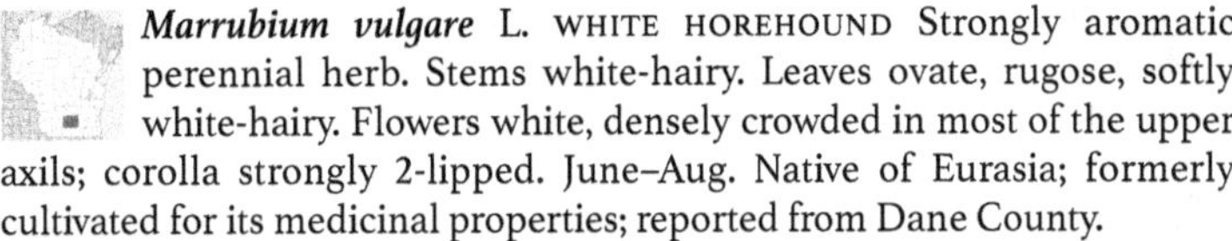

***Marrubium vulgare*** L. WHITE HOREHOUND Strongly aromatic perennial herb. Stems white-hairy. Leaves ovate, rugose, softly white-hairy. Flowers white, densely crowded in most of the upper axils; corolla strongly 2-lipped. June–Aug. Native of Eurasia; formerly cultivated for its medicinal properties; reported from Dane County.

## MENTHA *Mint*

Perennial herbs, spreading by rhizomes or stolons, with serrate leaves, and small, blue to lavender flowers borne in the axils of the leaves or in terminal spikes or heads. All species bloom in summer.

ADDITIONAL SPECIES ***Mentha aquatica*** L. (Water mint), introduced lemon-scented perennial, reported from e Wisc. ***Mentha* × *gracilis*** Sole (pro sp.), hybrid between *M. arvensis* and *M. spicata,* found across much of Wisc.

*Key to species*

1 Flowers in axillary whorls separated by internodes of normal length .......... ***M. arvensis***

1 Flowers in terminal spikes or heads, the internodes short ....... 2

2 Main leaves with petioles; peppermint-scented ..... ***M.* × *piperita***

2 Main leaves sessile or nearly so; spearmint-scented .... ***M. spicata***

***Mentha arvensis*** L. AMERICAN WILD MINT July–Sept. Wet meadows, marshes, swamps, thickets, streambanks, ditches, springs and other wet places.

***Mentha* × *piperita*** L. PEPPERMINT European origin; cultivated as an herb and commercially for its oil; escaped in wet soil.

***Mentha spicata*** L. SPEARMINT Native of Europe; commonly cultivated as a flavoring herb and escaped throughout southern Wisc.

## MONARDA *Beebalm*

Erect perennial herbs (ours). Leaves lance-shaped to ovate. Flowers conspicuous, densely aggregated into head-like clusters terminating the branches or also borne in the upper axils. Corolla strongly bilabiate.

*Key to species*

1 Flowers yellowish, dotted with purple; stamens and style not exserted .......... ***M. punctata***

1 Flowers lavender, white or scarlet; stamens and style strongly exserted beyond corolla; heads 2 or more and forming an interrupted spike .......... 2

2 Corolla lavender (rarely white) .......... *M. fistulosa*

2 Corolla bright scarlet .......... *M. didyma*

***Monarda didyma*** L. SCARLET BEEBALM, OSWEGO TEA Introduced. July–Sept. Moist woods and thickets. Often cultivated for ornament.

***Monarda fistulosa*** L. [•64] WILD BERGAMOT June–Sept. Upland woods, thickets, and prairies.

***Monarda punctata*** L. HORSE-MINT June–Sept. Sandy fields, sand dunes, open oak and pine woods, roadsides and disturbed areas.

## NEPETA *Catnip*

***Nepeta cataria*** L. CATNIP Perennial herb; stems, undersides of leaves, and inflorescences covered with grayish hairs. Leaves deltoid, coarsely crenate-dentate. Flower clusters continuous or interrupted, rather loosely many-flowered; corolla dull white, the lower lobe dotted with pink or purple. July–Oct. Native of se Europe and sw Asia, formerly cultivated for reputed medicinal properties, and now established in waste places, fencerows, and roadsides.

ADDITIONAL SPECIES ***Nepeta racemosa*** Lam. (Racemed catnip), introduced in s Wisc.

## PERILLA *Perilla-Mint*

***Perilla frutescens*** (L.) Britt. PERILLA-MINT Annual herbs, often purple or suffused with purple. Leaves ovate, coarsely serrate or incised. Flowers purple or white, borne singly in the axils of small bracteal leaves, forming a loose, elongate, spike-like raceme long, terminal and from the upper axils. Aug–Sept. Native of India; cultivated for its ornamental foliage; reported from Manitowoc County.

## PHYSOSTEGIA *False Dragonhead*

***Physostegia virginiana*** (L.) Benth. OBEDIENCE Perennial herb, spreading by rhizomes. Stems 4-angled. Leaves opposite, oval to oblong lance-shaped; margins with sharp teeth; sessile. Flowers in several racemes; petals pink-purple or white with purple spots. July–Sept. Sedge meadows, low prairie, shores, swamps, floodplain forests, thickets and ditches. Sometimes cultivated for its attractive flowers.

## PRUNELLA *Self-Heal*

***Prunella vulgaris*** L. SELF-HEAL Perennial herb. Stems 4-angled. Leaves opposite, lance-shaped to oval or ovate; margins entire or with a few small teeth. Flowers in dense spikes with obvious bracts; petals blue-violet (rarely pink or white). Subsp. *vulgaris,* introduced from Europe and found in mostly disturbed places, has broad leaves half as wide as long. The native subsp. *lanceolata* has narrower leaves, 1/3 as wide as long. June–Oct. Common in many types of wetlands (especially where disturbed):

swamps, wet forest depressions, wet trails, streambanks; also in drier forests, fields and lawns.

### PYCNANTHEMUM *Mountain-Mint*

Erect herbs, perennial from rhizomes. Leaves linear to ovate, entire to serrate. Flowers small, in crowded or head-like cymes terminating the stem and its branches, or also sessile or peduncled in the axils of the upper leaves. Corolla 2-lipped, purple to white, the lower lip commonly spotted with purple. Our species all bloom in summer.

*Key to species*

1 Stems and leaves glabrous ........................ ***P. tenuifolium***

1 Stems pubescent on the angles, leaves rough-to-touch .......... .................................................. ***P. virginianum***

***Pycnanthemum tenuifolium*** Schrad. NARROW-LEAF MOUNTAIN-MINT Introduced. Chiefly in dry soil of upland woods and prairies.

***Pycnanthemum virginianum*** (L.) T. Dur. & B.D. Jackson VIRGINIA MOUNTAIN-MINT July–Sept. Wet meadows, marshes, tamarack swamps, calcareous fens, low prairie.

### SALVIA *Sage*

Annual, biennial, or perennial herbs of diverse foliage and habit, the flowers in verticils subtended by usually much reduced bracteal leaves, forming a terminal, continuous or interrupted, loose or dense, spike-like raceme. Corolla 2-lipped, the lower lip 3-lobed. Several species cultivated for ornament.

ADDITIONAL SPECIES ***Salvia azurea*** Michx. ex Lam. (Blue sage), native of s USA, adventive in southern Wisc. ***Salvia splendens*** Sellow ex Roemer & J.A. Schultes (Scarlet sage), native of Brazil, introduced in southern Wisc; commonly cultivated for its showy scarlet corollas about 4 cm long.

*Key to species*

1 Main leaves linear to lance-shaped, to 12 mm wide, entire or nearly so .................................................. ***S. reflexa***

1 Main leaves ovate to ovate-oblong, 2 cm or more wide, margins with rounded teeth.................................................. 2

2 Leaves mostly at base of plant, long-petioled; corolla 15 mm or more long .................................................. ***S. pratensis***

2 Leaves along stems, petioles short or absent; corolla to 12 mm long .................................................. ***S. nemorosa***

***Salvia nemorosa*** L. WOODLAND SAGE June–July. Native of Europe and w Asia; adventive in fields and waste places.

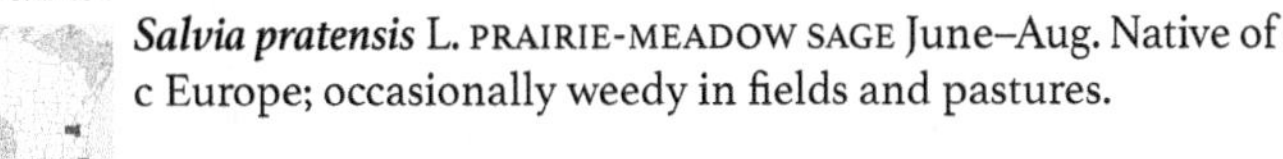

***Salvia pratensis*** L. PRAIRIE-MEADOW SAGE June–Aug. Native of s and c Europe; occasionally weedy in fields and pastures.

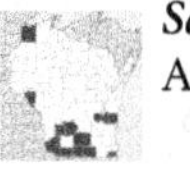

***Salvia reflexa*** Hornem. LANCE-LEAF SAGE Introduced. June–Sept. Adventive in dry sandy or gravelly soil of hillsides and prairies.

## SCUTELLARIA *Skullcap*

Perennial herbs, spreading by rhizomes. Stems 4-angled. Leaves opposite, ovate to lance-shaped, margins toothed. Flowers blue or blue with white markings, single on short stalks in axils of middle and upper leaves, or in racemes from leaf axils; corolla 2-lipped, pubescent on outer surface, upper lip hoodlike, lower lip more or less flat.

*Key to species*

1 Flowers single in leaf axils .................................... 2
1 Flowers in racemes from end of stem or leaf axils.............. 3

2 Corolla 15–20 mm long; leaves 2 or more times longer than wide; common ........................................ *S. galericulata*
2 Corolla to 10 mm long; leaves less than 2 times longer than wide; rare in Wisc ................................... *S. parvula*

3 Flowers in racemes from axils of stem leaves ........ *S. lateriflora*
3 Flowers mainly in terminal racemes, with smaller racemes from axils of upper leaves ........................................ 4

4 Leaves heart-shaped to truncate at base ................ *S. ovata*
4 Middle and upper stem leaves rounded or narrowed at base, with the blade extending along the actual petiole ............ *S. incana*

***Scutellaria galericulata*** L. HOODED SKULLCAP June–Sept. Shores, streambanks, marshes, wet meadows, swamps, thickets, bogs, ditches.

***Scutellaria incana*** Biehler HOARY SKULLCAP June–Aug. Moist or dry upland woods.

***Scutellaria lateriflora*** L. [•64] BLUE SKULLCAP July–Sept. Shores, streambanks, wet meadows, marshes, swamps, shaded wet areas.

***Scutellaria ovata*** Hill FOREST SKULLCAP June–July. Moist or dry woods.

***Scutellaria parvula*** Michx. LITTLE SKULLCAP *Endangered.* May–June. Upland woods, dry prairies, sandstone bluffs.

## STACHYS *Hedge-Nettle*

Perennial herbs, spreading by rhizomes; plants usually hairy. Stems 4-angled. Leaves opposite. Flowers in interrupted spikes at ends of stems, appearing whorled in more or less evenly spaced clusters; corolla 2-lipped, petals pink, often with purple spots or mottles.

*Key to species*

1 Plants glabrous; leaf petioles 8–25 mm long .......... *S. tenuifolia*
1 Plants pubescent, at least on the stem angles; leaves sessile or with petioles to 10 mm long ........................................ 2

2 Stems with stiff hairs on angles only; leaves with petioles to 10 mm long ................................................ *S. pilosa*
2 Stems with hairs on angles and sides; leaves sessile or nearly so . ................................................ *S. palustris*

*Clinopodium vulgare*
LAMIACEAE
*Agastache scrophulariifolia*
LAMIACEAE
*Galeopsis tetrahit*
LAMIACEAE

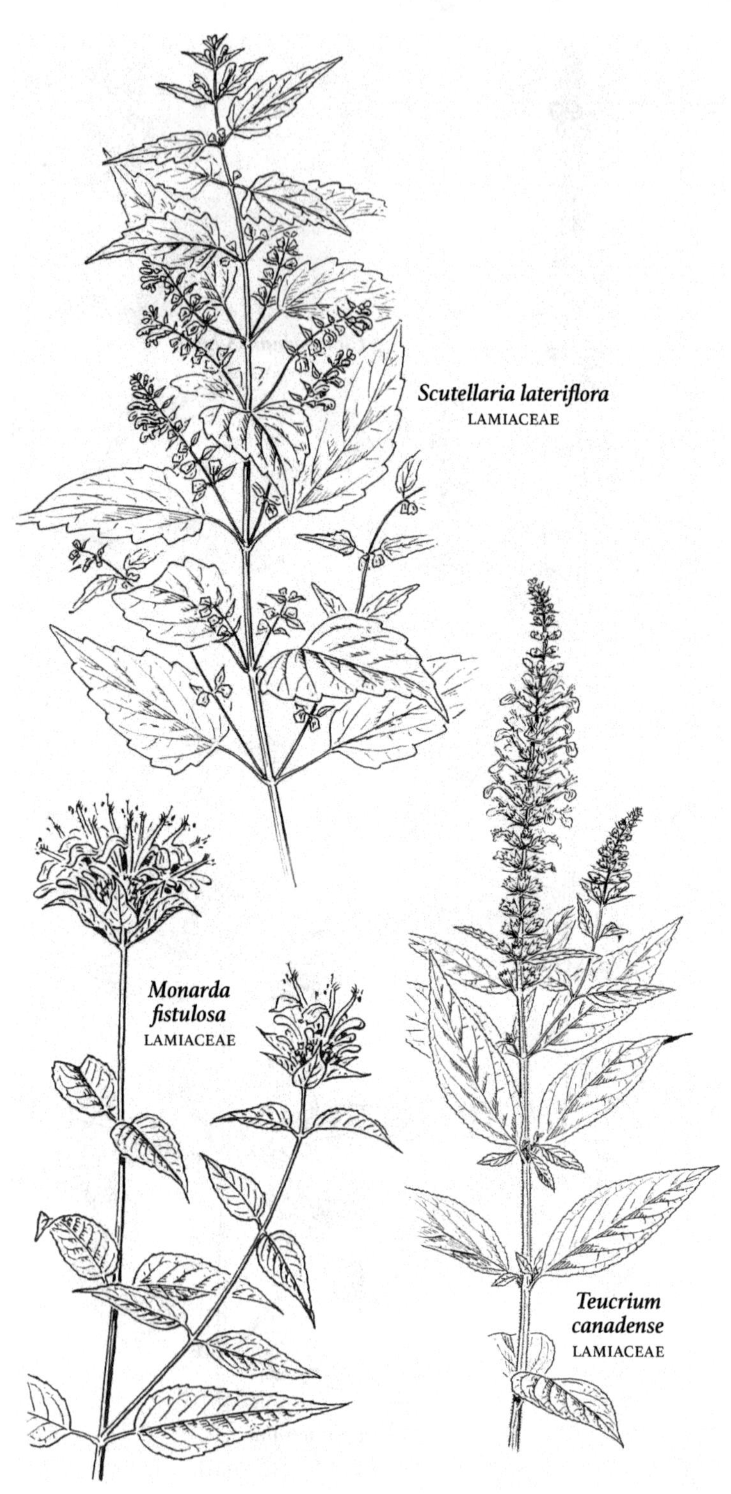
*Scutellaria lateriflora*
LAMIACEAE
*Monarda fistulosa*
LAMIACEAE
*Teucrium canadense*
LAMIACEAE

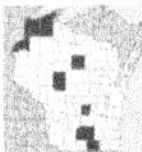

***Stachys palustris*** L. MARSH HEDGE-NETTLE Introduced. June–Aug. Marshes, wet meadows, ditches, thickets, shores, streambanks, openings in swamps.

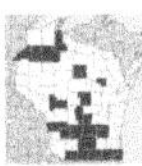

***Stachys pilosa*** Nutt. HEDGE-NETTLE July–Aug. Damp ground, ditch banks, beaches, and wet prairies.

***Stachys tenuifolia*** Willd. SMOOTH HEDGE-NETTLE July–Sept. Floodplain forests, shores, streambanks, thickets, wet meadows.

### TEUCRIUM *Germander*

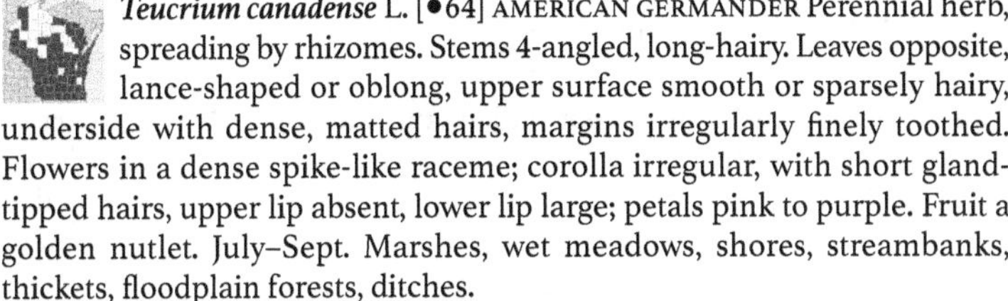

***Teucrium canadense*** L. [•64] AMERICAN GERMANDER Perennial herb, spreading by rhizomes. Stems 4-angled, long-hairy. Leaves opposite, lance-shaped or oblong, upper surface smooth or sparsely hairy, underside with dense, matted hairs, margins irregularly finely toothed. Flowers in a dense spike-like raceme; corolla irregular, with short gland-tipped hairs, upper lip absent, lower lip large; petals pink to purple. Fruit a golden nutlet. July–Sept. Marshes, wet meadows, shores, streambanks, thickets, floodplain forests, ditches.

### TRICHOSTEMA *False Pennyroyal*

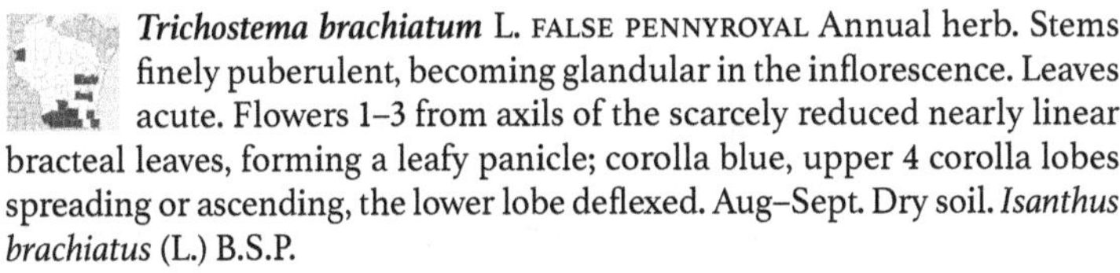

***Trichostema brachiatum*** L. FALSE PENNYROYAL Annual herb. Stems finely puberulent, becoming glandular in the inflorescence. Leaves acute. Flowers 1–3 from axils of the scarcely reduced nearly linear bracteal leaves, forming a leafy panicle; corolla blue, upper 4 corolla lobes spreading or ascending, the lower lobe deflexed. Aug–Sept. Dry soil. *Isanthus brachiatus* (L.) B.S.P.

## LENTIBULARIACEAE *Bladderwort Family*

Insectivorous herbs. Leaves in a basal rosette (*Pinguicula*), or floating, or in peat, muck, or wet soil (*Utricularia*). Flowers perfect, irregular, 2-lipped, sometimes with a spur, 1 to several on an erect stem. Fruit a capsule.

*Key to genera*

1 Leaves ovate or oval, in a basal rosette; flowers single on a bractless stalk .......................................... PINGUICULA

1 Leaves linear or dissected into narrow segments; flowers 1, or several in a raceme, each flower subtended by a bract ... UTRICULARIA

### PINGUICULA *Butterwort*

***Pinguicula vulgaris*** L. COMMON BUTTERWORT *Endangered.* Perennial herb. Leaves 3–6 in a basal rosette, upper surface sticky; margins inrolled. Flowers single atop a leafless stalk (scape); corolla violet-purple, spurred, 2-lipped. June–July. Mostly in rock crevices in cool sandstone cliffs along Lake Superior; usually occurring with Mistassini primrose (*Primula mistassinica*). Small insects are trapped by the sticky leaf surface.

## UTRICULARIA *Bladderwort*

Mostly aquatic, annual or perennial herbs. Leaves underwater, alternate, entire or dissected into many linear segments, some with bladders which trap tiny aquatic invertebrates; or leaves in wet soil and rootlike or absent. Flowers perfect, irregular, 1 to several in a raceme atop stalks raised above water or soil surface, each flower subtended by a small bract; corolla yellow or purple, similar to a snapdragon flower, 2-lipped, the upper lip erect, entire or slightly 2-lobed, lower lip entire or 3-lobed, the corolla tube extended backward into a sac or spur.

*Key to species*

1 Flowers purple or pink. . . . . . . . . . . . . . . . . . . . . . . . . . . . . . . . . . . . . . . . . 2
1 Flowers yellow . . . . . . . . . . . . . . . . . . . . . . . . . . . . . . . . . . . . . . . . . . . . . . 3

2 Flowers 2–5 atop a stout stalk; plants floating in water, masses of leaves present . . . . . . . . . . . . . . . . . . . . . . . . . . . . . . . . . . . . . *U. purpurea*
2 Flowers one atop a slender stalk; plants not free-floating, rooted in muck, appearing leafless . . . . . . . . . . . . . . . . . . . . . . . . . *U. resupinata*

3 Scapes appearing leafless; leaves simple or absent; plants of peat or moist sand or marl . . . . . . . . . . . . . . . . . . . . . . . . . . . . . . . . *U. cornuta*
3 Scapes with leaves at base, the leaves dissected and with bladderlike traps; plants mostly floating in water. . . . . . . . . . . . . . . . . . . . . . . . . . 4

4 Leaf divisions flat in cross-section. . . . . . . . . . . . . . . . . . . . . . . . . . . . . 5
4 Leaf divisions round in cross-section or threadlike . . . . . . . . . . . . 6

5 Bladders borne on leaves; smallest leaf divisions entire (visible with a 10x hand lens); flower with a sac or spur much shorter than lower lip . . . . . . . . . . . . . . . . . . . . . . . . . . . . . . . . . . . . . . . . . . . . . . . . . *U. minor*
5 Bladders on branches separate from leaves; smallest leaf divisions finely toothed, the teeth spine-tipped; flower with a spur as long as lower lip . . . . . . . . . . . . . . . . . . . . . . . . . . . . . . . . . . . . . . . . *U. intermedia*

6 Plants large; leaves floating; scapes 1 mm or more wide; flowers 13 mm or more long, 5 or more per head; larger bladders more than 2 mm wide . . . . . . . . . . . . . . . . . . . . . . . . . . . . . . . . . . . . *U. macrorhiza*
6 Plants smaller; leaves floating or creeping; scapes threadlike; flowers to 12 mm long, 1–3 per head; larger bladders mostly less than 2 mm wide . . . . . . . . . . . . . . . . . . . . . . . . . . . . . . . . . . . . . . . . . . . . . . . . . . 7

7 Plants forming tangled masses, creeping on bottom in shallow water, or on muck or drying pond edges; often with emergent scapes with at least 1 normal flower; cleistogamous flowers absent . . . . *U. gibba*
7 Plants forming a delicate mass of floating leaves; emergent scapes with normal flowers rare; cleistogamous flowers common, on stalks 4–8 mm long . . . . . . . . . . . . . . . . . . . . . . . . . . . . . . . . . *U. geminiscapa*

***Utricularia cornuta*** Michx. [•65] HORNED BLADDERWORT June–Sept. Acid lakes, shores, peatlands, calcareous pools between dunes, borrow pits.

***Utricularia geminiscapa*** Benj. [•65] HIDDEN-FRUIT BLADDERWORT July–Aug. Acid lakes, pools in open bogs.

***Utricularia gibba*** L. [•65] CREEPING BLADDERWORT July–Sept. Exposed shores, lakes, ponds, marshes, fens.

***Utricularia intermedia*** Hayne FLAT-LEAF BLADDERWORT June–Aug. Shallow water (usually alkaline), marly pools between dunes, calcareous fens, marshes, ponds and rivers. bogs and swamps.

***Utricularia macrorhiza*** Le Conte [•65] GREATER BLADDERWORT June–Aug. Shallow water of lakes, ponds, peatlands, marshes and rivers. Our most common bladderwort. *Utricularia vulgaris* L.

***Utricularia minor*** L. LESSER BLADDERWORT June–Aug. Fens, open bogs, sedge meadows and marshes; often in shallow water and where calcium-rich.

***Utricularia purpurea*** Walt. PURPLE BLADDERWORT July–Sept. Acid lakes and ponds in water to 1 m deep, peatlands, marshes.

***Utricularia resupinata*** B.D. Greene LAVENDER BLADDERWORT July–Aug. Shallow to deep water, wet lake and pond shores where sandy or mucky.

## LIMNANTHACEAE *Meadowfoam Family*

### FLOERKEA *False Mermaidweed*

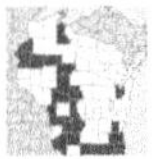

***Floerkea proserpinacoides*** Willd. FALSE MERMAIDWEED Annual herb. Stems weak, diffuse or decumbent. Leaves deeply divided into 3–7 linear, oblong lance-shaped, or narrowly elliptic lobes. Peduncles from the upper axils, at first about equaling the petiole, becoming much longer in fruit; petals white, about 2 mm long. April–May. Moist woods in rich soil. Distinct among our dicots in its completely 3-merous flowers, and distinguished from our monocots by the very deeply pinnately lobed leaves.

## LINACEAE *Flax Family*

### LINUM *Flax*

Annual or perennial herbs. Leaves simple, alternate or opposite, narrow, margins entire, petioles absent. Flowers regular, perfect, 5-parted. Petals yellow or blue. Fruit a 10-chambered capsule.

ADDITIONAL SPECIES ***Linum rigidum*** Pursh (Large-flower yellow flax); adventive in Wisc (Juneau and Pierce counties), native to Great Plains.

*Key to species*

1 Petals blue; pedicels becoming more than 1 cm long ............ 2

1 Petals yellow; pedicels to 1 cm long............................ 3

2 Margins of inner sepals fringed with short hairs, the sepal tips long-tapered ........................................ ***L. usitatissimum***

2 Margins of inner sepals entire, the sepal tips rounded or only short-pointed ........................................ ***L. perenne***

3 Leaves with a pair of rounded glands at their base .... *L. sulcatum*
3 Leaves without glands at base ........................ *L. medium*

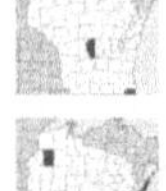
***Linum medium*** (Planch.) Britt. COMMON YELLOW FLAX June–Sept. Sandy, calcium-rich shores, moist places between dunes.

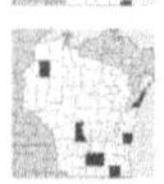
***Linum perenne*** L. WILD BLUE FLAX Introduced. May–July. Prairies and plains.

***Linum sulcatum*** Riddell GROOVED YELLOW FLAX June–July. Dry sandy soil, prairies.

***Linum usitatissimum*** L. CULTIVATED FLAX Summer. Of unknown origin; cultivated since prehistoric times for its fiber (linen) and more recently for its oil (linseed); sometimes escaped or adventive in fields and roadsides.

---

## LINDERNIACEAE *Lindernia Family*

### LINDERNIA *False Pimpernel*

***Lindernia dubia*** (L.) Pennell YELLOW-SEED FALSE PIMPERNEL Annual herb. Leaves opposite, ovate to obovate; margins entire or with small, widely spaced teeth; petioles absent. Flowers single, on slender stalks from leaf axils; corolla pale blue-purple, 2-lipped, the upper lip 2-lobed, the lower lip 3-lobed and wider than upper lip. June–Sept. Mud flats, sandbars, shores of temporary ponds and marshes, streambanks.

---

## LYTHRACEAE *Loosestrife Family*

Annual or perennial herbs, sometimes woody at base (*Decodon*). Leaves simple, opposite, or both opposite and alternate, or whorled, margins entire. Flowers 1 or several in leaf axils or in spike-like heads at ends of stems; perfect, regular or irregular; petals 4 or 6, separate, pink or purple. Fruit a dry, many-seeded capsule.

ADDITIONAL SPECIES ***Ammania robusta*** Heer & Regel, se Wisc in marshy ground and muddy flats, especially in wet disturbed areas with bare soil. Plants of this species and of *Rotala* are often strongly flushed with red; petals in both are quickly deciduous, but are a little larger in *Ammannia* than in *Rotala*.

*Key to genera*

1 Plants arching, woody near base; leaves with petioles and mostly whorled ........................................ DECODON
1 Plants annual or perennial herbs; leaves opposite, or if whorled, leaves without petioles........................................ 2

2 Plants perennial; flowers in spike-like heads at ends of stems; petals and sepals 6 ........................................ LYTHRUM
2 Plants annual; flowers from leaf axils; petals and sepals 4 or 5 (when present) ........................................ 3

3 Leaves lance-shaped, broadest at base, less than 3 mm wide .......................................................... DIDIPLIS

3 Leaves oval, widest near middle, larger leaves 3 mm or more wide .......................................................... ROTALA

## DECODON *Water-Willow*

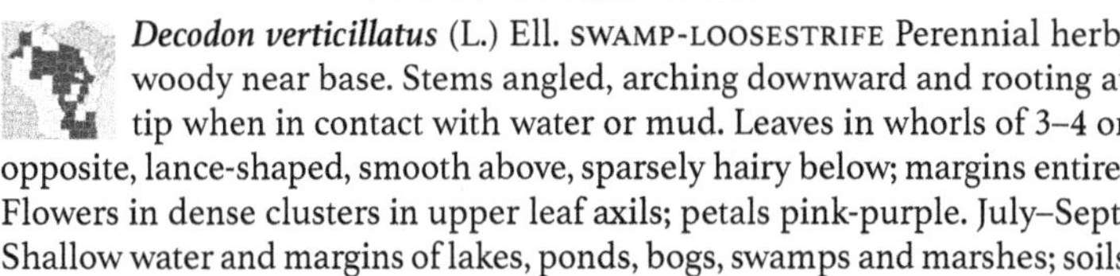

***Decodon verticillatus*** (L.) Ell. SWAMP-LOOSESTRIFE Perennial herb, woody near base. Stems angled, arching downward and rooting at tip when in contact with water or mud. Leaves in whorls of 3–4 or opposite, lance-shaped, smooth above, sparsely hairy below; margins entire. Flowers in dense clusters in upper leaf axils; petals pink-purple. July–Sept. Shallow water and margins of lakes, ponds, bogs, swamps and marshes; soils mucky.

## DIDIPLIS *Water-Purslane*

***Didiplis diandra*** (Nutt.) Wood WATER-PURSLANE Annual herb; plants underwater or on exposed shores. Leaves numerous, opposite; underwater leaves linear; emersed leaves shorter and wider; petioles absent. Flowers few, inconspicuous, green. Fruit a small round capsule. July–Aug. Shallow water and muddy pond margins. Plants somewhat resemble water-starwort (*Callitriche*), but in water-starwort the underwater leaves have a shallow notch at tip and a flattened capsule. *Peplis diandra* (Nutt.) ex DC.

## LYTHRUM *Loosestrife*

Perennial herbs. Stems erect, sometimes rather woody at base, upper stems 4-angled. Leaves opposite, alternate, or rarely whorled, lance-shaped, reduced to bracts in the head. Flowers in showy, spike-like heads, 1 to several in axils of upper leaves, regular or somewhat irregular; petals 6, purple.

*Key to species*

1 Flowers single in upper leaf axils; stamens usually 6 .... ***L. alatum***

1 Flowers many in spike-like heads at ends of stems; stamens usually 12 (6 long and 6 short) .............................. ***L. salicaria***

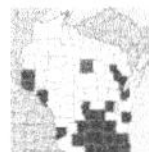

***Lythrum alatum*** Pursh [•66] WINGED LOOSESTRIFE June–Aug. Lakeshores, wet meadows, marshes, low prairie, calcareous fens, ditches; especially where sandy.

***Lythrum salicaria*** L. PURPLE LOOSESTRIFE June–Sept. Introduced from Europe and formerly planted as an ornamental, escaping to marshes, wet ditches, streambanks, cranberry bogs and shores, where a serious threat to our native flora and of little value to wildlife.

## ROTALA *Toothcup*

***Rotala ramosior*** (L.) Koehne LOWLAND TOOTHCUP Small annual herb. Stems 4-angled. Leaves opposite, linear to oblong, margins entire. Flowers single and stalkless in leaf axils; petals small, white to pink, 4, slightly longer than sepals. July–Oct. Muddy or sandy shores, marshes (especially those that dry during growing season), low spots in fields, ditches and other seasonally flooded places.

## MALVACEAE *Mallow Family*

Annual or perennial herbs with upright stems; trees in *Tilia.* Leaves alternate, entire to lobed or dissected, often round or kidney-shaped, palmately veined. Flowers single or in small, narrow clusters from leaf axils, with 5 united sepals (separate in *Tilia*) and 5 petals. Fruit a capsule.

*Key to genera*

1 Trees; with inflorescence apparently borne at the middle of a tongue-shaped bract ........ TILIA
1 Herbs or shrubs; inflorescences various, but never with a large, tongue-shaped bract ........ 2

2 Calyx subtended by a series of 2 or more bracts ........ 3
2 Calyx without involucral bracts ........ 7

3 Involucral bracts 6 or more ........ 4
3 Involucral bracts 5 or less ........ 6

4 Styles 5; involucral bracts narrow and linear ........ HIBISCUS
4 Styles many; involucral bracts triangular ........ 5

5 Upper stems more or less completely covered with felty hairs ........ ALTHAEA
5 Upper stems hairy but often with intermixed glabrous areas, the hairs not dense and felt-like ........ ALCEA

6 Flower petals straight across at tip, the tip finely fringed ........ CALLIRHOE
6 Petals obovate, rounded at tip, not fringed ........ MALVA

7 Leaves undivided ........ ABUTILON
7 Leaves deeply palmately dissected ........ 8

8 Petals white, to 9 mm long ........ NAPAEA
8 Petals pink, white, or red with a white spot at base, 15 mm or more long ........ CALLIRHOE

### ABUTILON *Velvetleaf*

***Abutilon theophrasti*** Medik. VELVETLEAF Annual herb, softly pubescent throughout with stellate hairs. Leaves cordate, toothed, on petioles of about the same length. Flowers yellow. July–Oct. Native of Asia; established as a weed in fields and waste places but more abundant south of Wisc.

### ALCEA *Hollyhock*

***Alcea rosea*** L. HOLLYHOCK Perennial herb. Leaves orbicular. Flowers vary from white to pink or purplish, on short peduncles from the upper axils, forming an elongate raceme-like inflorescence. June–Aug. Native perhaps to se Europe; cultivated and appearing near gardens, railroads, waste places. *Althaea rosea* (L.) Cav.

### ALTHAEA *Marsh-Mallow*

***Althaea officinalis*** L. COMMON MARSH-MALLOW Pubescent perennial herb. Leaves ovate, coarsely and irregularly serrate, commonly shallowly 3-lobed,

velvety-pubescent. Flowers several in a peduncled cluster from the axil of the upper leaves, pink. July–Sept. Native of Europe. The thick mucilaginous roots are used in confectionery.

## CALLIRHOE *Poppy-Mallow*

***Callirhoe triangulata*** (Leavenworth) Gray CLUSTERED POPPY-MALLOW Perennial herb; stems, leaves, pedicels, and calyx harshly stellate-pubescent. Leaves triangular, crenate, the upper much narrower. Flowers several, crowded at the end of axillary peduncles, the petals entire or inconspicuously erose. June–Aug. Dry sandy prairies.

ADDITIONAL SPECIES ***Callirhoe alcaeoides*** (Michx.) Gray; se Wisc, more common sw of state.

## HIBISCUS *Rose-Mallow*

Large perennial herbs. Leaves alternate, smooth or hairy, palmately divided. Flowers large and showy, pink to white. Fruit an ovate capsule.

*Key to species*

1 Annual herbs to 6 dm tall; petals pale yellow with purple center . . . . . . . . . . . . . . . . . . . . . . . . . . . . . . . . . . . . . . . . . . . . . . . . . . ***H. trionum***

1 Tall perennial herbs to 1 m tall or more; petals pink, 5–9 cm long **2**

2 Leaves and stems without hairs; many leaves lobed; along Mississippi and Wisconsin Rivers in sc Wisc . . . . . . . . . . . ***H. laevis***

2 Leaf undersides and upper stems with velvety hairs; leaves mostly unlobed; uncommon in sc Wisc . . . . . . . . . . . . . . . . . . ***H. moscheutos***

***Hibiscus laevis*** All. SMOOTH ROSE-MALLOW Aug–Sept. Marshes, muddy shores and shallow water.

***Hibiscus moscheutos*** L. SWAMP ROSE-MALLOW Aug–Sept. Marshes, streambanks and disturbed wet areas. Native to se USA, adventive in Walworth County in s Wisc.

***Hibiscus trionum*** L. FLOWER-OF-AN-HOUR July–Sept. Native of Europe; fields, roadsides, and waste places, often a troublesome weed.

## MALVA *Mallow*

Annual, biennial, or perennial herbs. Leaves broad, serrate, crenate, lobed, or parted. Flowers solitary or fascicled in the axils.

ADDITIONAL SPECIES ***Malva alcea*** L., introduced in Calumet County. ***Malva verticillata*** L., introduced in southeastern Wisc.

*Key to species*

1 Leaves deeply divided into 3–7 segments . . . . . . . . . . . ***M. moschata***

1 Leaf margins entire or only shallowly lobed . . . . . . . . . . . . . . . . . . . **2**

2 Petals purple or pink, more than 1.5 cm long; stems erect . . . . . . . . . . . . . . . . . . . . . . . . . . . . . . . . . . . . . . . . . . . . . . . . . . . . . ***M. sylvestris***

2 Petals white or purple-tinged, less than 1.5 cm long; stems ascending or prostrate . . . . . . . . . . . . . . . . . . . . . . . . . . . . . . . . . . . . . . . . . . . . . . . **3**

3 Stems glabrous or nearly so; petals only slightly longer than calyx lobes ........ *M. pusilla*

3 Stems pubescent; petals about 2 times longer than calyx lobes ........ *M. neglecta*

***Malva moschata*** L. MUSK MALLOW June–Sept. Native of Europe; escaped from cultivation along roadsides and in waste places.

***Malva neglecta*** Wallr. [•66] COMMON MALLOW May–Oct. Native of Eurasia and n Africa; common weed of gardens and waste places.

***Malva pusilla*** Sm. DWARF MALLOW Native of Europe; occasional as a weed. *Malva rotundifolia* L.

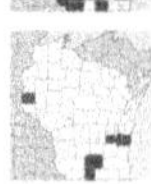

***Malva sylvestris*** L. HIGH MALLOW June–Aug. Native of Eurasia; occasionally escaped from cultivation.

### NAPAEA *Glade-Mallow*

***Napaea dioica*** L. GLADE-MALLOW Large perennial herb. Leaves round in outline, deeply 5–9 lobed, the lobes coarsely toothed; upper leaves smaller. Flowers either staminate or pistillate and on separate plants; many in large panicles at ends of stems; petals white. June–Aug. Moist floodplain forests, riverbanks.

### TILIA *Basswood; Linden*

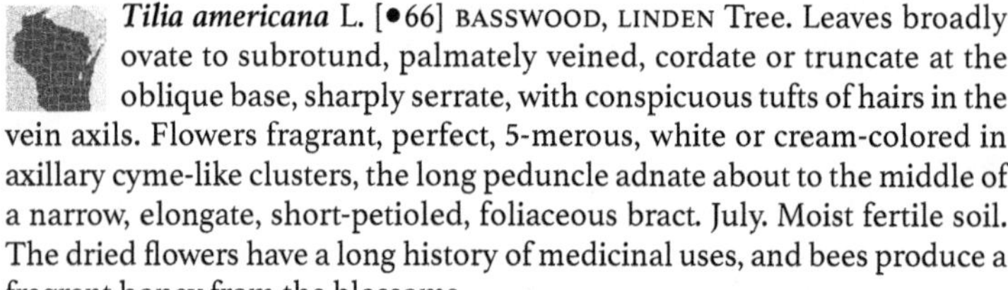

***Tilia americana*** L. [•66] BASSWOOD, LINDEN Tree. Leaves broadly ovate to subrotund, palmately veined, cordate or truncate at the oblique base, sharply serrate, with conspicuous tufts of hairs in the vein axils. Flowers fragrant, perfect, 5-merous, white or cream-colored in axillary cyme-like clusters, the long peduncle adnate about to the middle of a narrow, elongate, short-petioled, foliaceous bract. July. Moist fertile soil. The dried flowers have a long history of medicinal uses, and bees produce a fragrant honey from the blossoms.

## MELASTOMATACEAE *Melastome Family*

### RHEXIA *Meadow-Pitchers*

***Rhexia virginica*** L. [•66] WING-STEM MEADOW-PITCHERS Perennial herb, roots often with tubers. Stems 4-angled and 4-winged, with bristly hairs at nodes. Leaves ovate, margins finely toothed. Flowers perfect, regular, 4-merous, in cymes from ends of stems and upper leaf axils; petals purple. July–Sept. Open shores, moist meadows, thickets (often of *Aronia*); soils acidic, sandy or peaty. When growing in water, the stems are often covered toward the base by a spongy aerenchyma.

## MENISPERMACEAE *Moonseed Family*

### MENISPERMUM *Moonseed*

***Menispermum canadense*** L. CANADIAN MOONSEED Dioecious woody twiners. Leaves simple, alternate, broadly ovate to nearly orbicular, palmately veined,

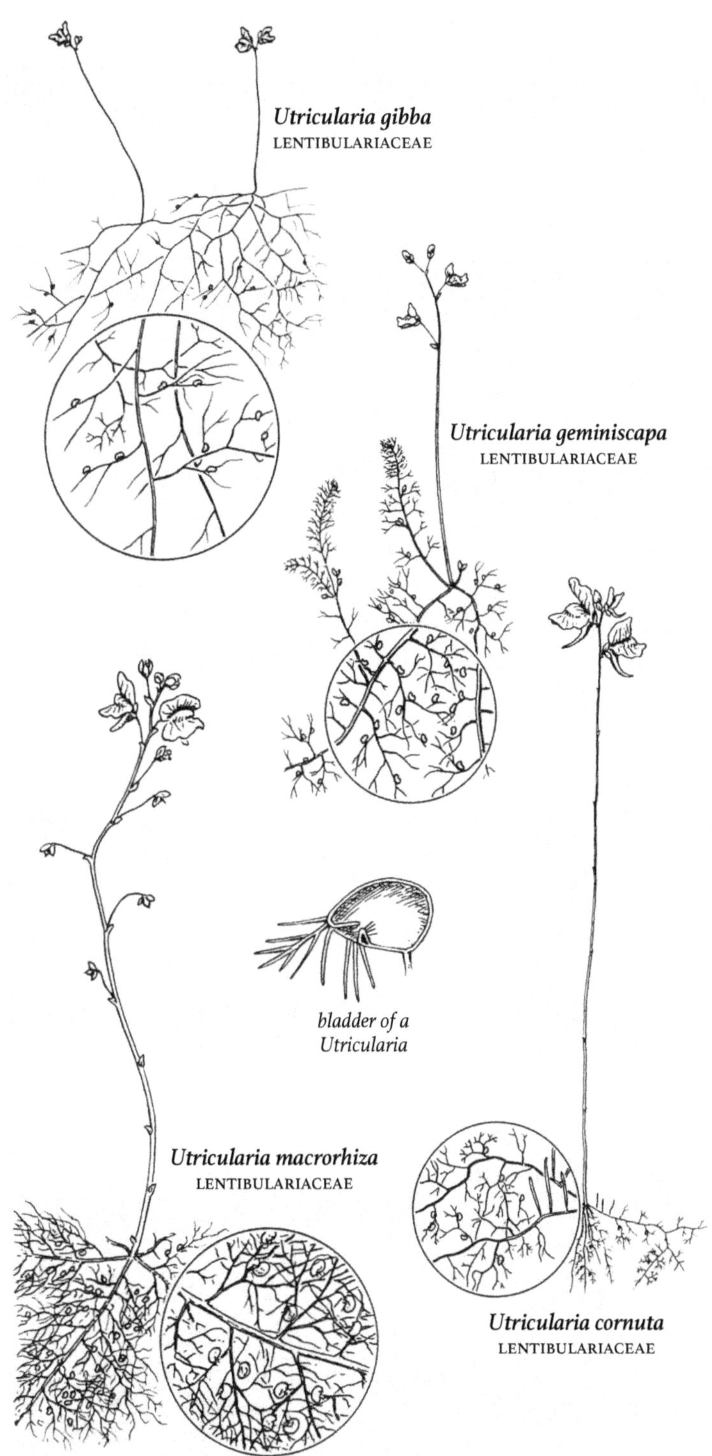
Utricularia gibba
LENTIBULARIACEAE
Utricularia geminiscapa
LENTIBULARIACEAE
bladder of a
Utricularia
Utricularia macrorhiza
LENTIBULARIACEAE
Utricularia cornuta
LENTIBULARIACEAE

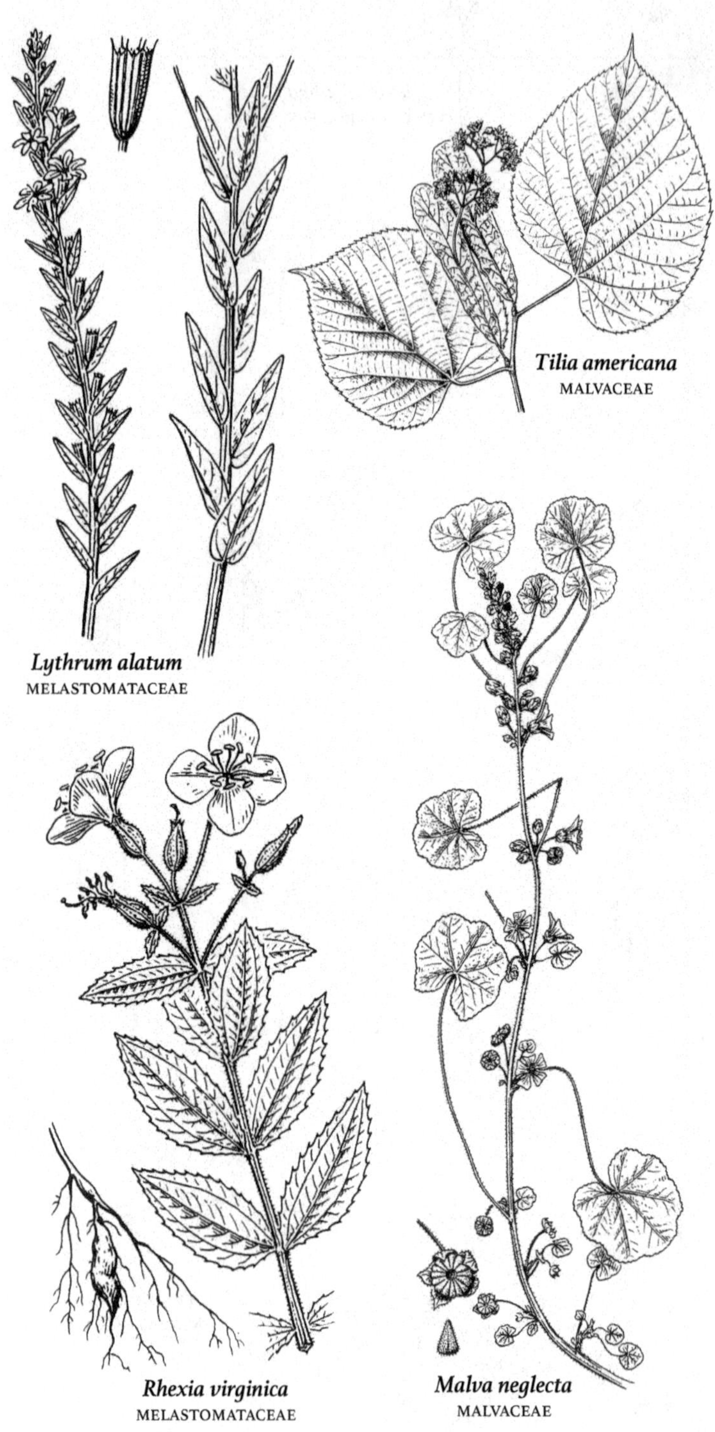
Tilia americana
MALVACEAE
Lythrum alatum
MELASTOMATACEAE
Rhexia virginica
MELASTOMATACEAE
Malva neglecta
MALVACEAE

shallowly 3–7-lobed to entire. Flowers small, unisexual, usually 3-merous, in racemes or panicles that arise just above the leaf-axils. Drupe bluish-black. June–July. Moist woods and thickets. The drupes, suspected to be toxic, resemble wild grapes.

## MENYANTHACEAE *Buckbean Family*

### MENYANTHES *Buckbean*

***Menyanthes trifoliata*** L. BUCKBEAN Perennial glabrous herb, with thick rhizomes covered with old leaf bases. Leaves alternate along rhizomes, palmately divided into 3 leaflets, the base of petiole expanded and sheathing stem. Flowers in racemes on leafless stalks; petals white, often purple-tinged, bearded with white hairs on inner surface. May–July. Open bogs and fens (especially in pools and outer moat), cedar swamps, wet thickets.

## MOLLUGINACEAE *Carpetweed Family*

### MOLLUGO *Carpetweed*

***Mollugo verticillata*** L. GREEN CARPETWEED Annual herb. Stems forming mats. Leaves in whorls of 3–8, narrowly to broadly oblong lance-shaped. Flowers perfect, 2–5 from each node; petals 5, pale green to white. June–Sept. Apparently native of tropical America; now a common weed in moist soil.

## MONTIACEAE *Montia Family*

The family Montiaceae includes our 2 genera formerly within Portulacaceae.

*Key to genera*

1 Leaves clustered at base of stem ............ PHEMERANTHUS

1 Leaves not all clustered at base of stem, in a single pair and opposite on stem ............................................ CLAYTONIA

### CLAYTONIA *Springbeauty*

Glabrous perennial herbs from rounded tubers (ours). Leaves one or few from the base and a single opposite pair on the stem below the loose terminal raceme. Petals 5, white or pale pink with pink veins. Flowers open from March to May, with stems and leaves withered by early summer.

*Key to species*

1 Leaves with distinct petiole; blades less than 5 times longer than wide; local in n Wisc ............................ ***C. caroliniana***

1 Petiole not distinct; blades 5 times or more longer than wide; widespread and common ......................... ***C. virginica***

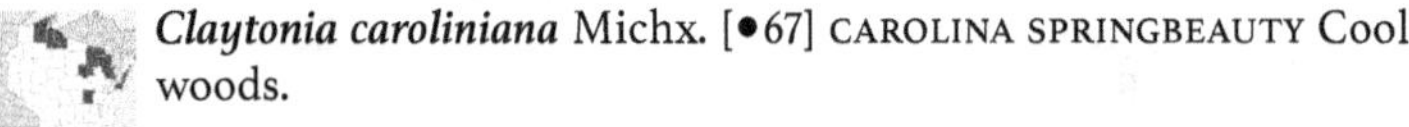

***Claytonia caroliniana*** Michx. [•67] CAROLINA SPRINGBEAUTY Cool woods.

***Claytonia virginica*** L. [•67] VIRGINIA SPRINGBEAUTY Damp woods and fields.

### PHEMERANTHUS *Fameflower*

***Phemeranthus rugospermus*** (Holz.) Kiger PRAIRIE FAMEFLOWER Glabrous, taprooted, perennial herb. Leaves crowded, succulent, terete 3–6 cm long, sessile. Flowers in long-peduncled bracted cymes; petals 5, pink. July–Aug. Thin soil overlying sandstone and on sand prairies. Flowers open in late afternoon only. *Talinum rugospermum* Holz.

---

## MORACEAE *Mulberry Family*

Trees (rarely herbs), juice milky or watery. Leaves alternate, simple or compound. Flowers small, crowded in dense clusters or heads; unisexual, the plants monoecious or dioecious. Corolla none. Fruits diverse.

ADDITIONAL SPECIES ***Fatoua villosa*** (Thunb.) Nakai, annual weed introduced from Asia, reported from Winnebago County.

*Key to genera*

1 Leaves entire; branches often with stout thorns; fruit large, 5 cm wide or more .......... MACLURA

1 Leaves toothed and often with 1 or several lobes; branches not thorny; fruit 2–3 cm wide .......... MORUS

### MACLURA *Osage-Orange*

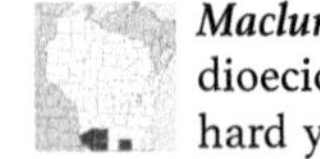

***Maclura pomifera*** (Raf.) Schneid. OSAGE-ORANGE Introduced tree; dioecious, with male and female flowers on different plants, with hard yellow wood and milky juice; branches armed with stout, straight, axillary spines 1–2 cm long. Leaves alternate, entire, ovate lance-shaped, shining. Staminate flowers numerous in loose axillary heads; pistillate flowers in dense, globose, axillary heads. Fruit hard, globose, yellowish-green, of numerous small drupes composed of the enlarged common receptacle and calyces, completely concealing the achenes. Rich moist soil; sometimes planted as a hedge, occasionally adventive or persistent but probably not fully established.

### MORUS *Mulberry*

Monoecious or dioecious trees. Leaves alternate, serrate or lobed, palmately veined. Flowers in cylindric catkins, the staminate longer and more loosely flowered than the pistillate. Fruit white or colored, edible, resembling a blackberry.

*Key to species*

1 Leaves pubescent on underside, often unlobed .......... *M. rubra*

1 Leaves glabrous or with scattered hairs on underside veins, often with 3–5 lobes .......... *M. alba*

***Morus alba*** L. WHITE MULBERRY Apparently native of Asia; long cultivated in Europe and America for its fruit or fiber or as food for the silkworm, and now used for ornament in several horticultural forms; escaped along roadsides, in vacant land, and open woods.

*Morus rubra* L. RED MULBERRY Rich woods.

## MYRICACEAE *Bayberry Family*

Monoecious or dioecious shrubs, with alternate simple leaves; leaves resinous-dotted and fragrant. Flowers unisexual, solitary in the axils of small bracts, aggregated into globose to cylindric catkins.

*Key to genera*

1 Leaves entire or nearly so; wet habitats ................ MYRICA
1 Leaves pinnately lobed; dry sandy habitats ....... COMPTONIA

### COMPTONIA *Sweet-Fern*

***Comptonia peregrina*** (L.) Coult. SWEET-FERN Shrub. Leaves linear-oblong, deeply pinnately lobed, resinous-dotted, more or less pubescent. Nutlets ellipsoid, subtended by elongate bractlets, the whole fruit bur-like. April–May. Dry, especially sandy soil.

### MYRICA *Bayberry*

***Myrica gale*** L. SWEET GALE Much-branched shrub; twigs hairy, dotted with glands. Leaves alternate, wedge-shaped, broadest above middle, dark green on upper surface, paler below, dotted with shiny yellow glands; fragrant when rubbed. Staminate and pistillate flowers separate and on different plants, appearing before or with unfolding leaves; staminate flowers in catkins with dark brown, shiny triangular scales; pistillate flowers in conelike, brown clusters. April–May. Lakeshores, marshes, swamps and bogs.

## NELUMBONACEAE *Lotus-Lily Family*

### NELUMBO *Lotus-Lily*

***Nelumbo lutea*** Willd. AMERICAN LOTUS-LILY Perennial aquatic herb, from a large, horizontal rootstock. Leaves large, shield-shaped, floating on water surface or held above water; petioles attached at center of blade. Flowers pale yellow; receptacle flat-topped, to 1 dm wide; seeds acornlike. July–Aug. Lakes, ponds, backwater areas, marshes; mostly near Mississippi River.

## NYCTAGINACEAE *Four-O'clock Family*

### MIRABILIS *Four-O'clock*

Perennial herbs, or woody at base. Leaves opposite. Flowers many in terminal panicles; petals rose to pink-purple, open in the morning, solitary or in clusters of 2–4. Fruit 5-ribbed, mucilaginous when wet.

*Key to species*

1 Leaves ovate, heart-shaped or truncate at base; inflorescence not glandular .......................................... *M. nyctaginea*

1 Leaves linear to lance-shaped, tapered to the base; inflorescence glandular hairy .......................................... *M. albida*

***Mirabilis albida*** (Walter) Heimerl HAIRY FOUR-O'CLOCK Summer. Dry prairies, hills, and barrens. *Mirabilis hirsuta* (Pursh) MacM.

***Mirabilis nyctaginea*** (Michx.) MacM. HEART-LEAF FOUR-O'CLOCK May–Aug. Dry soil, waste places.

## NYMPHAEACEAE *Water-Lily Family*

Aquatic, perennial herbs. Stems long and fleshy, from horizontal rhizomes rooted in bottom mud. Leaves large, leathery, mostly floating or emergent above water surface, heart-shaped to shield-shaped, notched at base. Flowers showy, single on long stalks and borne at or above water surface, perfect, white or yellow; petals numerous. Fruit a many-seeded, berrylike capsule, opening underwater when mature.

*Key to genera*

1 Flowers yellow, often red-tinged, sepals petal-like, true petals small; leaf blades oblong to oval or heart-shaped ............ **NUPHAR**

1 Flowers white (rarely pink), sepals green, true petals large and showy; leaf blades nearly round ................... **NYMPHAEA**

### NUPHAR *Yellow Water-Lily*

Aquatic herbs. Leaves mostly large and floating or emergent. Sepals 5–6, yellow and petal-like, forming a saucer-shaped flower; petals small and numerous.

*Key to species*

1 Disk at base of stigma red; anthers shorter than the filaments ... 2

1 Disk at base of stigma green or yellow; anthers longer than filaments ..................................................... 3

2 Leaf sinus 2/3 or more length of the midrib; sepals 5; anthers 1–3 mm long; n Wisc .............................. *N. microphylla*

2 Leaf sinus about 1/2 length of midrib; sepals 5 or 6; anthers 3–6 mm long; n Wisc ................................... *N. × rubrodisca*

3 Leaf blades mostly floating; petioles flattened and narrowly winged; fruit usually purplish; statewide .................... *N. variegata*

3 Leaf blades usually raised above water surface; petioles round or oval in cross-section; fruit usually green or yellowish; southern Wisc ....................................................... *N. advena*

***Nuphar advena*** (Ait.) Ait. f. YELLOW POND-LILY June–Sept. Shallow to deep water of slow-moving streams, lakes and ponds. *Nuphar lutea* (L.) Sm. subsp. *advena* (Ait.) Kartesz & Gandhi

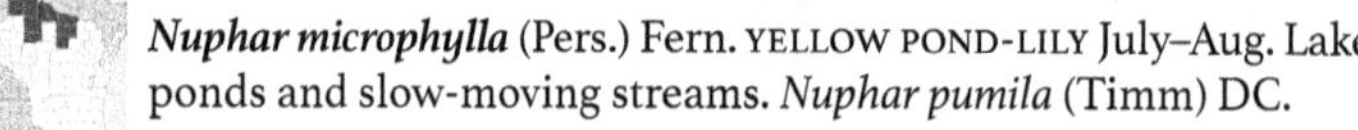

***Nuphar microphylla*** (Pers.) Fern. YELLOW POND-LILY July–Aug. Lakes, ponds and slow-moving streams. *Nuphar pumila* (Timm) DC.

***Nuphar × rubrodisca*** Morong YELLOW POND-LILY Summer. Considered a hybrid between *Nuphar microphylla* and *N. variegata*.

*Nuphar lutea* (L.) Sm. subsp. *rubrodisca* (Morong) Hellquist & Wiersema

***Nuphar variegata*** Dur. YELLOW POND-LILY June–Aug. Ponds, lakes, quiet streams.

### NYMPHAEA *Water-Lily*

***Nymphaea odorata*** Ait. [•67] WHITE WATER-LILY Aquatic perennial herb, rhizomes sometimes with knotty tubers. Leaves floating, round, 1–3 dm wide, with a narrow notch, green and shiny on upper surface, usually purple or red below. Flowers large and showy, white (rarely pink), usually fragrant, often opening in morning and closing in late afternoon (or remaining open on cool, cloudy days). June–Aug. Shallow water of ponds and lakes, quiet water of rivers. *Nymphaea tuberosa* Paine.

## OLEACEAE *Olive Family*

Trees or shrubs with opposite, simple or compound leaves. Flowers perfect or unisexual, regular. Calyx small or in some genera lacking. Corolla in our genera partially or wholly fused, or lacking (*Fraxinus*). Fruit a drupe, capsule, or samara.

*Key to genera*

1 Trees; leaves pinnately compound .................. **FRAXINUS**
1 Shrubs; leaves simple .................. 2

2 Flowers usually lilac-purple; leaves on petioles 1 cm or more long; fruit a capsule .................. **SYRINGA**
2 Flowers white; leaves nearly sessile or on short petioles to 1 cm long; fruit fleshy .................. **LIGUSTRUM**

### FRAXINUS *Ash*

Medium trees. Leaves deciduous, opposite, pinnately divided into leaflets. Flowers in clusters from axils of previous year's twigs, mostly single-sexed, staminate and pistillate flowers on different trees, rarely perfect, petals absent. Fruit a 1-seeded, winged samara.

*Key to species*

1 Twigs strongly 4-angled in cross-section, winged; rare in se Wisc .................. ***F. quadrangulata***
1 Twigs round or elliptic in cross-section, not winged .................. 2

2 Twigs densely hairy .................. 3
2 Twigs glabrous.................. 4

3 Lateral leaflets tapered at base to a short winged petiole, or the leaflets sessile .................. ***F. pennsylvanica***
3 Lateral leaflets rounded at base, short petioles present .................. ***F. americana***

4 Leaflets pale or waxy on underside, margins often entire; lateral leaflets on short petioles 5 mm or more long .................. ***F. americana***
4 Leaflets not waxy on underside, margins usually finely toothed; lateral leaflets sessile or on short petioles to 3 mm long .................. 5

5 Lateral leaflets usually number 8, sessile; body of fruit flat in cross-section ........ *F. nigra*

5 Lateral leaflets usually 4–6, on short petioles; body of fruit round in section ........ *F. pennsylvanica*

***Fraxinus americana*** L. [•68] WHITE ASH Rich moist woods; a valuable timber tree.

***Fraxinus nigra*** Marsh. [•68] BLACK ASH April–May. Floodplain forests, cedar swamps, wet depressions in forests.

***Fraxinus pennsylvanica*** Marsh. [•68] GREEN ASH April–May. Floodplain forests, swamps, shores, streambanks. Both smooth and hairy forms of *F. pennsylvanica* occur, with trees becoming less hairy as one moves w across the Great Lakes region.

***Fraxinus quadrangulata*** Michx. [•68] BLUE ASH *Threatened.* Rich, moist woods. The 4-angled twigs are distinctive.

### LIGUSTRUM *Privet*

***Ligustrum vulgare*** L. EUROPEAN PRIVET Shrub. Leaves simple, opposite, entire. Flowers white, in small panicles terminating the main axis and short lateral branches. Fruit a small black drupe. June. Native of Europe, sometimes grown as an ornamental hedge, and will flower and fruit abundantly if not heavily trimmed; may sometimes escape, reported from Dane County.

### SYRINGA *Lilac*

***Syringa vulgaris*** L. COMMON LILAC Deciduous shrub, spreading and forming thickets. Leaves opposite, simple, entire, ovate. Flowers fragrant, in dense panicles; 4-merous; corolla usually lilac. May. Native of se Europe, found near abandoned farms. Lilac apparently does not spread by seed but is very long-lived and persists indefinitely after planting.

---

## ONAGRACEAE *Evening-Primrose Family*

Annual or perennial herbs. Leaves opposite to alternate, simple to pinnately divided, stalkless or short-petioled. Flowers usually large and showy, perfect, regular, borne in leaf axils or in heads at ends of stems; petals 4, white, yellow, or pink to rose-purple. Fruit a 4-chambered capsule; seeds many, with or without a tuft of hairs (coma).

*Key to genera*

1 Petals 2, small, white; leaves opposite; fruit with bristly hairs ........ CIRCAEA

1 Petals 4 (rarely absent), white, pink, or yellow; leaves alternate or opposite; fruit without bristly hairs ........ 2

2 Hypanthium prolonged beyond ovary into a tube below the petals; leaves alternate........ 3

2 Hypanthium scarcely if at all prolonged beyond ovary, not tube-like; leaves alternate or opposite........ 5

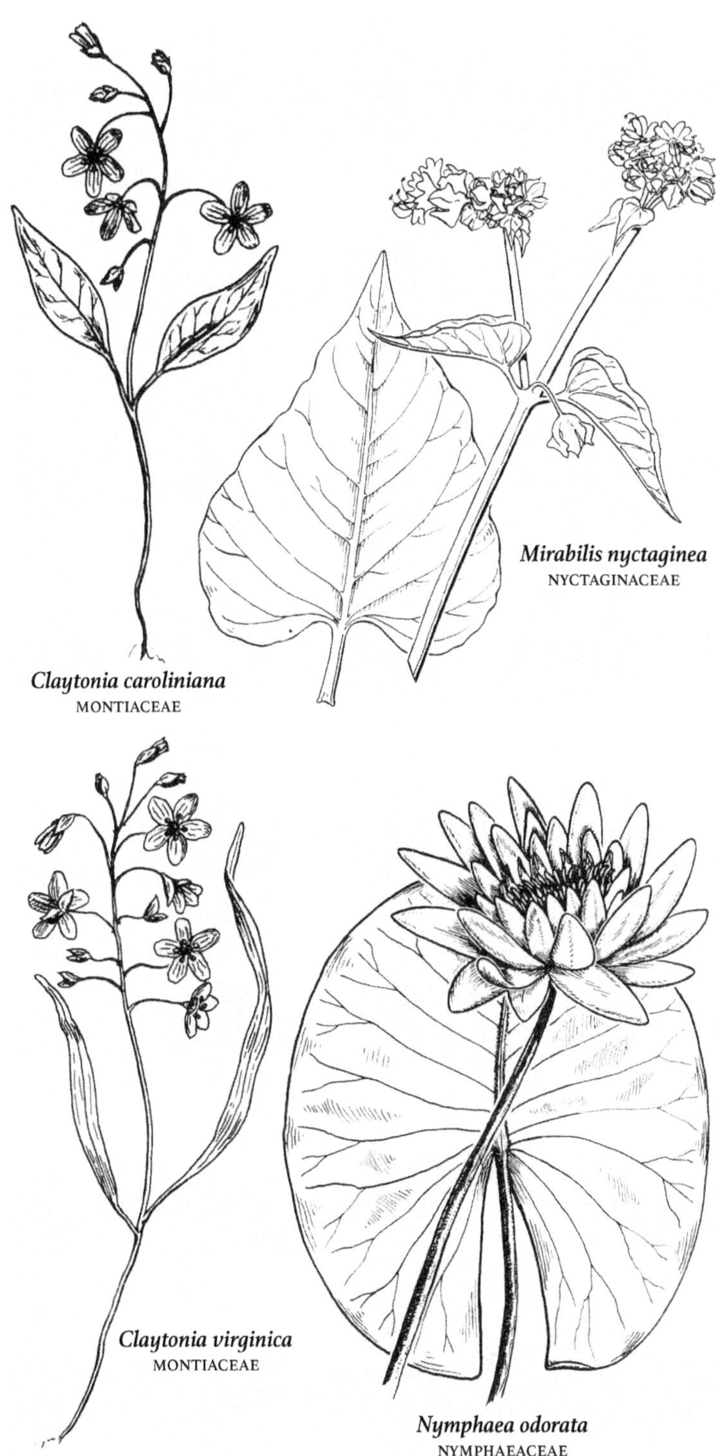

*Claytonia caroliniana*
MONTIACEAE

*Mirabilis nyctaginea*
NYCTAGINACEAE

*Claytonia virginica*
MONTIACEAE

*Nymphaea odorata*
NYMPHAEACEAE

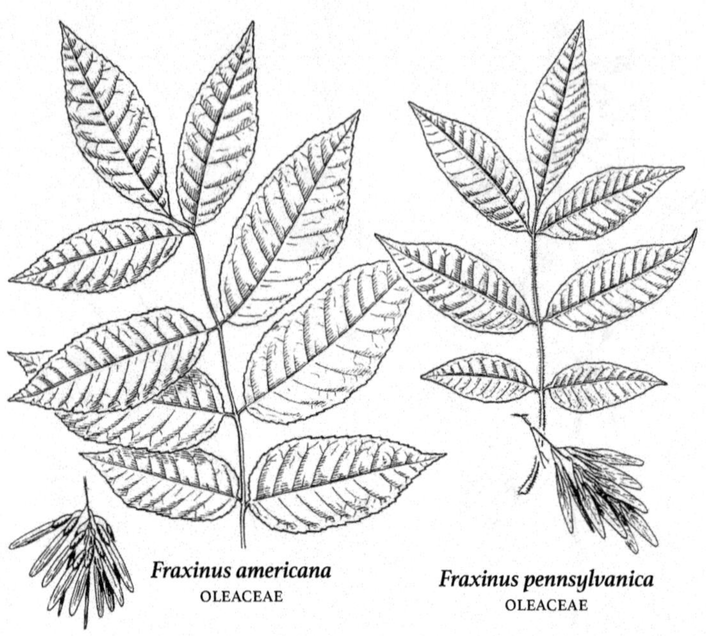

*Fraxinus americana*
OLEACEAE

*Fraxinus pennsylvanica*
OLEACEAE

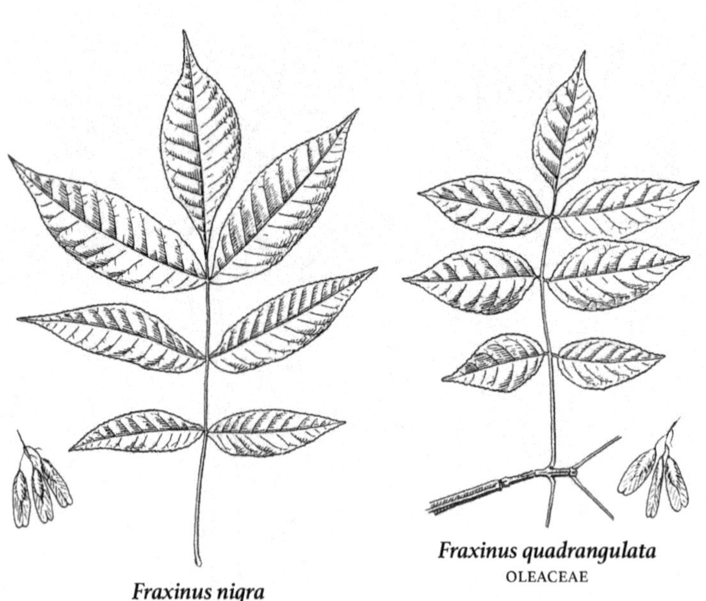

*Fraxinus nigra*
OLEACEAE

*Fraxinus quadrangulata*
OLEACEAE

3 Petals pink or white; fruit broadest in middle, tapering to each end, to 1 cm long .......................................... **GAURA**
3 Petals yellow (white in *Oenothera nuttallii*); fruit linear, usually more than 1 cm long .............................................. 4

4 Stigma 4-parted ................................ **OENOTHERA**
4 Stigma peltate ("umbrella-like," the style attached to center of stigma underside) ............................ **CALYLOPHUS**

5 Petals pink or rose-purple; seeds with a tuft of hairs (coma) ...... ............................................ **EPILOBIUM**
5 Petals yellow (or absent); seeds without a tuft of hairs **LUDWIGIA**

## CALYLOPHUS *Sundrops*

***Calylophus serrulatus*** (Nutt.) Raven YELLOW SUNDROPS Perennial herb. Leaves linear, entire or commonly sharply serrulate, glabrous to canescent. Flowers sessile in the axils of the upper leaves; petals yellow. June–July. Dry prairies. *Oenothera serrulata* Nutt.

## CIRCAEA *Enchanter's Nightshade*

Perennial herbs with opposite petioled leaves and small white flowers in one to few terminal racemes. Flowers 2-merous; petals obcordate or deeply notched. Fruit reflexed, beset with hooked bristles.

*Key to species*

1 Plants to 6 dm tall; flowers and fruit well-spaced on stalk; calyx lobes more than 1.5 mm long; leaves rounded at base, the margins very shallowly toothed ................................ ***C. lutetiana***
1 Plants smaller, to 3 dm tall; open flowers clustered near top of stem; calyx lobes less than 1.5 mm long; leaves usually heart-shaped at base, margins sharply toothed ........................ ***C. alpina***

***Circaea alpina*** L. [•69] ALPINE ENCHANTER'S NIGHTSHADE June–Aug. Cedar swamps (where often on rotting logs), low spots in forests.

***Circaea lutetiana*** L. [•69] COMMON ENCHANTER'S NIGHTSHADE June–Aug. Moist woods. *Circaea canadensis* (L.) Hill

## EPILOBIUM *Willow-Herb; Fireweed*

Perennial herbs, often producing leafy rosettes or bulblike offsets (turions) at base of stem late in growing season. Leaves simple, opposite, alternate, or opposite below and becoming alternate above. Flowers white to pink, single in axils of upper reduced leaves, or in spike or racemes at ends of stems; seeds tipped with a tuft of fine hairs (coma).

*Key to species*

1 Flowers showy in a terminal raceme; petals 8–16 mm long; stigma 4-parted ..................................... ***E. angustifolium***
1 Flowers single from upper leaf axils; petals 3–8 mm long........ 2

2 Stigma 4-parted ............................................ 3
2 Stigma entire, not 4-parted.................................. 4

3 Taprooted annual herb; leaves mostly alternate, linear .......... ............................................. *E. brachycarpum*

3 Perennial rhizomatous herb; leaves mostly opposite, lance-shaped and somewhat clasping ........................... *E. hirsutum*

4 Leaves entire or nearly so, the margins often revolute; stems round in cross-section, without lines of hairs on stem below base of each leaf........................................................ 5

4 Leaf margins conspicuously toothed; stems 4-angled in section, with lines of hairs on stems below leaf bases ................ 7

5 Stems with soft, straight hairs ....................... *E. strictum*

5 Stems finely hairy, the hairs appressed to stem................. 6

6 Upperside of leaves finely hairy ................ *E. leptophyllum*

6 Upper surface of leaves glabrous or nearly so ......... *E. palustre*

7 Tuft of hairs attached to tip of seeds (coma) white or nearly so, seeds with a broad, short beak; margins of stem leaves with mostly 10–30 teeth on a side ................................... *E. ciliatum*

7 Coma brown, seeds beakless; leaf margins with more than 30 teeth on a side ......................................... *E. coloratum*

***Epilobium angustifolium*** L. FIREWEED June–Sept. In a variety of habitats, preferring moist soils rich in humus; often abundant after fires. *Chamaenerion angustifolium* (L.) Scop.

***Epilobium brachycarpum*** K. Presl TALL ANNUAL WILLOWHERB Adventive from further west, Milwaukee County.

***Epilobium ciliatum*** Raf. [•69] AMERICAN WILLOWHERB July–Sept. Shores, streambanks, marshes, wet meadows, seeps, ditches and other wet places. *Epilobium glandulosum* Lehm.

***Epilobium coloratum*** Biehler PURPLE-LEAF WILLOWHERB July–Sept. Shores, seeps, swamps and wet woods, wet meadows, fens, ditches.

***Epilobium hirsutum*** L. HAIRY WILLOWHERB June–Sept. Introduced from Eurasia, established in e Wisc in marshes, shores, wet meadows, and ditches.

***Epilobium leptophyllum*** Raf. [•69] BOG WILLOWHERB July–Sept. Swamps, marshes, open bogs, sedge meadows, shores, streambanks and springs.

***Epilobium palustre*** L. MARSH WILLOWHERB July–Aug. Open bogs and swamps.

***Epilobium strictum*** Muhl. DOWNY WILLOWHERB July–Aug. Conifer swamps, sedge meadows, calcareous fens, marshes.

## GAURA *Gaura; Beeblossom*

Annual or perennial herbs, with alternate, entire, sinuate or dentate leaves, and long terminal spikes or racemes of small flowers. Flowers 4-merous, usually somewhat irregular.

ADDITIONAL SPECIES ***Gaura longiflora*** Spach (Long-flower beeblossom), native to central USA, considered adventive in Wisc.

*Key to species*

1 Leaves narrowly lance-shaped, less than 5 mm wide and to 3 cm long; fruit densely hairy, the hairs appressed; plants to 5 dm high ........ ***G. coccinea***

1 Leaves lance-shaped, wider and longer than in *G. coccinea;* fruit finely hairy, the hairs spreading; plants 5 dm or more high ...... **2**

2 Hairs on main stem mostly straight and widely spreading ....... ***G. biennis***

2 Hairs on main stem curled or appressed to stem .... ***G. longiflora****

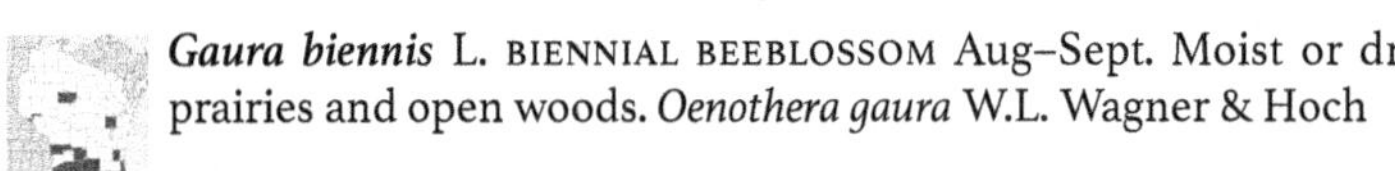

***Gaura biennis*** L. BIENNIAL BEEBLOSSOM Aug–Sept. Moist or dry prairies and open woods. *Oenothera gaura* W.L. Wagner & Hoch

***Gaura coccinea*** Nutt. ex Pursh SCARLET BEEBLOSSOM Introduced. May–Aug. Dry prairies and plains. *Oenothera suffrutescens* (Ser.) W.L. Wagner & Hoch

## LUDWIGIA *Primrose-Willow*

Perennial herbs. Stems floating, creeping, or upright. Leaves simple, opposite or alternate, entire. Flowers single in leaf axils; petals 4 (or absent), yellow or green; seeds without a tuft of hairs at tip (coma).

*Key to species*

1 Leaves opposite; stems floating, or creeping and rooting at nodes . ........ ***L. palustris***

1 Leaves alternate; stems erect or ascending .................... **2**

2 Stamens 8; leaf base extending downward on stem ... ***L. decurrens***

2 Stamens 4; leaves not extending downward on stem ........... **3**

3 Flower petals showy, yellow; flowers and fruit on stalks 3 mm or more long ................................ ***L. alternifolia***

3 Flower petals very small or absent; flowers and fruit stalkless .... ........ ***L. polycarpa***

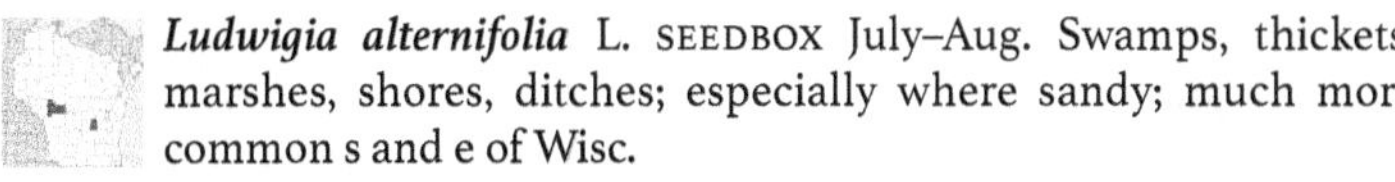

***Ludwigia alternifolia*** L. SEEDBOX July–Aug. Swamps, thickets, marshes, shores, ditches; especially where sandy; much more common s and e of Wisc.

***Ludwigia decurrens*** Walt. WING-STEM PRIMROSE-WILLOW July–Sept. Swamps and shallow water; Walworth County.

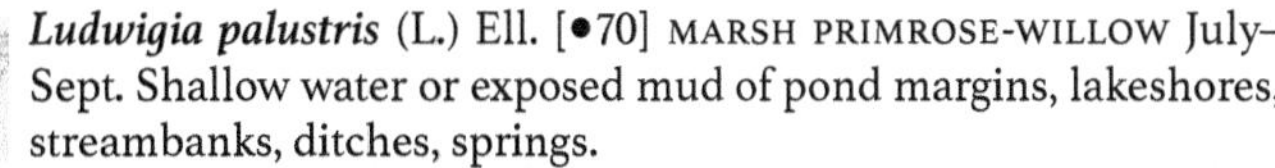

***Ludwigia palustris*** (L.) Ell. [•70] MARSH PRIMROSE-WILLOW July–Sept. Shallow water or exposed mud of pond margins, lakeshores, streambanks, ditches, springs.

***Ludwigia polycarpa*** Short & Peter MANY-FRUIT PRIMROSE-WILLOW July–Sept. Borders of swamps and marshes, muddy shores, wet depressions.

## OENOTHERA *Evening-Primrose*

Annual, biennial, or perennial herbs with alternate, mostly narrow leaves, and yellow, white, or pink flowers solitary in the axils or forming a terminal raceme. Flowers 4-merous. Five of our species can be considered part of an

*Oenothera biennis* "complex," including closely related (and sometimes difficult to distinguish) *O. biennis, O. glazioviana, O. oakesiana, O. parviflora,* and *O. villosa.*

*Key to species*

1 Flowers white, becoming pink . . . . . . . . *O. nuttallii*
1 Flowers yellow . . . . . . . . 2

2 Ovary round in cross-section or nearly so; fruit round or rounded 4-angled in cross-section, abruptly rounded at base . . . . . . . . 3
2 Ovary 4-angled; fruit sharply 4-angled or 4-winged, tapered to the base . . . . . . . . 10

3 Capsules linear, 2–3 mm wide and about same width throughout; stem leaves either pinnately lobed or linear to narrowly lance-shaped . . . . . . . . 4
3 Capsules tapered upward from base, 5–7 mm wide; stem leaves unlobed, lance-shaped, usually at least 1 cm wide . . . . . . . . 6

4 Leaves pinnately lobed; flowers few, from the axils of upper leaves . . . . . . . . *O. laciniata*
4 Stem leaves linear to lance-shaped, entire or nearly so; flowers many from leafy bracts in a terminal spike . . . . . . . . 5

5 Petals 1.5–3 cm long; stigma elevated above anthers when in full-flower . . . . . . . . *O. rhombipetala*
5 Petals to 1.5 cm long; stigma surrounded by anthers at full-flower . . . . . . . . *O. clelandii*

6 Petals 3.5 cm or more long . . . . . . . . *O. glazioviana*
6 Petals 1–2 cm long . . . . . . . . 7

7 Bases of sepals contiguous . . . . . . . . 8
7 Bases of awl-shaped sepals separate . . . . . . . . 9

8 Plant green in aspect, with mostly spreading long hairs and often shorter glandular ones . . . . . . . . *O. biennis*
8 Plant, especially the upper portion and inflorescence, gray in aspect, with dense appressed non-glandular hairs . . . . . . . . *O. villosa*

9 Calyx, ovary, capsule, and upper leaves or bracts densely pubescent with appressed whitish non-glandular hairs; largest leaves typically less than 15 mm wide, finely toothed . . . . . . . . *O. oakesiana*
9 Calyx, ovary, capsule, and other parts glabrate to sparsely pubescent, often with some long spreading hairs as well as shorter glandular hairs; largest leaves various, usually at least 15 mm wide, nearly entire . . . . . . . . *O. parviflora*

10 Flower petals 15–30 mm long; tip of stem erect . . . . . . . . *O. pilosella*
10 Petals to 8 mm long; tip of stem nodding when in bud . *O. perennis*

***Oenothera biennis*** L. [•70] COMMON EVENING-PRIMROSE July–Oct. Fields, roadsides, and waste places.

***Oenothera clelandii*** W. Dietr. Raven & W.L. Wagner LESSER FOUR-POINT EVENING-PRIMROSE June–Sept. Sandy fields and prairies.

***Oenothera glazioviana*** Micheli GARDEN EVENING-PRIMROSE Cultivated and sparingly escaped. Much like *O. biennis,* but with wider stem leaves and larger flowers.

***Oenothera laciniata*** Hill [•70] CUT-LEAF EVENING-PRIMROSE Dry sandy soil.

***Oenothera nuttallii*** Sweet WHITE-STEM EVENING-PRIMROSE Introduced. June–July. Prairies.

***Oenothera oakesiana*** (Gray) J.W. Robbins ex. S. Wats. & Coult. OAKES' EVENING-PRIMROSE July–Sept. Sandy or rocky shores, dunes, and clearings along the Great Lakes; occasionally inland along railroads, sandy shores, or disturbed places; the common evening-primrose of Lake Michigan sand dunes and beaches. Very similar to *O. parviflora, O. biennis* and *O. clelandii;* differs from *O. parviflora* by its inflorescence lacking gland-tipped hairs; differs from *O. biennis* in having sepals separate at base (in *O. biennis* sepals close together at base); differs from *O. clelandii* by its seeds not being pitted and the fruit being broad at the base. *Oenothera biennis* L. var. *oakesiana* A. Gray.

***Oenothera parviflora*** L. SMALL-FLOWER EVENING-PRIMROSE June–Oct. Shores, banks, marshy areas, also roadsides, fields and disturbed places. Similar to *O. laciniata* and *O. nuttallii* but plants usually smaller.

***Oenothera perennis*** L. [•70] LITTLE SUNDROPS June–Aug. Moist or dry soil, fields, meadows, and open woods.

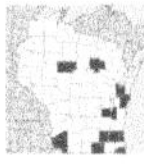

***Oenothera pilosella*** Raf. MIDWESTERN SUNDROPS Introduced. May–July. Moist soil, meadows, fields, and open woods.

***Oenothera rhombipetala*** Nutt. GREATER FOUR-POINT EVENING-PRIMROSE Introduced. June–Sept. Fields and prairies in sandy soil.

***Oenothera villosa*** Thunb. HAIRY EVENING-PRIMROSE July–August. Fields, shores, roadsides, railroads.

## OROBANCHACEAE *Broom-Rape Family*

Annual, biennial, or perennial herbs; some genera without green color, parasitic on the roots of other plants. Leaves opposite, alternate, or reduced to scales. Flowers mostly perfect, single or few from leaf axils, or numerous in clusters at ends of stems or leaf axils, usually with a distinct upper and lower lip; petals 4–5 (or sometimes absent). Fruit a several- to many-seeded 2-valved capsule. Now includes many former members of Scrophulariaceae.

*Key to genera*

1 Plants non-green, leaves converted to small, non-photosynthetic bracts .......... **2**

1 Plants green, leaves present and photosynthetic, green or sometimes strongly tinged with purple (often blackening upon drying) .... **4**

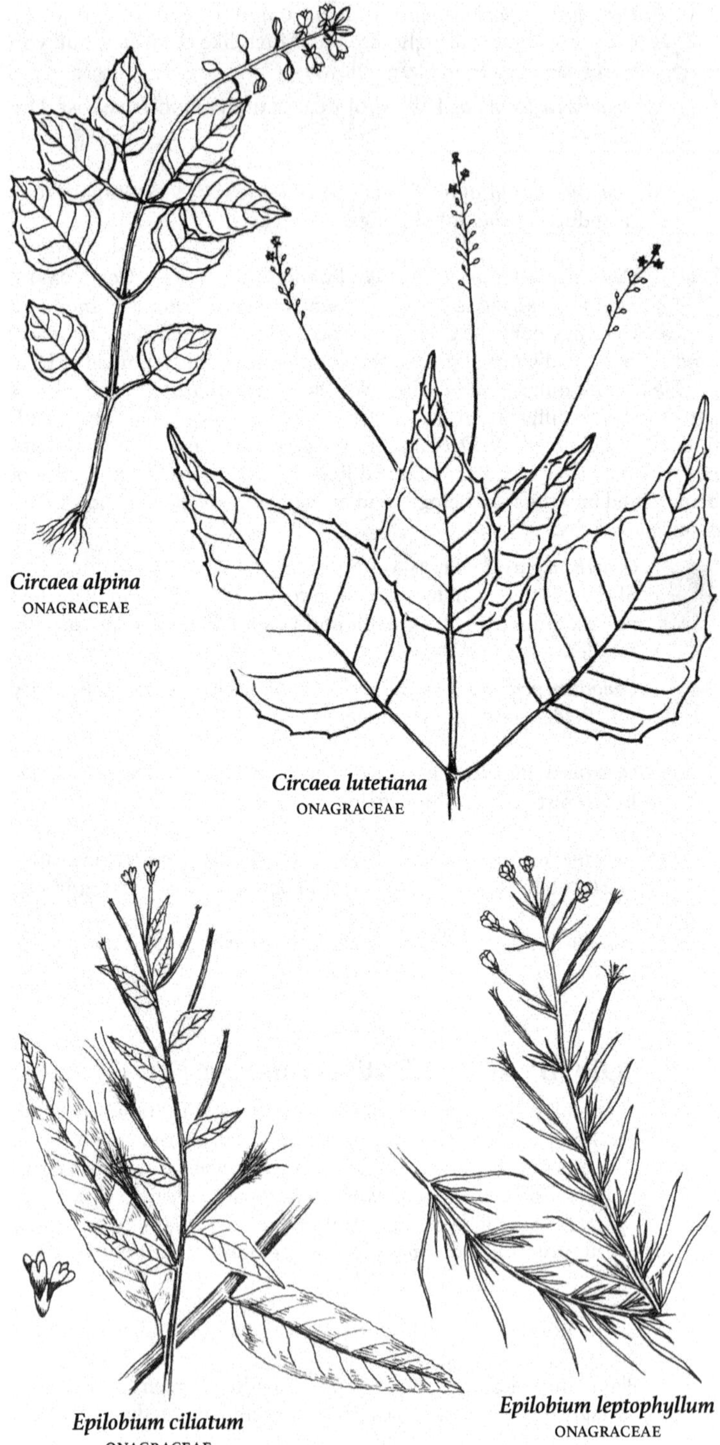
*Circaea alpina*
ONAGRACEAE
*Circaea lutetiana*
ONAGRACEAE
*Epilobium ciliatum*
ONAGRACEAE
*Epilobium leptophyllum*
ONAGRACEAE

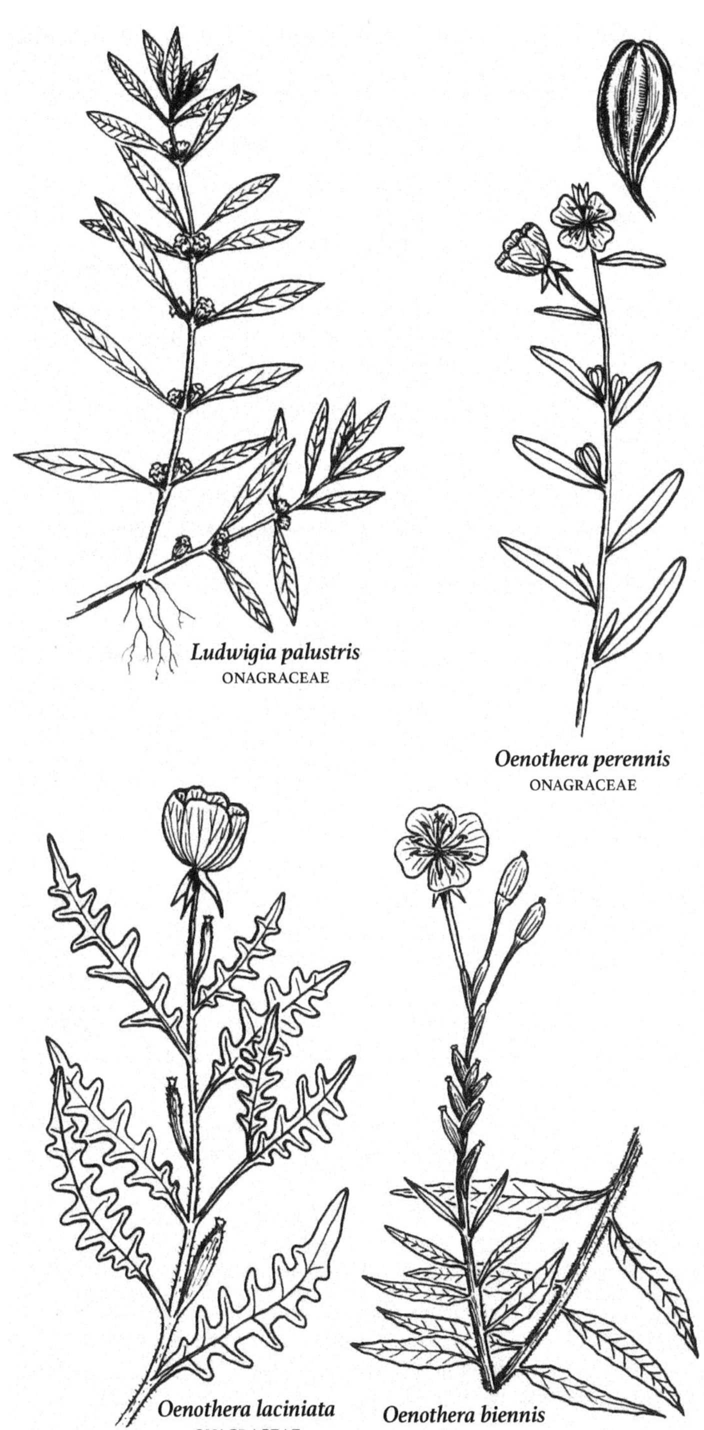
Ludwigia palustris
ONAGRACEAE
Oenothera perennis
ONAGRACEAE
Oenothera laciniata
ONAGRACEAE
Oenothera biennis
ONAGRACEAE

2 Stems well-branched; the flowers nearly sessile in spike-like racemes .......... **EPIFAGUS**

2 Stems unbranched or with only a few branches, the flowers single or few on long pedicels, or crowded into dense, spike-like racemes .......... 3

3 Stems thickened and pinecone-like; calyx deeply parted on both upper and lower sides .......... **CONOPHOLIS**

3 Stems not pinecone-like; calyx deeply parted on lower side .......... **OROBANCHE**

4 Stem leaves of fertile stems alternate .......... **CASTILLEJA**

4 Stem leaves of fertile stems all opposite or nearly so, sometimes alternate below the flowers .......... 5

5 Most leaves deeply pinnately divided; corolla cream-colored or yellow, 1.5–5 cm long .......... 6

5 Stem leaves toothed or entire, not deeply pinnately lobed; corolla various colors, in most species less than 1.5 cm long .......... 7

6 Flowers in dense racemes at ends of stems and branches; corolla strongly 2-lipped .......... **PEDICULARIS**

6 Flowers from leaf axils; corolla 5-lobed and only weakly 2-lipped Aureolaria

7 Calyx somewhat inflated in flower; conspicuously inflated and laterally compressed in fruit; flowers in a one-sided, leafy spike; leaf margins with large, forward-pointing teeth .......... **RHINANTHUS**

7 Calyx not inflated .......... 8

8 Flowers in the axils of alternate bracts, forming a terminal inflorescence .......... 9

8 Flowers solitary in the axils of opposite or whorled leaves or bracts .......... 10

9 Leaves 3–6 times longer than wide, pinnately veined, the midvein prominent .......... **ODONTITES**

9 Leaves less than 2 times longer than wide, palmately veined, with 3–5 prominent veins .......... **EUPHRASIA**

10 Corolla nearly regular, the 5 lobes similar .......... **AGALINIS**

10 Corolla irregular, 2-lipped, the upper lip 2-lobed, the lower lip 3-lobed .......... 11

11 Leaves much longer than wide, with short petioles **MELAMPYRUM**

11 Leaves less than 2 times longer than wide, sessile ... **EUPHRASIA**

## AGALINIS *False Foxglove*

Annual hemiparasitic herbs. Stems usually 4-angled. Leaves opposite, linear. Flowers showy, in clusters at ends of branches; petals pink to purple.

ADDITIONAL SPECIES ***Agalinis auriculata*** (Michx.) Blake, rare in prairies and open woods in s Wisc; separated from our other species by broader lanceolate leaves, the uppermost often with a pair of basal lobes, the terete stem, and the pubescent calyx.

*Key to species*

1 Pedicels less than 6 mm long, shorter than or equaling the calyx. 2

1 Longer pedicels more than 6 mm long, equaling or longer than the calyx ............................................... 3

2 Corolla 2–3 cm long; calyx lobes less than half length of calyx tube ............................................... ***A. purpurea***

2 Corolla less than 2 cm long; calyx lobes nearly as long as calyx tube ............................................... ***A. paupercula***

3 Plants yellow-green, remaining so when dried; leaves only about 1 mm wide; seeds yellow-brown............................... 4

3 Plants deep green, often tinged with purple, tending to darken when dried; leaves to 6 mm wide; seeds dark ........................ 5

4 Stems angled, with short rough hairs, unbranched or with short, upright branches ............................... ***A. skinneriana***

4 Stems round in cross-section, smooth, usually branched, the branches spreading ............................... ***A. gattingeri***

5 Flowers on slender, widely spreading pedicels; the calyx lobes v-shaped ............................................... ***A. tenuifolia***

5 Flowers on stout upright pedicels; the calyx lobes broadly rounded ............................................... ***A. aspera***

***Agalinis aspera*** (Dougl.) Britt. TALL FALSE FOXGLOVE Aug–Sept. Dry prairies.

***Agalinis gattingeri*** (Small) SMALL ROUND-STEM FALSE FOXGLOVE *Threatened.* Aug–Sept. Dry prairies and sparsely wooded glades and openings; sites usually with shallow soils and exposed bedrock.

***Agalinis paupercula*** (A.Gray) Britton SMALL-FLOWER FALSE FOXGLOVE Aug–Sept. Open, sandy wet places. Very similar to *A. purpurea* and perhaps best considered a variety of that species; *A. paupercula* has smaller flowers and longer calyx lobes than those of *A. purpurea. Gerardia purpurea* var. *paupercula* Gray.

***Agalinis purpurea*** (L.) Pennell [•71] PURPLE FALSE FOXGLOVE Aug–Sept. Wet meadows, fens, shores of Great Lakes and along inland lakes and ponds, moist areas between dunes, ditches; usually where sandy, often where calcium-rich. *Gerardia purpurea* L.

***Agalinis skinneriana*** (Wood) Britt. PALE FALSE FOXGLOVE *Endangered.* Aug–Sept. Dry, sandy, calcium-rich prairie, open woods, and barrens. The short-lived flowers open in on sunny days in the morning and the petals are often shed in the afternoon.

***Agalinis tenuifolia*** (Vahl) Raf. COMMON FALSE FOXGLOVE Aug–Sept. Wet meadows, low prairie, fens, shores, streambanks and ditches, usually where sandy. *Agalinis besseyana* (Britt.) Britt.; *Gerardia tenuifolia* var. *parviflora* Nutt.

## AUREOLARIA *False Foxglove*

Annual or perennial herbs; partially parasitic on roots of oaks. Principal leaves opposite, entire to deeply lobed, the upper ones reduced in size; the uppermost often irregularly alternate, subtending large, yellow, solitary, pediceled flowers.

*Key to species*

1 Stems and leaves glabrous and glaucous ................ *A. flava*
1 Stems and leaves long-hairy, the hairs often somewhat sticky.... 2

2 Corolla and pedicels glandular hairy on outside; stems and leaves with scattered, sticky glandular hairs .............. *A. pedicularia*
2 Corolla glabrous; pedicels, stems and leaves without glandular, sticky hairs ...................................... *A. grandiflora*

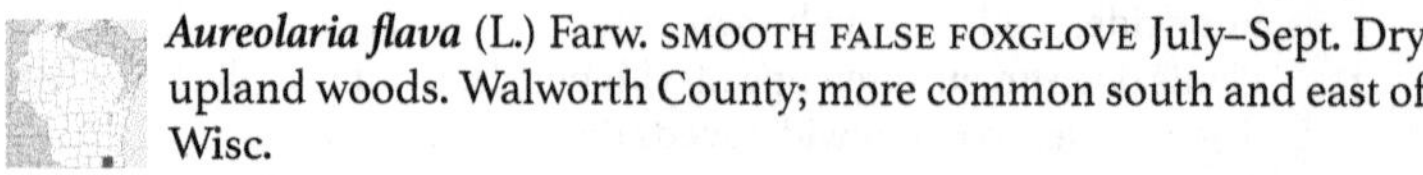

***Aureolaria flava*** (L.) Farw. SMOOTH FALSE FOXGLOVE July–Sept. Dry upland woods. Walworth County; more common south and east of Wisc.

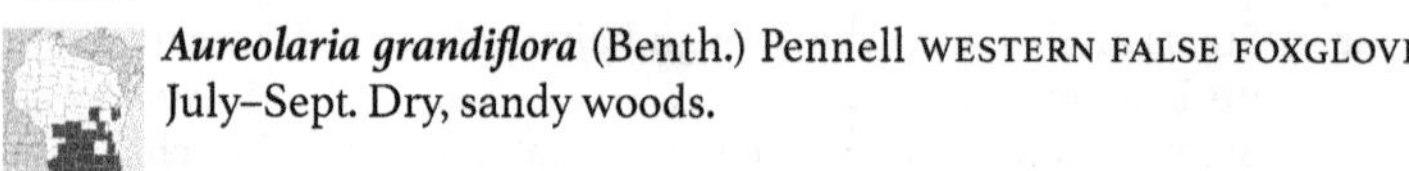

***Aureolaria grandiflora*** (Benth.) Pennell WESTERN FALSE FOXGLOVE July–Sept. Dry, sandy woods.

***Aureolaria pedicularia*** (L.) Raf. [•71] ANNUAL FALSE FOXGLOVE Aug–Sept. Dry upland woods.

## CASTILLEJA *Indian-Paintbrush*

Annual or perennial hemiparasitic herbs. Leaves alternate, often pinnatifid. Flowers in dense terminal spikes, each subtended by a large, entire or pinnatifid, sometimes brightly colored, bracteal leaf.

*Key to species*

1 Floral bracts green; plants perennial .............. *C. sessiliflora*
1 Floral bracts red, scarlet, orange or yellow; plants annual or biennial .................................................. *C. coccinea*

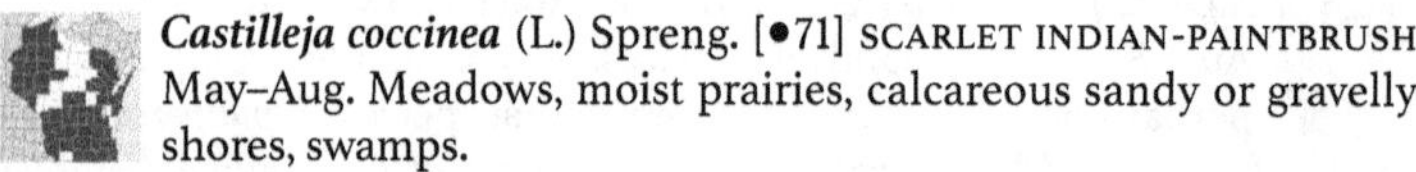

***Castilleja coccinea*** (L.) Spreng. [•71] SCARLET INDIAN-PAINTBRUSH May–Aug. Meadows, moist prairies, calcareous sandy or gravelly shores, swamps.

***Castilleja sessiliflora*** Pursh GREAT PLAINS INDIAN-PAINTBRUSH May–July. Dry prairies and plains.

## CONOPHOLIS *Squawroot*

***Conopholis americana*** (L.) Wallr. [•72] AMERICAN SQUAWROOT Unbranched perennial herb; parasitic on the roots of oaks (especially Quercus rubra). Stems entirely or mostly concealed by the numerous, fleshy, overlapping leaf-scales, pale brown or yellowish throughout; corolla tubular, curved downward. May–June. Deciduous or mixed forests with oak trees.

## EPIFAGUS *Beechdrops*

***Epifagus virginiana*** (L.) W. Bart. [•72] BEECHDROPS Freely branched annual herb; parasitic on the roots of beech trees, with small, scattered, alternate leaf scales and numerous, nearly sessile, solitary, axillary flowers, forming a large panicle. Stems pale brown, usually marked with fine brown-purple lines; the dead dry stems of the previous season persist through the winter and into the next summer. Leaf-scales triangular-ovate. Flowers dimorphic, the lower small, pistillate, fertile, the upper perfect but sterile; upper flowers white, commonly with two stripes of brown-purple. Aug–Sept. In forests with beech and usually other trees, such as sugar maple, hemlock, oak.

### EUPHRASIA *Eyebright*

***Euphrasia stricta*** D. Wolff ex J.F. Lehm. DRUG EYEBRIGHT Introduced annual herb. Stems usually freely branched below the middle. Leaves opposite, sessile, ovate, sharply 3–5-toothed on each margin, the teeth tipped with a hair-like bristle. Flowers small, sessile or nearly so; corolla bilabiate, lower lip 3-lobed, white with violet lines; upper lip shallowly 2-lobed or merely notched, suffused with purple. Dry fields, lawns, clearings, trails, roadsides, old railroad grades. *Euphrasia officinalis* L.

### MELAMPYRUM *Cow-Wheat*

***Melampyrum lineare*** Desr. [•72] AMERICAN COW-WHEAT Annual herb, partially parasitic on other plants, often red-tinged when in open habitats. Leaves opposite, lower leaves oblong lance-shaped, upper leaves linear or lance-shaped, often toothed near base. Flowers from upper leaf axils; corolla 2-lipped, the upper lip white, the lower pale yellow. June–Aug. In a wide variety of habitats, ranging from wet to dry forests and openings; in wetlands occasional in swamps and on hummocks in open fens.

### ODONTITES *Eyebright*

***Odontites vulgaris*** Moench EYEBRIGHT Annual herb, usually parasitic on the roots of other plants. Leaves opposite, lance-shaped, sessile with a broad base, roughly pubescent, with 2 or 3 blunt teeth on each margin. Flowers nearly sessile from the upper axils, forming a terminal, often secund spike or raceme; corolla tubular, pubescent, light red. July–Sept. Native of Europe; established as a weed in fields and waste places.

### OROBANCHE *Broom-Rape*

Plants of diverse aspect with white to yellowish or purple flowers in racemes or spikes, or apparently solitary. Corolla tubular, bilabiate.

*Key to species*

1 Flowers many, sessile or nearly so, in a dense spike ***O. ludoviciana***
1 Flowers 1 or several on long pedicels .......................... 2

2 Flowers solitary, white, cream or lilac; the leaves scale-like and glabrous ........................................... ***O. uniflora***
2 Flowers 2 or more, purple; the scale-like leaves pubescent ....... ............................................ ***O. fasciculata***

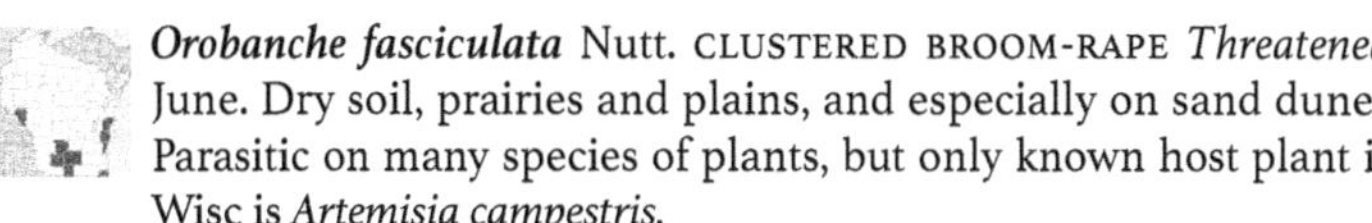

***Orobanche fasciculata*** Nutt. CLUSTERED BROOM-RAPE *Threatened.* June. Dry soil, prairies and plains, and especially on sand dunes. Parasitic on many species of plants, but only known host plant in Wisc is *Artemisia campestris.*

***Orobanche ludoviciana*** Nutt. [•72] PRAIRIE BROOM-RAPE *Endangered.* June–Aug. Sandy soil. Parasitic on members of Asteraceae, most commonly on *Artemisia* and *Ambrosia.*

***Orobanche uniflora*** L. NAKED BROOM-RAPE Moist woods and streambanks. Parasitic on numerous plant species.

### PEDICULARIS *Lousewort*

Perennial herbs (ours). Leaves either opposite, alternate, or scattered, sharply toothed to 2-pinnatifid. Flowers yellow or purple in terminal spikes or racemes, each subtended by a bracteal leaf.

*Key to species*

1 Stems glabrous or nealy so; leaves opposite; flowering in late summer . . . . . . . . . . . . . . . . . . . . . . . . . . . . . . . . . . . . . . . . . . ***P. lanceolata***

1 Stems usually long-hairy; leaves alternate; flowering in early summer . . . . . . . . . . . . . . . . . . . . . . . . . . . . . . . . . . . . . . . . . ***P. canadensis***

***Pedicularis canadensis*** L. WOOD BETONY April–June. Dry forests and savannas, moist hardwood forests, especially in openings; less often in conifer swamps, meadows, and grasslands.

***Pedicularis lanceolata*** Michx. [•72] SWAMP LOUSEWORT July–Sept. Wet meadows, calcareous fens, wetland margins, streambanks.

### RHINANTHUS *Yellow Rattle*

***Rhinanthus minor*** L. LITTLE YELLOW RATTLE Introduced annual herb. Stems pubescent on two sides, glabrous on the other two. Leaves narrowly oblong to lanceolate, conspicuously serrate. Corolla mostly yellow, often with pale or colored teeth on the upper lip or with dark markings on the lower. Roadsides and waste places; Door County. *Rhinanthus crista-galli* L.

---

## OXALIDACEAE *Wood-Sorrel Family*

### OXALIS *Wood-Sorrel*

Perennial herbs (ours). Leaves basal or alternate on the stem, 3-foliolate. Flowers solitary on axillary peduncles or in cymose or umbel-like clusters, 5-merous, white, yellow, pink, or purple.

*Key to species*

1 Plants with leaves from stems; flowers yellow . . . . . . . . . . . . . . . . . . 2

1 Plants without stems, the leaves and scapes all from plant base; flowers white to purple . . . . . . . . . . . . . . . . . . . . . . . . . . . . . . . . . . . . . . 3

2 Pubescence largely of septate hairs, the capsules glabrous or with only septate ± spreading hairs; pedicels remaining erect or ascending in fruit; stipules absent . . . . . . . . . . . . . . . . . . . . . . ***O. stricta***

2 Pubescence without or with very few septate hairs, the capsules with minute retrorse non-septate hairs; pedicels usually becoming ± strongly reflexed in fruit (but the capsules erect); stipules often evident, ± oblong, adnate to base of petiole . . . . . . . . . ***O. corniculata***

3 Peduncles with 1 flower . . . . . . . . . . . . . . . . . . . . . . . . . . . . ***O. montana***

3 Peduncles with 2 or more flowers . . . . . . . . . . . . . . . . . . . . ***O. violacea***

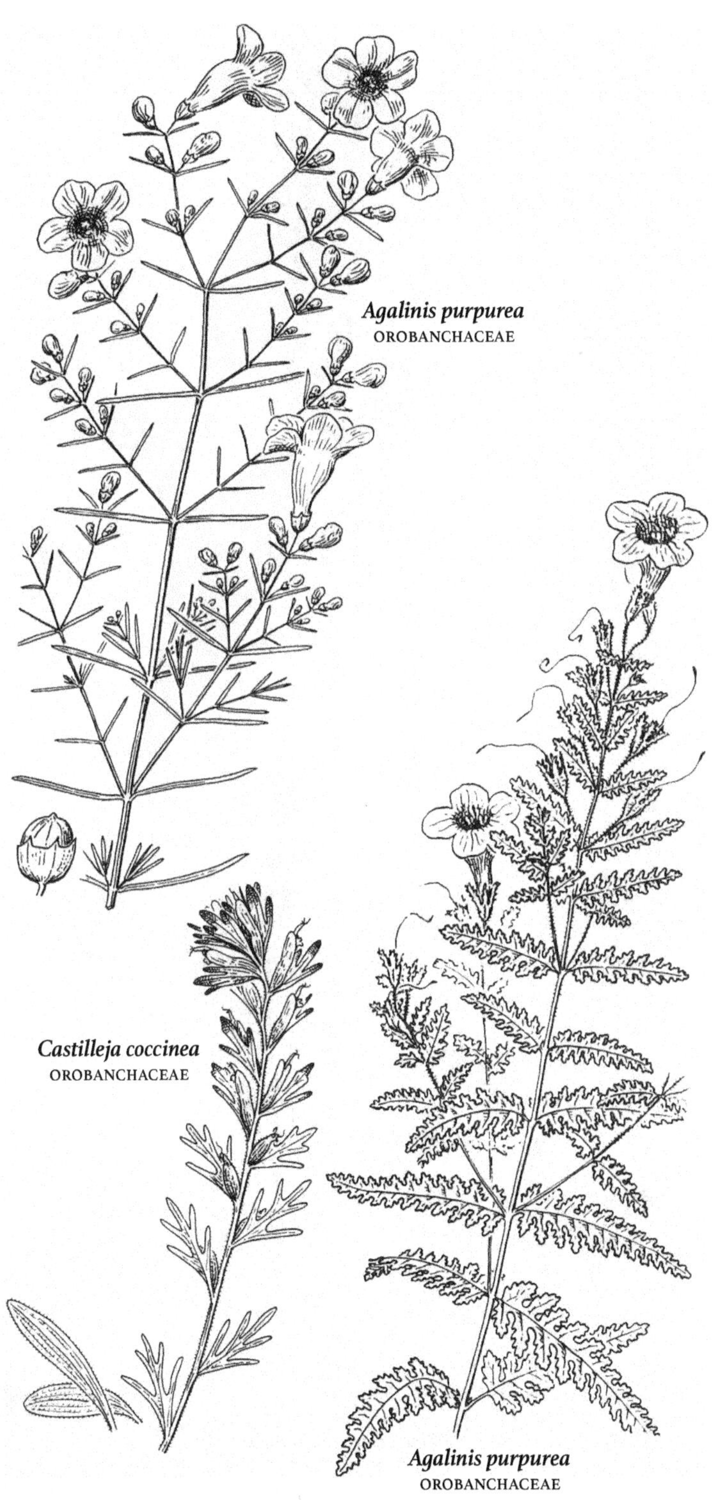
*Agalinis purpurea*
OROBANCHACEAE
*Castilleja coccinea*
OROBANCHACEAE
*Agalinis purpurea*
OROBANCHACEAE

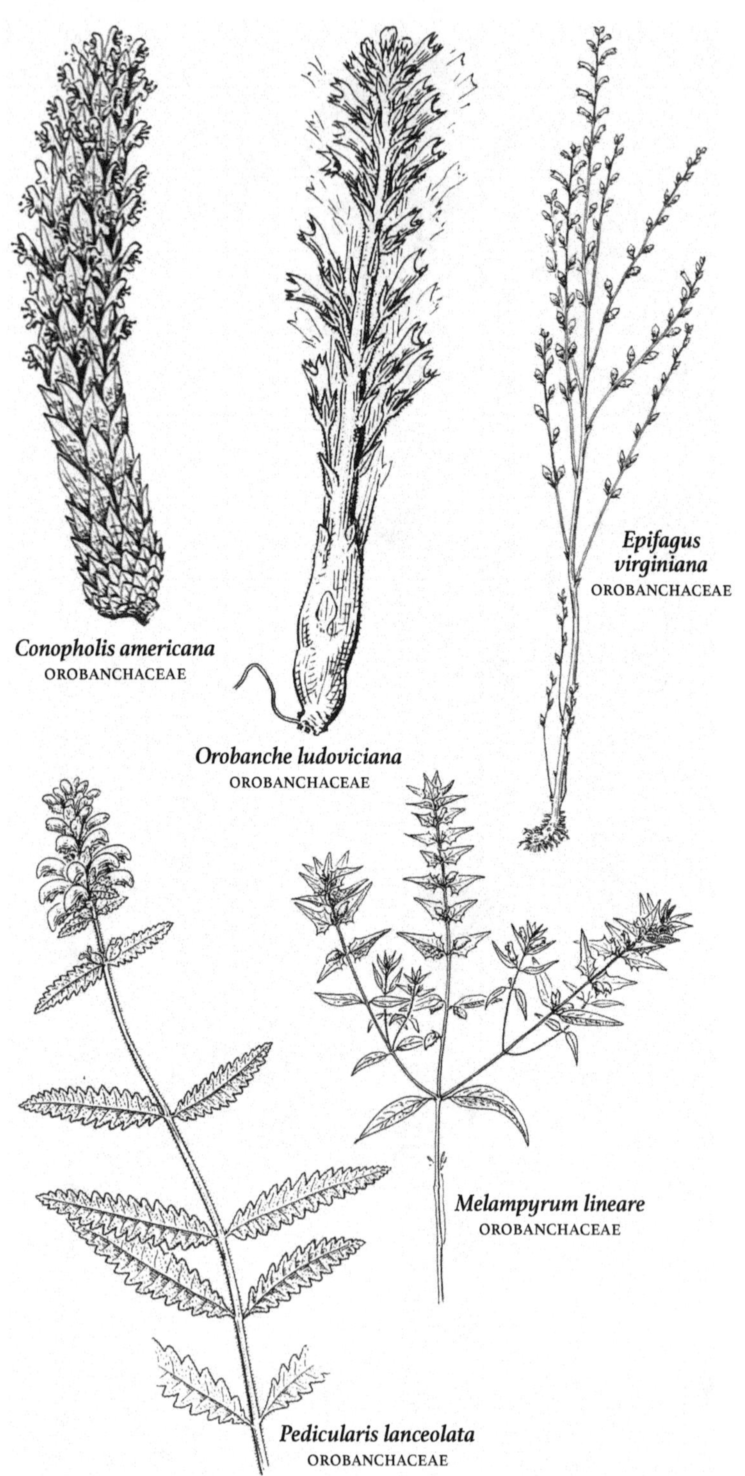
*Conopholis americana*
OROBANCHACEAE
*Orobanche ludoviciana*
OROBANCHACEAE
*Epifagus virginiana*
OROBANCHACEAE
*Melampyrum lineare*
OROBANCHACEAE
*Pedicularis lanceolata*
OROBANCHACEAE

***Oxalis corniculata*** L. CREEPING YELLOW WOOD-SORREL Introduced. Occasional in lawns and gardens.

***Oxalis montana*** Raf. NORTHERN WOOD-SORREL May–July. Hummocks in swamps, wet depressions in forests, moist wetland margins. *Oxalis acetosella* L.

***Oxalis stricta*** L. [•73] COMMON YELLOW WOOD-SORREL In many different habitats, mostly a common weed of roadsides, railroads, gardens, lawns, fields, and disturbed places; also in forests, especially along trails.

***Oxalis violacea*** L. [•73] VIOLET WOOD-SORREL April–June, and occasionally later. Dry upland woods and prairies.

## PAPAVERACEAE *Poppy Family*

Herbs or vines (*Adlumia*), with watery, milky, or colored juice. Leaves alternate or rarely opposite. Flowers regular, perfect. Sepals 2 or 3, early deciduous. Petals 4 or more (rarely absent), separate, conspicuous. Fruit a capsule, dehiscent by terminal valves or longitudinally (rarely otherwise). The Fumariaceae, now included as a subfamily of Papaveraceae, were previously recognized as a separate family, differing in bilateral symmetry of the flowers and watery juice. All of our members of Papaveraceae in the strict sense have colored juice (yellow to red-orange or milky).

*Key to genera*

1 Corolla bilaterally symmetrical; juice watery, clear ............. 2
1 Corolla regular, juice colored whitish to yellow or red-orange ... 6

2 Delicate vine ........................................ ADLUMIA
2 Upright herbaceous plants .................................... 3

3 Corolla white, with 2 spurs; leaves basal ........... DICENTRA
3 Corolla pink, yellow or red-purple, 1-spurred; leaves alternate... 4

4 Flowers red-purple; fruit globose .................. FUMARIA
4 Flowers yellow or pink with yellow tips; fruit a slender capsule.. 5

5 Flowers pink, yellow-tipped; seeds 1–1.5 mm wide; plants ± erect, the terminal inflorescences definitely surpassing the leaves ...... ............................................. CAPNOIDES
5 Flowers yellow or purple; seeds 1.8–2.2 mm wide; plants ± spreading or sprawling, the terminal inflorescences barely if at all surpassing the leaves ............................................ CORYDALIS

6 Leaf margins spiny-toothed ...................... ARGEMONE
6 Leaves not spiny ............................................. 7

7 Leaf 1 from base of plant; petals 8 or more ...... SANGUINARIA
7 Plants with leafy stems; petals 4 ............................. 8

8 Flowers yellow ............................. CHELIDONIUM
8 Flowers red, purple, or white ........................ PAPAVER

## ADLUMIA *Allegheny-Vine*

***Adlumia fungosa*** (Ait.) Greene ALLEGHENY-VINE Biennial climbing vine. During the first year acaulescent, with several ascending, non-prehensile, decompound leaves; climbing to 3 m high the second year. Flowers pearly pink, in drooping axillary panicles; corolla bilateral. June–Sept. Woods, rocky shores, thickets; sometimes where soil has been disturbed.

## ARGEMONE *Prickly Poppy*

***Argemone albiflora*** Hornem. WHITE PRICKLY POPPY Introduced annual herb, glaucous; stems, leaves, and sepals spiny. Flowers white, terminating the branches on peduncles; petals usually 4; stamens numerous, more than 100. Escape from cultivation on waste ground and roadsides.

## CAPNOIDES *Rock-Harlequin*

***Capnoides sempervirens*** (L.) Borkh. [•73] ROCK-HARLEQUIN Biennial herb, glaucous. Lower leaves petioled, upper leaves nearly sessile. Flowers in small panicles at the end of the branches; corolla pink, tipped with yellow. Dry or rocky woods, gravelly shores, especially where disturbed. *Corydalis sempervirens* (L.) Pers.

## CHELIDONIUM *Celandine*

***Chelidonium majus*** L. CELANDINE Biennial herb with saffron-colored juice. Stem leaves several, alternate, deeply pinnately parted (usually to the midrib) into 5–9 segments. Umbel peduncled, several-flowered; petals 4, yellow. April–Sept. Native of Eurasia; established in moist soil.

## CORYDALIS *Fumewort*

Annual or biennial herbs. Leaves cauline, alternate, 2-pinnately dissected. Flowers short-pediceled, in bracted racemes.

*Key to species*

1 Outer petals with well developed wings ............ ***C. micrantha***
1 Outer petals not winged ............................... ***C. aurea***

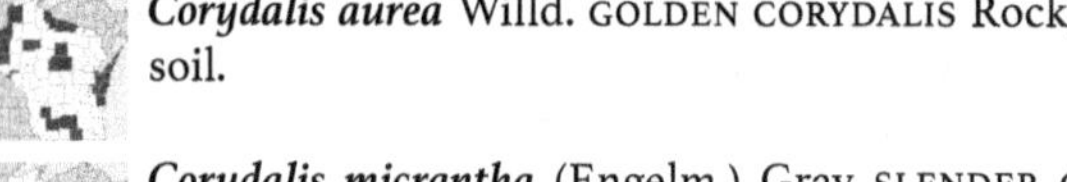

***Corydalis aurea*** Willd. GOLDEN CORYDALIS Rocky banks or sandy soil.

***Corydalis micrantha*** (Engelm.) Gray SLENDER CORYDALIS Moist, especially sandy soil.

## DICENTRA *Bleedinghearts*

Perennial herbs from rhizomes or a cluster of small tubers. Leaves basal or alternate, compound. Flowers white to red-purple, scapose or in axillary racemes or panicles.

*Key to species*

1 Corolla sac-like, with small rounded spurs about as long as wide; leaves waxy on underside ......................... ***D. canadensis***

1 Corolla with widely spreading spurs, these longer than wide; leaves green or only slightly waxy ........................ *D. cucullaria*

***Dicentra canadensis*** (Goldie) Walp. [•73] SQUIRREL-CORN May. Rich deciduous forests, rarely in swampy or dry forests. Very similar to *D. cucullaria* in foliage, size, and habit. Tubers fewer and about 2x larger; corolla narrowly ovate, the spurs short, broadly rounded, scarcely divergent.

***Dicentra cucullaria*** (L.) Bernh. DUTCHMAN'S-BREECHES May. Rich deciduous forests, occasionally in swampy or relatively dry woods.

### FUMARIA *Fumitory*

***Fumaria officinalis*** L. FUMITORY Introduced annual herb with the general aspect of Corydalis, with dissected foliage and small flowers in racemes. Racemes dense, many-flowered; corolla tube red-purple, the summit dark red. May–Sept. Occasional in waste places.

### PAPAVER *Poppy*

Annual herbs (ours), with colored juice and large, usually long-peduncled flowers terminating the stem and branches. Petals normally 4, white or colored; stamens numerous.

ADDITIONAL SPECIES ***Papaver orientale*** L. (Oriental poppy), introduced, reported from se Wisc.

*Key to species*

1 Stem leaves clasping at base ...................... ***P. somniferum***
1 Stem leaves not clasping .............................. ***P. rhoeas***

***Papaver rhoeas*** L. CORN POPPY Native of Eurasia and n Africa; introduced or escaped but seldom abundant.

***Papaver somniferum*** L. OPIUM POPPY June–Aug. Native of Eurasia, cultivated for ornament and sometimes escaped. The seeds are commonly used in baking; opium is derived from the milky juice of the capsule.

### SANGUINARIA *Bloodroot*

***Sanguinaria canadensis*** L. [•74] BLOODROOT Perennial herb with red juice, from a stout rhizome which sends up a single lobed leaf and a large white scapose flower. Leaves orbicular in outline, 3–9-lobed. Flowers white, varying rarely to pink; petals typically 8, but often more. March–April. Rich deciduous and floodplain forests. The leaves continue to expand after anthesis and may grow to 2 dm wide.

## PENTHORACEAE *Penthorum Family*

### PENTHORUM *Ditch-Stonecrop*

***Penthorum sedoides*** L. DITCH-STONECROP Perennial herb, spreading by rhizomes; plants often red-tinged. Stems smooth and round in section below, upper stem often angled and with gland-tipped hairs.

Leaves alternate, lance-shaped; margins with small, forward-pointing teeth. Flowers star-shaped, perfect, in branched racemes at ends of stems; sepals 5, green; petals usually absent. July–Sept. Muddy shores, ditches.

---

## PHRYMACEAE *Lopseed Family*

Perennial herbs. Calyx tubular, 5-lobed. Fruit a dehiscent capsule. Previously, this family was monotypic with only genus *Phryma;* now includes Wisc genera *Mazus* and *Mimulus.*

*Key to genera*

1 Flowers nearly sessile in pairs in terminal spike-like racemes, subtended by tiny bracts; upper calyx teeth bristle-like; fruit an achene, strongly reflexed .......... **PHRYMA**

1 Flowers peduncled, borne singly in the axils of opposite leaves or bracts with expanded blades or alternately in few-flowered bracteate racemes; calyx teeth not bristle-like; fruit a capsule, not strongly reflexed .......... **2**

2 Flowers borne singly in the axils of opposite leaves or bracts; bracts with expanded blades; statewide .......... **MIMULUS**

2 Flowers borne alternately in few-flowered racemes; bracts linear, often borne on the peduncle slightly above the node; Dane County .......... **MAZUS**

### MAZUS

***Mazus pumilus*** (Burm. f.) Steenis JAPANESE MAZUS Creeping perennial. Leaves chiefly basal; stem leaves much smaller. Flowers blue, alternate in a terminal raceme. Summer. Native of e Asia; cultivated for ornament and tending to escape in lawns; reported from Dane County.

### MIMULUS *Monkey-Flower*

Perennial herbs (ours). Leaves opposite, margins shallowly toothed. Flowers often large and showy, single on stalks from leaf axils or in leafy racemes at ends of stems; corolla 2-lipped, the upper lip 2-lobed, the lower lip 3-lobed, yellow or blue-violet; stamens 4, of 2 different lengths.

ADDITIONAL SPECIES ***Mimulus moschatus*** Dougl. ex Lindl., native plants with yellow corolla and lance-shaped calyx lobes; Bayfield County.

*Key to species*

1 Flowers blue to violet; leaves lance-shaped or oblong lance-shaped .......... ***M. ringens***

1 Flowers yellow; leaves nearly round .......... ***M. glabratus***

***Mimulus glabratus*** Kunth ROUND-LEAF MONKEY-FLOWER June–Aug. Cold springs, seeps, and banks of spring-fed streams; usually where calcium-rich.

***Mimulus ringens*** L. [•74] ALLEGHENY MONKEY-FLOWER July–Aug. Streambanks, oxbow marshes, swamp openings, floodplain forests, muddy shores, ditches; sometimes where disturbed.

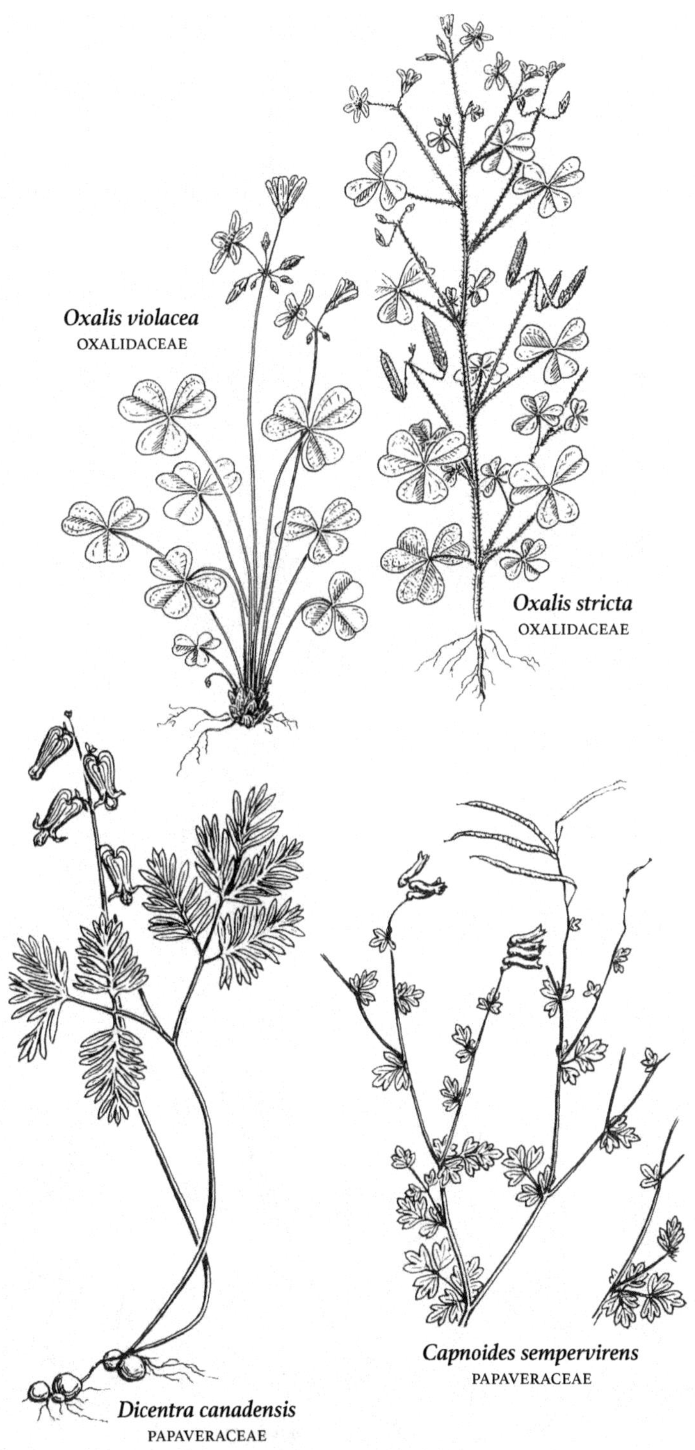
*Oxalis violacea*
OXALIDACEAE
*Oxalis stricta*
OXALIDACEAE
*Capnoides sempervirens*
PAPAVERACEAE
*Dicentra canadensis*
PAPAVERACEAE

***Mimulus ringens***
PHRYMACEAE

***Phryma leptostachya***
PHRYMACEAE

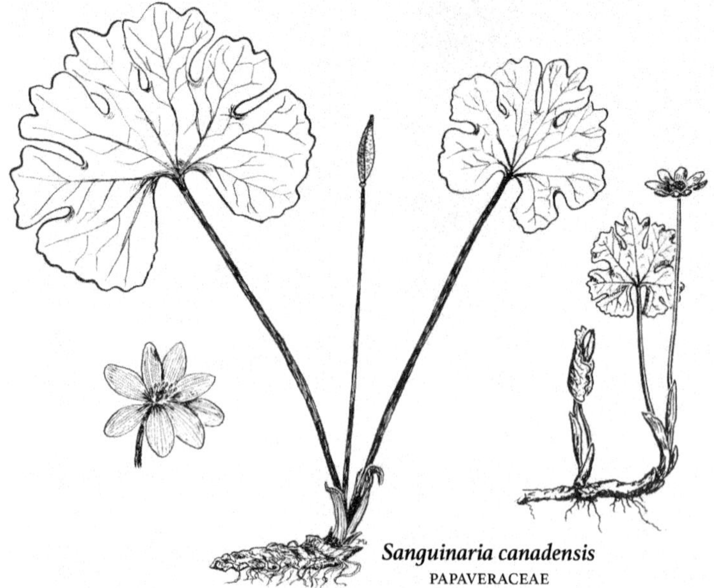

***Sanguinaria canadensis***
PAPAVERACEAE

### PHRYMA *Lopseed*

***Phryma leptostachya*** L. [•74] AMERICAN LOPSEED Perennial herb. Leaves opposite, ovate. Flowers pale purple to white, in elongate, long-peduncled, interrupted spike-like racemes terminating the stem and also from a few upper axils, opposite and horizontal. June–Aug. Rich deciduous forests, especially moist areas in beech-maple woods, but also in drier forests with oak and sometimes with conifers. Recognized by the distant paired flowers, reflexed fruit, and broad, opposite, petioled leaves.

## PHYTOLACCACEAE *Pokeweed Family*

### PHYTOLACCA *Pokeweed*

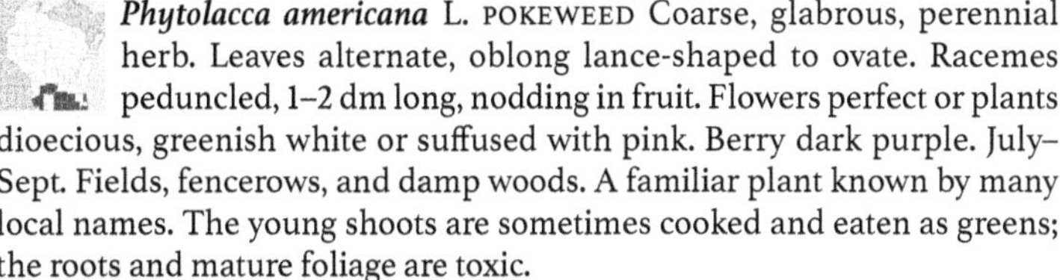

***Phytolacca americana*** L. POKEWEED Coarse, glabrous, perennial herb. Leaves alternate, oblong lance-shaped to ovate. Racemes peduncled, 1–2 dm long, nodding in fruit. Flowers perfect or plants dioecious, greenish white or suffused with pink. Berry dark purple. July–Sept. Fields, fencerows, and damp woods. A familiar plant known by many local names. The young shoots are sometimes cooked and eaten as greens; the roots and mature foliage are toxic.

ADDITIONAL SPECIES ***Phytolacca icosandra*** L. (Tropical pokeweed), reported from Dane County.

## PLANTAGINACEAE *Plantain Family*

Annual or perennial herbs. Leaves simple, entire, all from base of plant. Flowers perfect in a narrow spike (*Plantago*), or single-sexed, the staminate and pistillate flowers on same plant (*Littorella*); flower parts mostly in 4s. Fruit a capsule opening at tip. Plantaginaceae is the accepted name for the family that encompasses not only the plantains with their reduced flowers, but also the related larger-flowered genera formerly placed in the Scrophulariaceae, as well as highly reduced aquatics, such as *Hippuris* (Hippuridaceae) and *Callitriche* (Callitrichaceae).

ADDITIONAL SPECIES ***Antirrhinum majus*** L. (Garden snapdragon), familiar cultivated perennial; rarely escaping in se Wisc. ***Bacopa rotundifolia*** (Michx.) Wettst. (Disk water-hyssop), native perennial succulent herb, of mud flats and shallow water, spreading by stolons; flowers 1–2 from leaf axils, corolla white with a yellow throat, 2-lipped; Jefferson County. ***Digitalis purpurea*** L. (Common foxglove); cultivated for its showy flowers, occasional escape in Wisc.

*Key to genera*

1 Flowers tiny, lacking a corolla, or corolla regular and scarious . . . 2

1 Flowers usually conspicuous, with both calyx and corolla present, the corolla petal-like, usually conspicuously bilaterally symmetrical . 6

2 Leaves in a basal rosette . . . . . . . . . . . . . . . . . . . . . . . . . . . . . . . . . . . 3

2 Leaves opposite or whorled on an elongate stem . . . . . . . . . . . . . . . 4

3 Leaves terete (to ca. 3 mm thick at middle, then tapering to apex), at most 1-veined, glabrous; flowers unisexual (the staminate long-

stalked, the pistillate basal); fruit indehiscent; submersed or on moist shores .................................... LITTORELLA

3 Leaves flat, in most species with at least 3 prominent veins and/or pubescent; flowers bisexual (in heads or spikes); capsule circumscissile; dry or rarely wet habitats ............ PLANTAGO

4 Leaves in whorls of 6–12 (usually 9) .................. HIPPURIS

4 Leaves opposite ............................................. 5

5 Aquatic plants; flowers solitary in the leaf axils, corolla absent ... ............................................... CALLITRICHE

5 Introduced plants of dry places; flowers many in short spikes; corolla present ................................ *Plantago arenaria*

6 Stem leaves all or mostly alternate on fertile stems (lowermost leaves sometimes opposite and rosette of larger basal leaves sometimes present)............................................. 7

6 Stem leaves all or mostly opposite (rarely whorled) on fertile stems (may be alternate beneath flowers) .......................... 14

7 Corolla nearly regular, the lobes equaling or exceeding the tube . ............................................... VERONICA

7 Corolla bilaterally symmetrical, ± 2-lipped, the lobes distinctly shorter than the tube (including spur, if any)................... 8

8 Stem trailing or sprawling; leaf blades not over 1.5 times as long as broad; corolla with basal spur; capsules ± spherical, 3–4.5 mm in diameter ................................................ 9

8 Stem erect; leaf blades (or their principal lobe) over 1.5 times as long as broad; corolla spurred (*Chaenorrhinum, Linaria*) or not; capsules various, mostly longer than broad .......................... 10

9 Stem, leaves, and calyx glabrous; leaf blades palmately veined and lobed or scalloped; seeds with thin raised reticulate ornamentation ............................................... CYMBALARIA

9 Stems, leaves, and calyx pilose; leaf blades pinnately veined, usually hastate (1–2 pairs of small basal lobes); seeds covered with dense convolute-rounded ornamentation ................... KICKXIA

10 Corolla with a slender basal spur projecting back between the lower calyx lobes ................................................ 11

10 Corolla without spur (at most swollen or saccate at base) ...... 13

11 Flowers all solitary in axils of leaves (nearly to base of plant); corolla pale purple and white; leaves, calyx, and stem with ± dense gland-tipped hairs ............................ CHAENORRHINUM

11 Flowers in compact or elongate terminal inflorescences (half or less the height of the plant); corolla yellow, red, or blue; leaves, calyx, and usually stem glabrous and eglandular or nearly so ........ 12

12 Corolla 1.3–4 cm long (including spur), yellow (red-pink in a rare weedy annual); seeds strongly wrinkled, tuberculate, ridged, or winged ................................................ LINARIA

12 Corolla 0.6–1.1 cm long, blue; seeds smooth or weakly pebbled .. ............................................ NUTTALLANTHUS

13 Corolla half or more covered by the calyx ............ BESSEYA

13 Corolla much less than half covered by the calyx ............... ........................................ *Antirrhinum majus**

14 Inflorescence terminal and branched (± paniculate); stamens 4 fertile plus 1 staminodium ...................... PENSTEMON

14 Inflorescence a spike or raceme (no branched stalks), or flowers all axillary; stamens 2 or 4 fertile, in most genera with no staminodium (or only a very rudimentary one) .......................... 15

15 Sepals (at least at anthesis) fused one-third or more the length of the calyx; fertile stamens 4; historical record from Rock County, now likely absent from state ......................... *Collinsia verna*

15 Sepals separate nearly or quite to the base; fertile stamens 2 or 4 . 16

16 Corolla 2.3–3.5 cm long; sepals broadly ovate-orbicular, overlapping; stamens 4 fertile plus a filamentous elongate staminodium ................................. CHELONE

16 Corolla less than 1.5 cm long; sepals linear-lanceolate to somewhat ovate, not conspicuously overlapping; stamens 2 or 4 (including any staminodia) ............................................. 17

17 Corolla with a spur projecting back at the base; plant with ± dense gland-tipped hairs; leaves linear; fertile stamens 4 .............. ....................................... CHAENORHINUM

17 Corolla not spurred; plant glabrous or with eglandular hairs (or if with gland-tipped hairs, the leaves not linear); fertile stamens 2 (staminodia filamentous, reduced, or none) ................. 18

18 Leaves in whorls of 3–6, sharply toothed; inflorescence of 1–several dense elongate slenderly tapering spikes or spike-like racemes; corolla tube much longer than the lobes .... VERONICASTRUM

18 Leaves opposite, entire or toothed; inflorescence racemose or flowers solitary in axils of alternate or opposite bracts or leaves; corolla tube various ...................................... 19

19 Corolla 2-lipped, the tube much longer than the lobes; flowers solitary in axils of opposite leaves; sepals 5 .......... GRATIOLA

19 Corolla often nearly regular, the tube shorter than the lobes (usually a flat limb); flowers in axillary racemes or solitary in axils of bracts or leaves; sepals 4 ............................... VERONICA

## BESSEYA *Kitten-Tails*

***Besseya bullii*** (Eat.) Rydb. KITTEN-TAILS *Threatened.* Perennial herb. Scape unbranched, hirsute or villous, with several reduced leaves and a dense, terminal, spike-like raceme of small yellow flowers. Leaves in a basal rosette. Corolla strongly bilabiate, yellow. April–June. Sandy soil of prairies, open woods, bluffs, barrens, and hillsides.

## CALLITRICHE *Water-Starwort*

Small, perennial aquatic herbs with weak, slender stems. Leaves simple, opposite, all underwater or upper leaves floating; underwater leaves linear, 1-nerved, entire except for shallowly notched tip; floating leaves mostly in clusters at ends of stems, obovate to spatula-shaped, 3–5-nerved, rounded at tip. Flowers tiny, staminate and pistillate flowers usually separate on same plant, each flower with 1 stamen or 1 pistil; single and stalkless in middle and upper leaf axils, or 1 staminate and 1 pistillate flower in each axil.

ADDITIONAL SPECIES ***Callitriche marginata*** Torr., reported for Oneida County in northern Wisc.

*Key to species*

1 Plants terrestrial; fruit on a short pedicel ............. ***C. terrestris***
1 Plants aquatic; fruit sessile ................................... 2

2 Leaves all underwater, 1-veined, linear ........ ***C. hermaphroditica***
2 Leaves both underwater and floating; floating leaves 3-veined, spatula-shaped or obovate ................................... 3

3 Margins of fruit without wings; pits on fruit not in rows ......... ............................................ ***C. heterophylla***
3 Margins of fruit with small wings; fruit pitted in rows .......... 4

4 Fruit obovate, less than 1.5 mm long, slightly longer than wide ... ............................................ ***C. palustris***
4 Fruit nearly round in outline, 1.5–2 mm long and as wide ........ ............................................ ***C. stagnalis***

***Callitriche hermaphroditica*** L. AUTUMN WATER-STARWORT June–Sept. Shallow to deep water of lakes, ponds, marshes, ditches and streams.

***Callitriche heterophylla*** Pursh LARGE WATER-STARWORT *Threatened.* May–Aug. Shallow water or mud of springs, stream pools, ponds and wet depressions.

***Callitriche palustris*** L. [•75] VERNAL WATER-STARWORT June–Sept. Shallow water of lakes, ponds, streams; exposed mudflats.

***Callitriche stagnalis*** Scop. POND WATER-STARWORT Introduced. Cold ponds and streams.

***Callitriche terrestris*** Raf. TERRESTRIAL WATER-STARWORT Damp, usually shaded soil.

## CHAENORHINUM *Dwarf-Snapdragon*

***Chaenorhinum minus*** (L.) Lange DWARF-SNAPDRAGON Annual herb. Stems glandular-pubescent. Leaves linear. Corolla blue-purple, with yellow on the palate. June–Sept. Native of Europe; established in waste places, especially on railway ballast.

## CHELONE *Turtlehead*

***Chelone glabra*** L. WHITE TURTLEHEAD Perennial herb. Stems rounded 4-angled. Leaves opposite, lance-shaped, margins with sharp, forward-pointing teeth. Flowers in dense spikes at ends of stems; corolla white or light pink. Aug–Sept. Swamp openings, thickets, streambanks, shores, wet meadows, marshes, calcareous fens.

## COLLINSIA *Blue-Eyed Mary*

***Collinsia verna*** Nutt. SPRING BLUE-EYED MARY Annual herb, may form colonies. Stem leaves opposite, the uppermost often whorled. Flowers in 1–

3 whorls, also sometimes solitary from the axils of the upper stem leaves; upper lip normally white, varying to pale blue; lower lip bright blue. April–May, plants disappear in early summer after their seeds ripen. Rich moist woods, especially in alluvial soil. Historical record from Rock County; more common south and east of Wisc.

### CYMBALARIA *Kenilworth-Ivy*

***Cymbalaria muralis*** P.G. Gaertn. B. Mey. & Scherb. KENILWORTH-IVY Annual glabrous herb. Stems trailing, rooting at the nodes. Leaves alternate, nearly orbicular in outline with 3–7 shallow palmate lobes. Flowers solitary in the axils, on long slender pedicels; corolla bilabiate, distinctly spurred at base, blue with yellow palate. Summer. Native of Eurasia; commonly cultivated, escaped near gardens.

### GRATIOLA *Hedge-Hyssop*

Low annual or perennial herbs of shallow water and shores. Leaves opposite. Flowers on stalks from leaf axils; corolla white or yellow, 2-lipped, the upper lip entire or 2-lobed, the lower lip 3-lobed.

*Key to species*

1 Plants perennial, spreading by rhizomes; flowers bright yellow; leaves entire, widest at base .......... ***G. aurea***

1 Plants annual, rhizomes absent; flowers white; leaves toothed, widest near middle of blade .......... ***G. neglecta***

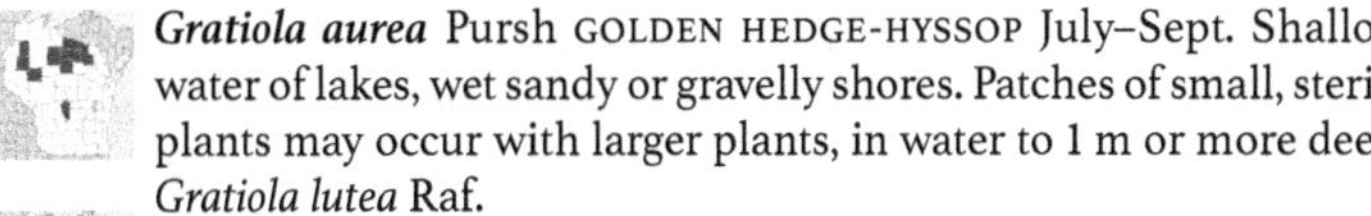

***Gratiola aurea*** Pursh GOLDEN HEDGE-HYSSOP July–Sept. Shallow water of lakes, wet sandy or gravelly shores. Patches of small, sterile plants may occur with larger plants, in water to 1 m or more deep. *Gratiola lutea* Raf.

***Gratiola neglecta*** Torr. [•75] CLAMMY HEDGE-HYSSOP June–Sept. Mud flats, shores of ponds and marshes.

### HIPPURIS *Mare's-Tail*

***Hippuris vulgaris*** L. COMMON MARE'S-TAIL Perennial herb, from large, spongy rhizomes. Stems unbranched, underwater and lax, or emersed and upright, densely covered by the closely spaced whorls of leaves. Flowers very small, perfect, stalkless and single in upper leaf axils, or often absent. June–Aug. Shallow water or mud of marshes, lakes, streams and ditches.

### KICKXIA *Cancerwort*

Our two species pubescent annuals, prostrate but not creeping. Leaves alternate, broadly ovate. Flowers small, yellow and purple, long pediceled and solitary in the axils; corolla bilabiate, spurred at the base.

*Key to species*

1 Leaves rounded-ovate, rounded to cordate at base; pedicels long-hairy .......... ***K. spuria***

1 Leaves triangular-ovate, the base straight; pedicels mostly glabrous .......... ***K. elatine***

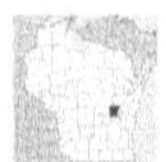

*Kickxia elatine* (L.) Dumort. SHARP-LEAF CANCERWORT June–Sept. Native of Eurasia; moist sandy soil, Winnebago County. Flowers essentially as in *K. spuria.*

*Kickxia spuria* (L.) Dumort. ROUND-LEAF CANCERWORT June–Sept. Native of s Eurasia and n Africa; moist sandy soil, Winnebago County.

## LINARIA *Toadflax*

Perennial herbs (ours), almost always glabrous, with erect flowering stems. Leaves numerous, narrow. Flowers several to many in terminal racemes. Corolla strongly bilabiate, spurred at base, the upper lip 2-lobed, the lower 3-lobed.

*Key to species*

1 Leaves 6 mm wide or more, ovate, the upper leaves clasping stem .. ............................................. ***L. dalmatica***

1 Leaves less than 5 mm wide, linear, sessile or with petioles ....... .............................................. ***L. vulgaris***

***Linaria dalmatica*** (L.) P. Mill. DALMATIAN TOADFLAX July–Aug. Native of Europe; roadsides and other disturbed sites.

***Linaria vulgaris*** P. Mill. [•76] BUTTER-AND-EGGS May–Sept. Native of Europe; fields, roadsides, and waste places.

## LITTORELLA

***Littorella uniflora*** (L.) Aschers. AMERICAN SHOREWEED Low perennial herb; plants clumped, often forming mats. Leaves bright green, linear, succulent; margins entire. Flowers only from emersed plants, single-sexed, staminate and pistillate flowers on same plant; staminate flowers stalks to 4 cm long; pistillate flowers stalkless among the leaves. July–Aug. Sandy or mucky lakeshores, or in water 1 m or more deep.

## NUTTALLANTHUS *Oldfield-Toadflax*

***Nuttallanthus canadensis*** (L.) D.A. Sutton OLDFIELD-TOADFLAX Annual herb. Leaves narrowly linear. Racemes congested at anthesis, later elongate; corolla blue, the lips much longer than the tube; lower lip with 2 short white ridges. May–Aug. Dry, open, sandy or rocky sterile ground; oak savanna, jack pine plains, dried lake beds. *Linaria canadensis* (L.) Chaz. Sometimes misidentified as the less common *Lobelia kalmii,* with which it shares narrow leaves and blue bilaterally symmetrical flowers, but *L. kalmii* is found in wetlands and has milky juice.

## PENSTEMON *Beardtongue*

Perennial herbs (ours), the erect stems rising from a rosette of petioled basal leaves, the stem leaves sessile and often clasping (the leaves of all our species are about alike and are of little value in distinguishing species). Flowers white to blue-violet or red-violet, in terminal clusters. Corolla tubular or trumpet-shaped, bilabiate, the upper lip 2-lobed, the lower 3-lobed.

*Key to species*

1 Upper leaves clasping stem, nearly round; stems glabrous and glaucous .......... ***P. grandiflorus***
1 Leaves sessile or with petioles, lance-shaped to oblong lance-shaped; stems various.......... 2

2 Inner surface of corolla covered with gland-tipped hairs; leaves entire or nearly so .......... ***P. tubiflorus***
2 Inner surface of corolla glabrous or somewhat hairy, the hairs not gland-tipped; leaves entire to toothed .......... 3

3 Throat of corolla inflated and broader than the tube; larger leaves often more than 1.5 cm wide .......... ***P. digitalis***
3 Throat of corolla not broader than tube; leaves mostly less than 1.5 cm wide.......... 4

4 Corolla pale to dark violet, the lower lip arched upward and nearly closing the throat; stems long-hairy, the hairs spreading or tangled .......... ***P. hirsutus***
4 Corolla white or pale violet, the lower lip not arched upward; stems finely hairy, the hairs somewhat appressed to stem .......... 5

5 Corolla white, often streaked with purple; peduncles often more than 1 cm long .......... ***P. pallidus***
5 Corolla pale violet; peduncles less than 1 cm long ...... ***P. gracilis***

***Penstemon digitalis*** Nutt. [•76] FOXGLOVE BEARDTONGUE Introduced (naturalized). May–July. Moist open woods and prairies.

***Penstemon gracilis*** Nutt. SLENDER BEARDTONGUE Prairies and open woods. The typical var. gracilis, has leaves glabrous on both sides. Var. *wisconsinensis* (Pennell) Fassett is known only from Wisc, ne Ill and nw Ind. Its leaves are puberulent on both sides and its stem is finely puberulent throughout. The pubescence suggests a possible origin through hybridization of var. *gracilis* with *P. pallidus.*

***Penstemon grandiflorus*** Nutt. LARGE-FLOWER BEARDTONGUE May–June. Dry prairies and barrens.

***Penstemon hirsutus*** (L.) Willd. [•76] HAIRY BEARDTONGUE May–July. Dry woods and fields.

***Penstemon pallidus*** Small PALE BEARDTONGUE April–June. Dry woods and fields.

***Penstemon tubiflorus*** Nutt. TUBE BEARDTONGUE Introduced. May–June. Prairies and moist woods.

## PLANTAGO *Plantain*

Perennial or annual herbs. Leaves all from base of plant, simple. Flowers small, perfect or single-sexed, green, more or less stalkless in axils of small bracts, grouped into crowded spikes.

*Callitriche palustris*
PLANTAGINACEAE
*Gratiola neglecta*
PLANTAGINACEAE

***Linaria vulgaris***
PLANTAGINACEAE

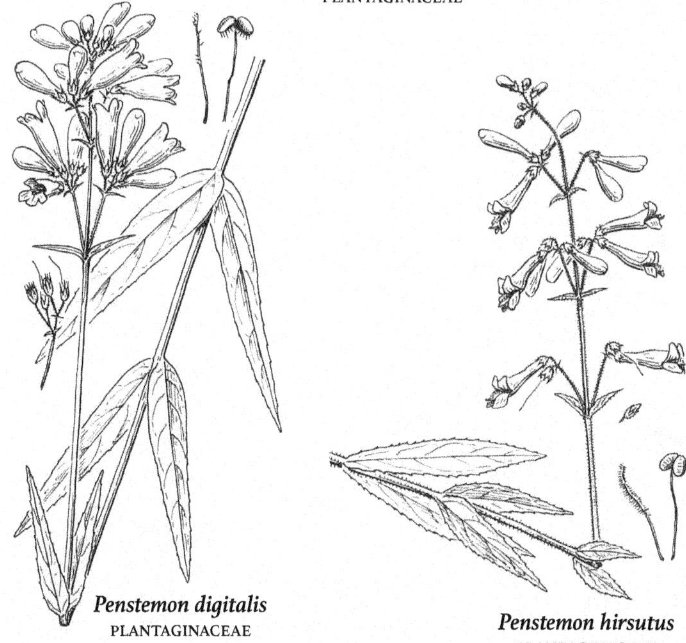

***Penstemon digitalis***
PLANTAGINACEAE

***Penstemon hirsutus***
PLANTAGINACEAE

*Key to species*

1 Leaves linear to lance-shaped, more than 5 times longer than wide ........ 2
1 Leaves narrowly ovate to ovate, less than 5 times longer than wide. 5

2 Leaves more than 5 mm wide ........ *P. lanceolata*
2 Leaves less than 5 mm wide ........ 3

3 Leaves opposite along stem or sometimes whorled; flowers on peduncles from upper leaf axils ........ *P. arenaria*
3 Leaves all from base of plant; flowers on a naked stalk ........ 4

4 Bracts subtending flowers elongate, 2–3 times longer than flowers ........ *P. aristata*
4 Bracts shorter, to 1.5 times longer than flowers ........ *P. patagonica*

5 Leaf blades glabrous, the lateral veins from the midrib; flower stalk hollow; rare in e Wisc ........ *P. cordata*
5 Leaf blades glabrous or hairy on upper surface, the veins originating from petiole; flower stalk not hollow ........ 6

6 Leaf petioles absent or less than 1.5 cm long ........ 7
6 Leaf petioles mostly more than 1.5 cm long ........ 8

7 Sepals and bracts with coarsely hairy ........ *P. virginica*
7 Sepals and bracts glabrous ........ *P. media*

8 Petioles green; bracts broadly ovate ........ *P. major*
8 Petioles red-tinged at base; bracts narrowly lance-shaped *P. rugelii*

***Plantago arenaria*** Waldst. & Kit. SAND PLANTAIN July–Aug. Native of Eurasia; a weed of waste places, roadsides, and railways.

***Plantago aristata*** Michx. [•77] LARGE-BRACT PLANTAIN Introduced (naturalized). Dry sterile or sandy soil. Depauperate forms much smaller, with filiform leaves and few-flowered spikes.

***Plantago cordata*** Lam. HEART-LEAF PLANTAIN *Endangered.* April–June. Floodplain forests, wooded streambanks, sometimes in shallow water; usually where calcium-rich. The large heart-shaped leaves are present only in summer; winter leaves are small and lance- or spatula-shaped; leaves that appear in the spring and fall are intermediate in size between summer and winter forms. Wisc at north edge of species' range.

***Plantago lanceolata*** L. [•77] ENGLISH PLANTAIN Native of the Old World; a common weed of lawns, roadsides, and waste places.

***Plantago major*** L. [•77] COMMON PLANTAIN Native of Eurasia; naturalized in lawns, roadsides, and waste places.

***Plantago media*** L. HOARY PLANTAIN Native of Eurasia, weed of waste places in the e states; in Wisc, known from La Crosse and Sheboygan counties. Resembles *P. lanceolata,* except for its broader leaves.

***Plantago patagonica*** Jacq. WOOLLY PLANTAIN Introduced (naturalized). Dry prairies. Sepals, petals, and seeds essentially as in *P. aristata.*

*Plantago rugelii* Dcne. [•77] AMERICAN PLANTAIN Lawns, gardens, roadsides, and waste places.

*Plantago virginica* L. [•77] PALE-SEED PLANTAIN Introduced. Dry or sandy soil, often weedy.

## VERONICA *Speedwell*

Annual or perennial herbs. Leaves opposite, or becoming alternate in the head. Flowers single or in racemes from leaf axils or at ends of stems; corolla blue or white, 4-lobed, somewhat 2-lipped. Fruit a flattened capsule, lobed or notched at tip; styles usually persistent on fruit.

ADDITIONAL SPECIES ***Veronica dillenii*** Crantz (Dillenius' speedwell), introduced and known from west-central Wisc. ***Veronica verna*** L. (Spring speedwell, see key), introduced, known from several Wisc locations.

### *Key to species*

1 Flowers in racemes from leaf axils, or leaves more than 4 cm long, or both; plants perennial .......... 2
1 Flowers single in axils of leafy bracts, or in terminal spikes; leaves less than 3 cm long; plants annual or perennial .......... 9

2 Stems glabrous or nearly so; leaves toothed or entire .......... 3
2 At least the upper stem pubescent; leaves toothed .......... 6

3 Leaves with short petioles .......... 4
3 Leaves sessile .......... 5

4 Leaves mostly widest near leaf base; styles 2.5–3.5 mm long .......... ***V. americana***
4 Leaves widest above leaf middle; styles to 2.2 mm long .......... ***V. beccabunga***

5 Upper leaves lance-shaped, with wide, clasping bases; rachis of raceme stout and straight; capsules swollen . ***V. anagallis–aquatica***
5 Leaves mostly linear, narrowed to a sessile base; rachis of raceme slender and zigzagged; capsules strongly flattened ... ***V. scutellata***

6 Flowers sessile or on short pedicels shorter than the bracts ..... 7
6 Flowers on long pedicels, the pedicels longer than the bracts ... 8

7 Leaves more than 5 cm long, widest near base, on petioles 1 cm or more long; flowers in dense terminal spikes; styles persistent on capsule .......... ***V. longifolia***
7 Leaves less than 5 cm long, widest near middle, sessile or nearly so; spikes from leaf axils, loosely flowered; styles deciduous .......... ***V. officinalis***

8 Racemes with incurved nonglandular hairs; leaves more than 2 times longer than wide; one pair of calyx lobes much shorter than other pair .......... ***V. austriaca***
8 Racemes with gland-tipped spreading hairs; leaves less than 2 times longer than wide; calyx lobes all about equal length ***V. chamaedrys***

9 Flowers and capsules on pedicels more than 4 mm long ....... 10
9 Flowers and capsules sessile or on short pedicels to 4 mm long . 11

10 Corolla much longer than sepals; capsules on pedicels more than 12 mm long .......... *V. persica*

10 Corolla about equaling the sepals; capsules on shorter pedicels to 12 mm long .......... *V. polita*

11 Middle stem leaves pinnately divided .......... *V. verna**

11 Leaves entire or toothed, not divided .......... 12

12 Plants matted perennial herbs; stems finely hairy; leaves glabrous or nearly so (except when young) .......... *V. serpyllifolia*

12 Plants erect annuals; stems and leaves glabrous or pubescent with mostly spreading hairs .......... 13

13 Flowers blue; the capsules only shallowly notched at tip; stems fleshy; plants glabrous .......... *V. peregrina*

13 Flowers white; the capsules deeply notched at tip; stems not fleshy; plants pubescent .......... *V. arvensis*

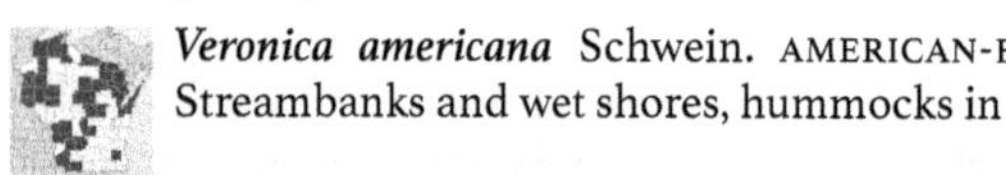

***Veronica americana*** Schwein. AMERICAN-BROOKLIME July–Sept. Streambanks and wet shores, hummocks in swamps, springs.

***Veronica anagallis-aquatica*** L. [•78] BLUE WATER SPEEDWELL June–Sept. Wet, sandy or muddy streambanks and ditches; often in shallow water.

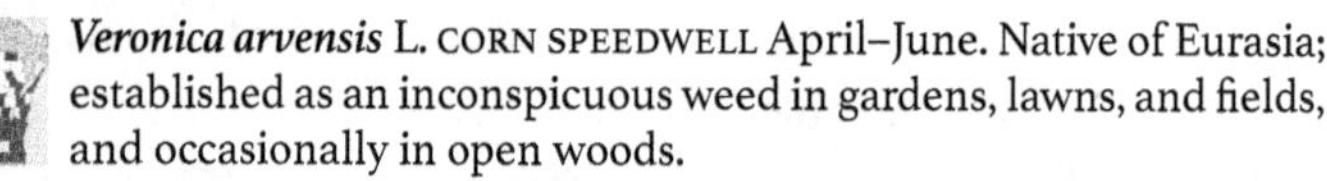

***Veronica arvensis*** L. CORN SPEEDWELL April–June. Native of Eurasia; established as an inconspicuous weed in gardens, lawns, and fields, and occasionally in open woods.

***Veronica austriaca*** L. BROAD-LEAF SPEEDWELL June–July. Native of Eurasia; occasionally escaped from cultivation in waste places, roadsides and meadows.

***Veronica beccabunga*** L. EUROPEAN SPEEDWELL, BROOKLIME Summer. Native of Eurasia, muddy shores and streambanks; Door County.

***Veronica chamaedrys*** L. GERMANDER SPEEDWELL May–June. Native of Europe; introduced in moist gardens, roadsides, and fields.

***Veronica longifolia*** L. LONG-LEAF SPEEDWELL June–Aug. Native of Europe; introduced in fields, roadsides, and waste places.

***Veronica officinalis*** L. COMMON SPEEDWELL Introduced. May–July. Dry fields and upland woods. At anthesis the main axis is often developed only slightly beyond the racemes, so that the inflorescence may appear terminal.

***Veronica peregrina*** L. [•78] PURSLANE SPEEDWELL May–July. Mud flats, shores, ditches, temporary ponds, swales; also weedy in cultivated fields, lawns and moist disturbed areas.

***Veronica persica*** Poir. [•78] BIRDSEYE SPEEDWELL April–Aug. Native of sw Asia; introduced in gardens, lawns, roadsides, and waste places.

***Veronica polita*** Fries GRAY FIELD SPEEDWELL May–July. Native of Eurasia; Dane County.

*Veronica scutellata* L. GRASS-LEAF SPEEDWELL June–Sept. Marshes, pond margins, hardwood swamps, thickets, springs, streambanks, wet depressions.

*Veronica serpyllifolia* L. THYME-LEAF SPEEDWELL May–July. Established in fields, meadows, and lawns, extending into moist open woods. Subsp. *serpyllifolia* introduced from Europe; subsp. *humifusa* native.

### VERONICASTRUM *Culver's-Root*

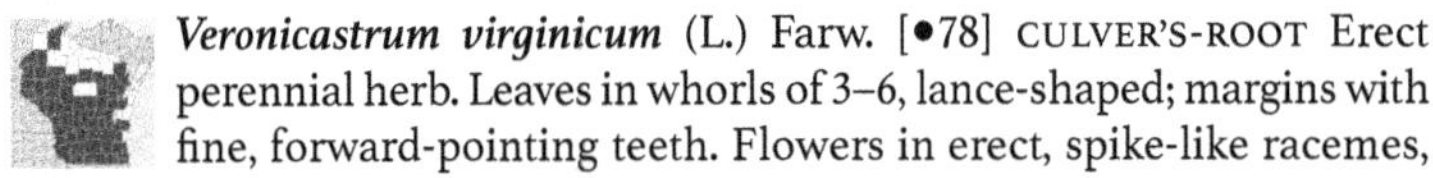

***Veronicastrum virginicum*** (L.) Farw. [•78] CULVER'S-ROOT Erect perennial herb. Leaves in whorls of 3–6, lance-shaped; margins with fine, forward-pointing teeth. Flowers in erect, spike-like racemes, the flowers crowded and spreading; corolla white, nearly regular; stamens 2, long-exserted from the corolla mouth. June–Aug. Moist to wet prairies, fens and streambanks; also in drier deciduous woods and sandy grasslands.

## PLATANACEAE *Planetree Family*

### PLATANUS *Sycamore; Planetree*

***Platanus occidentalis*** L. AMERICAN SYCAMORE Large tree; bark red-brown when young, soon breaking into thin, flat sections which fall away to expose white-green inner bark. Leaves alternate, divided into 3 or 5 shallow, sharp-pointed lobes, bright green and smooth on upper surface, underside paler. Staminate and pistillate flowers tiny, in dense clusters, separate but on same tree. Fruit a round, light brown head, on a long, drooping stalk. May. Riverbanks, floodplain forests and lakeshores. s Wisc; Wisc at n edge of species' range.

## POLEMONIACEAE *Phlox Family*

Perennial herbs (ours). Leaves opposite (Phlox) or pinnately divided (Polemonium). Flowers perfect, single or in clusters at ends of stems and from leaf axils; sepals and petals 5-parted and joined for part of length. Fruit a 3-chambered capsule, with usually 1 seed per chamber.

ADDITIONAL SPECIES ***Gilia achilleifolia*** Benth., uncommon garden escape, Manitowoc County. ***Ipomopsis rubra*** (L.) Wherry (Standing-Cypress), uncommon escape, Adam and Rock counties.

*Key to genera*

1 Leaves pinnately divided into leaflets .......... POLEMONIUM
1 Leaves undivided and entire.................................... 2

2 Leaves opposite or mostly so .......................... PHLOX
2 Leaves alternate .................................. COLLOMIA

### COLLOMIA *Mountain-Trumpet*

***Collomia linearis*** Nutt. NARROW-LEAF MOUNTAIN-TRUMPET Annual herb. Leaves alternate, linear to narrowly lance-shaped. Flowers in sessile cymes, several in the axils of the crowded upper leaves,

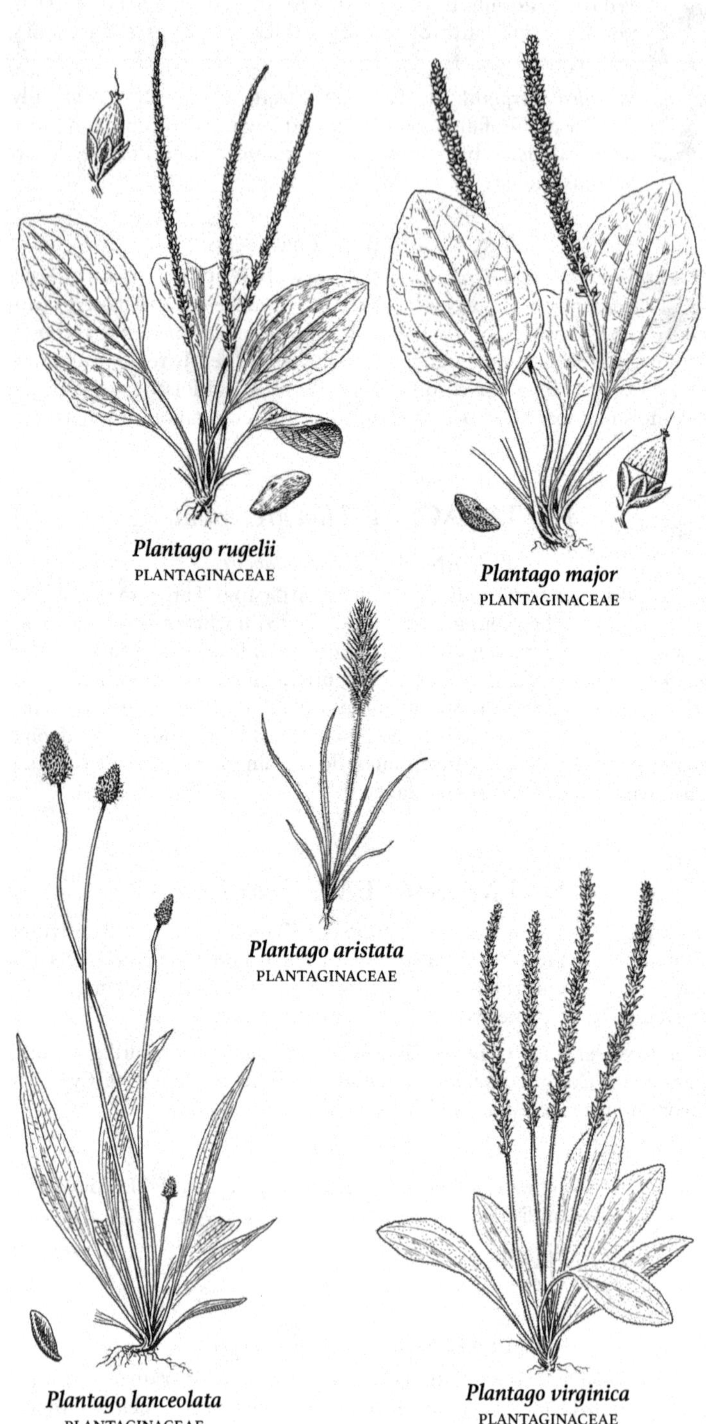

*Plantago rugelii*
PLANTAGINACEAE

*Plantago major*
PLANTAGINACEAE

*Plantago aristata*
PLANTAGINACEAE

*Plantago lanceolata*
PLANTAGINACEAE

*Plantago virginica*
PLANTAGINACEAE

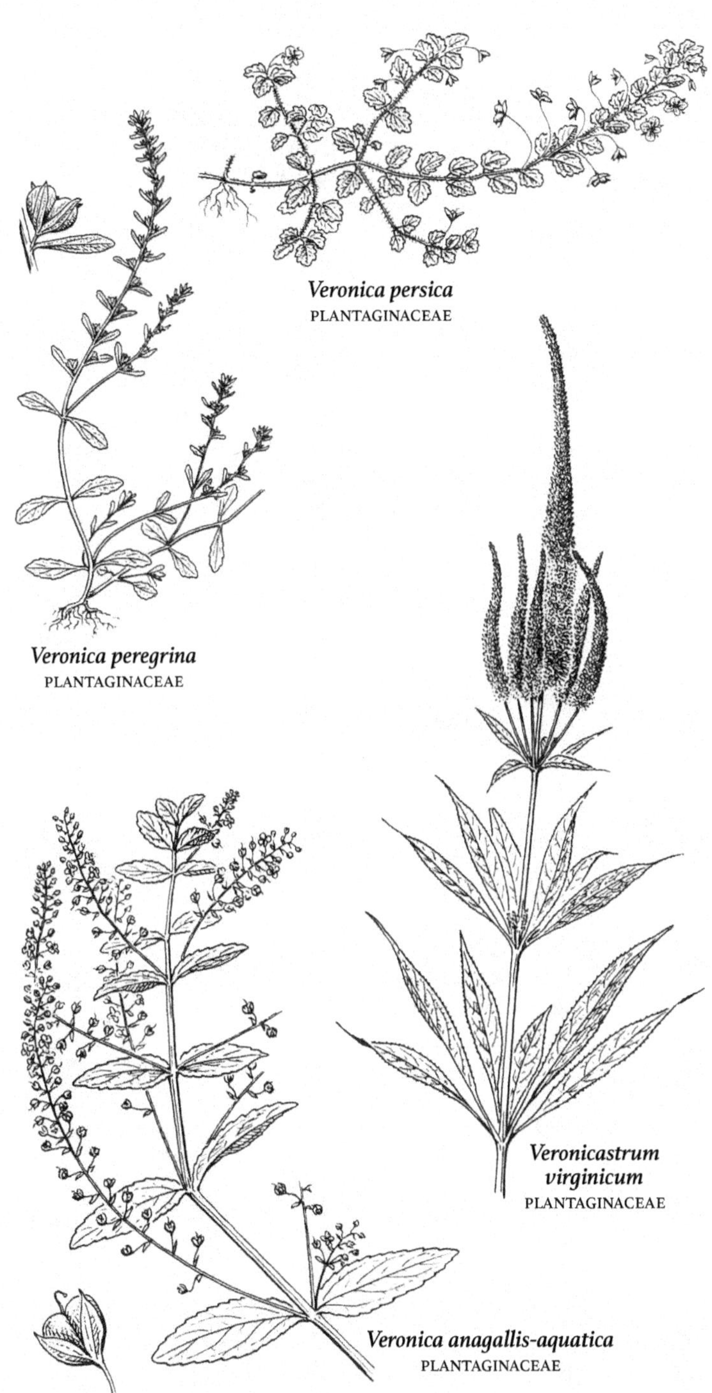

*Veronica persica*
PLANTAGINACEAE

*Veronica peregrina*
PLANTAGINACEAE

*Veronicastrum virginicum*
PLANTAGINACEAE

*Veronica anagallis-aquatica*
PLANTAGINACEAE

forming a dense headlike cluster; corolla narrowly trumpet-shaped, blue-purple to white. May–Aug. Dry sandy or gravelly grasslands, shores, and disturbed places. Native to western USA, adventive in Wisc.

## PHLOX *Phlox*

Erect perennial herbs. Leaves opposite, margins entire. Flowers pink, purple or rarely white, in stalked clusters at ends of stems and from upper leaf axils; corolla 5-lobed, tubelike but flared outward at tip.

*Key to species*

1 Stems somewhat woody, trailing on ground.................... 2
1 Stems herbaceous, upright .................................... 3

2 Petal lobes deeply notched, the notch half or more the petal length ..................................................... *P. bifida*
2 Petal lobes nearly entire or only slightly notched ..... ***P. subulata***

3 Leaves lance-shaped to ovate, less than 5 times longer than wide 4
3 Leaves linear to linear lance-shaped, more than 5 times longer than wide ................................................... 5

4 Stems with long soft hairs; flowering ends in mid- to late-June.... ............................................. *P. divaricata*
4 Stems glabrous below inflorescence; flowering begins in July .... ............................................. *P. paniculata*

5 Plants glabrous ................................ *P. glaberrima*
5 Plants long-hairy ................................... *P. pilosa*

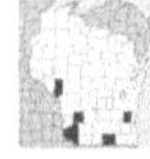

***Phlox bifida*** Beck SAND PHLOX April–May. Dry sandy soil and rock ledges.

***Phlox divaricata*** L. FOREST PHLOX April–June. Moist woods.

***Phlox glaberrima*** L. SMOOTH PHLOX *Endangered.* June–July. Wet to moist, calcium-rich meadows and prairies.

***Phlox paniculata*** L. [•79] FALL PHLOX Introduced (naturalized). July–Sept. Rich moist soil; cultivated in numerous horticultural varieties and escaped into roadsides and waste places.

***Phlox pilosa*** L. PRAIRIE PHLOX April–June. Upland woods and prairies.

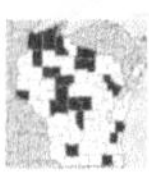

***Phlox subulata*** L. MOSS-PINK Introduced. April–May. Sandy or gravelly soil and rock-ledges; frequently cultivated and sometimes escaped.

## POLEMONIUM *Jacob's Ladder*

Perennial herbs. Leaves alternate, pinnately compound. Flowers blue, in terminal, panicle-like or thyrsoid clusters; corolla funnelform to broadly campanulate, 5-lobed.

ADDITIONAL SPECIES ***Polemonium caeruleum*** L. (Charity); cultivated and occasionally escaping, similar to the native and rare *P. occidentale;* sw Wisc.

*Key to species*

I Plants flowering in spring; inflorescence loosely flowered *P. reptans*

I Plants flowering in summer; inflorescence crowded . *P. occidentale*

***Polemonium occidentale*** Greene WESTERN JACOB'S LADDER *Endangered.* July. Cedar swamps and thickets. Disjunct in Florence County from main range of w North America.

***Polemonium reptans*** L. [•79] SPREADING JACOB'S LADDER May. Rich, moist woods.

## POLYGALACEAE *Milkwort Family*

### POLYGALA *Milkwort*

Annual, biennial, or perennial herbs. Leaves alternate or verticillate. Flowers perfect, in racemes; petals 3, all more or less united with each other and with the stamen-tube, the two upper ones similar, the lower one keel-shaped or boat-shaped with a fringe-like crest (in our species).

*Key to species*

I Flowers large, 13 mm long or more; stamens 6; leaves few, mostly near top of stem .................................. *P. paucifolia*

I Flowers smaller, mostly less than 10 mm long; stamens 8; leaves distributed along stem or mostly near base .................... 2

2 Corolla 7–10 mm long, about twice as long as the wings ........ ............................................ *P. incarnata*

2 Corolla shorter, less than 5 mm long, equaling or shorter than the wings.............................................. 3

3 Plants annual, the stems solitary from a small taproot .......... 4

3 Plants biennial or perennial................................. 7

4 Flowers in slender racemes, tapering to the tip................ 5

4 Flowers in head-like or rounded racemes...................... 6

5 Racemes elongate; wings about as long as the fruit .... *P. ambigua*

5 Racemes shorter; wings shorter than the fruit ...... *P. verticillata*

6 Leaves alternate .................................. *P. sanguinea*

6 Leaves whorled .................................... *P. cruciata*

7 Flowers white, in densely flowered spike-like racemes ... *P. senega*

7 Flowers rose-purple to white, in loose racemes ...... *P. polygama*

***Polygala ambigua*** Nutt. ALTERNATE MILKWORT June–Oct. Open woods and fields, especially in sandy soil; more common south and east of Wisc.

***Polygala cruciata*** L. DRUM-HEADS July–Sept. Sandy or mucky lakeshores, wet areas between dunes.

***Polygala incarnata*** L. PINK MILKWORT *Endangered.* Aug–Nov. Dry soil, prairies and open woods.

***Polygala paucifolia*** Willd. FRINGED POLYGALA May–June. Moist rich woods. Leaves usually pubescent only on the midrib and margin. *Triclisperma paucifolia* (Willd.) Nieuwl.

***Polygala polygama*** Walt. RACEMED MILKWORT Dry, sandy soil.

***Polygala sanguinea*** L. [•79] PURPLE MILKWORT July–Sept. Fields, meadows, and open woods.

***Polygala senega*** L. SENECA-SNAKEROOT May–June. Dry or moist woods and prairies.

***Polygala verticillata*** L. [•79] WHORLED MILKWORT July–Oct. Moist sandy soil, grasslands and woods.

---

## POLYGONACEAE *Buckwheat Family*

Annual or perennial herbs, plants sometimes vining. Leaves alternate, simple, sometimes wavy-margined, otherwise entire; the nodes usually enlarged. Stipules joined to form a membranous or papery sheath (ocrea) around stem at each node. Flowers in spike-like racemes or small clusters from leaf axils (*Persicaria, Polygonum*), or in crowded panicles at ends of stems (*Rumex*). Flowers small, perfect, regular, petals absent. In *Rumex* the sepals herbaceous, green to brown, in inner and outer groups, each group with 3 sepals, the 3 inner enlarging after flowering, becoming broadly winged, persisting to enclose the achene; in other genera of family, sepals more or less petal-like, white to pink or yellow, mostly 5 (sometimes 4). Polygonaceae recognized by presence of a stipular sheath (ocrea), which surrounds the stem above the attachment of each leaf. The similar reduced structure in the inflorescence is called an ocreola.

ADDITIONAL SPECIES ***Fagopyrum esculentum*** Moench (Buckwheat) occasional escape from cultivation or where seed spilled, but not persisting in our flora. ***Rheum rhabarbarum*** L. (Garden rhubarb), native of Asia; commonly cultivated; occasionally escaping.

*Key to genera*

1 Tepals 6, greenish or reddish, scarcely petaloid, the 3 inner (but not the outer) ones enlarging in fruit and concealing the achene; stigmas a feathery tuft; plants in some species dioecious or polygamous and hence some flowers entirely staminate **RUMEX**

1 Tepals 4–5, white to red and ± petaloid at least along the margins, uniform in size or the outer ones larger; stigmas usually not feathery and plants mostly with bisexual flowers . . . . . . . . . . . . . . . . . . . . . . . 2

2 Pedicels with a swollen joint near the middle (but not far above the sheathing ocreolae), solitary in each ocreola, the inflorescence thus composed of slender racemes, appearing jointed because of the overlapping ocreolae; leaves not over 1 mm wide; delicate-looking annual . . . . . . . . . . . . . . . . . . . . . . . . . . . . . . . . . . . **POLYGONELLA**

2 Pedicels usually jointed near the summit (if at all), often crowded, the inflorescence various; leaves at least 2 mm wide; annual or perennial, not delicate . . . . . . . . . . . . . . . . . . . . . . . . . . . . . . . . . . . . . 3

3 Stem and petioles with retrorse prickles; leaves hastate or sagittate (with acute basal lobes) .......................... **PERSICARIA**

3 Stem and petioles without prickles; leaves various ............. 4

4 Outer tepals winged or keeled in fruit, or plant somewhat twining or vine-like, or both; leaves ovate-cordate to broadly sagittate .... ...................................................... **FALLOPIA**

4 Outer tepals not winged or keeled; plant not twining; leaves various ........................................................... 5

5 Flowers 1–4 at a node, sessile or pediceled in the axils of foliage leaves or bracts; leaf blades jointed at the base, less than 2 (–2.4) cm broad; summit of ocrea silvery white, becoming lacerate-shredded; annuals ...................................... **POLYGONUM**

5 Flowers numerous in peduncled terminal or axillary spikes, racemes, or panicles, often densely crowded; leaves not jointed at base of blade, in some species over 2.5 cm broad; summit of ocrea tinged with brown, shattering at maturity but not shredding; annuals or perennials ............................ **PERSICARIA**

## FALLOPIA *Black-Bindweed*

Annual or perennial, twining or stout and erect and forming colonies. In the past, our species typically included in genus *Polygonum.*

*Key to species*

1 Stems twining and slender ................................... 2

1 Stems erect and stout, 1–3 m tall .............................. 4

2 Base of sheathing stipules with stiff, downward-pointing hairs ... ..................................................... *F. cilinodis*

2 Base of stipules not with stiff, downward-pointing hairs......... 3

3 Plants annual; styles united; achenes dull and finely roughened .. ................................................ *F. convolvulus*

3 Plants perennial; styles divergent; achenes shiny and smooth .... ..................................................... *F. scandens*

4 Leaf blades heart-shaped at base with rounded basal lobes, the blades often 20 cm or more long from leaf tip to lobe tip; flowers perfect ......................................... *F. sachalinense*

4 Leaf blades cut nearly straight across at base, the blades less than 20 cm long; flowers functionally either staminate or pistillate .... ..................................................... *F. japonica*

***Fallopia cilinodis*** (Michx.) Holub FRINGED BLACK BINDWEED July–Aug. Dry woods and thickets. Plants in open sun are often erect, with stouter red stems, the red color extending into the leaf veins. *Polygonum cilinode* Michx.

***Fallopia convolvulus*** (L.) Á. Löve [•80] BLACK BINDWEED June–Sept. Native of Europe; roadsides, railway tracks, and waste ground. *Polygonum convolvulus* L.

***Fallopia japonica*** (Houtt.) Ronse Decr. JAPANESE KNOTWEED Aug–Sept. Native of Japan; sometimes planted but often escaping to form large colonies. *Polygonum cuspidatum* Sieb. & Zucc.

***Fallopia sachalinensis*** (F. Schmidt) Ronse Decr. GIANT KNOTWEED Native of e Asia; occasionally planted and sometimes escaped and forming colonies. Closely resembling *F. japonica* in habit, flower, and fruit. Stems sometimes more than 4 m tall. Leaf blades ovate, cordate at base, the basal lobes broadly rounded. *Polygonum sachalinense* F. Schmidt.

***Fallopia scandens*** (L.) Holub FALSE BUCKWHEAT Aug–Sept. Moist woods, thickets, and roadsides. Variable in the shape of the wings and the length of the fruit. *Polygonum scandens* L.

## PERSICARIA *Lady's-Thumb; Smartweed*

Annual and perennial herbs. Flowers pink or sometimes white, in terminal spikes. The genus was formerly included in *Polygonum.*

*Key to species*

1 Tepals 4; styles elongate, persistent and becoming hard and stiff . . . . . . . . ***P. virginiana***
1 Tepals usually 5; styles short, not persistent nor becoming hard and stiff. . . . . . . . **2**

2 Stems with downward-pointing prickles on the stem angles . . . . . **3**
2 Stems smooth to hairy, but not prickly . . . . . . . . **4**

3 Basal lobes of leaves pointed downward; achenes 3-sided . . . . . . . . ***P. sagittata***
3 Basal lobes pointed outward; achenes 2-sided . . . . . . . . ***P. arifolia***

4 Perennial herbs from rhizomes or stolons . . . . . . . . **5**
4 Taprooted annual herbs . . . . . . . . **7**

5 Flowers in 1 or 2 terminal racemes . . . . . . . . ***P. amphibia***
5 Flowers in several to many terminal and axillary racemes . . . . . . . **6**

6 Perianth dotted with glands . . . . . . . . ***P. punctata***
6 Perianth not dotted with glands . . . . . . . . ***P. hydropiperoides***

7 Sheathing stipules (ocreae) fringed with bristles at tip . . . . . . . . . **8**
7 Ocreae entire or irregularly cut, not fringed with bristles . . . . . . . **13**

8 Perianth dotted with glands . . . . . . . . **9**
8 Perianth not dotted with glands. . . . . . . . **10**

9 Tepals usually 4; achenes dull . . . . . . . . ***P. hydropiper***
9 Tepals 5; achenes shiny . . . . . . . . ***P. punctata***

10 Leaves broadly ovate to heart-shaped, 5–10 cm wide . . ***P. orientalis***
10 Leaves linear to lance-shaped, usually less than 5 cm wide. . . . . . **11**

11 Upper stem and peduncles with gland-tipped hairs . . . . . ***P. careyi***
11 Upper stem and peduncles not with gland-tipped hairs . . . . . . . . **12**

12 Small stipules at base of each inflorescence (ocreolae) fringed with long hairs 2–3 mm long . . . . . . . . ***P. longiseta***
12 Small stipules at base of each inflorescence entire, or with a few short hairs to 1 mm long . . . . . . . . ***P. maculosa***

13 Outer sepals strongly 3-nerved, each nerve ending in an anchor shaped fork; racemes nodding to erect . . . . . . . . . . . . ***P. lapathifolia***
13 Outer sepals with faint, irregularly forked nerves; racemes erect . . . . . . . . ***P. pensylvanica***

***Persicaria amphibia*** (L.) Delarbre WATER SMARTWEED June–Sept. Ponds, lakes, marshes, bog pools, backwater areas, quiet streams. *Polygonum amphibium* L.

***Persicaria arifolia*** (L.) Haraldson HALBERD-LEAF TEARTHUMB July–Sept. Swamps, wet woods, streambanks and shores. Similar to arrow-leaf tearthumb (*P. sagittata*). *Polygonum arifolium* L.

***Persicaria careyi*** (Olney) Greene CAREY'S SMARTWEED July–Aug. Sandy lakeshores and streambanks, marshes, recently burned wetlands. *Polygonum careyi* Olney.

***Persicaria hydropiper*** (L.) Delarbre [•80] MILD WATER-PEPPER Introduced. July–Oct. Muddy shores, streambanks, floodplains, marshes, ditches and roadsides. *Polygonum hydropiper* L.

***Persicaria hydropiperoides*** (Michx.) Small SWAMP SMARTWEED July–Sept. Shallow water or wet soil; ponds, marshes, swamps, bogs and fens, streambanks, lakeshores and ditches. *Polygonum hydropiperoides* Michx.

***Persicaria lapathifolia*** (L.) Delarbre DOCK-LEAF SMARTWEED July–Sept. Marshes, wet meadows, shores, streambanks, ditches and cultivated fields. Common and weedy. *Polygonum lapathifolium* L.

***Persicaria longiseta*** (Bruijn) Kitag. BRISTLY LADY'S-THUMB Native of e Asia; waste places, preferably in moist soil. *Polygonum caespitosum* Blume.

***Persicaria maculosa*** Gray LADY'S-THUMB Introduced. July–Sept. Muddy shores, streambanks, ditches and cultivated fields, often weedy. *Polygonum persicaria* L.

***Persicaria orientalis*** (L.) Spach PRINCE'S FEATHER Native of India; escaped in waste places near gardens. *Polygonum orientale* L.

***Persicaria pensylvanica*** (L.) M. Gómez PINKWEED June–Sept. Streambanks, exposed shores, marshes, fens, ditches and cultivated fields. *Polygonum pensylvanicum* L.

***Persicaria punctata*** (Elliott) Small [•80] DOTTED SMARTWEED Aug–Sept. Floodplain forests, marshes, shores, streambanks and cultivated fields. *Polygonum punctatum* Ell.

***Persicaria sagittata*** (L.) H.Gross ARROW-LEAF TEARTHUMB July–Sept. Swamps, marshes, wet meadows and burned wetlands. *Polygonum sagittatum* L.

***Persicaria virginiana*** (L.) Gaertn. JUMPSEED Aug–Sept. Moist woods. *Polygonum virginianum* L.

## POLYGONELLA *Jointweed*

***Polygonella articulata*** (L.) Meisn. COASTAL JOINTWEED Annual herb. Leaves linear, revolute. Flowers perfect, or a few unisexual by abortion, in several racemes; sepals 5, petal-like, white or greenish to pink or red. Fruit a smooth, sharply 3-angled achene. July–Aug. Great Lakes sandy shores and dunes. *Polygonum articulatum* L.

## POLYGONUM *Smartweed; Knotweed; Tearthumb*

Annual herbs (ours). Stems erect to sprawling, often swollen at nodes. Leaves arrowhead-shaped to lance-shaped or oval; stipules joined to form a tubular sheath (ocrea) around the stem above each node; the ocreae membranous or papery, entire or with an irregular, jagged margin or fringed with bristles. Flowers from leaf axils; sepals usually 5, petal-like, green-white to pink. Fruit a brown to black achene, lens-shaped or 3-angled.

*Key to species*

1 Leaves pleated, with two lengthwise folds ............... ***P. tenue***
1 Leaves flat or revolute.......................................... 2

2 Perianth abruptly narrowed above achene ("bottle-shaped") .... 3
2 Perianth not narrowed above achene.......................... 4

3 Leaves yellow-green; fruiting perianth divided for about three-fourths of its length .................................. ***P. erectum***
3 Leaves blue-green; fruiting perianth divided for about one-third its length ............................................ ***P. achoreum***

4 Outer 3 tepals flat, shorter than or equaling inner 2 tepals ....... ................................................ ***P. aviculare***
4 Outer 3 tepals hood-like, longer than inner 2 tepals............. 5

5 Plants prostrate; leaves 2–4 times longer than wide ... ***P. aviculare***
5 Plants upright; leaves 4–12 times longer than wide .............. ............................................ ***P. ramosissimum***

***Polygonum achoreum*** Blake LEATHERY KNOTWEED Sandy and gravelly roadsides, barnyards, gardens, railroads.

***Polygonum aviculare*** L. YARD KNOTWEED Common weed of waste ground, streets, and lawns; also common on beaches and around salt marshes.

***Polygonum erectum*** L. ERECT KNOTWEED Common weed in waste ground.

***Polygonum ramosissimum*** Michx. YELLOW-FLOWER KNOTWEED Sandy fields and meadows, sandy or gravelly shores of the Great Lakes.

***Polygonum tenue*** Michx. PLEAT-LEAF KNOTWEED Dry, sandy open hills and old fields, roadsides.

## RUMEX *Dock; Sorrel*

Perennial, sometimes weedy herbs (annual in *R. fueginus*). Leaves large and clustered at base of plants, or leafy-stemmed; flat to wavy-crisped along margins. Membranous sheaths around stems present at nodes (ocreae). Flowers in crowded whorls in panicles at ends of stems; flowers small and numerous, green but turning brown; sepals in 2 series of 3, the inner 3 sepals (valves) enlarging, becoming winged and loosely enclosing the achene, giving the appearance of a 3-winged fruit, the midvein of the valve often swollen to produce a grainlike tubercle on the back. Fruit a brown, 3-angled achene, tipped with a short slender beak.

ADDITIONAL SPECIES ***Rumex stenophyllus*** Ledeb. (Narrow-leaf dock), introduced, reported from Douglas County.

*Key to species*

1 At least some of the leaves arrowhead-shaped, the basal lobes pointing backward or outward .................................. 2

1 Leaves not arrowhead-shaped with basal lobes .................. 3

2 Basal lobes pointing backward; valves with a conspicuous grain at base ............................................ ***R. acetosa***

2 Basal lobes pointing outward; valves without grains .. ***R. acetosella***

3 At least 1 of the valves with a prominent grain .................. 4

3 Valves without grains ........................................ 11

4 Margins of mature valves entire or shallowly lobed, not toothed . 5

4 Margins of mature valves with coarse or spine-tipped teeth .... 10

5 Flower pedicels without a large swollen joint; base of grain distinctly above base of valve .............................. ***R. britannica***

5 Flower stalks with a large swollen joint below the middle or near base; base of grain even with base of valve .................... 6

6 Fruit with 1 grain, the grain small, less than half as long as the valve ............................................ ***R. patientia***

6 Fruit with 1–3 grains, the larger grains at least half as long as the valve ................................................ 7

7 Fruit with 3 grains; flower pedicels 2–5 times longer than fruit ..... ............................................ ***R. verticillatus***

7 Fruit with 1–3 grains, the grains not projecting below the valves; flower stalks 1–2 times longer than fruit ...................... 8

8 Leaves crisp-margined (crinkled); grains two-thirds as wide as long ............................................ ***R. crispus***

8 Leaf margins flat; grains narrower, up to half as wide as long .... 9

9 Grains usually 1; leaves mostly less than 4 times longer than wide ............................................ ***R. altissimus***

9 Grains usually 3; leaves mostly more than 4 times longer than wide ............................................ ***R. triangulivalvis***

10 Plants annual from fibrous roots; grains 3 ............. ***R. fueginus***

10 Plants perennial from a stout taproot; grain 1 ....... ***R. obtusifolius***

11 Mature valves less than 1 cm wide .................. ***R. longifolius***

11 Mature valves 2–3 cm wide .......................... ***R. venosus***

***Rumex acetosa*** L. GREEN SORREL Native of Eurasia; occasionally cultivated for greens and sparsely naturalized.

***Rumex acetosella*** L. COMMON SHEEP SORREL Naturalized from Eurasia in fields, lawns, and waste places, soils acidic; often a troublesome weed.

***Rumex altissimus*** Wood PALE DOCK May–Aug. Marshes, shores, streambanks, ditches. Similar to willow-leaf dock (*R. triangulivalvis*).

***Rumex brittannica*** L. GREAT WATER-DOCK June–Aug. Marshes, fens, streambanks and ditches, often in shallow water. *Rumex orbiculatus* Gray.

***Rumex crispus*** L. [•80] CURLY DOCK Introduced (naturalized). July–Sept. Wet meadows, shores, ditches, old fields, and other wet and disturbed areas; weedy.

***Rumex fueginus*** Phil. GOLDEN DOCK July–Aug. Marshes, shores, streambanks and ditches, sometimes where brackish.

***Rumex longifolius*** DC. DOOR-YARD DOCK Native of Europe, disturbed places.

***Rumex obtusifolius*** L. BITTER DOCK Introduced (naturalized). June–Aug. Floodplain forests and openings, cultivated fields and disturbed areas.

***Rumex patientia*** L. PATIENCE DOCK Native of Europe; waste places.

***Rumex triangulivalvis*** (Danser) Rech. f. WILLOW-LEAF DOCK June–Aug. Wet meadows, marshes, shores, streambanks, ditches and other low areas, sometimes where brackish. *Rumex salicifolius* Weinm.

***Rumex venosus*** Pursh VEINY DOCK Introduced. Uncommon adventive in waste places and along railways.

***Rumex verticillatus*** L. SWAMP DOCK June–Sept. Marshes, swamps, wet forests, backwater areas and muddy shores, often in shallow water.

## PORTULACACEAE *Purslane Family*

### PORTULACA *Purslane*

Ours succulent annual herbs. Leaves cauline, mostly alternate, the uppermost crowded and forming an involucre to the flowers. Flowers ephemeral, opening only in the sunshine, solitary or glomerate at ends of the stems and branches. Petals 4–6, commonly 5.

*Key to species*

1 Plants prostrate, glabrous; leaves obovate; flowers yellow ........ ........ ***P. oleracea***

1 Plants upright or spreading, hairy at nodes; leaves terete; flowers bright red to yellow ........ ***P. grandiflora***

***Portulaca grandiflora*** Hook. MOSS-ROSE All summer. Native of Argentina; cultivated for ornament and occasionally escaped.

***Portulaca oleracea*** L. COMMON PURSLANE All summer. Reputedly native of w Asia, but now widely distributed as a familiar weed; sometimes cooked for greens.

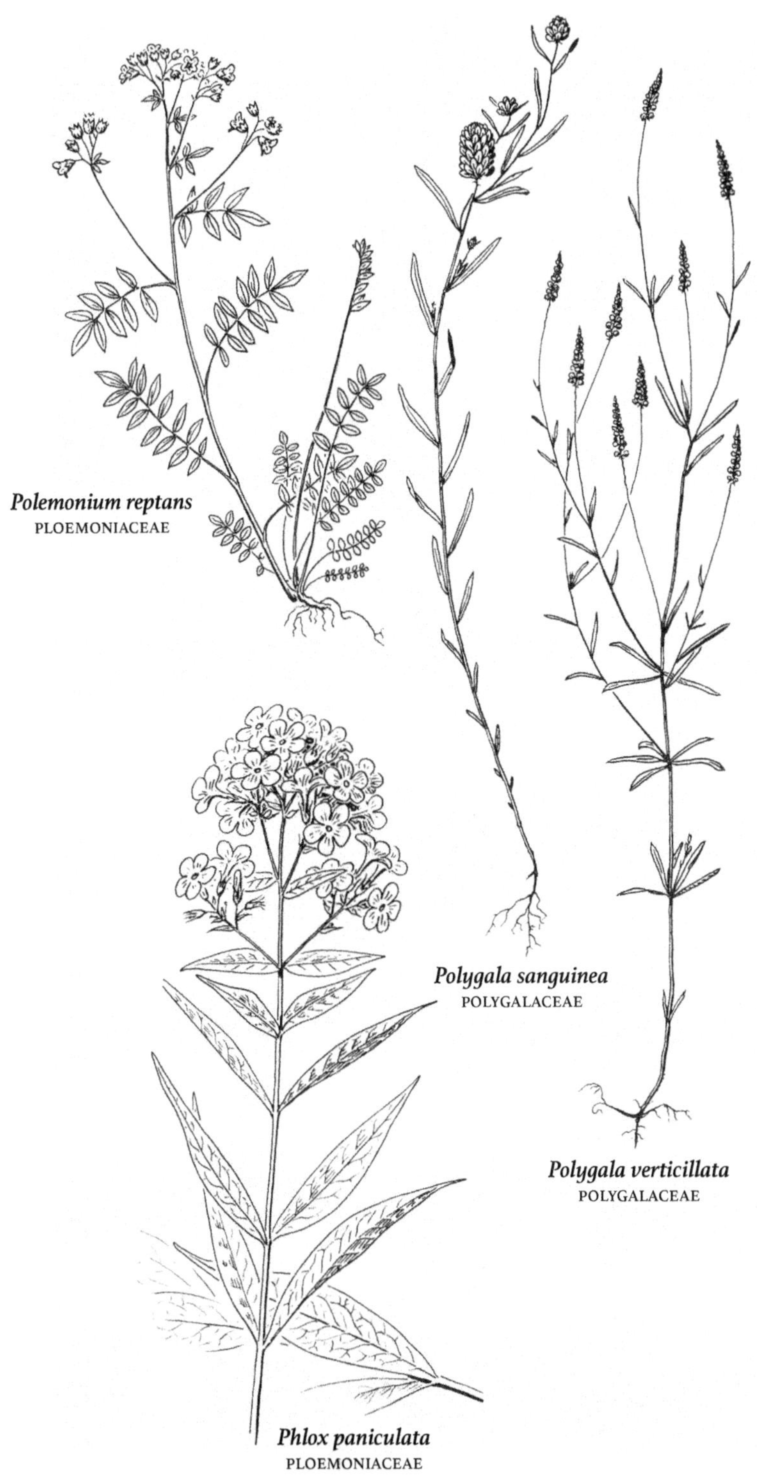
*Polemonium reptans*
PLOEMONIACEAE
*Polygala sanguinea*
POLYGALACEAE
*Polygala verticillata*
POLYGALACEAE
*Phlox paniculata*
PLOEMONIACEAE

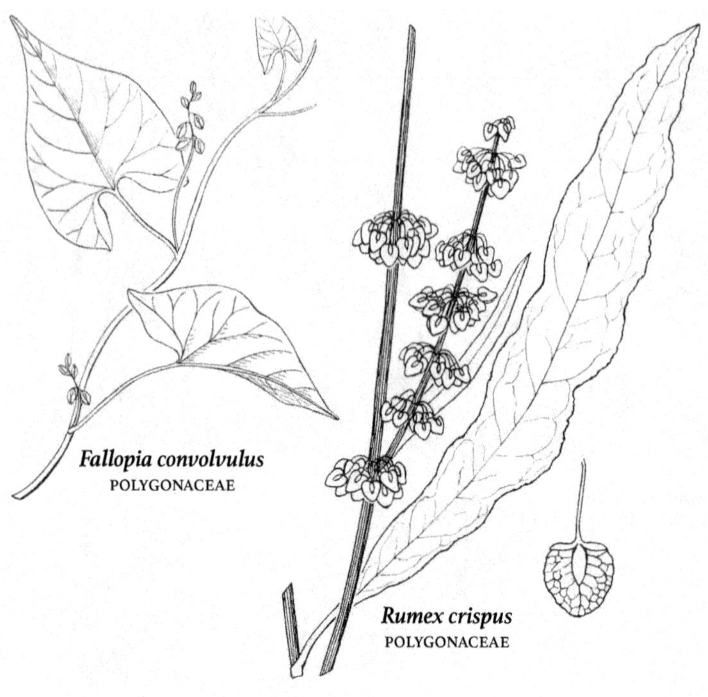
*Fallopia convolvulus*
POLYGONACEAE
*Rumex crispus*
POLYGONACEAE

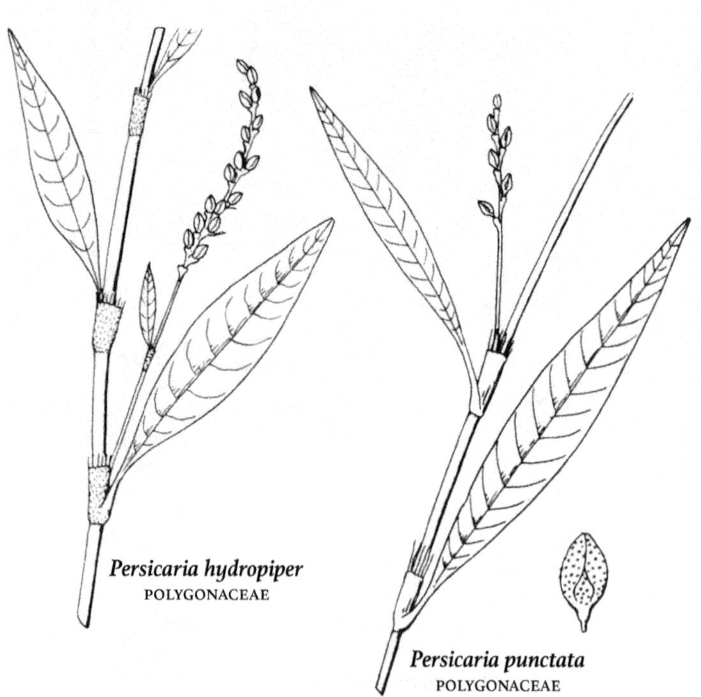
*Persicaria hydropiper*
POLYGONACEAE
*Persicaria punctata*
POLYGONACEAE

## PRIMULACEAE *Primrose Family*

Annual or perennial herbs. Leaves simple, opposite (sometimes whorled in Lysimachia), or leaves all basal. Flowers perfect, regular, single from leaf axils, or in clusters at ends of stems; petals mostly 5 (varying from 4–9), joined, tube-shaped below and flared above, deeply cleft to shallowly lobed at tip. Fruit a 5-chambered capsule.

*Key to genera*

1 Leaves all from base of plant, inflorescence an umbel at end of naked stalk .................................................. 2
1 Leaves from stem; inflorescence various ...................... 4

2 Lobes of corolla strongly reflexed ........... DODECATHEON
2 Lobes of corolla spreading or ascending ...................... 3

3 Corolla tube longer than the calyx ................. PRIMULA
3 Corolla tube shorter than the calyx ............. ANDROSACE

4 Leaves alternate ................................. SAMOLUS
4 Leaves opposite or whorled .................................. 5

5 Leaves in a single whorl near end of stem; flowers 7-merous ..... .............................................. TRIENTALIS
5 Leaves opposite or in several whorls; flowers 5–6-merous ....... 6

6 Perennial of mostly wetland habitats; flowers yellow LYSIMACHIA
6 Annual weed; flowers rusty red ................... ANAGALLIS

### ANAGALLIS *Pimpernel*

***Anagallis arvensis*** L. PIMPERNEL Annual herb. Stems usually much branched, 4-angled. Leaves opposite, entire, sessile. Flowers solitary in the axils, on slender pedicels, ascending at anthesis, recurved in fruit; corolla deeply 5-parted, usually scarlet or brick-red, varying to white, opening only in fair weather. May–Aug. Native of Eurasia, almost throughout the USA and s Canada on roadsides, lawns, gardens, and waste places.

### ANDROSACE *Rock-Jasmine*

***Androsace occidentalis*** Pursh WESTERN ROCK-JASMINE Small annual herb with a rosette of several basal leaves and one to several leafless scapes hearing a terminal umbel. Corolla salverform, white to pink; corolla tube inflated around the ovary, constricted at the throat, not exceeding the calyx. April. Dry, usually sandy soil.

### DODECATHEON *Shooting Star*

Perennial glabrous herbs. Leaves in a basal rosette. Flowers in a terminal bracted umbel of attractive nodding flowers on ascending or erect pedicels, atop a solitary, erect, leafless scape; corolla 5-cleft almost to the base, the lobes reflexed. Capsule ovoid to cylindric, erect, opening by 5 short terminal valves.

*Key to species*

1 Flowers deep rose-purple; plants slender, to 35 cm high .......... .............................................. *D. amethystinum*

1 Flowers white to lilac; plants stout, to 60 cm high ...... *D. meadia*

***Dodecatheon amethystinum*** (Fassett) Fassett WESTERN SHOOTING STAR April–June. Moist hillsides in the driftless area of sw Wisc. Very similar to *D. meadia* in foliage and habit, but usually smaller and blooming ca. 2 weeks later. *Dodecatheon radicatum* Greene; *Dodecatheon pulchellum* (Raf.) Merrill.

***Dodecatheon meadia*** L. [•81] EASTERN SHOOTING STAR May–June. Moist or dry woods and prairies.

## LYSIMACHIA *Loosestrife*

Perennial herbs, spreading by rhizomes. Leaves mostly opposite (sometimes appearing whorled), ovate or lance-shaped. Flowers 5-parted, single on stalks from leaf axils or in racemes or panicles; petals bright to pale yellow. Fruit a capsule.

ADDITIONAL SPECIES ***Lysimachia clethroides*** Duby (Goose-neck yellow-loosestrife), introduced in se Wisc; leaves alternate.

*Key to species*

1 Plants creeping; leaves opposite, nearly round .... *L. nummularia*
1 Plants upright; leaves opposite or whorled, longer than wide .... 2

2 Flowers in terminal racemes or panicles ..................... 3
2 Flowers solitary or in clusters or spikes from the leaf axils ...... 4

3 Plants pubescent .................................. *L. vulgaris*
3 Plants glabrous ................................... *L. terrestris*

4 Leaves rounded or heart-shaped at base; petioles 1–3 cm long, fringed with hairs ............................... *L. ciliata*
4 Leaves tapered to their base; petioles absent or short, smooth or fringed with hairs ........................................ 5

5 Flowers many in racemes from leaf axils; flowers mostly 6-merous ............................................... *L. thyrsiflora*
5 Flowers 1 to several from the leaf axils, 5-merous ............... 6

6 Leaves narrowly linear, to 5 mm wide ............. *L. quadriflora*
6 Leaves lance-shaped to ovate, usually more than 8 mm wide .... 7

7 Flowers in clusters of several from leaf axils .......... *L. punctata*
7 Flowers usually single from each leaf axil ..................... 8

8 Main leaves in whorls of 3 or more leaves; corolla lobes entire .... ............................................... *L. quadrifolia*
8 Leaves opposite or whorled; corolla lobes ragged-toothed at tip . 9

9 Calyx lobes ovate, usually with 5 distinct nerves ....... *L. hybrida*
9 Calyx lobes lance-shaped, with 1 distinct nerve ..... *L. lanceolata*

***Lysimachia ciliata*** L. FRINGED YELLOW-LOOSESTRIFE June–Aug. Usually shaded wet areas, such as shores, streambanks, wet meadows, ditches, floodplains, wet woods and thickets.

***Lysimachia hybrida*** Michx. LOWLAND YELLOW-LOOSESTRIFE July–Aug. Wet meadows, marshes, streambanks, ditches and shores, sometimes in shallow water.

***Lysimachia lanceolata*** Walt. LANCE-LEAF YELLOW-LOOSESTRIFE June–July. Moist or wet woods or prairies.

***Lysimachia nummularia*** L. [•81] CREEPING-JENNY Introduced (naturalized). June–Aug. Swamps, floodplain forests, streambanks, shores, meadows and ditches.

***Lysimachia punctata*** L. LARGE YELLOW-LOOSESTRIFE June–July. Native of se Europe and sw Asia; Sheboygan County.

***Lysimachia quadriflora*** Sims [•82] FOUR-FLOWER YELLOW-LOOSESTRIFE July–Aug. Wet meadows, pond and marsh margins, low prairie, calcareous fens; often where sandy and calcium-rich.

***Lysimachia quadrifolia*** L. [•81] WHORLED YELLOW-LOOSESTRIFE June–July. Moist or dry upland soil, chiefly in open woods.

***Lysimachia terrestris*** (L.) B.S.P. [•81] SWAMPCANDLES June–Aug. Marshes, fens, thickets, muddy shores, and ditches.

***Lysimachia thyrsiflora*** L. [•82] SWAMP LOOSESTRIFE June–Aug. Many types of wetlands: thickets, shores, fens and bogs, marshes, low places in conifer and deciduous swamps, often in shallow water.

***Lysimachia vulgaris*** L. GARDEN YELLOW-LOOSESTRIFE July–Sept. Native of Eurasia; escaped from cultivation, occasional on mudflats along rivers and in wet meadows.

## PRIMULA *Primrose*

***Primula mistassinica*** Michx. [•82] MISTASSINI PRIMROSE Perennial herb. Leaves all at base of plant, oblong lance-shaped; margins with outward pointing teeth. Flowers 2–10 in a cluster atop a leafless stalk; petals joined, tubelike and flared at ends, pink and sometimes with a yellow center. Fruit an upright capsule. May–June. Moist ledges near the Great Lakes and in St. Croix, Wisc and Kickapoo River valleys; often found with common butterwort (*Pinguicula vulgaris*).

## SAMOLUS *Brookweed*

***Samolus valerandi*** L. WATER-PIMPERNEL Perennial herb. Leaves in a basal cluster and alternate on stem, obovate. Flowers white, on slender spreading stalks, in loosely flowered racemes. Fruit a round capsule. June–Sept. Muddy and sandy streambanks, where often in shade; ditches and marshes. *Samolus floribundus* Kunth, *Samolus parviflorus* Raf.

## TRIENTALIS *Starflower*

***Trientalis borealis*** Raf. [•82] STARFLOWER Low perennial herb, with slender rhizomes. Stems usually with a small scale-leaf near the middle and at the summit a whorl of lance-shaped leaves, from the axils of which appear 1 or several white flowers on slender pedicels. Flowers ordinarily 7-merous; corolla rotate. Fruit a 5-valved, many-seeded capsule. May–June. Rich woods, hummocks in swamps and bogs. *Lysimachia borealis* (Raf.) U. Manns & A. Anderb.

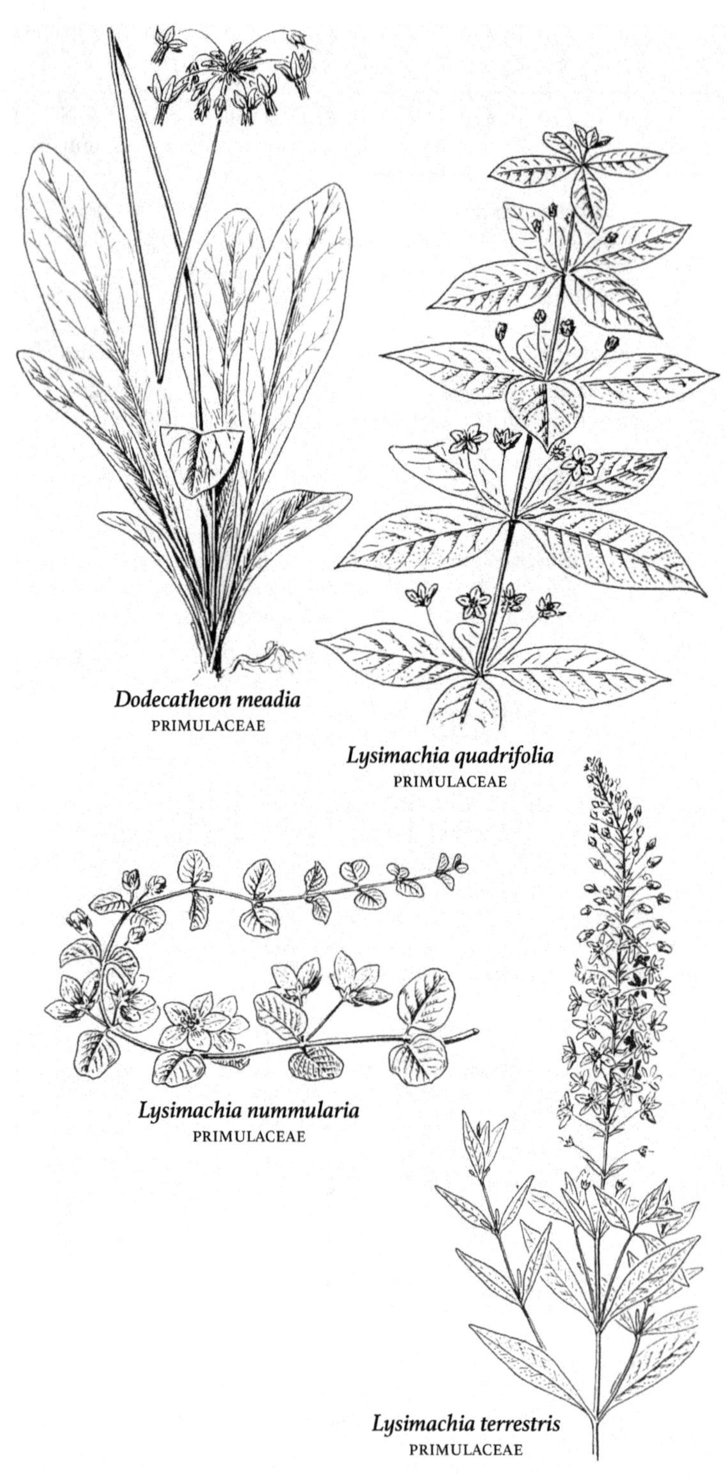

*Dodecatheon meadia*
PRIMULACEAE

*Lysimachia quadrifolia*
PRIMULACEAE

*Lysimachia nummularia*
PRIMULACEAE

*Lysimachia terrestris*
PRIMULACEAE

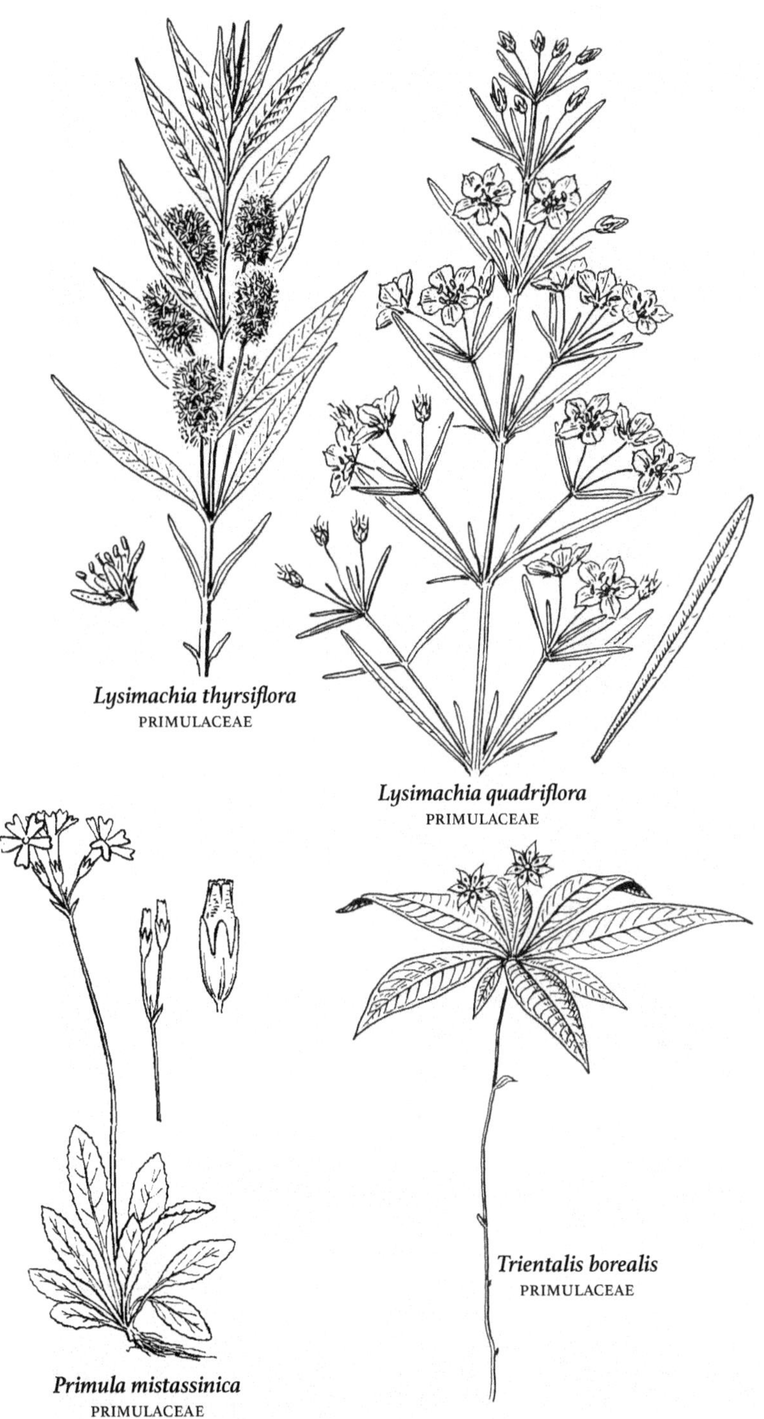

*Lysimachia thyrsiflora*
PRIMULACEAE

*Lysimachia quadriflora*
PRIMULACEAE

*Primula mistassinica*
PRIMULACEAE

*Trientalis borealis*
PRIMULACEAE

## RANUNCULACEAE *Buttercup Family*

Annual or perennial, aquatic or terrestrial herbs (or vines in *Clematis*). Leaves usually alternate, sometimes opposite or whorled, or all at base of plant. Flowers mostly white or yellow, usually with 5 (occasionally more) separate petals and sepals, or petals absent and then with petal-like sepals; sepals leafy and green or petal-like and colored; flowers perfect, stamens usually numerous; pistils several to many, ripening into beaked achenes or dry capsules (follicles).

ADDITIONAL SPECIES ***Ficaria verna*** Huds. (synonym *Ranunculus ficaria* L.; Lesser celadine), introduced, Walworth County.

*Key to genera*

1 Vines; leaves opposite; fruit with a long, feathery style . **CLEMATIS**
1 Herbs; leaves alternate or from base of plant; fruit not with a long, feathery style .......... **2**

2 Leaves linear, to 2 mm wide, all from base of plant; achenes in a spike-like cluster to 6 cm long .......... **MYOSURUS**
2 Leaves not linear, petioles usually distinct; achenes in round to short-cylindric heads .......... **3**

3 Flowers spurred or strongly irregular .......... **4**
3 Flowers regular and unspurred .......... **7**

4 Sepals petal-like, unequal, the upper sepal largest and helmet-shaped, unspurred; uncommon plant of driftless area of sw Wisc and ne Iowa .........***Aconitum columbianum*** subsp. ***columbianum***
4 Petal-like sepals equal or unequal, at least the upper sepal with a long spur .......... **5**

5 Flowers reddish .......... **AQUILEGIA**
5 Flowers blue with a single spur.......... **6**

6 Plants perennial; pistils 3 .......... **DELPHINIUM**
6 Plants annual; pistil 1 .......... **CONSOLIDA**

7 Stem leaves whorled .......... **ANEMONE**
7 Stem leaves alternate, or all leaves from base of plant .......... **8**

8 Flowers yellow, or leaves simple and not lobed, or plants aquatic **9**
8 Flowers not yellow; leaves compound or 3-lobed; plants not aquatic .......... **11**

9 Leaves all alike, unlobed; sepals yellow, large and petal-like; petals absent .......... **CALTHA**
9 Leaves usually of 2 types (stem leaves different from basal leaves), or leaves deeply lobed or divided; sepals green; petals yellow or white.......... **10**

10 Receptacle becoming long and cylindric; achenes covered with woolly hairs, with a sharp beak 2–3 mm long; small annual with pinnately dissected basal leaves .......... **CERATOCEPHALA**
10 Receptacle globose to short-cylindric; achenes neither woolly hairy nor sharp-beaked; leaves various .......... **RANUNCULUS**

11 Plants with a naked scape and solitary flowers, the leaves basal . **12**

11 Plants with leafy stems; flowers 1 to several.................... 13

12 Leaves glabrous, parted into 3 leaflets; sepals white ..... **COPTIS**
12 Leaves hairy, 3 lobed but not divided into leaflets; sepals white, pink, or blue ............................................ **HEPATICA**

13 Flowers several to many in terminal racemes or panicles....... 14
13 Flowers solitary or few, not in racemes or panicles............. 15

14 Fruit an achene ................................ **THALICTRUM**
14 Fruit berry-like, white or red ......................... **ACTAEA**

15 Leaves divided into leaflets; flowers on long slender peduncle; fruit a cluster of 4 follicles ............................... **ENEMION**
15 Leaves deeply palmately lobed; flower solitary on short stout peduncle; fruit a dense cluster of red berries ....... **HYDRASTIS**

## ACONITUM *Monkshood*

***Aconitum columbianum*** Nutt. subsp. ***columbianum*** NORTHERN BLUE MONKSHOOD *Threatened.* Perennial herb, from a thickened rootstock, toxic. Stems upright, to 1.5 m long. Leaves divided into 5–7 coarsely toothed segments. Flowers few in a raceme, showy, irregular, purple to blue, the upper sepal rounded, domelike, covering the petals. July–Sept. Ledges and bases of moist, shaded cliffs, often in cold air drainage pockets; driftless area of sw Wisc and ne Iowa. *Federally threatened. Aconitum noveboracense* Gray.

## ACTAEA *Baneberry*

Perennial herbs. Leaves 2–3x 3-partedly compound, the leaflets sharply toothed. Flowers small, white, in a dense, long-peduncled, terminal raceme; the raceme at anthesis short, the axis and pedicels elongating later, the pedicels becoming widely divergent. Sepals 3–5, petal-like. Fruit a several-seeded berry.

*Key to species*

1 Fruit red, on slender pedicels .......................... ***A. rubra***
1 Fruit white, on thicker pedicels ................... ***A. pachypoda***

***Actaea pachypoda*** Ell. WHITE BANEBERRY, DOLL'S EYES May–June. Rich woods.

***Actaea rubra*** (Ait.) Willd. RED BANEBERRY May–June. Rich woods.

## ANEMONE *Thimbleweed*

Perennial herbs. Basal leaves few to several, deeply palmately divided; stem erect with a whorl of 3 or more involucral leaves subtending one or more elongate peduncles. Flowers white to blue or red or greenish. Sepals 4–20, petal-like; petals absent. Pistils numerous, in a subglobose to cylindric head, pubescent. Achenes tipped with the persistent style.

*Key to species*

1 Styles elongating to 2 cm or more at maturity, becoming plumose ..................................................... ***A. patens***
1 Styles much shorter, not becoming plumose.................. 2

2 Achenes hidden by long, cottony hairs ........................ 3
2 Achenes nearly glabrous or only short-hirsute ................ 6

3 Involucral leaves sessile or nearly so ........................ 4
3 Involucral leaves with petioles ................................ 5

4 Sepals 4–9 stems from a stout caudex ................ *A. multifida*
4 Sepals 10–35; stems from a tuberous rhizhome ..... *A. caroliniana*

5 Involucral leaves 2 or 3; main leaflets thin and shallowly divided; achenes in an ovoid cluster ........................ *A. virginiana*
5 Involucral leaves 4 or more; main leaflets thickened and deeply divided; achenes borne in an cylindrical cluster ..... *A. cylindrica*

6 Involucral leaves with petioles; basal leaves absent or few ....... ........................................ *A. quinquefolia*
6 Involucral leaves without petioles; basal leaves usually many .... ........................................ *A. canadensis*

***Anemone canadensis*** L. ROUND-LEAF THIMBLEWEED May–Aug. Wet openings, streambanks, thickets, low prairie, ditches and roadsides.

***Anemone caroliniana*** Walt. CAROLINA THIMBLEWEED *Endangered.* April–May. Dry sandy and gravelly prairies, barrens, bluff tops.

***Anemone cylindrica*** Gray LONG-HEAD THIMBLEWEED June–Aug. Dry open woods and prairies.

***Anemone multifida*** Poir. CUT-LEAF ANEMONE *Endangered.* May–June. Limestone bluffs, open cliffs, hillside prairies, rocky calcareous soil.

***Anemone patens*** L. PASQUE-FLOWER April. Dry prairies. *Pulsatilla patens* (L.) P. Mill.

***Anemone quinquefolia*** L. [•83] WOOD-ANEMONE April–June. Moist woods.

***Anemone virginiana*** L. TALL THIMBLEWEED June–Aug. Rocky banks and open woods.

## AQUILEGIA *Columbine*

***Aquilegia canadensis*** L. RED COLUMBINE Perennial herb from a stout caudex-like rhizome. Stems with few to several large basal leaves. Leaves compound, stem leaves gradually reduced upward. Flowers nodding; sepals 5, red; petals 5, the blade yellow, prolonged backward from the base into an elongate red spur; stamens numerous, projecting in a column; pistils usually 5, each prolonged into a slender style. April–June. Dry woods, rocky cliffs and ledges.

ADDITIONAL SPECIES ***Aquilegia vulgaris*** L. (European columbine), resembling A. canadensis in foliage. Flowers nodding, blue, varying to purple, white, or pink; stamens not longer than the sepals; spurs much incurved. Native of Eurasia; cultivated for ornament and occasionally escaped.

## CALTHA *Marsh-Marigold*

Succulent perennial herbs. Leaves simple, heart-shaped, mostly from base of plant, becoming smaller upward. Flowers single at ends of stalks; sepals large and petal-like, bright yellow (*C. palustris*), to pink or white (*C. natans*); petals absent; stamens many. Fruit a follicle.

*Key to species*

1 Flowers bright yellow; stems upright; widespread in Wisc ....... ........................................................ ***C. palustris***

1 Flowers pink or white; stems floating; rare in nw Wisc .. ***C. natans***

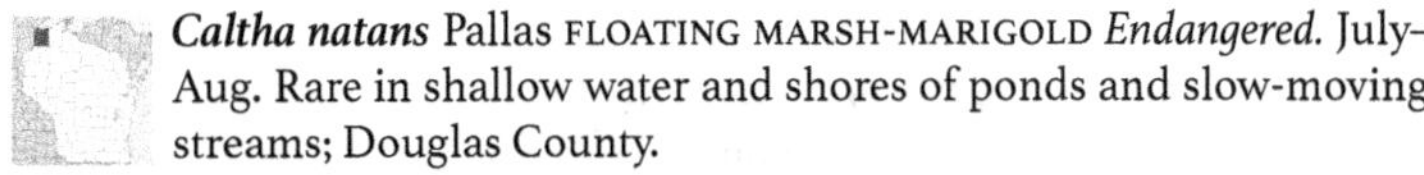

***Caltha natans*** Pallas FLOATING MARSH-MARIGOLD *Endangered.* July–Aug. Rare in shallow water and shores of ponds and slow-moving streams; Douglas County.

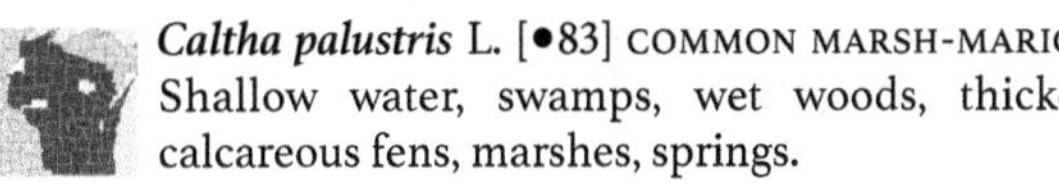

***Caltha palustris*** L. [•83] COMMON MARSH-MARIGOLD March–June. Shallow water, swamps, wet woods, thickets, streambanks, calcareous fens, marshes, springs.

## CERATOCEPHALA *Butterwort*

***Ceratocephala testiculata*** (Crantz) Bess. CURVESEED-BUTTERWORT Small, stemless annual, thinly silky-tomentose. Leaves 3-parted, the segments narrow. Petals pale yellow, fading to whitish with pink veins. Achenes forming a cylindric bur-like head. Spring ephemeral Eurasian weed, widespread in w USA and occasional in s Wisc. *Ranunculus testiculatus* Crantz.

## CLEMATIS *Virgin's Bower*

Herbaceous or woody plants, erect, or climbing by the prehensile leaf-rachis. Leaves opposite, simple or compound. Flowers solitary or panicled, usually dioecious. Sepals petal-like, commonly 4. Petals absent. Stamens numerous. Pistils numerous; style elongate. Fruit a flattened achene, terminated by the elongate persistent style.

*Key to species*

1 Sepals whitish, less than 1 cm long, in a branched inflorescence .. ...................................................... ***C. virginiana***

1 Sepals purple, 4–5 cm long, solitary .............. ***C. occidentalis***

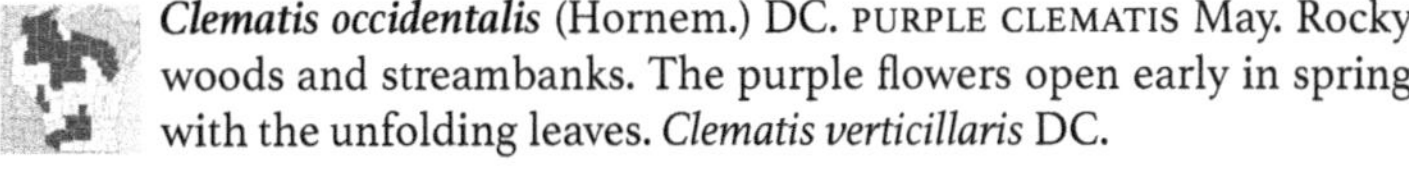

***Clematis occidentalis*** (Hornem.) DC. PURPLE CLEMATIS May. Rocky woods and streambanks. The purple flowers open early in spring with the unfolding leaves. *Clematis verticillaris* DC.

***Clematis virginiana*** L. [•83] VIRGIN'S BOWER July–Sept. Thickets, streambanks, moist to wet woods, rocky slopes.

## CONSOLIDA *Knight's-Spur*

***Consolida ajacis*** (L.) Schur DOUBTFUL KNIGHT'S-SPUR Annual herb. Leaves dissected into linear segments. Flowers blue, violet, pink, white, or intermediate colors, in an elongate raceme. Summer. Native of Europe; cultivated as a garden annual; occasionally escaped but probably never fully established.

## COPTIS *Goldthread*

***Coptis trifolia*** (L.) Salisb. [•83] THREE-LEAF GOLDTHREAD Perennial herb, with slender, bright yellow rhizomes. Leaves from base of plant on long petioles, evergreen, divided into 3-leaflets, the leaflets shallowly lobed, with rounded teeth tipped by an abrupt point. Flowers single, white, on a stalk from base of plant; sepals petal-like; petals absent. May–June. Wet conifer woods and swamps, often on mossy hummocks; especially n Wisc, mostly confined to bogs in s Wisc. *Coptis groenlandica* (Oeder) Fern.

## DELPHINIUM *Larkspur*

***Delphinium carolinianum*** Walt. CAROLINA LARKSPUR Perennial herb. Stems pubescent throughout. Leaves deeply dissected into linear segments. Flowers in an elongate terminal raceme; flowers irregular, prevailingly blue; upper sepal prolonged backward into a long spur. May–June. Dry woods, prairies, and sandhills.

ADDITIONAL SPECIES ***Delphinium elatum*** L. (Candle larkspur) introduced, Columbia County.

## ENEMION *False Rue-Anemone*

***Enemion biternatum*** Raf. [•84] EASTERN FALSE RUE-ANEMONE Perennial herb; roots bearing numerous small tuber-like thickenings. Basal leaves long-petioled, 2–3x 3-parted; stem leaves short-petioled or sessile, 1–2x 3-parted or the uppermost trifoliolate. Flowers white. April–May. Moist woods. *Isopyrum biternatum* (Raf.) Torr. & Gray.

## HEPATICA *Liverwort*

Perennial herbs, with several 1-flowered scapes bearing a calyx-like involucre of 3 entire bracts immediately below the flower. Leaves basal, simple, lobed. Sepals 5–12, petal-like. Petals none. The genus is very close to *Anemone* and often merged with it, differing only in the simple leaves and the position of the involucre. Leaves persist during the winter, the new leaves appearing after the very early blooming flowers.

*Key to species*

1 Leaves lobed nearly to middle of blade, the lobes rounded ...... **H. americana**

1 Leaves lobed to more than middle of blade, the lobes acute ...... **H. acutiloba**

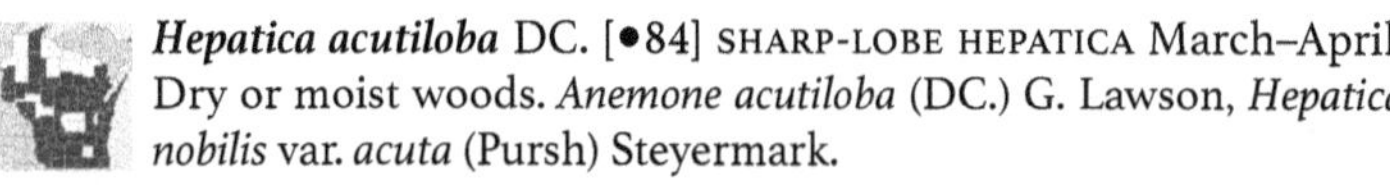

***Hepatica acutiloba*** DC. [•84] SHARP-LOBE HEPATICA March–April. Dry or moist woods. *Anemone acutiloba* (DC.) G. Lawson, *Hepatica nobilis* var. *acuta* (Pursh) Steyermark.

***Hepatica americana*** (DC.) Ker-Gawl. [•84] ROUND-LOBE HEPATICA March–April. Rich beech-maple forests, as for *H. acutiloba,* but more often found on drier sites with aspen, oak, hickory, pine; sometimes with spruce or cedar. Very similar to *H. acutiloba* except in leaves and bracts. Leaves averaging smaller, 3-lobed, the lobes broadly obtuse or rounded, the terminal one often wider than long. *Anemone americana* (DC.) Hara, *Hepatica nobilis* var. *obtusa* (Pursh) Steyermark, *Hepatica triloba* Chaix.

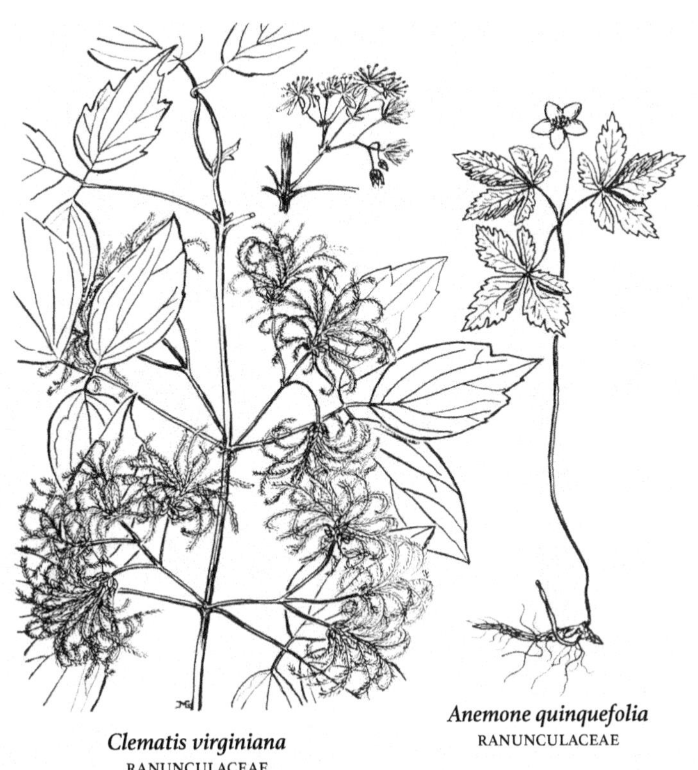

*Clematis virginiana*
RANUNCULACEAE

*Anemone quinquefolia*
RANUNCULACEAE

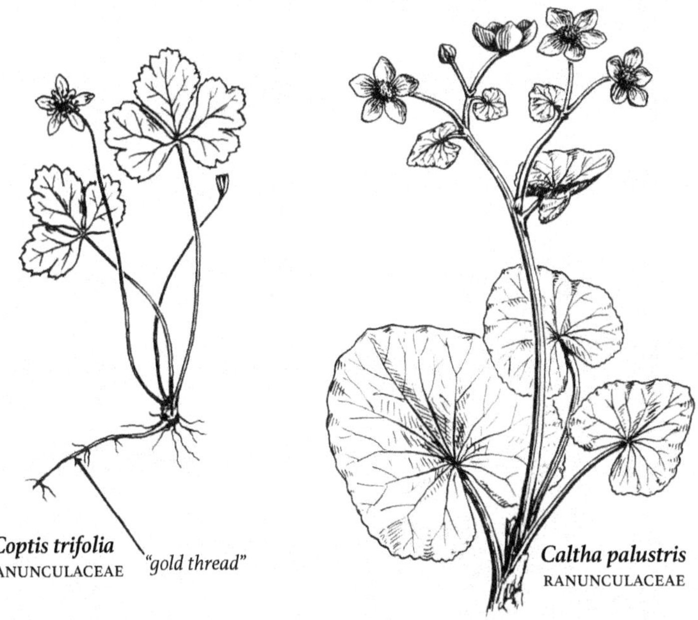


*Coptis trifolia*
RANUNCULACEAE

*Caltha palustris*
RANUNCULACEAE

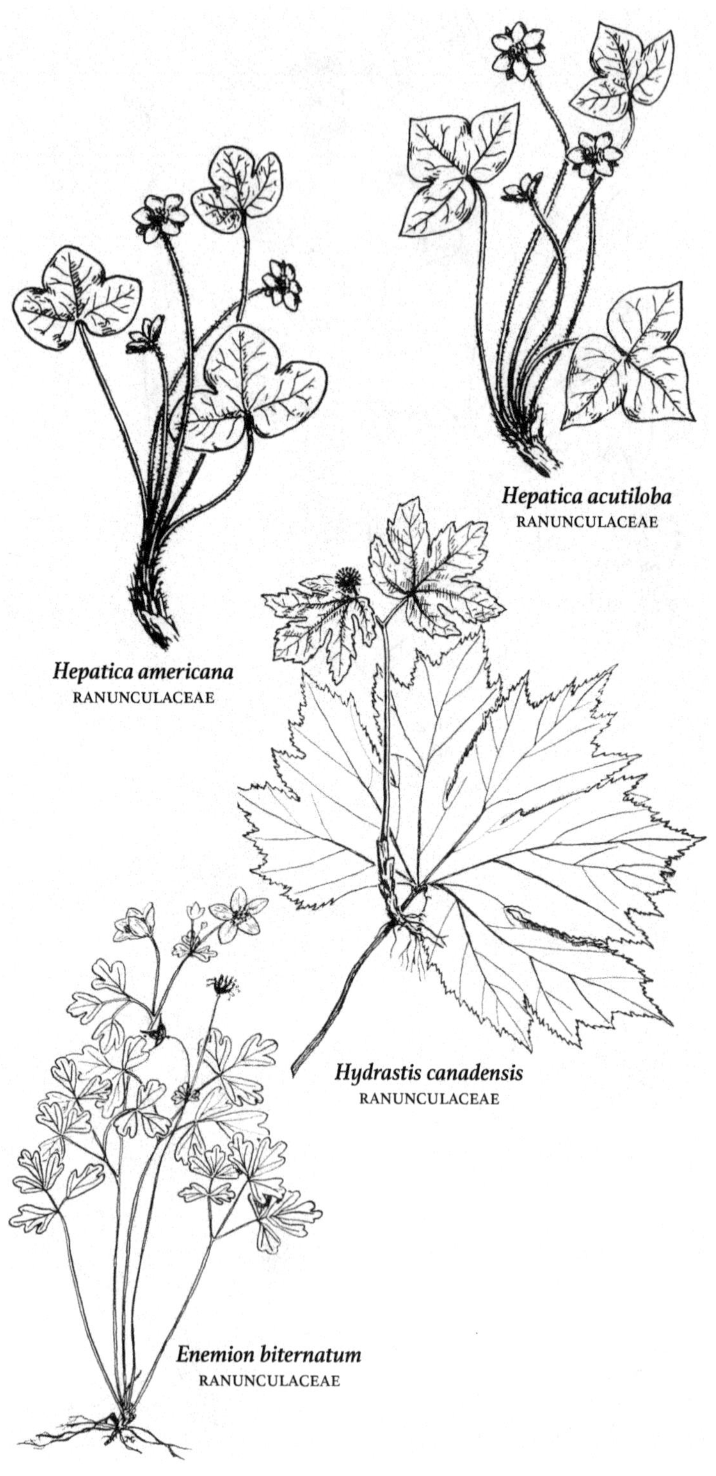
Hepatica acutiloba
RANUNCULACEAE
Hepatica americana
RANUNCULACEAE
Hydrastis canadensis
RANUNCULACEAE
Enemion biternatum
RANUNCULACEAE

## HYDRASTIS *Goldenseal*

***Hydrastis canadensis*** L. [•84] GOLDENSEAL Perennial herb. Stems pubescent, bearing one basal leaf and two stem leaves near the summit. Leaf blades broadly cordate in outline, 5-lobed and palmately 5-nerved. Flower solitary, terminal; sepals 3, petal-like, falling when the flower opens; petals none. Fruit a head of dark red, 1–2-seeded berries. April–May. Deep rich woods. The knotty yellow rhizomes are used in medicine and have been collected so extensively that the plant is scarce across its range (Wisc at northern edge of species' range).

## MYOSURUS *Mousetail*

***Myosurus minimus*** L. TINY MOUSETAIL Introduced. Inconspicuous annual herb. Leaves in a basal tuft, linear. Flowers in a spike atop a slender stalk; sepals 5, green; petals 5 or sometimes absent, white or pink. April–June. Wet to moist places such as streambanks and floodplains, sometimes temporarily in shallow water; also in disturbed drier areas.

## RANUNCULUS *Buttercup; Crowfoot; Spearwort*

Aquatic, semi-aquatic, or terrestrial annual and perennial herbs. Stems erect to sprawling, sometimes floating in water. Leaves simple, or compound and finely dissected, often variable on same plant; alternate on stem or all from base of plant; petioles short to long. Flowers borne above water surface in aquatic species; sepals usually 5, green; petals usually 5, yellow or white, usually with a small nectary pit covered by a scale near base of petal; stamens and pistils numerous. Achenes many in a round or cylindric head; achene body tipped with a straight or curved beak.

*Key to species*

1 Flowers white; leaves divided into linear or threadlike segments; plants typically aquatic ........................... ***R. aquatilis***

1 Flowers yellow; leaves simple to deeply lobed or divided into narrow segments; plants aquatic or emergent ........................ 2

2 Sepals 3 (rarely 4) ................................ ***R. lapponicus***

2 Sepals 5 (or rarely more) ................................... 3

3 All leaves simple and entire, or shallowly lobed with rounded teeth ............................................................ 4

3 All, or at least stem leaves, deeply lobed, divided, or compound . 5

4 Leaves ovate to round or kidney-shaped, shallowly lobed with rounded teeth; achenes with longitudinal ribs ..... ***R. cymbalaria***

4 Leaves oval to lance-shaped or linear, entire to sharp-toothed; achenes not ribbed .............................. ***R. flammula***

5 Basal and stem leaves distinctly different in shape, the basal leaves mostly entire or with rounded teeth, the stem leaves deeply divided ............................................................ 6

5 Basal and stem leaves similar, all deeply lobed, divided, or compound ................................................ 7

6 Flower petals 4–10 mm long, longer and wider than the sepals ... ................................................ ***R. rhomboideus***

6 Flower petals to 3 mm long, shorter and narrower than the sepals ........................................ *R. abortivus*

7 Achenes swollen, without a sharp-edged margin .............. 8
7 Achenes flattened, with a sharp or winglike margin ........... 10

8 Petals 2–4 mm long; achenes to 1.2 mm long, nearly beakless; plants terrestrial or in water only part of season ........... *R. sceleratus*
8 Petals 4–14 mm long; achenes 1.2–2.5 mm long, beaked; plants underwater or exposed later in season ...................... 9

9 Petals more than 7 mm long; achene body more than 1.6 mm long, achene margin thickened and white-corky below the middle .... ........................................ *R. flabellaris*
9 Petals less than 7 mm long; achene body less than 1.6 mm long, achene margin rounded but not thickened ............ *R. gmelinii*

10 Petals 2–5 mm long ........................................ 11
10 Petals 7–15 mm long........................................ 12

11 Beak of achene strongly hooked .................... *R. recurvatus*
11 Beak of achene straight or only slightly curved.. *R. pennsylvanicus*

12 Style short and outcurved; introduced and weedy species ...... 13
12 Style elongate and nearly straight; native, non-weedy species... 14

13 Stems creeping; terminal segment of the main leaves stalked .... ........................................ *R. repens*
13 Stems upright; terminal segment of the main leaves not stalked, usually with green tissue extending to the lateral segments ...... ........................................ *R. acris*

14 Main leaves mostly longer than wide; mature receptacle cone-shaped ........................................ *R. fascicularis*
14 Main leaves usually wider than long; mature receptacle often club-shaped and widest at tip .......................... *R. hispidus*

***Ranunculus abortivus*** L. KIDNEY-LEAF BUTTERCUP April–June. Wet to moist woods, floodplains, wet meadows, thickets, ditches; especially where soils disturbed or compacted.

***Ranunculus acris*** L. [•85] MEADOW-BUTTERCUP Introduced (naturalized). June–Aug. Common weed of fields, thickets, ditches and shores.

***Ranunculus aquatilis*** L. [•85] LONG-BEAK WATER-CROWFOOT May–Aug. Ponds, lakes, streams, rivers and ditches. Ranunculus longirostris Godr., Ranunculus trichophyllus Chaix. Our plants sometimes treated as 2 species as follows:

1 Styles (at least the longest) and achene beaks 0.6–1.1 mm long, more than 1/3 the length of the achene body ............. *R. longirostris*
1 Styles and achene beaks very short, less than 0.6 mm long, less than about 1/3 the length of the body ................. *R. trichophyllus*

***Ranunculus cymbalaria*** Pursh ALKALI BUTTERCUP *Threatened.* June–Sept. Wet meadows, streambanks, sandy or muddy shores, ditches and seeps, often in wet mud or sand; Lake Michigan shores; often where brackish.

***Ranunculus fascicularis*** Muhl. EARLY BUTTERCUP April–May. Prairies and dry woods.

***Ranunculus flabellaris*** Raf. GREATER YELLOW WATER BUTTERCUP May–July. Shallow water or muddy shores of ponds, quiet streams, swamps, woodland pools, marshes and ditches.

***Ranunculus flammula*** L. CREEPING SPEARWORT June–Aug. Sandy, gravelly, or muddy shores; shallow to deep water, water usually acid.

***Ranunculus gmelinii*** DC. LESSER YELLOW WATER BUTTERCUP *Endangered.* July–Aug. Muddy streambanks and lakeshores, cold springs, pools in swamps and bogs.

***Ranunculus hispidus*** Michx. [•85] NORTHERN SWAMP BUTTERCUP May–July. Wet woods, floodplains and swamps, thickets, lakeshores, wet meadows and fens.

***Ranunculus lapponicus*** L. LAPLAND BUTTERCUP *Endangered.* June–July. Cedar swamps and bogs. Plants resemble *Coptis trifolia* but Lappland buttercup leaves are lobed, not compound, lighter green and deciduous. *Coptidium lapponicum* (L.) Gandog.

***Ranunculus pensylvanicus*** L. f. BRISTLY CROWFOOT July–Aug. Marshes, wet meadows, ditches and streambanks, often in muck.

***Ranunculus recurvatus*** Poir. HOOKED CROWFOOT May–June. Moist deciduous forests (especially in openings), swamps; also in drier woods; southward also in partial shade in calcareous fens.

***Ranunculus repens*** L. CREEPING BUTTERCUP May–July. Native of Europe; introduced in fields, lawns, roadsides, and wet meadows. Plants vary in size and in kind and amount of pubescence.

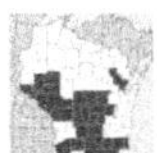
***Ranunculus rhomboideus*** Goldie LABRADOR BUTTERCUP April–May. Dry open woods and prairies.

***Ranunculus sceleratus*** L. [•86] CURSED CROWFOOT May–Sept. Muddy shores, streambanks, wet meadows, ditches, marshes and other wet places.

## THALICTRUM *Meadow-Rue*

Perennial herbs. Leaves alternate, compound. Staminate and pistillate flowers separate, in panicles on separate plants; sepals 4–5, green or petal-like but soon deciduous; petals absent; stamens numerous, the stalks (filaments) long and slender. Fruit a ribbed or nerved achene.

*Key to species*

1 Flowers few in an umbel ........................ ***T. thalictroides***

1 Flowers in racemes or panicles .............................. 2

2 Upper stem leaves with long petioles; leaflets glabrous and not glandular; flowering in April or May before leaves fully expanded; plants less than 1m tall ............................ ***T. dioicum***

2 Upper stem leaves sessile or nearly so (the 3 stalked leaflets appearing to arise together from the node); leaflets glabrous, or

hairy, or with small, short-stalked glands; flowering in summer after leaves expanded; plants often more than 1 m tall ............... 3

3 Leaflets 3-lobed, each lobe tipped with 1–3 teeth .... *T. venulosum*
3 Leaflets usually 3-lobed, the lobes usually not toothed.......... 4

4 Underside of leaflets with very short hairs (rarely smooth), not glandular; leaves odorless; widespread .......... *T. dasycarpum*
4 Underside of leaflets with small beads and hairs tipped with gray or amber exudate; leaves with strong odor when crushed; uncommon in se Wisc ........................... *T. revolutum*

***Thalictrum dasycarpum*** Fisch. & Avé-Lall. PURPLE MEADOW-RUE June–July. Wet to moist meadows, low prairie, swamps, thickets, streambanks.

***Thalictrum dioicum*** L. EARLY MEADOW-RUE Moist woods. Flowering with or before the expansion of leaves on deciduous trees.

***Thalictrum revolutum*** DC. [•86] WAXY-LEAF MEADOW-RUE June–July. Streambanks, thickets, moist meadows and prairies.

***Thalictrum thalictroides*** (L.) Eames & Boivin [•86] RUE-ANEMONE April–May. Dry to moist woods. Color of the sepals sometimes approaches red, blue, or green; double flowers are often observed. *Anemonella thalictroides* (L.) Spach.

***Thalictrum venulosum*** Trel. VEINY-LEAF MEADOW-RUE June–July. Streambanks, thickets and wet, calcium-rich shores of Lake Michigan.

## RESEDACEAE *Mignonette Family*

### RESEDA *Mignonette*

***Reseda lutea*** L. YELLOW UPRIGHT MIGNONETTE Introduced biennial herb. Leaves alternate, oblong lance-shaped in outline, deeply and irregularly pinnatifid above the middle into a few narrow segments. Flowers perfect, in dense terminal racemes; petals usually 6, unequal, the upper the largest, greenish yellow. Summer. Dry fields, roadsides, waste places.

ADDITIONAL SPECIES ***Reseda alba*** L. (White upright mignonette), introduced, Dane County.

## RHAMNACEAE *Buckthorn Family*

Shrubs, trees, or woody vines with simple, opposite or alternate leaves. Flowers perfect or unisexual, regular, 4–5-merous. Petals present or lacking, small, separate. Fruit a capsule or drupe.

*Key to genera*

1 Leaves 3-veined from base of leaf; flowers white in many-flowered, stalked clusters; fruit a capsule .................. **CEANOTHUS**

*Ranunculus acris*
RANUNCULACEAE
*Ranunculus hispidus*
RANUNCULACEAE
*Ranunculus aquatilis*
RANUNCULACEAE

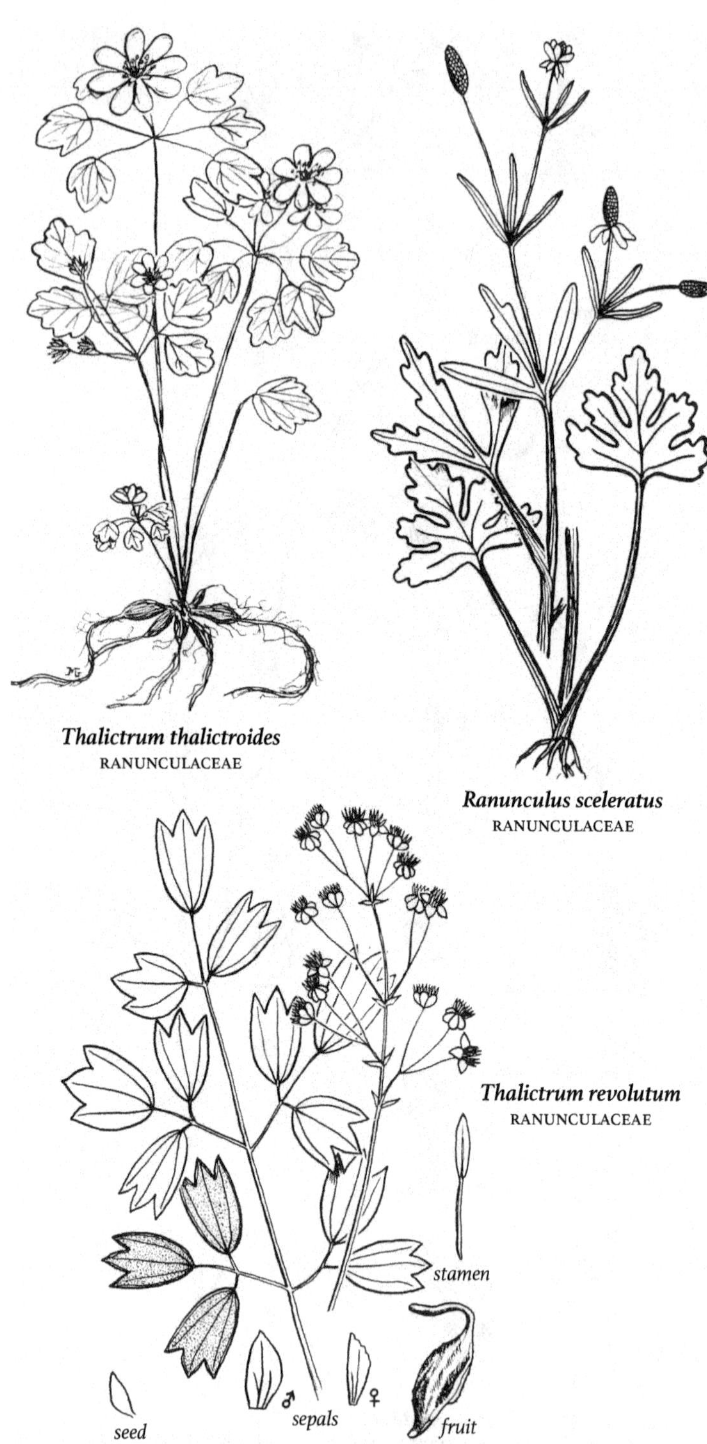
Thalictrum thalictroides
RANUNCULACEAE
Ranunculus sceleratus
RANUNCULACEAE
Thalictrum revolutum
RANUNCULACEAE
stamen
seed
♂
sepals
♀
fruit

1 Leaves not 3-veined from base; flowers greenish, single or few to a cluster; fruit a fleshy drupe .................................. 2

2 Leaf margins entire or nearly so ................. FRANGULA
2 Leaf margins toothed ........................... RHAMNUS

## CEANOTHUS *Buckbrush*

Low shrubs. Leaves 3-nerved, glandular-serrate. Flowers small, white, in sessile or short-peduncled umbels aggregated into terminal or axillary panicles. Many species highly ornamental. Characteristic are the leaves with 3 pairs of prominent parallel veins extending from the leaf base to the outer margins of the leaf tips.

*Key to species*

1 Leaves elliptic, less than 2 cm wide; inflorescences at ends of current year's shoots ..................................... *C. herbaceus*
1 Leaves ovate, mostly more than 2 cm wide; inflorescences from leaf axils .......................................... *C. americanus*

***Ceanothus americanus*** L. NEW JERSEY-TEA June–July. Upland woods, prairies, and barrens.

***Ceanothus herbaceus*** Raf. PRAIRIE REDROOT May–June. Sandy or rocky soil, prairies and plains.

## FRANGULA *False Buckthorn*

***Frangula alnus*** P. Mill. GLOSSY FALSE BUCKTHORN Shrub. Leaves usually obovate-oblong, entire or with a few marginal glands near the tip. Umbels sessile, 2–8-flowered. Flowers perfect. Fruit red, ripening to nearly black, 2–3-stoned. May–June. Native of Eurasia; escaped from cultivation, especially in wet soil. *Rhamnus frangula* L.

## RHAMNUS *Buckthorn*

Shrubs or small trees. Leaves simple, alternate or opposite, pinnately veined, usually with stipules. Flowers perfect, or staminate or pistillate, regular, single or few from leaf axils. Fruit a purple-black, berrylike drupe.

*Key to species*

1 Leaves with 2–4 obvious pairs of lateral veins ....... *R. cathartica*
1 Leaves with mostly 5 or more pairs of lateral veins.............. 2

2 Leaves less than 8 cm long; petals 4 ................ *R. lanceolata*
2 Leaves more than 8 cm long; petals absent ........... *R. alnifolia*

***Rhamnus alnifolia*** L'Hér. ALDER-LEAF BUCKTHORN May–June. Conifer swamps, thickets, sedge meadows, wet depressions in deciduous forests; usually where calcium-rich.

***Rhamnus cathartica*** L. EUROPEAN BUCKTHORN May–Aug. Conifer swamps, thickets, calcareous fens, lakeshores, moist to dry woods, especially where disturbed, heavily grazed, or cleared. Introduced from Eurasia; escaping from cultivation in ne and c North America.

 ***Rhamnus lanceolata*** Pursh LANCE-LEAF BUCKTHORN Calcareous fens.

## ROSACEAE *Rose Family*

Shrubs, and perennial, biennial, or annual herbs. Leaves evergreen or deciduous, mostly alternate and simple or compound. Flowers perfect, regular, with 5 sepals and petals; stamens numerous. Fruit an achene, capsule, or fleshy fruit with numerous embedded seeds (drupe), or a fleshy fruit with seeds within (pome).

ADDITIONAL SPECIES ***Rhodotypos scandens*** (Thunb.) Makino (Jetbead), ornamental deciduous shrub, introduced in se Wisc.

*Key to genera*

1 Plants trees, shrubs, or erect to trailing, thorny to bristly brambles. 2
1 Plants herbs (sometimes woody at base), not thorny or bristly . . 15

2 Leaves mostly compound; branches or stems often thorny or bristly . . . . . . . . 3
2 Leaves simple; branches and stems smooth or only with long stout spines . . . . . . . . 7

3 Stems biennial, prickly or bristly; leaves 3-parted or palmately compound; fruit a tight cluster of juicy drupelets; flowers usually white . . . . . . . . **RUBUS**
3 Stems perennial, smooth or thorny; leaves pinnately compound; fruit various but not a cluster of drupelets; flowers white, pink, or yellow. . . . . . . . 4

4 Flowers pink (rarely white or yellow), 2 cm or more wide; stems thorny; fruit fleshy, red to orange . . . . . . . . **ROSA**
4 Flowers white or yellow, mostly less than 2 cm wide; stems smooth; fruit various. . . . . . . . 5

5 Flowers solitary or few in an inflorescence, the petals yellow; leaflets entire . . . . . . . . ***Potentilla fruticosa***
5 Flowers many in a crowded inflorescence, the petals white; leaflets toothed. . . . . . . . 6

6 Colony-forming shrub, occasionally escaping from cultivation; inflorescence a panicle, much longer than wide; leaflets doubly-toothed (each main tooth with several smaller teeth) **SORBARIA**
6 Small trees; inflorescence much wider than long; leaflets not doubly toothed . . . . . . . . **SORBUS**

7 Style and ovary 1; fruit a drupe; leaves unlobed . . . . . . . . **PRUNUS**
7 Styles 2 or more (1 in *Crataegus monogyna*); fruit a pome, or a cluster of drupelets or dry fruits . . . . . . . . 8

8 Ovaries superior . . . . . . . . 9
8 Ovary inferior . . . . . . . . 10

9 Leaves mostly 3–5 lobed; bark shredding into long strips . . . . . . . . **PHYSOCARPUS**
9 Leaves not lobed; bark not shredding into long strips . . **SPIRAEA**

10 Leaves with red or black appressed glands along midrib of leaf upper surface .......................................... ARONIA

10 Leaves without glands on midrib .............................. 11

11 Branches never thorny; flower petals white, lance-shaped and usually more than 2 times longer than wide ... AMELANCHIER

11 Branches sometimes with stout spines; petals less than 2 times longer than wide ............................................ 12

12 Branches normally with spines; leaves toothed and often slightly lobed ........................................................ 13

12 Branches without spines; leaves toothed but not lobed......... 14

13 Spines shiny; bud scales glabrous; petals white; seeds within hard nutlets .......................................... CRATAEGUS

13 Spines dull; bud scales hairy; petals pinkish; seeds within papery carpels ............................................ MALUS

14 Young twigs densely hairy; petals pinkish; fruit an apple MALUS

14 Young twigs glabrous; petals white; fruit a pear .......... PYRUS

15 Leaves 3-parted or palmately compound....................... 16

15 Leaves pinnately compound or divided ........................ 21

16 Styles long, jointed near middle, the lower portion persistent on the achene as a long beak ................................. GEUM

16 Styles short, neither jointed nor persistent on the fruit ......... 17

17 Calyx with bractlets about as large as sepals, the calyx appearing 10-lobed ........................................................ 18

17 Calyx without bractlets between the sepals, the calyx 5-lobed .. 20

18 Petals white; fruit fleshy and red; leaflets 3 ........... FRAGARIA

18 Petals yellow or white; fruit dry; leaflets 3, 5, or 7............... 19

19 Flowers yellow, leaflets 3, 5, or 7, regularly toothed, deciduous .... ........................................... POTENTILLA

19 Flowers white; leaflets 3, entire except for a 3 (–5)-toothed apex, evergreen......................................... SIBBALDIA

20 Petals yellow; fruit an achene ................ *Geum fragarioides*

20 Petals white or pink; fruit fleshy drupelets ....... POTENTILLA

21 Leaves 2–3 times compound; inflorescence a large panicle of numerous spike-like racemes ...................... ARUNCUS

21 Leaves once-pinnate; inflorescence various, smaller ........... 22

22 Calyx 5-lobed, small bractlets absent; receptacle flat or concave 23

22 Calyx 10-lobed, small bractlets alternating with sepals; receptacle hemispherical or conical ...................................... 25

23 Petals pink; receptacle flat or nearly so; leaflets deeply lobed....... ......................................... FILIPENDULA

23 Petals yellow or absent; receptacle deeply concave; leaflets not lobed .............................................................. 24

24 Petals yellow; floral tube with hooked bristles at tip; inflorescence an elongate raceme ............................ AGRIMONIA

24 Petals absent; floral tube not bristly; inflorescence short ......... ........................................ SANGUISORBA

25 Styles elongating and becoming longer than achene, persistent as a beak atop the achene .................................. **GEUM**
25 Styles short, deciduous ........................................ **26**

26 Petals deep maroon to purple; sepals red tinged; stem usually decumbent, the lower portion in water or wet ground, rooting at nodes ............................................ **COMARUM**
26 Petals yellow or white; sepals green; stem usually erect, or with slender stolons; mostly upland .................................. **27**

27 Pubescence not glandular; petals deep yellow .... **POTENTILLA**
27 Pubescence glandular-viscid; petals white to pale yellow ......... ........................................ **DRYMOCALLIS**

## AGRIMONIA *Agrimony; Grooveburr*

Perennial herbs from stout rhizomes. Leaves pinnately compound, mostly below middle of stem. Flowers in long, interrupted, spike-like racemes. Hypanthium with hooked bristles and small resinous glands; petals 5, yellow. Fruit an achene.

*Key to species*

1 Leaves with 5 or more pairs of lateral leaflets (not including the much smaller leaflets), the leaflets mostly 3 or more times longer than wide ........................................ ***A. parviflora***
1 Leaves with up to 4 pairs of major leaflets, the leaflets to 2 times longer than wide ................................................ **2**

2 Inflorescence rachis covered with small glands, the pubsecence sparse or absent ................................ ***A. gryposepala***
2 Inflorescence rachis without glands or nearly so, but rachis covered with appressed to spreading hairs.............................. **3**

3 Leaflet underside velvety to touch, the hairs spreading ......... ...................................................... ***A. pubescens***
3 Leaflet underside smooth or rouch-to-touch, the hairs usually appressed ............................................ ***A. striata***

***Agrimonia gryposepala*** Wallr. [•87] TALL HAIRY AGRIMONY July–Aug. Moist or dry open woods.

***Agrimonia parviflora*** Ait. HARVESTLICE July–Aug. Streambanks, wet meadows, wet woods.

***Agrimonia pubescens*** Wallr. SOFT AGRIMONY July–Aug. Dry open woods.

***Agrimonia striata*** Michx. WOODLAND AGRIMONY July–Aug. Dry or moist woods.

## AMELANCHIER *Serviceberry*

Trees or shrubs, without thorns. Leaves simple, alternate, serrate. Flowers in short leafy racemes (except in *A. bartramiana*) terminating the branches of the season, opening with or before the leaves. Sepals 5, spreading to recurved. Petals 5, white. Stamens usually 20, shorter than the petals.

HYBRIDS ***Amelanchier* × *neglecta*** Eggl. ex G.N. Jones (pro sp.); cross between *A. bartramiana* and *A. laevis*, reported from several Wisc locations.

*Key to species*

1 Pedicels 1–3 in axils of leaves; petals less than twice as long as wide; leaves at least partly open and essentially glabrous (except margins and petioles) at flowering time, the blade tapering trough-like into raised petiole margins; petioles less than 8 (–15) mm long, n Wisc only ........................................ ***A. bartramiana***

1 Pedicels 4 or more (at least scars present if some have fallen with fruit), the inflorescence a raceme; petals at least twice as long as wide; leaves various (glabrous to tomentose) but the blade rounded or truncate to subcordate, not tapered at base; petioles usually longer than 8 mm; statewide ................................ 2

2 Tip of ovary glabrous; leaf blades short-acuminate, finely and closely serrate with 22–45 teeth per side ...................... 3

2 Tip of ovary tomentose; leaf blades variously shaped and toothed 4

3 Leaves just beginning to unfold at flowering time, densely white-tomentose beneath, otherwise green, retaining some of the pubescence on petioles and along midrib beneath into maturity . ....................................................... ***A. arborea***

3 Leaves mostly half-grown at flowering time, usually bronze-red, glabrous or nearly so, completely glabrous at maturity ... ***A. laevis***

4 Larger leaves with 25–50 fine teeth on a side (more than twice as many teeth as lateral veins), acute, at flowering time open though not fully grown and often glabrous or soon becoming so ***A. interior***

4 Larger leaves with fewer than 20 (–25) teeth on a side (no more than 2x as many teeth as lateral veins), the blades at flowering time ± folded and white-tomentose beneath, when mature the tip acute to rounded ................................................... 5

5 Most leaves coarsely toothed (2–5 teeth per cm toward tip when mature), the veins prominent and running to tips of the teeth (or a principal fork into the teeth) at least toward tip of blade; petals 10–20 mm long; plants typically solitary or in tall clumps with many stems ........................................ ***A. sanguinea***

5 Most leaves finely toothed at least toward apex (5–8 teeth per cm when mature), the veins anastomosing and becoming indistinct near the margin, at most with weak veinlets ending in the teeth; petals 5–9 mm long; plants typically spreading underground and forming colonies of low shrubs. ...................... ***A. spicata***

***Amelanchier arborea*** (Michx. f.) Fern. DOWNY SERVICEBERRY April–May. Usually in dry sandy open forests with red maple, aspen, oaks, or jack pine; sometimes in moist or swampy forests and along forest borders.

***Amelanchier bartramiana*** (Tausch) M. Roemer OBLONG-FRUIT SERVICEBERRY May–Aug. Conifer swamps, open bogs, thickets, old dune or rock ridges; borders of hardwood forests; plants may be low and sprawling on bare rock shores and ledges, otherwise a tall shrub.

***Amelanchier interior*** Nielsen INLAND SERVICEBERRY May–June. Sandy open savannas and dunes, shallow soil on rock outcrops and shores; sometimes at borders of hardwood forests and conifer swamps.

***Amelanchier laevis*** Wieg. SMOOTH SERVICEBERRY May. Most often in dry sandy open forests and savannas, rocky sites, sandy bluffs and shores; also on river banks and forest and bog margins.

***Amelanchier sanguinea*** (Pursh) DC. NEW ENGLAND SERVICEBERRY May–June. Dry, open, sandy savannas and clearings; sandy thickets, borders of forests, gravelly shores, and low dunes. *Amelanchier humilis* Wieg.

***Amelanchier spicata*** (Lam.) K. Koch RUNNING SERVICEBERRY May. Dry, sandy plains, dunes, and savannas, usually with jack pine or oaks, often little taller than the associated shrubby species of *Comptonia* and *Vaccinium.*

## ARONIA *Chokeberry*

***Aronia prunifolia*** (Marsh.) Rehder [•87] PURPLE CHOKEBERRY Shrub. Leaves alternate, oval or obovate, upper surface dark green and smooth (except for dark, hairlike glands along midveins), underside paler, smooth or hairy; margins with small, rounded, forward-pointing teeth, the teeth gland-tipped. Flowers in clusters of 5–15 at ends of stems and short, leafy branches; sepals usually glandular; petals white. Fruit a dark purple to nearly black, berrylike pome, not persisting into winter. May–June. Tamarack swamps, open bogs, thickets, marshes and shores.

ADDITIONAL SPECIES Completely glabrous plants are sometimes recognized as ***Aronia melanocarpa*** (Michx.) Elliott (black chokeberry).

## ARUNCUS *Goat's Beard*

***Aruncus dioicus*** (Walt.) Fern. GOAT'S BEARD Introduced, rhizomatous, perennial herbs. Leaves alternate; leaflets ovate lance-shaped to broadly ovate, coarsely doubly serrate. Flowers dioecious, 5-merous, in numerous racemes aggregated into a large terminal panicle; staminate flowers with petals about 1 mm long, stamens 15 or more; pistils rudimentary, 3–5; pistillate flowers with sepals and petals as in the staminate flowers but smaller. Rich woods; Dane and Iowa cos.

## COMARUM *Marshlocks*

***Comarum palustre*** L. MARSH CINQUEFOIL Perennial herb, from long, stout rhizomes. Stems ascending to sprawling or floating in shallow water, often rooting at nodes, more or less woody at base. Leaves all from stem, pinnately divided or nearly palmate, with 3–7 leaflets; underside waxy; margins with sharp, forward-pointing teeth. Flowers single or paired from leaf axils, or in open clusters; sepals dark red or purple; petals 5 (sometimes 10), very dark red; stamens about 25, dark red. June–Aug. Open bogs (especially in pools and wet margins), conifer swamps, shores. *Potentilla palustris* (L.) Scop.

## CRATAEGUS *Hawthorn*

Small trees or shrubs with usually spiny branches. Leaves simple, deciduous, alternate, serrate or dentate and otherwise entire or variously lobed. Flowers perfect, regular, in corymbs or rarely single or 2 or 3 together. Sepals 5. Petals 5, white or rarely pink. Fruit a globose pome, red or rarely yellow, blue, or black at maturity, with 1–5 bony nutlets. The leaves of sterile shoots or of the ends of branches (vegetative leaves) are often differently shaped and more deeply incised than those of the flowering branchlets.

Included in the key are the most common, well-defined species reported for Wisconsin; listed below are minor native and introduced hawthorns and hawthorn hybrids not treated in the Flora: *Crataegus* × *apiomorpha* Sarg., *C. beata* Sarg., *C. coccinea* L., *C. coccinioides* Ashe, *C.* × *desueta* Sarg., *C. distincta* Kruschke, *C. douglasii* Lindl., *C. holmesiana* Ashe, *C. intricata* Lange, *C. irrasa* Sarg., *C. jesupii* Sarg., *C.* × *lucorum* Sarg., *C. margaretta* Ashe, *C.* × *nitidula* Sarg., *C. pedicellata* Sarg., *C. prona* Ashe, *C. schuettei* Ashe (see key), *C. submollis* Sarg.

*Key to species*

1 Leaves with some of the primary lateral veins running to (or toward, forking just before) the sinuses as well as to the points of the lobes; blades ± deltoid in general outline or small and deeply lobed; thorns under 5 cm long; stamens ca. 20; introduced species and known from several Wisc locations ........................ ***C. monogyna***

1 Leaves with the primary lateral veins running only to (or toward) the points of the lobes (if any); blades, thorns, and stamens various; native species, widespread to uncommon in Wisc .............. 2

2 Nutlets with deep to shallow pits or depressions on their lateral surfaces; flowering in late May or June ........................ 3

2 Nutlets not pitted laterally; flowering in April–early June ....... 5

3 Leaves ± narrowly acute or short-acuminate in outline at the apex, completely glabrous beneath; nutlets 2–5, with rather shallow and irregular depressions on the ventral face, usually very strongly ridged and grooved dorsally; thorns mostly 3.5–4.5 (–5.5) cm long; inflorescences mostly glabrous ..................... ***C. brainerdii***

3 Leaves acute to rather broadly rounded or obtuse in outline at the apex, at least the midrib beneath often pubescent; nutlets 2–3, with a definite pit occupying most of each half of the ventral face, smoothly rounded or nearly so dorsally; thorns mostly 2.5–9.5 cm long; inflorescences often ± villous .......................... 4

4 Mature leaf blades thin, the veins (except sometimes for midrib) scarcely if at all impressed above, strigose above and usually pubescent beneath; inflorescences, branchlets of current year, and petioles all usually villous or lightly tomentose; thorns 2.5–5 cm long, often sparse or even absent; stamens ca. 20; late flowering (usually early June) ............................ ***C. calpodendron***

4 Mature leaf blades ± leathery, thickened at margins, the veins usually deeply impressed above, glabrous to pubescent on both surfaces; inflorescences, new branchlets, and petioles sparsely villous to glabrous (if inflorescence somewhat villous, at least the young branchlets nearly always glabrous, the veins deeply impressed, and/or the stamens ca. 10); thorns 2.5–9.5 cm long,

usually numerous; stamens ca. 20 or 10; mid-season flowering (mid- to late-May in s Wisc) .............................. *C. succulenta*

5 Blades of at least the floral leaves (in many species also the vegetative leaves) ± acute to broadly or (more commonly) narrowly tapered or cuneate at the base .............................. 6

5 Blades of both floral and vegetative leaves mostly broadly rounded, truncate, or subcordate at the base .............................. 9

6 Blades (especially of floral leaves) mostly obovate to oblong-elliptic, broadest above or rarely at the middle, unlobed or very obscurely lobed near the apex, mostly 1.5–3 or more times as long as broad, usually thick or even stiff and leathery .............................. 7

6 Blades (at least of floral leaves) mostly elliptic to ovate, broadest at or below the middle, often ± lobed, usually 1–1.5 times as long as broad, often thin .............................. 8

7 Leaves glossy above, the veins not (or only slightly) impressed; petioles mostly less than 1 cm long; styles and nutlets 1–3; inflorescence usually glabrous, stamens ca. 10 or 20 .. *C. crus-galli*

7 Leaves dull above, the veins rather conspicuously impressed; petioles mostly 1–2 cm long; styles and nutlets 3–5; inflorescence usually ± villous; stamens ca. 20 .............................. *C. punctata*

8 Stamens ca. 15–20 .............................. *C. pruinosa*

8 Stamens ca. 10 or fewer .............................. *C. chrysocarpa*

9 Inflorescence, calyx, and leaves (at least along main veins) densely tomentose; fruit short-hairy at least at the ends; stamens 20, the anthers white or yellow (rarely pink) .............................. *C. mollis*

9 Inflorescence, calyx, and leaves glabrous or pubescent; fruit glabrous; stamens ca. 10 or 20, anthers in most species pink to purple .............................. 10

10 Stamens 10 or fewer; anthers pink to purple; young leaves strigose above; inflorescences completely glabrous at flowering time .......................................... *C. macrosperma*

10 Stamens 15–20; anthers pink-purple or white to yellowish; young leaves glabrous to pubescent above .............................. 11

11 Leaves glabrous or nearly so on both sides when young; inflorescences glabrous; calyx lobes entire or weakly and sparsely serrate; ripe fruit with rather thin dry flesh .......... *C. pruinosa*

11 Leaves strigose above, at least when young; inflorescences glabrous; calyx lobes usually glandular-serrate; ripe fruit mellow or succulent .............................. *C. schuettei**

***Crataegus brainerdii*** Sarg. ROUGH HAWTHORN May–June; fruit ripe Sept. Thickets, pastures and roadsides.

***Crataegus calpodendron*** (Ehrh.) Medik. PEAR HAWTHORN May–June; fruit ripe Oct. Woods and thickets, often along small rocky streams.

***Crataegus chrysocarpa*** Ashe FIREBERRY HAWTHORN May; fruit ripe Sept–Oct. Sandy hillsides, stream and river banks, forest borders, roadsides, fields, pastures; sometimes in wet places.

***Crataegus crus-galli*** L. COCKSPUR HAWTHORN May–June; fruit ripe Oct. Borders of woods, thickets, and pastures. Variable, with many forms and varieties.

***Crataegus macrosperma*** Ashe BIG-FRUIT HAWTHORN May–June. Open woods, thickets, fields, and along river banks and rocky ridges.

***Crataegus mollis*** Scheele DOWNY HAWTHORN April–May; fruit ripe Aug–Oct. Open woods, usually in alluvial or fertile ground, and most common in limestone regions.

***Crataegus monogyna*** Jacq. ENGLISH HAWTHORN May–June; fruit ripe Sept. Thickets and pastures; native of Europe and w Asia, escaped and established.

***Crataegus pruinosa*** (Wendl. f.) K. Koch WAXY-FRUIT HAWTHORN May; fruit ripe Oct. Thickets and rocky woods, usually in dry or well drained ground.

***Crataegus punctata*** Jacq. DOTTED HAWTHORN May–June; fruit ripe Sept–Oct. Thickets and borders of woods, often in rocky ground.

***Crataegus succulenta*** Schrad. FLESHY HAWTHORN May–June; fruit ripe Sept. Thickets, pastures, and borders of woods, usually in dry or rocky ground.

## DRYMOCALLIS *Woodbeauty*

***Drymocallis arguta*** (Pursh) Rydb. TALL WOODBEAUTY Perennial from a stout rhizome, more or less viscid-pubescent throughout. Leaves pinnately compound; leaflets 7–11, or only 5 in the uppermost leaves. Flowers white, cream or pale yellow, crowded in a slender, elongate inflorescence. June–July. Dry woods and prairies. *Potentilla arguta* Pursh.

## FILIPENDULA *Queen-of-the-Prairie*

Perennial rhizomatous herbs. Leaves stipulate, pinnately compound. Flowers in large panicles. Flowers white to pink, 5–7-merous.

1 Petals white; leaves white-hairy on underside .......... ***F. ulmaria***
1 Petals rose-colored; leaves green on underside ........... ***F. rubra***

***Filipendula rubra*** (Hill) B.L. Robins. QUEEN-OF-THE-PRAIRIE June–July. Wet meadows and shores, calcareous fens; soils usually calcium-rich. Native to ne North America but considered adventive in Wisc.

***Filipendula ulmaria*** (L.) Maxim. QUEEN-OF-THE-MEADOW June–Aug. Native of Eurasia; occasionally planted for ornament and sometimes escaped. The leaves are usually pale beneath with a close tomentum.

## FRAGARIA *Strawberry*

Perennial herbs, usually spreading freely by runners and forming colonies. Leaves basal, 3-foliolate, serrate. Flowers several on peduncles. Petals white. Fruit of numerous minute achenes on the greatly enlarged, red, juicy receptacle, subtended by the persistent calyx and bracts.

*Key to species*

1 Terminal center tooth of leaflets smaller than the tooth on either side of it; calyx lobes appressed to fruit .............. ***F. virginiana***
1 Terminal center tooth of leaflets as large or larger than the tooth on either side of it; calyx lobes spreading away from fruit .... ***F. vesca***

***Fragaria vesca*** L. Thin-Leaved WILD STRAWBERRY April–June. Hardwood and mixed forests, cedar and tamarack swamps, shores and forest edges. In nearly all leaves, a line connecting the apices of the 2 uppermost lateral teeth passes across the projecting terminal tooth.

***Fragaria virginiana*** Duchesne THICK-LEAVED WILD STRAWBERRY April–June. Widespread in many habitats, in a diversity of deciduous, mixed, and coniferous forests, clearings, dry sandy forests, roadsides, and fields; more often in dry open sunny places than F. vesca. In nearly all leaves, a line connecting the tips of the 2 uppermost lateral teeth passes above the tip of the small terminal tooth.

## GEUM *Avens*

Perennial herbs. Lower leaves pinnately lobed or divided, upper leaves smaller, less divided or entire. Flowers yellow, white or purple; 1 to many in clusters at ends of stems; petals 5. Fruit an achene.

ADDITIONAL SPECIES ***Geum virginianum*** L. (Cream avens) native, southeastern Wisc, more common in eastern USA.

*Key to species*

1 Leaves all 3-foliolate and basal ................... ***G. fragarioides***
1 Leaves mostly pinnately compound or divided, if 3-foliolate, then cauline .................................................. **2**

2 Calyx bell-shaped; reddish; flowers nodding; petals yellow, tinged with purple ............................................. **3**
2 Calyx lobes spreading, green; flowers upright; petals white or yellow ...................................................... **4**

3 Plant of wetlands; terminal leaflet much larger than lateral leaflets; style with distinct joint near its middle, lengthening to less than 2x length of perianth ................................... ***G. rivale***
3 Plant of dry habitats; terminal leaflet barely larger than lateral leaflets; style not jointed, elongating to 2x or more longer than perianth ...................................... ***G. triflorum***

4 Plants fruiting by early June; calyx without small bractlets between the sepals; petals short, to 2 mm long ................ ***G. vernum***
4 Plants beginning to flower in June or July; calyx with small bractlets between the sepals; petals 3 mm or more long.................. **5**

5 Plants flowering ........................................ **6**
5 Plants fruiting........................................... **10**

6 Petals white to pale yellow ................................ **7**
6 Petals bright yellow ..................................... **8**

7 Petals equal to or longer than the sepals; stems glabrous or only sparsely hairy ................................ ***G. canadense***

7 Petals shorter than the sepals; stems densely hairy, the hairs spreading ........................................ *G. laciniatum*

8 Terminal leaflet of basal leaves much larger than lateral segments; lower portion of style with short-stalked glands . *G. macrophyllum*

8 Terminal leaflet various; lower portion of style without glands . . 9

9 Native species; petals 5 mm or more long; upper portion of style long-hairy ........................................ *G. allepicum*

9 Introduced weedy species; petals to 4 mm long; upper portion of style glabrous or nearly so ........................ *G. urbanum*

10 Receptacle glabrous or only sparsely hairy (remove a few achenes to check); plants with either the achene beak with short-stalked glands, or with the pedicels with dense long hairs over the much shorter hairs ........................................ 11

10 Receptacle densely hairy; achene beaks neither with glands nor the pedicels with dense long hairs ............................ 12

11 Achene beak not glandular; pedicels densely long-hairy ........ ............................................ *G. laciniatum*

11 Achene beak with short-stalked glands, especially near base; pedicels finely hairy and with only scattered long hairs .......... ........................................ *G. macrophyllum*

12 Stem leaves pinnately compound; achenes many (150 or more) in each head, the achene beak with long hairs at base . . *G. allepicum*

12 Stem leaves mostly 3-parted; achenes less than 100 in each head, the achene beak glabrous ................................ 13

13 Native species of natural habitats; upper segment of style long-hairy at base ........................................ *G. canadense*

13 Introduced species occasional in waste places; upper segment of style glabrous or nearly so ........................ *G. urbanum*

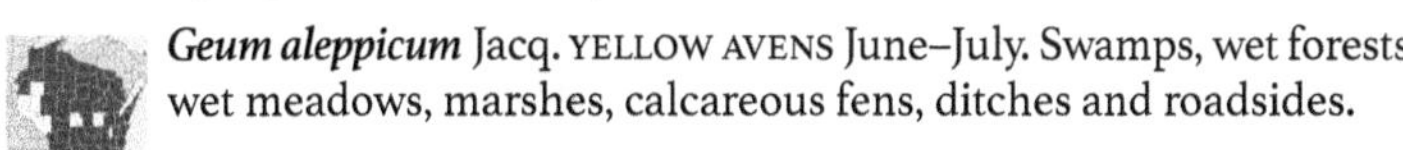

***Geum aleppicum*** Jacq. YELLOW AVENS June–July. Swamps, wet forests, wet meadows, marshes, calcareous fens, ditches and roadsides.

***Geum canadense*** Jacq. WHITE AVENS May–June. Dry or moist woods.

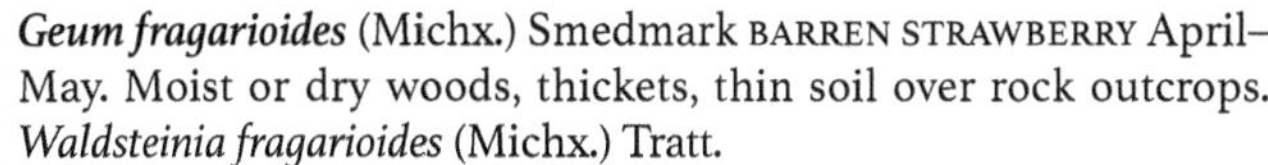

***Geum fragarioides*** (Michx.) Smedmark BARREN STRAWBERRY April–May. Moist or dry woods, thickets, thin soil over rock outcrops. *Waldsteinia fragarioides* (Michx.) Tratt.

***Geum laciniatum*** Murr. [•87] ROUGH AVENS May–June. Wet woods, floodplain forests, ditches.

***Geum macrophyllum*** Willd. BIG-LEAF AVENS May–July. Moist to wet forest openings, streambanks, wet meadows.

***Geum rivale*** L. PURPLE AVENS May–July. Conifer swamps, wet forests, bogs, fens, wet meadows; often where calcium-rich.

***Geum triflorum*** Pursh PRAIRIE SMOKE May–June. Dry woods and prairies.

*Geum urbanum* L. HERB-BENNET May–June. Native of Eurasia, weedy near yards and in disturbed places; Racine County.

*Geum vernum* (Raf.) Torr. & Gray SPRING AVENS April–May. Rich woods.

## MALUS *Apple*

Trees or shrubs, sometimes thorny, with simple, alternate, toothed or lobed leaves, and large flowers in simple umbels or umbel-like clusters on dwarf lateral branches (fruit-spurs). Fruit a fleshy pome. All bloom in April or May.

ADDITIONAL SPECIES *Malus sieboldii* (Regel) Rehder (Toringo Crab), native of Japan, rarely escaped, Kenosha and Richland counties; small tree; flowers pink or very deep rose in bud, finally nearly white.

### *Key to species*

1 Leaf margins toothed, the teeth sharp or rounded and not coarse 2

1 Leaf margins irregularly double-toothed, often coarsely so...... 3

2 Fruit 3 cm or more wide; pedicels shorter, less than 2.5 cm long; leaves never lobed .................................... *M. pumila*

2 Fruit less than 3 cm wide; pedicels longer, often more than 2.5 cm long; leaves sometimes lobed ........................ *M. sieboldii**

3 Leaves thick and leathery, the upperside veins sunken into leaf surface; petioles and pedicels densely woolly hairy; widespread in central and southern Wisc ............................ *M. ioensis*

3 Leaves thinner, the upperside veins not sunken; petioles and pedicels glabrous, or if hairy, the hairs thinning by end of flowering period; southeastern Wisc .......................... *M. coronaria*

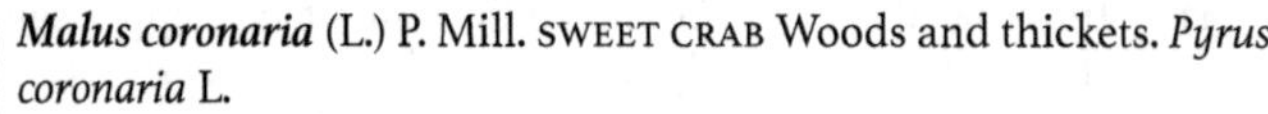

*Malus coronaria* (L.) P. Mill. SWEET CRAB Woods and thickets. *Pyrus coronaria* L.

*Malus ioensis* (Wood) Britt. PRAIRIE CRAB Woods and thickets. *Pyrus ioensis* (Wood) Bailey.

*Malus pumila* P. Mill. CULTIVATED APPLE Native probably of w Asia; long in cultivation and occasionally escaped. Many wild apples are persistent from planted trees, but grows from seed and appears in old fields and along fences and roads. *Pyrus malus* L., *Pyrus pumila* (P. Mill.) K. Koch.

## PHYSOCARPUS *Ninebark*

*Physocarpus opulifolius* (L.) Maxim. [•88] NINEBARK Much-branched shrub; bark of older stems shredding in long thin strips. Leaves alternate, ovate in outline, mostly 3-lobed; margins irregularly toothed. Flowers 5-parted, white; many in stalked, rounded clusters at ends of branches. Fruit a red-brown pod, in round clusters. June–July. Streambanks, lakeshores, swamps, rocky shores of Lake Superior.

## POTENTILLA *Cinquefoil*

Annual or perennial herbs, or woody in shrubby cinquefoil (*P. fruticosa*); stolons present in some species. Leaves pinnately or palmately divided, alternate or mostly from base of plant. Flowers perfect, regular; petals 5, yellow. Fruit a group of many small achenes, surrounded by the persistent hypanthium.

ADDITIONAL SPECIES **Potentilla inclinata** Vill., introduced in southeastern Wisc.

*Key to species*

1 Plants shrubs; leaflets 5–7, 1–2 cm long ............... *P. fruticosa*
1 Plants herbs (or woody only at base) .......................... 2

2 Flowers solitary on naked pedicels from nodes of creeping stems 3
2 Flowers few to many in cymes ................................ 5

3 Leaves pinnately compound; leaf underside densely white-hairy . ........................................................ **P. anserina**
3 Leaves palmately compound; leaf underside coarsely hairy ..... 4

4 Flowers 10–15 mm wide ............................ **P. simplex**
4 Flowers 18–25 mm wide ............................ **P. reptans**

5 Leaf undersides with long straight hairs or glabrous, but not woolly hairy ....................................................... 6
5 Leaf undersides woolly hairy ................................ 9

6 Main leaves below inflorescence 3-parted ..................... 7
6 Main leaves below inflorescence usually 5-parted or more ...... 8

7 Petals and sepals about same length; stamens usually 20; achenes ridged ............................................ **P. norvegica**
7 Petals much shorter than the sepals; stamens 5–10; achenes smooth ........................................................ **P. rivalis**

8 Plants erect, unbranched up to the inflorescence ......... **P. recta**
8 Plants often not erect, often much branched ........ **P. intermedia**

9 Stem leaves reduced in size; inflorescence with few branches .... ........................................................ **P. gracilis**
9 Stem leaves well developed; inflorescence much-branched..... 10

10 Leaves white-woolly on underside, the pubescence concealing the leaf surface ........................................ **P. argentea**
10 Leaves only thinly hairy on underside, the hairs mostly straight . ........................................................ **P. intermedia**

***Potentilla anserina*** L. SILVERWEED May–Sept. Wet meadows, marshes, sandy and gravelly shores and streambanks, Lake Michigan shoreline; soils often calcium-rich. *Argentina anserina* (L.) Rydb.

***Potentilla argentea*** L. SILVERY CINQUEFOIL Introduced (naturalized). May–Sept. Wet sandy beaches.

***Potentilla fruticosa*** L. SHRUBBY CINQUEFOIL June–Sept. Calcareous fens, lakeshores, open bogs, conifer swamps, wet meadows. *Dasiphora fruticosa* (L.) Rydb.

***Potentilla gracilis*** Dougl. SLENDER CINQUEFOIL Introduced. July–Aug. Prairies, rocky banks, and dry woods. *Potentilla flabelliformis* Lehm.

***Potentilla intermedia*** L. DOWNY CINQUEFOIL European; roadsides, waste places.

***Potentilla norvegica*** L. [•88] STRAWBERRY-WEED June–Aug. In a wide variety of moist or dry habitats, usually where somewhat disturbed; roadsides, railroads, fields, shores, meadows, rock outcrops, gardens.

***Potentilla recta*** L. [•88] SULPHUR CINQUEFOIL June–Aug. Native of Europe; weedy in dry soil, roadsides, fields, railroads, gravel pits; invading dry open forests. Sessile or short-stalked glands are usually present on the leaflet underside.

***Potentilla reptans*** L. CREEPING CINQUEFOIL June–Aug. Native of Eurasia; introduced in Racine County.

***Potentilla simplex*** Michx. OLDFIELD CINQUEFOIL April–June. Dry open sandy forests, fields, roadsides, and sandy barrens; also in moist thickets and deciduous forests, and on rocky ledges. Our most common *Potentilla* with solitary flowers and palmately compound leaves.

## PRUNUS *Plum; Cherry*

Trees or shrubs. Leaves alternate, simple, serrate, often with petiolar glands. Flowers umbellate or solitary from axillary buds or short lateral branches, or racemose and terminal. Petals 5, white to pink or red.

ADDITIONAL SPECIES ***Prunus persica*** (L.) Batsch (Peach) introduced, uncommon garden escape in s Wisc.

*Key to species*

1 Flowers 20 or more in elongate racemes; inflorescence bracts absent .......... 2
1 Flowers 1 to several, in umbel-like clusters .......... 3

2 Tree; leaves 2 times longer than wide, the margins with incurved teeth; fruit black .......... ***P. serotina***
2 Shrub; leaves less than 2 times longer than wide, the margins with sharp, outward pointing teeth; fruit dark red to purple ***P. virginiana***

3 Flowers in short, leafy-bracted corymb-like racemes; leaves about 1.5 times longer than wide, heart-shaped at base or nearly so, margins with rounded teeth; young twigs densely covered with small white hairs .......... ***P. mahaleb***
3 Plants not with the above combination of characters .......... 4

4 Plants in flower.......... 5
4 Plants in fruit and with fully developed leaves .......... 8

5 Sepals glabrous .......... 6
5 Sepals hairy, at least on upper surface near base .......... 7

6 Leaf margins mostly entire below middle; leaves widest above middle; flower pedicels mostly less than 1 cm long ..... ***P. pumila***

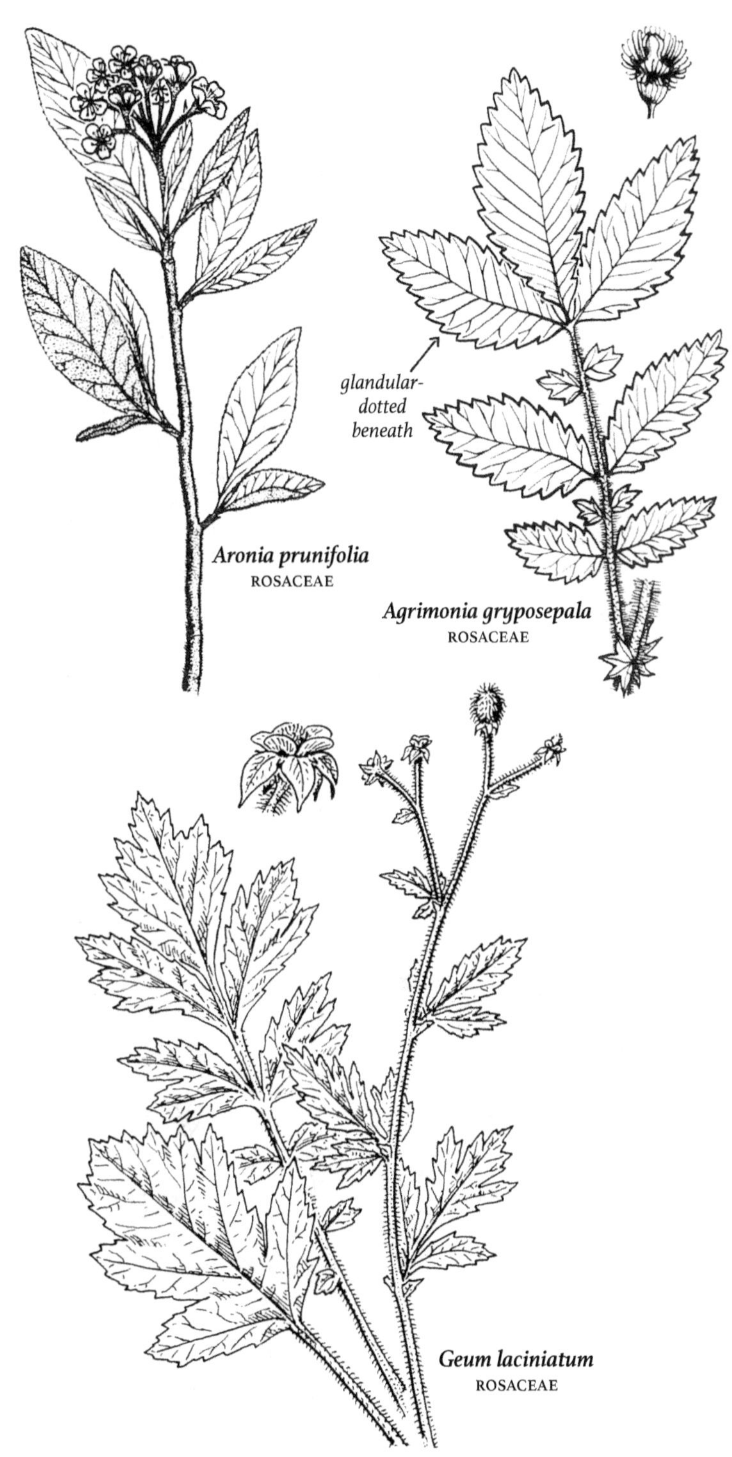
glandular-
dotted
beneath
Aronia prunifolia
ROSACEAE
Agrimonia gryposepala
ROSACEAE
Geum laciniatum
ROSACEAE

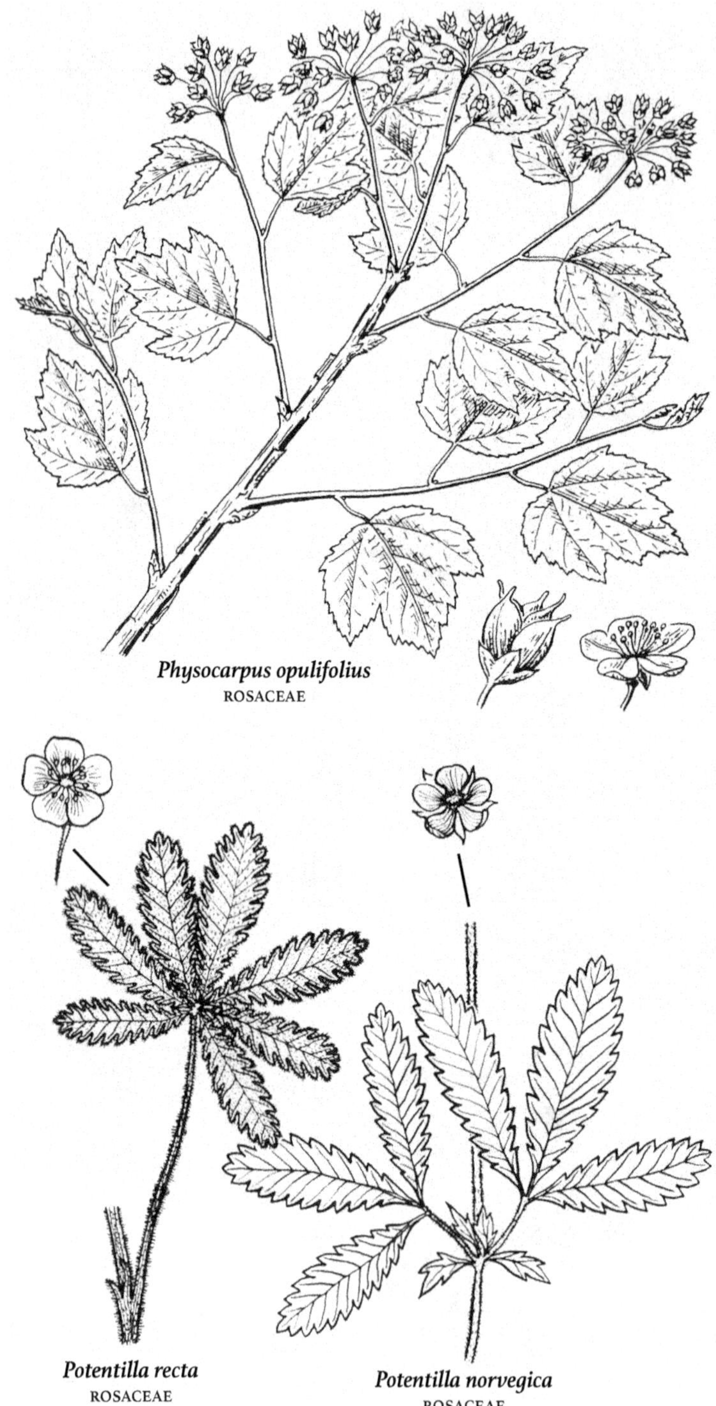
*Physocarpus opulifolius*
ROSACEAE
*Potentilla recta*
ROSACEAE
*Potentilla norvegica*
ROSACEAE

6 Leaf margins finely toothed for their entire length; leaves widest at or below middle; pedicels mostly more than 1 cm long ........................................................ *P. pensylvanica*

7 Sepals entire or with a few small teeth at tip ........ ***P. americana***
7 Sepal margins with gland-tipped teeth .................. ***P. nigra***

8 Leaves glabrous, obovate, distinctly widest above the middle; the margins entire or nearly so below the middle of the leaf ***P. pumila***
8 Leaves not as above ........................................ **9**

9 Leaves finely toothed, the teeth much less than 1 mm long; leaves more than 2 times longer than wide, widest below middle ....... ........................................... *P. pensylvanica*
9 Leaves coarsely toothed, the teeth 1 mm or more long; leaves less than 2 times longer than wide, widest at or above the middle... **10**

10 Margin teeth sharp, often tipped with a short spine .. ***P. americana***
10 Margin teeth rounded or tapered to a tip, sometimes gland-tipped ................................................... ***P. nigra***

***Prunus americana*** Marsh. WILD PLUM April–May. Moist woods, roadsides, and fencerows.

***Prunus mahaleb*** L. MAHALEB-CHERRY Native of Europe; occasionally planted and escaped.

***Prunus nigra*** Ait. CANADIAN PLUM May. Moist woods and thickets.

***Prunus pensylvanica*** L. f. PIN-CHERRY April–May. Dry or moist woods and forest clearings, often abundant after fires.

***Prunus pumila*** L. SAND-CHERRY May. Sand dunes and sandy soil, Great Lakes shores, dry or rocky woods.

***Prunus serotina*** Ehrh. WILD BLACK CHERRY May. Formerly a forest tree, now more common as a weedy tree of roadsides, waste land, and forest margins.

***Prunus virginiana*** L. [•89] CHOKE-CHERRY May. In a wide variety of habitats, from rocky hills and dunes to borders of swamps.

## PYRUS *Pear*

***Pyrus communis*** L. COMMON PEAR April–May. Native probably of e Europe and w Asia; long in cultivation and rarely escaped to fencerows and clearings. Pear leaves are shinier than those of apple, and have a more distinct pinnate venation.

## ROSA *Rose*

Shrubs or woody vines, usually thorny. Leaves pinnately compound with 3-11 serrate leaflets. Petals white to yellow or red. Stamens numerous.

ADDITIONAL SPECIES Several Eurasian roses occur, rarely escaped from cultivation in Wisc: ***Rosa centifolia*** L. (Cabbage rose), differing from *R. gallica* in its thin leaflets, nodding pedicels, and double flowers. ***Rosa gallica*** L. (French rose) Erect shrub about 1 m tall; stems with both stout thorns and numerous bristles, the latter usually gland-tipped; leaflets usually 5; flowers pink to red, about 5 cm wide. ***Rosa micrantha*** Borrer (Small-flower sweetbrier), very similar to *R. eglanteria;* leaflets elliptic; flowers smaller, about 3 cm wide. ***Rosa spinosissima*** L. (Scotch rose), shrub to 1 m tall, the stems with very numerous slender straight thorns; leaflets usually 7–9, usually less than 2 cm long; flowers solitary at the end of the branches; petals white, pink, or yellow.

*Key to species*

1 Styles joined to form a column, protruding from the hypanthium opening ........ 2

1 Styles distinct, styles not exserted from hypanthium opening, only the stigmas protruding........ 3

2 Leaflets 3 or 5, the leaflets 3–10 cm long; petals usually pink, 2–3 cm long; native species ........ ***R. setigera***

2 Leaflets 7 or 9, the leaflets to 3 cm long; petals usually white, 1–2 cm long; introduced species ........ ***R. multiflora***

3 Flowers solitary at end of branches, the pedicel not subtended by a bract; introduced species........ 4

3 Flowers solitary or in clusters; if solitary then the pedicel with a bract ........ 5

4 Leaflets 3–7, 2–6 cm long; petals deep pink, 3 cm or more long ........ ***R. gallica****

4 Leaflets 7–11, less than 2 cm long; petals white, to 2 cm long ........ ***R. spinosissima****

5 Sepals not all same shape and size, the outer sepals pinnately divided into lance-shaped segments; hypanthium opening small, 1 mm wide; styles short-exserted........ 6

5 Sepals entire; hypanthium opening 2–4 mm wide; styles not exserted ........ 7

6 Leaf underside glabrous or nearly so ........ ***R. canina***

6 Leaf underside with stalked glands ........ ***R. eglanteria***

7 Young twigs densely hairy; petals 3–5 cm long; introduced species ........ ***R. rugosa***

7 Young twigs glabrous or nearly so; petals 2–3 cm long; native species except for *R. cinnamomea* ........ 8

8 Pedicel and hypanthium with stalked glands; sepals spreading and then deciduous ........ 9

8 Pedicel and hypanthium glabrous; sepals persistent on fruit and typically upright ........ 10

9 Leaf margins with fine teeth; internodal prickles absent ........ ***R. palustris***

9 Leaf margins coarsely toothed; internodal prickles many ........ ***R. carolina***

10 Prickles at nodes below leaf stipules present and larger than internodal prickles........ 11

10 Prickles at nodes below leaf stipules absent or similar to internodal prickles .................................................. 12

11 Prickles at nodes straight and slender; flowers not double; native species .................................................. *R. woodsii*

11 Prickles at nodes wide at base, curved; flowers often double; introduced species .................................................. *R. cinnamomea*

12 Flowers from tips of currrent year's stems and also on lateral branches on stems from previous year; leaflets mostly 9 or 11 .... .................................................. *R. arkansana*

12 Flowers only on lateral branches of previous year's stems; leaflets mostly 5 or 7 .................................................. 13

13 Stems usually not prickly or bristly, or with slender prickles only on lower internodes .................................................. *R. blanda*

13 Stems densely prickly on most internodes .......... *R. acicularis*

***Rosa acicularis*** Lindl. BRISTLY ROSE Upland woods, hills, and rocky banks; statewide. *Rosa sayi* Schwein.

***Rosa arkansana*** Porter DWARF PRAIRIE-ROSE Prairies and plains, or in open or brushy sites.

***Rosa blanda*** Ait. SMOOTH ROSE Dry woods, hills, prairies, and dunes. Thorns may be entirely absent, or may extend a variable distance up the stem.

***Rosa canina*** L. DOG ROSE Native of Europe; occasionally planted and escaped.

***Rosa carolina*** L. PASTURE ROSE Upland woods, dunes, and prairies.

***Rosa cinnamomea*** L. CINNAMON ROSE Native of Eurasia; sometimes cultivated and rarely escaped; reported from Sheboygan County.

***Rosa eglanteria*** L. SWEETBRIER Native of Europe; cultivated and escaped. *Rosa rubiginosa* L.

***Rosa multiflora*** Thunb. MULTIFLORA ROSE Native of Asia; introduced as a hedgerow plant; sometimes escaping as the seeds are readily spread by birds eating the fruit.

***Rosa palustris*** Marsh. [•89] SWAMP ROSE July–Aug. Open bogs, conifer swamps, thickets, shores and streambanks; increasing in disturbed wetlands.

***Rosa rugosa*** Thunb. RUGOSA ROSE Native of e Asia; commonly cultivated and occasionally escaped; notable for its large "rose-hips" (fruit), rich in vitamin C.

***Rosa setigera*** Michx. CLIMBING PRAIRIE-ROSE Thickets and fencerows. Native of e USA but considered adventive in s Wisc.

***Rosa woodsii*** Lindl. WESTERN ROSE; WOOD'S ROSE Fields and openings.

## RUBUS *Blackberry; Raspberry; Dewberry*

Perennials, woody at least at base, usually with bristly stems. Stems biennial in some species, the first year's canes called primocanes, the second year's growth termed floricanes. Leaves alternate, palmately lobed or divided. Flowers 5-parted, usually perfect, white to pink or rose-purple. Fruit a group of small, 1-seeded drupes forming a berry.

*Key to species*

1 Stems without bristles or prickles .......... 2
1 Stems with bristles or prickles .......... 5

2 Leaves simple .......... 3
2 Leaves with 3–5 leaflets .......... 4

3 Flowers white; fruits orange to red, edible .......... ***R. parviflorus***
3 Flowers rose-purple; fruits pink to red, dry and inedible ***R. odoratus***

4 Stems trailing; flowers single or several in a cluster; fruit red .......... ***R. pubescens***
4 Stems erect or arching; flowers many in long, bracted clusters; fruit black .......... ***R. canadensis***

5 Leaves whitish or gray-hairy on underside; fruit separating easily from receptacle when ripe (*raspberries*) .......... 6
5 Leaves green on both sides, underside veins hairy; fruit falling with receptacle when ripe (*dewberries and blackberries*) .......... 7

6 Stems erect or spreading, with stiff straight bristles; fruit red .......... ***R. idaeus***
6 Stems arching, often rooting at tip, with broad-based, recurved prickles; fruit black .......... ***R. occidentalis***

7 Plants low and trailing (less than 0.5 m tall), often rooting at nodes; flowers 1 to several in a cluster; fruit red to red-purple (*dewberries*) 8
7 Plants tall, to 2 m; stems erect, neither rooting at nodes nor arching and rooting at tips; flowers numerous in elongate clusters; fruit black (*blackberries*) .......... 9

8 Stems with prickles, these hooked at tip and broad at base; leaves thin and deciduous; petals more than 1 cm long .......... ***R. flagellaris***
8 Stems with coarse hairs and slender bristles; leaves leathery and often evergreen; petals less than 1 cm long .......... ***R. hispidus***

9 Stems smooth or with scattered prickles; leaves glabrous .......... ***R. canadensis***
9 Stems glandular-hairy, with broad-based prickles or covered with spreading bristles; leaves hairy on underside veins .......... 10

10 Stems covered with stiff bristles, broad-based prickles absent .......... ***R. setosus***
10 Stems with bristles, gland-tipped hairs, and scattered broad-based prickles .......... 11

11 Petioles and pedicels covered with gland-tipped hairs .......... ***R. allegheniensis***
11 Petioles and pedicels without glandular hairs (or nearly so) .......... ***R. pensilvanicus***

***Rubus allegheniensis*** Porter COMMON BLACKBERRY Forests and forest edges, clearings, old fields, roadsides; usually on dry uplands, occasional in marshy or swampy ground. Our commonest tall blackberry.

***Rubus canadensis*** L. SMOOTH BLACKBERRY, DEWBERRY Woods, clearings, fields, roadsides; occasionally in moist soil.

***Rubus flagellaris*** Willd. WHIPLASH DEWBERRY May–June. Swamps, wetland margins; also in drier sandy woods, prairies and openings.

***Rubus hispidus*** L. [•89] BRISTLY DEWBERRY June–Aug. Conifer swamps, wet hardwood forests, thickets, wetland margins; usually where shaded.

***Rubus idaeus*** L. WILD RED RASPBERRY May–Aug. Thickets, moist to wet openings, streambanks; often where disturbed. Our native plants sometimes considered a variety (*R. idaeus* var. *strigosus*) of the cultivated red raspberry (*R. idaeus* L.) from Europe. *Rubus strigosus* Michx.

***Rubus occidentalis*** L. BLACK RASPBERRY May–June. Dry or moist woods, fields, and thickets. Often cultivated in many horticultural varieties.

***Rubus odoratus*** L. FLOWERING RASPBERRY June–Aug. Reported for Wisc, but presence not verified; common in New England (*no map*).

***Rubus parviflorus*** Nutt. THIMBLEBERRY May–July. Open woods and thickets. Where plentiful, harvested for jam-making.

***Rubus pensilvanicus*** Poir. PENNSYLVANIA BLACKBERRY Roadsides, fields, thickets, forests and forest borders; often in moist places such as borders of marshes and swamps. The coarse serration, commonly accentuated beyond the middle of the leaflets, is characteristic, but forms with simpler serration occur.

***Rubus pubescens*** Raf. [•89] DWARF RASPBERRY May–July. Conifer swamps, wet deciduous woods, rocky shores.

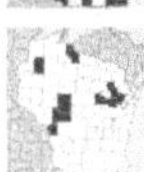
***Rubus setosus*** Bigelow BRISTLY BLACKBERRY June–Aug. Wetland margins, shores, occasional in open bogs; also in drier sandy prairie.

## SANGUISORBA *Burnet*

***Sanguisorba minor*** Scop. SALAD-BURNET Perennial herb. Stems 2–7 dm tall. Basal and lower leaves numerous, 1-pinnate; upper leaves progressively reduced. Heads densely flowered, several on elongate peduncles; flowers 4-merous, the lower staminate, the upper pistillate or perfect; petals none; stamens numerous, the long filaments drooping. May–June. Native of Eurasia, rarely established in roadsides, waste places, and fields. *Poterium sanguisorba* L.

## SIBBALDIA *Fivefingers*

***Sibbaldia tridentata*** (Aiton) Paule & Soják SHRUBBY-FIVEFINGERS Stems woody at base. Leaves mostly near the base, digitately compound; leaflets 3, 3-toothed at the truncate tip. Flowers several

in a flattened cyme, white. June–Aug. Open sandy places, dry savannas of jack pine and oak; rocky and gravelly shores, rock outcrops. *Potentilla tridentata* Ait., *Sibbaldiopsis tridenta* (Aiton) Rydb.

### SORBARIA *False Spiraea*

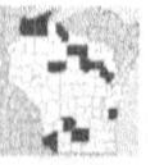

***Sorbaria sorbifolia*** (L.) A. Braun FALSE SPIRAEA Shrub, the younger parts covered with a flocculent, deciduous, stellate tomentum. Leaves 1-pinnate. Flowers 5-merous, in panicles; petals white. July. Native of e Asia; cultivated and escaped along roadsides and fencerows.

### SORBUS *Mountain-Ash*

Trees or shrubs. Leaves odd-pinnate with normally 11–17 serrate leaflets; flowers white, numerous, in repeatedly branched, round or flattened clusters. Fruit a small pome.

*Key to species*

1 Leaflets tapered to a tip, 3–5 times longer than wide; petals obovate, to 4 mm long; fruit 5–6 mm wide .................. ***S. americana***
1 Leaflets rounded at tip or abruptly tapered to a tip, 2–3 times longer than wide; petals orbicular, 4–5 mm long; fruit 8–10 mm wide... 2

2 Leaflets glabrous, pale on underside; inflorescence branches and pedicels glabrous or nearly so ........................ ***S. decora***
2 Leaflets soft-hairy on underside; inflorescence branches and pedicels with soft hairs .......................... ***S. aucuparia***

***Sorbus americana*** Marsh. [•90] AMERICAN MOUNTAIN-ASH May–June; fruit in late summer. In moist or wet soil; swamps (both cedar and deciduous), streambanks, forest borders. *Pyrus americana* (Marsh.) DC.

***Sorbus aucuparia*** L. EUROPEAN MOUNTAIN-ASH; rowan May–June. Native of Europe; planted for ornament and escaped into moist woods; often mistaken for a native plant. *Pyrus aucuparia* (L.) Gaertn.

***Sorbus decora*** (Sarg.) Schneid. NORTHERN MOUNTAIN-ASH May–June. Moist or dry, often rocky soil; wooded dunes and bluffs, forest margins. *Pyrus decora* (Sarg.) Hyl.

### SPIRAEA *Meadowsweet*

Shrubs with simple leaves and terminal or lateral clusters of white, pink, or purple flowers. Flowers 5-merous.

*Key to species*

1 Leaves glabrous on both sides; flowers white to pinkish ... ***S. alba***
1 Leaf underside densely covered with light brown, woolly hairs; flowers rose-pink .................................. ***S. tomentosa***

***Spiraea alba*** Du Roi [•90] MEADOWSWEET June–Aug. Wet meadows, streambanks, lakeshores, conifer swamps; soils often sandy.

***Spiraea tomentosa*** L. [•90] HARDHACK July–Sept. Open bogs, conifer swamps, thickets, lakeshores, wet meadows; soils often sandy.

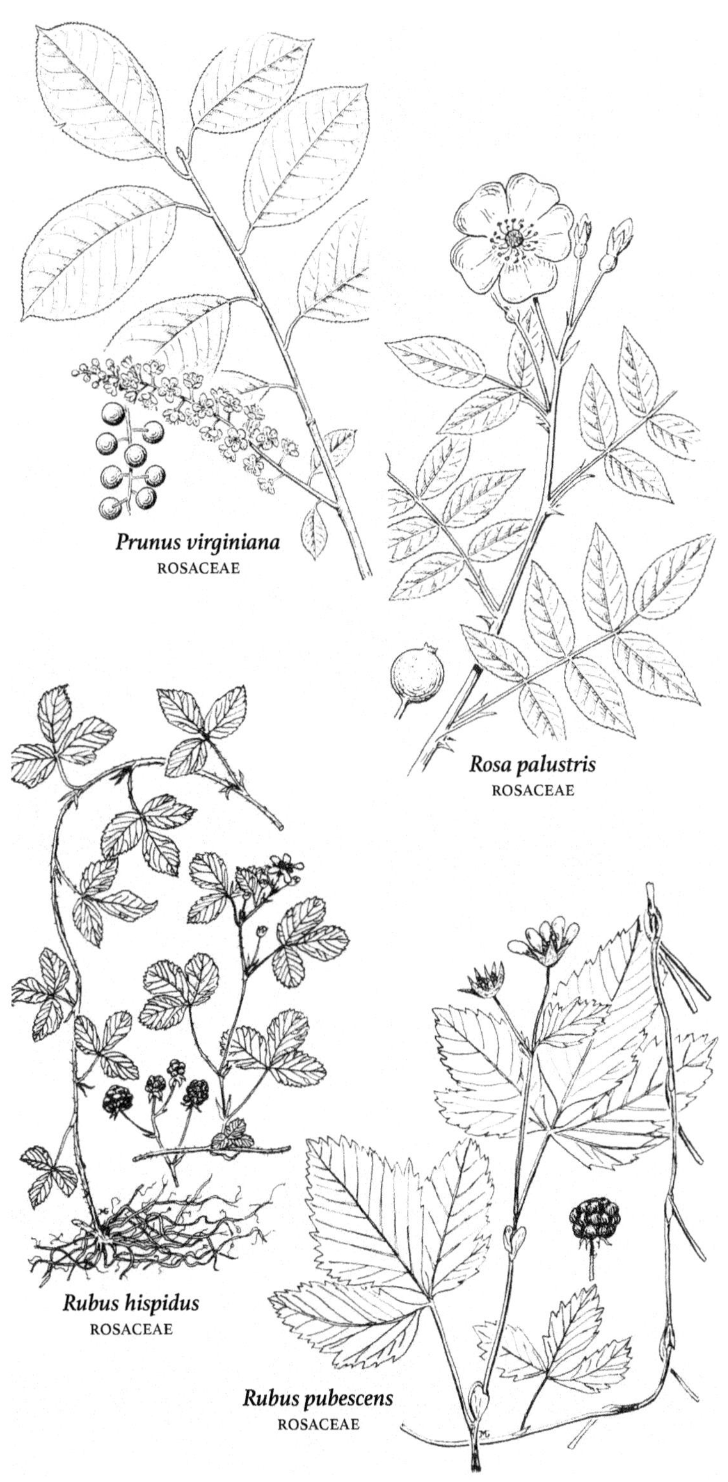
*Prunus virginiana*
ROSACEAE
*Rosa palustris*
ROSACEAE
*Rubus hispidus*
ROSACEAE
*Rubus pubescens*
ROSACEAE

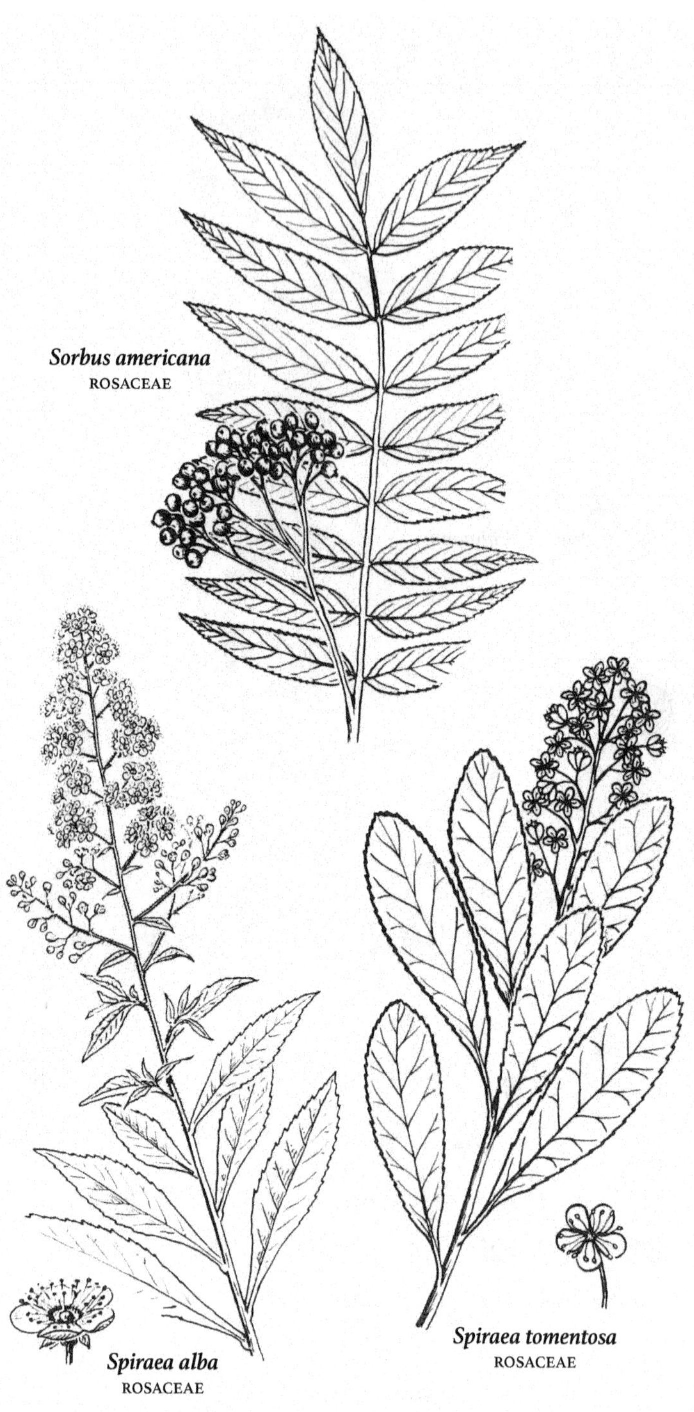
*Sorbus americana*
ROSACEAE
*Spiraea alba*
ROSACEAE
*Spiraea tomentosa*
ROSACEAE

## RUBIACEAE *Madder Family*

Shrubs (*Cephalanthus*) or herbs. Leaves simple, opposite or whorled. Flowers small, perfect, white to green, single or in loose or round clusters. Fruit a nutlet (*Cephalanthus, Diodia*), a capsule (*Galium, Houstonia*), or a berry (*Mitchella*).

*Key to genera*

1 Shrubs; flowers in spherical heads ........... CEPHALANTHUS
1 Herbs; flowers not in spherical heads.......................... 2

2 Leaves whorled ..................................... GALIUM
2 Leaves opposite ............................................ 3

3 Leaves evergreen; fruit a scarlet or white berry .... MITCHELLA
3 Leaves deciduous; fruit not a berry........................... 4

4 Stems spreading; flowers sessile; stipules with long bristles DIODIA
4 Stems usually erect; flowers on pedicels; stipules entire or nearly so ............................................... HOUSTONIA

### CEPHALANTHUS *Buttonbush*

***Cephalanthus occidentalis*** L. [•91] COMMON BUTTONBUSH Shrub or small tree. Leaves opposite or in whorls of 3, upper surface bright green and shiny, paler or finely hairy below; margins entire or slightly wavy; petioles grooved. Flowers small, in round, many-flowered heads, on long stalks at ends of stems or from upper leaf axils; petals 4, creamy white. Fruit a round head of brown, cone-shaped nutlets, tipped by 4 teeth of persistent sepals. June–Aug. Hardwood swamps, floodplain forests, thickets, streambanks, marshes, open bogs; often in standing water or muck.

### DIODIA *Buttonweed*

***Diodia teres*** Walt. BUTTONWEED Annual herb. Stems pubescent. Leaves stiff, sessile, linear to narrowly lance-shaped, scabrous, bristle-tipped. Flowers 4-merous, sessile in many of the upper axils. July–Sept. Dry or sandy soil, often weedy.

### GALIUM *Bedstraw*

Annual or perennial herbs, from slender rhizomes. Stems 4-angled, ascending to reclining, smooth or bristly. Leaves entire, in whorls of 4–6. Flowers 1 to several from leaf axils or in clusters at ends of stems; ovary 2-chambered and 2-lobed, maturing as 2 dry, round fruit segments which separate when mature.

*Key to species*

1 Fruit with bristly hairs ........................................ 2
1 Fruit smooth or nearly so ...................................... 7

2 Main leaves in whorls of 5 or more............................ 3
2 Leaves in whorls of 4 or less................................... 4

3 Annual herb; leaves in whorls of 7 or more; flowers white, blooming completed by early summer; stems very rough-to-touch ......... ................................................ *G. aparine*

3 Perennial herb; leaves in whorls of up to 6; flowers greenish, blooming beginning in early summer; stems rough or smooth ... ........ *G. triflorum*

4 Leaves linear to linear lance-shaped, usually less than 5 mm wide; flowers white in a large panicle ........ *G. boreale*

4 Leaves broader, often more than 5 mm wide; flowers greenish to purple, in few-flowered clusters........ 5

5 Stems hairy, at least on the angles ........ *G. circaezans*

5 Stems glabrous or nearly so ........ 6

6 Leaves lance-shaped, tapered to a tip, corolla becoming purple, glabrous ........ *G. lanceolatum*

6 Leaves ovate, rounded at tip; corolla greenish-yellow, pubescent ......... *G. circaezans*

7 Leaves tipped with a short spine or at least sharp-pointed ........ 8

7 Leaves rounded or blunt at tip ........ 11

8 Leaves linear, the margins revolute, hairy on underside; flowers yellow in elongate terminal panicles ........ *G. verum*

8 Leaves narrowly lance-shaped to ovate, glabrous or rough-hairy on underside; flowers white, the inflorescence various ........ 9

9 Leaves and stems with rough, downward-pointing hairs ......... *G. asprellum*

9 Leaves and stems smooth or with short, upward-pointing hairs . 10

10 Leaves in whorls of 7 or more, oblong lance-shaped; stems smooth or short-hairy ........ *G. mollugo*

10 Leaves in whorls of 6 or less, linear; stems usually rough-hairy on the angles ........ *G. concinnum*

11 Lobes of corolla 3, mostly wider than long ........ 12

11 Lobes of corolla 4, mostly longer than wide ........ 14

12 Leaves in whorls of 4; flowers and fruit on long, curved, rough-hairy pedicels ........ *G. trifidum*

12 Leaves usually in whorls of 5 or more; flowers and fruit on straight glabrous pedicels ........ 13

13 Pedicels 0.5–4 mm long and often curved at maturity, solitary or in pairs in leaf axils or at ends of branches but not on a common peduncle; corolla less than 1 mm wide; mature fruit to 1 mm long; leaves mostly 2.5–7 mm long ........ *G. brevipes*

13 Pedicels (at least the longest) 3–8 mm long and nearly always straight at maturity, often on a peduncle; corolla 1–1.8 mm wide; mature fruit 1–2 mm long; leaves mostly 5.5–14 (–22) mm long ... ........ *G. tinctorium*

14 Flowers in well-branched cymes; nodes of stems glabrous ......... *G. palustre*

14 Cymes only once or twice branched; nodes short-hairy ........ 15

15 Leaves linear, bent downward, less than 2 mm wide ......... *G. labradoricum*

15 Leaves linear to oblong, spreading but not angled downward, mostly more than 2 mm wide ........ *G. obtusum*

***Galium aparine*** L. [•92] STICKY-WILLY, CLEAVERS May–June. Damp ground, usually in shade. Variable.

***Galium asprellum*** Michx. ROUGH BEDSTRAW July–Sept. Swamps, streambanks, thickets, marshes, wet meadows, calcareous fens.

***Galium boreale*** L. NORTHERN BEDSTRAW June–Aug. Streambanks, shores, thickets, swamps, moist meadows; also in drier woods and fields.

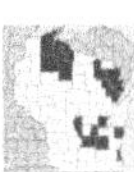

***Galium brevipes*** Fern. & Weig. LIMESTONE SWAMP BEDSTRAW July–Aug. Marshes, thickets; exposed calcareous shores, interdunal hollows, ditches. *Galium trifidum* subsp. *brevipes* (Fernald & Wiegand) A. & D. Löve.

***Galium circaezans*** Michx. [•91] FOREST BEDSTRAW June–July. Dry woods and thickets.

***Galium concinnum*** Torr. & Gray SHINING BEDSTRAW June–Aug. Dry woods.

***Galium labradoricum*** (Wieg.) Wieg. NORTHERN BOG BEDSTRAW June–July. Conifer swamps, sphagnum bogs, fens, sedge meadows.

***Galium lanceolatum*** Torr. [•91] LANCE-LEAF WILD LICORICE June–July. Dry woods and thickets.

***Galium mollugo*** L. FALSE BABY'S-BREATH Introduced. May–July. Meadows, fields, roadsides, and lawns.

***Galium obtusum*** Bigelow BLUNTLEAF BEDSTRAW May–July. Wet deciduous forests, wet meadows, streambanks, thickets, floodplains, moist prairie.

***Galium palustre*** L. COMMON MARSH BEDSTRAW June–Aug. Wet soil. Adventive in ne Wisc; more common in ne USA.

***Galium tinctorium*** (L.) Scop. STIFF MARSH BEDSTRAW July–Sept. Plants similar to G. trifidum and sometimes considered a variety of that species. Conifer swamps, open bogs, fens, thickets, wet shores and marshes. *Galium trifidum* L. subsp. *tinctorium* (L.) Hara.

***Galium trifidum*** L. NORTHERN THREE-LOBED BEDSTRAW June–Sept. Lakeshores, streambanks, swamps, marshes, bogs, springs.

***Galium triflorum*** Michx. [•92] SWEET-SCENTED BEDSTRAW June–Aug. Moist to wet woods, hummocks in cedar swamps, wetland margins and shores, clearings.

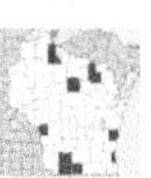

***Galium verum*** L. YELLOW SPRING BEDSTRAW Introduced. June–Sept. Fields and roadsides, usually in dry soil.

### HOUSTONIA *Bluets*

Perennial herbs. Leaves small, opposite. Flowers 4-merous, blue, purple, or nearly white, on terminal peduncles or in terminal cymes.

*Key to species*

1 Leaves oblong-elliptic to spatula-shaped, at least the lower leaves with petioles; peduncles with 1 flower; stamens not exserted .... ........ ***H. caerulea***

1 Leaves linear to oblong lance-shaped, sessile; peduncles with 2–several flowers; stamens exserted .................... ***H. longifolia***

***Houstonia caerulea*** L. QUAKER-LADIES May–July. Moist sandy meadows. *Hedyotis caerulea* (L.) Hook.

***Houstonia longifolia*** Gaertn. LONG-LEAF SUMMER BLUETS June–Aug. Dry to sometimes moist, sandy or gravelly soil; shallow soil over limestone; sandy fields. *Hedyotis longifolia* (Gaertn.) Hook.

### MITCHELLA *Partridge-Berry*

***Mitchella repens*** L. [•91] PARTRIDGE-BERRY Creeping perennial herb, forming mats. Leaves evergreen. Flowers 4-merous, in pairs; corolla white. Fruit a scarlet berry, edible but insipid, persistent through the winter. May–July. Dry or moist woods.

## RUTACEAE *Rue Family*

Mostly trees or shrubs with alternate, simple or compound leaves and small flowers. Flowers perfect or unisexual. Fruit commonly separating into segments, in some genera a capsule, drupe, or berry. Most parts of the plant contain oil-glands; those of the leaves appear as translucent dots. The most important economic genus is *Citrus* L.; the two Wisc species are the northernmost members of the family.

*Key to genera*

1 Leaflets 3 ........................................ PTELEA

1 Leaflets 5–11 .............................. ZANTHOXYLUM

### PTELEA *Hop-Tree*

***Ptelea trifoliata*** L. COMMON HOP-TREE Deciduous shrub or small tree, without spines. Leaves alternate, 3-foliolate, long-petioled. Flowers small, greenish white or yellowish white, in terminal cymes. Fruit a thin, flat, circular samara, reticulately veined, with the odor of hops. May–June. Moist or rich woods and thickets.

### ZANTHOXYLUM *Prickly Ash*

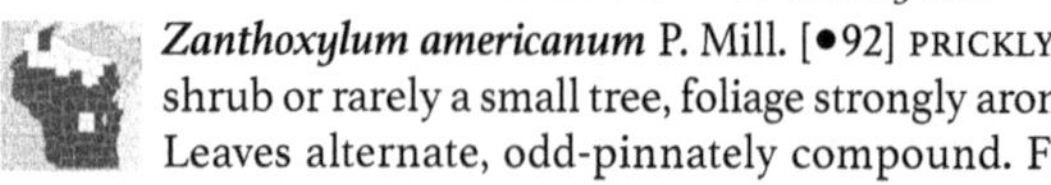

***Zanthoxylum americanum*** P. Mill. [•92] PRICKLY ASH Tall dioecious shrub or rarely a small tree, foliage strongly aromatic. Stems thorny. Leaves alternate, odd-pinnately compound. Flowers greenish or whitish, in short-peduncled, sessile, axillary clusters on branches of the previous year. April–May. Moist woods and thickets.

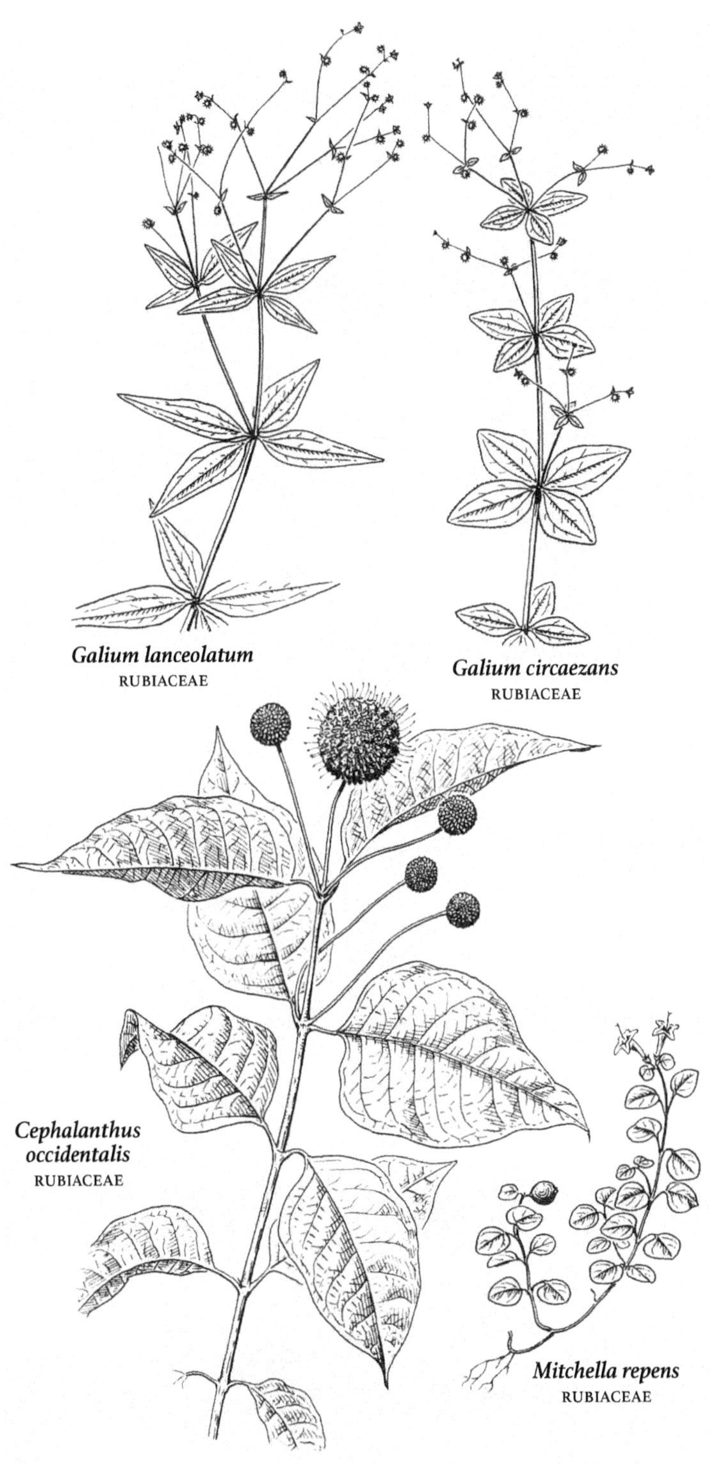
Galium lanceolatum
RUBIACEAE
Galium circaezans
RUBIACEAE
Cephalanthus occidentalis
RUBIACEAE
Mitchella repens
RUBIACEAE

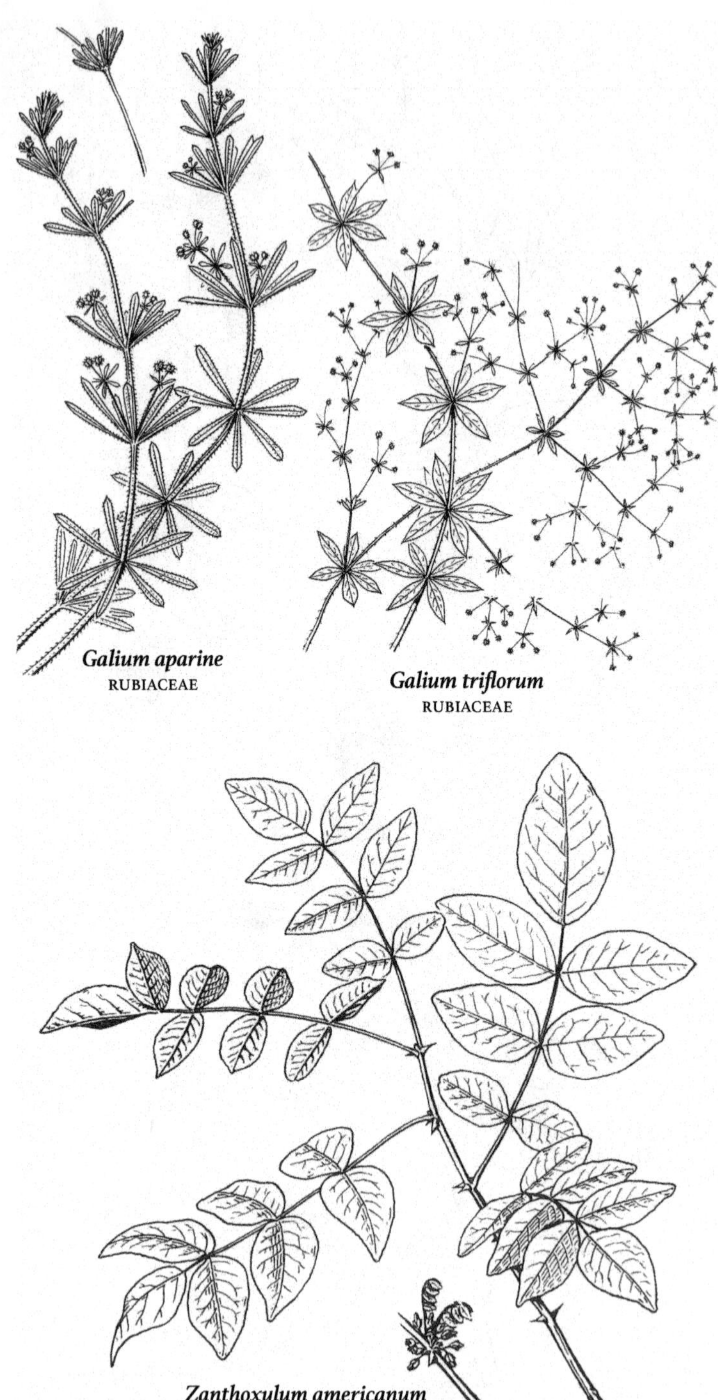

***Galium aparine***
RUBIACEAE

***Galium triflorum***
RUBIACEAE

***Zanthoxylum americanum***
RUTACEAE

## SALICACEAE *Willow Family*

Deciduous trees or shrubs. Leaves alternate. Flowers borne in catkins near ends of branches. Flowers imperfect, the staminate and pistillate flowers on separate plants, usually appearing before leaves open, or in a few species after leaves open; flowers without petals or sepals, each flower with either 1 or 2 enlarged basal glands (*Salix*) or a cup-shaped disk (*Populus*). Fruit a dry, many-seeded capsule; seeds small, covered with long, silky hairs.

*Key to genera*

1 Large trees; leaves heart-shaped to ovate, mostly less than 2 times longer than wide; buds often sticky and covered by 2 or more overlapping scales; catkins drooping, flowers subtended at base by a cup-shaped disk; stamens many, 12–80 ............ **POPULUS**

1 Shrubs and trees; leaves ovate, lance-shaped or linear, 2 or more times longer than wide; buds covered by 1 scale; catkins upright or drooping, flowers subtended by 1 or 2 enlarged glands; stamens 2–8 ............ **SALIX**

### POPULUS *Aspen; Poplar; Cottonwood*

Trees with deciduous, ovate to triangular leaves. Flowers in drooping catkins that develop and mature before and with leaves in spring.

*Key to species*

1 Leaf petioles round in section, leaf underside often stained brown from resin ............ ***P. balsamifera***

1 Leaf petioles strongly flattened, leaf underside not stained brown 2

2 Leaf underside and petioles densely woolly hairy ......... ***P. alba***

2 Leaf underside and petioles glabrous............ 3

3 Leaves strongly triangular in shape ............ 4

3 Leaves ovate to nearly round ............ 5

4 Leaf blades about as long or longer than wide, often with glands at tip of petiole; trees with broad crowns ............ ***P. deltoides***

4 Leaf blades wider than long, never with glands on petiole; trees narrow and spire-like ............ ***P. nigra***

5 Leaf margins coarsely wavy-toothed; leaves 7–13 cm long ............ ***P. grandidentata***

5 Leaf margins finely sharp-toothed; leaves less than 7 cm long ............ ***P. tremuloides***

***Populus alba*** L. WHITE POPLAR Native of Eurasia, commonly planted and spreading by root sprouts.

***Populus balsamifera*** L. BALSAM-POPLAR April–May. Swamps, floodplain forests, shores, streambanks, forest depressions, moist dunes.

***Populus deltoides*** Bartr. PLAINS COTTONWOOD April–May. Floodplains, streambanks and bars, shores, wet meadows, ditches.

***Populus grandidentata*** Michx. [•93] BIG-TOOTH ASPEN Dry or moist soil; common northward where it usually grows in drier soil than *P. tremuloides.*

***Populus nigra*** L. LOMBARDY POPLAR Native of Eurasia; often planted and occasionally escaped.

***Populus tremuloides*** Michx. [•93] QUAKING ASPEN Dry or moist soil, especially in cut-over land.

## SALIX *Willow*

Shrubs and trees. Leaves variable in shape, petioles glandular in some species; stipules early deciduous or persistent, sometimes absent. Catkins (aments) stalkless or on leafy branchlets, usually shed early in season. Staminate and pistillate flowers on separate plants. Fruit a 2-chambered, stalked or stalkless capsule.

ADDITIONAL SPECIES AND HYBRIDS ***Salix cinerea*** L., introduced, Racine County. ***Salix* × *conifera*** Wangenh. (pro sp.), *S. discolor* and *S. humilis,* s Wisc. ***Salix* × *pendulina*** Wenderoth, *S. babylonica* and *S. euxina,* nearly statewide. ***Salix* × *smithiana*** Willd. (pro sp.), *S. cinerea* and *S. viminalis,* s Wisc.

### *Key to species*

1 Leaves opposite or nearly so; young branches often dark purple .. .......... ***S. purpurea***

1 Leaves alternate; branches various colors .......... 2

2 Leaf petioles with glands at or near base of blade .......... 3

2 Petioles without glands .......... 11

3 Trees, usually with a single trunk; leaves narrow .......... 4

3 Small trees or shrubs, usually with several to many stems; leaves broader .......... 7

4 Leaves often curved sideways (scythe-shaped), tapered to a long, slender tip; vigorous shoots with large stipules; native species .... .......... ***S. nigra***

4 Leaves not curved sideways, tapered to a short tip; stipules small, early deciduous; introduced .......... 5

5 Twigs drooping, slender .......... ***S. babylonica***

5 Twigs not drooping, thicker .......... 6

6 Leaves glabrous; twigs easily broken at base .......... ***S. fragilis***

6 Leaf underside usually silky-hairy; twigs not easily broken at base .......... ***S. alba***

7 Leaves waxy-coated on underside .......... 8

7 Leaves not waxy-coated on underside .......... 10

8 Leaf tips rounded or with a short point; leaf base heart-shaped or rounded; young leaves translucent; buds and leaves with a balsamlike scent .......... ***S. pyrifolia***

8 Leaves tapered to tip; leaf base blunt or rounded; young leaves not translucent; buds and leaves not balsam-scented .......... 9

9 Young leaves sparsely hairy; margins with small forward-pointing teeth; flowering in early summer .......... ***S. amygdaloides***

9 Young leaves without hairs; margins with small, gland-tipped, forward-pointing teeth; flowering summer or fall ..... *S. serissima*

10 Leaves short-tapered to tip; twigs and young leaves resin-scented; introduced species ................................ *S. pentandra*

10 Leaves long-tapered to tip; twigs and young leaves not resin-scented; native species ........................................ *S. lucida*

11 Mature leaves hairy, at least on underside .................. 12

11 Mature leaves without hairs (sometimes hairy on petiole and midvein) ................................................ 21

12 Leaves linear or narrowly lance-shaped ..................... 13

12 Leaves broadly lance-shaped, oblong, or ovate ................ 16

13 Underside of leaves with felt-like covering of white tangled hairs; young twigs white-hairy; plant of peatlands, often where calcium-rich ............................................... *S. candida*

13 Leaves not with felt-like hairs; twigs smooth or sparsely hairy .. 14

14 Leaf margins entire and somewhat revolute; leaf underside pubescent ........................................... *S. pellita*

14 Leaf margins with gland-tipped teeth; leaf underside sparsely hairy ............................................................ 15

15 Leaf margins with widely spaced sharp teeth; petioles 1–5 mm long; colony-forming shrub of sandy banks ................ *S. interior*

15 Leaf margins with small teeth at least above middle of blade; petioles 3–10 mm long; stems clustered but not forming large colonies ........................................ *S. petiolaris*

16 Leaves rounded or heart-shaped at base; margins toothed; stipules present and persistent .................................... 17

16 Leaves tapered to base; margins entire or toothed; stipules usually falling early ............................................ 18

17 Leaves oblong lance-shaped, tapered to a long tip; young leaves reddish ...................................... *S. eriocephala*

17 Leaves obovate to oblong, tapered to a short tip; young leaves not reddish .......................................... *S. cordata*

18 Leaves narrowly to broadly lance-shaped, more than 5 times longer than wide, underside velvety-hairy with shiny white hairs *S. pellita*

18 Leaves obovate or elliptic, less than 5 times longer than wide, underside hairs not shiny ................................. 19

19 Small branches widely spreading; young leaves with white hairs; catkins appearing with leaves in spring; catkin bracts yellow or straw-colored; capsules on pedicels 2–5 mm long ..... *S. bebbiana*

19 Small branches not widely spreading; young leaves with some red or copper-colored hairs; catkins appearing before leaves in spring; catkin bracts dark brown to black; capsules on pedicels 1–3 mm long ............................................................ 20

20 Leaf upperside smooth or the veins slightly raised, underside sparsely hairy; twigs often shiny ..................... *S. discolor*

20 Leaf upperside somewhat wrinkled, the veins sunken, the leaf underside densely woolly hairy; twigs dull ............. *S. humilis*

21 Leaves green on both sides or slightly paler on underside, not glaucous or white-waxy below .............................. 22
21 Leaves glaucous or white-waxy on underside ................ 24

22 Leaves ovate to oblong-ovate ......................... *S. cordata*
22 Leaves narrower, linear to linear lance-shaped ............... 23

23 Many-stemmed, colony-forming shrub ............... *S. interior*
23 Single-stemmed tree ..................................... *S. nigra*

24 Leaf margins entire to shallowly lobed or with irregular teeth, sometimes revolute ........................................ 25
24 Leaf margins distinctly toothed .............................. 29

25 Leaf margins entire and somewhat revolute ................ 26
25 Leaf margins irregularly toothed, the teeth sharp or rounded .. 27

26 Stems upright, to 1 m tall, or creeping and rooting in moss; upper surface of leaves with raised, net-like veins; catkins appearing with leaves; capsules not hairy ......................... *S. pedicellaris*
26 Stems upright, 1–4 m tall; leaf veins not net-like; catkins appearing before leaves; capsules hairy ....................... *S. planifolia*

27 Leaves dull green above, wrinkled below; catkins appearing with leaves; bracts of pistillate catkins green-yellow to straw-colored .. ....................................................... *S. bebbiana*
27 Leaves dark green and shiny above; catkins appearing before leaves; bracts of pistillate catkins dark brown to black ................ 28

28 Stipules large on vigorous shoots; capsules on distinct stalks 2 mm or more long ...................................... *S. discolor*
28 Stipules lance-shaped and soon deciduous, or absent; capsules sessile or on short stalks less than 2 mm long ......... *S. planifolia*

29 Leaves narrowly to broadly lance-shaped, long- or short-tapered to tip ........................................................ 30
29 Leaves broadly elliptic, ovate, or obovate, rounded or somewhat abruptly short-tapered to tip ............................. 33

30 Leaves more or less equally tapered from middle of blade to tip and base ....................................................... 31
30 Leaves unequally tapered, the tip tapered to a point; base usually rounded or heart-shaped ................................. 32

31 Twigs from previous year glabrous and shining, often also somewhat waxy; young leaves with few to many coppery hairs mixed with white hairs; mature leaves smooth on both sides or hairy above; lateral veins not prominent; capsules lance-shaped . ....................................................... *S. petiolaris*
31 Twigs from previous year finely hairy (or at least with some patches of fine hairs); young leaves densely silky-hairy, without copper-colored hairs; underside of mature leaves silky-hairy with conspicuous, riblike, lateral veins; capsules ovate and rounded at tip ....................................................... *S. sericea*

32 Young twigs glabrous; stipules small or absent; bracts of pistillate catkins pale yellow and soon deciduous .......... *S. amygdaloides*
32 Young twigs gray-hairy; stipules large; bracts of pistillate catkins dark brown to black, persistent .................. *S. eriocephala*

33 Leaves balsam-scented (especially dried), underside net-veined, stipules tiny or absent; catkins appearing with leaves, on leafy or leafless branches .................................. *S. pyrifolia*

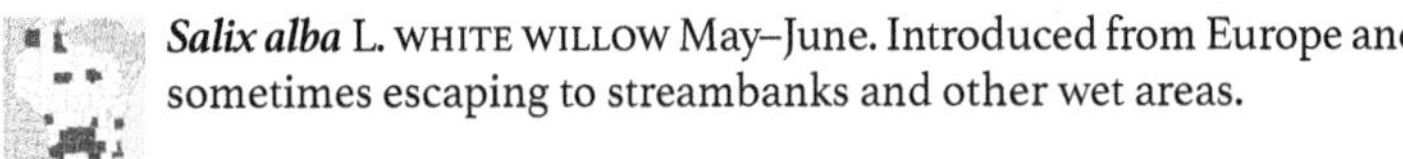

33 Leaves neither balsam-scented nor net-veined; stipules large on vigorous shoots; catkins appearing before or with leaves, sessile or on short, leafy branches ........................... *S. myricoides*

***Salix alba*** L. WHITE WILLOW May–June. Introduced from Europe and sometimes escaping to streambanks and other wet areas.

***Salix amygdaloides*** Anderss. [•93] PEACH-LEAF WILLOW May–June. Floodplains, streambanks, lake and pond borders.

***Salix babylonica*** L. WEEPING WILLOW April–May. Early introduction from Europe, especially for cemeteries, and sparingly escaped. Perhaps best treated as *Salix* × *sepulcralis* Simonkai as true *S. babylonica* not cold-hardy.

***Salix bebbiana*** Sarg. [•93] BEBB'S WILLOW, BEAKED WILLOW May–June. Swamps, thickets, wet meadows, streambanks, marsh borders.

***Salix candida*** Flueggé [•94] SAGE WILLOW May–June. Fens, bogs, open swamps, streambanks, usually where calcium-rich.

***Salix cordata*** Michx. HEART-LEAF WILLOW *Endangered.* Open sand dunes and sandy shores.

***Salix discolor*** Muhl. PUSSY-WILLOW April–May. Swamps, fens, streambanks, floodplains, marsh borders. Common.

***Salix eriocephala*** Michx. MISSOURI WILLOW April–May. Shores, streambanks, floodplains, ditches and wet meadows, especially along major rivers. *Salix cordata* Muhl., *Salix rigida* Muhl.

***Salix* × *fragilis*** L. CRACK WILLOW April–May. Introduced to North America from Europe in colonial times for ornament, shade, and gunpowder charcoal; common in farmyards and pastures and sometimes escaped. Considered of hybrid origin from *S. alba* and *S. euxina.*

***Salix humilis*** Marsh. [•94] UPLAND WILLOW March–April. Open woodlands, dry barrens, and prairies.

***Salix interior*** Rowlee [•93] SANDBAR WILLOW May–June. Shores, streambanks, sand and mud bars, ditches and other wet places; often colonizing exposed banks. *Salix exigua* Nutt. subsp. *interior* (Rowlee) Cronq.

***Salix lucida*** Muhl. [•94] SHINING WILLOW May. Swamps, shores, wet meadows, moist sandy areas.

***Salix myricoides*** Muhl. BLUELEAF WILLOW May. Dune hollows and sandy shorelines, fens, mostly near Great Lakes; inland on wet, calcium-rich sites. *Salix glaucophylloides* Fern.

***Salix nigra*** Marsh. BLACK WILLOW May. Streambanks, lakeshores and wet depressions; not tolerant of shade.

***Salix pedicellaris*** Pursh BOG WILLOW May–June. Bogs, fens, sedge meadows, interdunal wetlands.

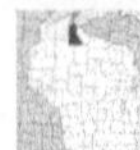

***Salix pellita*** (Anderss.) Anderss. SATINY WILLOW *Endangered.* May. Streambanks, sandy shores and rocky shorelines.

***Salix pentandra*** L. BAY-LEAF WILLOW Native of Europe, cultivated and occasionally escaped.

***Salix petiolaris*** Sm. [•94] MEADOW WILLOW May. Wet meadows, fens, streambanks, shores, open bogs, floating sedge mats, ditches. Common.

***Salix planifolia*** Pursh TEA-LEAF WILLOW *Threatened.* May. Rocky lakeshores, cedar swamps, black spruce bogs, streambanks, margins of sedge meadows.

***Salix purpurea*** L. PURPLE WILLOW, BASKET WILLOW May–June. Introduced from Europe, occasionally escaping to lakeshores and streambanks; mostly near Lake Michigan.

***Salix pyrifolia*** Anderss. BALSAM WILLOW May–June. Conifer swamps, bogs, rocky shores.

***Salix sericea*** Marsh. SILKY WILLOW May. Moist sandy or gravelly riverbanks and shores, sometimes in shallow water; however may no longer be present in Wisc.

***Salix serissima*** (Bailey) Fern. AUTUMN WILLOW Late May–July (our latest blooming willow). Fens, cedar and tamarack swamps, marshes, floating sedge mats, streambanks and shores, often where calcium-rich.

---

## SANTALACEAE *Sandalwood Family*

Herbs (ours); usually root-parasites. Leaves simple, alternate or opposite. Flowers perfect or unisexual, in terminal or axillary clusters, or solitary. Fruit a nut or drupe. Both *Comandra* and *Geocaulon,* though bearing green leaves, are hemiparasitic, and are apparently always attached (by means of modified roots, or haustoria) to some other plant. Both species also serve as alternate hosts for the canker-producing *Comandra* blister rust fungus (*Cronartium comandrae*), which in Wisc infects trees of jack pine.

*Key to genera*

1 Plant an essentially leafless, non-green parasite on the branches of coniferous trees .......... ARCEUTHOBIUM

1 Plant leafy, green, terrestrial .......... 2

2 Flowers green-purple, 2–3 from leaf axils; fruit a juicy orange to red drupe .......... GEOCAULON

2 Flowers white, numerous in a terminal inflorescence; fruit a dry green or yellowish drupe .......... COMANDRA

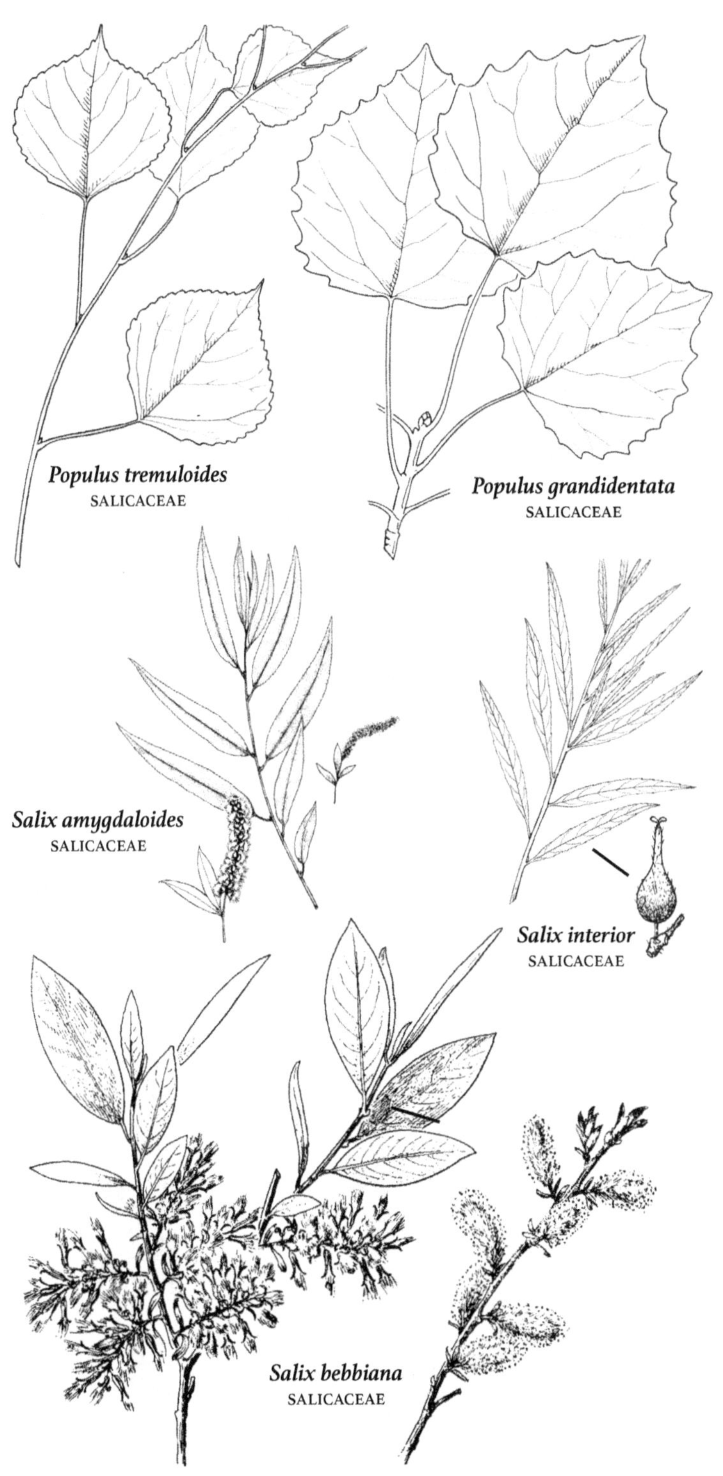
*Populus tremuloides*
SALICACEAE
*Populus grandidentata*
SALICACEAE
*Salix amygdaloides*
SALICACEAE
*Salix interior*
SALICACEAE
*Salix bebbiana*
SALICACEAE

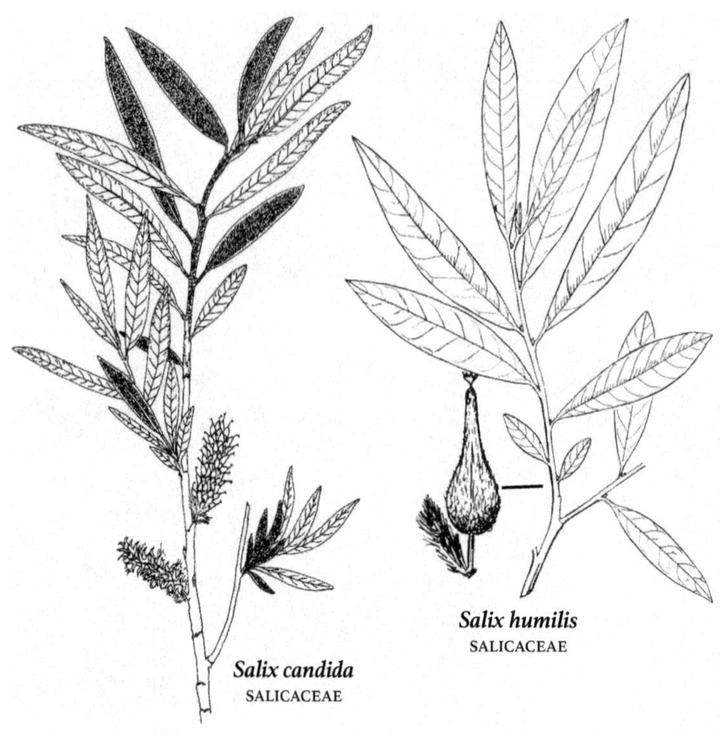

*Salix candida*
SALICACEAE
*Salix humilis*
SALICACEAE

*Salix petiolaris*
SALICACEAE
*Salix lucida*
SALICACEAE
*pine cone gall*

### ARCEUTHOBIUM *Dwarf-Mistletoe*

***Arceuthobium pusillum*** Peck EASTERN DWARF-MISTLETOE June–July. Chiefly on trees of black spruce, rarely on tamarack or white spruce; reported on white pine. On spruce it often produces witches' brooms. Usually a single host tree supports only one sex of the mistletoe.

### COMANDRA *Bastard Toadflax*

***Comandra umbellata*** (L.) Nutt. BASTARD TOADFLAX May–July. Prairies, shores, upland woods, and rock bluffs.

### GEOCAULON *False Toadflax*

***Geocaulon lividum*** (Richards.) Fern. FALSE TOADFLAX *Endangered.* Cedar swamps, open bogs; more commonly in sandy conifer woods and forested dune edges; Door Peninsula. *Comandra livida* Richards.

---

## SAPINDACEAE *Soapberry Family*

Soapberry Family now includes former members of Aceraceae and Hippocastanaceae.

*Key to genera*

1 Leaves simple and lobed or pinnately compound; fruit a 2-winged samara .......... ACER
1 Leaves palmately compound; fruit a large capsule.... AESCULUS

### ACER *Maple*

Trees or shrubs. Leaves opposite, simple or compound. Staminate and pistillate flowers borne on same or separate plants. Flowers with 5 sepals and 5 petals (sometimes absent), clustered into a raceme or umbel. Fruit a samara with 2 winged achenes joined at base.

ADDITIONAL SPECIES ***Acer ginnala*** Maxim. (Amur maple), a shrubby Asian species grown for its brilliant fall color, occasionally escaped throughout Wisc.

*Key to species*

1 Leaves pinnately compound .......... ***A. negundo***
1 Leaves simple .......... 2

2 Leaf sinuses between main leaf lobes sharp at their base.......... 3
2 Leaf sinuses rounded at their base .......... 5

3 Leaves deeply lobed to middle of blade or below, the lobes long and narrow .......... ***A. saccharinum***
3 Leaf lobes shorter and wider .......... 4

4 Leaves with downy white hairs on underside, tips of twigs with appressed hairs; shrubs or small trees; fruit persistent on plants until autumn .......... ***A. spicatum***
4 Leaves not downy-hairy on underside; twigs glabrous; medium trees; fruit shed in early summer .......... ***A. rubrum***

5 Leaf undersides and petioles covered with downy hairs; margins of the lobes undulate or entire; blades drooping at their edges; base of petioles with stipules .............................. *A. nigrum*

5 Leaf undersides and petioles glabrous or nearly so; margins of the lobes coarsely toothed; blades not drooping at their edges; base of petioles without stipules .................................. 6

6 Leaf blades large, the margins finely doubly toothed; bark with vertical white stripes; small trees or shrubs .....*A. pensylvanicum*

6 Leaf blades smaller, the margins only coarsely toothed; bark not vertically white-striped; medium to large trees ................ 7

7 Leaf petioles exuding milky juice when broken; twigs stout; samara wings widely divergent; bark becoming closely fissured, not scaly ........................................... *A. platanoides*

7 Leaf petioles not exuding milky juice when broken; twigs slender; samara wings less divergent; bark becoming deeply furrowed and plate-like ........................................ *A. saccharum*

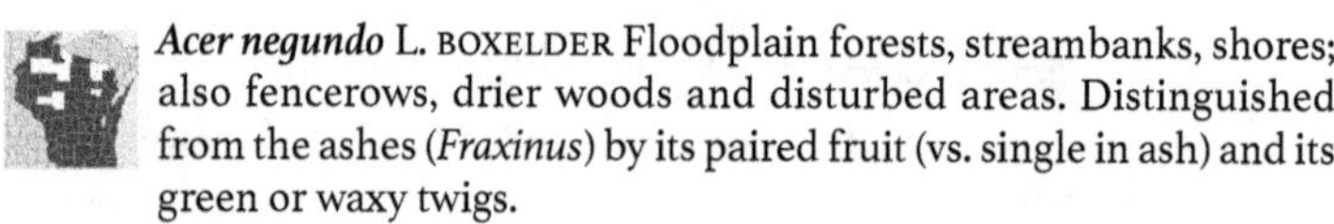

***Acer negundo*** L. BOXELDER Floodplain forests, streambanks, shores; also fencerows, drier woods and disturbed areas. Distinguished from the ashes (*Fraxinus*) by its paired fruit (vs. single in ash) and its green or waxy twigs.

***Acer nigrum*** Michx. f. BLACK MAPLE In moist soil, often associated with Acer saccharum. *Acer saccharum* Marsh. var. *nigrum* (Michx. f.) Britt.

***Acer pensylvanicum*** L. STRIPED MAPLE May–June. Moist woods; Door County.

***Acer platanoides*** L. NORWAY MAPLE April–May. Native of Europe; planted as a shade tree and established as a weedy tree in vacant lots.

***Acer rubrum*** L. [•95] RED MAPLE Floodplain forests, swamps; also common in drier forests. Distinguished from silver maple (*Acer saccharinum*) by its shallowly lobed leaves vs. the deeply lobed leaves of silver maple.

***Acer saccharinum*** L. SILVER MAPLE Floodplain forests, swamps, streambanks, shores, low areas in moist forests.

***Acer saccharum*** Marsh. SUGAR MAPLE April–May. Rich woods, especially in calcareous soils.

***Acer spicatum*** Lam. [•95] MOUNTAIN MAPLE June. Moist woods.

### AESCULUS *Buckeye*

***Aesculus glabra*** Willd. [•95] OHIO-BUCKEYE Small to medium tree; bark and leaves foul-smelling when bruised. Leaves opposite, palmately compound into usually 5 (rarely 7) leaflets. Flowers perfect or either staminate or pistillate on same tree, yellow-green, appearing after leaves unfold in spring, in panicles at ends of branches; petals 4, pale yellow. Fruit a prickly red-brown capsule. April–May. Floodplain forests,

streambanks, thickets. Local in Wisc (and considered adventive), where reaching its northern range limit.

## SARRACENIACEAE *Pitcherplant Family*

### SARRACENIA *Pitcherplant*

***Sarracenia purpurea*** L. [•96] PITCHERPLANT Perennial insectivorous herb. Flower stalks leafless. Leaves clumped, hollow and vaselike, green or veined with red-purple, winged, smooth on outside, upper portion of inside with downward-pointing hairs. Flowers large and nodding, single at ends of stalks. Fruit a 5-chambered capsule. May–July. Sphagnum bogs, floating bog mats, occasionally in calcium-rich wetlands.

## SAXIFRAGACEAE *Saxifrage Family*

Perennial herbs. Leaves alternate, opposite or basal. Flowers perfect, regular, single on stalks or in narrow heads. Sepals and petals 5 (4 in *Chrysosplenium*).

*Key to genera*

1 Leaves all from stem; petals absent, the flowers 4-merous ........ CHRYSOSPLENIUM

1 Leaves all (or nearly all) from base of plant; the flowers 5-merous 2

2 Leaves entire or sometimes the margins somewhat undulate .... MICRANTHES

2 Leaf margins with rounded or sharp teeth .................... 3

3 Petioles densely long-hairy, the hairs spreading; flowers small and irregular, in a panicle ........................... HEUCHERA

3 Plants not with the above combination of characters ........... 4

4 Petioles glabrous; inflorescence a panicle ........ SULLIVANTIA

4 Petioles with stiff, white, appressed hairs, or only finely hairy, or with tiny glandular hairs; inflorescence a raceme............... 5

5 Petals entire; flowers and fruit on slender pedicels mostly more than 5 mm long; scapes without any leaves ............... TIARELLA

5 Petals deeply fringed and comb-like; flowers and fruit on short thick pedicels less than 5 mm long; scape naked or with a single pair of leaves ......................................... MITELLA

### CHRYSOSPLENIUM *Golden-Saxifrage*

***Chrysosplenium americanum*** Schwein. [•96] AMERICAN GOLDEN-SAXIFRAGE Small, perennial herb, often forming large mats. April–June. Springs, shallow streams, shady wet depressions; soils mucky.

### HEUCHERA *Alumroot*

***Heuchera richardsonii*** R. Br. PRAIRIE ALUMROOT Perennial herb. Stems more or less hirsute, becoming glandular in the inflorescence. Leaves broadly cordate-ovate. Flowers in relatively narrow and congested panicles. Prairies and dry woods.

### MICRANTHES *Saxifrage*

***Micranthes pensylvanica*** (L.) Haw. [•96] SWAMP SAXIFRAGE Perennial herb. Stems stout, erect, with sticky hairs. Leaves all from base of plant. Flowers small, in clusters atop stem. May–June. Swamps, wet deciduous forests, marshes, moist meadows and low prairie; often where calcium-rich. *Saxifraga pensylvanica* L.

### MITELLA *Mitrewort; Bishop's Cap*

Perennial rhizomatous herbs. Leaves basal or alternate from the rhizome, the flowering stems bearing a terminal raceme of small white, greenish, or purple flowers.

*Key to species*

1 Plants small, the scape naked; the basal leaves not lobed or only slightly so; flowers green-yellow ........................ ***M. nuda***

1 Plants larger, with a pair of nearly sessile leaves on the scape below the inflorescence; the basal leaves clearly 3-lobed; flowers white . ................................................ ***M. diphylla***

***Mitella diphylla*** L. [•96] TWO-LEAF MITREWORT May–June. Rich woods. The follicles at dehiscence diverge and open widely, exposing the shiny seeds.

***Mitella nuda*** L. NAKED MITREWORT June–July. Hummocks in swamps and alder thickets, ravines, seeps, moist mixed conifer and deciduous forests.

### SULLIVANTIA *Coolwort*

***Sullivantia sullivantii*** (Torr. & Gray) Britt. SULLIVANT'S COOLWORT Perennial herb. Leaves basal, long-petioled, with numerous shallow, rounded lobes each terminating in 2 or 3 short teeth. Flowers small, in an open panicle atop a scape-like flowering stem. Petals white. June–July. Moist shaded cliffs, usually of sandstone.

### TIARELLA *Foam-Flower*

***Tiarella cordifolia*** L. FOAM-FLOWER *Endangered.* Perennial herb with long stolons. Flowering stems glandular-puberulent. Leaves basal, broadly cordate-ovate, shallowly 3–5-lobed. Flowers in a raceme atop an erect, usually leafless stem. Petals white. May–early June. Rich, mesic hardwood forests (sometimes with trees of hemlock present).

## SCROPHULARIACEAE *Figwort Family*

Annual, biennial, or perennial herbs. Leaves mostly opposite or alternate (*Verbascum*). Flowers single or few from leaf axils, or numerous in clusters at ends of stems or leaf axils, perfect, usually with a distinct upper and lower lip; sepals and petals 4–5 (petals sometimes absent). Fruit a several- to many-seeded capsule. Formerly a much larger family, many of our genera now segregated into other families, especially Orobanchaceae and Plantaginaceae.

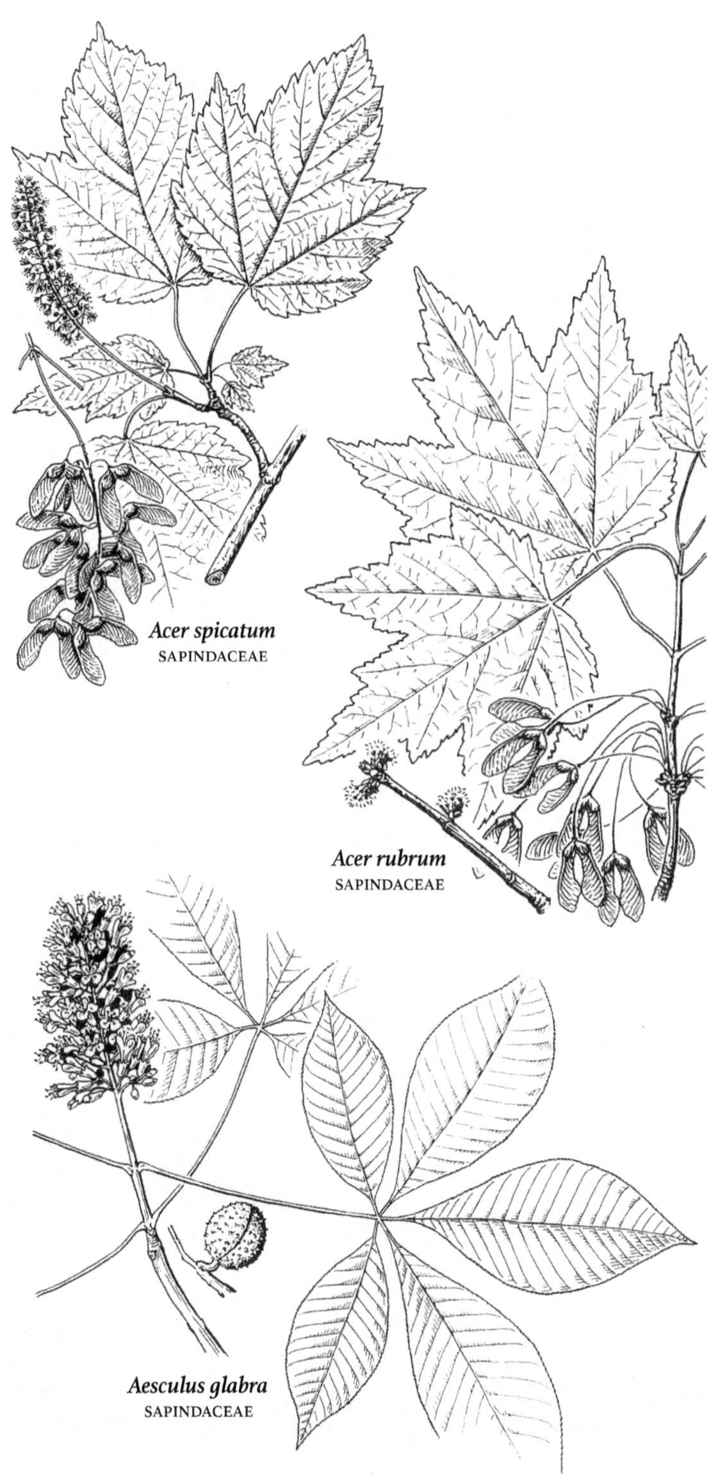
Acer spicatum
SAPINDACEAE
Acer rubrum
SAPINDACEAE
Aesculus glabra
SAPINDACEAE

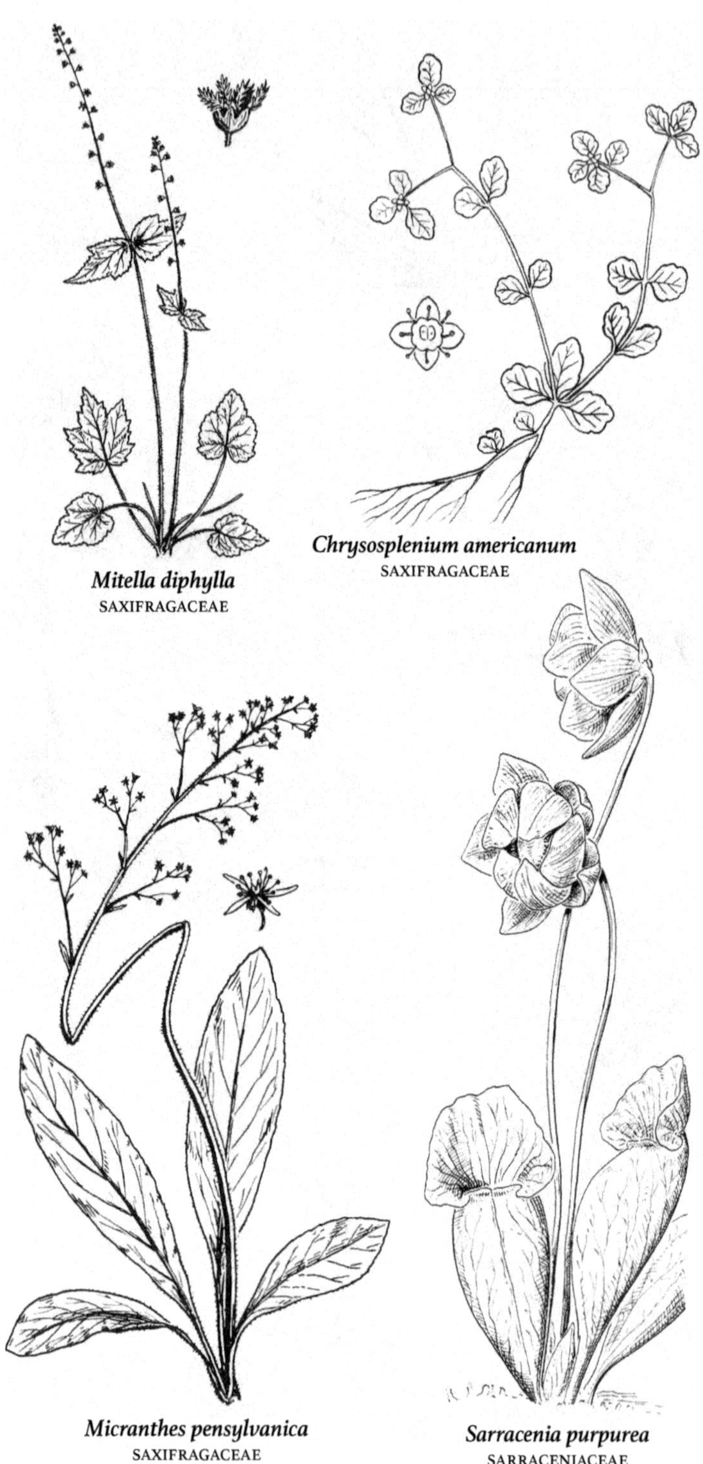

*Mitella diphylla*
SAXIFRAGACEAE

*Chrysosplenium americanum*
SAXIFRAGACEAE

*Micranthes pensylvanica*
SAXIFRAGACEAE

*Sarracenia purpurea*
SARRACENIACEAE

*Key to genera*

1 Stem leaves of fertile stems mostly alternate (a basal rosette may be present and the lower leaves may sometimes be opposite) ....... .......................................... VERBASCUM

1 Stem leaves of fertile stems all opposite or nearly so, sometimes alternate below the flowers .................................. 2

2 Inflorescence terminal, branched ............ SCROPHULARIA

2 Inflorescence leafy and spike-like; the flowers from the leaf axils . ........................................... DASISTOMA

## DASISTOMA *Mullein-Foxglove*

***Dasistoma macrophylla*** (Nutt.) Raf. MULLEIN-FOXGLOVE June–Aug. Moist rich woods; more common south of Wisc. *Seymeria macrophylla* Nutt.

## SCROPHULARIA *Figwort*

Perennial herbs. Stems 4-angled. Leaves opposite. Flowers 2-lipped, in open terminal clusters. Corolla brownish.

*Key to species*

1 Main leaves coarsely toothed or cleft, the teeth mostly long-tapered to their tip; leaf base tapered and extending downward along the petiole; sterile stamen green-yellow ................ ***S. lanceolata***

1 Main leaves evenly toothed, the teeth rounded at tip or abruptly tapered to a short point; leaf base rounded to heart-shaped, not extending downward along petiole; sterile stamen brown to purple ........................................... ***S. marilandica***

***Scrophularia lanceolata*** Pursh [•97] LANCE-LEAF FIGWORT Late May–July. Roadsides, railroads, old roads; forests, especially in clearings and edges; fields, fencerows, shores, swamp borders.

***Scrophularia marilandica*** L. CARPENTER'S-SQUARE July–Aug, its flowering period rarely overlapping that of *S. lanceolata.* Less common than *S. lanceolata* but in similar habitats: riverbank thickets and floodplains; open woods, especially in clearings and on margins; roadsides. Not always easily distinguished from *S. lanceolata,* though often taller and more branched.

## VERBASCUM *Mullein*

Biennial herbs (ours), producing a rosette of leaves the first year, from which the tall flowering stem rises the following season. Leaves alternate, entire, crenate, or rarely deeply toothed. Flowers yellow, white, or blue, in one to many spike-like racemes.

ADDITIONAL SPECIES ***Verbascum densiflorum*** Bertol. (Dense-flower mullein); introduced, Ozaukee and Washington counties. ***Verbascum nigrim*** L. (Black mullein); introduced, Portage and Sauk counties.

*Key to species*

1 Stems glabrous, or often glandular-hairy on upper stem and inflorescence; flowers yellow or white ............... ***V. blattaria***

1 Stems densely woolly hairy; flowers yellow .................... 2

2 Leaves not extending downward on stem (or only very shortly so) ................................................ *V. phlomoides*

2 Leaves extending downward on stem to about next lower leaf . . . 3

3 Flowers 1–2.5 cm wide; common throughout Wisc ..... *V. thapsus*

3 Flowers 2.5–4.5 cm wide; se Wisc ................. *V. densiflorum**

***Verbascum blattaria*** L. [•97] WHITE MOTH MULLEIN June–Oct. Eurasian weed of fields, roadsides, and waste places.

***Verbascum phlomoides*** L. ORANGE MULLEIN June–Sept. Native of Europe; locally established as a weed, but never as common as *V. thapsus.*

***Verbascum thapsus*** L. GREAT MULLEIN June–Sept. European weed of fields, roadsides, and disturbed places.

## SIMAROUBACEAE *Quassia-Wood Family*

### AILANTHUS *Tree-of-Heaven*

***Ailanthus altissima*** (P. Mill.) Swingle TREE-OF-HEAVEN Rapidly growing, weedy tree. Leaves odd-pinnate; leaflets 11–41. Dioecious, with staminate and pistillate flowers borne on different individuals (sometimes flowers perfect). Flowers greenish or greenish yellow, in large terminal pyramidal panicles. Native of e Asia; introduced as a shade tree and now established in vacant lots, roadsides, and occasionally in open woodlands.

## SOLANACEAE *Potato Family*

Herbs or shrubs, rarely climbing. Leaves alternate or appearing opposite. Flowers perfect, almost always 5-merous, regular (in most of our genera) or irregular. Fruit a capsule or berry. A large family, most numerous in tropical America, and with many plants, such as tomato, potato, eggplant, and peppers, of economic importance.

ADDITIONAL SPECIES ***Hyoscyamus niger*** L. (Black henbane), plants clammy-pubescent, corolla greenish yellow with purple veins; introduced, Sheboygan County.

*Key to genera*

1 Plants woody (at least at base), sprawling or climbing vines; fruit a red berry ...................................................... 2

1 Plants herbs; fruit various .................................. 3

2 Leaves unlobed; stems mostly woody .................. LYCIUM

2 Main leaves 3–4 lobed; stems woody near base ...... SOLANUM

3 Plants in flower ............................................... 4

3 Plants in fruit .................................................. 8

4 Corolla with short tube and widely flared upper portion ........ 5

4 Corolla tube longer, the lobes joined for most of their length.... 6

5 Anthers opening at tip .......... SOLANUM
5 Anthers opening along sides .......... LEUCOPHYSALIS

6 Calyx tubular, more than 3 cm long .......... DATURA
6 Calyx less than 3 cm long .......... 7

7 Corolla blue .......... NICANDRA
7 Corolla yellow or white .......... PHYSALIS

8 Calyx or capsule spiny .......... 9
8 Calyx and capsule not spiny.......... 10

9 Stems and leaves spiny; calyx spiny; berry smooth ... SOLANUM
9 Plants smooth apart from spiny capsule .......... DATURA

10 Calyx inflated, covering the berry .......... 11
10 Calyx not inflated, the berry visible .......... SOLANUM

11 Calyx strongly inflated, not filled by the berry ........ PHYSALIS
11 Calyx only slightly inflated, filled within by the berry .......... LEUCOPHYSALIS

## DATURA *Jimsonweed*

Coarse annual herbs (ours). Leaves ovate, petioled. Flowers large, white to violet. Fruit a many-seeded capsule. All species toxic if eaten.

*Key to species*

1 Plants glabrous or nearly so; flowers less than 10 cm long .......... *D. stramonium*
1 Plants covered with soft hairs; flowers more than 10 cm long .......... *D. inoxia*

***Datura inoxia*** P. Mill. DOWNY THORN-APPLE July–Oct. Native of the sw states; occasionally escaped from cultivation in waste places.

***Datura stramonium*** L. JIMSONWEED Introduced (naturalized). Widely distributed weed of fields, barnyards, and waste places.

## LEUCOPHYSALIS *False Ground-Cherry*

***Leucophysalis grandiflora*** (Hook.) Rydb. LARGE FALSE GROUND-CHERRY Annual herb; thinly villous and more or less viscid. Flowers commonly 2–4 from the upper nodes; corolla white with pale yellow center, rotate. June–Aug. Dry sandy soil. *Physalis grandiflora* Hook.

## LYCIUM *Matrimony-Vine*

***Lycium barbarum*** L. MATRIMONY-Vine Shrub with long slender branches to 3 m long, arched or recurved, sometimes climbing, often spiny at the nodes, bearing alternate leaves on the young shoots, later bearing at each node a tuft of leaves of various sizes, often subtended by a thorn. Flowers in clusters of 1–4; corolla dull pinkish violet, short-lived and fading the second day the. Berry scarlet. May–Aug. Native of s Europe; cultivated in old gardens and now escaped in vacant lots, roadsides, fencerows, and at the edge of woods.

## NICANDRA *Apple-of-Peru*

***Nicandra physalodes*** (L.) Gaertn. APPLE-OF-PERU Glabrous annual. Leaves ovate, coarsely and unevenly toothed. Corolla blue, campanulate, shallowly 5-lobed. Berry dry, many-seeded, enclosed by the calyx. July–Sept. Native of Peru; cultivated for ornament and escaped on roadsides and waste places.

## PHYSALIS *Ground-Cherry*

Annual or perennial herbs. Leaves alternate or falsely opposite. Flowers solitary or few at the nodes, white, greenish yellow, or yellow, often with a darker center. Berry many-seeded, pulpy.

ADDITIONAL SPECIES ***Physalis grisea*** (Waterfall) M. Martinez (Strawberry-tomato), adventive in s and c Wisc. ***Physalis hispida*** (Waterfall) Cronq. (Prairie ground-cherry), introduced, Grant County in sw Wisc.

*Key to species*

1 Colony-forming perennial herbs, spreading by rhizomes; corolla 1–2 cm long .......... 2
1 Taprooted annual herbs; corolla less than 1 cm long (except in *P. philadelphica*) .......... 5

2 Corolla white; distinctly 5-lobed; mature calyx bright red-orange .......... ***P. alkekengi***
2 Corolla yellow, only slightly lobed; mature calyx green or brown 3

3 Upper stems with soft, spreading hairs .......... ***P. heterophylla***
3 Upper stems with short, stiff, appressed hairs or glabrous .......... 4

4 Leaves sparsely hairy on both sides; calyx tube with stiff hairs to 1.5 mm long .......... ***P. virginiana***
4 Leaves nearly glabrous, hairs if present mostly along main veins; calyx tube with very short, appressed hairs, the hairs less than 0.5 mm long .......... ***P. longifolia***

5 Upper stems glabrous or with very short appressed hairs .......... ***P. philadelphica***
5 Upper stems with long, soft, spreading hairs .......... 6

6 Leaves grayish due to hairs, leaves also with sessile glands .......... ***P. grisea****
6 Leaves greenish, the pubescence less dense; sessile glands absent .......... ***P. pubescens***

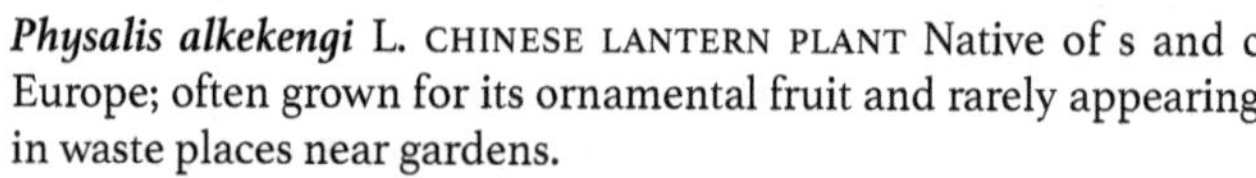

***Physalis alkekengi*** L. CHINESE LANTERN PLANT Native of s and c Europe; often grown for its ornamental fruit and rarely appearing in waste places near gardens.

***Physalis heterophylla*** Nees CLAMMY GROUND-CHERRY June–Sept. In dry or sandy soil, upland woods and prairies, our most abundant *Physalis.*

***Physalis longifolia*** Nutt. LONGLEAF GROUND-CHERRY July–Aug. Moist or dry fields, open woods, and prairies.

***Physalis philadelphica*** Lam. TOMATILLO Native of Mexico; rarely adventive.

***Physalis pubescens*** L. DOWNY GROUND-CHERRY May–Sept. Moist soil. Sometimes cultivated for its fruit, and persisting around gardens.

***Physalis virginiana*** P. Mill. VIRGINIA GROUND-CHERRY Dry or moist fields, sandy upland woods, prairies.

### SOLANUM *Nightshade*

Herbs or vines. Corolla rotate or broadly campanulate. Fruit a many-seeded berry. Our species bloom in summer, often continuing into the fall.

ADDITIONAL SPECIES ***Solanum physalifolium*** Rusby (Ground-cherry nightshade), introduced in several Wisc locations.

*Key to species*

1 Plants with spines or prickles . . . . . . . . . . 2
1 Plants without spines or prickles . . . . . . . . . . 3

2 Corolla bright yellow; leaves 1–2 times pinnately compound; calyx covered with spines . . . . . . . . . . ***S. rostratum***
2 Corolla pale violet or white; leaves entire or toothed or lobed; calyx not spiny . . . . . . . . . . ***S. carolinense***

3 Plants climbing or trailing vines, woody at base; flowers usually light blue or violet . . . . . . . . . . ***S. dulcamara***
3 Plants upright herbs; flowers white . . . . . . . . . . 4

4 Plants sicky-hairy; calyx enlarging to cover lower half of berry . . . . . . . . . . ***S. physalifolium****
4 Plants glabrous to pubescent, the hairs appressed and not gland-tipped; calyx covering only bottom of berry . . . . . . ***S. ptychanthum***

***Solanum carolinense*** L. [•98] HORSE-NETTLE Introduced (naturalized). Fields and waste places, especially in sandy soil.

***Solanum dulcamara*** L. CLIMBING NIGHTSHADE Native of Eurasia; moist thickets.

***Solanum ptychanthum*** Dunal WEST INDIAN NIGHTSHADE Both native and Eurasian forms present; a cosmopolitan weed. *Solanum nigrum* L.

***Solanum rostratum*** Dunal [•98] BUFFALO-BUR Introduced (naturalized). Dry prairies.

## STAPHYLEACEAE *Bladdernut Family*

### STAPHYLEA *Bladdernut*

***Staphylea trifolia*** L. [•98] AMERICAN BLADDERNUT Erect shrub with striped bark. Leaves opposite, 3-foliolate. Inflorescence a terminal drooping panicle. Flowers regular, perfect, 5-merous. Fruit a 3-lobed, inflated capsule. May. Moist deciduous forests and thickets, especially on riverbanks and floodplains.

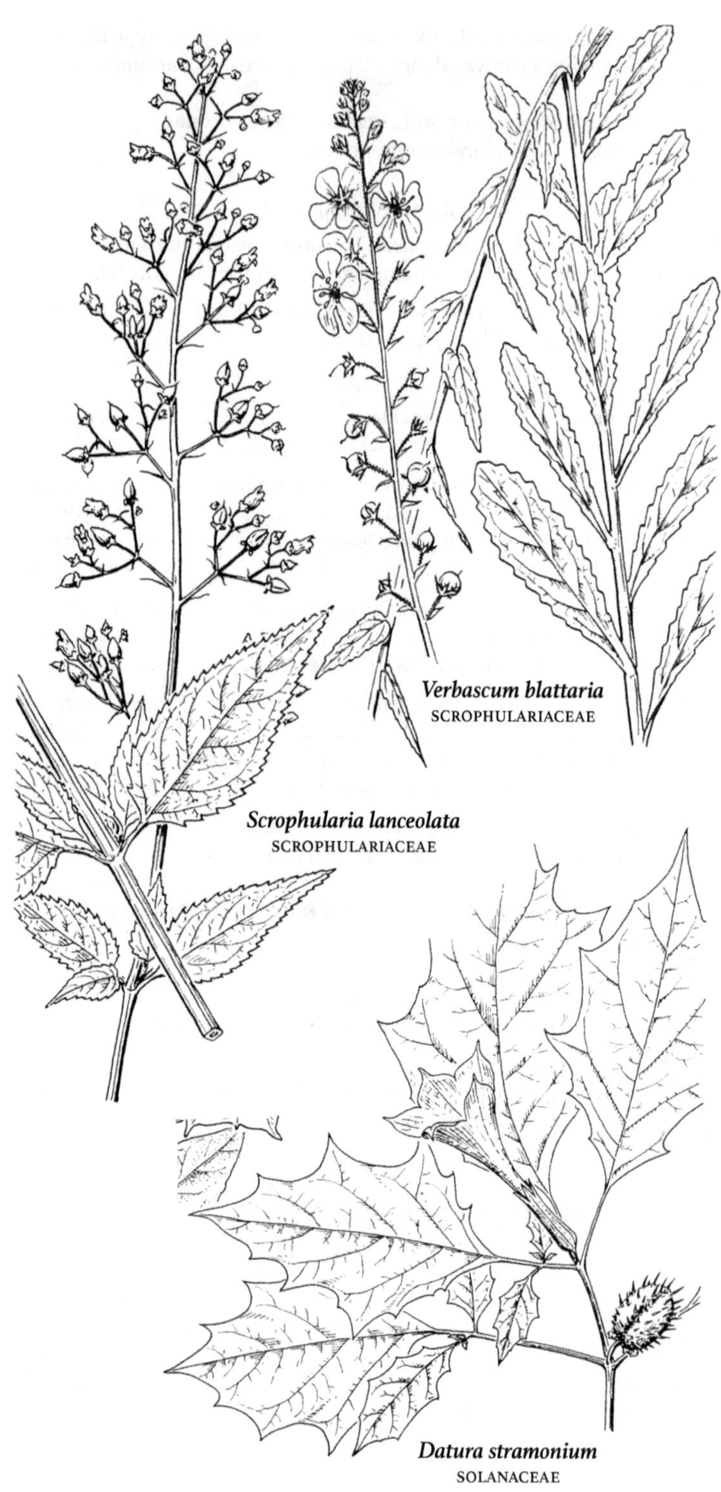

*Verbascum blattaria*
SCROPHULARIACEAE

*Scrophularia lanceolata*
SCROPHULARIACEAE

*Datura stramonium*
SOLANACEAE

*Staphylea trifolia*
STAPHYLEACEAE

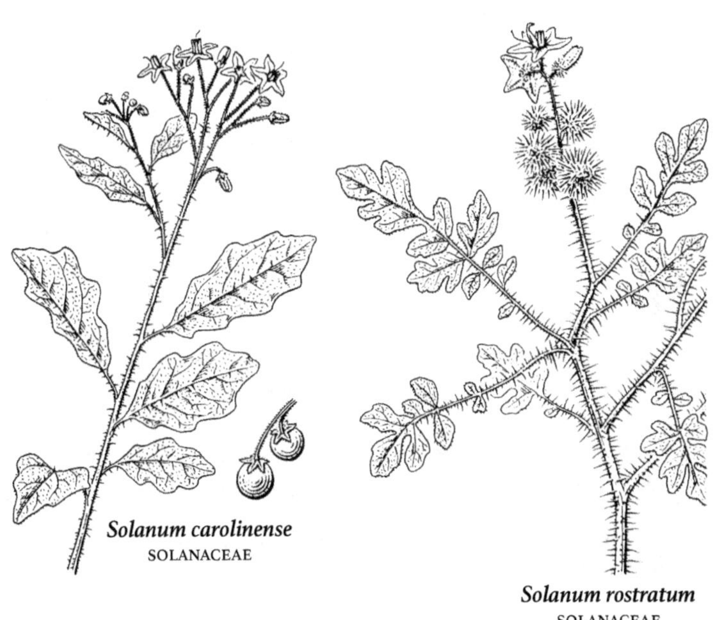

*Solanum carolinense*
SOLANACEAE

*Solanum rostratum*
SOLANACEAE

## THYMELAEACEAE *Mezereum Family*

### DIRCA *Leatherwood*

***Dirca palustris*** L. [•99] EASTERN LEATHERWOOD Branched shrub, bark very tough and pliable, twigs jointed. Leaves alternate, entire. Flowers perfect, regular, pale yellow, subtended by hairy bud scales in early spring before the leaves appear. Fruit an ellipsoid drupe. Spring. Rich, moist deciduous woods.

ADDITIONAL SPECIES ***Thymelaea passerina*** (L.) Coss. & Germ. (Mezereon), introduced annual herb, Milwaukee County.

## ULMACEAE *Elm Family*

### ULMUS *Elm*

Trees. Leaves alternate, simple, inequilateral; margins usually doubly serrate. Flowers perfect, in short racemes or, by abbreviation of the axis, in fascicles. Fruit a flat, 1-seeded samara. Several exotic elm species are sometimes planted as street trees. Hackberry (*Celtis*), a former member of this family, now included in Cannabaceae.

*Key to species*

1 Leaf blades small, 3–7 cm long . . . . . . . . . . *U. pumila*

1 Leaf blades larger, 7–18 cm long. . . . . . . . . . 2

2 Leaves smooth on each side; branches with corky wing-like ridges, the lowermost branches short and strongly drooping; main trunk usually not dividing into several large limbs . . . . . . . . . . *U. thomasii*

2 Leaves roughened on one or both sides; branches not with corky wings, the lowermost branches longer and not strongly drooping; main trunk usually dividing into several large limbs, tree vase-shaped in outline . . . . . . . . . . 3

3 Leaves usually rough only on upperside; bark gray, deeply fissured . . . . . . . . . . *U. americana*

3 Leaves usually rough on both sides; bark dark red-brown, shallowly fissured . . . . . . . . . . *U. rubra*

***Ulmus americana*** L. [•99] AMERICAN ELM Floodplain forests, streambanks and moist, rich woods; less common now than formerly due to losses from Dutch elm disease.

***Ulmus pumila*** L. SIBERIAN ELM Introduced (invasive) Escaping from cultivation to waste places, roadsides, fencerows. Distinguished from our other elms by its singly serrate leaf margins.

***Ulmus rubra*** Muhl. [•99] SLIPPERY ELM Moist woods. The inner bark is very mucilaginous and has some medicinal value.

***Ulmus thomasii*** Sarg. ROCK ELM Rich upland woods. Its thick, corky-winged branches are distinctive.

## URTICACEAE *Nettle Family*

Annual or perennial herbs with watery juice, sometimes with stinging hairs. Leaves alternate or opposite, simple, with petioles. Flowers small, green, in simple or branched clusters from leaf axils, staminate and pistillate flowers usually separate, on same or separate plants. Fruit an achene, often enclosed by the sepals which enlarge after flowering.

*Key to genera*

1 Leaves alternate ........ 2
1 Leaves opposite ........ 3

2 Plants large and coarse, with stiff stinging hairs, the leaves sharply toothed ........ LAPORTEA
2 Plants smaller, without stinging hairs; leaves entire PARIETARIA

3 Plants with stinging hairs; leaves lance-shaped ........ URTICA
3 Plants without stinging hairs; leaves ovate ........ 4

4 Stems translucent and fleshy; flowers in dense short clusters from leaf axils; achene equal or longer than sepals ........ PILEA
4 Stems neither translucent nor fleshy; flowers in cylindric spikes from leaf axils; achene shorter than and hidden by the sepals ........ BOEHMERIA

### BOEHMERIA *False Nettle*

***Boehmeria cylindrica*** (L.) Sw. [•99] SMALL-SPIKE FALSE NETTLE Perennial, nettle-like herb, stinging hairs absent. Leaves opposite, rough-textured, ovate to broadly lance-shaped, with 3 main veins; margins coarsely toothed. Flowers tiny, green, staminate and pistillate flowers usually on separate plants, in small spikelike clusters along unbranched stalks from upper leaf axils. July–Aug. Floodplain forests, swamps, marshes and bogs.

### LAPORTEA *Wood-Nettle*

***Laportea canadensis*** (L.) Weddell CANADIAN WOOD-NETTLE Perennial herb, spreading by rhizomes. Stems somewhat zigzagged. Leaves alternate, ovate, with small stinging hairs, margins coarsely toothed. Flowers small, green, staminate and pistillate flowers separate but borne on same plant; staminate flowers in branched clusters from lower leaf axils; pistillate flowers in open, spreading clusters from upper axils. July–Sept. Floodplain forests, rich moist woods, low places in hardwood forests.

### PARIETARIA *Pellitory*

***Parietaria pensylvanica*** Muhl. PENNSYLVANIA PELLITORY Annual pubescent herb. Leaves alternate, entire, lance-shaped, 3-nerved. Flowers green, from the middle and upper axils. June–Sept. Moist to dry forests, gravelly shores, disturbed sites.

### PILEA *Clearweed*

Annual herbs, sometimes forming colonies from seeds of previous year. Stems erect to sprawling, smooth and watery. Leaves opposite, stinging hairs absent, thin and translucent, ovate, with 3 major veins from base of leaf, margins toothed. Flowers green, staminate and pistillate flowers separate, borne on same or different plants, in clusters from leaf axils.

*Key to species*

1 Achenes 1–1.5 mm wide, olive-green to dark purple with a narrow pale margin, covered with low bumps ................. *P. fontana*

1 Achenes to 1 mm wide, green to yellow, often marked with purple spots, smooth ........................................ *P. pumila*

***Pilea fontana*** (Lunell) Rydb. LESSER CLEARWEED Aug–Sept. Lakeshores, riverbanks, swamps, marshes and springs.

***Pilea pumila*** (L.) Gray [•100] CANADIAN CLEARWEED July–Sept. Swampy woods (often on logs), wooded streambanks, floodplain forests, wet depressions, rocky hollows; usually in partial shade.

### URTICA *Stinging Nettle*

***Urtica dioica*** L. STINGING NETTLE Perennial herb, often forming dense patches from spreading rhizomes, with stinging hairs on stems and leaves. Leaves opposite, ovate to lance-shaped; margins coarsely toothed. Flowers small, green, staminate and pistillate flowers separate but mostly on same plants; flower clusters branched and spreading from leaf axils, all of one sex or a mix of staminate and pistillate flowers, the pistillate clusters usually above the staminate clusters when both present on a plant. July–Sept. Moist woods, thickets, ditches, streambanks and disturbed areas.

---

## VERBENACEAE *Verbena Family*

Perennial herbs with 4-angled stems. Leaves opposite, toothed. Flowers small, numerous, perfect, in branched or unbranched spikes or heads at ends of stems or from upper leaf axils, the spikes elongating as flowers open upward from the base. Fruit dry, enclosed by the sepals, splitting lengthwise into 2 or 4 nutlets when mature.

*Key to genera*

1 Flowers in dense rounded heads or short spikes, at ends of long, solitary peduncles from leaf axils; corolla 2-lipped ....... PHYLA

1 Flowers in short to long racemes or spikes, terminal or from upper leaf axils; corolla regular or nearly so ........................ 2

2 Corolla usually small, less than 1 cm wide, not showy . VERBENA

2 Corolla large, 1 cm or more wide, showy ....... GLANDULARIA

### GLANDULARIA *Mock Vervain*

***Glandularia canadensis*** (L.) Nutt. ROSE MOCK VERVAIN Introduced perennial herb. Leaves ovate or narrower, incised or pinnatifid. Flowers in peduncled spikes, elongating in fruit; corolla blue to purple or white. March–Oct. Various habitats, often in disturbed soil. *Verbena canadensis* (L.) Britt.

## PHYLA *Fogfruit*

***Phyla lanceolata*** (Michx.) Greene NORTHERN FOGFRUIT Perennial herb, sometimes forming mats. Stems slender, weak, 4-angled, creeping to ascending, often rooting at nodes, the stem tips and lateral branches upright. Leaves opposite, ovate to oblong lance-shaped, bright green; margins with coarse, forward-pointing teeth to below middle of blade. Flowers small, crowded in spikes from leaf axils; corolla pale blue or white, 4-lobed and 2-lipped. June–Sept. Margins of lakes, ponds, streams, ditches, mud flats; often where seasonally flooded. *Lippia lanceolata* Michx.

## VERBENA *Vervain*

Annual or perennial herbs. Leaves usually opposite, simple, entire to somewhat lobed. Inflorescence of terminal spikes, usually many-flowered, often flat-topped, sometimes elongate with scattered flowers. Fruit mostly enclosed by the mature calyx, separating at maturity into 4 nutlets.

*Key to species*

1 Plants spreading; leaves incised and often somewhat 3-lobed .... .................................................. ***V. bracteata***

1 Plants erect; leaves unlobed or lobed only near base ............ 2

2 Leaves narrowly lance-shaped, less than 1.5 cm wide, tapered at base to an indistinct petiole; plants glabrous or with scattered appressed hairs ...................................... ***V. simplex***

2 Leaves mostly ovate, 2 cm or more wide, with petioles or sessile; plants usually with at least some hairs......................... 3

3 Plants densely gray-hairy; leaves sessile ................. ***V. stricta***

3 Plants not densely hairy; leaves with petioles .................. 4

4 Flowers blue to purple, densely overlapping on spike; leaves often lobed at their base .................................. ***V. hastata***

4 Flowers white, not overlapping on spike; leaves not lobed ........ ............................................... ***V. urticifolia***

***Verbena bracteata*** Lag. & Rodr. CARPET VERVAIN April–Oct. Prairies, fields, roadsides, and waste places.

***Verbena hastata*** L. [•100] COMMON VERVAIN July–Sept. Marshes, wet meadows, shores, streambanks, openings in swamps, ditches.

***Verbena simplex*** Lehm. NARROW-LEAF VERVAIN June–Aug. Dry soil of woods, fields, rocky places, and roadsides.

***Verbena stricta*** Vent. [•100] HOARY VERVAIN June–Sept. Prairies, barrens, fields, and roadsides.

***Verbena urticifolia*** L. WHITE VERVAIN June–Oct. Thickets, meadows, waste places.

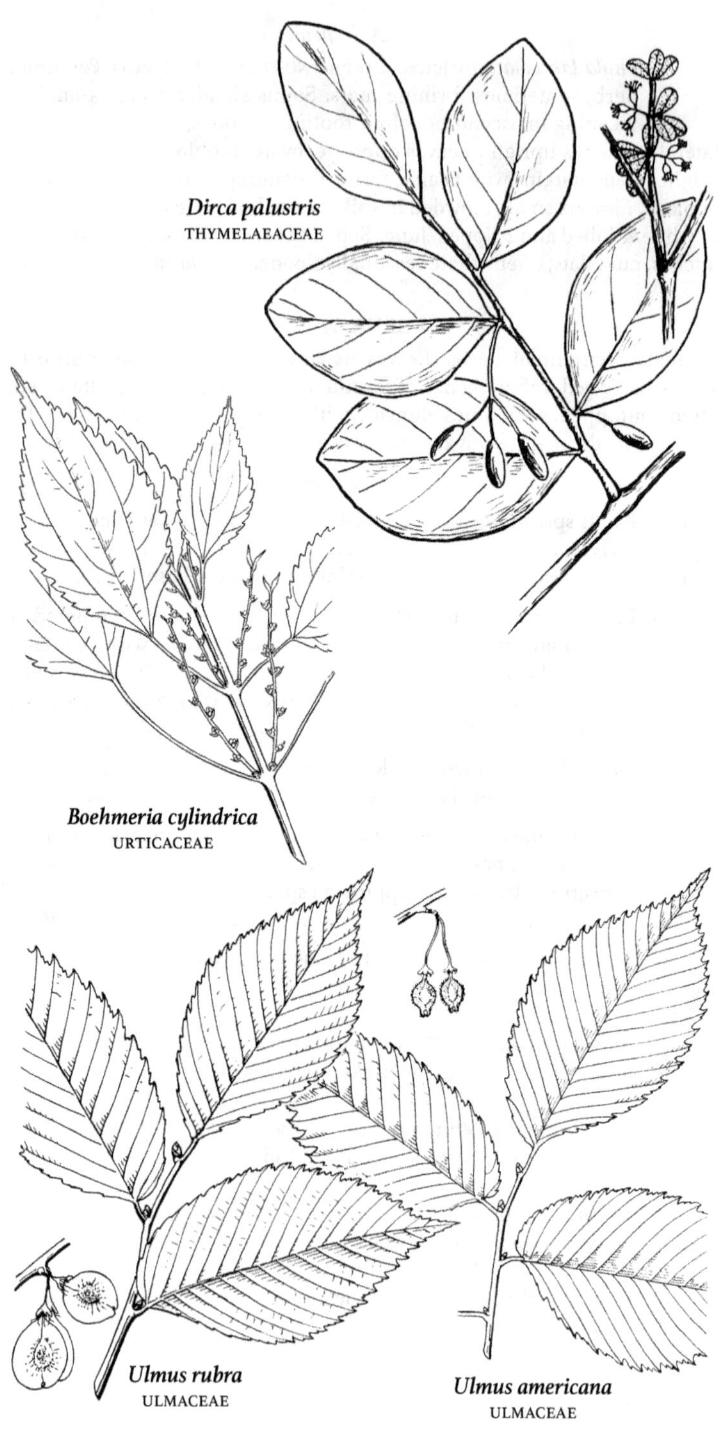
Dirca palustris
THYMELAEACEAE
Boehmeria cylindrica
URTICACEAE
Ulmus rubra
ULMACEAE
Ulmus americana
ULMACEAE

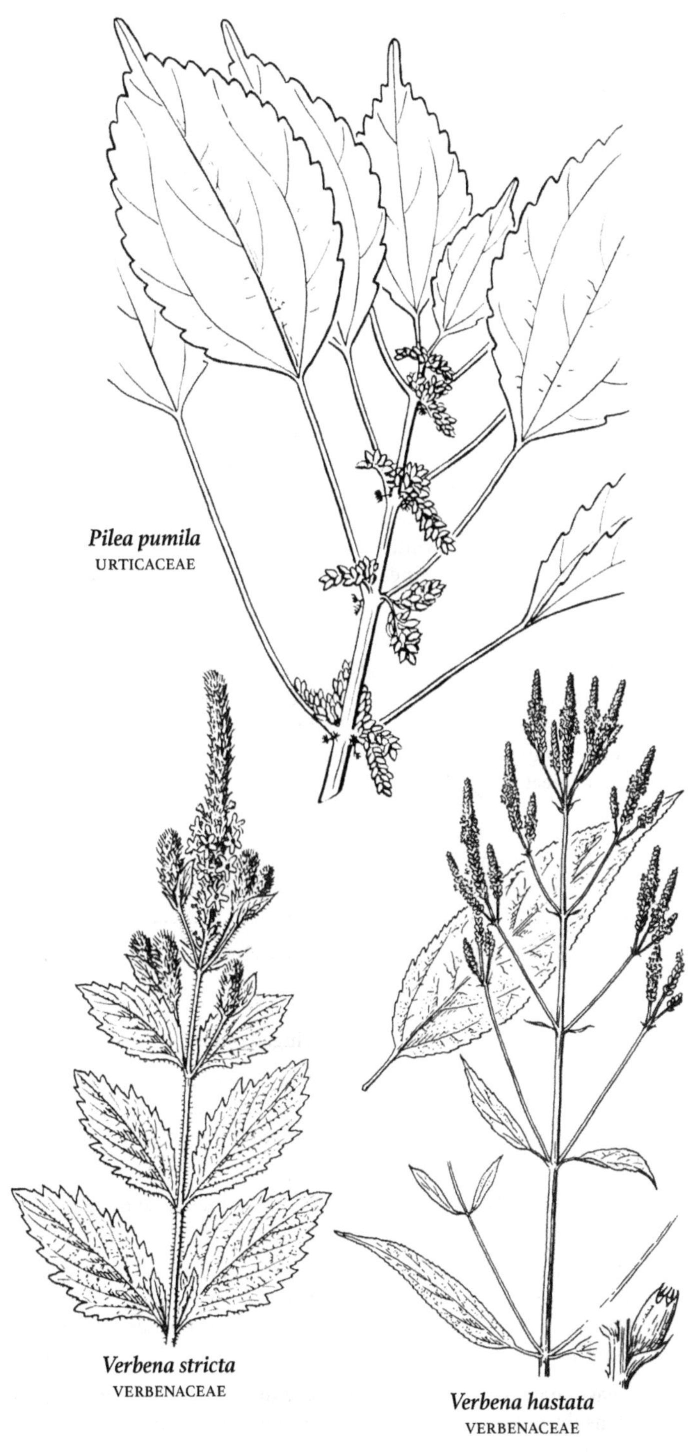
*Pilea pumila*
URTICACEAE
*Verbena stricta*
VERBENACEAE
*Verbena hastata*
VERBENACEAE

# VIOLACEAE *Violet Family*

Herbs. Leaves simple, alternate or rarely opposite. Flowers perfect, 5-merous, usually irregular, axillary or basal, usually nodding. Lower petal usually spurred or larger than the others. Fruit a capsule.

*Key to genera*

1 Leaves all from stem, 10 or more; plants 3–10 dm tall; corolla green or greenish-white .............................. **HYBANTHUS**

1 Leaves either all basal, or if from stem, then numbering less than 10; plants often less than 3 dm tall; corolla not green or greenish-white .................................................. **VIOLA**

## HYBANTHUS *Green-Violet*

***Hybanthus concolor*** (T.F. Forst.) Spreng. EASTERN GREEN-VIOLET Perennial herb. Stems solitary to several in a cluster, leafy, arising from a crown of fibrous roots, more or less pubescent. Leaves alternate, broadly elliptic to ovate-oblong, abruptly acuminate, more or less pubescent. Flowers greenish white, on strongly recurved axillary peduncles. May–June. Rich woods and ravines.

## VIOLA *Violet*

Perennial herbs, with or without leafy stems. Leaves all at base of plant or alternate on stems. Flowers perfect, nodding, and single at ends of stems, with 5 unequal sepals, 2 upper petals, 2 lateral, bearded petals, and 1 lower petal prolonged into a nectar-holding spur at its base. Fruit an ovate capsule which splits to eject the seeds.

*Key to species*

1 Plants with stems; leaves and flowers borne on the upright stems 2

1 Plants without stems; leaves and flowers borne directly from rootstock ....................................................... 9

2 Corolla solid yellow, or white with a yellow center; stipules entire or jagged-tooth on margins .................................. 3

2 Corolla creamy-white to yellow-orange, or lavender to blue, with or without a yellow center; stipules fringed or deeply lobed ....... 4

3 Corolla yellow; stipules ovate, widened above base before tapering to tip .............................................. *V. pubescens*

3 Corolla white with yellow center; stipules long-tapered from base to tip ............................................ *V. canadensis*

4 Stipules deeply lobed near base into long oblong segments ..... 5

4 Stipules fringed with short, slender segments.................. 6

5 Petals shorter than or about equal to the sepals; the 5 petals cream-colored on upper half, all about same length .......... *V. arvensis*

5 Petals longer than the sepals; the upper pair of petals dark blue or purple on upper half, and longer than the other 3 petals *V. tricolor*

6 Leaves narrowly ovate to triangular, tapered to a rounded tip, often densely pubescent, the hairs tiny; margins entire or nearly so; corolla dark blue ..................................... *V. adunca*

6 Leaves ovate to kidney-shaped, glabrous or only slightly hairy; the margins with rounded or sharp teeth; corolla creamy-white, lavender or light blue .......................................... 7

7 Corolla solid creamy-white; leaf margins regularly toothed with small rounded teeth ................................. *V. striata*

7 Corolla light blue or lavender; leaf margins not regularly rounded-toothed .................................................. 8

8 Corolla solid light blue; lateral petals bearded on inner surface; stem leaves broadly ovate or kidney-shaped; margins with low, rounded teeth ................................................ ............................................ *V. labradorica*

8 Corolla light blue to lavender with a darker purple spot; lateral petals not bearded; stem leaves ovate to oblong-ovate; margins with scattered sharp teeth ................................ *V. rostrata*

9 Style tipped by a slender, recurved hook; stolons green and cord-like; introduced species mostly of lawns and parks .... *V. odorata*

9 Style slightly expanded into a spatula-like tip, not hooked; stolons slender and pale or absent; native............................ 10

10 Corolla blue; petals not bearded within; spur more than 2 times longer than wide .................................. *V. selkirkii*

10 Corolla white or purple; petals sometimes bearded within; spur less than 2 times longer than wide................................ 11

11 Corolla white ................................................ 12

11 Corolla purple; lateral petals bearded within; stolons absent ... 16

12 Leaf blades more than 1.5 times longer than wide.............. 13

12 Leaf blades often wider than long............................ 14

13 Leaves lance-shaped, tapered to a narrow base ...... *V. lanceolata*

13 Leaves broader, ovate and narrowly heart-shaped at base ........ ............................................ *V. primulifolia*

14 Leaves dull, upper and lower surface without hairs, lower surface not paler than upper; margins nearly entire or with low rounded teeth; petioles often with long, soft hairs ........... *V. macloskeyi*

14 Leaves shiny and smooth on upper surface, or dull and hairy on either upper or lower surface; underside paler than upper surface; margins with sharp teeth...................................... 15

15 Plants with stolons and horizontal rhizomes; upper and lower surface of leaves sparsely to densely hairy with short hairs less than 1 mm long .............................................. *V. blanda*

15 Plants without stolons, rhizomes turned upright; leaves often shiny and smooth on upper surface, or densely hairy on upperside with hairs about 1–2 mm long and smooth on underside .... *V. renifolia*

16 Leaf blades lobed or divided................................. 17

16 Leaf blades toothed on margin, not lobed or divided........... 19

17 Leaf blades deeply divided into slender linear segments ......... ............................................ *V. pedatifida*

17 Leaf blades less deeply lobed or divided, the segments triangular to ovate .................................................. 18

18 Leaf blades much longer than wide; the spurred petal densely bearded within .......................................... *V. sagittata*

18 Leaf blades about as long as wide; the spurred petal glabrous within or with only a few hairs .................................... *V. pedata*

19 Leaf blades distinctly longer than wide........................ **20**

19 Leaf blades as wide as or wider than long...................... **23**

20 Plants of wet places; leaves glabrous or nearly so; sepal margins not fringed with hairs.................................................. **21**

20 Plants of dry or rocky habitats; leaves sparsely to densely hairy; sepal margins usually fringed with hairs...................... **22**

21 Lateral petals with long, threadlike hairs on inner surface; spurred petal densely hairy within .......................... *V. affinis*

21 Lateral petals with short, knob-tipped hairs on inner surface; spurred petal without hairs ........................ *V. cucullata*

22 Hairs on leaves over 1 mm long; sepals ovate, rounded at tip ..... ...................................... *V. novae-angliae*

22 Hairs on leaves shorter, less than 1 mm long; sepals long-tapered to a sharp tip ......................................... *V. sagittata*

23 Sepals long-tapered to a sharp tip; lateral petals with short, knob-tipped hairs on inner surface; spurred petal without hairs ....... ...................................................... *V. cucullata*

23 Sepals oblong to broadly lance-shaped, rounded at tip; lateral petals with long, threadlike hairs on inner surface................... **24**

24 Flowers held above the leaves; leaves and stems without hairs, leaves rounded at tip, margins with rounded teeth; the spurred petal densely hairy within; plants of wetlands ......... *V. nephrophylla*

24 Flowers overtopped by leaves; leaves and stems usually hairy, leaves tapered to a pointed tip, margins with sharp, forward-pointing teeth; spurred petal glabrous to only slightly hairy within; plants of moist forests ............................................. *V. sororia*

***Viola adunca*** Sm. HOOK-SPURRED VIOLET Dry sandy open places, often with jack pine and oaks; crevices in rock outcrops.

***Viola affinis*** Le Conte SAND VIOLET April–May. Swamps, floodplain forests, streambanks and lakeshores, low prairie.

***Viola arvensis*** Murr. EUROPEAN FIELD-PANSY Native of Europe; mostly in cultivated or abandoned fields.

***Viola blanda*** Willd. SWEET WHITE VIOLET April–May. Hummocks in swamps and bogs, low wet areas in deciduous and conifer forests.

***Viola canadensis*** L. [•101] TALL WHITE VIOLET Mesic deciduous woods.

***Viola cucullata*** Ait. MARSH BLUE VIOLET April–June. Swamps, sedge meadows, shady seeps; occasionally in bogs and low areas in forests.

***Viola labradorica*** Schrank ALPINE VIOLET April–June. Swamps, streambanks, moist hardwood forests. *Viola adunca* Sm. var. *minor* (Hook.) Fern.

***Viola lanceolata*** L. [•101] STRAP-LEAF VIOLET April–June. Open bogs, sedge meadows; soils sandy or mucky. The lance-shaped leaves are distinctive.

***Viola macloskeyi*** Lloyd WILD WHITE VIOLET April–July. Marshes, sedge meadows, open bogs and swamps, alder thickets; sometimes in shallow water. *Viola pallens* (Banks) Brainerd.

***Viola nephrophylla*** Greene NORTHERN BOG VIOLET May, sometimes again flowering in Aug or Sept. Wet meadows, calcareous fens, low areas between dunes, streambanks, rocky shores.

***Viola novae-angliae*** House NEW ENGLAND BLUE VIOLET Gravelly and sandy shores and in rock crevices along streams. *Viola sororia* var. *novae-angliae* (House) McKinney.

***Viola odorata*** L. SWEET BLUE VIOLET Native of Europe; commonly cultivated in many forms (both single-flowered and double-flowered), and sometimes escaped from gardens to waste places or woods.

***Viola pedata*** L. BIRD-FOOT VIOLET Dry fields and sandy open woods of oak and jack pine. Leaves and flowers variable in shape and color.

***Viola pedatifida*** G. Don CROW-FOOT VIOLET Prairies. *Viola palmata* var. *pedatifida* (G. Don) Cronq.

***Viola primulifolia*** L. PRIMROSE-LEAF VIOLET May. Wet meadows and bogs, often in sphagnum moss; sandy streambanks; soil sandy or peaty, acidic.

***Viola pubescens*** Ait. [•101] YELLOW FOREST VIOLET Mesic woods.

***Viola renifolia*** Gray KIDNEY-LEAF WHITE Violet May–July. Cedar swamps, sphagnum hummocks in peatlands.

***Viola rostrata*** Pursh LONG-SPUR VIOLET Shady slopes and moist woods, usually in deep humus.

***Viola sagittata*** Ait. [•101] ARROWHEAD VIOLET *Endangered.* April–June. Open, dry pine and oak woods, usually where sandy. Var. *ovata* (Nutt.) Torr. & Gray, found mostly in c Wisc, and more pubescent than the typical variety, is listed as state endangered.

***Viola selkirkii*** Pursh GREAT-SPUR VIOLET Deciduous woods and shady ravines; preferring calcareous soils.

***Viola sororia*** Willd. [•p. 432] HOODED BLUE VIOLET April–June. Moist hardwood forests; occasionally in swamps, floodplain forests and along rocky streambanks. Wisc state flower.

***Viola striata*** Ait. STRIPED CREAM VIOLET April–June. Floodplain forests, moist deciduous woods, streambanks, thickets. Distinguished from the more

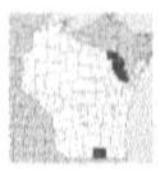

common white-flowered *V. canadensis* by the lack of blue color on the outside of the petals (normally conspicuous in *V. canadensis*), and the green, fringed stipules (rather than scarious, entire ones).

***Viola tricolor*** L. JOHNNY-JUMP-UP Native of the Old World; escaped from gardens in waste places, only rarely persistent. The cultivated pansy is the product of long cultivation and hybridization of *V. tricolor* with several allied European species.

---

## VITACEAE *Grape Family*

Mostly woody vines, climbing by tendrils. Leaves alternate, simple or compound; tendrils and flower clusters produced opposite the leaves. Flowers regular, 4–5-merous, perfect or unisexual. Fruit a berry.

ADDITIONAL SPECIES ***Ampelopsis glandulosa*** (Wallich.) Momiy., an Asian introduction, is an escape in southern Wisc; plants are vines with prominently 3-lobed leaves.

*Key to genera*

1 Stems brown-pithy inside, the bark shredding into strips; leaves simple . . . . . . . . VITIS
1 Stems white-pithy, the bark tight, not shredding; leaves simple or palmately compound . . . . . . . . 2

2 Leaves palmately compound, leaflets 5 . . . . . PARTHENOCISSUS
2 Leaves simple, the margins shallowly to deeply 3–5-lobed . . . . . . . . AMPELOPSIS

### PARTHENOCISSUS *Creeper*

Woody vines, trailing or climbing by tendrils. Leaves palmately compound with typically 5 leaflets (ours). Flowers in panicles borne opposite the leaves or aggregated into terminal clusters.

*Key to species*

1 Plants often climbing trees by means of adhesive disks on the tendrils; inflorescence with central axis . . . . . . . . . . . ***P. quinquefolia***
1 Plants not climbing, without adhesive disks on tendrils; inflorescence branched, without an evident central axis . ***P. inserta***

***Parthenocissus inserta*** (Kerner) Fritsch THICKET-CREEPER June. Moist soil. Very similar to *P. quinquefolia* in habit and foliage. *Parthenocissus vitacea* (Knerr) A.S. Hitchc.

***Parthenocissus quinquefolia*** (L.) Planch. [•102] VIRGINIA-CREEPER June. Moist soil, thickets, swamps.

### VITIS *Grape*

Woody vines, climbing by tendrils. Flowers actually or functionally unisexual, 5-merous. Calyx essentially none. Petals cohering at the tip, separating at the base, falling early. Fruit a juicy berry. Our species bloom in May or June.

*Key to species*

1 Leaf underside with felt-like covering of white or rust-colored hairs ........ ***V. labrusca***

1 Leaf underside glabrous or white hairy, the hairs not felt-like and soon disappearing or only remaining in small patches as leaves mature ........ **2**

2 Leaves white hairy on underside when young, becoming smooth and strongly blue-green waxy when fully developed, with persistent hairs on the prominent veins ........ ***V. aestivalis***

2 Leaves smooth and greenish on underside when fully developed, not waxy; often with persistent tufts of hairs in leaf axils ***V. riparia***

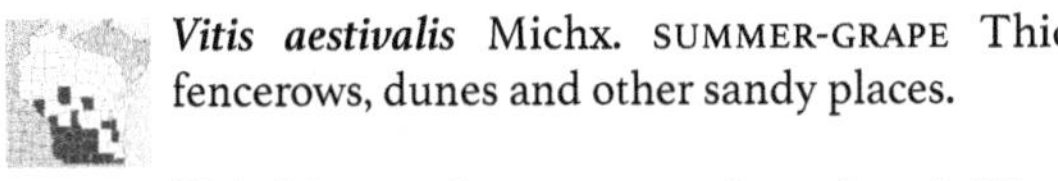

***Vitis aestivalis*** Michx. SUMMER-GRAPE Thickets, dry forests, fencerows, dunes and other sandy places.

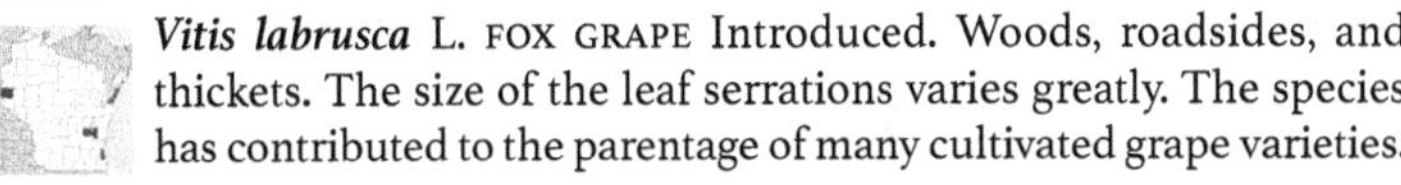

***Vitis labrusca*** L. FOX GRAPE Introduced. Woods, roadsides, and thickets. The size of the leaf serrations varies greatly. The species has contributed to the parentage of many cultivated grape varieties.

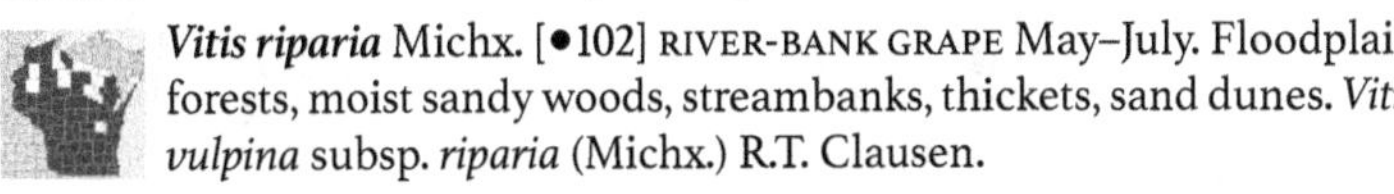

***Vitis riparia*** Michx. [•102] RIVER-BANK GRAPE May–July. Floodplain forests, moist sandy woods, streambanks, thickets, sand dunes. *Vitis vulpina* subsp. *riparia* (Michx.) R.T. Clausen.

## ZYGOPHYLLACEAE *Creosote-Bush Family*

### TRIBULUS *Puncturevine*

***Tribulus terrestris*** L. PUNCTUREVINE Recognized by its prostrate habit, hairy stems and leaves, the leaves usually with 6–7 pairs of small leaflets, and the spiny nutlets. All summer. Native of Europe; dry or sandy soil, along railroads and highways, in dry lawns, parking lots, and disturbed places.

***Viola sororia***
VIOLACEAE

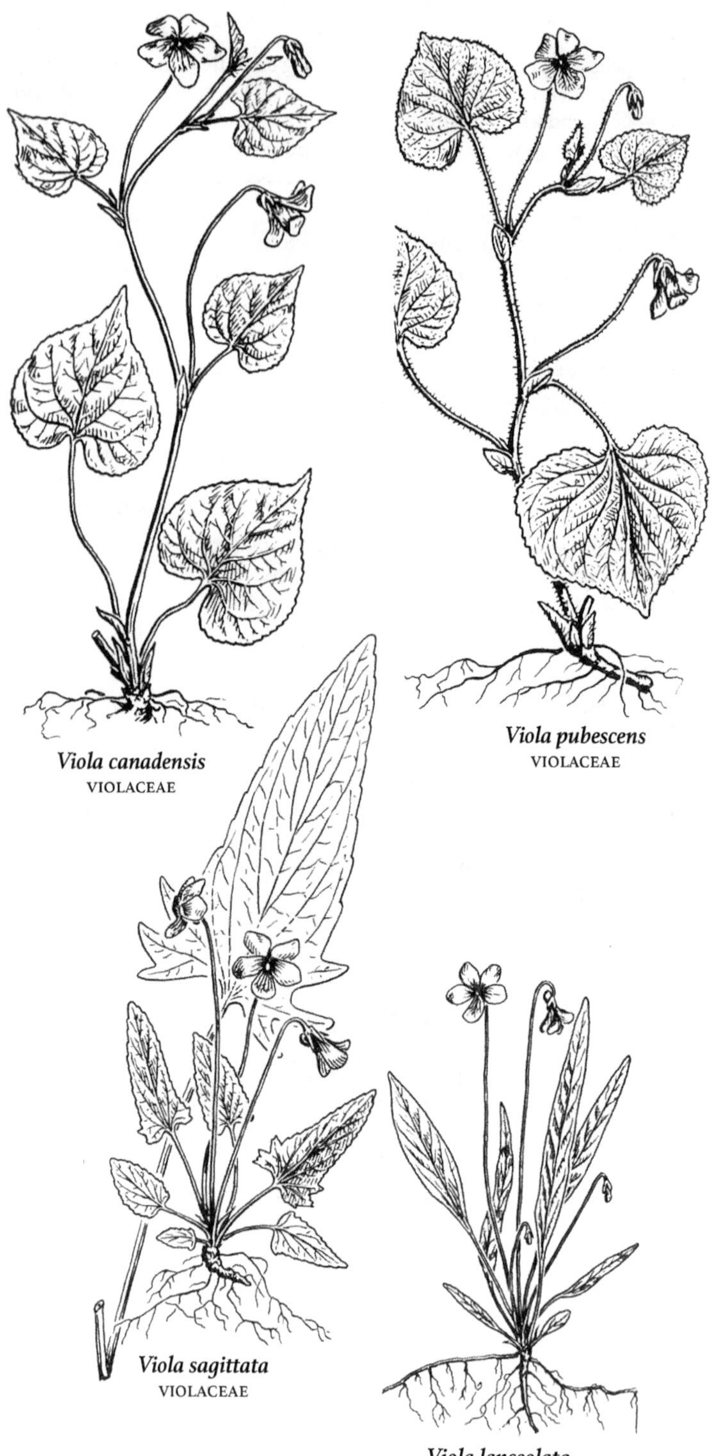

*Viola canadensis*
VIOLACEAE

*Viola pubescens*
VIOLACEAE

*Viola sagittata*
VIOLACEAE

*Viola lanceolata*
VIOLACEAE

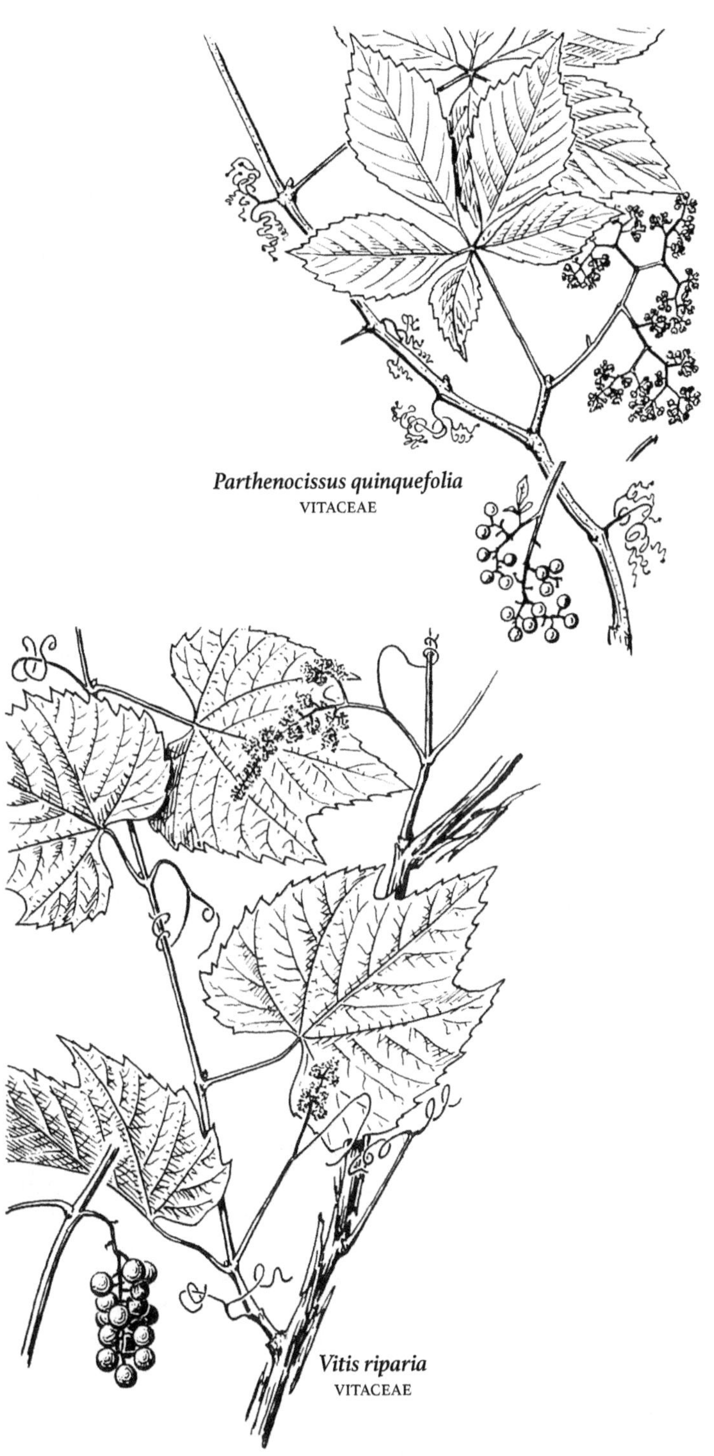
*Parthenocissus quinquefolia*
VITACEAE
*Vitis riparia*
VITACEAE

# ACORACEAE *Calamus Family*

## ACORUS *Sweetflag; Calamus*

Perennial herbs of wetlands; rhizomes and leaves pleasantly scented. Leaves sword-shaped, bright green, with 1–6 prominent veins parallel along length of leaf. Inflorescence a solitary spadix, borne from near midway of leaf; true spathe absent. Flowers bisexual; tepals 6, light brown. Fruit light brown to reddish berry with darker streaks. Seeds embedded in a mucilagenous jelly.

*Key to species*

1 Leaves lacking a single prominent raised midvein, but with 2–several clearly separate veins of ± equal strength along with numerous fainter veins, the leaf thus not obviously ridged to the naked eye; fruit maturing........................ ***A. americanus***

1 Leaves with a single prominent raised midvein (best observed at about mid-portion of the leaf) that is much more conspicuous than any other vein, appearing as a ridge to the naked eye; fruit not maturing ........................................... ***A. calamus***

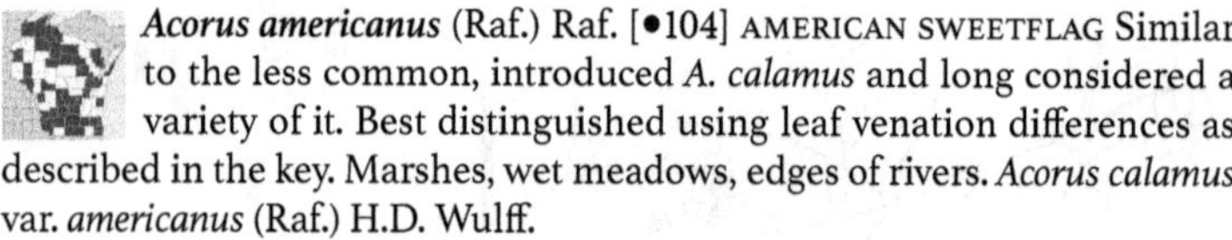

***Acorus americanus*** (Raf.) Raf. [•104] AMERICAN SWEETFLAG Similar to the less common, introduced *A. calamus* and long considered a variety of it. Best distinguished using leaf venation differences as described in the key. Marshes, wet meadows, edges of rivers. *Acorus calamus* var. *americanus* (Raf.) H.D. Wulff.

***Acorus calamus*** L. SWEETFLAG Introduced (naturalized). June–July. Marshes (often with cat-tails), bogs, streambanks.

# ALISMATACEAE *Water-Plantain Family*

Perennial, aquatic or emergent herbs; plants swollen and tuberlike at base. Leaves all from base of plant and clasping an erect stem; underwater leaves often ribbonlike; emergent leaves broader. Flowers perfect or imperfect, in racemes or panicles at ends of stems, with 3 sepals and 3 petals. Fruit a compressed achene, usually tipped by the persistent style.

*Key to genera*

1 Leaves often arrowhead-shaped; pistils or achenes in several series around a large, round receptacle, and forming a dense, round head .............................................. SAGITTARIA

1 Leaves never arrowhead-shaped; achenes in a single whorl on a small, flat receptacle or few in a small loose head.............. 2

2 Plants small, less than 10 cm high; achenes about 1 mm long, plump ........................................... ECHINODORUS

2 Plants larger, more than 10 cm high; achenes larger, flattened .... ................................................. ALISMA

## ALISMA *Water-Plantain*

Perennial herbs, from cormlike rootstocks. Leaves emersed or floating, ovate to lance-shaped, never arrowhead-shaped; underwater leaves sometimes ribbonlike (in *A. gramineum*).

*Key to species*

1 Leaves lance-shaped to oval, or if underwater, leaves long and ribbonlike; flower stalks rarely much longer than leaves; petals somewhat pink-tinged .......................... ***A. gramineum***

1 Leaves ovate, rounded to truncate or heart-shaped at base; flower stalks usually much longer than leaves; petals white........... **2**

2 Flowers larger; petals about 4 mm long and to 4 mm wide ....... ....................................................... ***A. triviale***

2 Flowers smaller; petals to 2.5 mm long and 2 mm wide .......... ............................................... ***A. subcordatum***

***Alisma gramineum*** Lej. NARROW-LEAF WATER-PLANTAIN July–Sept. Shallow, often brackish water, muddy shores, streambanks; in Wisc, near Mississippi River. *Alisma geyeri* Torr.

***Alisma subcordatum*** Raf. [•103] AMERICAN WATER-PLANTAIN July–Sept. Shallow water marshes, shores, ditches. *Alisma plantago-aquatica* L. var. *parviflorum* (Pursh) Torr.

***Alisma triviale*** Pursh NORTHERN WATER-PLANTAIN June–Sept. Marshes, ponds, and streams. *Alisma brevipes* Greene, *Alisma plantago-aquatica* L. var. *americanum* J.A. Schultes.

## ECHINODORUS *Burhead*

***Echinodorus berteroi*** (Spreng.) Fassett UPRIGHT BURHEAD Aquatic perennial; scape erect, simple or branched, bearing an inflorescence of 1–many verticils of flowers. Leaves broadly ovate to lance-shaped, commonly cordate at base. Flowers perfect, in whorls of 3–8; petals 3, white. Swamps and ditches.

## SAGITTARIA *Arrowhead*

Perennial or annual herbs, with fleshy or tuberous rootstocks. Leaves sheathing, all from base of plant, variable in shape and size. Emersed and floating leaves usually arrowhead-shaped with large lobes at base, or sometimes ovate to oval and without lobes; underwater leaves often linear in a basal rosette, normally absent by flowering time. Flowers in a raceme of mostly 3-flowered whorls; upper flowers usually staminate, lower flowers usually pistillate or sometimes perfect. Fruit a crowded cluster of achenes in more or less round heads, the achenes flattened and winged, beaked with a persistent style.

*Key to species*

1 Emersed leaves not arrowhead-shaped, basal lobes absent...... **2**

1 Emersed leaves all or mostly arrowhead-shaped, with large basal lobes ................................................... **3**

2 Pistillate flowers and fruiting heads sessile or nearly so .. ***S. rigida***

2 Pistillate flowers and fruiting heads obviously stalked. ***S. graminea***

3 Plants annual, rhizomes absent; sepals appressed to fruiting heads; stalks of fruiting heads stout .................... *S. montevidensis*

3 Plants perennial, with rhizomes; sepals reflexed on fruiting heads; stalks of fruiting heads slender ............................. 4

4 Bracts below flowers mostly less than 1 cm long; achene beak projecting horizontally from tip of achene ............ *S. latifolia*

4 Bracts below flowers usually more than 1 cm long; achene beak erect or ascending ............................................ 5

5 Achene beak short, erect, to 0.4 mm long; basal lobes of leaves mostly shorter than terminal lobe .................... *S. cuneata*

5 Achene beak larger, curved and ascending, 0.5 mm or more long; basal lobes of leaves usually equal or longer than terminal lobe . ................................................ *S. brevirostra*

***Sagittaria brevirostra*** Mackenzie & Bush SHORT-BEAK ARROWHEAD July–Sept. Shallow water and muddy shores. *Sagittaria engelmanniana* subsp. *brevirostra* (Mack. & Bush) Bogin.

***Sagittaria cuneata*** Sheldon ARUM-LEAF ARROWHEAD June–Sept. Shallow water, lakeshores and streambanks.

***Sagittaria graminea*** Michx. [•103] GRASS-LEAF ARROWHEAD June–Sept. Shallow water and shores. *S. graminea* includes *S. graminea* Michx. var. *cristata* (Engelm.) Bogin, sometimes treated as a separate species, *S. cristata* Engelm., and distinguished as follows:

1 Achenes with beak 0.4–0.6 mm long; anthers clearly shorter than the filaments ........................................ *S. cristata*

1 Achenes with beak minute, scarcely discernable, ca. 0.2 mm long; anthers as long as or longer than filaments ........... *S. graminea*

***Sagittaria latifolia*** Willd. DUCK-POTATO July–Sept. Shallow water, shores, marshes and pools in bogs.

***Sagittaria montevidensis*** Cham. & Schltdl. MISSISSIPPI ARROWHEAD July–Sept. Muddy shores. *Sagittaria calycina* Engelm.

***Sagittaria rigida*** Pursh SESSILE-FRUIT ARROWHEAD June–Sept. Shallow water, shores and streambanks.

## ARACEAE *Arum Family*

Perennial herbs with alternate, simple or compound, often fleshy leaves. Flowers small and numerous, mostly single-sexed, staminate flowers usually above pistillate, crowded in a cylindric or rounded spadix subtended by a leaflike spathe. Fruit a usually fleshy berry, containing 1 to few seeds, or the entire spadix ripening as a fruit. Now included in the Araceae are the duckweeds (*Lemna, Spirodela,* and *Wolffia*), aquatic genera formerly treated as their own family (Lemnaceae). These are small perennial herbs, floating at or near water surface, single or forming colonies. Plants thallus-like (not differentiated into stems and leaves).

*Key to genera*

1 Plants tiny floating or submerged aquatic species less than 1 mm long, without differentiation into leaves or stems .............. 2

1 Plants large, with clearly differentiated normal leaves and with rhizomes or tubers .......................................... 4

2 Roots absent; leaves thickened, less than 1.5 mm long . **WOLFFIA**

2 One to several roots present on leaf underside; leaves flat, mostly more than 1.5 mm long ...................................... 3

3 Each leaf with 1 root; leaf underside green or purple-tinged ...... ....................................................... **LEMNA**

3 Each leaf with 3 or more roots; leaf underside solid purple ....... .................................................. **SPIRODELA**

4 Leaves compound ................................ **ARISAEMA**

4 Leaves simple, not divided .................................. 5

5 Leaves arrowhead-shaped, lobes equaling 1/3 or more length of blade ........................................... **PELTANDRA**

5 Leaves heart-shaped or rounded at base....................... 6

6 Leaves broadly heart-shaped, abruptly tapered to a tip; spathe white, long-stalked; flowering in spring ................. **CALLA**

6 Leaves rounded ovate, tapered to a rounded tip; spathe green-yellow to purple-brown, short-stalked or stalkless; flowering in late winter to early spring ............................ **SYMPLOCARPUS**

## ARISAEMA *Jack-in-the-Pulpit*

Perennial herbs. Leaves compound. Flowers either staminate or pistillate, on same or different plants; staminate flowers with 2–5, more or less stalkless stamens, above the pistillate flowers on a fleshy spadix, the spadix subtended by a green or purple-brown spathe. Fruit a cluster of round, red berries.

*Key to species*

1 Leaflets 7–13; spadix longer than spathe ............ ***A. dracontium***

1 Leaflets usually 3; spathe arching over spadix ....... ***A. triphyllum***

***Arisaema dracontium*** (L.) Schott [•104] GREENDRAGON May–July. Wet woods and floodplain forests.

***Arisaema triphyllum*** (L.) Schott [•104] JACK-IN-THE-PULPIT April–July. Moist forests, cedar swamps.

## CALLA *Water-Arum*

***Calla palustris*** L. WATER-ARUM Perennial herb, from thick rhizomes, the rhizomes creeping in mud or floating in water. Leaves broadly heart-shaped, abruptly tapered to a tip. Flowers perfect or the uppermost staminate, on a short-cylindric spadix, 1.5–3 cm long, shorter than the spathe; the spathe white, ovate, tipped with a short, sharp point. Fruit a fleshy, few-seeded berry, turning red when ripe. May–July. Bog pools, swamps, shores and wet ditches.

## LEMNA *Duckweed*

Small perennial floating herbs, with 1 root per frond (or roots sometimes absent on oldest and youngest leaves). Blades single or 2 to several and joined in small colonies, floating on water surface or underwater (*L. trisulca*), varying from round, ovate, to obovate or oblong, tapered to a long point (petiole) in *L. trisulca;* green or often red-tinged. Reproduction mostly by budding of new leaves from the reproductive pouches.

*Key to species*

1 Fronds denticulate toward tip, tapered to a slender stipitate base, the stipe often as long as the main body and commonly attached to the parent frond; colonies star-shaped, usually submersed .. ***L. trisulca***

1 Fronds entire on the margin, nearly rounded and not obviously stipitate at the base, solitary or in tight colonies, these not star-shaped, floating on the water surface or stranded on mud ...... **2**

2 Root sheath winged at the base; root tip sharply pointed; roots not longer than 3 cm; fronds completely green .................... **3**

2 Root sheath not winged at the base; root tip mostly rounded; roots often longer than 3 cm; fronds often red-tinged beneath or with red spots on either surface ........................................ **4**

3 Fronds very often with 2-3 papillae in a row on the upper surface above the node (the level at which daughter fronds attach); seeds whitish, with 35-60 faint ribs, not escaping the fruit wall when ripening ........................................ ***L. perpusilla***

3 Fronds with only 1 prominent papilla above the node; seeds brownish, with 8-22 prominent ribs, falling out of the fruit wall when ripening ................................ ***L. aequinoctialis***

4 Fronds with several about equal sized, small papillae on the upper surface from the midline to the tip (often obscure), very often red-tinged on the lower surface, forming small, obovate to orbicular, rootless, dark green to brown turions under unfavorable conditions, these sinking to the bottom of the water ***L. turionifera***

4 Fronds lacking papillae or with one prominent papilla at the tip and another just above the node and with smaller papillae between them, rarely forming turions; if formed, the turionlike fronds have short roots and are slowly forming daughter fronds ............ **5**

5 Papilla at tip of the frond very prominent; fronds often red beneath ................................................ ***L. obscura***

5 Papilla at tip of the frond not very prominent; fronds never red beneath ............................................ ***L. minor***

***Lemna aequinoctialis*** Welw. LESSER DUCKWEED Very similar to *L. perpusilla* except with only 1 prominent papilla above the node on the upper surface. Main range s USA; in Wisc, known from Grant County (*no map*).

***Lemna minor*** L. [•104]] COMMON DUCKWEED July–Sept. Quiet or stagnant water of ponds, oxbows, shores, slow-moving rivers, ditches.

***Lemna obscura*** (Austin) Daubs LITTLE DUCKWEED Much less common than *L. minor,* known from several locations in w Wisc.

***Lemna perpusilla*** Torr. MINUTE DUCKWEED Quiet water of ponds and ditches.

***Lemna trisulca*** L. IVY-LEAF DUCKWEED Ponds, streams, ditches.

***Lemna turionifera*** Landolt TURION DUCKWEED Quiet water of ponds and lakes.

## PELTANDRA *Arrow-Arum*

***Peltandra virginica*** (L.) Schott GREEN ARROW-ARUM Perennial herb, with thick, fibrous roots. Leaves all from base of plant on long petioles, bright green, oblong to triangular in outline; leaf base with a pair of lobes. Flowers in a white to orange spadix about as long as the spathe, atop a curved stalk. Fruit a head of green-brown berries, the berries with 1–3 seeds surrounded by a jellylike material. June–July. Shallow water, shores, bog pools; often where shaded.

## SPIRODELA *Greater Duckweed*

***Spirodela polyrrhiza*** (L.) Schleid. [•104]] GREATER DUCKWEED Perennial herb, floating on water surface; roots 5–12 per frond. Fronds usually in clusters of 2–5, flat, round to obovate, upper surface green, underside red-purple. Stagnant or slow-moving water of lakes, ponds, marshes and ditches, often with *Lemna.*

## SYMPLOCARPUS *Skunk-Cabbage*

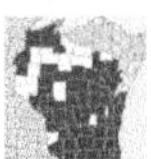

***Symplocarpus foetidus*** (L.) Salisb. [•104]] SKUNK-CABBAGE Perennial, foul-smelling herb, from thick rootstocks. Leaves all from base of plant, ovate to heart-shaped. Flowers appearing before leaves in late winter or early spring, perfect; the spathe ovate, curved over spadix, green-purple and often mottled. Feb–May (our earliest flowering native plant). Floodplain forests, swamps, streambanks, calcareous fens, moist wooded slopes.

## WOLFFIA *Watermeal*

Tiny perennial herbs, floating at or just below water surface, sometimes abundant and forming a granular scum across surface, usually mixed with other aquatic species of this family, roots absent. Leaves single or often paired, globe-shaped or ovate, flat or rounded on upper surface. Flowers uncommon, consisting of 1 stamen (staminate flower) and 1 pistil (pistillate flower) in the pouch. Watermeal is the world's smallest flowering plant. The blades feel granular or mealy and tend to stick to the skin. The 3 species in Wisc often occur together.

*Key to species*

1 Leaves rounded on upper surface, not brown-dotted ........ ***W. columbiana***

1 Leaves flattened on upper surface, brown-dotted (under 10x magnification) ........ **2**

2 Leaves rounded at tip, with a wartlike bump in center of upper surface ........ ***W. brasiliensis***

2 Leaves with an upturned point at tip, wartlike bump absent ...... .......... *W. borealis*

***Wolffia borealis*** (Engelm.) Landolt [•104] NORTHERN WATERMEAL Quiet water of ponds, marshes and ditches, often with other species of *Wolffia* and *Lemna*. *Wolffia punctata* auct. non Griseb.

***Wolffia brasiliensis*** Weddell BRAZILIAN WATERMEAL Quiet ponds and ditches. *Wolffia papulifera* C.H. Thompson, *Wolffia punctata* Griseb.

***Wolffia columbiana*** Karst. COLUMBIAN WATERMEAL Stagnant water of ponds and marshes.

## BUTOMACEAE *Flowering Rush Family*

### BUTOMUS *Flowering-Rush*

***Butomus umbellatus*** L. FLOWERING-RUSH Introduced (invasive) Perennial herb, from creeping rhizomes. Leaves all from base of plant, erect when emersed, or floating when in deep water, linear, to 1 m long, parallel-veined. Flowers pink, perfect, in a many-flowered umbel, borne on a stalk 1–1.2 m tall; sepals 3, petal-like; petals 3. Fruit a dry capsule. June–Aug. Marshes and shores.

## COMMELINACEAE *Spiderwort Family*

Herbs, often succulent. Leaves parallel-veined, dilated at base into a tubular sheath. Flowers cymose or rarely solitary, from the axils of foliaceous bracts or spathes; 3-merous.

*Key to genera*

1 Corolla irregular, with 2 blue petals and 1 smaller petal; inflorescence subtended by a wide bract less than 3 cm long...... .......... **COMMELINA**

1 Corolla regular; the petals all alike; inflorescence subtended by long, narrow, leaflike bracts 4–20 cm long ..... **TRADESCANTIA**

### COMMELINA *Day-Flower*

Annual or perennial herbs with succulent stems. Leaves alternate, linear to lance-shaped. Inflorescence a small cyme, closely subtended by a folded, heart-shaped spathe from which the pedicels protrude; flowers attractive but short-lived. Our species bloom in summer and early autumn.

*Key to species*

1 Plants erect; leaf blades lance-shaped or linear, 7–10 times longer than wide .......... *C. erecta*

1 Plants prostrate or creeping; leaf blades ovate, 2–4 times longer than wide .......... *C. communis*

***Commelina communis*** L. COMMON DAY-FLOWER Introduced (naturalized). Moist or shaded ground, often a weed in gardens.

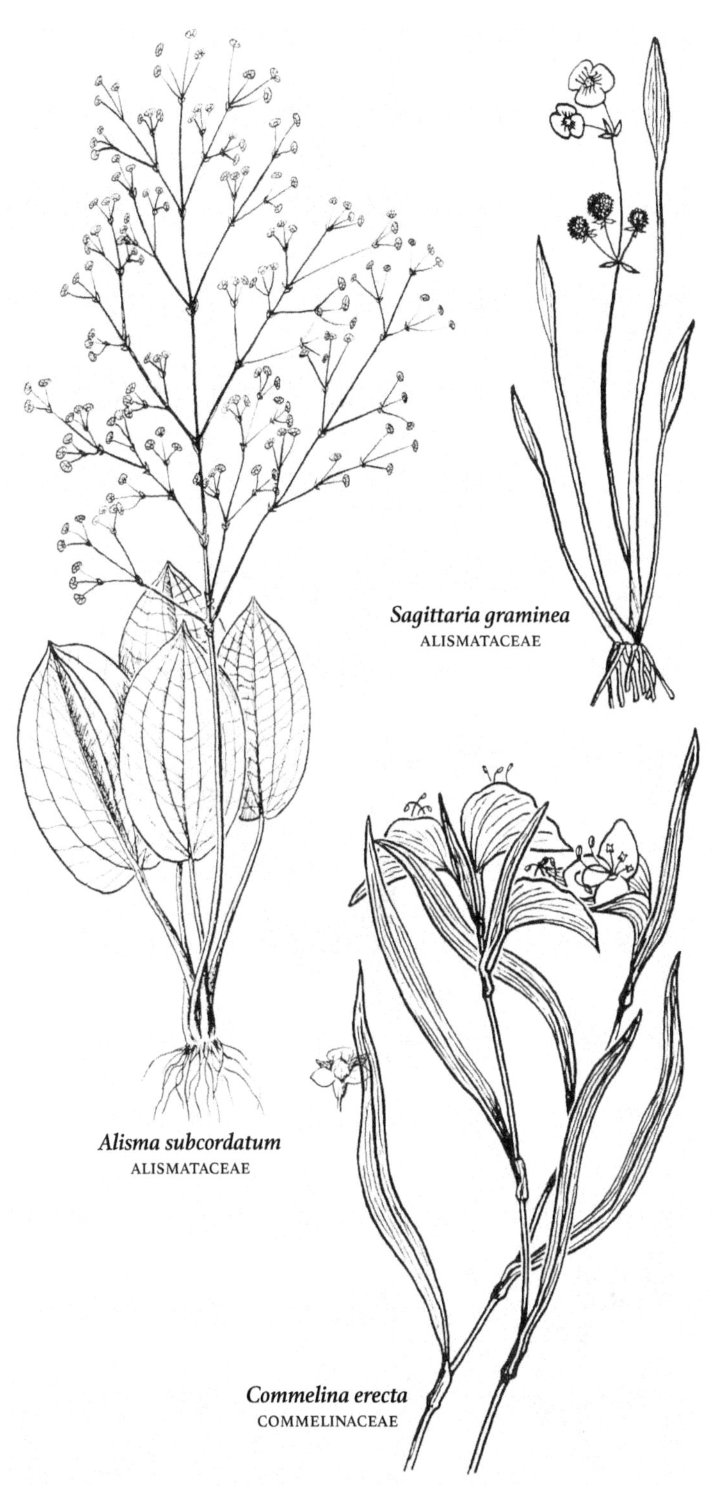
*Sagittaria graminea*
ALISMATACEAE
*Alisma subcordatum*
ALISMATACEAE
*Commelina erecta*
COMMELINACEAE

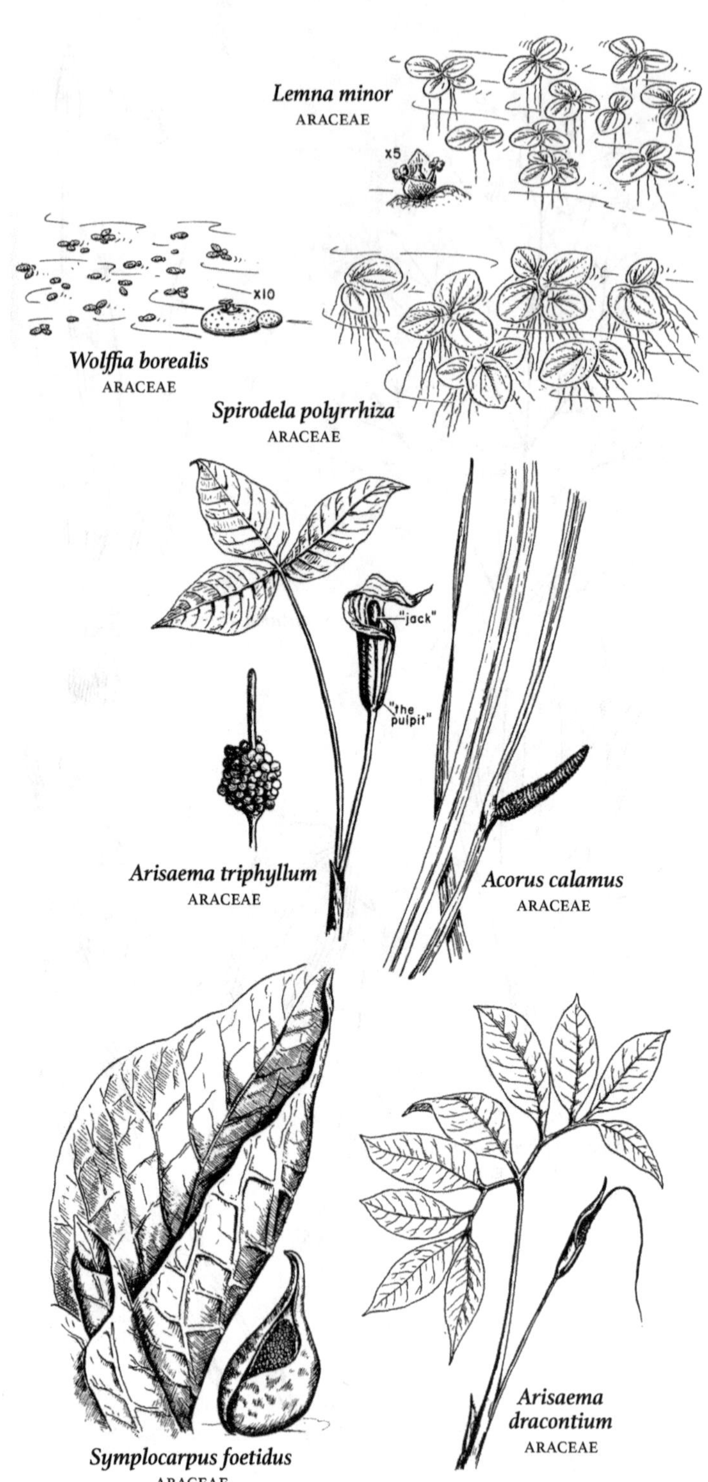

*Lemna minor*
ARACEAE

*Wolffia borealis*
ARACEAE

*Spirodela polyrrhiza*
ARACEAE

*Arisaema triphyllum*
ARACEAE

*Acorus calamus*
ARACEAE

*Symplocarpus foetidus*
ARACEAE

*Arisaema dracontium*
ARACEAE

***Commelina erecta*** L. [•103] ERECT DAY-FLOWER Dry, usually sandy soil. Variable in the size of the leaves.

### TRADESCANTIA *Spiderwort*

Perennial herbs, usually somewhat succulent. Leaves alternate, elongate, linear to lance-shaped, dilated into conspicuous basal sheaths. Flowers in umbel-like cymes of several to many flowers subtended by elongate foliaceous bracts.

*Key to species*

1 Sepals and pedicels glabrous or with only a few hairs at either tip or base of the sepal .................................. ***T. ohiensis***

1 Sepals and pedicels distinctly pubescent ...................... **2**

2 Hairs not glandular .............................. ***T. virginiana***

2 Hairs mostly glandular-sticky ................................ **3**

3 Sepals and pedicels densely soft-hairy, with both gland-tipped and nonglandular hairs, the hairs often more than 1 mm long; sepals 10–15 mm long ...................................... ***T. bracteata***

3 Sepals and pedicels sparsely hairy, the hairs all gland-tipped and shorter, to only 0.5 mm long; sepals to 10 mm long . . ***T. occidentalis***

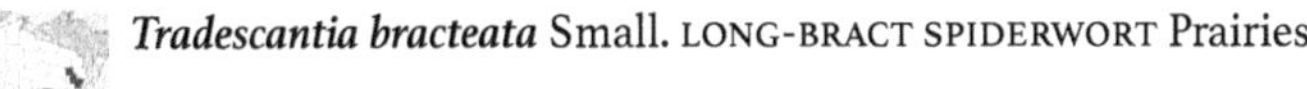

***Tradescantia bracteata*** Small. LONG-BRACT SPIDERWORT Prairies.

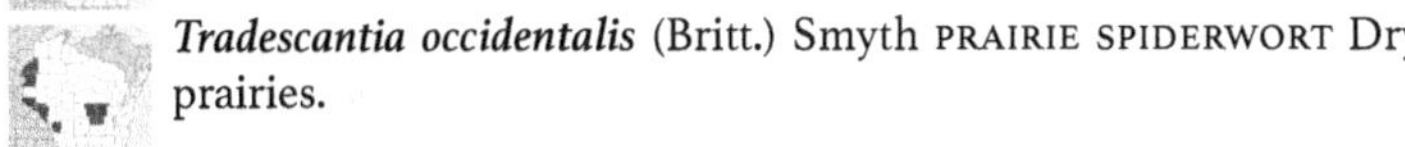

***Tradescantia occidentalis*** (Britt.) Smyth PRAIRIE SPIDERWORT Dry prairies.

***Tradescantia ohiensis*** Raf. BLUEJACKET Moist prairies and meadows.

***Tradescantia virginiana*** L. VIRGINIA SPIDERWORT Moist woods and prairies; Buffalo County in w Wisc, more common in e and se USA.

## CYPERACEAE *Sedge Family*

Mostly perennial, grasslike, rushlike or reedlike plants. Stems 3-angled, or more or less round in section, solid or pithy. Leaves 3-ranked or reduced to sheaths at base of stem; leaf blades, when present, grasslike, parallel-veined, often keeled; sheaths mostly closed around the stem. Flowers small, perfect, or single-sexed, each flower subtended by a bract (scale); perianth of 1 to many (often 6) small bristles, or a single perianth scale, or absent; stamens usually 3; ovary contained in a saclike covering (perigynium) in *Carex*, maturing into an achene, stigmas 3 or 2. Flowers arranged in spikelets (termed spikes in *Carex*), the spikelets single as a terminal or lateral spike, or several to many in various types of heads, the head often subtended by 1 to several bracts.

*Key to genera*

1 Achenes enclosed in a closed sac (perigynium) subtended by a scale, the style protruding through the apex; flowers strictly unisexual (sedges with exclusively staminate flowers should be keyed here) . ............................................................ **CAREX**

1 Achenes not enclosed in a closed sac, naked beside the subtending scale; at least some flowers bisexual (except in *Scleria*) .......... 2

2 Achenes white, hard (bone-like), ± spherical; flowers all unisexual ......................................................... **SCLERIA**

2 Achenes yellow, brown, or black, rarely whitish, not spherical; at least some flowers bisexual ................................... 3

3 Scales of spikelets 2-ranked; spikelets ± flattened in cross-section and always more than one per inflorescence ................. 4

3 Scales of spikelets spirally arranged (or if 2-ranked, the spikelet solitary); spikelets round or several-angled in cross-section, solitary or several to many per inflorescence .......................... 5

4 Stems usually ± angled, solid; inflorescences terminal; achenes without subtending bristles .......................... **CYPERUS**

4 Stems round, hollow; inflorescences in the axils of stem leaves; achenes with subtending bristles................. **DULICHIUM**

5 Perianth bristles 6, 3 slender and 3 with an expanded, ± spongy, spoon-like portion at the tip ......................... **FUIRENA**

5 Perianth bristles absent or 1 to many, all slender ............... 6

6 Spikelet or cluster of spikelets borne on one side of the stem at the base of a single ± erect to somewhat angled or curved involucral bract that appears to be a continuation of the stem ............. 7

6 Spikelet or spikelets terminating the stem or borne both terminally and laterally; if more than one spikelet, the inflorescence with (1–) 2 to several spreading to reflexed, leaflike involucral bracts...... 8

7 Stems less than 0.5 mm thick; plants tiny, less than 10 cm tall ............................................................ **LIPOCARPHA**

7 Stems thicker than 0.5 mm; plants usually much taller than 10 cm ........................................ **SCHOENOPLECTUS**

8 Spikelet solitary and terminal on the stem (very rarely a few smaller accessory spikelets occur at the base of the terminal spikelet in the bladeless genus Eleocharis) ................................. 9

8 Spikelets several to many on the stem, terminal or lateral ...... 12

9 Sheaths totally bladeless or at most with an apical tooth up to 1 mm long; achenes usually with an apical tubercle formed by the expanded and persistent base of the style ......... **ELEOCHARIS**

9 Upper sheaths with short green blades 0.3–12 cm long; achenes blunt at apex, tubercle absent .............................. 10

10 Achenes subtended by 1–8 bristles less than twice as long as the achenes, or bristles absent ................. **TRICHOPHORUM**

10 Achenes subtended by conspicuous silky, white or tawny, hair-like bristles many times as long as the achenes ................... 11

11 Bristles numerous, (12–) 15–50 or more; rhizomes erect, very short ............................................... **ERIOPHORUM**

11 Bristles 6; rhizomes horizontal and short-creeping ............... ......................................... ***Trichophorum alpinum***

12 Achenes subtended by (12–) 15–50 conspicuous, silky, white or tawny, hair-like bristles many times as long as the achenes ....... ............................................ **ERIOPHORUM**

12 Achenes subtended by 1–8 bristles, or bristles absent .......... 13

13 Leaves flat or folded; with a definite, ± keeled midrib .......... 14
13 Leaves inrolled and wiry; rounded on the back and without a definite midrib .................................................. 17

14 Achenes with a conspicuous tubercle formed by the expanded, persistent style base ........................ **RHYNCHOSPORA**
14 Achenes blunt at apex, without a tubercle; style base, if expanded, not persistent to maturity .................................... 15

15 Widest leaves 4–15 mm wide; achenes subtended by 1–8 bristles 16
15 Widest leaves 0.5–3 mm wide; achenes lacking bristles .......... ................................................ **FIMBRISTYLIS**

16 Spikelets (10–) 15–36 mm long; achenes 3–5 mm long, including apiculus; anthers 4–5 mm long; stems sharply 3-angled nearly or quite to the base; colonial from rhizomes with large corm-like thickenings .............................. **BOLBOSCHOENUS**
16 Spikelets 2–10 (–12) mm long, achenes 0.9–1.2 mm long; anthers 0.5–1.3 mm long; stems terete, obtusely 3-angled, or sharply 3-angled only toward summit; tufted or with rhizomes lacking corm-like enlargements ........................................ **SCIRPUS**

17 Styles 2-cleft; achenes subtended by slender bristles ............ ........................................ **RHYNCHOSPORA**
17 Styles 3-cleft; achenes lacking bristles ........................ 18

18 Achenes 2.2–3.5 mm long; perennials 4–11 dm tall; colonial from elongated rhizomes ................................ **CLADIUM**
18 Achenes 0.7–0.8 mm long; tufted annuals 0.2–4 dm tall .......... ........................................... **BULBOSTYLIS**

## BOLBOSCHOENUS *Club-Rush*

***Bolboschoenus fluviatilis*** (Torr.) Soják RIVER CLUB-RUSH Perennial, spreading by rhizomes and often forming large colonies. Stems stout, erect, 6–15 dm long, sharply 3-angled, the sides more or less flat. June–Aug. Shallow water of streams, ditches, marshes, lakes and ponds; sometimes where brackish. *Schoenoplectus fluviatilis* (Torr.) M.T. Strong, *Scirpus fluviatilis* (Torr.) Gray.

## BULBOSTYLIS *Hair Sedge*

***Bulbostylis capillaris*** (L.) Kunth HAIR SEDGE Stems threadlike, to 3 dm tall. Cymes capitate or umbel-like; longest bract usually exceeding the spikelets; spikelets few-flowered; scales purple-brown with strong green midvein, minutely pubescent and ciliate. Achenes trigonous, straw-colored, capped with a tiny tubercle Wet sandy or muddy soil.

## CAREX *Sedge*

Perennial grasslike plants. Stems mostly 3-angled. Leaves 3-ranked, margins often finely toothed. Flowers either staminate or pistillate, with both sexes in same spike, or in separate spikes on same plant, or the staminate and pistillate flowers on different plants. Staminate flowers with 3 or rarely 2 stamens; pistillate flowers with style divided into 2 or 3 stigmas. Achenes lens-shaped or flat on 1 side and convex on other (in species with 2 stigmas), or achenes 3-angled or nearly round (in species with 3 stigmas), enclosed in a sac called the perigynium (singular) or perigynia (plural). *Carex* is the largest genus of plants in Wisconsin. To aid in identification, a grouping of most sedges by their typical habitat is included, followed by a key to 42 *Carex* sections containing closely related species (section divisions largely follow those of Reznicek et al. (2011) and Hipp (2008).

*Quick entry key to Carex Sections*

- Spike 1 per stem, all flowers attached to main stem in terminal spike couplet .......... 2
- Spikes 2 or more per stem, all flowers staminate couplet .......... 10
- Spikes 2 or more per stem, at least some flowers pistillate, stigmas 2, achenes flat to biconvex in cross- section couplet .......... 13
- Spikes 2 or more per stem, at least some flowers pistillate, stigmas usually 3, achenes usually more or less 3-angled in cross-section, body of perigynium pubescent couplet .......... 33
- Spikes 2 or more per stem, at least some flowers pistillate, stigmas usually 3, achenes usually more or less 3-angled in cross-section, body of perigynium glabrous couplet .......... 44

*Key to Sections of Carex*

1 Spike solitary, terminal (entirely staminate, entirely pistillate, or mixed) .......... 2
1 Spikes 2 or more, sometimes crowded but distinguishable by the lobed appearance of inflorescence or protruding bracts or visible short segments of rachis between spikes .......... 9

2 Styles 2-cleft; achenes 2-sided (lenticular); basal sheaths brown . 3
2 Styles 3-cleft; achenes 3-sided (or nearly terete); basal sheaths brown or purple-red .......... 4

3 Plants with slender rhizomes; perigynia obscurely or not at all serrate, plump (usually at least as convex on upper face as on the lower), the lowermost tending to be remote (as much as 1 mm apart at points of attachment); spikes without empty basal scales; anthers to 2.5 (–3) mm long .......... *Carex sect.* **PHYSOGLOCHIN**
3 Plants densely tufted, not rhizomatous; perigynia minutely but strongly and regularly serrate on upper portion and beak, ± flattened, crowded; spikes usually with 1–2 empty basal scales; anthers 2–3.5 mm long .......... *Carex sect.* **STELLULATAE** (*C. exilis*)

4 Spikes unisexual (either staminate or pistillate); perigynia pubescent .......... *Carex sect.* **ACROCYSTIS**
4 Spikes containing both staminate and pistillate flowers; perigynia usually glabrous .......... 5

5 Perigynia minutely pubescent .......... *Carex sect.* **ACROCYSTIS**
5 Perigynia glabrous .......... 6

6 Spikes staminate at base, pistillate toward tip, densely flowered, mostly 1 cm or more thick; perigynia inflated (much larger than the included achene), abruptly contracted to a long, very slender beak .......... *Carex sect.* **SQUARROSAE**

6 Spikes pistillate at base, staminate above, more slender and sparsely flowered (fewer than 10 perigynia); perigynia various (but neither inflated nor with a long slender beak) .......................... **7**

7 Lower pistillate scale leaflike, at least on most spikes, much exceeding the perigynium; perigynia distinctly beaked, the body plump and filled by the mature achene ...... *Carex sect.* **PHYLLOSTACHYAE**

7 Lower pistillate scale not leaflike, scarcely if at all exceeding perigynium; perigynia essentially beakless or linear-lanceolate (tapering into an indistinct beak) ............................ **8**

8 Perigynia slender, linear-lanceolate (more than 5 times as long as thick), strongly reflexed at maturity *Carex sect.* **LEUCOGLOCHIN**

8 Perigynia broad, less than 5 times as long as thick, appressed-ascending ...................... *Carex sect.* **LEPTOCEPHALAE**

9 All spikes staminate ...................................... **10**

9 At least some spikes bisexual or pistillate ..................... **12**

10 Plants with long-creeping rhizomes ............................ ............................ *Carex sect.* **DIVISAE** (*C. praegracilis*)

10 Plants densely tufted ...................................... **11**

11 Leaves flat, lax and spreading; usually in swamps and marshes ... ......................... Carex sect. **DEWEYANAE** (*C. bromoides*)

11 Leaves channeled, stiff and erect; fens and other calcareous open wetlands .......................... *Carex sect.* **STELLULATAE**

12 Styles 2-cleft; achenes 2-sided ............................... **13**

12 Styles 3-cleft; achenes 3-sided (or nearly terete) ............... **33**

13 Lateral spikes peduncled, or if sessile, then elongate; terminal spike often entirely staminate ................................... **14**

13 Lateral spikes sessile, short, often crowded; terminal spike at least partly pistillate (rarely staminate) ........................... **16**

14 Plants slender, the stems to 3 dm tall and less than 1 mm thick (excluding leaf bases) even near the base; terminal (staminate or sometimes mixed) spike solitary, ca. 1 cm long; lowermost bract usually with a short sheath 2–7 mm long; perigynia white-pulverulent or golden-yellow at maturity .................... **15**

14 Plants coarse, the stems over (3–) 5 dm tall and usually over 1 mm thick, at least toward base; staminate spikes often 2 or more, mostly 2.5–7 cm long; lowermost bract essentially sheathless (rarely with very short sheath); perigynia neither white-pulverulent nor golden-yellow ............................. *Carex sect.* **PHACOCYSTIS**

15 Lowermost pistillate spike sessile or nearly so (except rarely one arising from near base of plant); terminal spike staminate; perigynia green or slightly glaucous, crowded ... *Carex sect.* **PHACOCYSTIS**

15 Lowermost pistillate spike nearly always peduncled; terminal spike often pistillate near apex, or the pistillate spikes ± loosely flowered; fresh perigynia white-pulverulent or golden-yellow ............ .......................................... *Carex sect.* **BICOLORES**

16 Stems arising mostly singly from rhizome or stolon ........... **17**

16 Stems tufted, the tufts with or without connecting rhizomes ... **22**

17 Perigynia plumply plano-convex to nearly terete in cross-section, not winged or sharply margined; plants of sphagnum bogs, cedar swamps, etc .................................................. 18

17 Perigynia strongly flattened, with distinctly winged or sharply edged margins; plants mostly of wet or dry open habitats ............ 19

18 Scales pale-hyaline with green midrib; perigynia apiculate or with very small beak; at least the lower few-flowered spikes ± separated; plants clumped from short, slender rhizomes ................... ................................... *Carex sect.* **GLAREOSAE**

18 Scales rich brown; perigynia with distinct beak ca. 0.5 mm long; spikes crowded as if in a single head; stems arising from axils of old decumbent stems (stolons) ...... *Carex sect.* **CHORDORRHIZAE**

19 Perigynia mostly over 2 mm wide; staminate flowers only at the base of some or all spikes ......................... *Carex sect.* **OVALES**

19 Perigynia mostly not over 2 mm wide; staminate flowers not restricted to base of spikes .................................... 20

20 Mature perigynia with the body ± narrowly wing-margined above and the beak bidentate (firm teeth 0.5 mm long); rhizome slender (1–1.5 mm in diameter), with brownish fibrous sheaths; spikes often dissimilar, some largely or entirely staminate or pistillate, others mixed ........................... *Carex sect.* **AMMOGLOCHIN**

20 Mature perigynia distinctly 2-edged but not winged, the beak with short weak teeth; rhizome stout (2–3 mm in diameter), with black fibrous sheaths; spikes mostly similar (each one staminate apically and pistillate basally; in section Holarrhenae the upper sometimes largely staminate) .......................................... 21

21 Sheaths of upper leaves green-nerved ventrally, usually not covering the inconspicuous nodes ........... *Carex sect.* **HOLARRHENAE**

21 Sheaths of upper leaves with broad white-hyaline stripe on ventral side covering the included nodes ............ *Carex sect.* **DIVISAE**

22 Staminate flowers at the base of some or all spikes, not at the apex (note especially the terminal spike) .......................... 23

22 Staminate flowers at the apex of some or all spikes (even when anthers have fallen, protruding filaments are usually visible)... 28

23 Perigynia with thin-winged margins, at least narrowly so along apical part of body and basal part of beak, strongly flattened and scale-like (in some species elongate), ± appressed and overlapping (or in some species spreading at the tips) .................... 24

23 Perigynia at most with a ridge along the margin, not winged, the achene plumply filling at least the apical part of the body all the way to the margins .......................................... 25

24 Bracts not resembling the leaves, narrower than 2 mm most or all their length and not over twice as long as the inflorescence; perigynia various .......................... *Carex sect.* **OVALES**

24 Bracts leaflike, the broadest 2–4 mm wide, many times exceeding the spikes (which are crowded in a dense head); perigynia very narrowly lanceolate, not over 1 mm wide ....... *Carex sect.* **CYPEROIDEAE**

25 Body of perigynium elliptic or nearly so (except in *C. arcta*) with at most a very short beak, and with rounded or slightly margined edges, nearly or entirely filled by the achene .... *Carex sect.* **GLAREOSAE**

25 Body of perigynium ovate or lanceolate or prominently beaked, sharp-edged, only 1/2 to 2/3 filled by achene (very spongy around and below base of achene) .................................. **26**

26 Mature perigynia loosely to strongly appressed-ascending, 4–5.7 mm long; anthers 1.3–2.6 mm long ..... *Carex sect.* **DEWEYANAE**

26 Mature perigynia strongly spreading to reflexed, 2–3.6 mm long; anthers 0.8–2 mm long .................................... **27**

27 Spikes 7–15, usually crowded, except sometimes the lowest, the inflorescence axis mostly concealed; beaks not bidentate ........ ............................ *Carex sect.* **GLAREOSAE** (*C. arcta*)

27 Spikes 3–8, not usually crowded, inflorescence axis clearly visible; beaks clearly bidentate with teeth 0.1–0.4 mm long ............ .................................. *Carex sect.* **STELLULATAE**

28 Stems stout (often 1.5 mm thick at ca. 3 cm below inflorescence) and very sharply angled (or even narrowly winged), ± soft and easily compressed (flattened in pressing); wider leaves 5–10 mm broad, with rather loose sheaths; perigynia spongy-thickened basally, on short slender stalks; anthers 1.3–2.6 mm long .... *Carex sect.* **VULPINAE**

28 Stems slender (not over 1.5 mm thick at ca. 3 cm below inflorescence, or rarely so in some species), firm, not wing-angled nor easily compressed (hence, not flattened in pressing); leaves, perigynia, and anthers various .............................. **29**

29 Spikes 10 or fewer, usually greenish at maturity, crowded or remote in a simple inflorescence (one spike, no branches, at each node of it) ...................................................... **30**

29 Spikes numerous (10–many), yellowish or brownish at maturity; inflorescence tending to be compound, at least its lower nodes with 2 or more spikes crowded on a lateral branch ................ **32**

30 Perigynia elliptic, essentially beakless, very plump (nearly terete) and filled by the achene; at least the lower spikes well separated, containing 1–5 perigynia ................ *Carex sect.* **DISPERMAE**

30 Perigynia ± ovate, beaked, plano-convex or lenticular; spikes various ........................................................... **31**

31 Mature perigynia brownish; some spikes (especially terminal) entirely or mostly staminate or staminate at their bases only ..... .................................... *Carex sect.* **STELLULATAE**

31 Mature (not over-ripe) perigynia generally greenish; no spikes entirely or mostly staminate (a few may have stamens at their base in addition to their apex) ...... *Carex sect.* **PHAESTOGLOCHIN**

32 Pistillate scales terminating in a distinct rough awn; bracts, at least lower ones, very slender and exceeding spikes or branches; ventral surface of leaf sheaths usually transversely wrinkled or puckered (very rarely smooth) ................ *Carex sect.* **MULTIFLORAE**

32 Pistillate scales acute or minutely cuspidate; bracts mostly short, inconspicuous, or absent; leaf sheaths smooth ventrally ......... ................................ *Carex sect.* **HELEOGLOCHIN**

33 Perigynia at least sparsely puberulent, pubescent, hispidulous, or scabrous ........................................................ 34
33 Perigynia glabrous (in some species, papillose or granular, but not even sparsely puberulent or scabrous) ........................ 43

34 Perigynia 12–18 mm long, in 1–2 short-oblong to spherical spikes 2–3.5 cm wide ............................ *Carex sect.* **LUPULINAE**
34 Perigynia 2–11 mm long, in 2–5 ± elongate, cylindrical spikes less than 2 cm wide ........................................ 35

35 Perigynia with distinct and definite slender beak and/or the apex with 2 firm teeth .......................................... 36
35 Perigynia beakless or merely apiculate ("beak" not over 0.4 mm long) and the apex not toothed ................................ 42

36 Leaves hairy ................................................ 37
36 Leaves glabrous (often rough or scabrous, but not hairy) ...... 38

37 Beak of perigynium with minute, scarcely visible teeth; body of perigynium strongly 3-angled, closely enveloping the achene, essentially nerveless, tapered to a stalk-like base; stems pubescent .................................... *Carex sect.* **HIRTIFOLIAE**
37 Beak of perigynium with strong spreading teeth 0.8 mm or more long; body of perigynium ± rounded, loosely enveloping achene (especially at summit), strongly ribbed, ± rounded (not cuneate-tapered) at base; stems glabrous ............. *Carex sect.* **CAREX**

38 Pistillate spikes not over 10 mm long (occasionally 12 mm in *C. communis*); achenes mostly with very convex or rounded sides (the angles thus obscured), at least apically, very tightly enveloped by the perigynium, especially on the apical half; anthers 1.5–3.7 mm long; plants of dryish habitats ......... *Carex sect.* **ACROCYSTIS**
38 Pistillate spikes mostly over 10 mm long; achenes with flattish to slightly concave sides (the angles thus ± evident), the summit (especially around base of style) ± loosely enveloped by the perigynium; anthers 2.5–4.7 mm long; plants of dry to wet habitats ....................................................... 39

39 Perigynium beak usually more than half as long as the body, the apex not or weakly and obscurely toothed; perigynia scabrous or with short stiff ascending hairs ............................ 40
39 Perigynium beak less than half as long as the body, with two firm apical teeth; perigynia ± densely short-hairy .................. 41

40 Perigynia conspicuously 6–8 nerved; spikes densely flowered, with 20–75 perigynia; basal sheaths pale brown *Carex sect.* **ANOMALAE**
40 Perigynia 2-ribbed, otherwise nerveless; spikes very loosely flowered with only 3–6 (–8) perigynia; basal sheaths reddish purple ....... ............ *Carex sect.* **HYMENOCHLAENAE** (*C. assiniboinensis*)

41 Perigynia 6–11 mm long, beak teeth 1.2–2.3 mm long, inner band of upper sheaths strongly purple-red tinged and thickened at apex, the thickened reddish portion opaque, smooth *Carex sect.* **CAREX**
41 Perigynia 2.5–6.5 mm long, beak teeth 0.2–0.8 mm long, inner band of upper sheaths whitish to brown, brown- or purple-dotted, but not uniformly colored, not strongly opaque-thickened at apex, often scabrous .............................. *Carex sect.* **PALUDOSAE**

42 Leaf sheaths (and usually the blades) ± pubescent, especially toward base of plant; terminal spike pistillate toward apex, staminate toward base .......................... *Carex sect.* **POROCYSTIS**

42 Leaf sheaths and blades glabrous; terminal spike staminate toward apex or entirely staminate .............. *Carex sect.* **DIGITATAE**

43 Leaf sheaths (at least at apex) finely pubescent; blades often also pubescent or at least strongly hispidulous, especially toward base of plant .................................................. 44

43 Leaf sheaths and blades completely glabrous (though sometimes scabrous) ................................................ 47

44 Beak of perigynium with firm teeth ca. 1.5–3 mm long; perigynia ca. 8–10 mm long, in spikes 4–12 cm long .......................... ............................... *Carex sect.* **CAREX** (*C. atherodes*)

44 Beak of perigynium with teeth scarcely 0.5 mm long or absent; perigynia less than 6 mm long, in spikes less than 3 cm long ... 45

45 Basal sheaths pale brown, leaf blades and stems glabrous or scabrous, perigynia with ca. 50 fine, impressed nerves .......... .......................... *Carex sect.* **GRISEAE** (*C. hitchcockiana*)

45 Basal sheaths reddish purple tinged, leaf blades and stems pubescent, perigynia 5–12 nerved ........................... 46

46 Pistillate spikes laxly spreading or drooping on slender peduncles, the lowest (20–) 25–60 mm long (including portion inside sheath, if any); perigynia tapering to distinct beak ...................... ............................ *Carex sect.* **HYMENOCHLAENAE**

46 Pistillate spikes erect or ascending, sessile, short-peduncled or on stiff, erect peduncles less than 20 (–25) mm long; perigynia beakless .................................. *Carex sect.* **POROCYSTIS**

47 Perigynia ± rounded to broadly tapered at summit, beakless or essentially so (the tiny beak or apiculus less than 0.5 mm long if distinct, or up to 0.8 mm long if vaguely defined, often strongly bent or curved); beak or apiculus (if present) never toothed (or teeth scarcely 0.1 mm long) ..................................... 48

47 Perigynium abruptly contracted or more gradually tapering to a definite slender beak 0.5 mm or more long, or to an indistinct tapering beak 1 mm or more long; beak in some species with short apical teeth ................................................ 59

48 Leaf blades not over 0.5 mm broad, linear-filiform; perigynia dark brown or nearly black at maturity, 2 mm or less long, in few-flowered spikes, of which at least the upper ones are on peduncles usually surpassing the sessile staminate spike.. *Carex sect.* **ALBAE**

48 Leaf blades 0.5 mm or more broad; perigynia and spikes various (but not as above) ........................................ 49

49 Bract of lowest pistillate spike sheathless (at most with a thin scarious sheath 1–3 mm long) .............................. 50

49 Bract of lowest pistillate spike with a sheath 4 mm or more long 53

50 Terminal spike partly pistillate; pistillate spikes nearly or quite sessile and erect or ascending; roots glabrous or nearly so ....... .................................. *Carex sect.* **RACEMOSAE**

50 Terminal spike normally entirely staminate; spikes and roots various .......................................................... 51

51 Pistillate spikes mostly drooping at maturity on slender peduncles; species of wet peat lands with roots with dense felt-like pubescence .......................................... *Carex sect.* **LIMOSAE**

51 Pistillate spikes erect or ascending, sessile or peduncled; roots glabrous ........................................................ 52

52 Perigynia 2.5–3.4 mm wide; leaves involute, 0.5–2.5 mm wide ..... ...................... *Carex sect.* **VESICARIAE** (*C. oligosperma*)

52 Perigynia 1–2.5 mm wide; leaves flat or folded, 1–35 mm wide .... ................................................ *go to couplet 56*

53 Terminal spike bearing some perigynia (very rarely a few individuals with one entirely staminate); plants very strongly reddish tinged at base ........................................ 54

53 Terminal spike entirely staminate; plants reddish or not at base 55

54 Staminate flowers at apex of terminal spike, pistillate flowers at base; cauline sheaths bladeless or with rudimentary blades to 2 (rarely 4) cm long; pistillate spikes short-cylindric, bearing fewer than 10 perigynia, very long-peduncled, some elongate peduncles usually arising from base of plant ............................ ......................... *Carex sect.* **DIGITATAE** (*C. pedunculata*)

54 Staminate flowers at base of terminal spike, pistillate flowers at apex; cauline sheaths with well-developed blades; pistillate spikes linear-cylindric, bearing more than 10 perigynia, on peduncles about as long as the spike or shorter, all arising from the upper part of the stem .................. *Carex sect.* **HYMENOCHLAENAE**

55 Perigynia concave- or at least cuneate-tapering toward the base, ± 3-angled and often somewhat broadly spindle-shaped ........ 56

55 Perigynia convex-rounded toward the base, nearly or quite circular in cross-section (or very obscurely triangular), ellipsoid-cylindric to nearly spherical ................................................ 57

56 Plants with elongate deep or shallow rhizomes and very slender, firm stems; leaf blades 1–4 mm wide ...... *Carex sect.* **PANICEAE**

56 Plants without elongate rhizomes, the stems sharply triangular, sometimes nearly wing-margined, rather weak and easily compressed, soon shriveling after maturity of the fruit; leaf blades usually more than 4 (and up to 35) mm wide................... .................................. *Carex sect.* **LAXIFLORAE**

57 Larger perigynia 4–5 mm long, the nerves not raised above the surface at maturity ........................ *Carex sect.* **GRISEAE**

57 Larger perigynia 2–3.5 mm long; nerves various .............. 58

58 Perigynia with the nerves not raised above the surface, usually ± impressed; staminate spike usually long-peduncled; plants not strongly rhizomatous nor with any pistillate spikes on basal peduncles ..................... *Carex sect.* **GRISEAE** (*C. conoidea*)

58 Perigynia with the nerves slightly raised above the surface; staminate spike nearly or quite sessile or, if long-peduncled, the plants strongly rhizomatous and with basal pistillate spikes ...... ................................. *Carex sect.* **GRANULARES**

59 Body of perigynium obovoid or obconic, ± truncately contracted into a distinct long slender beak; terminal spike often mostly pistillate (staminate at base only) ...... *Carex sect.* **SQUARROSAE**

59 Body of perigynium ovoid to lanceolate or ellipsoid, tapered or contracted into the beak; terminal spike usually staminate, at least apically .................................................. **60**

60 Lower pistillate scales leaflike or bract-like, much exceeding the perigynia; achenes abruptly constricted to a short thick base; body of perigynium nearly terete, essentially nerveless except for 2 ribs; anthers 0.5–1.6 mm long ......... *Carex sect.* **PHYLLOSTACHYAE**

60 Lower pistillate scales scarcely if at all exceeding the perigynia; achenes not abruptly constricted at the base; perigynia and anthers various .................................................. **61**

61 Perigynia in densely crowded spherical to very short-cylindric spikes, spreading and with the lowermost usually reflexed, usually strongly few-ribbed; at least the uppermost pistillate spikes ± sessile and often crowded; the terminal spike (staminate or partly pistillate) often sessile or short-peduncled ............................ **62**

61 Perigynia in elongate or long-peduncled spikes or both, all ascending, 2-ribbed or variously many-nerved; inflorescences various, but the upper spikes often not crowded and the terminal spike often long peduncled ................................ **63**

62 Perigynia 2–6.2 mm long; basal sheaths brown .................................................. *Carex sect.* **CERATOCYSTIS**

62 Perigynia 11–18 mm long; basal sheaths reddish purple tinged ... ........................................ *Carex sect.* **LUPULINAE**

63 Bract of lowest pistillate spike sheathless (or pistillate spikes all crowded at base of plant in *Carex tonsa*); check several stems; rarely, a pistillate spike will be borne abnormally low on the stem and this spike may then have a sheath, which should be disregarded in keying.................................................. **64**

63 Bract of lowest pistillate spike consistently with sheath 4 mm or more long .................................................. **70**

64 Pistillate scales subtending at least some of the perigynia terminated by a distinct slender scabrous awn; perigynia 3–9 mm long .................................................. **65**

64 Pistillate scales smooth-margined and awnless or very short-awned, or at most with a scabrous margin toward an acuminate (sometimes inrolled) apex (occasionally a long rough awn in species with perigynia more than 9 mm long); perigynia (4–) 4.5–18 mm long **66**

65 Scales toward apex of pistillate spikes merely acuminate or with awns shorter than their bodies (the latter easily visible, about half as long as perigynia or longer); staminate spikes 2 or more; body of perigynium rather gradually tapered to a beak 1.5 mm long, including the short (not over 0.8 mm) teeth ........ *Carex sect.* **PALUDOSAE**

65 Scales toward apex of pistillate spikes ordinarily with awns (as on the other pistillate scales) nearly or fully as long as their bodies (the latter small and mostly hidden among the bases of the densely crowded perigynia); staminate spike solitary (or very rarely a second

smaller one present); body of perigynium tapered or strongly contracted into a beak 1.2–3.5 mm long, including teeth up to 2.2 mm long ............................ *Carex sect.* **VESICARIAE**

**66** Basal sheaths pale brown; perigynia very narrowly lanceolate, 4–6.5 times as long as wide and not over 3 mm wide, many-nerved, tapering to apex (not strongly contracted into a beak); staminate spike solitary (pistillate spikes may be staminate at apex) ........ ........................................ *Carex sect.* **ROSTRALES**

**66** Basal sheaths reddish purple tinged, at least on the youngest shoots; perigynia lanceolate or broader, less than 4 times as long as wide, or more than 3 mm wide, or strongly contracted into a conspicuous beak (or all of these); staminate spikes solitary or 2 or more .... **67**

**67** Perigynia strongly inflated, not tight around the achene, 2–8 mm wide ........................................................ **68**

**67** Perigynia not inflated, ± tightly enclosing achene, 1–1.6 mm wide ............................................................ **69**

**68** Perigynia 4–12 mm long, 6–12 (–15)-nerved ..................... ....................................... *Carex sect.* **VESICARIAE**

**68** Perigynia 12–17 (–18) mm long, 15–20-nerved ................... ........................................ *Carex sect.* **LUPULINAE**

**69** Pistillate spikes linear-cylindric, drooping or curving on slender peduncles; perigynia (somewhat twisted) and achenes strongly angled, the latter with concave sides; tall plants (stems over 3 dm high) with scattered thin leaves ............................... .................. *Carex sect.* **HYMENOCHLAENAE** (*C. prasina*)

**69** Pistillate spikes short, thick, and few-flowered, often crowded at base of plant; perigynia and achenes very convex-sided; low plants (stems less than 1 dm high) with crowded, very stiff leaves ....... ............................ *Carex sect.* **ACROCYSTIS** (*C. tonsa*)

**70** Perigynia (6–) 9–17 (–18) mm long; beak teeth usually conspicuous and stiff ........................................ *go to couplet 64*

**70** Perigynia 2–6.5 (–9) mm long; beak teeth absent or weak and inconspicuous.................................................. **71**

**71** Perigynia with several to many conspicuous fine nerves on each side ............................................................ **72**

**71** Perigynia with 2 (–3) main ribs, the sides otherwise nerveless or with much less prominent nerves.................................. **75**

**72** Nerves of perigynia very numerous (20–65) and impressed, giving a longitudinally corrugated appearance; awns of pistillate scales rough or even ciliate ...................... *Carex sect.* **GRISEAE**

**72** Nerves of perigynia several to many (5–40) and slightly raised; awns of pistillate scales absent, smooth, or rough.................. **73**

**73** Awns rough and/or summit of pistillate scales minutely ciliate; lower spikes drooping on long very thin peduncles; beak slightly bidentate at maturity; plants strongly reddish at base ........... ........................... *Carex sect.* **HYMENOCHLAENAE**

**73** Awns of pistillate scales usually smooth or absent; lower spikes mostly not drooping; beak not bidentate; plants pale, brown, or reddish at base ............................................ **74**

74 Perigynia ± sharply triangular with flattish sides, short-tapering at the base; stems bluntly trigonous, firm and not easily compressed; anthers mostly 3–4.5 mm long or lower pistillate spikes on elongate filiform spreading or drooping peduncles *Carex sect.* **CAREYANAE**

74 Perigynia ± rounded-triangular with swollen sides, long-tapering to a ± stalk-like base; stems sharply triangular to nearly wing-margined, easily compressed; anthers mostly 1.5–3 mm long and lower pistillate spikes usually on erect or ascending peduncles ... ........................................ *Carex sect.* **LAXIFLORAE**

75 Lowermost pistillate spikes erect or ascending at maturity ..... 76

75 Lowermost pistillate spikes drooping on long slender peduncles at maturity.................................................... 77

76 Staminate spike well-peduncled; perigynia ± convex-sided toward the base; bracts with poorly developed blades; plants mat-forming from long-creeping rhizomes . *Carex sect.* **PANICEAE** (*C. vaginata*)

76 Staminate spike sessile or nearly so; perigynia tapered-cuneate toward the base; bracts with well-developed blades; plants tufted. ....................... *Carex sect.* **LAXIFLORAE** (*C. leptonervia*)

77 Pistillate spikes not over 15 mm long ............................ ................................ *Carex sect.* **CHLOROSTACHYAE**

77 Pistillate spikes mostly 20 mm or more long ..................... ............................... *Carex sect.* **HYMENOCHLAENAE**

## CAREX SECTION ACROCYSTIS

First sedges to flower each year, fruits maturing in April or May and soon shed. Basal leaf sheaths in most species becoming fibrous with age. Perigynium beaks bidentate, less than 0.5 mm long. Most common in dry woods, prairies, and open sandy places; less common in mesic woods or wetlands. Similar to section Digitatae but basal sheaths not becoming fibrous.

### *Key to species*

1 Pistillate spikes on stems of varying length, at least some of the stems short (up to 5 cm long) and partly hidden among the tufted leaf bases; anthers 1.5–2 mm long ........................... 2

1 Pistillate spikes all on elongate stems (none borne on short basal peduncles); anthers various .................................. 4

2 Bract of the lowest non-basal pistillate spike leaflike, equaling or exceeding the tip of the staminate spike; remnants of old leaves only slightly breaking into fibrous shreds at the base ........ *C. deflexa*

2 Bract of the lowest non-basal pistillate spike scale-like or bristle-like, not exceeding the staminate spike (or all spikes often on short basal stems, but foliage and stems stiffer and much more scabrous than in *C. deflexa,* which nearly always have some elongate stems); remnants of old leaves breaking into copious fibrous shreds at the base .... 3

3 Perigynia 3.2–4 mm long, the beak 1.2–1.6 (–2) mm, about half as long as the body or longer ............................ ***C. tonsa***

3 Perigynia 2.5–2.9 mm long, the beak 0.4–0.9 mm, about 1/4–1/3 as long as the body .................................. ***C. umbellata***

4 Main body of perigynium, not including spongy-tapered base or beak, orbicular to short-obovoid, about the same diameter as length; anthers 2.1–3.7 mm long; plants either with the widest leaves 3–8 mm broad or with elongate shallow rhizomes .............. 5

4 Main body of perigynium ± elliptic (to slightly obovoid or oblong), definitely longer than thick; anthers 1.3–2.5 mm long; plants with mostly narrow leaves and lacking stout elongate rhizomes (and otherwise not fitting either lead of couplet 5) .................. 8

5 Widest leaves (at least the oldest dry ones) 3–5 mm broad; cauline leaves above base of plant (when present on stem) usually with the ligule longer than the width of the leaf; bract subtending the middle (and sometimes the lowest) pistillate spike(s) ± scarious-lobed at base, blade awn-like to leaflike, usually green, arising from between the lobes; staminate spike ca. 1–2 (–2.5) mm thick; plants without elongate rhizomes ................................ *C. communis*

5 Widest leaves 1.5–3 mm broad; cauline leaves with ligule no longer than the width; bracts subtending middle pistillate spikes tapered to apex, without an elongate awn-like or leaflike blade (the lowermost bract often green but seldom lobed); staminate spike 2–3.5 (–5) mm thick; plants with stout, shallow elongate rhizomes with fibrous sheaths ............................................ 6

6 Larger perigynia 1.7–2.2 mm wide ...................... *C. inops*

6 Larger perigynia 1.2–1.7 mm wide.............................. 7

7 Beak of perigynium 1–1.6 mm, half or more as long as the body .. ................................................ *C. lucorum*

7 Beak of perigynium 0.2–0.8 mm, much less than half as long as the body ..................................... *C. pensylvanica*

8 Widest leaves (at least the oldest dry ones) 3–5 mm broad; bract subtending the middle (and sometimes also the lowest) pistillate spike(s) ± scarious-lobed at base, the blade awn-like or leaflike, usually green, arising from between the lobes ....... *C. communis*

8 Widest leaves not over 3 mm broad; bracts either scale-like or leaflike and lacking a scarious-lobed base .................... 9

9 Lower two pistillate spikes 7.5–22 mm distant; lowest inflorescence bracts 18–35 mm long, 3/4 as long to exceeding inflorescence; loosely mat-forming from delicate, ascending rhizomes .. *C. novae-angliae*

9 Lower two pistillate spikes mostly close together, up to 7 mm distant; lowest inflorescence bracts rarely more than 17 mm long, often less than 3/4 as long the inflorescence; ± tufted .................. 10

10 Bodies of mature perigynia about as long as their scales or even slightly shorter; beak of perigynium 0.5–1.4 mm long .. *C. albicans*

10 Bodies of mature perigynia mostly distinctly exceeding their scales; beak of perigynium ca. 0.4–0.7 mm long ..................... 11

11 Perigynia 2–3 mm long, minutely puberulent to short-hairy; stems very slender (seldom over 0.4 mm thick) and mostly surpassed by the leaves ........................................... *C. deflexa*

11 Perigynia 3–4 mm long, definitely short-hairy; stems usually 1 mm or more in thickness and surpassing the leaves .......... *C. peckii*

***Carex albicans*** Willd. WHITE-TINGE SEDGE Wet to moist sandy woods.

***Carex communis*** Bailey [•105] FIBROUS-ROOT SEDGE Common in forests of n Wisc; occasional in mesic forests of s Wisc.

***Carex deflexa*** Hornem. NORTHERN SEDGE June–Aug. Moist woods and swamps, wetland margins, often where sandy or in sphagnum moss.

***Carex inops*** Bailey LONG-STOLON SEDGE Sandy woods and fields. *Carex heliophila* Mackenzie, *Carex pensylvanica* Lam. var. *digyna* Boeckl.

***Carex lucorum*** Willd ex Link BLUE RIDGE SEDGE Dry woods. *Carex pensylvanica* Lam. var. *distans* Peck.

***Carex novae-angliae*** Schwein. NEW ENGLAND SEDGE Woodlands.

***Carex peckii*** Howe PECK'S SEDGE Open woods. *Carex nigromarginata* Schwein. var. *elliptica* (Boott) Gleason

***Carex pensylvanica*** Lam. [•105] PENNSYLVANIA SEDGE Common in a wide range of dry to mesic woods and prairies.

***Carex tonsa*** (Fern.) Bickn. SHAVED SEDGE Dry sandy fields and open woods. *Carex rugosperma* Mackenzie.

***Carex umbellata*** Schkuhr PARASOL SEDGE Dry, often calcareous fields and prairies.

### CAREX SECTION ALBAE

One member of the section in Wisc. Rhizomes elongate, the plants forming mats. Leaf blades involute, wiry. Perigynia becoming dark in age, beaks short, white-tipped.

***Carex eburnea*** Boott BRISTLE-LEAF SEDGE Dry sand prairies, and rarely in fens.

### CAREX SECTION AMMOGLOCHIN

One member of the section in Wisc.

***Carex siccata*** Dewey DRY-SPIKE SEDGE Dry sandy prairies and woods.

### CAREX SECTION ANOMALAE

One member of the section in Wisc. Upper surface of leaf blade and perigynia are scabrous.

***Carex scabrata*** Schwein. [•105] EASTERN ROUGH SEDGE May–Aug. Low shaded areas in forests, streambanks, seeps.

## CAREX SECTION BICOLORES

Plants short, colonial, loosely tufted, shoots arising singly or few in a clump; rhizomes elongate; bases brown. Terminal spike staminate or gynecandrous, hidden by the crowded lateral spikes. Perigynia plump, golden to whitish, weakly veined; margins and apex rounded, beakless to short-beaked. Stigmas 2. Calcium-rich sites, often where somewhat disturbed.

*Key to species*

1 Mature perigynia golden-orange when fresh (drying dark brown or, especially if immature, ± white); terminal spikes mostly all staminate (occasionally with a very few perigynia); pistillate scales ± loosely spreading, distinctly shorter than the mature perigynia (usually averaging 3/4 or less as long), most of them acute to cuspidate .................................................. *C. aurea*

1 Mature perigynia white-pulverulent when fresh; terminal spikes usually staminate at base only, with several to numerous perigynia apically; pistillate scales ± appressed, nearly (averaging about 3/4) to quite as long as the perigynia, most of them blunt to acute .... .................................................. *C. garberi*

***Carex aurea*** Nutt. GOLDEN-FRUIT SEDGE May–July. Moist to wet meadows, low prairie, swales, wet woods and along sandy or gravelly shores; often where calcium-rich.

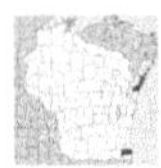

***Carex garberi*** Fern. ELK SEDGE *Threatened.* Wet sandy, gravelly, or marly shores, limestone pavements, interdunal flats, and edges of cedar thickets, typically near Lake Michigan in Door and Racine counties.

## CAREX SECTION CAREX

Plants typically colonial; rhizomes elongate. Vegetative stems prominent. Perigynia long-beaked with prominent beak teeth.

*Key to species*

1 Perigynia covered with hairs .................................. 2

1 Perigynia smooth and hairless .................................. 3

2 Inner band of the uppermost leaf sheaths red to purple and thickened at the summit, glabrous; native, in wetlands across much of Wisc .................................. *C. trichocarpa*

2 Inner band of leaf sheaths not colored, pubescent; introduced at several locations in Wisc .................................. *C. hirta*

3 Inner band of the uppermost leaf sheaths red to purple and thickened at the summit, glabrous ................. *C. trichocarpa*

3 Inner band of leaf sheaths pale or brown, not thickened at the summit, glabrous or pubescent .................................. 4

4 Vegetative stems hollow, easily flattened; inner band of the leaf sheaths pubescent, rarely glabrous, not obviously veined; basal leaf sheaths ladder fibrillose .......................... *C. atherodes*

4 Vegetative stems solid; inner band of the leaf sheaths strongly veined, glabrous or the veins scabrous; upper and lower leaf sheaths ladder fibrillose .................................. *C. laeviconica*

***Carex atherodes*** Spreng. SLOUGH SEDGE June–Aug. Marshes, wet meadows, prairie swales, stream and pond margins, usually in shallow water where may form dense colonies.

***Carex hirta*** L. HAMMER SEDGE Introduced from Europe; reported from 2 disturbed sites in Wisc.

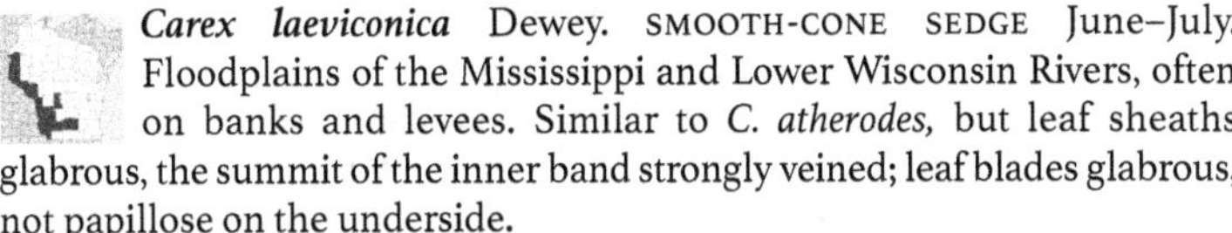

***Carex laeviconica*** Dewey. SMOOTH-CONE SEDGE June–July. Floodplains of the Mississippi and Lower Wisconsin Rivers, often on banks and levees. Similar to *C. atherodes,* but leaf sheaths glabrous, the summit of the inner band strongly veined; leaf blades glabrous, not papillose on the underside.

***Carex trichocarpa*** Muhl. [•106] HAIRY-FRUIT SEDGE May–Aug. Riverbanks and old river channels, marshes, wet meadows, low prairie. Similar to slough sedge (*C. atherodes*) but sheaths strongly purple-tinged at tip, the leaf blades not hairy on underside, and the perigynia with short white hairs (vs. smooth in *C. atherodes*).

## CAREX SECTION CAREYANAE

Resembling section Laxiflorae in appearance, but stems generally firm. Perigynia acutely angled, tightly enclosing the achene; veins more than 40, impressed in fresh plants, raised when dried.

*Key to species*

1 Larger leaf blades (especially on vegetative shoots) mostly 12–40 mm wide (or if largest blades as narrow as 8 mm, the bases of plants and the staminate scales strongly reddish) .................... 2
1 Widest leaf blades less than 12 mm wide; plant bases brownish . 4

2 Bases of plants and staminate scales pale or brownish, not reddish; perigynia to 4 mm long .......................... ***C. platyphylla***
2 Bases of plants and staminate scales strongly tinged with reddish; perigynia 4–6.5 mm long .................................. 3

3 Sheaths of cauline bracts and leaves bladeless or nearly so; perigynia 4–5 mm long ......................... ***C. plantaginea***
3 Sheaths of cauline bracts and leaves with flat green blades; perigynia 5–6.5 mm long ........................... ***C. careyana***

4 Pistillate spikes (except sometimes the upper ones) with 1–2 staminate flowers at the base; leaves green or glaucous, 4–12 mm wide ............................................ ***C. laxiculmis***
4 Pistillate spikes without staminate flowers at the base; leaf blades deep or bright green, the wider ones 2.5–4 mm wide ... ***C. digitalis***

***Carex careyana*** Torr. CAREY'S SEDGE *Threatened.* May–June. Rich moist woods.

***Carex digitalis*** Willd. SLENDER WOODLAND SEDGE Rich moist woods.

***Carex laxiculmis*** Schwein. SPREADING SEDGE Sugar maple forests.

*Carex plantaginea* Lam. [•106] PLANTAIN-LEAF SEDGE Mesic forests.

*Carex platyphylla* Carey BROAD-LEAF SEDGE Forests of sugar maple and beech. Leaves strongly glaucous.

## CAREX SECTION CERATOCYSTIS

Plants tufted; rhizomes short; bases brown. Terminal spike staminate, occasionally androgynous. Lateral spikes pistillate, densely flowered, globose to oblong. Perigynia strongly veined, abruptly beaked; beak toothed, generally reflexed. Stigmas 3. Usually where wet and calcareous.

*Key to species*

1 Larger perigynia ca. 2–3 mm long, horizontally spreading, the beak about 1/4 to nearly 1/2 as long as the body ............ *C. viridula*

1 Larger perigynia (3–) 3.5–6.2 mm long, at least the beaks becoming conspicuously reflexed on lower half of spike, the beak nearly or fully half as long as the body .................................. 2

2 Pistillate scales at maturity strongly flushed with shiny brown or reddish color, hence conspicuous in the spike; widest leaves 3–5 mm wide .................................................. *C. flava*

2 Pistillate scales greenish or yellowish, the same color as the perigynia and essentially invisible in the spikes; widest leaves 1.5–4 mm wide ................................................ *C. cryptolepis*

*Carex cryptolepis* Mackenzie NORTHEASTERN SEDGE June–Aug. Wet meadows and marshy areas, peatlands, swamp margins; often where calcium-rich. Similar to *C. flava.*

*Carex flava* L. YELLOW-GREEN SEDGE May–Aug. Wet, peaty meadows, often where calcium-rich.

*Carex viridula* Michx. LITTLE GREEN SEDGE May–Aug. Wet meadows, sandy lake margins, fens and seeps; often where calcium-rich.

## CAREX SECTION CHLOROSTACHYAE

One member of the section in Wisc. Plants small, densely tufted, with fibrous basal leaf sheaths and small beadlike perigynia borne in slender spikes on threadlike stalks.

*Carex capillaris* L. HAIR-LIKE SEDGE June-July. Alder thickets, wetland margins, usually in shade.

## CAREX SECTION CHORDORRHIZAE

One member of the section in Wisc. Plants stoloniferous, the stolons arching and rooting.

*Carex chordorrhiza* Ehrh. ROPE-ROOT SEDGE May–Aug. Open floating mats around lakes and ponds, fens, conifer swamps, interdunal hollows.

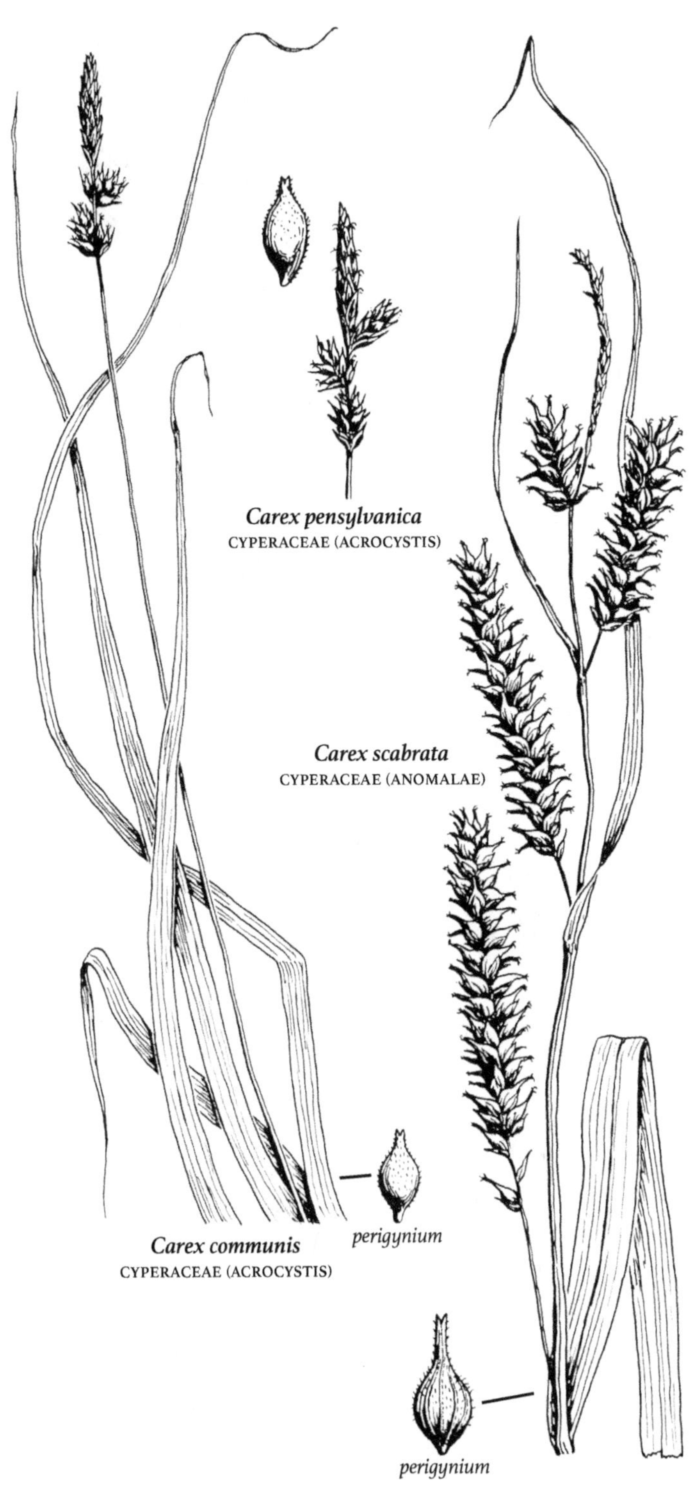


*Carex pensylvanica*
CYPERACEAE (ACROCYSTIS)

*Carex scabrata*
CYPERACEAE (ANOMALAE)

*Carex communis*
CYPERACEAE (ACROCYSTIS)

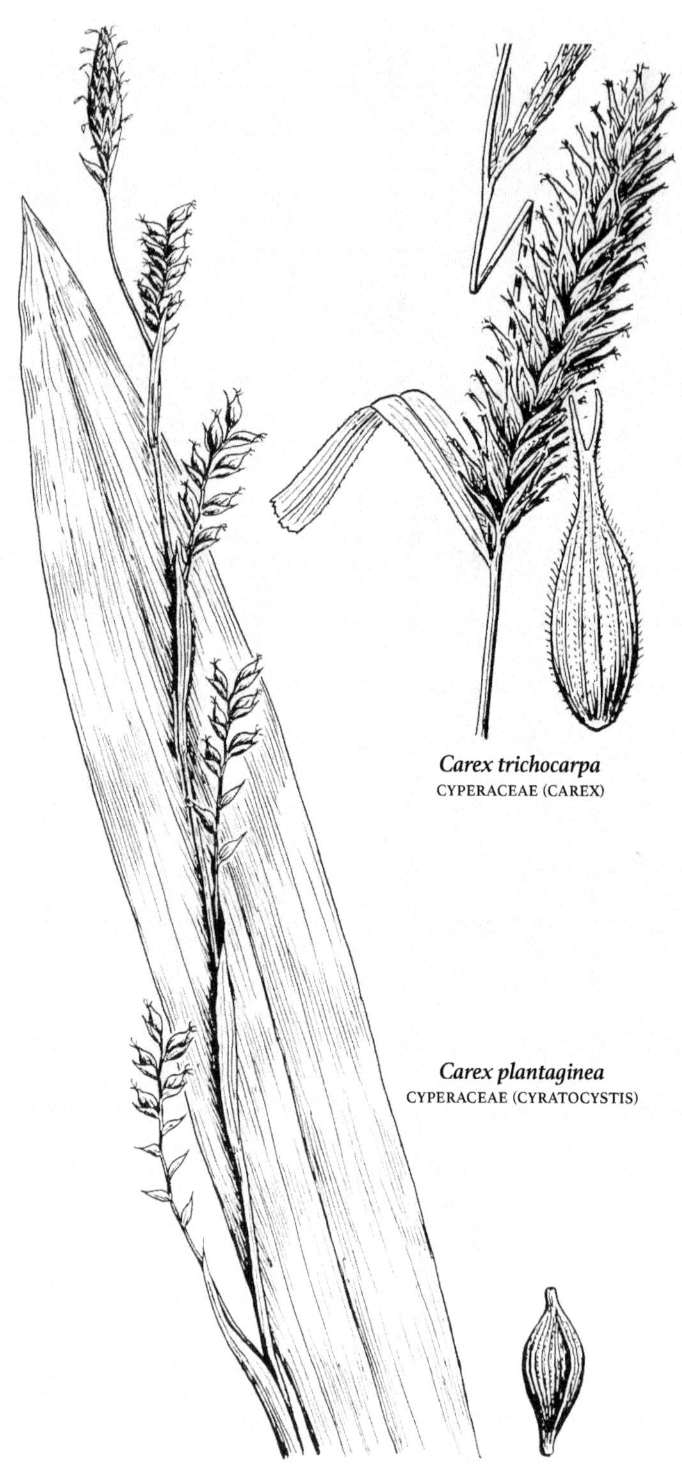

*Carex trichocarpa*
CYPERACEAE (CAREX)

*Carex plantaginea*
CYPERACEAE (CYRATOCYSTIS)

## CAREX SECTION CYPEROIDEAE

One member of the section in Wisc. Similar to the Ovales and sometimes placed within that section; distinguished by leafy bracts more than 3 times as long as the inflorescence and very long-tapering perigynia, the beak often twice as long as the body.

***Carex sychnocephala*** Carey MANY-HEAD SEDGE June–Aug. Wet meadows, sandy lakeshores, marshes.

## CAREX SECTION DEWEYANAE

Two members of the section in Wisc. Plants tufted; rhizomes mostly short; bases brown. Inflorescence slender, open, at least the lowest spike(s) distinct; bracts setaceous. Spikes mostly gynecandrous, lateral spikes sometimes pistillate, mixed, or (rarely) staminate. Perigynia appressed to ascending, ovate to lanceolate, plano-convex, slender; base spongy; beak distinct, margins serrate, tip bidentate. Achenes mostly filling the perigynium body. Usually in moist to wet shaded places.

*Key to species*

1 Perigynia 0.8–1.2 mm wide and 4–5 times as long as wide, conspicuously nerved on dorsal face, weakly to strongly nerved on ventral face ........................................ ***C. bromoides***

1 Perigynia 1.3–1.6 mm wide and usually 3–3.5 times as long as wide, faintly nerved or nerveless on both faces ............ ***C. deweyana***

***Carex bromoides*** Schkuhr BROME-LIKE SEDGE April–July. Floodplain forests, old river channels, swamps.

***Carex deweyana*** Schwein. DEWEY'S SEDGE May–Aug. Thickets, swamps, and moist to dry woods.

## CAREX SECTION DIGITATAE

Basal sheaths not fibrous. Bracts reduced to bladeless sheaths. Perigynium beaks untoothed, mostly less than 0.5 mm long. Similar to section Acrocystis but basal sheaths not fibrous.

*Key to species*

1 Terminal spike pistillate at base; basal spikes usually present, on long very thin peduncles; pistillate scales abruptly truncate and awned; anthers 2–3 mm long ...................... ***C. pedunculata***

1 Terminal spike usually entirely staminate; basal spikes not present; pistillate scales not awned; anthers various ...................... **2**

2 Staminate spike 4–6 (–8) mm long; pistillate spikes less than 10 mm long; pistillate scales obtuse, minutely ciliate, distinctly shorter than the perigynia; anthers 1–1.5 mm long ................. ***C. concinna***

2 Staminate spike 10–22 mm long; pistillate spikes (often staminate at their tips) (8–) 10 mm long; pistillate scales mostly acute to acuminate, glabrous, and equaling or exceeding the perigynia; anthers 2–3.5 mm long .......................... ***C. richardsonii***

***Carex concinna*** R. Br. LOW NORTHERN SEDGE *Threatened.* Open, moist, sandy places, usually underlain by dolomitic limestone. Ashland and Door counties.

*Carex pedunculata* Muhl. LONG-STALK SEDGE Rich, mesic forests.

*Carex richardsonii* R. Br. RICHARDSON'S SEDGE Dry sandy prairies and barrens; rarely in fens.

## CAREX SECTION DISPERMAE

One member of the section in Wisc. Plants slender, shoots arising singly or in small bunches from pale, slender rhizomes; spikes few-flowered, androgynous; perigynia spreading, darkening at maturity, plump.

*Carex disperma* Dewey SOFT-LEAF SEDGE May–July. Hummocks in conifer swamps and alder thickets, wetland margins; usually where shaded.

## CAREX SECTION DIVISAE

One member of the section in Wisc. Plants strongly rhizomatous, unisexual. Not native in most of e North America but spreading, especially along expressways, where tolerant of road salt.

*Carex praegracilis* W. Boott CLUSTERED FIELD SEDGE May–June. Wet to moist meadows, shores, streambanks and ditches; most common along salted highways. Native of w USA, considered adventive in Wisc.

## CAREX SECTION GLAREOSAE

Tufted sedges of wetlands, soils often peaty. Spikes distinct, mostly nonoverlapping (except Carex arcta which has spikes overlapping, the upper not separated), mostly or all gynecandrous, lateral spikes sometimes pistillate. Perigynia ascending to spreading; margins rounded in most species, smooth or finely serrate, often finely papillose.

*Key to species*

1 Lowest bract bristle-like, several times as long as its spike; perigynia mostly 2.8–3.8 (–4) mm long, including very short smooth beak; spikes widely separated, containing 1–5 perigynia each .......... *C. trisperma*

1 Lowest bract absent or at most about twice as long as its spike (if rarely prolonged, the perigynia smaller and often with serrulate beak); perigynia and spikes various .......... 2

2 Perigynia broadest near the base of the body, with a conspicuous beak 0.7–1.1 mm long; spikes mostly 7–15, usually ± overlapping or crowded into an ovoid to narrowly pyramidal head 2–4.5 cm long .......... *C. arcta*

2 Perigynia broadest at or near the middle of the body; beak essentially absent or less than 0.6 mm long; spikes 2–8, at least the lower spikes well separated or, if crowded, the inflorescence only 0.6–2 cm long .......... 3

3 Spikes 2–4, crowded into a short inflorescence 0.6–2 cm long; perigynia 2.5–3.5 mm long, beak often smooth-margined .......... *C. tenuiflora*

3 Spikes 4–8 (–10), remote or ± crowded, but total inflorescence over 2 cm long; perigynia 1.7–2.6 mm long, beak serrulate usually minutely or scabrous . . . . . . . . . . . . . . . . . . . . . . . . . . . . . . . . . . . . . . . . 4

4 Perigynia 3–9 per spike (occasionally one or two spikes on a plant, especially terminal one, with as many as 15), loosely spreading, becoming rich brown in age; largest leaves 1–2 mm wide; foliage and perigynia green when fresh. . . . . . . . . . . . . . . . . . . *C. brunnescens*

4 Perigynia mostly 10–many per spike, appressed-ascending, greenish or dull brown in age; largest leaves 2–2.7 (–3.7) mm wide; foliage and perigynia glaucous or gray-green at least when fresh. . *C. canescens*

***Carex arcta*** Boott NORTHERN CLUSTER SEDGE June–Aug. Floodplain forests, old river channels, swamps and wetland margins.

***Carex brunnescens*** (Pers.) Poir. [•107] BROWNISH SEDGE June–Aug. Wet forests and swamps, peatland margins.

***Carex canescens*** L. HOARY SEDGE May–July. Peatlands (including hummocks in patterned fens), tamarack swamps, floating mats, swamps, alder thickets, wet forest depressions. Similar to *C. brunnescens* but leaves waxy blue-green rather than green and spikes somewhat larger and silver-green vs. brown.

***Carex tenuiflora*** Wahlenb. SPARSE-FLOWER SEDGE June–Aug. Hummocks in peatlands, floating mats, conifer swamps; mostly confined to tamarack swamps in s part of Wisc.

***Carex trisperma*** Dewey [•107] THREE-SEED SEDGE May–Aug. Forested wetlands and conifer swamps, alder thickets.

## CAREX SECTION GRANULARES

Plants tufted or shoots arising singly from elongate rhizomes. Pistillate spikes oblong to narrowly oblong, densely packed with perigynia. Pistillate scales and perigynia dotted or finely streaked with red. Perigynia more than 25 per pistillate spike; veins 25–40, raised.

*Key to species*

1 Staminate spike long-peduncled, elevated above summit of uppermost pistillate spikes; lowest pistillate spike usually on a separate basal peduncle; stems mostly solitary from elongate rhizomes; widest leaves 1.5–4 mm broad . . . . . . . . . . . . . . . . *C. crawei*

1 Staminate spike sessile or nearly so; lowest pistillate spike not on a basal peduncle; stems clumped, without elongate rhizomes; widest leaves 4.5–10 mm broad . . . . . . . . . . . . . . . . . . . . . . . . . . . . *C. granularis*

***Carex crawei*** Dewey CRAWE'S SEDGE May–July. Wet to moist meadows and prairies, marly lakeshores, ditches, especially where calcium-rich.

***Carex granularis*** Muhl. LIMESTONE-MEADOW SEDGE May–July. Wet to moist meadows and swales, streambanks and pond margins, especially where calcium-rich.

## CAREX SECTION GRISEAE

Perigynia round or obtusely angled in cross-section, many-veined; veins impressed on both fresh and dried plants. Pistillate scales awned.

*Key to species*

1 Perigynia contracted to a distinct beak 0.5–1.3 mm long ......... 2
1 Perigynia essentially beakless ................................ 3

2 Leaf sheaths strongly hispidulous; perigynia 4–5.5 mm long; plants brownish at base ............................. ***C. hitchcockiana***
2 Leaf sheaths glabrous; perigynia 3.5–4 mm long; plants reddish at base ........................................... ***C. oligocarpa***

3 Peduncles of lateral spikes finely scabrous; staminate spike long-peduncled; perigynia 2.5–3.6 (–4) mm long, usually more than 20 per spike ............................................ ***C. conoidea***
3 Peduncles of lateral spikes smooth; staminate spike sessile or nearly so; perigynia 4–5 mm long, usually fewer than 15 per spike ***C. grisea***

***Carex conoidea*** Schkuhr OPEN-FIELD SEDGE May–July. Wet calcareous prairies, sedge meadows; also in drier old fields.

***Carex grisea*** Wahlenb. INFLATED NARROW-LEAF SEDGE Mesic to wet deciduous forests, roadside ditches.

***Carex hitchcockiana*** Dewey HITCHCOCK'S SEDGE Mesic woods.

***Carex oligocarpa*** Schkuhr RICHWOODS SEDGE Mesic woods.

## CAREX SECTION HELEOGLOCHIN

Plants densely tufted; bases brown. Stems narrowing toward the tip, typically arching at maturity. Inner band of the leaf sheaths smooth, pigmented toward the summit. Leaf blades less than 3 mm wide (ours). Spikes androgynous, the lower branched. Perigynia plano-convex to biconvex, darkening at maturity, mostly less than 3 mm long; beak short-triangular, scabrous on the margin, bidentate. Wetlands, primarily in peaty soils.

*Key to species*

1 Leaf sheaths whitish or pale ventrally except for purplish dots; inflorescence ± crowded, the lowermost spike (or branch) usually at least slightly overlapping the next above it (occasionally separated by a distance no more than its total length); perigynia tending to spread at maturity, therefore not concealed by the scales ................................................... ***C. diandra***
1 Leaf sheaths strongly tinged with copper color toward their summits ventrally; inflorescence ± interrupted, the lowermost spikes (or branches) often well separated or even peduncled; perigynia ± appressed at maturity, nearly or completely concealed by the large scales .................................. ***C. prairea***

***Carex diandra*** Schrank LESSER TUSSOCK SEDGE May–July. Wet meadows, ditches, calcareous peatlands, floating mats.

*Carex prairea* Dewey PRAIRIE SEDGE May–July. Wet meadows, calcareous fens, marshes, tamarack swamps and peaty lakeshores.

## CAREX SECTION HIRTIFOLIAE

One member of the section in Wisc; recognized by the soft pubescence covering the entire plant, including the distinctly beaked, 2-ribbed perigynia.

*Carex hirtifolia* Mackenzie PUBESCENT SEDGE Rich mesic woods.

## CAREX SECTION HOLARRHENAE

One member of the section in Wisc; resembling those of sections Divisae and Ammoglochin but distinguished by the green-veined inner band of its leaf sheaths.

*Carex sartwellii* Dewey SARTWELL'S SEDGE May–July. Wet to moist meadows, marshes, fens and shores, often where calcium-rich.

## CAREX SECTION HYMENOCHLAENAE

Includes nearly all of the forest understory sedges with long, nodding pistillate spikes. Superficially similar to section Gracillimae but plants more delicate. Terminal spike wholly staminate. Perigynia 8–45 per spike (fewer in *Carex assiniboinensis*), narrow and long-tapering to the beak. Woodlands and wetlands in mostly northern Wisc.

*Key to species*

1 Terminal spike gynecandrous; sheaths ± softly pubescent or perigynia essentially beakless (except *C. prasina*) . . . . . . . . . . . . . . 2

1 Terminal spike staminate; sheaths glabrous (except *C. castanea*) and perigynia conspicuously beaked . . . . . . . . . . . . . . . . . . . . . . . . . . . . 5

2 Perigynia strongly angled, gradually tapering into a beak 1–1.5 mm long; bract of lowest pistillate spike sheathless or with sheath up to 1.2 cm long; terminal spike mostly staminate, with at most a few perigynia at apex . . . . . . . . . . . . . . . . . . . . . . . . . . . . . . . . . . . ***C. prasina***

2 Perigynia obscurely angled or nearly terete, essentially beakless or beak less than 0.5 mm long; bract of lowest spike with sheath 1.5–8 cm or more in length; terminal spike staminate at base, pistillate toward apex . . . . . . . . . . . . . . . . . . . . . . . . . . . . . . . . . . . . . . . . . . . . 3

3 Perigynia 1.3–1.6 mm wide, beakless; sheaths and blades glabrous . . . . . . . . . . . . . . . . . . . . . . . . . . . . . . . . . . . . . . . . . . . . . . ***C. gracillima***

3 Perigynia 1.7–2.5 mm wide, abruptly contracted to a short beak; sheaths and leaf blades ± softly pubescent, at least below (sometimes very sparsely so) . . . . . . . . . . . . . . . . . . . . . . . . . . . . . . . . 4

4 Upper pistillate scales with a distinct prolonged awn greater than 0.5 mm long, often nearly equaling or exceeding the perigynia; lateral spikes entirely pistillate . . . . . . . . . . . . . . . . . . . . . . . ***C. davisii***

4 Upper pistillate scales merely acute (or at most with a tip less than 0.5 mm long), distinctly shorter than the perigynia; lateral spikes usually with a few staminate flowers at base . . . . . . . . . . ***C. formosa***

5 Perigynia pubescent . . . . . . . . . . . . . . . . . . . . . . . . . . ***C. assiniboinensis***

5 Perigynia glabrous . . . . . . . . . . . . . . . . . . . . . . . . . . . . . . . . . . . . . . . . . . 6

6 Leaf sheaths and blades (at least toward the base) ± hairy; pistillate spikes 1–2.5 cm long ................................ *C. castanea*

6 Leaf sheaths and blades glabrous (at most the lowermost bladeless sheaths minutely hispidulous); pistillate spikes mostly (2–) 2.5–6.5 cm long ................................................ 7

7 Basal sheaths reddish purple for at least several cm above the base; perigynia clearly nerved between the 2 ribs .................. 8

7 Basal sheaths brown, lacking any trace of reddish purple color (at most a small trace on the smaller sheaths in *C. prasina*); perigynia 2-ribbed, but otherwise nerveless or faintly nerved............. 9

8 Perigynia short-stalked, the achene within sessile or nearly so; broadest leaves (5–) 6–10 (–12) mm wide; pistillate scales mostly awned or cuspidate ................................ *C. arctata*

8 Perigynia sessile but the achene within on a definite short stalk 0.5–1 mm long; broadest leaves 2.5–4.5 (–5.5) mm wide; pistillate scales mostly not awned .................................. *C. debilis*

9 Perigynia (somewhat twisted) gradually tapering to a poorly defined conical beak, the cylindrical apical portion only 0.2–0.5 mm long ................................................... *C. prasina*

9 Perigynia (symmetrical) tapering to abruptly contracted into a well developed beak ca. 1.2–4.5 mm long ................ *C. sprengelii*

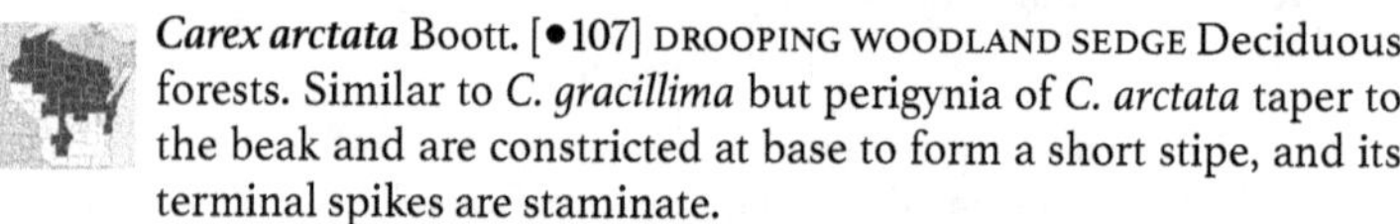

***Carex arctata*** Boott. [•107] DROOPING WOODLAND SEDGE Deciduous forests. Similar to *C. gracillima* but perigynia of *C. arctata* taper to the beak and are constricted at base to form a short stipe, and its terminal spikes are staminate.

***Carex assiniboinensis*** W. Boott ASSINIBOIA SEDGE Rich mesic woods.

***Carex castanea*** Wahlenb. CHESTNUT-COLOR SEDGE June–July. Swamps, moist openings, wetland margins and ditches.

***Carex davisii*** Schwein. & Torr. DAVIS' SEDGE May–June. Floodplain forests, moist woods.

***Carex debilis*** Michx. WHITE-EDGE SEDGE May–Aug. Wet woods (usually under conifers), swamp margins, wet sandy ditches.

***Carex formosa*** Dewey HANDSOME SEDGE *Threatened.* Rich mesic forests, usually where soils calcareous.

***Carex gracillima*** Schwein. [•107] GRACEFUL SEDGE Mesic to wet forests, sometimes in drier oak woods.

***Carex prasina*** Wahlenb. DROOPING SEDGE May–June. Springs, seeps and low areas in deciduous woods, shaded streambanks.

***Carex sprengelii*** Dewey LONG-BEAK SEDGE Mesic and floodplain forests, sometimes where disturbed, not tolerant of heavy shade.

## CAREX SECTION LAXIFLORAE

Plants tufted; bases pale to brown or occasionally reddish. Stems weak, ascending to decumbent, sharply triangular in cross-section, angles sometimes winged. Perigynia triangular in cross-section with rounded edges, 25–40-veined (except *Carex leptonervia*); beak (in our species) abrupt, short, often bent. Woodland species.

*Key to species*

1 Sides of perigynia with at most 1 main nerve, otherwise nerveless or each with up to 6 obscure nerves; perigynium with a straightish or slightly bent short beak ........................ ***C. leptonervia***

1 Sides of perigynia each with 7 or more conspicuous nerves; perigynium with straightish or strongly bent beak ............. 2

2 Angles of bract sheaths smooth or nearly so (granular-papillose in *C. ormostachya*); beak of perigynium usually straight or slightly bent ........................................................... 3

2 Angles of bract sheaths minutely ciliate-serrulate; beak of perigynium strongly bent ..................................... 4

3 Perigynia mostly more than twice as long as wide, tapered to the straightish beak .................................. ***C. laxiflora***

3 Perigynia mostly twice as long as wide, or shorter, abruptly contracted to a very short bent beak ............. ***C. ormostachya***

4 Widest leaves 8 mm or more broad; pistillate scales broadly obtuse or truncate, at most scarcely toothed at apex; staminate spike sessile or nearly so ..................................... ***C. albursina***

4 Widest leaves often less thann 8 mm broad; pistillate scales acuminate, awned, or cuspidate; staminate spike sessile or peduncled .............................................. 5

5 Staminate spike sessile or at most short-peduncled, exceeded by one or more of the bracts; upper pistillate spikes ± crowded; scales not (or only slightly) flushed with orange-brown; bases of plants not red tinged............................................... ***C. blanda***

5 Staminate spike peduncled, elevated above pistillate spikes and (usually) ends of the bracts; pistillate spikes scattered; staminate and sometimes pistillate scales often strongly flushed with orange-brown; bases of plants red tinged, at least on a few bladeless basal sheaths ........................................... ***C. gracilescens***

***Carex albursina*** Sheldon WHITE BEAR SEDGE Mesic forests.

***Carex blanda*** Dewey EASTERN WOODLAND SEDGE Mesic to wet deciduous forests, sometimes in moist open places, tolerant of disturbance.

***Carex gracilescens*** Steud. SLENDER LOOSE-FLOWER SEDGE Mesic deciduous woods. Occurrences in Wisc are considered historical and the species may no longer be present in the state.

***Carex laxiflora*** Lam. BROAD LOOSE-FLOWER SEDGE Mostly in beech woods near Lake Michigan.

***Carex leptonervia*** (Fern.) Fern. NERVELESS WOODLAND SEDGE Woodlands.

***Carex ormostachya*** Wieg. NECKLACE SPIKE SEDGE Woodlands.

## CAREX SECTION LEPTOCEPHALAE

One member of the section in Wisc. Plants soft, very slender, rhizomatous; spike androgynous (with staminate flowers at tip, pistillate flowers below); perigynia few.

***Carex leptalea*** Wahlenb. BRISTLY STALK SEDGE May–July. Swamps, alder thickets, open bogs, calcareous fens; usually in partial shade.

## CAREX SECTION LEUCOGLOCHIN

One member of the section in Wisc. Spike solitary; perigynia several.

***Carex pauciflora*** Lightf. [•108] FEW-FLOWER SEDGE June–July. Open peatlands and floating mats in sphagnum moss, true bogs.

## CAREX SECTION LIMOSAE

Plants loosely tufted or stems arising singly, strongly rhizomatous; bases reddish. Roots covered in a dense yellow felt-like tomentum. Vegetative shoots becoming decumbent, behaving like stolons, producing shoots at the nodes. Pistillate spikes pendulous on slender stalks. Perigynia pale, short-beaked, papillose. Stigmas 3. Common in northern bogs and fens.

*Key to species*

1 Pistillate scales nearly or quite as broad as the perigynia and often only slightly if at all longer; staminate spike (12–) 15–30 (–50) mm long; plants strongly stoloniferous .................... ***C. limosa***

1 Pistillate scales distinctly narrower than perigynia, generally with narrowly acuminate tips much exceeding them; staminate spike 5–12 (–15) mm long; plants loosely clumped ........ ***C. magellanica***

***Carex limosa*** L. [•108] MUD SEDGE May–July. Open bogs and floating mats. Common northward, less common in s where mostly confined to calcareous fens. Poor sedge (*C. magellanica*) similar but has scales much narrower than perigynia; *C. buxbaumii* also similar but lacks yellow roots.

***Carex magellanica*** Lam. POOR SEDGE July–Aug. Open bogs, partly shaded peatlands, floating mats, cedar swamps and thickets, usually in sphagnum moss. *Carex paupercula* Michx.

## CAREX SECTION LUPULINAE

Distinctive sedges of wet forests; recognized by the strongly inflated, ribbed perigynia, 1–2 cm long.

*Key to species*

1 Pistillate spikes spherical or nearly so, scarcely if at all longer than wide; sheath of uppermost leaf absent or less than 1.5 (–2.5) cm; style straight or sinuous or contorted (especially in *C. intumescens*) just below or at the middle; beak of perigynium much shorter than the body ........................................................ **2**

1 Pistillate spikes cylindrical or short-oblong, usually definitely longer than broad; sheath of uppermost leaf usually 1.7 cm or longer; style strongly bent and contorted immediately above the body of the achene; beak of perigynium nearly or quite as long as the body . 3

2 Perigynia (7–) 10–31 per spike, radiating in all directions, narrowed at the base to a ± broad cuneate stalk, sometimes hispidulous basally; pistillate spikes 1–2 (–3) ........................ *C. grayi*

2 Perigynia 2–8 (–12) per spike, mostly spreading-ascending, rounded at the base, glabrous (and often very shiny); pistillate spikes (1–) 2–5 ............................................ *C. intumescens*

3 Body of achene with broadly diamond-shaped sides, mostly 2.4–3.4 mm wide, at most 0.5 mm longer than wide, the angles each with a prominent swollen knob ........................ *C. lupuliformis*

3 Body of achene with somewhat diamond-shaped to ± elliptic or ovate sides, 1.7–2.6 (–2.8) mm wide, usually 1 mm or more longer than wide, the angles obscurely if at all knobbed ...... *C. lupulina*

***Carex grayi*** Carey GRAY'S SEDGE June–Sept. Floodplain forests and backwater areas (as along Mississippi River).

***Carex intumescens*** Rudge [•108] GREATER BLADDER SEDGE May–Aug. Mixed and deciduous moist forests, kettle wetlands in woods, swamps and alder thickets.

***Carex lupuliformis*** Sartwell FALSE HOP SEDGE *Endangered.* July–Sept. Low areas in forests (floodplains and seasonally wet depressions), swamps and marshes; often in shallow water. Similar to *C. lupulina* but much less common.

***Carex lupulina*** Muhl. HOP SEDGE June–Aug. Wet woods, swamps, wet meadows and marshes, ditches and shores. Shining bur sedge (*C. intumescens*) is similar but differs from hop sedge by having fewer, uncrowded perigynia which are olive-green and glossy.

## CAREX SECTION MULTIFLORAE

Plants tufted; bases fibrous, brown or pale. Inner band of the leaf sheaths hyaline, corrugated. Inflorescence compound, cylindrical, densely flowered, stiff. Bracts setaceous. Spikes androgynous, at least the lowest branched. Perigynia plano-convex, weakly or inconspicuously spongy at the base. Primarily in wetlands. Characterized by the corrugated inner band of the leaf sheaths; firm, narrow stems; and densely flowered, straight, compound inflorescence.

*Key to species*

1 Perigynium abruptly contracted into a beak mostly 0.25–0.5 times as long as body; larger perigynia 1.5–2.3 mm wide with broadly ovate to ± orbicular bodies, fully mature (orange-brown) early to midsummer ..................................... *C. annectens*

1 Perigynium tapering or contracted into a beak 0.5–1 times as long as body; larger perigynia 1.1–1.9 mm wide with ovate bodies, fully mature (greenish to dull yellow or brown) late summer .......... .............................................. *C. vulpinoidea*

*Carex annectens* (Bickn.) Bickn. YELLOW-FRUIT SEDGE Dry fields.

*Carex vulpinoidea* Michx. COMMON FOX SEDGE May–Aug. Wet to moist meadows, marshes, lakeshores, streambanks, roadside ditches.

## CAREX SECTION OVALES

In general, Ovales are characterized by a tufted habit, brownish basal sheaths, and sterile shoots with both nodes and internodes; this is in contrast to the sterile shoots of most species of *Carex*, where the stem-like portion is formed only of overlapping leaf sheaths, and nodes and internodes are absent. Mature perigynia (and often a dissecting microscope) are often needed for accurately identifying species in this large group. Considering the preferred moisture regime of Wisconsin Ovales may help narrow the list of possible species:

Wetlands: *C. bebbii, C. crawfordii, C. cristatella, C. longii, C. muskingumensis, C. normalis, C. projecta, C. scoparia, C. straminea, C. suberecta, C. tenera, C. tribuloides.*

Non-wetlands: *C. adusta, C. bicknellii, C. brevior, C. cumulata, C. festucacea, C. foenea, C. merritt-fernaldii, C. molesta.*

ADDITIONAL SPECIES *Carex ovalis* Goodenough (Oval sedge), adventive in Wisc in Bayfield and Dane counties. Plants densely tufted; inflorescence erect, the lower spikes usually separate, otherwise all spikes overlapping. The perigynia with terete, entire-margined beak tips are unique in our Ovales. *Carex leporina* L., a Eurasian species, has been incorrectly applied to this species.

*Key to species*

1 Pistillate scales about or fully as long as the perigynia and nearly the same width as the beaked portion (not necessarily the body), so that the tip of each perigynium is largely concealed; anthers 1.5–3 mm long ........ 2

1 Pistillate scales (or most of them) both shorter and narrower than beaks of perigynia, so the mature perigynia are largely exposed at the tip; anthers various........ 3

2 Inflorescence stiff, the spikes close together, mostly overlapping; pistillate scales nearly as wide as the bodies of the perigynia, almost concealing them ........ *C. adusta*

2 Inflorescence ± lax or flexuous, the lowermost spikes usually remote; pistillate scales distinctly narrower than bodies of perigynia (the wings of which clearly protrude at maturity) ....... *C. foenea*

3 Pistillate scales in the middle or lower portions of the spikes acuminate with a subulate tip or awned ........ 4

3 Pistillate scales obtuse, acute or acuminate, sometimes inconspicuous in the spikes ........ 7

4 Perigynia 2.6–4 times as long as wide, the bodies lanceolate; 0.9–2 mm wide ........ 5

4 Perigynia less than 2.5 times as long as wide, the bodies lance-ovate, ovate, broadly elliptic, orbicular, or obovate; 1.8–3.9 mm wide .. 6

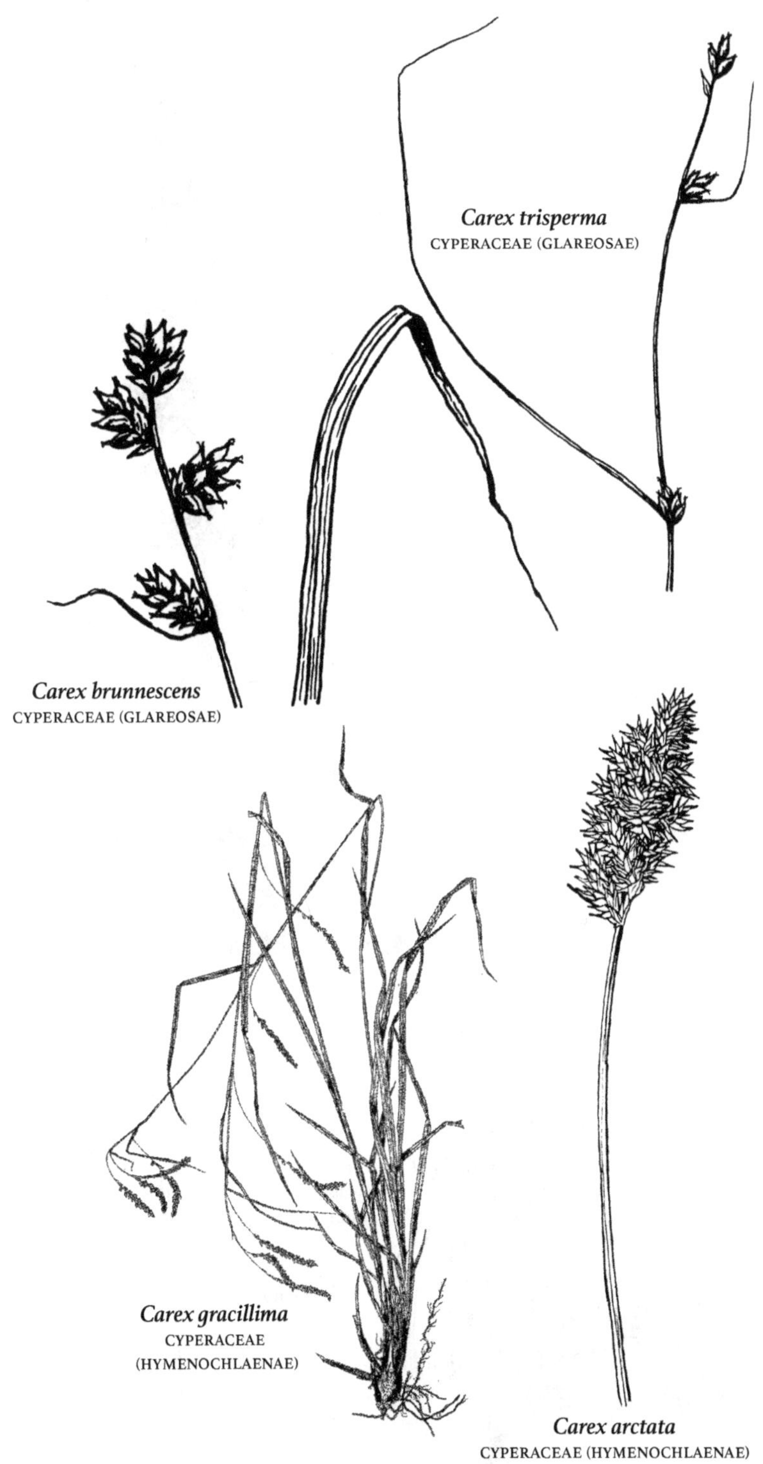

*Carex trisperma*
CYPERACEAE (GLAREOSAE)

*Carex brunnescens*
CYPERACEAE (GLAREOSAE)

*Carex gracillima*
CYPERACEAE
(HYMENOCHLAENAE)

*Carex arctata*
CYPERACEAE (HYMENOCHLAENAE)

*Carex vulpinoidea*
CYPERACEAE (MULTIFLORAE)
*Carex limosa*
CYPERACEAE (LIMOSAE)
*Carex intumescens*
CYPERACEAE
(LUPULINAE)
*Carex pauciflora*
CYPERACEAE
(LEUCOGLOCHIN)

5 Perigynia 0.9–1.2 mm wide; achenes 0.6–0.8 mm wide; inflorescences dense, lowest inflorescence internodes 2–3 (–5) mm long .......................................... *C. crawfordii*

5 Perigynia 1.2–2.0 mm wide; achenes 0.7–1.1 mm wide; inflorescences dense to open or flexuous, lowest internodes 2–17 mm long ...... .............................................. *C. scoparia*

6 Perigynium bodies cuneately tapering to the base, the base therefore subacute and the body ± diamond shaped; inflorescences ± compact, stiffly erect, with 3–5 spikes .............. *C. suberecta*

6 Perigynium bodies rounded to the base, the body therefore elliptic, orbicular or weakly obovate; inflorescences elongate and nodding, the larger with 5–7 spikes ......................... *C. straminea*

7 Perigynia (6.5–) 7–9 mm long; larger spikes on each stem 1.5–2.5 cm long and 1/5–1/3 as thick in the middle, tapered to both ends ..... ...................................... *C. muskingumensis*

7 Perigynia shorter; spikes usually shorter and a third or more as thick as long ................................................ **8**

8 Mature perigynia more than 2 mm broad at widest part......... **9**

8 Mature perigynia not over 2 mm broad ...................... **19**

9 Perigynium bodies obovate, widest above the middle; leaf sheaths green-nerved ventrally nearly to the summit with at most a narrow V-shaped hyaline area .................................. **10**

9 Perigynium bodies lanceolate, ovate, elliptic, or orbicular, widest at or below the middle; leaf sheaths various, some with prominent hyaline band near the apex ventrally ........................ **11**

10 Perigynia nerveless on ventral face; broadest leaves 3–6 (–8) mm wide, their sheaths truncate at summit and extending 0.3 mm above base of leaf blade ................................ *C. cumulata*

10 Perigynia nerved on ventral face; broadest leaves 2–4 mm wide, their sheaths concave at summit and not prolonged above base of leaf blade ............................................ *C. longii*

11 Leaf sheaths green-nerved ventrally nearly to the summit; perigynia cuneately tapering to the base, the body therefore ± diamond-shaped ........................................ *C. suberecta*

11 Leaf sheaths with a white hyaline area ventrally; perigynia rounded to the base, the bodies therefore ovate, elliptic, or orbicular .... **12**

12 Perigynium bodies narrowly to broadly ovate, greenish, gradually tapered to the beak; pistillate scales with a green midstripe and hyaline or pale margins, rarely brown tinged; leaves 2.5–6.5 mm wide, the sheaths green-mottled, the mouth of sheaths truncate and prolonged up to 2 mm above the base of leaf blades ... *C. normalis*

12 Perigynium bodies broadly ovate, broadly elliptic, or orbicular, yellowish to tan brown, often abruptly contracted to the beak; pistillate scales greenish to dark brown; leaves 1.5–4 (–5) mm wide, the sheaths evenly colored, the mouth of sheaths concave (prolonged above base of leaf blades in *C. merritt-fernaldii*) ..... **13**

13 Leaf sheaths finely papillose at high magnification (30–40×), especially near leaf base .................................... 14
13 Leaf sheaths smooth ........................................ 17

14 Perigynia strongly and evenly 4–8-nerved over the achene on the ventral face, (4.5–) 5.1–6.7 mm long, pistillate scales usually (1–) 1.4–2.3 mm shorter than the perigynia; anthers 2.8–4.2 mm long ..... ............................................... *C. bicknellii*
14 Perigynia nerveless or faintly and irregularly 0–5 (–6)-nerved over the achene on the ventral face, 2.5–5.5 mm long, pistillate scales 0.2–1.3 mm shorter than the perigynia; anthers 1.3–2.6 mm long .... 15

15 Pistillate scales dark rust or brown; leaves of fertile shoots 2–4, the leaf sheaths with ventral hyaline area sometimes puckered or cross-corrugated ............................................. *C. tincta*
15 Pistillate scales greenish to yellowish; leaves of fertile shoots 3–6, the leaf sheaths not puckered ............................. 16

16 Perigynia 2.5–3.4 mm wide; achenes 1.3–1.5 mm wide ........... ......................................... *C. merritt-fernaldii*
16 Perigynia 2–2.5 mm wide; achenes (0.9–) 1–1.3 mm wide ......... ............................................... *C. festucacea*

17 Spikes on larger inflorescences 2–4 (rarely more), rounded at the base, the terminal one lacking a conspicuous staminate base; inflorescences mostly 1.3–3 cm long (the lowest internodes generally 1.5–6 mm long); perigynium bodies elliptic to ovate (rarely orbicular), 1–1.6 times as long as wide ................ *C. molesta*
17 Spikes on larger inflorescences (4–) 5–7 or more, tapered at the base, the terminal one with a conspicuous staminate base; inflorescences typically 2.5–4.5 (–6) cm long (the lowest internodes generally 5–13 mm long); perigynium bodies broadly ovate to orbicular, (0.7–) 0.9–1.2 times as long as wide .................................. 18

18 Larger perigynia 2.5–3.5) mm wide, the ventral face usually nerveless; larger achenes 1.4–1.8 mm wide ............ *C. brevior*
18 Larger perigynia 1.5–2.5 mm wide, the ventral face mostly 2–4-nerved; larger achenes 0.9–1.3 mm wide ............ *C. festucacea*

19 Perigynia thin, ± scale-like, often not winged to the base; leaf sheaths somewhat expanded towards apex and bearing narrow wings continuous with midrib and edges of leaf blade, blades 3–7 mm wide; vegetative shoots tall, conspicuous, and with numerous leaves spaced along upper 1/2 of stem ....................... 20
19 Perigynia thicker, plano-convex, winged to the base; leaf sheaths with ± rounded edges, not distinctly expanded towards apex, blades 1–4.5 mm wide (except in *C. normalis*); vegetative shoots usually inconspicuous, with leaves relatively few and clustered at apex . 22

20 Perigynia stiffly spreading or recurved; spikes ± spherical; pistillate scales hidden, 1.6–2.3 mm long .................... *C. cristatella*
20 Perigynia loosely spreading or appressed ascending; spikes nearly spherical to ovate-oblong; pistillate scales evident, 2–3 mm long . 21

21 Inflorescences stiff, spikes overlapping; perigynia usually more than 40, beaks appressed-ascending; leaf sheaths firm at summit ...... ............................................... *C. tribuloides*

21 Inflorescences flexuous, the lower spikes usually separated; perigynia usually 15–40, the beaks spreading; leaf sheaths firm or friable at summit ........ *C. projecta*

22 Perigynia 2.6–4 times longer than wide, the bodies lanceolate, the distance from beak tip to top of achene 2.2–5.0 mm (as little as 1.8 mm long in *C. crawfordii* with perigynia less than 1.2 mm wide) . **23**

22 Perigynia less than 2.5 times longer than wide, the bodies obovate, orbicular, or ovate, the distance from beak tip to top of achene 0.8–2.2 mm ........ **25**

23 Perigynia 0.9–1.2 mm wide; achenes 0.6–0.8 mm wide; inflorescences dense, lowest inflorescence internodes 2–3 (–5) mm long ........ *C. crawfordii*

23 Perigynia 1.2–2 mm wide; achenes 0.7–1.1 mm wide; inflorescences dense to open or even flexuose, lowest internodes 2–17 mm long **24**

24 Inflorescences dense or open, spikes usually overlapping; perigynia usually ascending ........ *C. scoparia*

24 Inflorescences nodding or flexuous, spikes separated; perigynia spreading ........ *C. echinodes*

25 Perigynium bodies obovate, widest above the middle of the body ........ *C. longii*

25 Perigynium bodies ovate, elliptic, or orbicular, widest at or below middle of the body ........ **26**

26 Inflorescences on tallest stems compact, 1.5–3 times as long as wide, erect, the spikes overlapping; lowest inflorescence internodes 1–6 (–7.5) mm long, 1/12–1/5 (–1/4) the total length of the inflorescence . **27**

26 Inflorescences on tallest stems elongate, ± open proximally, (2.5–) 3–5.1 times as long as wide, often arching or nodding; lowest inflorescence internodes (5–) 7–19 mm long, mostly 1/5–1/3 (–1/2) the total length of the inflorescence ........ **30**

27 Achenes 0.6–0.9 mm wide; perigynia nerveless or with 1–3 faint or basal nerves on the ventral face; inflorescences less than 3 cm long ........ *C. bebbii*

27 Achenes 0.9–1.3 mm wide; perigynia often with 3 or more well-defined ventral nerves; inflorescences 1–6 cm long ........ **28**

28 Perigynium body broadly elliptic, or nearly orbicular, with wing margin 0.4–0.8 mm wide, conspicuous ventral nerves 0–6 ........ *C. molesta*

28 Perigynium body ovate to broadly ovate, their wing margins 0.2–0.4 mm wide, ventral nerves 4–7 ........ **29**

29 Sheaths smooth, whitish mottled, the inner band not corrugated; perigynia greenish at maturity ........ *C. normalis*

29 Sheaths finely papillose (30–40×, most easily seen near the leaf base) not whitish mottled, the inner band sometimes corrugated; perigynia pale brown at maturity ........ *C. tincta*

30 Perigynium body orbicular, widest at middle and abruptly contracted into the beak ........ *C. festucacea*

30 Perigynium body narrowly to broadly ovate, widest below the middle and tapering to contracted into the beak ........ **31**

31 At least some sheaths papillose near the collar (30–40×), not prominently whitish mottled; perigynium beaks appressed or ascending in the spikes, exceeding pistillate scales by 0–0.8 mm; beaks and shoulders of perigynia stramineous to reddish brown at maturity ........................................ *C. tenera*

31 Sheaths totally smooth, often whitish mottled; perigynium beaks spreading, mostly exceeding pistillate scales by 0.7–1.6 mm; beaks and shoulders of perigynia greenish to yellowish or greenish brown at maturity ........................................ 32

32 Inflorescences erect to somewhat bent, the lowest internodes mostly 6–10long, the rachis stiff; leaves 2.2–6.5 mm wide; larger perigynia mostly 3.1–3.8 mm long and 1.8–2.2 times longer than wide; plants forming small, ± erect clumps often with fewer than 20 stems .... ........................................ *C. normalis*

32 Inflorescences arching or nodding, the lowest internodes (6–) 10–21 mm long, the rachis usually thin and wiry; leaves 1.5–3.5 mm wide; larger perigynia mostly 3.5–4.6 mm long, 2.1–3 times longer than wide; plants often forming large, spreading clumps of more than 30 stems ........................................ *C. echinodes*

***Carex adusta*** Boott LESSER BROWN SEDGE Dry soil.

***Carex bebbii*** Olney [•109] BEBB'S SEDGE June–Aug. Wet to moist meadows, marshes, streambanks, ditches and other wet places; calcareous fens in s Wisc.

***Carex bicknellii*** Britt. BICKNELL'S SEDGE Dry soil.

***Carex brevior*** (Dewey) MACKENZIE SHORT-BEAK SEDGE Open places.

***Carex crawfordii*** Fern. CRAWFORD'S SEDGE July–Sept. Moist openings and wetland margins, sandy shorelines.

***Carex cristatella*** Britt. CRESTED SEDGE June–Aug. Wet meadows, ditches, floodplains, marshy shores and streambanks.

***Carex cumulata*** (Bailey) Fern. CLUSTERED SEDGE Rocky woods and sandy places.

***Carex festucacea*** Schkuhr FESCUE SEDGE Woodlands.

***Carex foenea*** Willd. BRONZE-HEAD OVAL SEDGE Dry open sandy places, roadsides, cut-over forests. *Carex aenea* Fernald.

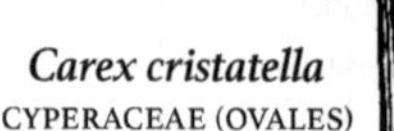

*Carex cristatella*
CYPERACEAE (OVALES)

***Carex longii*** Mackenzie LONG'S SEDGE May–Aug. Open wetlands and wetland margins, often where sandy. La Crosse County, more common southward in USA.

***Carex merritt-fernaldii*** Mackenzie MERRITT FERNALD'S SEDGE Dry woodlands.

***Carex molesta*** Mackenzie TROUBLESOME SEDGE Woods.

***Carex muskingumensis*** Schwein. MUSKINGUM SEDGE June–Aug. Floodplain forests (as along Mississippi River), wet woods.

***Carex normalis*** Mackenzie GREATER STRAW SEDGE June–Aug. Moist to wet deciduous woods, floodplain forests, alder thickets, marshes, pond margins.

***Carex projecta*** Mackenzie NECKLACE SEDGE June–Aug. Floodplain forests, swamps, thickets, wet openings, shaded slopes. Similar to *C. tribuloides* but the perigynia tips spreading rather than erect as in *C. tribuloides.*

***Carex scoparia*** Schkuhr POINTED BROOM SEDGE May–July. Wet meadows and openings, low prairie, swamps and sandy lakeshores.

***Carex straminea*** Willd. EASTERN STRAW SEDGE May–July. Marshes and wetland margins.

***Carex suberecta*** (Olney) Britt. PRAIRIE STRAW SEDGE May–July. Calcareous swamps, marshes, wet meadows, low prairie, calcareous fens and shores.

***Carex tenera*** Dewey [•109] QUILL SEDGE June–Aug. Wet to moist meadows, streambanks, floodplains and moist woods.

ADDITIONAL TAXON ***Carex tenera*** var. ***echinodes*** sometimes treated as separate species ***Carex echinodes*** (Fernald) P. Rothr., found in mesic to wet forests, in mostly s Wisc. Inflorescences nodding, similar to those of *C. tenera* var. *tenera,* but the perigynium tips arch outward and are often 1 mm or more longer than the pistillate scales.

***Carex tincta*** (Fern.) Fern. TINGED SEDGE Woodlands. Ashland County, disjunct from main range of northeastern North America.

***Carex tribuloides*** Wahlenb. [•109] BLUNT BROOM SEDGE June–July. Floodplain forests, shady low areas in woods, pond and lake margins, marshes, low prairie.

## CAREX SECTION PALUDOSAE

Mostly slender, long-rhizomatous plants, with red basal leaf sheaths (and ladder-fibrillose in all but *C. houghtoniana*), and pubescent perigynia. *Carex lacustris* is somewhat different, having glabrous perigynia.

*Key to species*

1 Perigynia glabrous .................................... ***C. lacustris***

1 Perigynia pubescent ............................................ **2**

2 Perigynia 4.5–6.5 mm long, sparsely hairy, the strong nerves of perigynium and even cellular detail of body therefore evident; plants usually of dry and sandy habitats ......... *C. houghtoniana*

2 Perigynia 3–4.5 (–5.2) mm long, densely pubescent, nerving of perigynium and cellular detail therefore obscured; plants usually of wetlands ............................................. 3

3 Leaf blades involute to triangular-channeled, 0.7–2 (–2.2) mm wide, those of vegetative shoots especially long-prolonged into a curled, filiform tip; leaves and lowermost bracts with the midvein low, rounded, and forming an inconspicuous keel (at least proximally) ............................................. *C. lasiocarpa*

3 Leaf blades flat or folded into an M-shape except at the base and near the tip, (2–) 2.2–4.5 (–6.5) mm wide, not prolonged into a long filiform tip; leaves and lowest bract with the midvein forming a prominent and sharply pointed keel for much of the length ......... *C. pellita*

***Carex houghtoniana*** Torr. ex Dewey HOUGHTON'S SEDGE Open sandy or rocky soil.

***Carex lacustris*** Willd. [•110] LAKEBANK SEDGE May–Aug. Swamps, marshes, kettle wetlands, wetland margins, usually in shallow water; low areas in tamarack swamps.

***Carex lasiocarpa*** Ehrh. [•110] SLENDER SEDGE June–Aug. Peatlands and wet peaty soils, open bogs, pond margins (where a pioneer mat-former). *Carex lanuginosa* Michx. var. *americana* (Fern.) Boivin.

***Carex pellita*** Muhl. ex Willd. [•110] WOOLLY SEDGE Swamps. Carex lanuginosa auct. non Michx., *Carex lasiocarpa* Ehrh. var. *latifolia* (Boeckl.) Gilly.

## CAREX SECTION PANICEAE

Plants colonial, shoots arising singly or few together. Leaf blades typically stiff. Terminal spike staminate, typically raised above the uppermost pistillate spike. Calciphiles, growing mostly in wet soils (but *C. meadii* common in dry calcareous prairies). The section is fairly distinctive and easy to recognize, apart from *C. vaginata,* which is morphologically distinct.

*Key to species*

1 Perigynium with a beak 1 mm long .................. *C. vaginata*

1 Perigynium beakless, indistinctly beaked, or contracted to beak less than 0.5 mm ............................................. 2

2 Perigynia strongly ascending, beakless or tapering to an erect, very short straight beak; leaves stiff, thick, channeled, strongly glaucous ............................................. *C. livida*

2 Perigynia ascending to spreading, tapering to a bent apex; leaves relatively thin and flexible, flat or folded, green to somewhat glaucous ............................................. 3

3 Bladeless basal sheaths and proximal leaf sheaths strongly tinged with reddish purple; plants forming loose clumps to extensive closed colonies of vegetative shoots from superficial rhizomes; perigynia ± 2-ranked; plants of rich forests .............. *C. woodii*

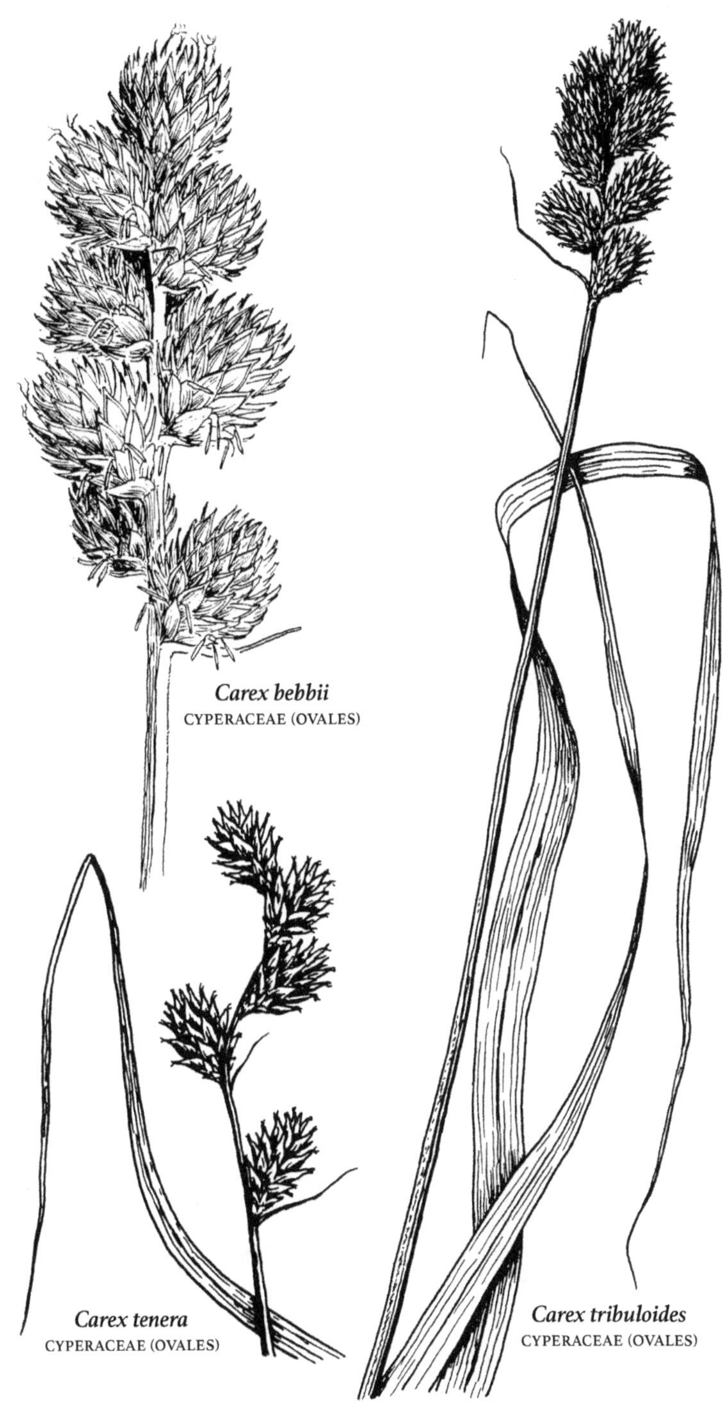

*Carex bebbii*
CYPERACEAE (OVALES)

*Carex tenera*
CYPERACEAE (OVALES)

*Carex tribuloides*
CYPERACEAE (OVALES)

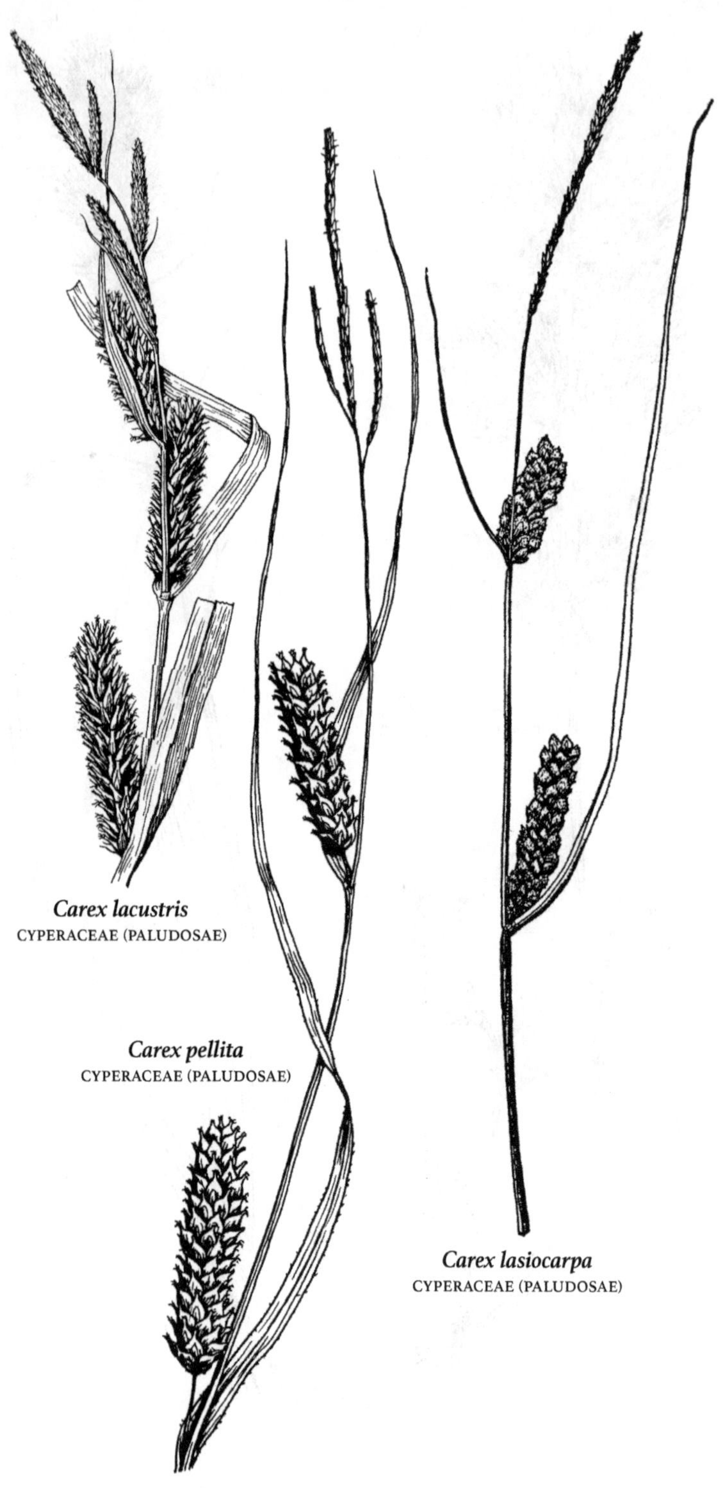

*Carex lacustris*
CYPERACEAE (PALUDOSAE)

*Carex pellita*
CYPERACEAE (PALUDOSAE)

*Carex lasiocarpa*
CYPERACEAE (PALUDOSAE)

3 Bladeless basal sheaths and proximal leaf sheaths brownish, green, or faintly, irregularly tinged with reddish purple; plants usually with vegetative shoots widely scattered and inconspicuous from deep rhizomes; perigynia 3–6-ranked; plants of moist, usually sunny habitats .................................................. 4

4 Largest achenes 1.7–2.2 (–2.5) mm wide; longest anthers (3.5–) 4–4.6 mm long; ligules mostly 0.4–1.2 times as long as wide ... *C. meadii*

4 Largest achenes 1.2–1.7 mm wide; longest anthers 2.8–3.5 (–3.8) mm long; ligules mostly (0.8–) 1–2 times as long as wide .... *C. tetanica*

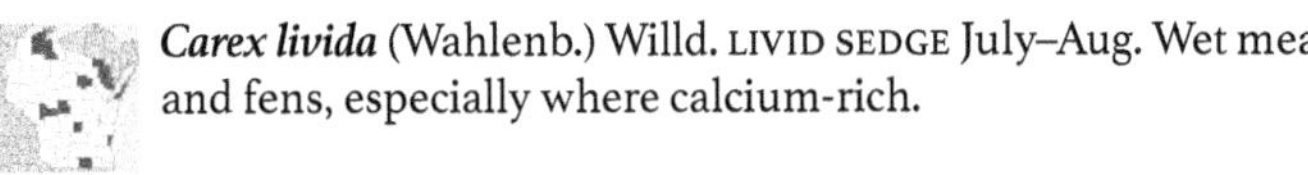

***Carex livida*** (Wahlenb.) Willd. LIVID SEDGE July–Aug. Wet meadows and fens, especially where calcium-rich.

***Carex meadii*** Dewey MEAD'S SEDGE Prairies.

***Carex tetanica*** Schkuhr RIGID SEDGE May–July. Wet meadows and openings, low prairies, marshy areas.

***Carex vaginata*** Tausch SHEATHED SEDGE June–Aug. Swamps and thickets, especially where calcium-rich.

***Carex woodii*** Dewey PRETTY SEDGE Dry woodlands.

## CAREX SECTION PHACOCYSTIS

Plants often cespitose; rhizomes short or long. Lower leaf sheaths brown to red, fibrous in some species. Terminal spike typically staminate, ascending. Lateral spikes pistillate or androgynous, ascending to nodding or drooping, elongate. Perigynia biconvex with distinct marginal veins. Stigmas 2. Mostly common in Wisconsin wetlands, ranging from floodplains and wet forests (*Carex crinita, C. gynandra, C. emoryi*), to sedge meadows (*C. stricta*), wet prairies (*C. haydenii*), bogs and marshes (*C. aquatilis*), and wet roadsides and ditches.

ADDITIONAL SPECIES ***Carex nebrascensis*** Dewey (Nebraska sedge), native of w USA, known in Wisc from an old Milwaukee County collection; pistillate scales purplish or brownish black with 1–3-nerved lighter center and sometimes hyaline margins. Perigynia 30–150 to a spike, leathery, straw-colored, red-dotted, granular; beak bidentate. ***Carex nigra*** (L.) Reichard, reported from wetlands in Douglas and Manitowoc counties; distinguished by dark-spotted perigynia 2–3.5 mm long and black pistillate scales.

*Key to species*

1 Pistillate spikes on ± lax peduncles, at length drooping, the scales prominently awned; body of achene with an irregular notch, constriction, or wrinkle on one side .......................... 2

1 Pistillate spikes erect, often sessile, the scales acute or acuminate, not awned; body of achene smooth and ± regular ............. 3

2 Sheaths smooth; bodies of most if not all pistillate scales shallowly lobed at summit (on each side of base of the awn) ...... *C. crinita*

2 Sheaths scabrous-hispidulous; bodies of most or all pistillate scales on lower part of spike truncate or tapered at summit . *C. gynandra*

3 Fertile stems of current year with conspicuous bladeless sheaths at base, not surrounded by dried-up bases of the previous year's leaves but arising laterally; lowest bract usually shorter than to approximately equaling the inflorescence .................... 4

3 Fertile stems of current year mostly lacking bladeless sheaths at base, arising centrally from tufts of dried-up bases of previous years leaves; lowest bract usually conspicuously longer than the inflorescence............................................... 6

4 Perigynia suborbicular to obovoid, 2–2.3 mm long at maturity, broadest at or slightly above middle, rather abruptly contracted to a minute apiculus, at least the lower ones in a spike much exceeded by the spreading scales; lower leaf sheaths not or only slightly tearing to form a ladder like arrangement of fibers, the intact sheaths smooth ventrally; ligule longer than width of leaf blade; plants with short, ascending rhizomes ................ ***C. haydenii***

4 Perigynia elliptic to ovate, mostly 2.2–2.7 (–3.3) mm long at maturity, broadest at or slightly below the middle, ± tapered to apex, as long as or longer than the scales (rarely exceeded by scales); ventral surface of lower leaf sheaths tearing to form a ladder-like arrangement of fibers (ladder-fibrillose) or if not, then the ligule shorter than width of leaf blade; plants with long horizontal rhizomes ............................................... 5

5 Ligule longer than width of leaf blade (deeply inverted V-shaped); ventral surface of lower leaf sheaths tearing to form a ladder-like arrangement of fibers and usually minutely scabrous and red-dotted, especially near the tip ......................... ***C. stricta***

5 Ligule shorter than width of leaf blade (often nearly horizontal); ventral surface of lower leaf sheaths not tearing to form a ladder like arrangement of fibers, smooth and whitish ......... ***C. emoryi***

6 Perigynia essentially nerveless, except sometimes at the base only; staminate spikes usually 2 or more .................. ***C. aquatilis***

6 Perigynia conspicuously few-ribbed on both sides; staminate spike usually 1............................................... 7

7 Plants densely tufted, without long rhizomes; scales with a broad central green portion about as wide as the darker margins; leaves mostly overtopping spikes ........................ ***C. lenticularis***

7 Plants colonial from elongated rhizomes; scales with very narrow green portion much narrower than the broad, dark margins, scarcely if at all broader than the midrib; leaves mostly shorter than stems ............................................ ***C. nigra***

***Carex aquatilis*** Wahlenb. WATER SEDGE May–Aug. Wet meadows, marshes, shores, streambanks, kettle lakes, ditches and fens.

***Carex crinita*** Lam. FRINGED SEDGE May–July. Swamps and alder thickets, wet openings, ditches and potholes. Similar to *C. gynandra* but with smooth sheaths, lower pistillate scales rounded at tip, and perigynia round in section and inflated.

***Carex emoryi*** Dewey EMORY'S SEDGE May–July. Shores, streambanks, wet meadows and floodplain forests, sometimes forming pure stands, especially along rivers.

***Carex gynandra*** Schwein. NODDING SEDGE June–July. Wet openings and swamps. Similar to *C. crinita,* but with finely hairy sheaths, lower pistillate scales tapered to an awned tip, and perigynia somewhat flattened and not inflated.

***Carex haydenii*** Dewey CLOUD SEDGE May–July. Wet to moist meadows and swales, marshes and streambanks; often with *Carex buxbaumii.*

***Carex lenticularis*** Michx. LAKESHORE SEDGE *Threatened.* June–Sept. Rocky and sandy lakeshores, rock pools along Lake Superior, shallow ponds, sedge mats.

***Carex stricta*** Lam. TUSSOCK SEDGE May–July. Often dominant sedge of wet meadows, marshes, fens, shores, streambanks, ditches.

## CAREX SECTION PHAESTOGLOCHIN

Plants tufted; rhizomes short or inconspicuous; bases pale to brown, occasionally reddish. Inner band of the leaf sheaths hyaline, corrugated or smooth. Spikes all or mostly androgynous, simple in most taxa, the lower branched in some species. Perigynia mostly plano-convex, beaks typically bidentate. Mostly upland species of forests and open, sometimes disturbed habitats.

ADDITIONAL SPECIES ***Carex aggregata*** Mackenzie, native, Lafayette County, more common southward in USA; open, disturbed habitats and dry prairies in s Wisc; similar to *C. gravida,* but inner band of leaf sheaths are yellow or brown and thickened at top. ***Carex spicata*** Huds. (Prickly sedge), introduced from Asia, known from sw Wisc; stems tinged red to purple-tinged at base, inflorescence elongate.

*Key to species*

1 Leaf sheaths loose, white with green veins or mottled green and white on back; wider blades 5–10 mm broad (or rarely only 3 mm in *C. gravida* and *C. aggregata* with very slender elongate stigmas) 2

1 Leaf sheaths ± tight and slender and uniform green or whitish on back (or sometimes mottled in the slender and narrow-leaved *C. leavenworthii*); wider blades 0.9–4.3 mm broad ................ 5

2 Pistillate scales with narrowly acuminate or awned tips reaching over the bases or all the way to the ends of the beaks of the perigynia they subtend; anthers 1.1–2.4 mm long; stigmas quite elongate and slender, when intact and well developed, protruding 1.5 mm or more from the perigynia; spikes crowded in a dense inflorescence ..... 3

2 Pistillate scales with short-acuminate, slightly cuspidate, acute, or obtuse tips almost or not at all reaching the bases of the beaks of the perigynia they subtend; anthers 0.7–1.1 (–1.3) mm long; stigmas shorter and stouter, protruding slightly from perigynia; spikes crowded or the lower (in *C. sparganioides*) becoming well separated ........................................................ 4

3 Ventral surface of leaf sheath strongly concave and thickened at the summit, usually intact on specimens; face of mature perigynia green at maturity; rare, sw Wisc ......................... ***C. aggregata****

3 Ventral surface of leaf sheaths thin or slightly thickened at the summit, easily broken; face (over achene) of mature perigynia mostly yellow-brown; common in all but n Wisc ....... *C. gravida*

4 Spikes close together, the lower not separated more than their length, usually ± overlapping; perigynia 3.6–4.5 mm long, 2–3 times longer than wide, the bodies not wing-margined; widest leaf blades 5–7 (–8) mm wide .............................. *C. cephaloidea*

4 Spikes well separated below, the lower ones ± remote; perigynia 3–4.1 mm long, 1.3–2 times as long as wide, the bodies ± narrowly thin-winged; widest leaf blades 5.5–10 mm wide .. *C. sparganioides*

5 Pistillate scales with the sides brown or purple, acuminate-awned; larger perigynia mostly 3.5–5.5 mm long ............. *C. spicata**

5 Pistillate scales whitish or greenish, obtuse to acuminate; larger perigynia 2.6–4.1 mm long ................................ **6**

6 Perigynia mostly widely spreading at maturity, conspicuously spongy-thickened at their bases and there puckered in drying, the wire-like margin above the base tending to turn inward......... **7**

6 Perigynia mostly ascending and not widely spreading, at most with thin spongy area at base not conspicuously puckered in drying (unless immature), the margin above flat or slightly incurved ... **8**

7 Wider leaf blades mostly 0.9–1.8 (very rarely 2.5) mm broad; stigmas reddish to dark brown, slender and elongate (when intact), often protruding 1–1.5 mm or more, often reflexed but otherwise straight or slightly sinuous .................................. *C. radiata*

7 Wider leaf blades mostly (1.5–2.7 mm broad; stigmas very dark reddish brown, comparatively short and stout, strongly curled ... .................................................... *C. rosea*

8 Inflorescence crowded to oblong and interrupted (the lower spikes overlapping but distinct); leaf blades densely papillose above (at 20×–30×; bodies of scales more (often much more) than half as long as bodies of the perigynia they subtend; larger perigynia in spike 3–4.1 mm long, 2–2.6 mm wide ................... *C. muehlenbergii*

8 Inflorescence densely crowded, ± ovoid, the spikes in a close head and nearly indistinguishable except by the slightly protruding setaceous bracts; leaf blades smooth above the collar or the cellular outlines conspicuous, but only rarely some leaves papillose; bodies of scales usually about or only slightly more than half as long as bodies of the perigynia; perigynia 2.5–3.2 mm long, 1.5–2 mm wide........... **9**

9 Beak of perigynium uniformly strongly serrulate; anthers 0.7–1 mm long; perigynia broadest at or near the middle of the orbicular to broadly elliptic body .......................... *C. cephalophora*

9 Beak of perigynium smooth or sparsely serrulate at very base, usually near the junction with body; anthers 0.8–1.7 mm long; perigynia broadest toward the base of the very broadly ovoid body ............................................. *C. leavenworthii*

*Carex cephaloidea* (Dewey) Dewey THIN-LEAF SEDGE Woods.

*Carex cephalophora* Muhl. OVAL-LEAF SEDGE Dry woodlands.

*Carex gravida* Bailey HEAVY SEDGE Prairies.

*Carex leavenworthii* Dewey LEAVENWORTH'S SEDGE Dry woodlands. Native to s USA, considered adventive in Wisc.

*Carex muehlenbergii* Schkuhr ex Willd. MUHLENBERG'S SEDGE Sand hills and dry places.

*Carex radiata* (Wahlenb.) Small EASTERN STAR SEDGE Dry woods.

*Carex rosea* Schkuhr ROSY SEDGE Dry woodlands. *Carex convoluta* Mackenzie.

*Carex sparganioides* Muhl. BUR-REED SEDGE Dry woods.

### CAREX SECTION PHYLLOSTACHYAE

Plants tufted, bases brown. Bracts lacking. Lateral spikes absent or basal, pistillate or androgynous. Terminal spike androgynous. Lowest pistillate scale foliose, suggesting the lowest bract of the inflorescence in most other sections, exceeding the tip of the spike. Perigynia 2-ribbed, beak untoothed.

*Key to species*

1 Pistillate scales much wider than the perigynia, embracing and nearly concealing them; staminate scales about 3 (usually hidden by upper perigynia); wider leaf blades (3–) 3.5–5.5 mm broad; anthers 1.3–1.6 mm long ................................ *C. backii*

1 Pistillate scales mostly narrower than the perigynia, not concealing them; staminate scales about 4 or more; wider leaf blades 2–2.6 (–3.5) mm broad; anthers 0.5–1.1 mm long ................ *C. jamesii*

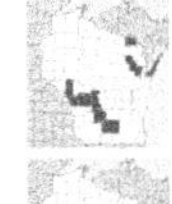

*Carex backii* Boott Back's Sedge Dry woods.

*Carex jamesii* Schwein. James' Sedge Dry woods.

### CAREX SECTION PHYSOGLOCHIN

One member of the section in Wisc; sphagnum moss peatlands.

*Carex gynocrates* Wormsk. NORTHERN BOG SEDGE June–July. Conifer swamps and open peatlands, usually in sphagnum and wet, peaty soils. *Carex dioica* L.

### CAREX SECTION POROCYSTIS

Plants tufted. Leaves and stems usually pubescent, at least sparsely. Pistillate spikes erect to spreading, ovoid to oblong-cylindrical. Perigynia beakless or very short-beaked, glabrous or pubescent. Stigmas 3. The cylindrical pistillate spikes of *C. pallescens* and *C. swanii* resemble the spikes of *C. granularis* and relatives.

ADDITIONAL SPECIES ***Carex hirsutella*** Mackenzie, native, Sheboygan County, more common in s and e USA. ***Carex torreyi*** Tuckerman (Torrey's sedge); Great Plains species known from woodlands in w and se Wisc, similar to *C. pallescens,* but terminal spike rarely gynecandrous, and perigynia strongly many-nerved and short-beaked.

*Key to species*

1 Perigynia pubescent; terminal spike pistillate at apex, staminate at base ........ ***C. swanii***
1 Perigynia glabrous; terminal spike staminate or pistillate at tip .. 2

2 Terminal spike entirely staminate; perigynia ellipsoid-cylindric, rather faintly nerved ........ ***C. pallescens***
2 Terminal spike pistillate at tip; perigynia obovoid-orbicular, ± compressed, strongly nerved (especially on dorsal face)........ 3

3 Pistillate scales longer than the perigynia, with an awn 0.5–2 mm long ........ ***C. bushii***
3 Pistillate scales shorter than the perigynia, awnless, or with an awn not more than 0.5 mm long ........ ***C. hirsutella****

***Carex bushii*** Mackenzie BUSH'S SEDGE Dry meadows and banks. Native of e and s USA, known from Iowa County, considered adventive in Wisc.

***Carex pallescens*** L. PALE SEDGE Dry banks and meadows.

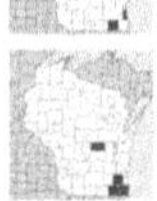

***Carex swanii*** (Fern.) Mackenzie SWAN'S SEDGE Dry woods.

## CAREX SECTION RACEMOSAE

Plants loosely to densely tufted; rhizomes variable in length; bases dark red, generally fibrous; roots not clothed with yellow felt. Terminal spike gynecandrous (in our species). Pistillate scales dark, often black. Perigynia pale, often greenish, very short-beaked to beakless, smooth or papillose, 2-ribbed, inconspicuously veined (in our species). Stigmas 3.

*Key to species*

1 Pistillate scales mostly awned or narrowly acuminate, exceeding the perigynia; ventral surface of lower leaf sheaths tearing into fibers; throughout Wisc ........ ***C. buxbaumii***
1 Pistillate scales obtuse or acute, equaling or shorter than the perigynia; ventral surface of lower sheaths not tearing to form fibers; uncommon in sw Wisc ........ ***C. media***

***Carex buxbaumii*** Wahlenb. [•111] BROWN BOG SEDGE May–Aug. Wet meadows and fens, shallow marshes, low prairie, hollows in patterned peatlands.

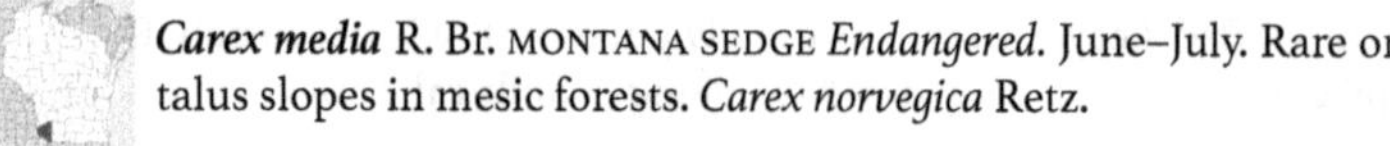

***Carex media*** R. Br. MONTANA SEDGE *Endangered.* June–July. Rare on talus slopes in mesic forests. *Carex norvegica* Retz.

## CAREX SECTION ROSTRALES

Plants tufted; rhizomes elongate; bases brown, not reddish or purplish. Staminate spike solitary, the base lower than or roughly equaling the apex of the uppermost pistillate spike. Pistillate spikes 2–6, approximately as long as thick. Perigynia ≤20 per spike, divergent or the lowermost reflexed, somewhat inflated, narrow, tapering continuously to the apex, generally 4–7x as long as wide, beakless, subtly bidentate. Species in this section superficially resemble the more common *Carex intumescens*, but differ in their narrower perigynia.

*Key to species*

1 Broadest leaf blades 5–17 mm wide; sheaths of bracts with a ± prolonged lobe at mouth; staminate spike usually peduncled, its tip projecting well above the pistillate spikes ........... ***C. folliculata***

1 Broadest leaf blades 1.5–3.5 mm wide; sheaths of bracts concave at mouth; staminate spike sessile or very short-peduncled, scarcely if at all projecting above the pistillate spikes ........ ***C. michauxiana***

***Carex folliculata*** L. NORTHERN LONG SEDGE June–Aug. Wet woods and cedar swamps.

***Carex michauxiana*** Boeckl. MICHAUX'S SEDGE *Threatened.* June–Aug. Wet meadows, sphagnum peatlands, ditches and swales.

## CAREX SECTION SQUARROSAE

The obconic perigynia of this section are highly distinctive, widest at the apex and abruptly narrowed to the beak; perigynia in all of the other "bladder" and "bottlebrush" sections taper more gradually to the beak. *Carex typhina* is the only one of our "bladder" and "bottlebrush" sedges to occasionally produce plants with a single spike.

***Carex typhina*** Michx. [•111] CAT-TAIL SEDGE June–Sept. Floodplain forests of large rivers (especially Mississippi and St. Croix), often occurring with *Carex muskingumensis* and *C. grayi;* marshy areas.

## CAREX SECTION STELLULATAE

Plants tufted; rhizomes short; bases brown, not fibrous. Inflorescence mostly open, spikes readily distinguished from each other, the lowest in our more common species not overlapping; bracts inconspicuous or lacking. Spikes 2–10 (solitary in *Carex exilis*), gynecandrous (unisexual in *C. sterilis*). Perigynia spreading to reflexed, typically plano-convex, widest at the base, generally chestnut brown to dark brown or even blackish at maturity; margins acute; base spongy; beak generally bidentate, margins finely serrate. Achenes much smaller than the perigynia. Wetlands.

The distinctions between species in this section are subtle; however, the species have habitat preferences that help with field identification. When examining perigynia, view the lowest 2–3 perigynia in the spike; the upper perigynia are very similar in all of our species.

*Key to species*

1 Spikes solitary; leaves involute; anthers 2–3.6 mm long ... ***C. exilis***

1 Spikes 2–8; leaves flat or plicate; anthers 0.6–2.2 mm long ....... **2**

2 Terminal spikes entirely staminate ........................ *C. sterilis*

2 Terminal spikes partly or wholly pistillate ........................ 3

3 Terminal spikes without a distinct clavate base of staminate scales, staminate portion less than 1 mm long.................. *C. sterilis*

3 Terminal spikes with a distinct clavate base of staminate scales mostly 1–8 mm long ........................................ 4

4 Lower perigynia mostly 2.9–3.6 mm long, 1.8–3.6 times as long as wide; beaks 0.9–2 mm long, mostly 0.5–0.8 times as long as the body ................................................... *C. echinata*

4 Lower perigynia mostly 1.9–3 mm long, 1–2 times as long as wide; beaks 0.4–0.9 mm long, mostly 0.2–0.5 times as long as body ..... ................................................... *C. interior*

***Carex echinata*** Murr. STAR SEDGE July–Sept. Swamp margins, wet sandy lakeshores, hummocks in peatlands.

***Carex exilis*** Dewey COASTAL SEDGE *Threatened.* June–Aug. Sphagnum peatlands, interdunal wetlands near Great Lakes; coastal disjunct.

***Carex interior*** Bailey INLAND SEDGE May–Aug. Swamps, tamarack bogs, alder thickets, wet meadows and wetland margins.

***Carex sterilis*** Willd. [•111] DIOECIOUS SEDGE April–June. Spring-fed calcareous fens, calcium-rich wet meadows. Similar to *Carex interior.* *Carex muricata* L. var. *sterilis* (Willd.) Gleason.

## CAREX SECTION VESICARIAE

Includes typical bottlebrush sedges of former section Pseudocypereae, with pistillate spikes tightly packed with perigynia, and pistillate scales with scabrous awns conspicuous between the perigynia; and also former section Vesicariae, with pistillate spikes often narrower, longer, less densely packed with perigynia in some species, and pistillate scales mostly not awned, hidden by the perigynia.

*Key to species*

1 Pistillate scales with a prominent, scabrous awn; often the body also ciliate ........................................................ 2

1 Pistillate scales smooth-margined, obtuse to acuminate, awnless (rarely the lowermost awned in *C. rostrata* and *C. utriculata*) ..... 6

2 Perigynia ± reflexed at maturity, hard-walled, uninflated, flattened-triangular in cross-section, strongly and closely nerved with most nerves separated by less than three times their width; longest beak teeth 0.7–2.2 mm long ........................................ 3

2 Perigynia spreading to ascending, thin-textured, ± inflated, ± round in cross-section; many nerves separated by more than three times their width; longest beak teeth 0.3–0.7 mm long ............... 4

3 Spikes 12–18 mm thick; beak teeth strongly outcurved, the longest 1.3–2.1 mm long ................................... *C. comosa*

3 Spikes 9–12 mm thick; beak teeth straight or slightly outcurved, the longest 0.7–1.2 mm ............................ *C. pseudocyperus*

4 Staminate scales (except sometimes the lowermost) acute to acuminate, essentially smooth-margined except at the very tip; plants extensively colonial from elongate, creeping rhizomes; perigynia 7–11-nerved . . . . . . . . . . . . . . . . . . . . . . . . . . . *C. schweinitzii*

4 Staminate scales (at least some) with a distinct, scabrous awn and sometimes also ciliate-margined; plants densely to loosely tufted, rhizomes connecting individual stems in a clump not more than 10 cm long; perigynia 7–25-nerved . . . . . . . . . . . . . . . . . . . . . . . . . . . . . 5

5 Perigynia 15–20-nerved, the nerves (except for the two prominent lateral nerves) fusing together and becoming indistinguishable from about the middle of the beak to the apex, bodies ellipsoid, 1.4–2.2 mm wide; achenes smooth . . . . . . . . . . . . . . . . . . . . *C. hystericina*

5 Perigynia 7–12-nerved, the nerves separate nearly to the beak tip, the bodies broadly ellipsoid to ± spherical, 2–3.5 mm wide; achenes rough-papillate . . . . . . . . . . . . . . . . . . . . . . . . . . . . . . . . . . . . . . . *C. lurida*

6 Leaf blades and bracts involute-filiform, wiry, 1–3 mm wide; stems round or obtusely 3-angled in cross-section, smooth; pistillate spikes 3–15-flowered, nearly spherical or short-oblong (not over 2 cm long); staminate spike usually solitary . . . . . . . . . *C. oligosperma*

6 Leaf blades and bracts flat, U-, V-, or W-shaped in cross-section, 1.5–12 mm wide; stems round to 3-angled, often scabrous-angled; pistillate spikes usually more than 15-flowered, oblong to long-cylindric; staminate spikes normally 2 or more (often 1 in *C. retrorsa*) . . . . . . . . . . . . . . . . . . . . . . . . . . . . . . . . . . . . . . . . . . . . . . . . . . . 7

7 Perigynia 4–7 mm thick; achenes with a deep notch or constriction on one angle . . . . . . . . . . . . . . . . . . . . . . . . . . . . . . . . . . . *C. tuckermanii*

7 Perigynia 2.5–3.5 mm thick; achenes symmetrical, not notched on one angle. . . . . . . . . . . . . . . . . . . . . . . . . . . . . . . . . . . . . . . . . . . . . . . 8

8 Lowest pistillate bract 3–9 times as long as the entire inflorescence; mature perigynia 7–12 mm long, at least the lower reflexed or widely spreading; staminate spike often 1, its base (or base of lowest staminate spike if more than one) slightly if at all elevated above summit of the crowded pistillate spikes (rarely lower spike remote) . . . . . . . . . . . . . . . . . . . . . . . . . . . . . . . . . . . . . . . . . . . . . . *C. retrorsa*

8 Lowest pistillate bract less than 3 times as long as inflorescence; perigynia 4–7.5 mm long, ascending or spreading; staminate spikes mostly 2–4, generally well elevated above the pistillate spikes . . . 9

9 Leaves strongly papillose on upper surface, U-shaped in cross-section, glaucous, widest leaves 1.5–4.5 (–7.5) mm wide; stems round . . . . . . . . . . . . . . . . . . . . . . . . . . . . . . . . . . . . . . . . . . . . . . *C. rostrata*

9 Leaves smooth or scabrous on upper surface, flat or folded, pale to dark green, widest leaves 3–12 mm wide; stems triangular, often scabrous below the inflorescence . . . . . . . . . . . . . . . . . . . . . . . . . . 10

10 Colonial from long-creeping rhizomes; widest leaves 5–12 mm wide; ligules about as long as wide; basal sheaths usually spongy-thickened with little or no red tingeing; perigynia (at least those on lower portion of fully mature spike) ± widely spreading; stems bluntly triangular and sparsely and irregularly scabrous below the inflorescence . . . . . . . . . . . . . . . . . . . . . . . . . . . . . . . . *C. utriculata*

10 Tufted; widest leaves 3–6 mm wide; ligules longer than wide; basal sheaths not spongy-thickened and often tinged with reddish purple; perigynia ascending; stems sharply triangular and scabrous-angled below the inflorescence ............................ *C. vesicaria*

***Carex comosa*** Boott BEARDED SEDGE June–Aug. Marshes, wetland margins, floating mats, ditches.

***Carex hystericina*** Muhl. PORCUPINE SEDGE May–July. Swamps, alder thickets, wet meadows and ditches; calcareous fens in s Wisc.

***Carex lurida*** Wahlenb. SHALLOW SEDGE June–Aug. River floodplains, swamps, open bogs, fens and wet meadows, especially in Wisc and Black River valleys.

***Carex oligosperma*** Michx. FEW-SEED SEDGE June–Aug. Open bogs and swamps, floating mats, pioneer mat-former along pond margins. Sometimes a dominant sedge in poor fens in n Wisc, less common southward.

***Carex pseudocyperus*** L. CYPRESS-LIKE SEDGE June–Aug. Marshy lake margins, swamps, fens, wet ditches; in Minn, an indicator of calcium-rich fens in the Red Lake peatland. Similar to *C. comosa.*

***Carex retrorsa*** Schwein. RETRORSE SEDGE June–Aug. Floodplain forests, swamps, thickets and marshes.

***Carex rostrata*** Stokes SWOLLEN BEAKED SEDGE July–Sept. Peat mats or shallow water. Similar to *Carex utriculata,* but much less common, and with the leaves waxy blue and dotted with fine bumps on upper surface, v-shaped in section or inrolled, and only 2–4 mm wide.

***Carex schweinitzii*** Dewey SCHWEINITZ'S SEDGE *Endangered.* May–July. Shaded streambanks.

***Carex tuckermanii*** Dewey TUCKERMAN'S SEDGE June–Aug. Swamps, alder thickets, low areas in forests, pond margins.

***Carex utriculata*** Boott [•112] NORTHWEST TERRITORY Sedge June–Aug. Wet meadows, marshes, fens, swamps and lakeshores. Long confused with *C. rostrata,* a boreal species with waxy blue leaves to only 4 mm wide and which has numerous small bumps on upper leaf surface. *Carex rostrata* Stokes var. *utriculata* (Boott) Bailey.

***Carex vesicaria*** L. [•112] LESSER BLADDER SEDGE June–Aug. Wet meadows, marshes, forest depressions and shores.

## CAREX SECTION VULPINAE

Plants tufted; bases generally pale. Inner band of the leaf sheaths hyaline, in other regards various: corrugated or smooth, thickened or fragile at the summit, sparsely purple-dotted or lacking pigmentation, and combinations of the above. Stems thick, spongy, weak, the angles narrowly winged, scabrous. Inflorescence ovate to cylindrical. Bracts setaceous. Spikes densely flowered, the lower branched, mostly or all androgynous (the terminal always androgynous). Perigynia plano-convex, bases spongy (not spongy in

*Carex alopecoidea*). Wetlands. The thick, spongy stems, branched lower spikes, and spongy perigynium bases (except in *C. alopecoidea*) are characteristic.

*Key to species*

1 Perigynia 6.5–8 mm long, enlarged below with a spongy disc-like base much wider than the rest of the body; beak 2x or more longer than perigynium body; ventral surface of leaf sheaths with many tiny purplish dots ................................ ***C. crus-corvi***

1 Perigynia 3–6.2 mm long, corky below but without a distinct a disc-like base; beak slightly longer or shorter than perigynium body; ventral surface of leaf sheaths dotted or not .................. 2

2 Perigynia contracted into a beak no longer than the body, 3–4.5 mm long, essentially nerveless ventrally; ventral surface of leaf sheaths sparsely to strongly dotted with purplish, especially toward the tip ........................................... ***C. alopecoidea***

2 Perigynia somewhat contracted or ± cuneately tapered into the beak (this then difficult to define, but about equaling or slightly exceeding the body, if the latter is measured from the base of perigynium to tip of achene), 4–6.2 mm long, with at least a few nerves ventrally; ventral surface of leaf sheaths not dotted with purplish ............................................ 3

3 Sheaths thickened (or even ± cartilaginous) at the concave or truncate mouth, smooth and unwrinkled ventrally; perigynia 4.7–6.2 mm long ........................................ ***C. laevivaginata***

3 Sheaths thin (usually broken) at the prolonged (when intact) mouth, rather strongly puckered or cross-wrinkled ventrally, very rarely nearly or quite smooth; perigynia 4–5 mm long ........ ***C. stipata***

***Carex alopecoidea*** Tuckerman FOX-TAIL SEDGE May–July. Swamps and floodplain forests, streambanks, swales and moist fields.

***Carex crus-corvi*** Shuttlw. RAVEN-FOOT SEDGE *Endangered.* June–July. Floodplains, marshes, edges of seasonally wet forest depressions.

***Carex laevivaginata*** (Kükenth.) Mackenzie SMOOTH-SHEATH SEDGE *Endangered.* May–July. Swamps, marshy areas and streambanks.

***Carex stipata*** Muhl. [•112] STALK-GRAIN SEDGE May–July. Floodplain forests and swamps, thickets, wet meadows, wetland margins and ditches; usually not in sphagnum bogs.

## CLADIUM *Saw-Grass*

***Cladium mariscoides*** (Muhl.) Torr. SMOOTH SAW-GRASS Grasslike perennial, spreading by rhizomes and forming colonies. Stems single or in small groups, stiff, slender. Leaves 1–3 mm wide, upper portion round in section, middle portion flattened. Flowers in spikelets, in branched clusters (umbels) at end of stem and also with 1–2 clusters on slender stalks from leaf axils. June–Aug. Shallow water, sandy or mucky shores, floating bog mats, calcium-rich wet meadows, seeps, fens and low prairie.

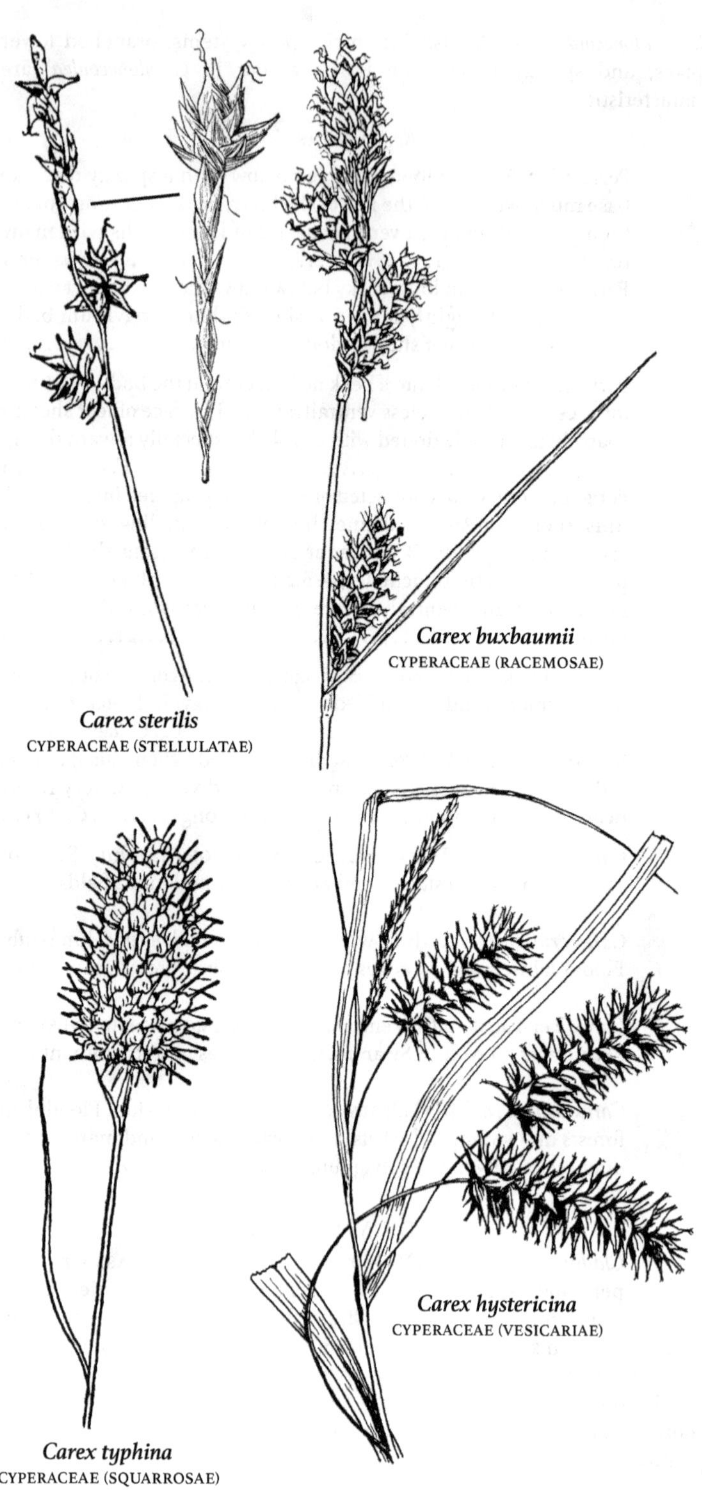
Carex buxbaumii
CYPERACEAE (RACEMOSAE)
Carex sterilis
CYPERACEAE (STELLULATAE)
Carex hystericina
CYPERACEAE (VESICARIAE)
Carex typhina
CYPERACEAE (SQUARROSAE)

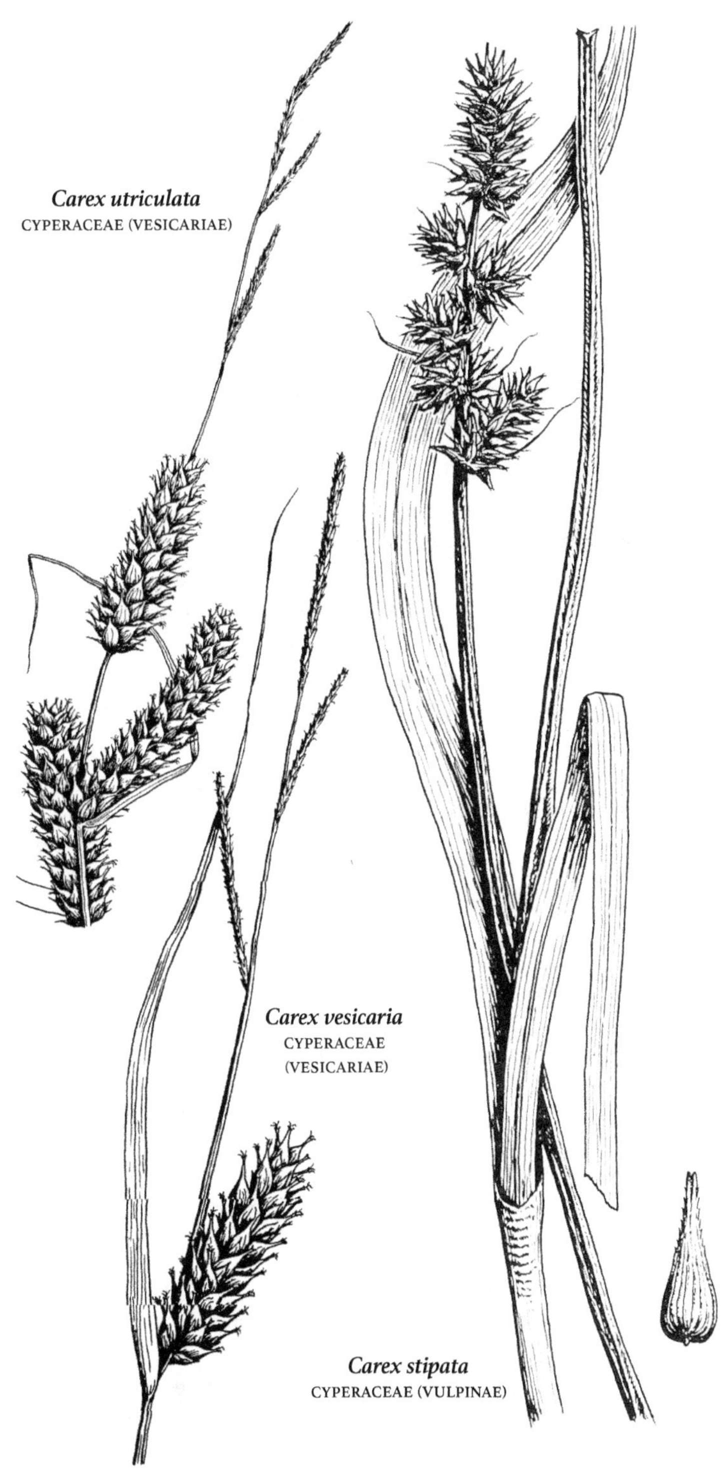
*Carex utriculata*
CYPERACEAE (VESICARIAE)
*Carex vesicaria*
CYPERACEAE
(VESICARIAE)
*Carex stipata*
CYPERACEAE (VULPINAE)

## CYPERUS *Flat Sedge*

Annual or perennial, grasslike plants. Stems often clumped, sharply 3-angled. Leaves mostly from base of plants, with 1 or more leaflike bracts near top of stems, the blades flat or folded along midvein. Flower heads in umbels at ends of stems; the spikelets many, grouped in 1 to several rounded or cylindric spikes.

*Key to species*

1 Achenes lens-shaped; stigmas 2 .......... 2
1 Achenes 3-angled; stigmas 3 .......... 3

2 Style cleft to slightly below middle .......... *C. bipartitus*
2 Style cleft almost to base .......... *C. diandrus*

3 Rachilla of spikelets continuous, scales gradually deciduous, falling from base of rachilla to apex .......... 4
3 Rachilla of spikelets articulated; scales persistent and then falling all at once from the rachilla .......... 5

4 Scales to 2 mm long; achenes 0.6–1.2 mm long ... *C. erythrorhizos*
4 Scales mostly 2–3 mm long; achenes 1–2 mm long ... *C. esculentus*

5 Rachilla articulating at the base of each scale .......... *C. odoratus*
5 Rachilla not separating into joints (short segments) .......... 6

6 Scales with strongly recurved acuminate tips .......... *C. squarrosus*
6 Scales with incurved or straight blunt tips .......... 7

7 Rachilla of spikelet with wings to 1.2 mm wide .......... *C. strigosus*
7 Rachilla of spikelet essentially wingless .......... 8

8 Culms scabrous, rarely smooth; mucro of the scales 0.3–1.5 mm long; achenes 2.2–2.6 mm long .......... *C. schweinitzii*
8 Culm smooth; mucro of the scale very short, to 0.3 mm long; achenes 1.4–2.2 mm long .......... 9

9 Involucral leaves strongly ascending, slightly scabrous or smoothish toward base; stem leaves smooth or nearly so .......... *C. houghtonii*
9 Involucral leaves mostly widely spreading or recurved at maturity, the margins of these and of the stem leaves scabrous . *C. lupulinus*

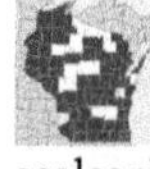

***Cyperus bipartitus*** Torr. [•113] SHINING FLAT SEDGE July–Sept. Wet, sandy, gravelly or muddy shores, streambanks, wet meadows, ditches. Very similar to umbrella flatsedge (*C. diandrus*), but the scales shiny and the styles not as deeply divided (vs. dull scales and the styles cleft nearly to base in *C. diandrus*). *Cyperus niger* Ruiz & Pavón var. *rivularis, Cyperus rivularis* Kunth, *Pycreus rivularis* (Kunth) Palla.

***Cyperus diandrus*** Torr. UMBRELLA FLAT SEDGE July–Sept. Sandy or muddy shores, streambanks, wet meadows.

***Cyperus erythrorhizos*** Muhl. RED-ROOT FLAT SEDGE July–Sept. Sandy or muddy shores, streambanks, exposed mud flats, ditches; often with rusty flatsedge (*Cyperus odoratus*). *Cyperus halei* Torr.

***Cyperus esculentus*** L. CHUFA July–Sept. Sandy or muddy shores, streambanks, marshes, ditches and other wet places; weedy in wet or moist cultivated fields.

***Cyperus houghtonii*** Torr. HOUGHTON'S FLAT SEDGE Dry, especially sandy soil.

***Cyperus lupulinus*** (Spreng.) Marcks GREAT PLAINS FLAT SEDGE Dry woods and fields.

***Cyperus odoratus*** L. RUSTY FLAT SEDGE July–Sept. Sandy or muddy shores, ditches, wet cultivated fields. *Cyperus engelmannii* Steud., *Cyperus ferruginescens* Boeckl., *Cyperus speciosus* Vahl.

***Cyperus schweinitzii*** Torr. SAND FLAT SEDGE Dry or moist sandy soil. *Cyperus × mesochoreus* Geise, *Mariscus schweinitzii* (Torr.) T. Koyama.

***Cyperus squarrosus*** L. [•113] AWNED FLAT SEDGE July–Sept. Wet, sandy or muddy lakeshores, streambanks, mud and gravel bars, wet meadows. *Cyperus aristatus* Rottb., *Cyperus inflexus* Muhl.

***Cyperus strigosus*** L. STRAW-COLOR FLAT SEDGE July–Sept. Wet, sandy or muddy shores, streambanks, marshes, wet meadows, ditches, cultivated fields.

## DULICHIUM *Three-Way Sedge*

***Dulichium arundinaceum*** (L.) Britt. [•113] THREE-WAY SEDGE Grasslike perennial, spreading by rhizomes and often forming large colonies. Stems stout, erect, jointed, hollow, rounded in section. Leaves 3-ranked, flat, short; lower leaves reduced to sheaths. Flower heads from leaf axils, in linear clusters of 5–10 spikelets. July–Sept. Shallow marshes, wet meadows, shores, bog margins.

## ELEOCHARIS *Spike-Rush*

Rushlike plants, perennial from rhizomes, or annual, often forming large, matlike colonies. Stems round, flattened, or angled in section. Leaves reduced to sheaths at base of stems. Flower head a single spikelet at tip of stem. Achenes rounded on both sides or 3-angled.

ADDITIONAL SPECIES ***Eleocharis mamillata*** (H.Lindb.) H.Lindb. (Soft-stem spike-rush), a boreal species reported from Ashland and Douglas counties in n Wisc; often included within *E. palustris.* ***Eleocharis tenuis*** (Willd.) J.A. Schultes (Slender spike-rush), native tufted perennial, spreading by rhizomes; stems slender, 4–8-angled; sheaths red-purple at base; achenes rounded 3-angled, tubercle cone-shaped. ***Eleocharis wolfii*** (Gray) Gray ex Britt. (Wolf's spike-rush), endangered in Wisc (Juneau and Marinette counties); plants tufted, stems flattened, 2-edged, often twisted; spikelets narrowly ovate, wider than stem; achenes gray; tubercle cone-shaped, constricted at base where joins achene; wet meadows.

*Key to species*

1 Mature spikelet scarcely if at all thicker than main portion of stem; scales persistent; stems quadrangular, triangular, or terete and cross-partitioned .......... 2

1 Mature spikelet decidedly thicker than stem; scales usually deciduous; stems terete (or sometimes flattened or many-ridged), not cross-partitioned .......... 4

2 Stems terete, cross-partitioned, appearing as if jointed (with narrow, light-colored bands) .............................. *E. equisetoides*

2 Stems angled, not clearly partitioned or appearing jointed ..... 3

3 Stems sharply 4-angled, stout (3–5 mm thick); spikelets mostly 2–5 cm long ..................................... *E. quadrangulata*

3 Stems 3-angled, not over 2 mm thick; spikelets mostly 1–2 cm long .................................................. *E. robbinsii*

4 Tubercle a slender or tiny conical continuation of the body of the achene, slightly differentiated in texture or color, not separated by a constriction or shaped as a distinct apical cap; stigmas 3; tip of leaf sheath without a prominent tooth ........................... 5

4 Tubercle differentiated in shape as well as texture, and usually separated from body of achene by a narrow constriction to form a distinct apical cap; stigmas 2 or 3; leaf sheaths sometimes with a prominent tooth at tip ...................................... 7

5 Fertile stems 20–70 cm tall, flattened, stout; vegetative stems often as long or longer and rooting at their tips; spikelets 9–17 mm long ................................................ *E. rostellata*

5 Fertile stems to 35 cm tall, but often all tufted and less than 20 cm tall, very slender; stems not rooting at tips; spikelets 2–7 mm long .... 6

6 Plants less than 5 cm tall; achenes ca. 1–1.3 mm long, including tiny tubercle; spikelets 2–3 mm long ...................... *E. parvula*

6 Plants mostly over 5 cm tall; achenes 2–2.5 mm long; spikelets 4–7 mm long ......................................... *E. quinqueflora*

7 Achenes 3-sided (the angles sharp, or obscure and the achene plumply rounded); styles 3-cleft; surface of achene normally ridged, reticulate, roughened, or in a few species only minutely punctate 8

7 Achenes 2-sided (lenticular or biconvex); styles 2- or 3-cleft; surface of achene smooth, usually ± shiny .......................... 12

8 Achenes white or pearly, with prominent longitudinal ridges connected by numerous tiny cross-bars; basal scales of spikelet fertile ............................................ *E. acicularis*

8 Achenes greenish, yellow, golden, brown, black (rarely whitish), and reticulate, smooth, or roughened; basal scales of spikelets sterile. 9

9 Plants tufted, without rhizomes; achenes whitish, greenish, olive, or black, smooth to finely reticulate ............... *E. intermedia*

9 Plants with very stout rhizomes; achenes yellow, golden, or brown, the surface strongly papillate-roughened or honeycombed .... 10

10 Stems very strongly flattened and often ± twisted, with obscure ridges; scales at middle of spikelet reddish brown with narrow, deeply bifid scarious whitish tips to 1 mm long ...... *E. compressa*

10 Stems slightly or not at all flattened, prominently ridged; scales at middle of spikelet deep reddish brown to nearly black, with short, entire, lacerate, or bifid tips mostly less than 0.6 mm long ...... 11

11 Stems usually 10–50 cm tall, 0.4–0.8 mm wide, scales 2–3.4 mm long ................................................ *E. elliptica*

11 Stems ca. 3–10 (–15) cm tall, ca. 0.2–0.3 mm wide; scales 1–1.4 mm long ................................................ *E. nitida*

12 Top of leaf sheaths thin and membranous, cleft on one side, usually whitish; achene olive green to brown, ca. 1–1.5 mm long, including the green tubercle; anthers to 1 mm long ........... *E. flavescens*

12 Top of leaf sheaths thin to firm, truncate, not split (sometimes with a tooth); achenes and anthers various ........................ 13

13 Plants perennial, with stiff stems and rhizomes; scales acute to acuminate at tip (or somewhat obtuse); achenes 1.5–2.8 mm long, including tubercle; anthers ca. 1–3 mm long .......... ***E. palustris***

13 Plants annual, with soft, easily compressed, densely tufted stems; scales broadly rounded at tip; achenes 1.1–1.5 mm long, including the strongly flattened tubercle; anthers to 0.7 mm long ........ 14

14 Base of tubercle less than 2/3 as wide as the broadest part of the mature achene; scales purplish brown .................. ***E. ovata***

14 Base of tubercle at least 2/3 as wide as broadest part of mature achene; scales brown or reddish brown (rarely flushed with purple) ..... 15

15 Tubercle very depressed, not over 1/4 of the total length of the achene, nearly or quite as wide as the truncate achene body, on which it sits as a flattish cap ...................... ***E. engelmannii***

15 Tubercle broadly triangular, more than 1/4 the total length of the achene and nearly as wide as the broadest part of the body ...... ................................................. ***E. obtusa***

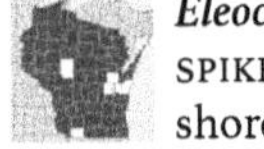

***Eleocharis acicularis*** (L.) Roemer & J.A. Schultes [•114] NEEDLE SPIKE-RUSH May–Sept. Shallow water, exposed muddy or sandy shores, marshes and streambanks.

***Eleocharis compressa*** Sullivant [•114] FLAT-STEM SPIKE-RUSH May–Aug. Low calcareous prairie, wet meadows, swamps, ditches. *Eleocharis elliptica* var. *compressa* (Sullivant) Drapalik & Mohlenbrock.

***Eleocharis elliptica*** Kunth ELLIPTIC SPIKE-RUSH Wet, stony or sandy places, often where marly; shores, marshes, swamps.

***Eleocharis engelmannii*** Steud. ENGELMANN'S SPIKE-RUSH Moist sandy soil.

***Eleocharis equisetoides*** (Ell.) Torr. HORSETAIL SPIKE-RUSH Sandy or mucky lakeshores and pond margins, sometimes in shallow water. *Eleocharis elliottii* A. Dietr.

***Eleocharis flavescens*** (Poir.) Urban YELLOW SPIKE-RUSH Shallow water, sandy or muddy shores, mud flats; sometimes where calcium-rich. *Eleocharis flaccida* var. *olivacea* (Torr.) Fern. & Grisc., *Eleocharis olivacea* Torr.

***Eleocharis intermedia*** J.A. Schultes INTERMEDIATE SPIKE-RUSH June–Sept. Wet, sandy or mucky shores, streambanks, mud flats.

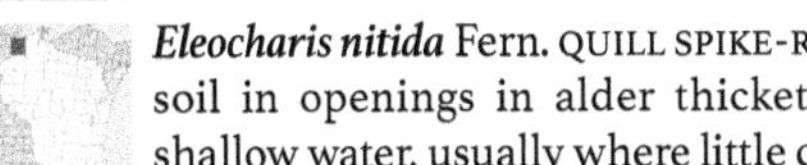

***Eleocharis nitida*** Fern. QUILL SPIKE-RUSH *Endangered.* May–June. Wet soil in openings in alder thickets and marshes, sometimes in shallow water, usually where little competing vegetation; disturbed moist places such as ditches and wheel ruts; crevices in rocks along Lake Superior shoreline. A boreal species, known in Wisc from Douglas County.

***Eleocharis obtusa*** (Willd.) J.A. Schultes OVOID SPIKE-RUSH June–Sept. May form large colonies, especially on exposed mud flats and drying shores of receding lakes.

***Eleocharis ovata*** (Roth) Roemer & J.A. Schultes OVOID SPIKE-RUSH Muddy or sandy shores; likely occurs nearly statewide (*no map*). *Eleocharis obtusa* var. *ovata* (Roth) Drapalik & Mohlenbrock.

***Eleocharis palustris*** (L.) Roemer & J.A. Schultes [•114] COMMON SPIKE-RUSH May–Aug. Shallow water of marshes, wet meadows, muddy shores, bogs, ditches, streambanks and swamps. A variable and common species known by a number of synonyms including *Eleocharis erythropoda* Steud. and *Eleocharis smallii* Britt.

***Eleocharis parvula*** (Roemer & J.A. Schultes) Link LITTLE-HEAD SPIKE-RUSH July–Sept. Wet saline or alkaline flats and shores.

***Eleocharis quadrangulata*** (Michx.) Roemer & J.A. Schultes SQUARE-STEM SPIKE-RUSH *Endangered.* Shallow water and wet, sandy or mucky shores; sedge meadows near lakeshores.

***Eleocharis quinqueflora*** (F.X. Hartmann) Schwarz [•114] FEW-FLOWER SPIKE-RUSH June–Aug. Wet, sandy or gravelly shores and flats marshes and fens; often where calcium-rich. *Eleocharis bernardina* Munz & Johnston.

***Eleocharis robbinsii*** Oakes ROBBINS' SPIKE-RUSH July–Aug. Wet, sandy or mucky lake and pond shores, marshes, exposed flats.

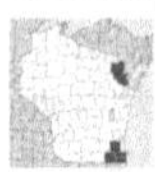

***Eleocharis rostellata*** (Torr.) Torr. BEAKED SPIKE-RUSH *Threatened.* July–Sept. Shores, wet meadows, calcareous fens and mud flats; typically where calcium-rich and often associated with mineral springs.

## ERIOPHORUM *Cotton-Grass*

Grasslike perennials. Stems clumped or single, round to rounded 3-angled in section. Leaves mostly at base of plant, the blades flat, folded or inrolled; upper leaves often reduced to bladeless sheaths. Flower heads at ends of stems, with 1 or several spikelets; spikelets resemble cottonballs when mature.

*Key to species*

1 Head a single spikelet at end of stem; leaflike bracts absent ..... 2
1 Head of 2 or more spikelets; leaflike bracts present ............. 3

2 Plants forming colonies from rhizomes ........... ***E. chamissonis***
2 Plants densely clumped, rhizomes absent .......... ***E. vaginatum***

3 Leaves 1–2 mm wide; leaflike bract 1, erect, the head appearing lateral from side of stem ..................................... 4
3 Leaves 3 mm or more wide; leaflike bracts 2 or 3, the head appearing terminal ................................................ 5

4 Blade of uppermost stem leaf much shorter than its sheath ...... ............................................... ***E. gracile***
4 Blade as long or longer than its sheath .............. ***E. tenellum***

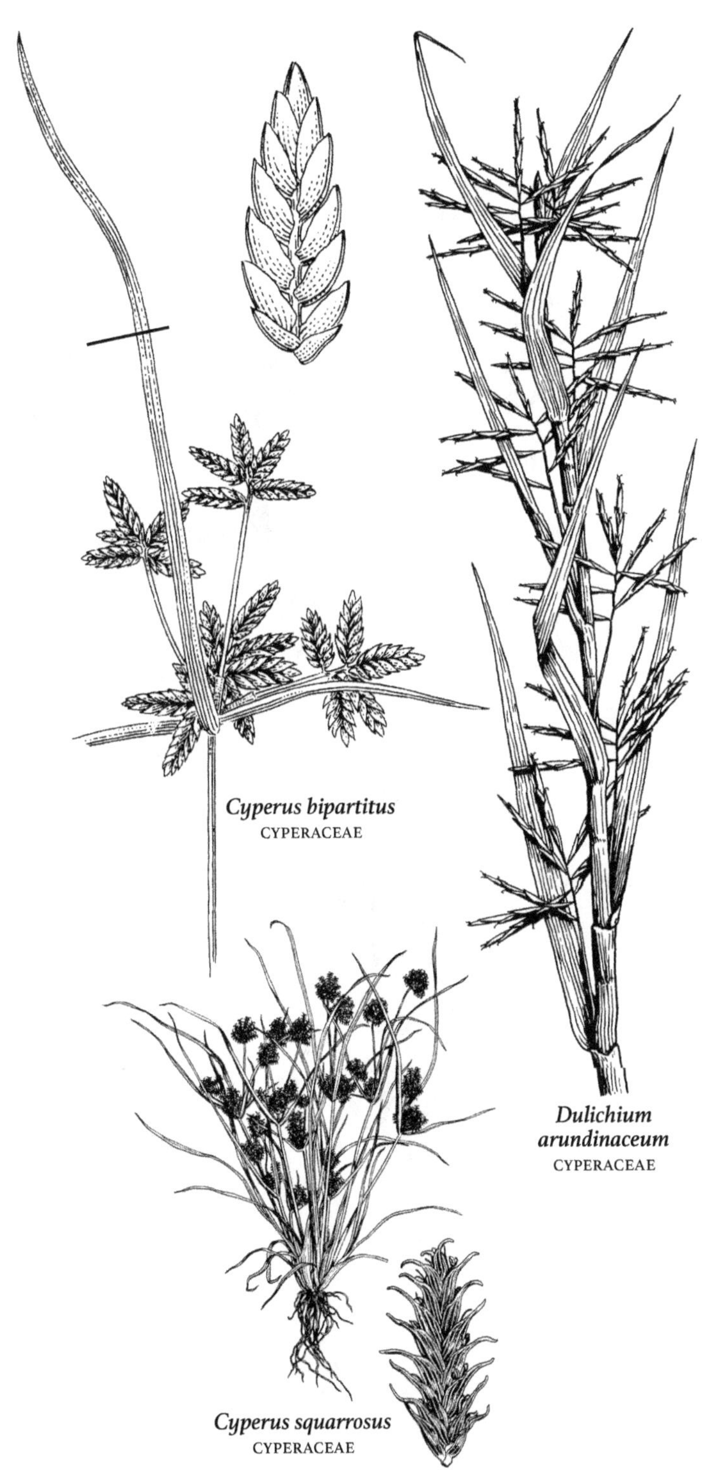
*Cyperus bipartitus*
CYPERACEAE
*Dulichium arundinaceum*
CYPERACEAE
*Cyperus squarrosus*
CYPERACEAE

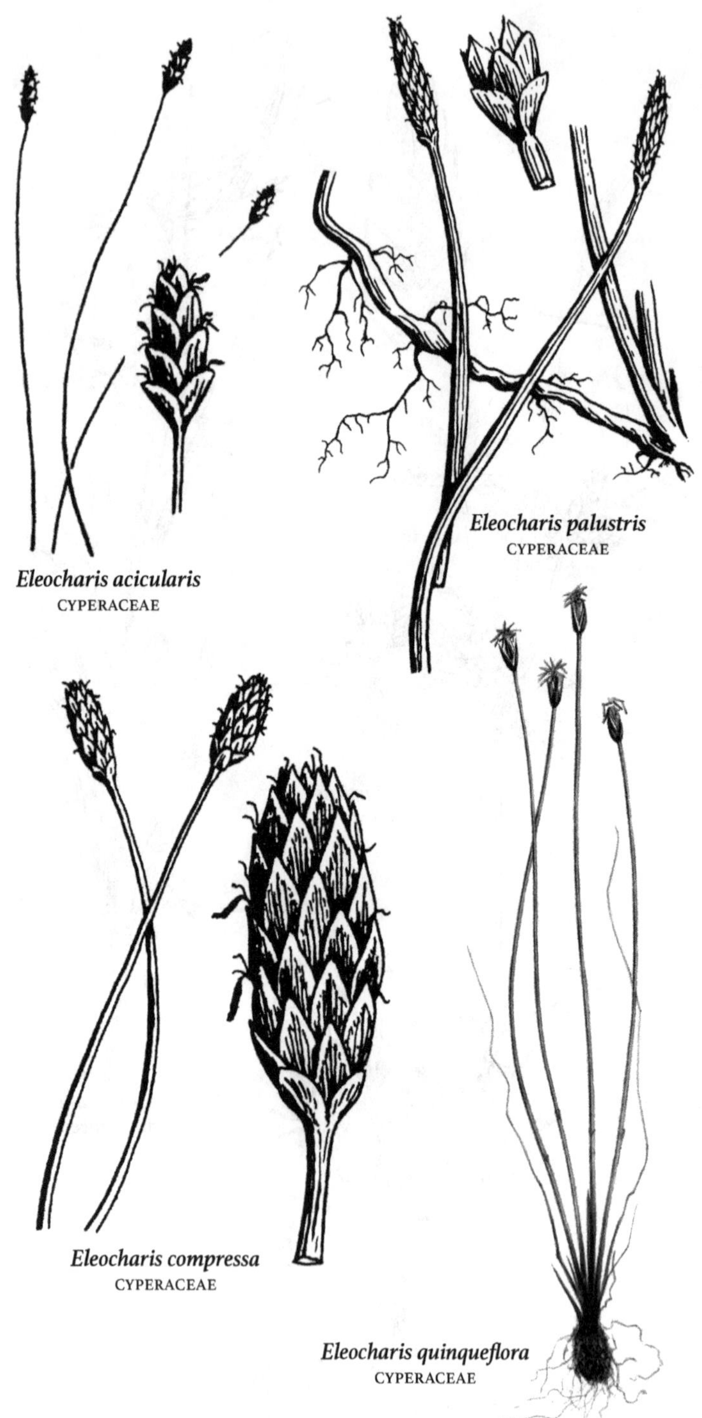

*Eleocharis acicularis*
CYPERACEAE

*Eleocharis palustris*
CYPERACEAE

*Eleocharis compressa*
CYPERACEAE

*Eleocharis quinqueflora*
CYPERACEAE

5 Scales 3–7-nerved, copper-brown on sides .......... *E. virginicum*
5 Scales with 1 nerve, sides olive-green to nearly black ........... 6

6 Midvein of scale slender, fading before reaching tip of scale ...... ...................................................... *E. angustifolium*
6 Midvein of scale widening toward tip of scale and reaching scale tip ................................................. *E. viridicarinatum*

***Eriophorum angustifolium*** Honckeny [•115] THIN-SCALE COTTON-GRASS May–July. Bogs, calcareous fens, wet meadows. Similar to *E. viridicarinatum. Eriophorum polystachion* L.

***Eriophorum chamissonis*** C.A. Mey. CHAMISSO'S COTTON-GRASS June–July. Bogs.

***Eriophorum gracile*** W.D.J. Koch [•115] SLENDER COTTON-GRASS May–July. Fens and bogs.

***Eriophorum tenellum*** Nutt. FEW-NERVE COTTON-GRASS Bogs and conifer swamps.

***Eriophorum vaginatum*** L. TUSSOCK COTTON-GRASS June. Sphagnum bogs and tamarack swamps. *Eriophorum spissum* Fern.

***Eriophorum virginicum*** L. [•115] TAWNY COTTON-GRASS July–Aug. Sphagnum moss peatlands.

***Eriophorum viridicarinatum*** (Engelm.) Fern. DARK-SCALE COTTON-GRASS May–July. Bogs and open conifer swamps. Similar to *E. angustifolium,* but usually with more spikelets, the scale midvein extending to tip of scale, and the leaf sheaths not dark-banded at tip.

## FIMBRISTYLIS *Fimbry*

Annual or perennial grasslike plants. Stems slender, clumped or single. Leaves mostly at base of plants, narrowly linear, flat to inrolled. Spikelets many-flowered, in umbel-like clusters at ends of stems.

*Key to species*

1 Plants annual; achenes 3-angled; styles 3-parted .... *F. autumnalis*
1 Plants perennial; achenes lens-shaped; styles 2-parted *F. puberula*

***Fimbristylis autumnalis*** (L.) Roemer & J.A. Schultes SLENDER FIMBRY July–Sept. Sandy or mucky shores (especially where seasonally flooded and then later exposed), streambanks, wet meadows, ditches.

***Fimbristylis puberula*** (Michx.) Vahl CHESTNUT FIMBRY *Endangered.* June–Sept. Wet meadows, shores, and low prairie, often where sandy and calcium-rich; also in drier prairies; Wisc at northern edge of species' range.

## FUIRENA *Umbrella Sedge*

***Fuirena pumila*** (Torr.) Spreng. DWARF UMBRELLA SEDGE *Endangered.* Tufted, grasslike annual. Leaves flat, the margins hairy; lower sheaths hairy. Spikelets

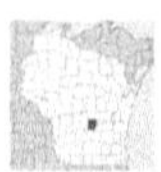

stalkless, in clusters of mostly 1–3, the spikelets ovate. Achenes 3-angled. July–Sept. Sandy or mucky shores, especially where seasonally flooded; floating sedge mats.

### LIPOCARPHA *Halfchaff Sedge*

Small tufted grasslike annuals, the hairlike stems bearing 2 hair-like leaves at the base, only the upper leaves with a short blade. Spikelets very small, 1-3 in a stalkless cluster surpassed by 2 or 3 hairlike involucral bracts.

*Key to species*

1 Main outer scales with a long curved awn almost equaling the length of the scale itself; small inner scale equaling and partly inclosing the achene; mature achene obovate, compressed, black ........ ***L. drummondii***

1 Main outer scales with a short extended tip much shorter than the length of the scale; small inner scale much shorter than the achene, often absent; mature achene cylindrical, brown ..... ***L. micrantha***

***Lipocarpha drummondii*** (Nees) G. Tucker DRUMMOND'S HALFCHAFF SEDGE Aug–Sept. Wet sandy margins of ponds and streams. *Hemicarpha drummondii* Nees.

***Lipocarpha micrantha*** (Vahl) G. Tucker SMALL-FLOWER HALFCHAFF SEDGE Aug–Sept. Sandy or muddy shores and streambanks, usually where seasonally flooded. *Hemicarpha micrantha* (Vahl) Pax.

### RHYNCHOSPORA *Beak Sedge*

Grasslike perennials (annual in *R. scirpoides*), clumped or spreading by rhizomes. Stems erect, leafy, usually 3-angled or sometimes round. Leaves flat or rolled inward. Spikelets clustered in dense heads.

*Key to species*

1 Plants annual; spikelets with many perfect flowers, scales not empty; bristles subtending the achene absent ............... ***R. scirpoides***

1 Plants mostly perennial; spikelets few-flowered, the lower scales empty; bristles present ........................ 2

2 Spikelets white to tan; bristles 8 or more .................. ***R. alba***

2 Spikelets brown, dark olive-green or nearly black; bristles 5–6... 3

3 Scales dark olive-green to black; bristles with upward-pointing barbs, at least some of the bristles longer than the tubercle ....... ***R. fusca***

3 Scales brown; bristles with downward pointing barbs (rarely smooth), the bristles shorter to as long as the tubercle .......... 4

4 Stems narrow and threadlike; achene margins not translucent, achene body less than half as wide as long ........... ***R. capillacea***

4 Stems stout; achene with translucent margins, body more than half as wide as long ..................................... ***R. capitellata***

***Rhynchospora alba*** (L.) Vahl [•116] WHITE BEAK SEDGE June–Sept. Bogs, open conifer swamps of black spruce and tamarack, fens.

***Rhynchospora capillacea*** Torr. NEEDLE BEAK SEDGE June–Aug. Calcareous fens, interdunal flats, wet sandy or gravelly shores, seeps; usually where calcium-rich.

***Rhynchospora capitellata*** (Michx.) Vahl [•116] BROWNISH BEAK SEDGE June–Sept. Wet sandy or mucky shores and flats, wet meadows, bogs, calcareous fens, ditches. *Rhynchospora glomerata* var. *capitellata* (Michx.) Kük.

***Rhynchospora fusca*** (L.) Ait. f. BROWN BEAK SEDGE Wet sandy shores, interdunal wetlands, sedge meadows, bog mats.

***Rhynchospora scirpoides*** (Torr.) Gray LONG-BEAK BEAK SEDGE *Threatened.* Aug–Sept. Wet sandy or mucky shores and mudflats. *Psilocarya scirpoides* Torr.

## SCHOENOPLECTUS *Club-Rush*

Perennial or annual, tufted or rhizomatous herbs. Stems cylindric to strongly 3-angled, smooth, spongy with internal air cavities. Leaves basal, rarely 1(–2) on stem; blades well-developed to rudimentary. Inflorescences terminal, head-like to openly paniculate; spikelets 1–100 or more. Our 3 annual species, *S. hallii, S. purshianus,* and *S. smithii,* are sometimes placed in genus *Schoenoplectiella.*

ADDITIONAL SPECIES ***Schoenoplectus hallii*** (Gray) S.G. Sm., endangered in Wisc, reported from Dane County.

*Key to species*

1 Spikelets (at least several of them) distinctly pediceled (sometimes congested in S. acutus); stems terete, often over 1 m tall ......... 2

1 Spikelets 1-few, crowded, sessile or nearly so (rarely one on a short pedicel); stems 3-angled or terete (if terete, then slender, soft, and not over 1 m tall) ........................................ 4

2 Styles 3-cleft; achenes 3-sided; perianth bristles 2–4 (–5); scales glabrous on the back; mature achenes ca. 2.5 mm long, including short apiculus; stems firm ...................... ***S. heterochaetus***

2 Styles 2-cleft; achenes plano-convex (flat on one side and rounded on the other) or biconvex; perianth bristles mostly 6; scales puberulent on back; achenes shorter, or stems soft and easily compressed ............................................. 3

3 Stems firm and dark olive-green when fresh; spikelets ovoid to cylindrical (often 2.5 or more times as long as wide), usually in a stiffer, sometimes condensed, inflorescence; scales dull, pale or whitish brown, the midrib not strongly contrasting, the margins often more copiously ciliate than in S. tabernaemontani, and the backs copiously flecked with shiny red dots, often puberulent; mature achenes ca. 2.2–2.7 mm long, including apiculus, completely hidden by the scales ................................ ***S. acutus***

3 Stems rather soft and easily compressed, pale blue-green when fresh; spikelets ovoid (about twice as long as wide, or shorter), in an open, lax inflorescence; scales ± shiny, rich orange-brown, often with prominent greenish midrib, the margins ciliate but the backs essentially glabrous (puberulence and swollen red flecks, if any, limited to region of midrib); mature (dark gray or lead-colored)

achenes ca. 1.6–2.1 (–2.4) mm long, including apiculus, barely covered by the scale ........................ *S. tabernaemontani*

4 Spikelet 1, strongly ascending, the involucral bract surpassing its tip by not more than 15 (–20) mm; leaves normally many, hair-like, submersed; stem seldom over 1 mm thick; anthers (2.1–) 2.5–3.5 mm long; achenes 3-sided, the body ca. 2.5–3 mm long S. subterminalis

4 Spikelets usually more than 1 and the involucral bract surpassing them by more than 15 mm (except in smallest plants of some populations); leaves stiff and stems thicker; anthers and achenes various................................................ 5

5 Plants annual, with soft, terete or obscurely 3-angled, tufted stems; anthers 0.3–0.7 mm long ................................... **6**

5 Plants perennial, with elongate rhizomes; stems sharply 3-angled, at least distally; anthers 1–3 mm long ......................... **8**

6 Achene covered with prominent transverse ridges; perianth bristles none ............................................. *S. hallii**

6 Achene smooth or obscurely pitted; perianth bristles present or absent............................................... **7**

7 Taller shoots with stems (base of plant to inflorescence) more than 3/4 as long as height of the plant, including the involucral bract; achenes thickly and asymmetrically biconvex (inner face slightly but clearly convex, outer faces forming a clear angle) .... *S. purshianus*

7 Taller shoots with stems to 3/4 as long as height of the plant; achenes flattened-plano-convex (inner face essentially flat, the outer faces gently rounded) ............................... *S. smithii*

8 Midrib of scale ± greenish, excurrent as a short (not over 0.5 mm) tip extending beyond the tapered (sometimes very slightly notched) apex of the scale; bristles slightly exceeding body of achene; rhizome soft; achene with apiculus 0.5 mm or more in length; styles 3-cleft and achenes 3-sided; leaves more than half as tall as the stems ............................................. *S. torreyi*

8 Midrib of scale brown, excurrent as a long (0.5–1 mm) tip equaling or exceeding lobes; bristles shorter than body of achene; rhizome firm and hard; achene with apiculus shorter than 0.5 mm; styles usually 2-cleft and achenes biconvex to plano-convex (occasionally some styles 3-cleft and achenes 3-sided in a spikelet); leaves less than half as tall as the stems ......................... *S. pungens*

***Schoenoplectus acutus*** (Muhl.) A. & D. Löve [•116] HARDSTEM CLUB-RUSH May–Aug. Usually emergent in shallow to deep water (1–2 m deep) of marshes, ditches, ponds and lakes; sometimes where brackish. *Scirpus acutus* Muhl.

***Schoenoplectus heterochaetus*** (Chase) Soják PALE GREAT CLUB-RUSH June–Aug. Emergent in shallow to deep water (1–2 m deep) of marshes, ponds and lakes, ditches. *Scirpus heterochaetus* Chase.

***Schoenoplectus pungens*** (Vahl) Palla [•116] COMMON THREESQUARE May–Sept. Shallow water (to about 1 m deep), wet sandy, gravelly or mucky shores, streambanks, wet meadows, ditches, seeps and other wet places. *Scirpus americanus* var. *pungens* (Vahl) Barros & Osten, *Scirpus pungens* Vahl

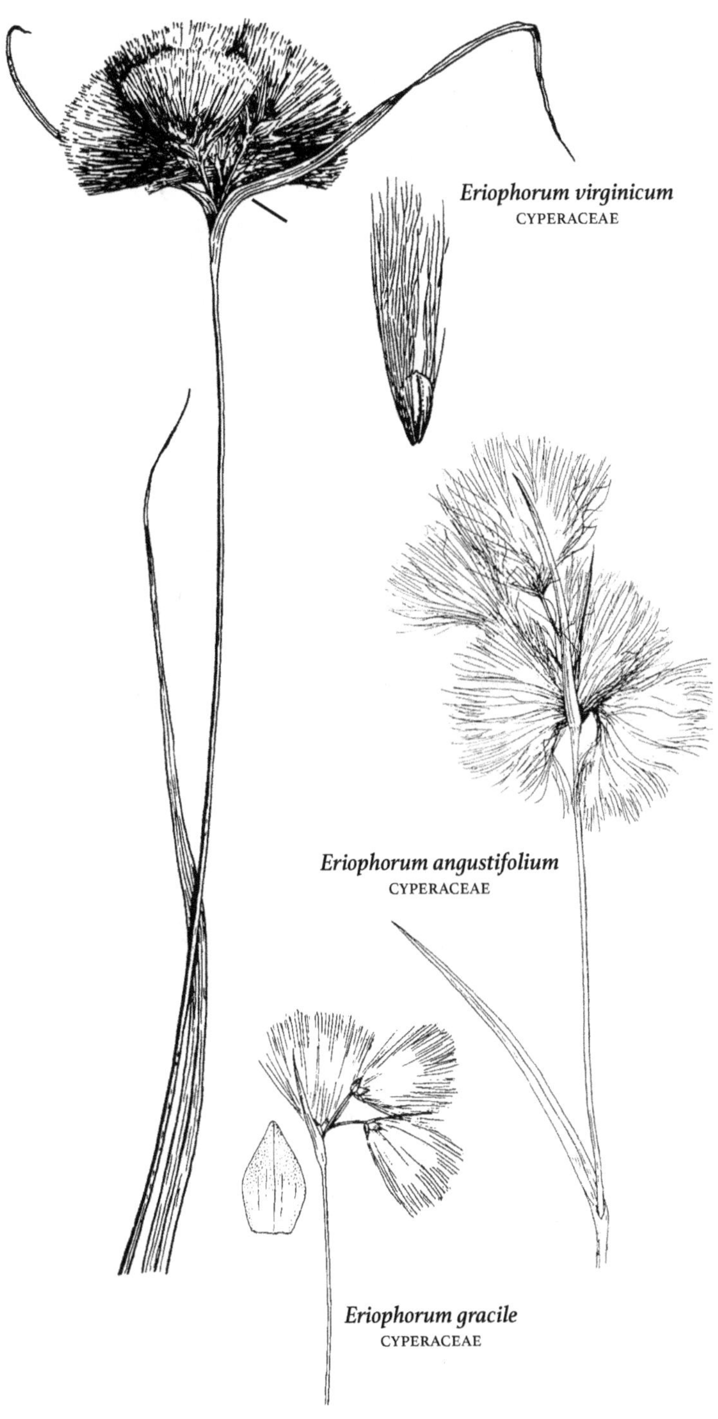
*Eriophorum virginicum*
CYPERACEAE
*Eriophorum angustifolium*
CYPERACEAE
*Eriophorum gracile*
CYPERACEAE

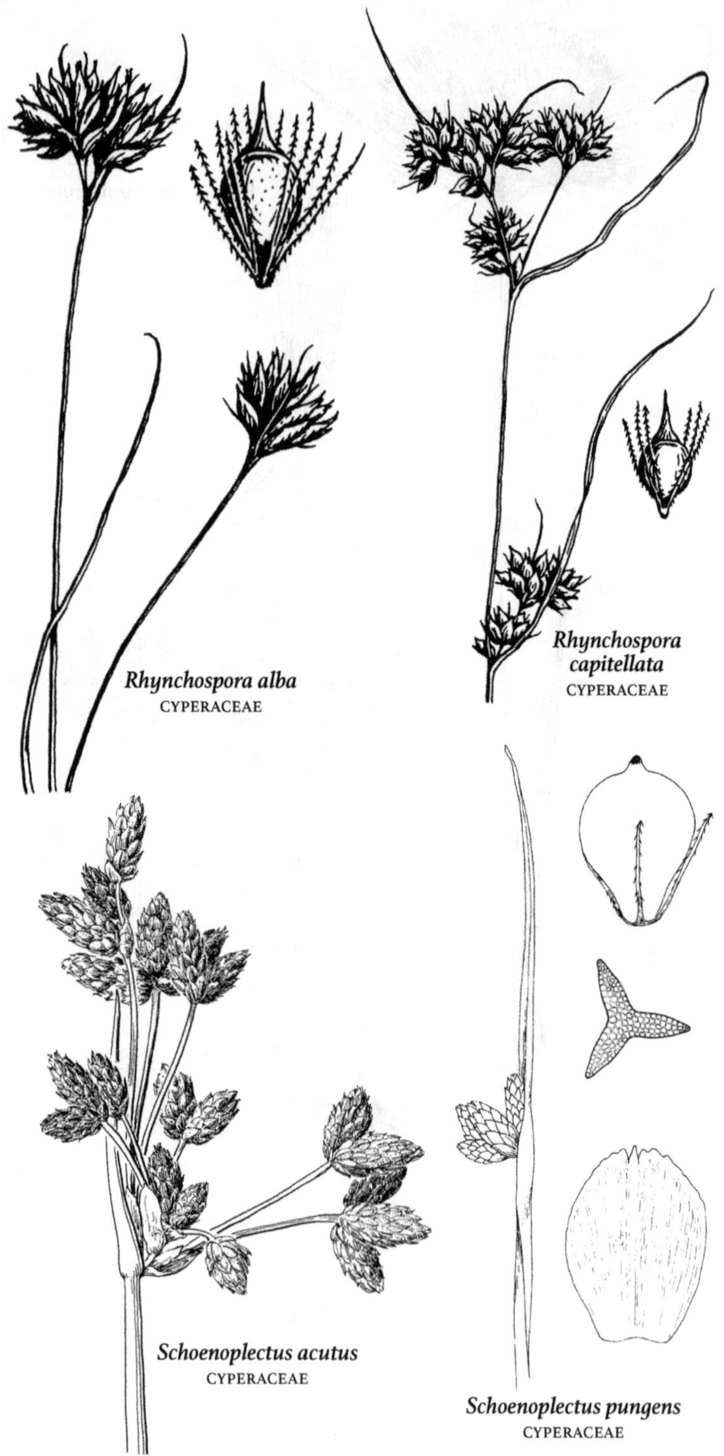
*Rhynchospora alba*
CYPERACEAE
*Rhynchospora capitellata*
CYPERACEAE
*Schoenoplectus acutus*
CYPERACEAE
*Schoenoplectus pungens*
CYPERACEAE

***Schoenoplectus purshianus*** (Fern.) M.T. Strong WEAK-STALK CLUB-RUSH Sandy to mucky, sometimes marly, shores, especially where water levels have receded.

***Schoenoplectus smithii*** (Gray) Soják SMITH'S CLUB-RUSH July–Aug. Sandy, gravelly or mucky shores, floating mats, bogs. *Scirpus smithii* Gray.

***Schoenoplectus subterminalis*** (Torr.) Soják SWAYING CLUB-RUSH July–Aug. In water to about 1 m deep of lakes, ponds and bog margins. *Scirpus subterminalis* Torr.

***Schoenoplectus tabernaemontani*** (K. C. Gmel.) Palla [•117] SOFT-STEM CLUB-RUSH June–Aug. Shallow water and shores of lakes, ponds, marshes, streams, and ditches. Similar to hardstem club-rush (*S. acutus*) but the stems easily crushed between the fingers, plants generally smaller and more slender, and the head more open. *Scirpus tabernaemontani* K.C. Gmel., *Scirpus validus* Vahl

***Schoenoplectus torreyi*** (Olney) Palla TORREY'S CLUB-RUSH June–Aug. Shallow water, wet sandy or mucky shores. *Scirpus torreyi* Olney.

## SCIRPUS *Bulrush*

Stout, rushlike perennials, mostly spreading by rhizomes. Stems unbranched, 3-angled or round in section, solid or pithy. Leaves broad and flat, to narrow and often folded near tip, or reduced to sheaths at base of stems; involucral bracts several and leaflike, or single and appearing like a continuation of the stem. Spikelets single, or in panicle-like or umbel-like clusters at ends of stems, or appearing lateral from the stem.

*Key to species*

1 Lower sheaths red-tinged ........................ ***S. microcarpus***
1 Sheaths green or brown .................................... 2

2 Spikelets many in dense, more or less round heads; bristles about as long as achene or shorter ...................... ***S. atrovirens***
2 Spikelets few in open clusters; bristles much longer than achene 3

3 Mature bristles equal or only slightly longer than scales, spikelets not woolly ........................................ ***S. pendulus***
3 Mature bristles longer than scales, giving spikelets woolly appearance .................... ***S. cyperinus*** complex (see desc.)

***Scirpus atrovirens*** Willd. DARK-GREEN BULRUSH June–Aug. Wet meadows, shores, ditches, streambanks, swamps, springs and other wet places. Here, includes *S. hattorianus* Makino and *S. pallidus* (Britt.) Fern.

***Scirpus cyperinus*** (L.) Kunth [•117] WOOL-GRASS July–Sept. Common in wet meadows, marshes, swamps, ditches, bog margins, thickets; where wet or in very shallow standing water. The *Scirpus cyperinus* complex, including this species, *S. atrocinctus* Fern., and *S. pedicellatus* Fern., is often regarded as one highly variable species. Alternately, the 3 taxa can be separated as follows:

1 Spikelets all or mostly all sessile in clusters of (2–) 3–7 or more ... ............................................... ***S. cyperinus***

1 Spikelets mostly pediceled, the ultimate branches of the inflorescence typically bearing 1 central, sessile spikelet with 2–3 pediceled ones ........................................ 2

2 Scales and bases of bracts dark blackish green; plants slender with leaves 2–5 mm wide .............................. ***S. atrocinctus***

2 Scales and bases of bracts brown or gray-brown; plants more robust with leaves 3–10 mm wide ........................ ***S. pedicellatus***

*Scirpus atrocinctus* flowers and fruits earlier than the other two species, often with inflorescences fully developed by late June, and achenes ripe by late July. *S. atrocinctus* readily hybridizes with *S. cyperinus* to form hybrid swarms. Scales of *S. atrocinctus* are usually distinctly blackened, at least near the tip, while those of *S. pedicellatus* have no black pigment or only slightly so. *S. pedicellatus* is paler and larger than *S. atrocinctus,* the spikelets greenish to pale brown.

***Scirpus microcarpus*** J. & K. Presl [•117] RED-TINGE BULRUSH June–July. Streambanks, wet meadows, marshes, wet shores, thickets, swamps, springs; not in dense shade.

***Scirpus pendulus*** Muhl. RUFOUS BULRUSH June–Aug. Marshes, wet meadows, streambanks, swamp openings and ditches.

## SCLERIA *Nut-Rush*

Annual or perennial sedge-like herbs, tufted or spreading by short rhizomes. Stems slender, 3-angled. Leaves narrow, shorter than the stem. Flowers either staminate or pistillate, borne in separate spikelets on the same plant. Fruit a hard, white achene.

*Key to species*

1 Achenes smooth, ca. 3 mm long, including the whitish, foam-like basal disc; larger leaves 4–7.5 mm wide; mature anthers 2.5–4 mm long ........................................... ***S. triglomerata***

1 Achenes papillose-roughened or wrinkled, 1–2 mm long, including basal disc (not a foam-like crust); larger leaves not over 2.5 mm wide; anthers ca. 1–2.5 mm long .................................. 2

2 Flowers in clusters at ends of stems or from leaf axils S. reticularis

2 Flowers 1 to few in stalkless heads along a spike ..... ***S. verticillata***

***Scleria reticularis*** Michx. NETTED NUT-RUSH *Endangered.* July–Sept. Sandy marshes, wet sand flats, fluctuating lakeshores.

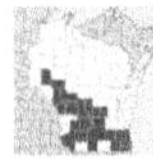

***Scleria triglomerata*** Michx. WHIP NUT-RUSH Moist sandy soil.

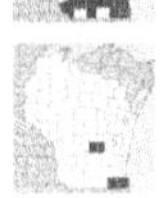

***Scleria verticillata*** Muhl. [•118] LOW NUT-RUSH July–Sept. Sandy or gravelly shores, interdunal flats, wet meadows, marshes.

## TRICHOPHORUM *Leafless-Bulrush*

Tufted perennials. Stems 3-angled or terete. Leaves basal or nearly so; sheaths bladeless or with very short blades less than 1 cm long and to 1 mm wide. Inflorescences terminal; spikelets 1.

*Key to species*

1 Stems more or less round in section, smooth ...... *T. caespitosum*
1 Stems 3-angled, rough on angles ........................... 2

2 Perianth bristles ciliate, slightly or not at all exceeding the blunt achene; scales of spikelet not more than 7; achenes ca. 1.6–1.8 mm long; not in wetlands ................................ *T. clintonii*
2 Perianth bristles smooth, several times as long as the apiculate achene at maturity; scales of spikelets slightly more than 7; body of achene less than 1.5 mm long; wetland species ........ *T. alpinum*

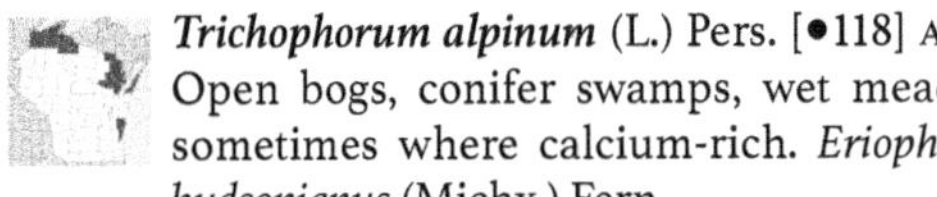

***Trichophorum alpinum*** (L.) Pers. [•118] ALPINE LEAFLESS-BULRUSH Open bogs, conifer swamps, wet meadows, wet sandy shores; sometimes where calcium-rich. *Eriophorum alpinum* L., *Scirpus hudsonianus* (Michx.) Fern.

***Trichophorum caespitosum*** (L.) Hartman TUFTED LEAFLESS-BULRUSH *Threatened.* Open bogs, cedar swamps, calcareous fens, wet dune swales; also Lake Superior rocky shores. *Scirpus caespitosus* L.

***Trichophorum clintonii*** (Gray) S.G. Sm. CLINTON'S LEAFLESS-BULRUSH Dry woods. *Scirpus clintonii* Gray.

## DIOSCOREACEAE *Yam Family*

### DIOSCOREA *Yam*

***Dioscorea villosa*** L. WILD YAM, COLIC-ROOT Perennial dioecious herb. Stems twining, to 5 m long. Leaves alternate, cordate-ovate. Flowers regular; unisexual; small, white to greenish yellow. Fruit a 3-winged capsule. June–July. Moist to dry woods, thickets, pond and marsh borders, river bottoms, roadsides and railroads.

## ERIOCAULACEAE *Pipewort Family*

### ERIOCAULON *Pipewort*

***Eriocaulon aquaticum*** (Hill) Druce [•118] SEVEN-ANGLE PIPEWORT Perennial, spongy at base, with fleshy roots. Stems usually single, leafless, slightly twisted, 5–7-ridged. Leaves grasslike, in a rosette at base of plant, thin and often translucent, with conspicuous cross-veins. Flowers either staminate or pistillate, grouped together in a single, more or less round head at end of stem, the heads white-woolly. July–Sept. Shallow water, sandy or peaty shores. *Eriocaulon septangulare* Withering.

## HYDROCHARITACEAE *Tape-Grass Family*

Aquatic herbs. Stems leafy, the leaves opposite (*Najas*), whorled (*Elodea*), or plants stemless with clusters of long, linear, ribbonlike leaves (*Vallisneria*). Flowers usually either staminate or pistillate and borne on separate plants, small and stalkless, or in a spathe at end of a stalk. Fruit several-seeded, maturing underwater.

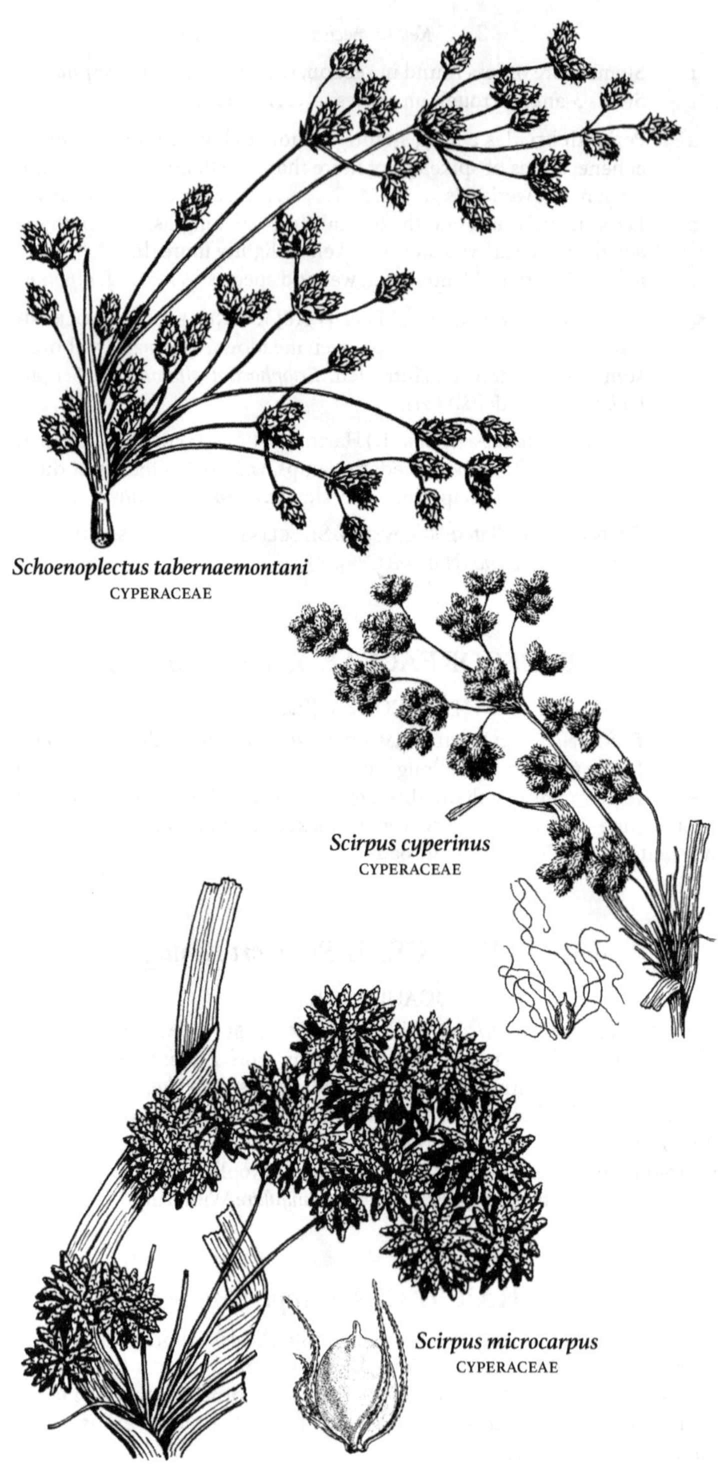

*Schoenoplectus tabernaemontani*
CYPERACEAE

*Scirpus cyperinus*
CYPERACEAE

*Scirpus microcarpus*
CYPERACEAE

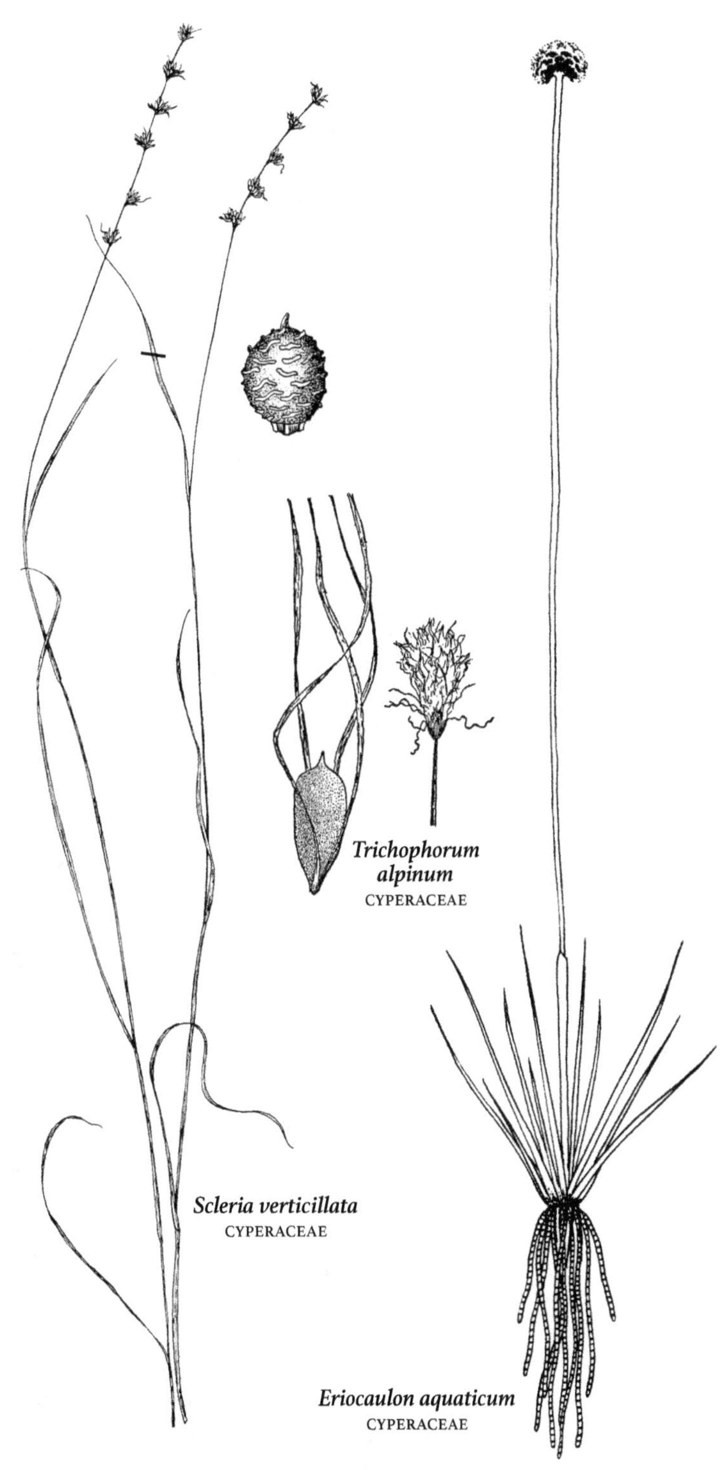
*Trichophorum alpinum*
CYPERACEAE
*Scleria verticillata*
CYPERACEAE
*Eriocaulon aquaticum*
CYPERACEAE

*Key to genera*

1 Leaves very long and ribbon-like (mostly 3–11 mm wide), in a basal rosette .......... **VALLISNERIA**
1 Leaves to 6 (–12) cm long, opposite or whorled .......... 2

2 Leaves whorled, entire .......... **ELODEA**
2 Leaves opposite, minutely denticulate to visibly toothed .. **NAJAS**

## ELODEA *Waterweed*

Aquatic perennial herbs, rooting from lower nodes or free-floating. Stems slender, leafy, branched. Leaves crowded near tip of stem, mostly in whorls of 3–4, or opposite; margins finely sharp-toothed. Flowers either staminate or pistillate and on separate plants, tiny, single in upper leaf axils.

*Key to species*

1 Leaves mostly 2 mm or more wide; staminate flowers long-stalked in a spathe, the spathe more than 7 mm long, extended to water surface by a long, threadlike hypanthium .......... ***E. canadensis***
1 Leaves to 1.5 mm wide; staminate flowers stalkless in a spathe, the spathe 2–4 mm long, breaking free to float to water surface at flowering time .......... ***E. nuttallii***

***Elodea canadensis*** Michx. [•119] CANADIAN WATERWEED June–Aug. Shallow to deep water of lakes, streams and ditches. *Anacharis canadensis* (Michx.) Planch.

***Elodea nuttallii*** (Planch.) St. John [•119] WESTERN WATERWEED June–Aug. Shallow to deep water of lakes, streams and ditches. Similar to *E. canadensis* but less common, smaller and more delicate overall, the leaves narrower, paler green, and not closely overlapping at stem tips, and the staminate flowers not elevated on a long slender stalk. *Anacharis nuttallii* Planch.

## NAJAS *Waternymph*

Aquatic annual herbs, roots fibrous, rhizomes absent. Stems wavy, with slender branches. Leaves simple, opposite or in crowded whorls, abruptly widened at base to sheath the stem; margins toothed to nearly entire, the teeth sometimes spine-tipped. Flowers either staminate or pistillate, separate on same plant or on different plants, tiny, single in leaf axils.

*Key to species*

1 Leaves coarsely toothed and spine-tipped (spines visible without a lens), bright green; midvein of leaf underside and stems between nodes often prickly .......... ***N. marina***
1 Leaves nearly entire or toothed (if spine-tipped, the spines, except in *N. minor,* not visible without a lens), often olive green; leaf surface and stems between leaves smooth .......... 2

2 Base of leaves lobed .......... 3
2 Leaves tapered to base, not lobed .......... 4

3 Leaves somewhat stiff, curved downward near tip; base of leaf truncate-lobed, the lobes herbaceous; seed coat pitted, the pits wider than long and arranged in regular, ladderlike rows ***N. minor***

3 Leaves slender, not stiff, not curved downward near tip; base of leaf auriculate and scarious; seed coat pitted, the pits longer than wide ........................................... *N. gracillima*

4 Seeds smooth and glossy, widest above middle ......... *N. flexilis*

4 Seeds rough and pitted, widest at middle and tapered to ends .... ........................................... *N. guadalupensis*

***Najas flexilis*** (Willd.) Rostk. & Schmidt WAVY WATERNYMPH July–Sept. Ponds, lakes, streams.

***Najas gracillima*** (A. Braun) Magnus [•119] SLENDER WATERNYMPH Shallow water, usually in muck; intolerant of polluted water.

***Najas guadalupensis*** (Spreng.) Magnus SOUTHERN WATERNYMPH July–Sept. Shallow to deep water of lakes, ponds and sometimes rivers; often with *Najas flexilis* but less common.

***Najas marina*** L. ALKALINE WATERNYMPH July–Sept. Shallow water (to 1 m deep) of lakes and marshes.

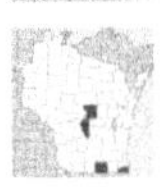

***Najas minor*** Allioni [•119] EUTROPHIC WATER-NYMPH Introduced. Marshes, lakes, ponds.

### VALLISNERIA *Eel-Grass*

***Vallisneria americana*** Michx. [•119] AMERICAN EEL-GRASS Submerged perennial herb, fibrous rooted, spreading by stolons and often forming large colonies. Stems absent. Leaves long and ribbonlike, in tufts from a small crown, rounded at tip, margins smooth. Flowers either staminate or pistillate and on separate plants; staminate flowers released singly from the spathe and floating to water surface where they open; pistillate flowers single in a spathe, on long slender stalks that extend to water surface, the stalk contracting and coiling after flowering to draw the fruit underwater. July–Sept. Shallow to sometimes deep water of lakes and streams.

## HYPOXIDACEAE *Liliid Monocot Family*

### HYPOXIS *Star-Grass*

***Hypoxis hirsuta*** (L.) Coville [•120] EASTERN YELLOW STAR-GRASS Low perennial herb, from a small corm. Leaves from base of plant, linear, hairy. Flowers 1–6 (usually 2), yellow, in racemes at ends of stems, tepals hairy on outside, persistent. May–July. Wet meadows, shores, moist prairie; often where calcium-rich.

## IRIDACEAE *Iris Family*

Perennial herbs with rhizomes, bulbs, or fibrous roots. Leaves parallel-veined, narrow, 2-ranked, the margins joined to form an edge facing the stem (equitant). Flowers perfect, with 6 petal-like segments, single or in clusters at ends of stem. Fruit a 3-chambered capsule.

*Key to genera*

1 Flowers more than 2 cm wide; stems not winged; leaves more than 6 mm wide .......... **IRIS**

1 Flowers to 2 cm wide; stems winged; leaves to 6 mm wide .......... **SISYRINCHIUM**

## IRIS *Iris; Flag*

Perennial herbs, spreading by thick rhizomes. Stems erect. Leaves swordlike, erect or upright, the margins joined to form an edge facing the stem. Flowers 1 or several at ends of stems; yellow or blue-violet. Fruit an oblong capsule.

ADDITIONAL SPECIES ***Iris pumila*** L. (Dwarf iris), introduced and escaped, Columbia and Ozaukee counties; plants small, usually less than 10 cm tall.

*Key to species*

1 Plants dwarf, the flowering stems less than 15 (usually less than 10) cm tall .......... 2

1 Plants more than 15 cm tall .......... 3

2 Sepals with a prominent beard above; rhizomes stout (much more than 5 mm thick) .......... ***I. pumila****

2 Sepals beardless; rhizomes slender (less than 5 mm thick most of their length) .......... ***I. lacustris***

3 Styles club-shaped, not concealing the stamens; sepals and petals alike; seeds round and shiny-black .......... ***I. domestica***

3 Style branches broad and petal-like, concealing the stamens; sepals and petals similar (the sepals larger, ± recurved, though petaloid in color and texture); seeds dull brown, flattened in common species 4

4 Sepals with a prominent median beard above .......... ***I. germanica***

4 Sepals without a prominent beard, at most minutely pubescent . 5

5 Flowers yellow, mature capsules spreading or pendant .......... ***I. pseudacorus***

5 Flowers blue (white in albinos), capsules erect .......... 6

6 Base of expanded portion of sepal with a bright yellow spot, finely pubescent with hairs as long as the thickness of the sepal; outer spathe bracts of uniform texture and color; seeds round to D-shaped, irregularly (but shallowly) pitted .......... ***I. virginica***

6 Base of expanded portion of sepal at most with a greenish yellow spot, with papillae shorter than thickness of the sepal; outer spathe bracts with the margins generally darker and more shiny than the rest of the dull surface; seeds D-shaped, with a ± regularly pebbled surface .......... ***I. versicolor***

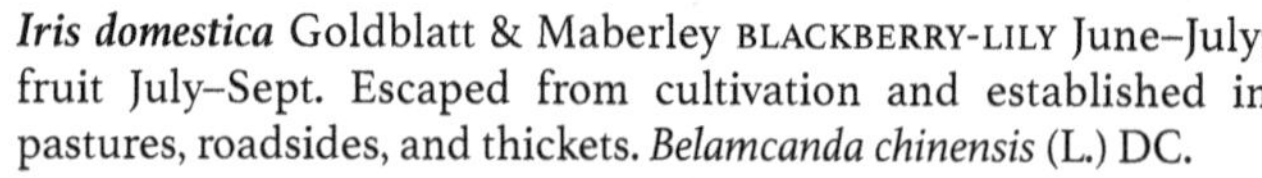

***Iris domestica*** Goldblatt & Maberley BLACKBERRY-LILY June–July; fruit July–Sept. Escaped from cultivation and established in pastures, roadsides, and thickets. *Belamcanda chinensis* (L.) DC.

***Iris germanica*** L. GERMAN IRIS May–June. Introduced from Europe; occasional along roadsides, in waste places, and around abandoned houses where it persists after cultivation.

***Iris lacustris*** Nutt. DWARF LAKE IRIS *Threatened.* May. Gravelly shores in calcareous soil, sandy beach ridges and stabilized dunes near Lake Michigan.

***Iris pseudacorus*** L. [•120] PALE-YELLOW IRIS Introduced. May–June. Lakeshores, streambanks, marshes, ditches.

***Iris versicolor*** L. NORTHERN BLUEFLAG June–July. Marshes, shores, wet meadows, open bogs, swamps, thickets, forest depressions; often in shallow water.

***Iris virginica*** L. [•120] SOUTHERN BLUEFLAG May–July. Swamps, thickets, shores, streambanks, marshes, ditches.

## SISYRINCHIUM *Blue-Eyed-Grass*

Tufted perennial herbs, from fibrous roots. Stems slender, leafless, flattened or winged. Flowers in an umbel at end of stem, above a pair of erect green bracts (spathe), blue-violet or rarely white, with 6 spreading segments, the segments joined only at base. Fruit a rounded capsule.

*Key to species*

1 Spathes on peduncles arising from a leaflike bract, usually more than one, the upper portion of the stem thus appearing branched ..... 2

1 Spathes sessile or nearly so at the end of a simple stem ......... 4

2 Stems 1.5–2.5 mm wide; spathe 16–22 mm long, the bracts subequal; pedicels (at least some of them) wing-margined basally more than half their length; capsule pale whitish tan (sometimes purplish at tip) ........................................ *S. strictum*

2 Stems 0.5–5.5 mm wide, if less than 2 mm then the inner spathe bract less than 15 mm long; pedicels merely 2-edged or slightly winged basally; capsules dark ............................... 3

3 Stems 2.5–3.5 mm wide, with broad wings, each wider than the central portion; inner bract of spathe 13–22 mm long, the outer bract longer or even leaflike; margins of outer bract fused for 3–6 mm; stems usually very minutely denticulate on margins (at 30–40×); plants turning dark green in drying .............. *S. angustifolium*

3 Stems 0.5–2 mm wide, the wings distinctly narrower than the central portion; inner bract of spathe 10–13 mm long, the outer scarcely longer; margins of outer bract fused for 2.5–3.5 (–4.5) mm; stems completely smooth-margined; plants drying pale green ... ........................................ *S. atlanticum*

4 Spathes 2, surrounded at base by an outer leaflike involucral bract with margins not fused beyond the wing (if any) of stem ......... ........................................ *S. albidum*

4 Spathe 1, the outer (usually ± leaflike) bract with the margins slightly to moderately fused at the base beyond wing of stem (except in *S. campestre*) ........................................ 5

5 Margins of outer spathe distinct nearly to the base, spathes greenish, slightly or not at all purple tinged; stems less than 2 mm wide ........................................ *S. campestre*

5 Margins of outer spathe fused for 1–5 mm, spathes often strongly purple tinged or stems often more than 2 mm wide (or both) ... 6

6 Plants very slender, the stem usually 1 mm or less wide and barely margined or narrowly winged; largest leaves to 1.5 mm wide; capsule 2.5–4 mm long .......................... *S. mucronatum*

6 Plants stout, the stem usually 2–2.5 mm wide, winged; largest leaves 2–3 mm wide; mature capsule 5–7 mm long ......... *S. montanum*

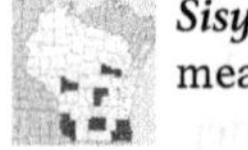

***Sisyrinchium albidum*** Raf. WHITE BLUE-EYED-GRASS Prairies, meadows, grassy or sandy places, and open woodlands.

***Sisyrinchium angustifolium*** P. Mill. [•120] NARROW-LEAF BLUE-EYED-GRASS Meadows, fields, and open woods.

***Sisyrinchium atlanticum*** Bickn. EASTERN BLUE-EYED-GRASS Spring–early summer. Sandy, moist fields and moist sandy shores.

***Sisyrinchium campestre*** Bickn. PRAIRIE BLUE-EYED-GRASS Prairies, meadows, sandy places, and open woodlands.

***Sisyrinchium montanum*** Greene STRICT BLUE-EYED-GRASS May–July. Wet meadows, shores, thickets, ditches, swales; also in drier woods and fields.

***Sisyrinchium mucronatum*** Michx. NEEDLE-TIP BLUE-EYED-GRASS May–June. Wet meadows, calcareous fens.

***Sisyrinchium strictum*** E.P. Bicknell BLUE-EYED-GRASS June–July. Openings with gravelly and sandy soil.

## JUNCACEAE *Rush Family*

Distinguished from grasses and sedges by the presence of a true perianth of 6 tepals and a 3–many-seeded capsule rather than a 1-seeded grain (grasses) or achene (sedges). No ligule (as in the grasses) is present at junction of leaf blade and sheath; however an auricle (an ear-like appendage) may occur at top of leaf sheath.

*Key to genera*

1 Foliage completely glabrous; capsules usually many-seeded ........................................................... JUNCUS

1 Foliage ± hairy, at least toward summit of sheaths; capsules 3-seeded ........................................................... LUZULA

### JUNCUS *Rush*

Clumped or rhizomatous rushes, mostly perennial (annual in Juncus bufonius). Stems erect and unbranched. Leaves from base of plant or along stem, alternate, round in section, flat to involute, or reduced to sheaths at base of stem. Flowers perfect, regular, in compact to open clusters of few to many flowers, subtended by 1 or several leaflike involucral bracts.

ADDITIONAL SPECIES ***Juncus squarrosus*** L., European native, adventive in USA in Ashland County, northern Wisc.

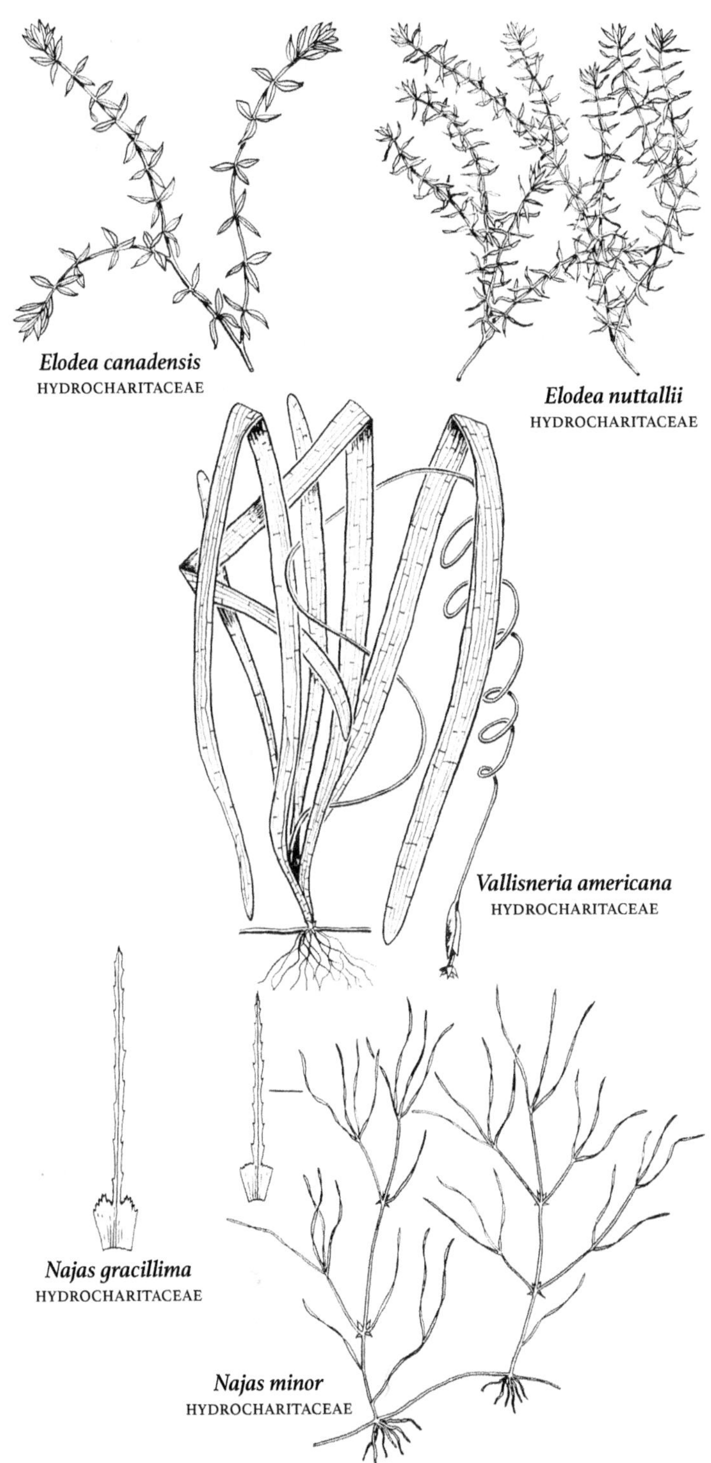
*Elodea canadensis*
HYDROCHARITACEAE
*Elodea nuttallii*
HYDROCHARITACEAE
*Vallisneria americana*
HYDROCHARITACEAE
*Najas gracillima*
HYDROCHARITACEAE
*Najas minor*
HYDROCHARITACEAE

*Sisyrinchium angustifolium*
IRIDACEAE

*Hypoxis hirsuta*
HYPOXIDACEAE

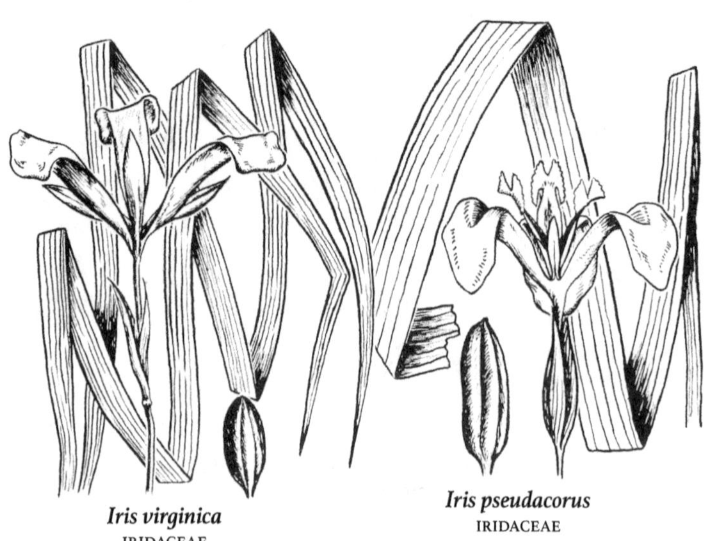

*Iris virginica*
IRIDACEAE

*Iris pseudacorus*
IRIDACEAE

*Key to species*

1 Head from side of stem, the involucral bract erect, round in section and appearing to be a continuation of stem; basal and stem leaves absent . . . . . 2
1 Head at end of stem; basal leaves present, stem leaves present or absent . . . . . 4
2 Involucral bract more than half height of plant . . . . . *J. filiformis*
2 Involucral bract less than half height of plant . . . . . 3
3 Stems densely clumped; stamens 3 . . . . . *J. effusus* (*J. pylaei*)
3 Stems single from rhizomes, the stems often in rows; stamens 6 . . . . . *J. balticus*
4 Leaves flat or somewhat channeled . . . . . 5
4 Leaves round, inrolled, or narrowly channeled . . . . . 15
5 Flowers in dense heads; leaves 2 mm or more wide . . . . . 6
5 Flowers single on branches of head; leaves usually less than 1.5 mm wide . . . . . 7
6 Leaves folded and joined near tip, with 1 edge facing the stem (as in Iris); nw Wisc only . . . . . *J. ensifolius*
6 Leaves not folded or joined, the leaf surface facing the stem . . . . . *J. marginatus*
7 Plants annual, to 10 cm tall . . . . . *J. bufonius*
7 Plants perennial, more than 10 cm tall . . . . . 8
8 Larger stem leaves more than 3 mm wide, with 5 main veins . . . . . *J. marginatus*
8 Stem leaves to 3 mm wide, with less than 5 main veins . . . . . 9
9 Stems clumped; leaves with spreading blades extend from base to middle of stem . . . . . *J. compressus*
9 Stems not clumped; leaves with spreading blades only near base of stem . . . . . 10
10 Mature capsule to as long as the sepals; stamens nearly equaling the tepals; anthers 1 mm or more; longest involucral bract usually shorter than or ± equaling the inflorescence . . . . . *J. gerardii*
10 Mature capsule strongly surpassmg the sepals; stamens to half as long as the tepals; anthers 1 mm or less long; longest involucral bract often greatly exceeding inflorescence . . . . . *J. compressus*
11 Leaves flat or inrolled but open for entire length; seeds less than 0.5 mm long . . . . . 12
11 Leaves round in section, closed for more or less entire length; seeds more than 0.7 mm long . . . . . 14
12 Auricles flaplike, prolonged into a membranaceous or scarious projection 3-5 mm long . . . . . *J. tenuis*
12 Auricles shorter, not flaplike, prolonged up to 2 mm beyond the sheath, submembraneous or cartilaginous . . . . . 13
13 Auricles cartilaginous, dull to shiny yellow, very rigid . . . *J. dudleyi*
13 Auricles submembraneous, not rigid . . . . . *J. interior*
14 Ends of seeds with white "tails" about half as long as the slender body; sepals ca. (3.5–) 4 mm long; longest involucral bract 1–6 cm long (often less than 3 cm) . . . . . *J. vaseyi*
14 Ends of seeds without "tails" or these at most half the width of the

plump body; sepals 3–4 mm long; longest involucral bract to 21 cm long .......... *J. greenei*

15 Leaf blades without cross-partitions at regular intervals . *J. stygius*

15 Leaf blades with cross-partitions at regular intervals .......... 16

16 Heads with 1–2 flowers; the head less than 1/4 total height of plant .......... *J. pelocarpus*

16 Heads with 3–13 flowers; the head more than 1/4 total height of plant .......... 17

17 Leaf sheath distinctly ribbed; seeds 1–2 mm long, with short to long white tails .......... 18

17 Leaf sheath smooth or veined; seeds less than 1 mm long, without white tails or with very short dark tails .......... 20

18 Petals short-tapered to a rounded tip; seeds with a short tail (about 1/10 of body length) .......... *J. brachycephalus*

18 Petals short-tapered to a pointed tip; seeds with long or short tails .......... 19

19 Head narrowly cylindric in outline, the branches erect; heads with 2–5 flowers .......... *J. brevicaudatus*

19 Head ovate in outline, the branches ascending or spreading; heads with 5–40 flowers .......... *J. canadensis*

20 Flowers in dense round heads, the flowers radiating in all directions 21

20 Heads various but the flowers not radiating in all directions ... 23

21 Heads to 10 mm wide; most capsules longer than sepals *J. nodosus*

21 Heads more than 10 mm wide; capsules shorter to nearly equaling the sepals .......... 22

22 Stamens 6; capsules lance-shaped, about as long as sepals *J. torreyi*

22 Stamens 3; capsules ovate, shorter to about as long as sepals .......... *J. acuminatus*

23 Branches of head ascending; petals shorter than sepals .......... *J. alpinoarticulatus*

23 Branches of head spreading; petals as long or slightly longer than sepals .......... *J. articulatus*

***Juncus acuminatus*** Michx. KNOTTY-LEAF RUSH June–Aug. Wet sandy shores, streambanks and ditches; not in open bogs.

***Juncus alpinoarticulatus*** Chaix NORTHERN GREEN RUSH June–Sept. Sandy or gravelly shores, streambanks, fens; often where calcium-rich. *Juncus alpinus* Vill.

***Juncus articulatus*** L. JOINT-LEAF RUSH July–Sept. Sandy, gravelly or mucky shores, streambanks and springs.

***Juncus balticus*** Willd. [•121] SMALL-HEAD RUSH May–Aug. Wet sandy or gravelly shores, interdunal wetlands near Lake Michigan, meadows, ditches, marshes, seeps. *Juncus arcticus* Willd.

***Juncus brachycephalus*** (Engelm.) Buch. SMALL-HEAD RUSH June–Sept. Sandy or gravelly shores, streambanks, open bogs, calcium-rich springs.

***Juncus brevicaudatus*** (Engelm.) Fern. [•121] NARROW-PANICLE RUSH Aug–Sept. Wet meadows, marshes, sandy and rocky lakeshores.

*Juncus bufonius* L. TOAD RUSH June–Aug. Sandy or silty shores, mud flats, streambanks, wet compacted soil of trails and wheel ruts.

*Juncus canadensis* J. Gay [•121] CANADIAN RUSH July–Sept. Sandy, muddy or mucky shores, marshes, streambanks, thickets, ditches.

*Juncus compressus* Jacq. ROUND-FRUIT RUSH Introduced. June–Aug. Wet meadows, disturbed wet areas, ditches along highways where forming dark green colonies, often where saline.

*Juncus dudleyi* Wieg. DUDLEY'S RUSH Damp to drier open places, including lakeshores, marsh margins, ditches. Juncus tenuis var. dudleyi (Wiegand) F.J. Herm.

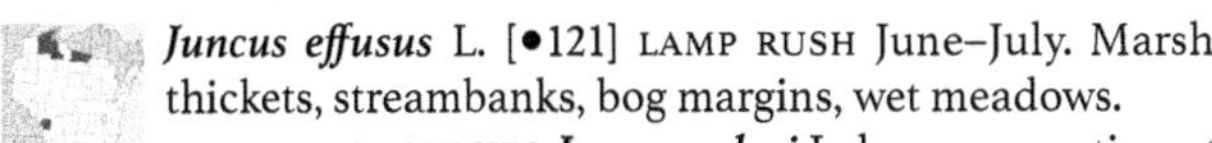

*Juncus effusus* L. [•121] LAMP RUSH June–July. Marshes, shores, thickets, streambanks, bog margins, wet meadows.

ADDITIONAL SPECIES ***Juncus pylaei*** Laharpe, sometimes treated as a species or as *J. effusus* var. *pylaei* (Laharpe) Fern. & Weig., and apparently the common taxon in Wisc. *J. effusus* has soft, broad stems rather smooth because the ridges are tiny and numerous and readily flattened by pressing; *J. pylaei* has fewer, prominent ridges; and stems that are narrower, sturdier, and generally not flattened by pressing.

*Juncus ensifolius* Wikstr. DAGGER-LEAF RUSH July–Sept. Margins of streams, ponds and springs. Disjunct from main range of western USA in Ashland County (adventive) and Michigan Upper Peninsula.

*Juncus filiformis* L. THREAD RUSH Sandy, mucky, or gravelly shores, streambanks, thickets.

*Juncus gerardii* Loisel. BLACK-GRASS Salt marsh species of ne USA; considered adventive in Wisc, especially where salt is applied (as along highways).

*Juncus greenei* Oakes & Tuckerman GREENE'S RUSH Moist to dry sandy open places: shores, swales, fields, clearings, dunes.

*Juncus interior* Wieg. INLAND RUSH Differs from *J. tenuis* in having shorter auricles.

*Juncus marginatus* Rostk. GRASS-LEAF RUSH June–Aug. Sandy shores and streambanks, wet meadows, marshes, low prairie, springs.

*Juncus nodosus* L. [•121] INLAND RUSH July–Sept. Sandy, gravelly or clayey shores and streambanks, wet meadows, fens, ditches, springs; often where calcium-rich.

*Juncus pelocarpus* E. Mey. BROWN-FRUIT RUSH July–Aug. Shallow water, sandy or mucky shores, bog margins.

*Juncus stygius* L. MOOR RUSH *Endangered.* Open bogs, marshes, and shallow water.

*Juncus tenuis* Willd. POVERTY RUSH June–July. Wet meadows, shores, streambanks, springs, common in disturbed places (often where soils compacted) such as trails, roadsides, ditches; also in drier woods and meadows.

ADDITIONAL SPECIES ***Juncus anthelatus*** (Wiegand) R. E. Brooks, like a larger plant of *J. tenuis* with a lax, large inflorescence, treated as valid species or as *J. tenuis* var. *anthelatus* Weig. Plants mostly 60–90 cm tall with inflorescences 7–20 cm long; *J. tenuis* rarely more than 60 cm tall, the inflorescence 1–10 cm.

***Juncus torreyi*** Coville [•121] TORREY'S RUSH June–Sept. Sandy shores, streambanks, wet meadows, marsh borders, springs, ditches.

***Juncus vaseyi*** Engelm. VASEY'S RUSH July–Aug. Wet meadows, sandy shores. *Juncus greenei* Oakes & Tuckerman var. *vaseyi* (Engelm.) Boivin.

### LUZULA *Wood-Rush*

Perennial grasslike herbs, with narrow, flat, more or less pubescent leaves often involute toward the tip, and an umbel-like or spike-like inflorescence, the flowers solitary, rarely paired, or glomerulate. Distinguished from Juncus by the presence of few to many long hairs on the leaves, especially toward the base of the blades.

ADDITIONAL SPECIES ***Luzula luzuloides*** (Lam.) Dandy & Wilmott (Oak-forest wood-rush), native of Europe; introduced in Oneida and Walworth counties. Stems loosely tufted; leaves narrow, to only 3–5 mm wide; flowers in terminal clusters of 3–6; perianth segments white or tinged with rose.

*Key to species*

1 Flowers grouped in heads or short dense spikes; capsule usually no longer than the perianth ........................... ***L. multiflora***

1 Flowers single at ends of inflorescence branches; capsule at maturity slightly longer than perianth ............. ***L. acuminata***

***Luzula acuminata*** Raf. [•122] HAIRY WOOD-RUSH May–June. Moist or dry woods and forest openings, meadows, streambanks, hillsides.

***Luzula multiflora*** (Ehrh.) Lej. [•122] COMMON WOOD-RUSH Forests, sometimes with *L. acuminata;* also in swamps and moist grassy areas. *Luzula campestris* var. *multiflora* (Ehrh.) Celak.

## JUNCAGINACEAE *Arrow-Grass Family*

### TRIGLOCHIN *Arrow-Grass*

Grasslike perennial herbs, clumped from creeping rhizomes, often in brackish habitats. Stems slender, leafless. Leaves all from base of plant, slender, linear, round or somewhat flattened in section. Flowers perfect, regular, on short stalks in a spike-like raceme at end of stem.

*Key to species*

1 Plants generally small and slender; stigmas 3; fruit linear, clublike toward tip ............................................ ***T. palustris***

1 Plants larger, usually 3 dm or more tall; stigmas 6; fruit short-cylindric .......... *T. maritima*

***Triglochin maritima*** L. [•122] SEASIDE ARROW-GRASS June–Aug. Sandy, gravelly, or marly lakeshores and streambanks; marshes, brackish wetlands. Plants larger than marsh arrow-grass (*Triglochin palustris*) and the fruit ovate rather than linear.

***Triglochin palustris*** L. MARSH ARROW-GRASS June–Sept. Sandy, gravelly, or marly lakeshores and streambanks, calcareous fens, marshes, interdunal swales; often where calcium-rich.

## LILIACEAE *Lily Family*

Perennial herbs, from corms, bulbs or rhizomes. Stems leafy or leafless. Leaves linear to ovate, usually from base of plant, sometimes along stem, alternate to opposite or whorled. Flowers perfect, regular; sepals and petals of 6 petal-like tepals in 2 series of 3. Fruit a capsule or round berry.

Under the Angiosperm Phylogeny Group III system (APG III), genera in the Liliaceae have been placed into various new familes. However, the family designations are still in a state of flux, and may change in the future. As a convenience, our genera are retained within the traditional Lily Family grouping, with the proposed new family name noted in parentheses after each genus name.

ADDITIONAL SPECIES ***Narcissus poeticus*** L. (Poet's narcissus, Amaryllidaceae), introduced, Bayfield County. ***Yucca flaccida*** Haw. (Adam's-needle, Asparagaceae), native of se USA, considered adventive in s Wisc on dry sandy soil. Leaves stiff, fibrous along the margins, prolonged at tip into a short stout spine; inflorescence a panicle of showy, creamy white flowers.

*Key to genera*

1 Sepals and petals of quite different color and/or texture, the former green or brownish .......... **TRILLIUM**

1 Sepals and petals both colored and petal-like, usually similar in shape (tepals) or the sepals (in Iris) of different size and shape... 2

2 Ovary inferior (flowers bisexual) .......... 3

2 Ovary superior (or flowers unisexual) .......... 5

3 Ovary clearly inferior; uncommon garden escape ... **NARCISSUS**

3 Ovary half-inferior, part of it adnate to the perianth, glabrous (at most granular-roughened); native .......... 4

4 Plants with at least some principal leaves clearly cauline; inflorescences paniculate (except in depauperate individuals) .... .......... **ANTICLEA**

4 Plants with the principal leaves all basal (at ground level) or nearly so, cauline leaves absent, reduced to bracts, or at most much smaller or fewer than basal leaves; inflorescences racemose .... **ALETRIS**

5 Flowers or inflorescences lateral, arising from the axils of alternate cauline leaves or scales .......... 6

5 Flowers or inflorescences terminal on scapes or leafy (simple or branched) stems .......... 7

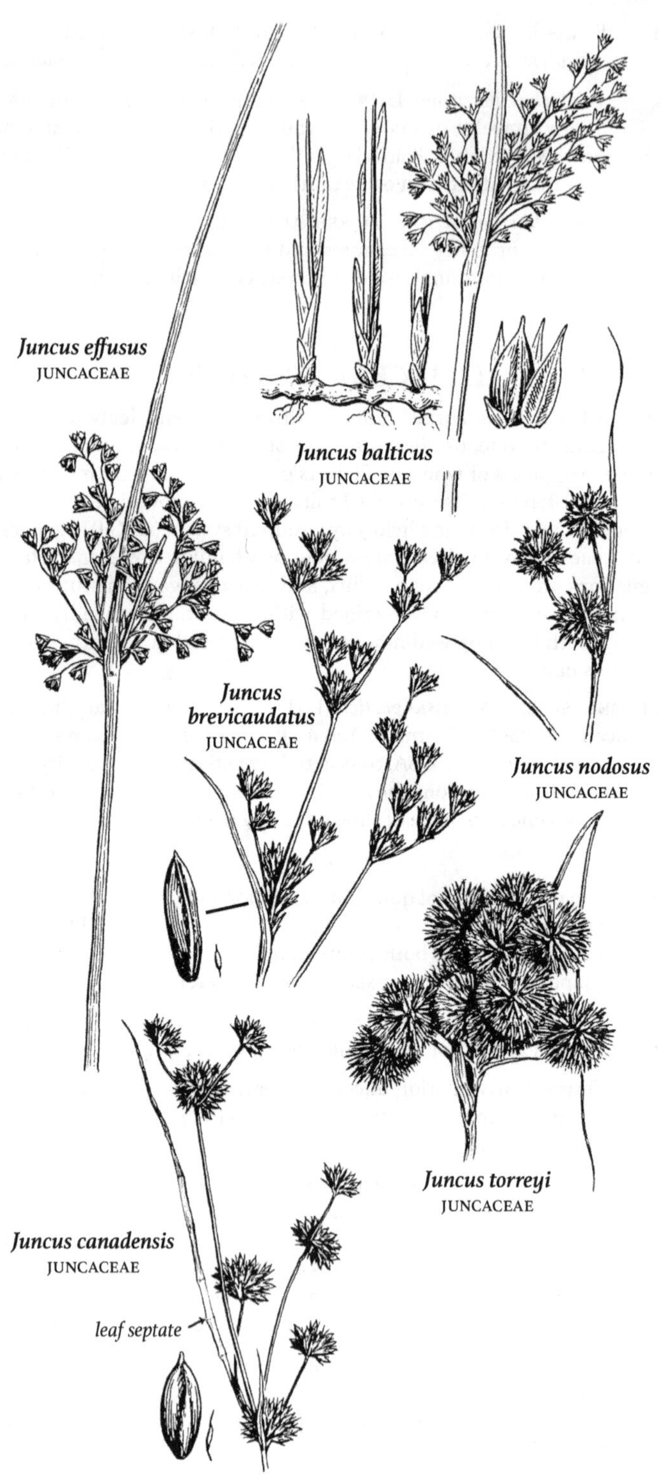
Juncus effusus
JUNCACEAE
Juncus balticus
JUNCACEAE
Juncus brevicaudatus
JUNCACEAE
Juncus nodosus
JUNCACEAE
Juncus torreyi
JUNCACEAE
Juncus canadensis
JUNCACEAE
leaf septate

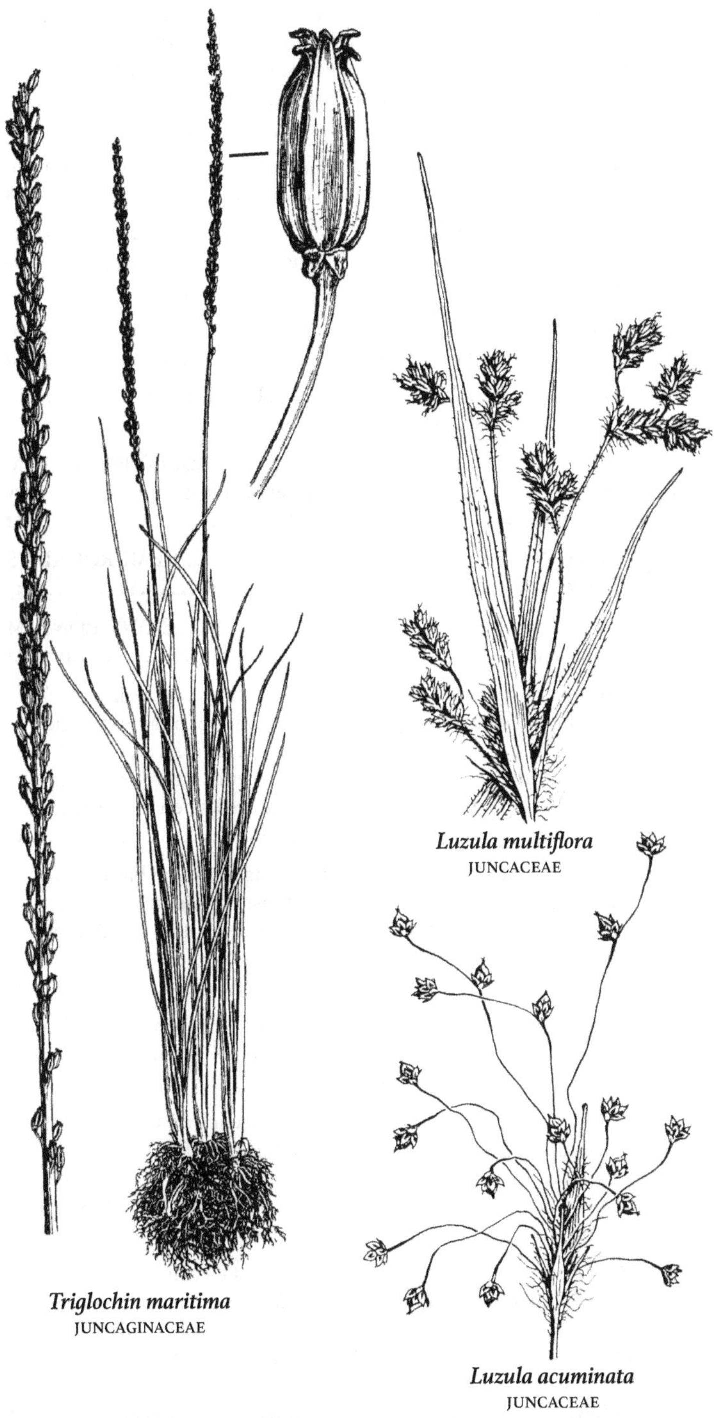

*Triglochin maritima*
JUNCAGINACEAE

*Luzula multiflora*
JUNCACEAE

*Luzula acuminata*
JUNCACEAE

6 Leaves scale-like, mostly brownish or yellowish, those on the much-branched upper portion of the plant subtending short green filiform branches (often mistaken for leaves) ....... **ASPARAGUS**

6 Leaves broad, flat, green (scale-like leaves or bracts may be present in addition to normal leaves) ........................ *go to lead 29*

7 Leaves all withering before plant flowers .. **ALLIUM** (*A. tricoccum*)

7 Leaves present at flowering time .............................. 8

8 Leaves all in one or two whorls on the stem ......... **MEDEOLA**

8 Leaves alternate or basal, or if in whorls these more than 2 or some alternate leaves also present ................................ 9

9 Flowers more than 3.5 cm long .............................. 10

9 Flowers less than 3.5 cm long ............................... 14

10 Leaves perfoliate .................... **UVULARIA** (*U. grandiflora*)

10 Leaves not perfoliate ........................................ 11

11 Principal leaves cauline, not crowded toward base of plant **LILIUM**

11 Principal leaves basal or nearly so; stem leafless above or with very small bracts ............................................... 12

12 Flowers 2–many ........................... **HEMEROCALLIS**

12 Flowers solitary ............................................ 13

13 Leaves basal ................................ **ERYTHRONIUM**

13 Leaves cauline........................................... **LILIUM**

14 Flowers in a many (7–60 or more) flowered umbel on an unbranched stem or scape; plants with odor of onion or garlic ... ........................................................ **ALLIUM**

14 Flowers solitary or in a raceme, panicle, or corymb, on a simple or branched stem (if in a few-flowered umbel, then either the stem branched or forked or flowers less than 7)..................... 15

15 Plants with principal leaves clearly along the stem; basal leaves (at least of current season) absent or at most apparently one ...... 16

15 Plants with the principal leaves all basal (at ground level) or nearly so; stem leaves absent, reduced to bracts, or much smaller or fewer than basal leaves ........................................... 20

16 Stem forked or branched; perianth over 12 mm long .. *go to lead 29*

16 Stem unbranched (above the ground and below the inflorescence); perianth usually less than 12 mm long........................ 17

17 Ovary with 1 style; fruit a berry; inflorescence a raceme (tepals up to 7 mm long) or if a panicle, the tepals less than 3 mm long; leaves ovate ........................................... *go to lead 29*

17 Ovary with 3 styles (one on each lobe); fruit a capsule; inflorescence a panicle; tepals 5–13 mm long; leaves very elongate .......... 18

18 Plants with at least some principal leaves clearly along the stem; inflorescences paniculate (except in depauperate individuals) .... ...................................................... **ANTICLEA**

18 Plants with the principal leaves all basal (at ground level) or nearly so, stem leaves absent, reduced to bracts, or at most much smaller or fewer than basal leaves; inflorescences racemose............ 19

19 Tepals united for half or more of their length .......... ALETRIS
19 Tepals completely separate or united at base only; peduncle and pedicels glandular-sticky .......................... TRIANTHA

20 Flower solitary.............................. ERYTHRONIUM
20 Flowers 2 or more in an inflorescence ........................ 21

21 Tepals united for half or more of their length .................. 22
21 Tepals completely separate or united at base only.............. 24

22 Perianth blue, less than 6 mm long; leaves linear; plants bulbous ................................................ MUSCARI
22 Perianth white, 5–10 mm long when mature; leaves lanceolate to elliptic; plants not bulbous ................................ 23

23 Leaves elliptic, the widest 2–6 cm broad; stems up to 35 cm tall, about equaling or shorter than the leaves; flowers nodding, on pedicels longer than the subtending bracts; perianth smooth on outside; fruit a berry ......................... CONVALLARIA
23 Leaves narrowly lanceolate or oblanceolate, the widest less than 2 cm broad; stems over 40 cm tall, much surpassing the leaves; flowers ascending on pedicels shorter than the subtending bracts; perianth granular-roughened on the outside; fruit a capsule .... ALETRIS

24 Ovary with 3 styles (1 on each lobe) in bisexual or pistillate flowers (or flowers all staminate) ................................... 25
24 Ovary with a single style; flowers bisexual ..................... 27

25 Plants with at least some principal leaves clearly cauline; inflorescences paniculate ......................... ANTICLEA
25 Plants with the principal leaves all basal or nearly so, stem leaves absent, reduced to bracts, or at most much smaller or fewer than basal leaves; inflorescences racemose ......................... 26

26 Tepals united for half or more of their length .......... ALETRIS
26 Tepals completely separate or united at base only; peduncle and pedicels glandular-sticky ......................... TRIANTHA

27 Leaves broad (over 3 cm wide); perianth yellow; fruit a blue berry. ................................................ CLINTONIA
27 Leaves long and narrow (less than 1.5 cm wide); perianth white to blue; fruit a capsule ........................................ 28

28 Tepals white, with green median stripe on outer side; filaments broad and flat (winged), wider at the middle than the anthers; flowers fewer than 20, in a raceme or corymb ORNITHOGALUM
28 Tepals white to blue, without green stripe; filaments slender or somewhat flattened toward base only; flowers usually more than 20, in an elongate raceme ............................ CAMASSIA

*Lead 29. Fruit red, blue, or black berries, except in Uvularia, which has a capsule.*

29 Leaves all in 1 or 2 whorls on the stem .............. MEDEOLA
29 Leaves alternate or basal ...................................... 30

30 Plants with the leaves all basal or nearly so; stem leaves absent, reduced to bracts, or at most much smaller or fewer than the basal leaves ........................................................ 31
30 Plants with leaves clearly along the stem ...................... 32

31 Flowers yellow, in an umbel; fruit blue ............. **CLINTONIA**

31 Flowers white, in a raceme; fruit (rarely produced) orange or red . ............................................ **CONVALLARIA**

32 Plant unbranched ............................................. 33

32 Plant branched (above the ground)............................ 34

33 Flowers in a terminal raceme or panicle, white, tepals separate; ripe fruit red, with dark stripes in one species .... **MAIANTHEMUM**

33 Flowers in the axils of leaves, greenish, greenish white, or yellowish, tepals united most of their length; ripe fruit blue to black ....... ............................................ **POLYGONATUM**

34 Perianth pale to deep yellow; fruit a glabrous capsule; stem and pedicels glabrous ................................. **UVULARIA**

34 Perianth greenish, rose-purple, white, or creamy, fruit a pubescent to glabrate or tuberculate red berry; stem (at least when young) and pedicels often pubescent ....................... **STREPTOPUS**

## ALETRIS *Colicroot* (NARTHECIACEAE)

***Aletris farinosa*** L. WHITE COLICROOT Perennial herb, with a basal rosette of narrow leaves, an erect stem bearing bract-like leaves, and a terminal spike-like raceme. Flowers perfect, white, tubular. Fruit a capsule, tipped by the persistent style and enclosed by the persistent withered perianth. June, July. Moist or dry sandy soil, open woods and barrens.

## ALLIUM *Onion* (AMARYLLIDACEAE)

Biennial or perennial herbs from a coated bulb, with a strong odor of onion or garlic, the leaves usually narrow, basal, or on the lower part of the stem, the scape-like stem erect, terminated by an umbel.

ADDITIONAL SPECIES ***Allium cepa*** L. (Cultivated onion), probably native of sw Asia; cultivated and rarely adventive. ***Allium sativum*** L. (Cultivated garlic), introduced culinary herb; perhaps adventive in Wisc. ***Allium tuberosum*** Rottl. ex Spreng. (Chinese chives), introduced in s Wisc.

### *Key to species*

1 Leaves usually over 2 cm wide, flat, petiolate, withering before the plant flowers ......................................... ***A. tricoccum***

1 Leaves linear, flat or terete, less than 2 cm wide (usually less than 1 cm), not petiolate, present at flowering time .................. 2

2 Umbel nodding, on bent or reflexed tip of scape; leaves flat ...... ............................................... ***A. cernuum***

2 Umbel erect on straight tip of scape; leaves flat or terete ........ 3

3 Leaf blades terete, hollow, at least most of their length (else flattened where pressed in drying, but the base of blade, just above summit of sheath, will not show 2 distinct surfaces) ................... 4

3 Leaf blades flat (sometimes keeled) .......................... 6

4 Stem stout, over 5 mm wide for most or all of its length, distinctly inflated below the middle .............................. ***A. cepa****

4 Stem 5 mm in diameter or less (very rarely to 7 mm), without

inflated section ............................................5

5 Pedicels equaling or shorter than the flowers; inflorescence without bulblets; filaments unappendaged..............*A. schoenoprasum*

5 Pedicels longer than mature flowers; umbels bearing bulblets in addition to 0–many flowers; inner filaments with a long slender appendage on each side ..............................*A. vineale*

6 Umbels bearing bulblets (flowers few or none).................7

6 Umbels not bearing bulblets .................................8

7 Involucral spathe composed of one bract with a beak usually 2–10 cm long; outer coverings of bulb membranous .......*A. sativum**

7 Involucral spathe composed of 2–3 bracts, with beaks less than 1.5 cm long; outer coats of bulb strongly fibrous .......*A. canadense*

8 Flowers white; umbels flat-topped to hemispherical, flowers all facing up in bloom ...............................*A. tuberosum**

8 Flowers pink, umbels ± spherical, flowers facing in all directions ..............................................*A. stellatum*

***Allium canadense*** L. MEADOW GARLIC May–June. Moist or dry open woods and prairies.

***Allium cernuum*** Roth NODDING ONION July–Aug. Dry woods, rocky banks, and prairies.

***Allium schoenoprasum*** L. WILD CHIVES June–July. Subsp. *schoenoprasum* is the cultivated chives, common in gardens.

***Allium stellatum*** Nutt. AUTUMN ONION July–Sept. Prairies, barrens, and rocky hills.

***Allium tricoccum*** Ait. WILD LEEK Rich woods, often in large colonies. The leaves develop in early spring and disappear before the flowers appear in June and July.

***Allium vineale*** L. FIELD-GARLIC June. Native of Europe; a weed in lawns, fields, and meadows; reported from Wisc and likely to occur as the species is common throughout Illinois (*no map*).

### ANTICLEA *Death-Camas* (MELANTHIACEAE)

***Anticlea elegans*** (Pursh) Rydb. MOUNTAIN DEATH-CAMAS Perennial herb, from an ovate bulb; plants waxy, especially when young. July–Aug. Sandy or rocky shores of Great Lakes, open bogs, calcareous fens. Highly toxic if eaten. *Zigadenus elegans* Pursh.

### ASPARAGUS (ASPARAGACEAE)

***Asparagus officinalis*** L. ASPARAGUS Native of Europe; cultivated in gardens and escaped in waste places.

### CAMASSIA *Camas* (ASPARAGACEAE)

***Camassia scilloides*** (Raf.) Cory [•123] ATLANTIC CAMAS *Endangered.* Perennial from bulbs, with several linear basal leaves, an erect scape, and a loose, terminal, bracted raceme of conspicuous, white to blue or violet flowers. April–May. Prairies, roadsides, along railroads, moist open woods.

### CLINTONIA *Bluebead-Lily* (LILIACEAE)

***Clintonia borealis*** (Ait.) Raf. [•123] YELLOW BLUEBEAD-LILY Perennial herbs from a rhizome, bearing 2–4 ample basal leaves, the bases sheathing a leafless erect scape bearing a few-flowered umbel of conspicuous flowers. Fruit a blue berry. May–June. Rich moist woods, swamps.

### CONVALLARIA *Lily-of-the-Valley* (ASPARAGACEAE)

***Convallaria majalis*** L. EUROPEAN LILY-OF-THE-VALLEY Perennial herb from a rhizome, the short stem bearing a few leafless sheaths and 2 or 3 broad leaves, the scape terminating in a bracted raceme. Fruit a red berry. May. Widely distributed in n Eurasia, commonly cultivated, and occasionally escaped near gardens.

### ERYTHRONIUM *Trout-Lily; Fawn-Lily* (LILIACEAE)

Perennial from a deep corm, the slender stem about half underground; leaves borne near middle of the stem and therefore appearing basal, usually mottled with brown; scape bearing a single nodding flower. Our species grow in colonies, producing numerous 1-leaved sterile plants and a few 2-leaved fertile ones.

*Key to species*

1 Flowers white ........................................ ***E. albidum***
1 Flowers yellow .................................... ***E. americanum***

***Erythronium albidum*** Nutt. SMALL WHITE FAWN-LILY April–May. Moist woods.

***Erythronium americanum*** Ker-Gawl. TROUT-LILY April–May. Moist woods.

### HEMEROCALLIS *Day-Lily* (XANTHORRHOEACEAE)

Tall perennial herbs, with numerous, elongate, linear, basal leaves and leafless scapes bearing a terminal cluster of large flowers, each lasting a single day.

*Key to species*

1 Flowers orange ........................................... ***H. fulva***
1 Flowers yellow ................................ ***H. lilioasphodelus***

***Hemerocallis fulva*** (L.) L. ORANGE DAY-LILY Introduced (invasive). Flowers tawny-orange. June–July. Long in cultivation and freely escaped.

***Hemerocallis lilioasphodelus*** L. YELLOW DAY-LILY Introduced. Flowers lemon-yellow. May–June. Commonly cultivated and occasionally escaped along roadsides. *Hemerocallis flava* (L.) L.

## LILIUM *Lily* (LILIACEAE)

Tall perennial herbs from a scaly bulb, in our species the erect stem bearing numerous narrow leaves, either alternate or whorled, and at the summit 1 to many, large, erect or nodding, yellow to red flowers. Fruit a more or less 3-angled capsule.

*Key to species*

1 Flowers erect; tepals narrowed at the base to a slender claw; leaves to 8 (–14) mm wide ........................... *L. philadelphicum*

1 Flowers nodding (fruit becoming erect); tepals narrowed gradually toward base, not clawed; widest leaves 8–35 mm wide .......... 2

2 Tepals glabrous within; stem glabrous above; leaves mostly whorled ............................................ *L. michiganense*

2 Tepals with pubescent strip basally within; stems with cobwebby pubescence, especially above; leaves alternate but crowded ...... ............................................ *L. lancifolium*

***Lilium lancifolium*** Thunb. TIGER LILY July–August. Native of e Asia; cultivated and escaped around dwellings and on roadsides.

***Lilium michiganense*** Farw. MICHIGAN LILY July–Aug. Wet meadows and low ground.

***Lilium philadelphicum*** L. WOOD-LILY June–July. Wet meadows, low prairie, fens and open bogs, seeps, ditches; also in drier meadows, prairies and woods.

## MAIANTHEMUM *False Solomon's-Seal* (ASPARAGACEAE)

Perennial herbs from a slender creeping rhizome, the erect or ascending stems bearing few to many alternate, sessile or nearly sessile leaves, and a terminal raceme or panicle of small white flowers. Fruit a globose berry.

*Key to species*

1 Perianth of 4 parts; leaves 3 or fewer (very rarely 4, usually 2), sometimes pubescent beneath .................... *M. canadense*

1 Perianth of 6 parts; leaves often more than 3 (1–4 in one species, where completely glabrous) ................................ 2

2 Inflorescence a panicle; perianth 1–2.5 mm long, the stamens up to 3 mm long ..................................... *M. racemosum*

2 Inflorescence a raceme; perianth 2.5–9 mm long, exceeding the stamens ............................................... 3

3 Stem leaves more than 6, finely pubescent beneath (rarely almost glabrous); uppermost leaves surpassing the top of the inflorescence ............................................ *M. stellatum*

3 Stem leaves 1–4, completely glabrous; inflorescence almost always overtopping leaves ................................ *M. trifolium*

***Maianthemum canadense*** Desf. [•123] FALSE LILY-OF-THE-VALLEY May–July. Common in moist to dry woods; also on hummocks in swamps, open bogs and thickets.

***Maianthemum racemosum*** (L.) Link [•123] FALSE SOLOMON'S-SEAL May–June. Rich woods. *Smilacina racemosa* (L.) Desf.

***Maianthemum stellatum*** (L.) Link [•123] STARRY FALSE SOLOMON'S-SEAL May–June. Moist, especially sandy soil of woods, shores, and prairies. *Smilacina stellata* (L.) Desf.

***Maianthemum trifolium*** (L.) Sloboda THREE-LEAF FALSE SOLOMON'S-SEAL May–June. Open bogs, conifer swamps, thickets. *Smilacina trifolia* (L.) Desf.

## MEDEOLA *Cucumber-Root* (LILIACEAE)

***Medeola virginiana*** L. INDIAN CUCUMBER-ROOT Perennial herb from a thick, tuber-like, horizontal rhizome, the slender stem bearing 2 whorls of leaves and a sessile, few-flowered, terminal umbel of greenish yellow flowers. Fruit a dark purple berry. May–July. Rich woods.

## MUSCARI *Grape-Hyacinth* (ASPARAGACEAE)

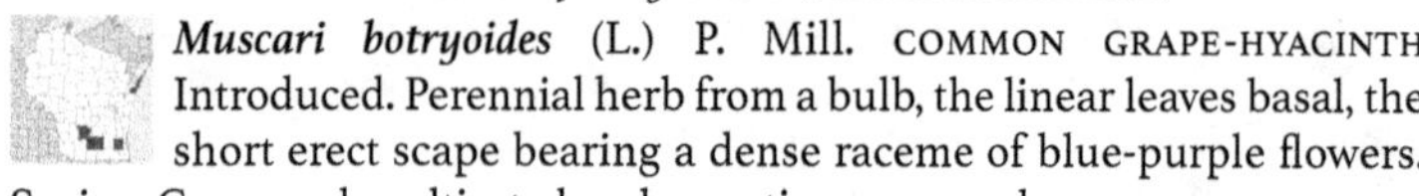
***Muscari botryoides*** (L.) P. Mill. COMMON GRAPE-HYACINTH Introduced. Perennial herb from a bulb, the linear leaves basal, the short erect scape bearing a dense raceme of blue-purple flowers. Spring. Commonly cultivated and sometimes escaped.

ADDITIONAL SPECIES ***Muscari neglectum*** Guss. (Starch grape-hyacinth), introduced, reported to occur in Wisc as a weed in fields and waste ground. In contrast to *M. botryoides,* its leaves are only 1–2 mm wide and nearly terete rather than flat.

## ORNITHOGALUM *Star-of-Bethlehem* (ASPARAGACEAE)

***Ornithogalum umbellatum*** L. STAR-OF-BETHLEHEM Perennial herb, with linear basal leaves, an erect scape, and a short bracted raceme of white flowers. May–June. Native of Europe; uncommon escape from cultivation in gardens, roadsides, and occasionally fields and woods.

## POLYGONATUM *Solomon's-Seal* (ASPARAGACEAE)

Perennial herbs from a horizontal knotty rhizome, the stem erect or arching, bearing in the upper portion numerous alternate leaves in two ranks and short, axillary, 1–15-flowered peduncles with pendent, white to greenish or yellow flowers. Fruit a dark blue or black berry.

*Key to species*

1 Leaves completely glabrous ........................ ***P. biflorum***
1 Leaves finely pubescent on the veins beneath ........ ***P. pubescens***

***Polygonatum biflorum*** (Walt.) Ell. KING SOLOMON'S-SEAL May–July. Moist woods and thickets.

***Polygonatum pubescens*** (Willd.) Pursh [•124] HAIRY SOLOMON'S-SEAL May–July. Moist woods and thickets.

## STREPTOPUS *Twisted Stalk* (LILIACEAE)

Perennial herbs from a rhizome, with alternate sessile or clasping leaves, and small, greenish white to purple, solitary or paired, axillary flowers. Fruit a red berry.

*Key to species*

1 Leaves entire or minutely denticulate, strongly clasping at the base, glaucous beneath; nodes and upper internodes glabrous (lower internodes sometimes hispid); tepals spreading or curving from near the middle; flowers whitish green. . . . . . . . . . . . ***S. amplexifolius***

1 Leaves prominently ciliate on the margins, the cilia usually visible to the naked eye, sessile or slightly clasping (the larger ones subtending branches more strongly clasping), sometimes paler but not glaucous beneath; nodes and upper internodes ± pubescent or sparsely hispidulous; tepals spreading or recurved only at the tips; flowers usually pinkish (or even maroon) . . . . . . . . . . . ***S. lanceolatus***

***Streptopus amplexifolius*** (L.) DC. CLASPING TWISTED STALK June–July. Rich moist woods.

***Streptopus lanceolatus*** (Ait.) Reveal [•124] LANCE-LEAF TWISTED STALK May–July. Rich woods. *Streptopus roseus* Michx.

## TRIANTHA *False Asphodel* (TOFIELDIACEAE)

***Triantha glutinosa*** (Michx.) Baker STICKY FALSE ASPHODEL *Threatened.* Perennial herb, from a bulb. Stems covered with sticky hairs. Leaves 2–4 from base of plant, linear, hairy, sometimes with 1 bractlike leaf near middle of stem. Flowers white, on sticky-hairy stalks, in a raceme. June–July. Sandy or gravelly shores, interdunal wetlands, calcareous fens, rocky shores of Lake Superior. *Tofieldia glutinosa* (Michx.) Pers.

## TRILLIUM *Trillium; Wake-Robin* (MELANTHIACEAE)

Perennial herbs from a stout rhizome, the erect stem bearing a single whorl of 3 ample leaves and a single, large, terminal, sessile or peduncled flower. Fruit a many-seeded berry. A few species are cultivated, especially *T. grandiflorum.*

*Key to species*

1 Flower sessile; petals maroon or yellow; leaves usually mottled when fresh . . . . . . . . . . . . . . . . . . . . . . . . . . . . . . . . . . . . . . . . ***T. recurvatum***

1 Flower peduncled; petals maroon to white (or green in aberrant forms); leaves not mottled . . . . . . . . . . . . . . . . . . . . . . . . . . . . . . . . . . . . 2

2 Ovary 3-angled to obscurely 3-lobed; leaves definitely petiolate . . . . . . . . . . . . . . . . . . . . . . . . . . . . . . . . . . . . . . . . . . . . . . . . . . . . . ***T. nivale***

2 Ovary strongly 6-angled or -winged and the leaves sessile or subsessile (except in aberrant forms with petioled leaves and/or 3-lobed ovary; these usually also have the petals ± marked with green and are obvious sports) . . . . . . . . . . . . . . . . . . . . . . . . . . . . . . . . . . . . . 3

3 Petals white to pink (never maroon), 3.5–9 cm long, distinctly longer than the sepals, ± obtuse (occasional small plants with shorter petals – though still longer than sepals – may be recognized by the straight styles and broad obovate petals); stigmatic styles straight (though sometimes spreading) or slightly curved at very tip, uniform in diameter; peduncles held above the leaves . . . . . . . . ***T. grandiflorum***

3 Petals white to maroon, usually less than 3.5 cm long (if longer, maroon and/or narrowly acute at tip), seldom much longer than sepals; stigmatic styles spreading, thick at base, tapering, and recurved; peduncles in white-flowered plants (and often also in maroon ones) usually reflexed and held below the leaves ....... 4

4 Petals white (rarely rosy), less than 2.5 (rarely to 3.5) cm long; f ilaments about as long as the anthers or occasionally as short as half as long; anthers 3–7 mm long, usually pink when fresh *T. cernuum*

4 Petals white to maroon, 2–3 (–4) cm long; f ilaments very short, almost always less than 2 mm long and less than 1/4 the length of the anthers; anthers 6–15 mm long, yellowish (to pink in maroon-flowered forms) ....................................... *T. flexipes*

***Trillium cernuum*** L. [•124] WHIP-POOR-WILL-FLOWER May–June. Conifer swamps, bog margins, moist or wet mixed forests, often with paper birch; thickets along streams; less often in rich hardwood forests.

***Trillium flexipes*** Raf. NODDING WAKEROBIN May. Rich deciduous forests, wet floodplain woods. Plants with purplish or maroon flowers occur.

***Trillium grandiflorum*** (Michx.) Salisb. LARGE-FLOWER WAKEROBIN Moist to rather dry deciduous forests, forming large colonies in beech-maple forests; less common in oak-hickory woods, swamps, mixed conifer-hardwoods.

***Trillium nivale*** Riddell SNOW-TRILLIUM *Threatened.* March–April, sometimes blooming through snow. Rich moist woods, floodplain forests, often adjacent to rivers or streams.

***Trillium recurvatum*** Beck BLOODY-BUTCHER Moist, rich deciduous woods. April–May.

## UVULARIA *Bellwort* (COLCHICACEAE)

Perennial herbs from a slender rhizome, the erect stem forked above the middle, the lower portion bearing a few bladeless sheaths and up to 4 leaves; leaves sessile or perfoliate; flowers yellow or greenish yellow, terminal, but appearing axillary by prolongation of the branches, nodding.

*Key to species*

1 Leaves perfoliate, finely puberulent (rarely almost glabrous) and usually light or dark green but not glaucous beneath; rhizome short with many crowded roots; mature capsule less than usually 8–10 mm long ........................................ *U. grandiflora*

1 Leaves sessile, glaucous but glabrous beneath; rhizome elongate, bearing scattered small roots; mature capsule over 15 mm long ... ........................................ *U. sessilifolia*

***Uvularia grandiflora*** Sm. [•124] LARGE-FLOWER BELLWORT April–May. Rich woods, preferring calcareous soil.

***Uvularia sessilifolia*** L. SESSILE-LEAF BELLWORT April–May. Dry or moist woods.

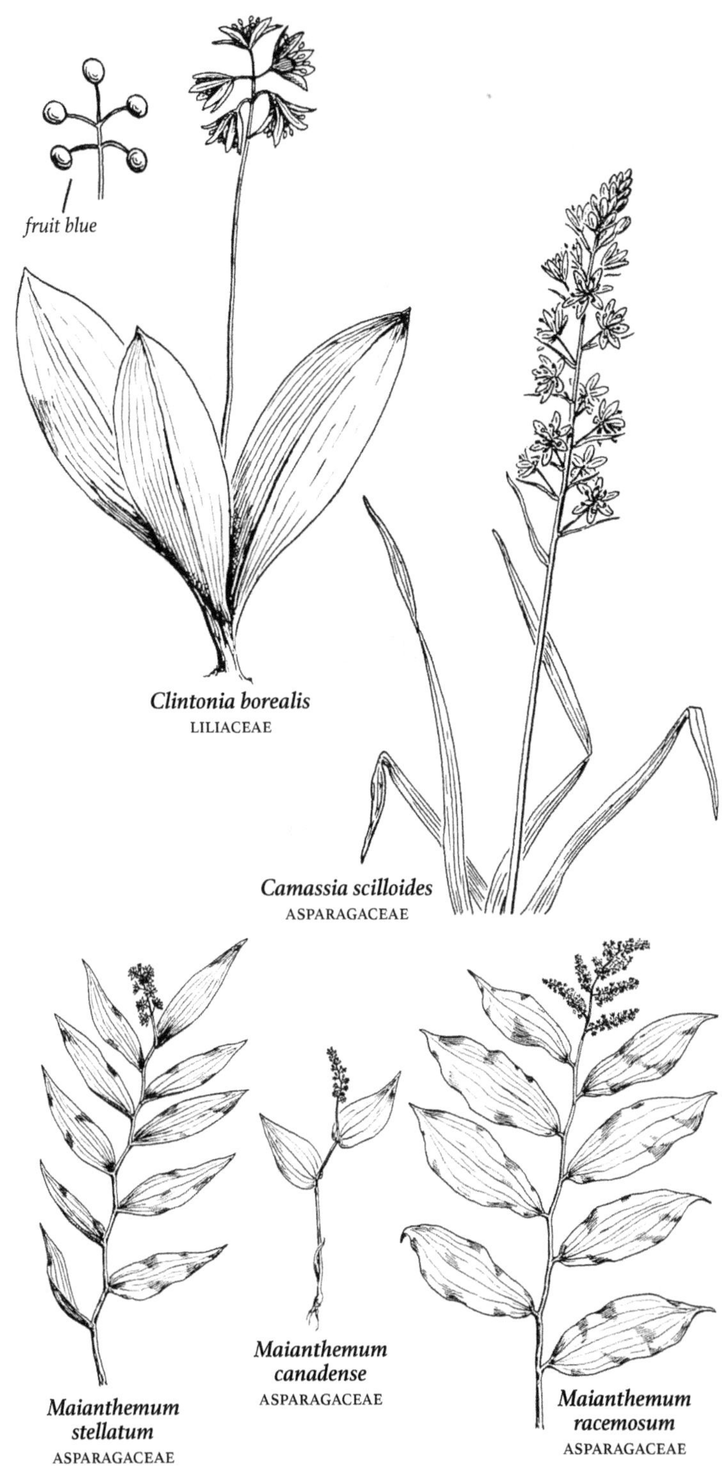
fruit blue
Clintonia borealis
LILIACEAE
Camassia scilloides
ASPARAGACEAE
Maianthemum stellatum
ASPARAGACEAE
Maianthemum canadense
ASPARAGACEAE
Maianthemum racemosum
ASPARAGACEAE

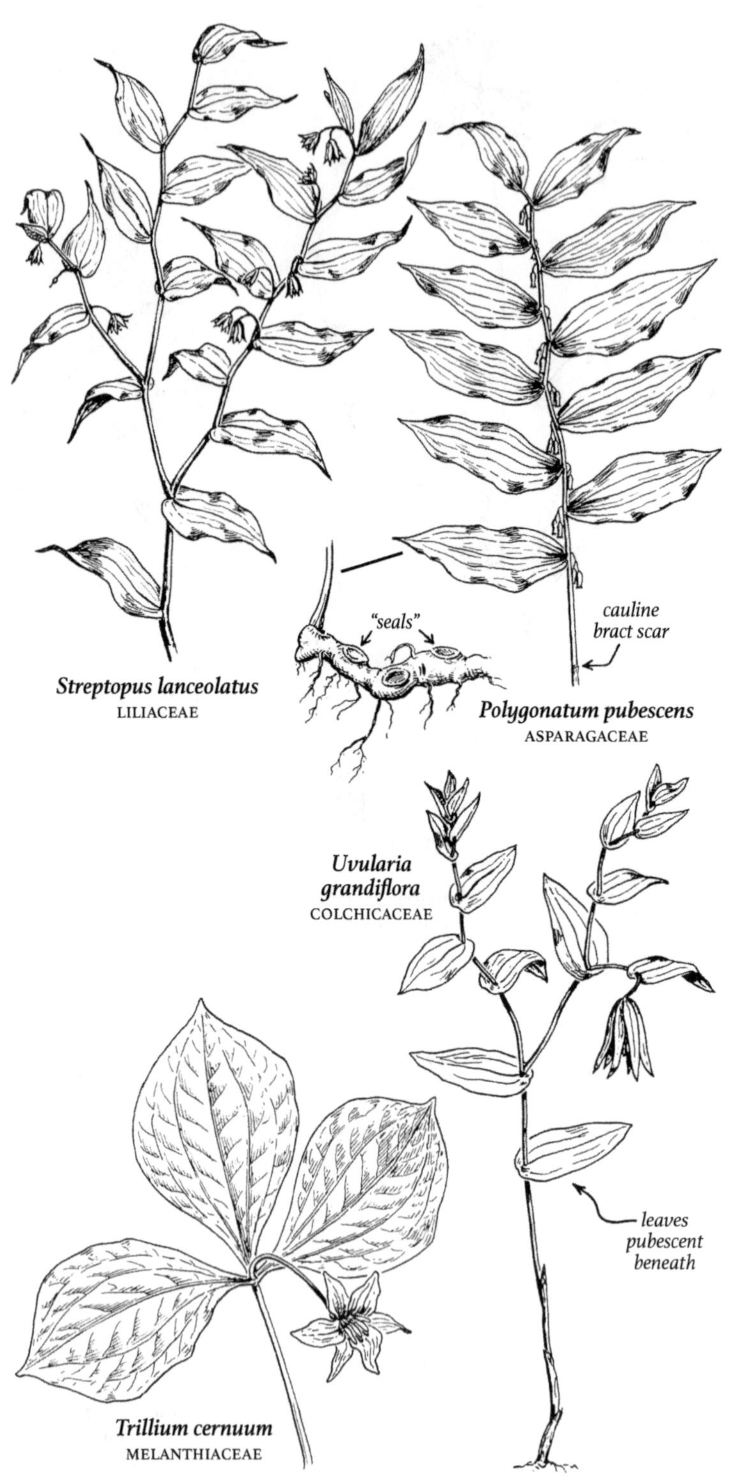


***Streptopus lanceolatus***
LILIACEAE

***Polygonatum pubescens***
ASPARAGACEAE

***Uvularia grandiflora***
COLCHICACEAE

***Trillium cernuum***
MELANTHIACEAE

## ORCHIDACEAE *Orchid Family*

Perennial herbs, from fleshy or tuberous roots, corms, or bulbs. Leaves simple, along the stem and alternate, or mostly at base of plant, stalkless and usually sheathing the stem, parallel-veined, often somewhat fleshy. Flowers perfect, irregular, showy in some species, in heads of 1 or 2 flowers at ends of stems, or with several to many flowers in a spike, raceme or panicle, each flower usually subtended by a bract. Fruit a many-seeded capsule, opening by 3 or sometimes 6 longitudinal slits, but remaining closed at tip and base; seeds very small. One of the world's largest families of vascular plants, with over 900 genera and an estimated 25,000–30,000 species, most of which occur in the tropics.

*Key to genera*

1 Lip a showy inflated pouch 1–5 cm long ...................... 2

1 Lip showy or inconspicuous, but not an inflated pouch with a small opening, usually ± flat with or without a slender basal spur (or if somewhat saccate, hardly showy and less than 1 cm long) ...... 4

2 Plants with leafy stems; lip a closed pouch (i.e., open only at base above) ...................................... **CYPRIPEDIUM**

2 Plants with leaves basal; lip split down middle above or open at base about half its length ........................................ 3

3 Basal leaf single, petiolate, the blade less than 7 cm long, produced in late summer and withering after the plant blooms the following spring; lip ca. 1.5–2 cm long; plants less than 20 cm tall **CALYPSO**

3 Basal leaves 2, longer, tapered to sheathing bases and not distinctly petiolate, present throughout the summer (but not winter); lip ca. 4–5 cm long; plants more than 20 cm tall ..... ***Cypripedium acaule***

4 Flower solitary (rarely plants with 2 flowers in a population of 1-flowered ones) ........................................ 5

4 Flowers 2 or more on one plant ............................ 6

5 Leaf linear, at most up to 7 (very rarely 10) mm wide, often poorly developed at flowering time, ± folded or plicate longitudinally, sheathing stem at base; plant from a small bulbous corm ........ ............................................ **ARETHUSA**

5 Leaf ± elliptic or lanceolate, usually over 7 mm wide and well developed at flowering time, flat, arising near middle of stem, sessile but not sheathing at base; plant from slender roots and rhizome.. ............................................ **POGONIA**

6 Lip prolonged into a distinct (usually slender and elongate) spur at base 2–40 mm long (pouch-like and only 2–3 mm in *Coeloglossum*) 7

6 Lip at most somewhat swollen or saccate (but not with a spur 2 mm or more long) ........................................ 11

7 Leaves cauline .............................................. 8

7 Leaves all basal or nearly so, or absent at flowering time (bracts subtending flowers may be leaflike) .......................... 9

8 Spur a thick pouch 2–3 mm long, much shorter than the lip ..... ........................................ **COELOGLOSSUM**

8 Spur slender, sometimes ± clavate, 7–40 mm long, ± equaling (at most slightly shorter than) to much longer than the lip .................................................. **PLATANTHERA**

9 Flowers entirely white and/or green, the lip lanceolate to narrowly linear, entire; lateral petals free ................ **PLATANTHERA**

9 Flowers with white lip (spotted or not) broadly ovate to oblong, often crenate or lobed; lateral petals connivent or fused with dorsal sepal to form a pink to purple hood .......................... 10

10 Leaves 1; lip less than 1 cm long, spotted, notched at apex and with a lateral lobe on each side ...................... **AMERORCHIS**

10 Leaves normally 2; lip over 1 cm long, unspotted, not lobed ...................................................... **GALEARIS**

11 Plants lacking green color (except sometimes in fruit), leafless with red, yellow, brown, or purplish stems arising from a coralloid rhizome ................................... **CORALLORHIZA**

11 Plants with green color, bearing leaves at some time in the year (if leaves absent at flowering time or plants apparently lacking green, arising from tubers, corms, or short rhizomes, not a coralloid mass) ........................................................ 12

12 Leaves a single opposite pair, definitely cauline, not at all sheathing the stem ........................................ **NEOTTIA**

12 Leaves solitary, alternate, absent, or basal (or almost basal, with sheathing bases) ........................................ 13

13 Stem leafy, with 4 or more conspicuous broadly ovate-lanceolate to elliptic leaves; perianth ca. 7–10 mm long; flowers greenish, at least the petals suffused with pink; upper part of stem and axis of inflorescence finely pubescent ..................... **EPIPACTIS**

13 Stem with the leaves fewer than 4, narrow, and/or basal (or absent); perianth various, but if pinkish then 10 mm or more long and the vegetative parts completely glabrous ........................ 14

14 Perianth 10–12 mm long, white or creamy; inflorescence dense, spike-like ................................. **SPIRANTHES**

14 Perianth longer or shorter, or not whitish and the inflorescence not spike-like .................................................. 15

15 Perianth 10 mm or more long, at least in part usually with some shade of pink or purple (yellowish in a form of *Aplectrum*) ..... 16

15 Perianth less than 10 mm long, greenish, white, or yellowish, with no trace of pink or purple ................................. 19

16 Flowers 2–3 cm or more wide, the lip uppermost, bearded with a tuft of yellow-tipped hairs; leaf solitary (rarely 2), several times longer than wide ............................. **CALOPOGON**

16 Flowers less than 1.5 cm broad, the lip lowermost and not bearded; blade of leaf not over 3.5 times longer than wide .............. 17

17 Leaves cauline, sessile and clasping, very seldom over 1.5 cm long ..................................................... **TRIPHORA**

17 Leaves basal or absent at flowering time, sheathing at base or petioled, much larger (usually 7–15 cm long) ................. 18

18 Leaf solitary, petioled, developing in fall and overwintering, usually withered before plant flowers .................... **APLECTRUM**

18 Leaves 2, sheathing at base, developing in current season and present at flowering .................................. **LIPARIS**

19 Leaves 1 or 2, sheathing at the base, the scape naked to the inflorescence; flowers on short pedicels, the raceme glabrous and not 1-sided nor noticeably twisted .......................... **20**

19 Leaves 3 or more (or withering at flowering time), the stem above them bearing small bracts or scales; flowers sessile or almost so in a narrow spike-like inflorescence, which is 1-sided or spirally twisted, or pubescent (or both) .......................... **21**

20 Leaves 1 (very rarely 2); perianth less than 4 mm long . **MALAXIS**

20 Leaves 2; perianth over 4 mm long ..................... **LIPARIS**

21 Leaves ovate to elliptic, basal or nearly so, present and firm at flowering time, the midvein and/or other veins margined in white or pale green (not always visible in dry plants); lip pouched or saccate at the base................................ **GOODYERA**

21 Leaves ovate-elliptic to linear and grass-like, sometimes cauline, often withering at flowering time (in wider-leaved species), not marked with whitish; lip not pouched ............ **SPIRANTHES**

## AMERORCHIS *Round-Leaf Orchid*

***Amerorchis rotundifolia*** (Banks) Hultén ROUND-LEAF ORCHID *Threatened.* Perennial herb, roots few from a slender rhizome. Leaves single from near base of plant, oval, usually with 1–2 bladeless sheaths below. Flowers 4 or more, in a raceme; sepals white to pale pink; petals white to pink or purple-tinged; lip white, with purple spots. June–July. Conifer swamps (on moss under cedar, tamarack, or black spruce); southward in cold conifer swamps of balsam fir, black spruce and cedar; usually found over underlying limestone and where sphagnum mosses not predominant. *Orchis rotundifolia* Banks.

## APLECTRUM *Adam-and-Eve*

***Aplectrum hyemale*** (Muhl.) Torr. [•125] ADAM-AND-EVE, PUTTY-ROOT Perennial from globose corms, which produce a single leaf in the late summer and a bracted scape the following spring; the corms are connected by a slender rhizome. Leaf single, basal; leaf blade elliptic. Scape with a few linear-oblong sheathing bracts. Flowers 7–15 in a loose terminal raceme; sepals and petals similar, purplish toward the base, brown toward the summit; lip white, marked with violet. May–June. Rich woods.

## ARETHUSA *Dragon's-Mouth*

***Arethusa bulbosa*** L. DRAGON'S-MOUTH Perennial herb; roots few, fibrous, from a corm. Stems leafless. Leaves 1, linear, small and bractlike at flowering time, later expanding; lower stem with 2–4 bladeless sheaths. Flowers single at ends of stems, sepals rose-purple; petals joined and more or less hoodlike over the column; lip pink, streaked with rose-purple. June–July. Open bogs and conifer swamps (in sphagnum moss), floating mats around bog lakes, calcareous fens; often with grass-pink (*Calopogon tuberosus*) and rose pogonia (*Pogonia ophioglossoides*).

## CALOPOGON *Grass-Pink*

***Calopogon tuberosus*** (L.) B.S.P. GRASS-PINK Perennial herb, from a corm. Stems leafless. Leaves 1 near base of plant, linear. Flowers pink to purple, 2–15 in a loose raceme; the lip located above the lateral petals, bearded on inside with yellow-tipped bristles. Open bogs and floating mats, openings in conifer swamps, calcareous fens near Great Lakes shoreline. Distinguished from swamp-pink (*Arethusa bulbosa*) and rose pogonia (*Pogonia ophioglossoides*) by having a raceme of several flowers vs. single flowers in *Arethusa* and *Pogonia*.

ADDITIONAL SPECIES ***Calopogon oklahomensis*** D.H. Goldman (Oklahoma grass-pink, native, reported from several s Wisc locations; typically in somewhat drier situations than the more common *C. tuberosus*. Our 2 species can be separated as follows:

1 Flowers opening sequentially; dilated upper portion of middle lip lobe usually much wider than long, typically anvil shaped; corms globose to elongate, not forked ..................... ***C. tuberosus***

1 Flowers opening nearly simultaneously; dilated upper portion of middle lip lobe usually much narrower than long, triangular to broadly rounded; corms elongate, forked ......... ***C. oklahomensis***

## CALYPSO *Fairy-Slipper Orchid*

***Calypso bulbosa*** (L.) Oakes CALYPSO, FAIRY-SLIPPER *Threatened.* Perennial herb, from a corm. Stems with 2–3 bladeless sheaths on lower portion. Leaves ovate, single from the corm. Flowers 1, nodding at end of stem; sepals and lateral petals similar, pale purple to pink; lip white to pink, streaked with purple, the lip extended to form a white "apron" with several rows of yellow bristles. May–June. Mature conifer forests or mixed forests of conifers and deciduous trees (such as balsam fir, hemlock, and paper birch), usually in shade; soils rich in woody humus. The single leaf of calypso appears in late August or September, persists through the winter, and withers after flowering in spring. Between fruiting in June and July and the emergence of the new leaf in late summer of fall, no aboveground portions of the plant may be visible.

## COELOGLOSSUM *Bracted Orchid*

***Coeloglossum viride*** (L.) Hartman BRACTED ORCHID Lowest 1 or 2 leaves reduced to bladeless sheaths; principal foliage leaves obovate, the upper progressively narrower and shorter and passing gradually into the bracts. Flowers greenish, often tinged with purple. June–Aug. Moist woods. *Habenaria viridis* (L.) R. Br.

## CORALLORHIZA *Coral-Root*

Yellow, brown, or purplish saprophytic herbs, lacking in chlorophyll, and parasitic on fungi inhabiting their characteristic coral-like rhizomes.

*Key to species*

1 Lip with a small lobe or elongate tooth on each side near the base (sometimes difficult to see in dried specimens) ................ **2**

1 Lip entire, or merely denticulate or erose .......................... 3

2 Sepals and petals 3-nerved; summit of ovary with a low protuberance (like a rudimentary spur) usually visible below the base of the lip; lip 4.5–7 mm long .................... *C. maculata*

2 Sepals and petals 1-nerved (or the latter rarely weakly 3-nerved); summit of ovary without visible protuberance; lip 2.5–4.5 mm long .......................................................... *C. trifida*

3 Sepals and petals 3–5-nerved, 8–15 mm long, conspicuously striped with purple, the lip solid purplish apically ............ *C. striata*

3 Sepals and petals 1-nerved (or faintly 3-nerved), less than 6 mm long, not conspicuously striped .............................. 4

4 Perianth 3–4.5 mm long, purplish; lip white, spotted with purplish ................................................ *C. odontorhiza*

4 Perianth 4–5.5 mm long, yellowish; lip unspotted white (or rarely spotted) .............................................. *C. trifida*

***Corallorhiza maculata*** (Raf.) Raf. SPOTTED CORAL-ROOT July–Sept. Woods.

***Corallorhiza odontorhiza*** (Willd.) Poir. AUTUMN CORAL-ROOT Aug–Sept. Open woods.

***Corallorhiza striata*** Lindl. STRIPED CORAL-ROOT May–July. Moist or dry woods.

***Corallorhiza trifida*** Chatelain YELLOW CORALROOT May–June. Moist to wet, mostly conifer woods, swamps (often under white cedar); usually where shaded.

## CYPRIPEDIUM *Lady's-Slipper*

Erect perennial herbs, from coarse, fibrous roots. Stems unbranched, often clumped, hairy. Leaves 2 or more at base of plant or along stem. Flowers 1 or 2, large and mostly showy at ends of stems, white, pink or yellow.

HYBRIDS ***Cypripedium × andrewsii*** A.M. Fuller, hybrid between *C. candidum* and *C. parviflorum;* s Wisc.

*Key to species*

1 Lip pouch pink to purple; leaves 2 at base of stem ....... *C. acaule*

1 Lip pouch yellow or white; leaves 3 or more on stem............ 2

2 Pouch yellow, sometimes brown- or purple-dotted ............. 3

2 Pouch white to pink, or pink with white patches ............... 4

3 Sepals and petals red-brown; lateral petals strongly twisted, brown-purple; pouch less than 4 cm long .... *C. parviflorum* var. *makasin*

3 Sepals and petals yellow to brown-green; lateral petals wavy, green with red-brown streaks; pouch more than 4 cm long ............ ................................... *C. parviflorum* var. *pubescens*

4 Pouch projected downward into a cone-shaped spur . *C. arietinum*

4 Pouch not spurred............................................ 5

5 Sepals and lateral petals white; lip 3–5 cm long ........ *C. reginae*
5 Sepals and lateral petals green; lip 1.5–2 cm long .... *C. candidum*

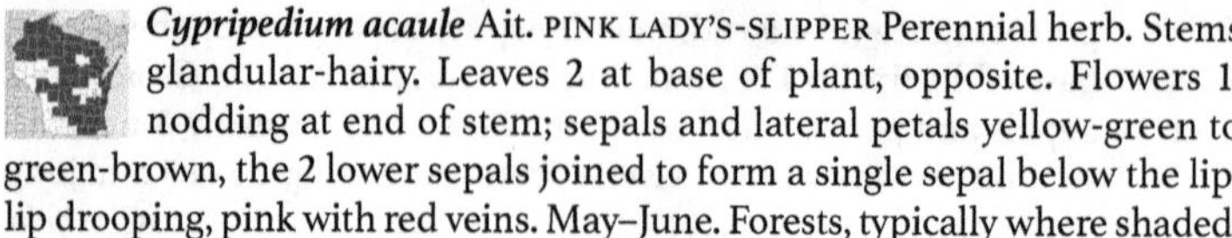

***Cypripedium acaule*** Ait. PINK LADY'S-SLIPPER Perennial herb. Stems glandular-hairy. Leaves 2 at base of plant, opposite. Flowers 1, nodding at end of stem; sepals and lateral petals yellow-green to green-brown, the 2 lower sepals joined to form a single sepal below the lip; lip drooping, pink with red veins. May–June. Forests, typically where shaded, acidic, and nutrient-poor; sometimes on hummocks in conifer swamps.

***Cypripedium arietinum*** Ait. f. RAM'S-HEAD LADY'S-SLIPPER *Threatened.* Perennial herb. Stems thinly hairy. Leaves 3–5, above middle of stem, oval, often folded, finely hairy. Flowers 1 or sometimes 2 at ends of stems; sepals and lateral petals similar, green-brown; lip an inflated pouch, white or pink-tinged, with prominent red-veins. Late May–June. Conifer swamps, wet forest openings (often with white cedar); also in drier, sandy, conifer and mixed conifer-deciduous forests, and on low dunes under conifers near shores of Great Lakes (usually with tamarack near Lake Mich). Our smallest and rarest lady's-slipper.

***Cypripedium candidum*** Muhl. SMALL WHITE LADY'S-SLIPPER *Threatened.* Perennial herb. Stems hairy. Leaves 2–4, upright, alternate along upper stem, oval, sparsely glandular-hairy. Flowers 1 at end of stems, the subtending bract leaflike, erect; sepals and lateral petals green-yellow, often streaked with purple; lip a small inflated pouch, white with faint purple veins. May–June. Calcium-rich wet meadows, low prairie, wet shores along Great Lakes, calcareous fens (often with shrubby cinquefoil (*Potentilla fruticosa*); usually where open and sunny.

***Cypripedium parviflorum*** Salisb. var. ***makasin*** (Farw.) Sheviak YELLOW LADY'S-SLIPPER Perennial herb. Stems glandular-hairy. Leaves 2–5, alternate along stem, oval, sparsely hairy. Flowers 1 (rarely 2) at ends of stems; sepals purple-brown; lateral petals linear, purple-brown, spirally twisted; lip an inflated pouch, yellow, often with purple veins and spots near opening. May–July. Conifer swamps, wet meadows, fens, and moist forests (often under cedar); sphagnum mosses are usually sparse; sites are shaded or sunny, with organic or mineral, often calcium-rich soil; s in our region also in open, calcium-rich swales. *Cypripedium calceolus* var. *parviflorum* (Salisb.) Fernald.

Our two varieties may be distinguished by the size of the pouch (lip) and the color of the sepals and petals: in var. *makasin,* the lip is mostly 2–3 cm long, and the sepals and petals are dark red; in var. *pubescens,* the lip is mostly 3–6 cm long and the sepals and petals are yellow-green; however, intermediate forms may occur.

***Cypripedium parviflorum Salisb.*** var. ***pubescens*** (Willd.) Knight YELLOW LADY'S-SLIPPER Perennial herb. Stems glandular-hairy. Leaves 3–6, alternate along stem, ovate to oval, sparsely hairy. Flowers 1 (rarely 2) at ends of stems; sepals yellow-green; lateral petals linear, yellow-green, often streaked with red-brown, usually spirally twisted; lip an inflated pouch, yellow, often with purple veins near opening. May–July. Conifer swamps, bogs, fens, prairies, especially where soils derived from limestone; also in wetter hardwood forests. *Cypripedium calceolus* var. *pubescens* (Willd.) Correll.

***Cypripedium reginae*** Walt. [•125] SHOWY LADY'S-SLIPPER Perennial herb. Stems strongly glandular-hairy. Leaves broadly oval, nearly smooth to hairy. Flowers 1 or often 2 at ends of stems; sepals and lateral petals white; lip an inflated pouch, white, streaked and spotted with pink or purple.June–July. Conifer and hardwood swamps (especially balsam fir-cedar-tamarack swamps), bogs, calcareous fens, sedge meadows, floating mats, wet openings, wet clayey slopes, ditches; especially where open and sunny; most abundant in openings in wet forests and swamps not dominated by sphagnum mosses. Showy lady's-slipper is our largest lady's-slipper. Avoid touching plants as the hairs can be irritating.

## EPIPACTIS *Helleborine*

***Epipactis helleborine*** (L.) Crantz HELLEBORINE Perennial herb. Leaves alternate, sessile and clasping, ovate to lance-shaped. Flowers in a terminal, many-flowered raceme; sepals and lateral petals dull green, strongly veined with purple; lip greenish and purple. July–Aug. Native of Europe; established and spreading in deciduous and mixed woods.

## GALEARIS *Showy Orchid*

***Galearis spectabilis*** (L.) Raf. SHOWY ORCHID Perennial herb. Leaves 2, rather fleshy. Sepals and lateral petals pink to pale purple; lip white. May–June. Rich woods. *Orchis spectabilis* L.

## GOODYERA *Rattlesnake-Plantain*

Perennial herbs from a short rhizome, plants glandular-pubescent on the scape, bracts, ovary, and sepals. Leaves in a basal cluster, commonly reticulated with white, narrowed to a broad, petiole-like base. Flowers in a spike-like raceme of white or greenish flowers, atop an erect scape with several scale-like bracts.

*Key to species*

1 Leaf blades with only the midvein outlined above in white or pale green, the largest blades usually 4–6 cm long; plants 20–50 cm tall; perianth 6–9 mm long .......... ***G. oblongifolia***

1 Leaf blades with white or pale green reticulation ± throughout (sometimes not on the midvein), the largest blades often less than 4 cm long; plants 5–30 cm tall; perianth 2.5–5.5 (rarely to 6.5) mm long .......... 2

2 Stem with usually 7–10 cauline bracts (undeveloped leaves); beak of lip (beyond the large pouch) less than 1 mm long, about 1/4 the total length of the lip or usually less; inflorescence ± densely flowered on all sides .......... ***G. pubescens***

2 Stem with 2–5 cauline bracts; beak of lip 1–2 mm long, about 1/2 the total length of the lip; pouch shallow or deep; inflorescence strongly one-sided or ± loosely flowered on all sides .......... 3

3 Lip deeply pouched, the pouch about as deep as long, the beak often strongly turned downward at maturity; plants mostly 10–20 cm tall; largest leaf blades mostly 1–2 cm long; cauline bracts 2–4 (usually 3) .......... ***G. repens***

3 Lip shallowly pouched, the pouch longer than deep, the beak horizontal or slightly recurved; plants usually 17–25 cm tall; largest leaf blades mostly 2–4 cm long; cauline bracts 4–5 ..... ***G. tesselata***

***Goodyera oblongifolia*** Raf. GREEN-LEAF, WESTERN RATTLESNAKE-PLANTAIN July–Aug. Dry woods.

***Goodyera pubescens*** (Willd.) R. Br. DOWNY RATTLESNAKE-PLANTAIN July–Aug. Dry woods.

***Goodyera repens*** (L.) R. Br. [•125] DWARF RATTLESNAKE-PLANTAIN July–Aug. Dry woods.

***Goodyera tesselata*** Lodd. CHECKERED RATTLESNAKE-PLANTAIN July–Aug. Dry woods.

## LIPARIS *Wide-Lip Orchid*

Low perennial herbs from a solid bulb. Leaves at base few and scalelike and with a pair of larger, shining leaves on the stem. Flowers atop a naked scape in a loose raceme.

*Key to species*

1 Lip ca. 10 mm long, purplish; capsules equaling or shorter than pedicels ........................................ ***L. liliifolia***

1 Lip 4–6 (–6.5) mm long, yellow-green; capsules longer than pedicels ................................................ ***L. loeselii***

***Liparis liliifolia*** (L.) L.C. Rich. BROWN WIDE-LIP ORCHID June–July. Rich woods.

***Liparis loeselii*** (L.) L.C. Rich. [•125] FEN-ORCHID June–Aug. Conifer swamps, fens, floating mats, streambanks, sandy shores, ditches; soils peaty to mineral, acid to calcium-rich.

## MALAXIS *Adder's-Mouth Orchid*

Small perennial herbs. Leaves 1–5 from base of plant or single along stem. Flowers green-white, spaced or crowded in slender or cylindric racemes at ends of stems.

*Key to species*

1 Flowers evenly spaced in a raceme 5–11 cm long .. ***M. monophyllos***

1 Flowers crowded near top of raceme, the raceme 2–5 cm long .... ................................................ ***M. unifolia***

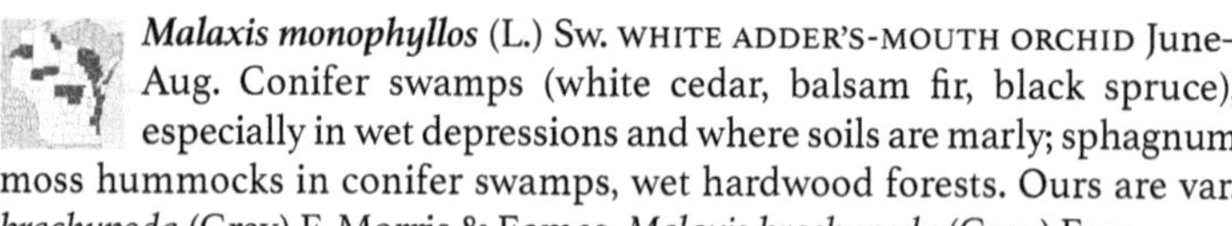

***Malaxis monophyllos*** (L.) Sw. WHITE ADDER'S-MOUTH ORCHID June–Aug. Conifer swamps (white cedar, balsam fir, black spruce), especially in wet depressions and where soils are marly; sphagnum moss hummocks in conifer swamps, wet hardwood forests. Ours are var. *brachypoda* (Gray) F. Morris & Eames. *Malaxis brachypoda* (Gray) Fern.

***Malaxis unifolia*** Michx. GREEN ADDER'S-MOUTH ORCHID June–Aug. Sphagnum moss hummocks in swamps, sedge meadows, thickets; also in drier forests including pine plantations.

## NEOTTIA *Twayblade*

Perennial herbs. Stems with a pair of opposite leaves near middle, stems smooth below leaves, hairy above. Leaves broad, stalkless. Flowers small, green to purple, in a raceme at end of stem, the lip 2-lobed or deeply parted. Formerly considered part of genus *Listera.*

*Key to species*

1 Lip 3–5 mm long, divided to about middle into 2 narrow segments ........................................................ ***N. cordata***
1 Lipp 7–12 mm long, shallowly notched or divided 1/3 of length, the segments broad ................................................ 2

2 Lip wide at base, with a pair of auricles ............. ***N. auriculata***
2 Lip narrowed to base, auricles absent ........... ***N. convallarioides***

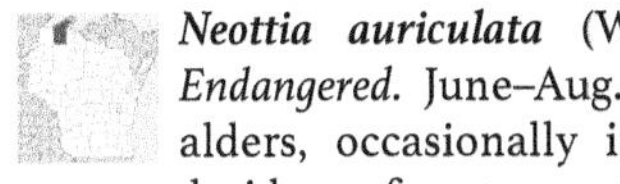

***Neottia auriculata*** (Wiegand) Szlach. AURICLED TWAYBLADE *Endangered.* June–Aug. Alluvial sand along rivers, often under alders, occasionally in moist conifer or mixed conifer and deciduous forests; usually where shaded. *Listera auriculata* Wieg.

***Neottia convallarioides*** (Sw.) Rich. BROAD-LIP TWAYBLADE *Threatened.* July–Aug. Seeps in forests, cedar swamps, wet, mixed conifer-deciduous woods, streambanks. *Listera convallarioides* (Sw.) Nutt.

***Neottia cordata*** (L.) Rich. [•126] HEART-LEAF TWAYBLADE June–July. Bogs and conifer swamps, where usually on sphagnum moss hummocks; hemlock groves. *Listera cordata* (L.) R. Br.

## PLATANTHERA *Rein-Orchid*

Perennial herbs, from a cluster of fleshy roots. Stems erect, smooth. Leaves mostly along the stem, upright, reduced to sheaths at base and upward on stem; leaves basal in *P. orbiculata.* Flowers white or green, several to many in a spike or raceme.

HYBRIDS ***Platanthera × andrewsii*** (M. White) Luer [syn: *Platanthera lacera* var. *terrae-novae* (Fern.) Luer], hybrid between *P. lacera* and *P. psycodes,* known from several c and n Wisc locations.

*Key to species*

1 Lip prominently ciliate or fringed ............................ 2
1 Lip entire or toothed, but not fringed ......................... 4

2 Flowers pink-purple; divisions of the lip broadly fan-shaped, copiously lacerate-fringed, but the fringe usually cut less than half the distance to the base of the division of the lip ....... ***P. psycodes***
2 Flowers yellowish, cream, or greenish; at least the lateral divisions of the lip more narrowly cuneate, mostly cut into a long fringe more than half their length ........................................ 3

3 Sepals 6–9 mm long; lateral petals broadly obovate, erose or denticulate at tip ................................ ***P. leucophaea***
3 Sepals 3.5–5 mm long; lateral petals linear-oblong or almost lanceolate, usually essentially entire .................... ***P. lacera***

4 Leaves all basal, the stem at most with reduced bracts .......... 5
4 Leaves cauline (along the stem)................................ 8

5 Leaves about twice as long as wide, or longer; spur less than 12 mm long . . . . . . 6
5 Leaves less than twice as long as broad, orbicular or almost so; spur 16–40 mm long . . . . . . 7

6 Lip with truncate 3-toothed or crenate tip; spur 7–11 mm long, much exceeding the lip . . . . . . *P. clavellata*
6 Lip tapered to a pointed or rounded, untoothed tip; spur about equaling lip or at most ca. 2 mm longer . . . . . . *P. obtusata*

7 Scape naked (rarely with a bract); spurs (16–24 mm long) tapered ± evenly to rounded tip; lip yellowish green, tending to turn upward near the end . . . . . . *P. hookeri*
7 Scape with 1–6 bracts between leaves and inflorescence; spurs parallel-sided or even somewhat club-shaped toward tip; lip whitish green, tending to turn downward . . . . . . *P. orbiculata*

8 Lip truncate and 2–3-toothed or -lobed at tip . . . . . . *P. clavellata*
8 Lip tapered, rounded (or almost truncate and obscurely crenulate) but not 2–3-toothed at tip . . . . . . 9

9 Lip much shorter than the spur, broadly rounded (or almost truncate) at tip, with an erect tubercle near the base and a lateral tooth or projection on each side near the base . . . . . . *P. flava*
9 Lip 1–2 mm shorter than, about equaling, or slightly longer than the spur, tapered to narrow tip, with neither a tubercle nor lateral teeth (at most, broadly widened basally) . . . . . . 10

10 Flowers pure white, lip strongly expanded basally . . . . . . *P. dilatata*
10 Flowers green, greenish yellow or greenish white, lip cuneate to strap-shaped, not or only slightly widened at base . . . . . . 11

11 Anther sacs essentially in contact above the rounded stigma (separated at tip by less than 0.3 mm); lips 2.5–5 mm long . . . . . . *P. aquilonis*
11 Anther sacs separated at tip by ca. 0.4 mm or more, stigma pointed; lips 4–8 mm long . . . . . . *P. huronensis*

***Platanthera aquilonis*** Sheviak BOG ORCHID May–Aug. Moist to wet including moist forests, cedar swamps, riverbanks, wet meadows, fens, ditches and borrow pits. The *Platanthera hyperborea* complex, including *P. dilatata, P. aquilonis,* and *P. huronensis,* are often difficult to separate; living rather than dried plants are easiest to identify.

***Platanthera clavellata*** (Michx.) Luer GREEN WOODLAND ORCHID July–Aug. Acid bogs and wet soils, especially in sphagnum. *Habenaria clavellata* (Michx.) Spreng.

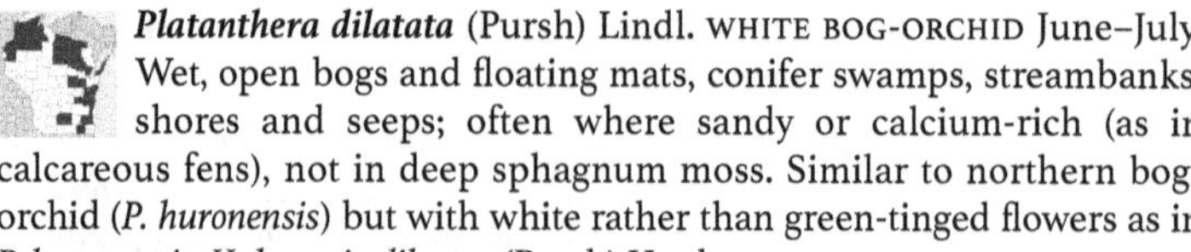

***Platanthera dilatata*** (Pursh) Lindl. WHITE BOG-ORCHID June–July. Wet, open bogs and floating mats, conifer swamps, streambanks, shores and seeps; often where sandy or calcium-rich (as in calcareous fens), not in deep sphagnum moss. Similar to northern bog-orchid (*P. huronensis*) but with white rather than green-tinged flowers as in *P. huronensis. Habenaria dilatata* (Pursh) Hook.

***Platanthera flava*** (L.) Lindl. PALE GREEN ORCHID *Threatened.* June–July. Wet depressions in hardwood swamps, alder thickets, sedge meadows, moist sand prairies; often where calcium-rich, sometimes where disturbed. Ours are var. *herbiola* (R. Br.) Luer. *Habenaria flava* (L.) R. Br.

***Platanthera hookeri*** (Torr.) Lindl. HOOKER'S ORCHID June–July. Coniferous or mixed forests, wooded dunes, soils often sandy. *Habenaria hookeri* Torr.

***Platanthera huronensis*** (Nutt.) Lindl. NORTHERN BOG-ORCHID June–Aug. Moist to wet forests and swamps, thickets, streambanks, wet meadows, wet sand along Great Lakes shoreline, ditches. *Habenaria hyperborea* (L.) R. Br.

***Platanthera lacera*** (Michx.) G. Don Green FRINGED ORCHID June–Aug. Hummocks in open sphagnum bogs, conifer bogs, swamps, wet meadows, sandy prairie, thickets, ditches. *Habenaria lacera* (Michx.) R. Br.

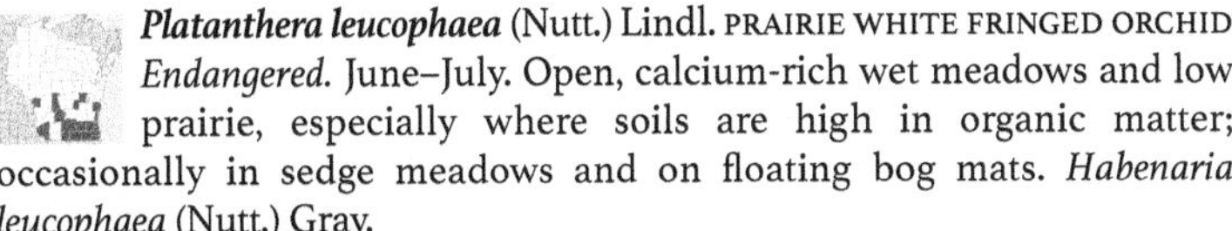

***Platanthera leucophaea*** (Nutt.) Lindl. PRAIRIE WHITE FRINGED ORCHID *Endangered.* June–July. Open, calcium-rich wet meadows and low prairie, especially where soils are high in organic matter; occasionally in sedge meadows and on floating bog mats. *Habenaria leucophaea* (Nutt.) Gray.

***Platanthera obtusata*** (Banks) Lindl. BLUNT-LEAF ORCHID June–Aug. Shaded hummocks in conifer swamps (especially under cedar, black spruce or balsam fir), wet mixed conifer-deciduous forests, alder thickets. *Habenaria obtusata* (Banks) Richards.

***Platanthera orbiculata*** (Pursh) Lindl. ROUND-LEAF ORCHID Late June–Aug. Shaded conifer swamps of white cedar, balsam fir, and black spruce, especially where underlain by marl; also in drier pine forests. *Habenaria orbiculata* (Pursh) Torr., *Platanthera macrophylla* (Goldie) P.M. Brown.

***Platanthera psycodes*** (L.) Lindl. [•126] LESSER PURPLE FRINGED ORCHID July–Aug. Wetland margins, shores, wet forests, wet meadows, low prairie, roadside ditches; typically not on sphagnum moss. *Habenaria psycodes* (L.) Spreng.

## POGONIA *Snake-Mouth*

***Pogonia ophioglossoides*** (L.) Ker-Gawl. [•126] ROSE POGONIA, SNAKE-MOUTH ORCHID Perennial herb, spreading by surface runners (stolons) which send up a stem every 10 cm or more apart. Leaves single, attached about halfway up stem. Flowers pink to purple, usually 1 at end of stems; lip pink with purple veins, fringed at tip, bearded with yellow bristles. June–July. Conifer swamps and open bogs in sphagnum moss, floating sedge mats, sedge meadows, sandy interdunal wetlands.

## SPIRANTHES *Ladies'-Tresses*

Perennial herbs, from a cluster of tuberous roots. Stems slender, erect. Leaves largest at base of plant, becoming smaller upward on stem, the stem leaves erect and sheathing. Flowers small, white or creamy, spirally twisted in a densely flowered, spike-like raceme.

ADDTIONAL SPECIES & HYBRIDS ***Spiranthes ovalis*** Lindl.; stems slender, flowers very late blooming, beginning late Sept through Oct; moist fields, shrub thickets, open woods; Grant County. ***Spiranthes* × *simpsonii*** Catling & Sheviak, a cross between *S. lacera* and *S. romanzoffiana,* Brown County.

*Key to species*

1 Leaves widely spreading or lying flat in a basal rosette, short-petioled, sometimes withered at flowering time, their blades less than 4.5 cm long, about 2/5 as wide as long or wider; perianth 2.5–5.5 mm long .......... ***S. lacera***

1 Leaves ascending, not distinctly petioled, usually present at flowering time (except in S. magnicamporum, with flowers 9–11 mm long), their blades (non-sheathing portion) over 4.5 cm long and less than 2/5 as wide as long; flowers usually in 2 or more rows in a ± crowded spike (sometimes one-sided); perianth larger than 5.5 mm long .......... **2**

2 Plants flowering in June and early July; largest leaves (non-sheathing portion) about 5–10x longer than wide; lip bright yellow or yellowish orange; leaves all basal (rarely 1 on the stem), 4.5–10 cm long, the stem with 1–2 bracts (including reduced leaf, if present) .......... ***S. lucida***

2 Plants flowering in mid- to late summer and fall; largest leaves commonly over 10x longer than wide, or leaves absent; lip white or creamy, sometimes the central portion pale yellow; leaves often present on lower portion of stem, the stem bracts and leaves numbering 3–6 .......... **3**

3 Lower flowers with perianth 3–7.5 mm long (most easily measured using the dorsal sepal); flowers in one row, usually in a loose spiral because of the twisted rachis .......... **4**

3 Lower flowers with perianth 7–11 mm long (most easily measured using the dorsal sepal); flowers in 2 or more rows, often tightly spiraled .......... **5**

4 Perianth 5.5–7.5 mm long; rachis and bracts conspicuously capitate-glandular .......... ***S. casei***

4 Perianth 3–5 mm long; rachis sparsely glandular, bracts glandular only on margins and base .......... ***S. ovalis****

5 Lip fiddle-shaped, strongly constricted behind expanded tip; lateral sepals united for at least half their length with dorsal sepal and lateral petals, forming a hood .......... ***S. romanzoffiana***

5 Lip oblong, often erose-margined but not strongly constricted; at least the lateral sepals free (or easily separated if connivent when young) .......... **6**

6 Plant leafless when flowering; upper stem bracts usually overlapping; lateral sepals in the fresh flowers curved and spreading .......... ***S. magnicamporum***

6 Plant with leaves usually present at flowering time, upper stem bracts not or barely overlapping; lateral sepals in the fresh flowers appressed .......... ***S. cernua***

***Spiranthes casei*** Catling & Cruise CASE'S LADIES'-TRESSES Aug–Sept. Sandy acidic soil, often with bracken fern (*Pteridium*).

***Spiranthes cernua*** (L.) L.C. Rich. [•126] WHITE NODDING LADIES'-TRESSES Aug–Oct. Open, usually sandy wetlands such as wet meadows, lakeshores, moist prairies, ditches and roadsides.

***Spiranthes lacera*** (Raf.) Raf. NORTHERN SLENDER LADIES'-TRESSES Aug–Sept. Dry sandy soil, often with blueberry and bracken fern in open woods of jack pine, red pine, and oak; moist aspen groves, conifer thickets along shores and on dunes.

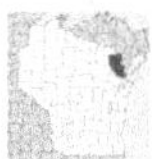

***Spiranthes lucida*** (H. H. Eat.) Ames SHINING LADIES'-TRESSES June–July. In Wisc, known only from several locations on rocky banks of the Menominee River, Marinette Co.

***Spiranthes magnicamporum*** Sheviak GREAT PLAINS LADIES'-TRESSES Sept–Oct. Calcareous prairies, fens, moist to mesic prairies.

***Spiranthes romanzoffiana*** Cham. HOODED LADIES'-TRESSES July–Sept. Open wetlands including wet meadows, fens, lakeshores, open swamps, ditches, seeps; usually in neutral or calcium-rich habitats.

### TRIPHORA *Three Birds Orchid*

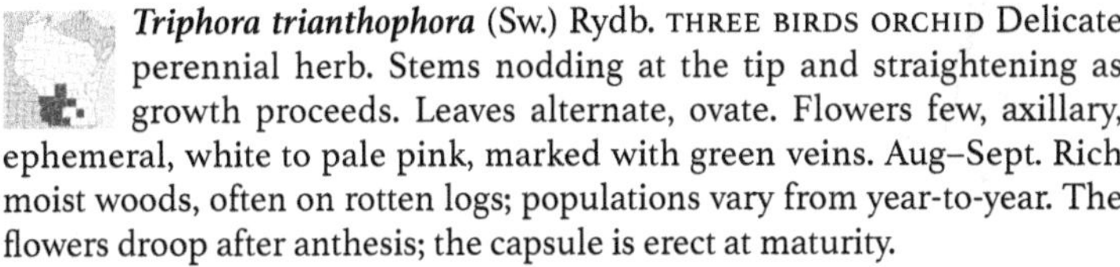

***Triphora trianthophora*** (Sw.) Rydb. THREE BIRDS ORCHID Delicate perennial herb. Stems nodding at the tip and straightening as growth proceeds. Leaves alternate, ovate. Flowers few, axillary, ephemeral, white to pale pink, marked with green veins. Aug–Sept. Rich moist woods, often on rotten logs; populations vary from year-to-year. The flowers droop after anthesis; the capsule is erect at maturity.

## POACEAE *Grass Family*

Perennial or annual herbs, clumped or spreading by rhizomes. Stems (culms) usually hollow, with swollen, solid nodes. Leaves linear, parallel-veined, alternate in 2 ranks or rows, sheathing the stem, the sheaths usually split vertically, sometimes joined and tubular as in brome (*Bromus*) and mannagrass (*Glyceria*); with a membranous or hairy ring (ligule) at top of sheath between blade and stem, or the ligule sometimes absent; a pair of projecting lobes (auricles) sometimes present at base of blade.

Flowers (florets) small, usually perfect, or sometimes either staminate or pistillate, the staminate and pistillate flowers separate on the same or different plants. Florets grouped into spikelets, each spikelet with 1 to many florets, the florets stalkless and alternate along a small stem or axis (rachilla), with a pair of small bracts (glumes) at base of each spikelet (the glumes rarely absent); the glumes usually of different lengths, the lowermost (or first) glume usually smaller, the upper (or second) glume usually longer. Within the spikelet, each floret subtended by 2 bracts, the larger one (lemma) containing the flower, the smaller one (palea) covering the flower; the lemma and palea often enclosing the ripe fruit (grain or caryopsis); stamens usually 3 or sometimes 6, usually exserted when flowering; ovary superior, never enclosed in a sac (as in sedges); styles 2–3-parted, the stigmas often feathery.

Spikelets grouped in a variety of heads, most commonly in branching heads (panicles), or stalked along an unbranched stem (rachis) in a raceme, or the spikelets stalkless along an unbranched stem in a spike; spikelets

*Goodyera repens*
ORCHIDACEAE

*Cypripedium reginae*
ORCHIDACEAE

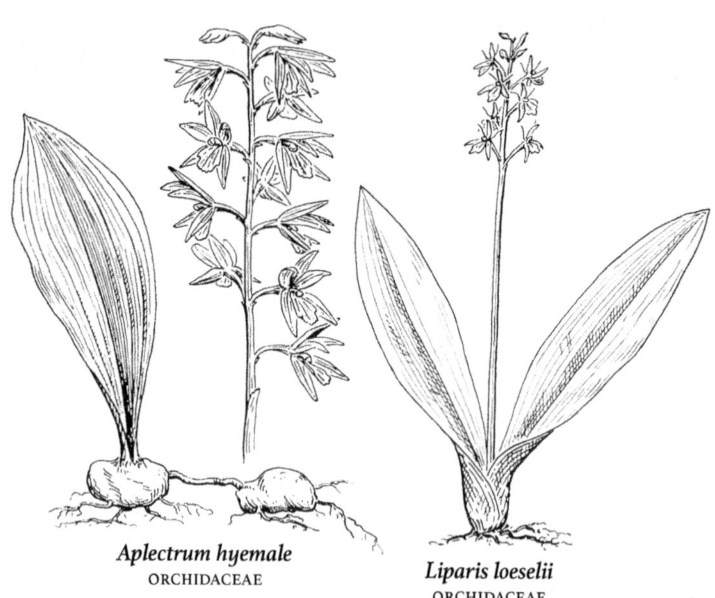

*Aplectrum hyemale*
ORCHIDACEAE

*Liparis loeselii*
ORCHIDACEAE

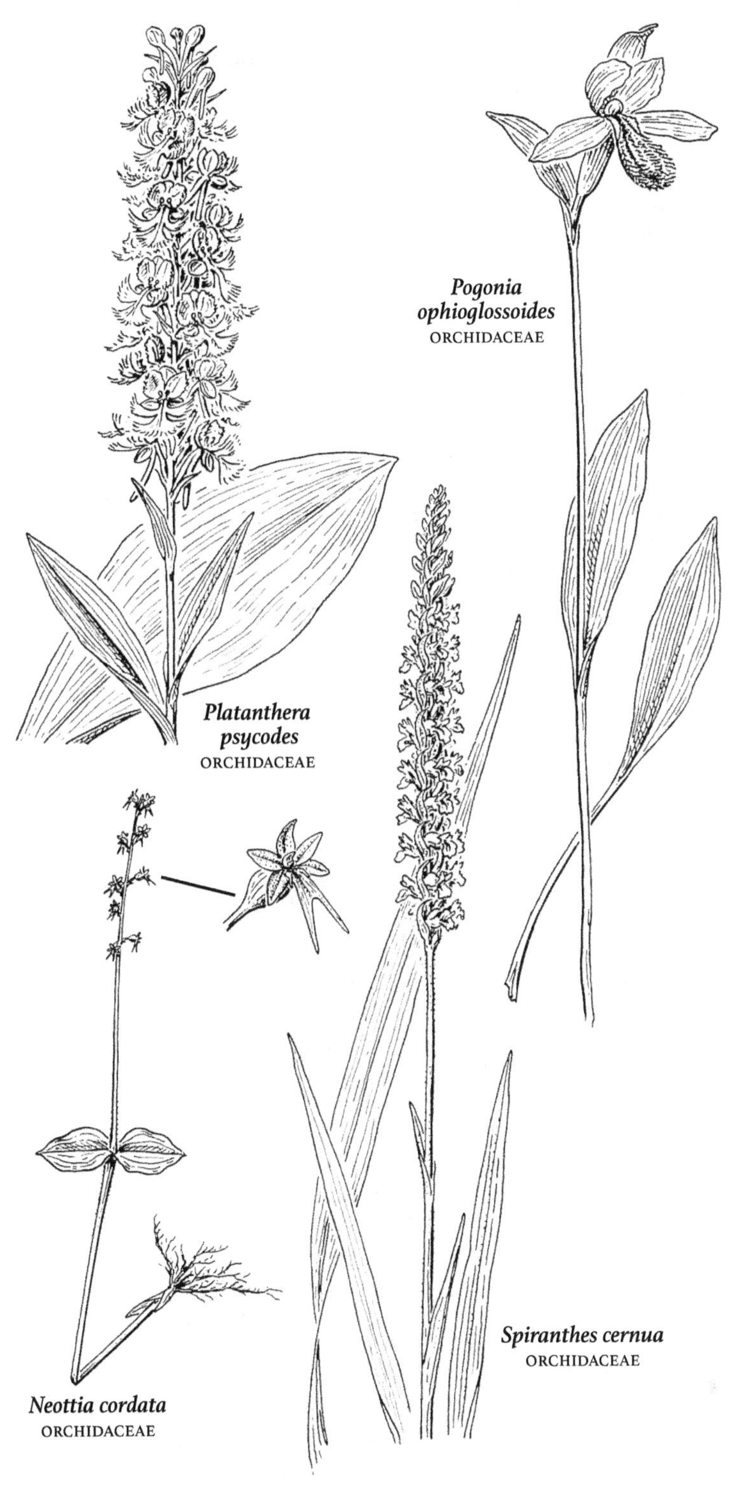

*Pogonia ophioglossoides*
ORCHIDACEAE

*Platanthera psycodes*
ORCHIDACEAE

*Spiranthes cernua*
ORCHIDACEAE

*Neottia cordata*
ORCHIDACEAE

breaking (disarticulating) either above or below the glumes when mature, the glumes remaining in the head if falling above the glumes, or the glumes falling with the florets if disarticulation is below the glumes.

ADDITIONAL SPECIES

***Apera interrupta*** (L.) Beauv. (Dense silky-bent), introduced, Oconto County.

***Briza media*** L. (Perennial quaking-grass), introduced in Ashland County.

***Buchloe dactyloides*** (Nutt.) Engelm. (Buffalo grass), native of Great Plains, adventive in sw Wisc.

***Catabrosa aquatica*** (L.) Beauv. (Brook-grass), endangered in Wisc and known from Adams and St. Croix counties; plants perennial, stems often horizontal, branching and rooting at nodes in mud or water; leaves flat; head an open panicle; spikelets mostly 2-flowered or sometimes mostly 1-flowered.

***Chloris verticillata*** (Windmill grass), native of s USA becoming weedy in n Illinois; in Wisc, known from Grant County. Plants resemble a pale green, prickly crabgrass due to finger-like spikes of awned spikelets.

***Distichlis spicata*** (L.) Greene (Coastal salt-grass), adventive in La Crosse and Richland counties from w USA; plants short, perennial grass, spreading by scaly rhizomes and forming patches, especially where brackish; the staminate and pistillate flowers on separate plants.

***Molinia caerulea*** (L.) Moench (Purple moor grass), native of Europe, introduced in Ashland County; plants densely tufted with long flat blades and elongated upper internodes, bearing a slender panicle.

***Nassella viridula*** (Trin.) Barkworth (Green needlegrass), an important grass of the Great Plains, adventive in s Wisc; panicle contracted, the ascending branches each bearing 2–several spikelets; lemma awn 2–3 cm long, weakly twice geniculate below the middle.

***Sclerochloa dura*** (L.) Beauv. (Fairgrounds grass), Eurasian grass, tolerant of heat, salt, and trampling; Rock and Walworth counties.

***Ventenata dubia*** (Leers) Coss. (North African grass), introduced in Wisc.

***Zea mays*** L. (Corn), escape from cultivation from scattered seed but not persisting.

*Key to Poaceae groups*

1 Tip of plant with a large "tassel" or spike-like raceme bearing staminate florets in pairs, the pistillate florets either sunken in hardened joints of rachis below the staminate portion or in separate "ears" lower on the plant .................................. **KEY 1**

1 Tip of plant not as above; upper portion of inflorescence bearing pistillate or bisexual florets, or rarely staminate florets in short one-sided spikes .................................................. 2

2 Spikelets concealed within ± globular, hard, bur-like structures . 3

2 Spikelets exposed, not concealed within globular, bur-like structures .................................................. 4

3 Bur-like clusters strongly spiny, the spines radiating out in all directions; plants tufted fibrous rooted annuals...... **CENCHRUS**

3 Bur-like clusters not spiny, the projections all pointing upward; plants perennial and colonial from stolons.... **BUCHLOE** (female plants)

4 Spikelets all unisexual, segregated into different and dissimilar parts of the inflorescence or on different plants .................... 5

4 Spikelets perfect, or if unisexual, then scattered among bisexual spikelets .................................................. 6

5 Plants small, terrestrial; staminate and pistillate spikelets on different plants .................................. BUCHLOE

5 Plants large, typically in shallow water; staminate and pistillate spikelets on the same plant .......................... ZIZANIA

6 Spikelets forming a simple spike or spikes, directly sessile or subsessile on main axis of inflorescence or at most on secondary branches .................................................. 7

6 Spikelets not forming simple spikes as above; pediceled and/or on tertiary or further branches of the inflorescence; in some species congested and hence spike-like, but not directly sessile or subsessile (reduced panicle branches usually visible upon removal of some spikelets) .................................................. 9

Spike solitary, terminal (its rachis a continuation of the culm), the spikelets on opposite sides of rachis ..................... KEY 2

7 Spikes several, one-sided (spikelets in two rows on one side of rachis).................................................. 8

Glumes keeled and ± equal (or the smaller half or more as long as the larger) .............................................. KEY 3

8 Glumes rounded on the back (not keeled), very unequal ... KEY 4

9 Spikelets with only 1–2 florets .............................. 10

9 Spikelets with 3 or more florets, including any sterile ones ..... 16

10 Spikelets with an involucre consisting of long subtending bristles .................................................. KEY 4

10 Spikelets without an involucre of bristles (although glumes or lemmas may be awned) .................................. 11

11 Glumes or lemmas (or both) ± laterally compressed or keeled (lateral nerves, if present, less prominent than midnerve) ........ ............................................ *go to couplet 16*

11 Glumes and lemmas rounded on back, not keeled (nerves, if present, about equally prominent) .......................... 12

12 Glumes very unequal in length, one of them minute, or absent, or at most about half as long as the spikelet. (Note: In *Key 4* species, a sterile lemma is present that closely resembles the large second glume opposite it and might easily be misinterpreted as a glume; the true first glume is a small, sometimes minute and membranous, even deciduous, scale at the very base of the spikelet; a reduced palea, often associated with the sterile lemma, will also help to identify the latter as part of a sterile floret and not a glume) .... 13

12 Glumes ± equal in length, neither of them much reduced nor absent .................................................. 14

13 Spikelets disarticulating below the glumes (except in *Setaria italica*), ± elliptic (less than 3 times as long as wide); a sterile lemma resembling the larger glume present ..................... KEY 4

13 Spikelets disarticulating above the glumes, ± lanceolate (3–10 times as long as wide); no sterile lemma present ................ KEY 5

14 Spikelets paniculate, all (or mostly) 1-flowered, bisexual, the florets all alike (no sterile lemmas or separate sterile pedicels present); spikelets less than 4 mm long, except in *Hesperostipa* with awns over 5 cm long .......... KEY 5

14 Spikelets paniculate or racemose, basically 2-flowered (the lower floret staminate or sterile with often suppressed palea); spikelets 3 mm or more long .......... 15

15 Spikelets all alike, not paired with a pedicel bearing a rudimentary, staminate, or no floret .......... ***Panicum virgatum***

15 Spikelets in pairs, usually of two kinds (except in *Miscanthus*): one sessile and with a bisexual floret, the other a hairy pedicel with or without a staminate or rudimentary floret (rarely 2 stalked florets with 1 sessile one) .......... KEY 1

16 Spikelets all or mostly containing 1 bisexual floret and no sterile or vestigial ones below it .......... 17

16 Spikelets all or mostly containing 2–several florets, the lower ones sometimes staminate or rudimentary (scale-like or reduced to tiny hairy appendages) .......... 18

17 Glumes both completely absent; spikelets strongly flattened, appressed and ± overlapping, the lemmas scabrous or hispid-ciliate .......... LEERSIA

17 Glumes (one or both) usually present; spikelets various but not as above .......... KEY 5

18 Glumes shorter than the lowest floret (excluding awns if present); awn of lemma none, terminal, arising from between terminal teeth and not twisted, or at most subterminal .......... KEY 7

18 Glumes (at least one of them, not necessarily the first glume) longer than lowest floret; awn of lemma none or arising from between terminal teeth and strongly twisted below, or (the usual condition) inserted on the middle or lower part of the lemma .......... 19

19 Spikelets containing one bisexual awnless floret (the lemma sometimes membranaceous) with two additional, often dissimilar (sometimes awned) staminate, sterile, or vestigial lemmas below it .......... KEY 8

19 Spikelets usually containing 2 or more bisexual florets (staminate or sterile florets, if present, above the fertile one and/or fertile lemma awned) .......... KEY 9

**KEY 1.** Tribe SACCHAREAE. Subtribe Andropogoninae (*Andropogon, Schizachyrium*); Subtribe Saccharinae (*Miscanthus*); Subtribe Sorghinae (*Sorghastrum, Sorghum*); Subtribe Tripsacinae (*Zea*)

1 Summit of plant with a large "tassel" or spike-like raceme bearing staminate florets in pairs, the pistillate florets either sunken in thick, hard joints of rachis below the staminate portion or in separate leafy bracted spikes (ears) lower on the plant .......... ***Zea mays****

1 Summit of plant not as above; upper portion of inflorescence bearing pistillate or sterile and pistillate florets; pistillate florets not in leafy bracted spikes nor enclosed in thick, hard joints of the rachis .......... 2

2 Inflorescence an open to contracted panicle .................... 3
2 Inflorescence of 1 or several narrow or spike-like simple racemes (some spikelets sessile and some pediceled) .................... 4

3 Stalked spikelet absent or represented by a sterile hairy pedicel closely resembling the segments of the panicle axis; ligule stiff, ± cartilaginous, glabrous or at most with minute hairs Sorghastrum
3 Stalked spikelet present, sterile or staminate; ligule at least in part of evident soft hairs .................... **SORGHUM**

4 Spike-like simple racemes solitary at the ends of the branches ... .................... **SCHIZACHYRIUM**
4 Spike-like simple racemes 2–ca. 20 at the ends of the branches .. 5

5 Racemes 10–20, not disarticulating; spikelets with a dense basal tuft of silky hairs longer than the glumes, the pedicel glabrous or very short hairy .................... **MISCANTHUS**
5 Racemes 2–6 (–10), disarticulating readily; spikelets lacking a dense basal tuft of silky hairs, the pedicel, however, hairy and sometimes with an apical tuft of hairs just below the articulation with the spikelet .................... **ANDROPOGON**

KEY 2. Tribe HORDEEAE. Subtribe Hordeinae (*Agropyron, Elymus, Hystrix, Hordeum, Leymus, Pascopyrum, Secale*); Subtribe Triticinae (*Triticum*). Tribe POEAE. Subtribe Loliinae (*Lolium*)

1 Lemmas smooth and glabrous, except for a spiny-ciliate keel and exposed margin, tapering into a long awn .................... **SECALE**
1 Lemmas smooth to scabrous or pubescent, but not simply with spiny-ciliate keel and margin, awned or awnless .................... 2

2 Larger glumes 3.3–6.5 mm broad with at least 3 prominent nerves, the keel or midnerve not centered .................... 3
2 Larger glumes less than 2.5 mm broad, variously nerved (or glumes absent) .................... 4

3 Glumes glabrous, or pubescent toward the base on nerves and margins (rarely pubescent throughout), the larger ones 4–6.5 mm wide, less than 3 times as long (excluding awns if present); lemmas awned or awnless .................... **TRITICUM**
3 Glumes softly hairy or glabrous throughout, 3.3–4.2 mm wide, ca. 6–10 times as long; lemmas awnless .................... **LEYMUS**

4 Spikelets mostly 2–3 at each node of the rachis; glumes usually 4–6 (the spikelet arrangement may be obscured by reduction or asymmetric positions of some spikelets, but the basic structure is revealed by the presence of usually 4–6 glumes subtending the entire group of spikelets); glumes usually vestigial or absent in *Elymus hystrix* with mostly 2 easily recognized narrow spikelets at each node .................... 5
4 Spikelets clearly 1 at each node (or most nodes) of the rachis; glumes variously arranged (or absent), but not more than 2 .................... 8

5 Spikelets 2 at each node of the rachis (or at some nodes, only 1, rarely 3, but total number of glumes (awn-like or broader) developed at a node not more than 4) .................... **ELYMUS**

5 Spikelets basically 3 at each node of the rachis (the lateral 2 in commonest species reduced to bristles), this arrangement most easily recognized by the presence of 6 awn-like or narrowly lanceolate and awn-tipped glumes at a node .................. 6

6 Body of larger lemmas ca. 3.5–6 mm long; rachis of spike readily disintegrating at maturity .......................... **HORDEUM**

6 Body of larger lemmas ca. 8–12 mm long; rachis not disintegrating ........................................................ 7

7 Awn of lemmas much stouter than awn of glumes, ± straight ..... ................................................ ***Hordeum vulgare***

7 Awn of lemmas as slender as awn of glumes, spreading to recurved at maturity ........................................... **ELYMUS**

8 Glumes 1 (except terminal spikelet with 2), the narrow edge of the spikelet against the rachis and lacking a glume ......... **LOLIUM**

8 Glumes 2 on all spikelets...................................... 9

9 Flowering stems 3–20 cm; inflorescences 1–3.5 cm long; ± prostrate annual ................................... **SCLEROCHLOA***

9 Flowering stems 20–150 cm tall; inflorescences (3–) 5–20 cm long; plants erect, perennial (except *Agropyron*).................... 10

10 Body of the glumes 2.7–4 mm long, the midrib asymmetrically positioned and formed into a prominent keel; spike densely crowded and strongly 2-ranked with middle internodes ca. 1–2 mm long ........................................... **AGROPYRON**

10 Body of the glumes more than 4 mm long; midrib symmetrically positioned and not raised into a prominent keel; spikes with the middle internodes ca. 3–15 mm long ......................... 11

11 Lemmas with awns strongly divergent or recurved at maturity ... .......................................... ***Elymus trachycaulus***

11 Lemmas with awns ± straight or absent ...................... 12

12 Lemmas densely hairy ...................... ***Elymus lanceolatus***

12 Lemmas glabrous (rarely slightly pubescent), smooth or scabrous ........................................................ 13

13 Leaf blades mostly broad and flat, slightly or not at all involute when dry, no more deeply grooved above than below between the numerous fine nerves (not strongly scabrous, usually with scattered long hairs above); glumes completely lacking cilia toward the base; cartilaginous belt at upper nodes ..................... **ELYMUS**

13 Leaf blades strongly involute when dry, deeply grooved above between the prominent raised nerves (and usually strongly scabrous above); glumes mostly with margin minutely ciliate toward base; cartilaginous belt (sharply defined, usually darker, non-green zone) at upper nodes of culm usually less than half as long as its diameter .................................... **PASCOPYRUM**

**KEY 3.** Tribe CHLORIDEAE. Subtribe Boutelouinae (*Bouteloua, Buchloe*); Subtribe Eleusininae (*Eleusine, Leptochloa*). Tribe POEAE. Subtribe Poinae (*Beckmannia*); Tribe ZOYSIEAE. Subtribe Sporobolinae (*Spartina*)

1 Spikelets concealed within ± globular, hard, bur-like structures .. .......................................... **BUCHLOE*** (*female plants*)

1 Spikelets exposed, not concealed within bur-like structures .... 2

2 Spikes radiating from summit of culm (i.e., umbellate or nearly so) or at least the lower ones whorled (solitary in depauperate individuals); anthers not over 1.5 mm long ........... **ELEUSINE**

2 Spikes all racemose or panicled; anthers various ............... 3

3 Glumes equal, ca. 2–3 mm long, deeply pouch-like and largely covering the floret, the spikelet strongly flattened and about as wide as long; ligule membranous, not ciliate; anthers 0.5–1 mm long .. ............................................. **BECKMANNIA**

3 Glumes unequal, the longer ones ca. 2.5–11 mm long, not pouch-like, equaling or shorter than florets, the spikelet not strongly flattened, much narrower than long; ligules and anthers various 4

4 Ligule membranous, not ciliate, ca. 4–6 mm long, becoming shredded; anthers up to ca. 0.5 mm long; end of rachis of spike spikelet-bearing, not prolonged; spikelets short-pediceled ....... ............................................. **LEPTOCHLOA**

4 Ligule ciliate (entirely of hairs, or hairs longer than any membranous portion), up to 3.5 mm long (a few hairs rarely to 4.5 mm); anthers 1.8–6 mm long; end of rachis of spike prolonged into a naked projection 0.5–14 mm beyond the last spikelet (except in *Bouteloua gracilis*); spikelets sessile or subsessile ................ 5

5 Spikes short peduncled, peduncles glabrous to hispid, 0.5–20 mm long ............................................... **SPARTINA**

5 Spikes ± sessile or on finely hispidulous or short-pubescent peduncles up to 3 mm long ................................. 6

6 Inflorescences with 20–40 (–55) spikes; end of rachis of spike prolonged into a stout naked projection 3–14 mm beyond the last spikelet ................................ ***Bouteloua curtipendula***

6 Inflorescences with 1–4 spikes; end of rachis of spike not prolonged into a naked projection or this only ca. 0.5–1 mm long .......... 7

7 Spikes 15–40 mm long, with ca. 17–50 spikelets; plants in small clumps ..................................... ***Bouteloua gracilis***

7 Spikes 0.6–1.5 mm long, with ca. 5–10 spikelets; plants colonial from stolons ............................. **BUCHLOE*** (*male plants*)

**KEY 4.** Tribe PANICEAE. Subtribe Anthephorinae (*Digitaria*); Subtribe Boivinellinae (*Echinochloa*); Subtribe Cenchrinae (*Cenchrus, Setaria*); Subtribe Melinidinae (*Eriochloa*); Subtribe Panicinae (*Dicanthelium*, Panicum). Tribe PASPALEAE. Subtribe Paspalinae (*Paspalum*)

1 Spikelets with an involucre consisting of a spiny bur or of long subtending bristles ........................................... 2

1 Spikelets without an involucre (although glumes or lemmas may be awned) .................................................... 3

2 Involucre a spiny bur enclosing much or all of the spikelets, the whole readily disarticulating ....................... **CENCHRUS**

2 Involucre of long slender bristles subtending but not concealing the spikelet, remaining attached to pedicels when spikelets disarticulate .................................................. **SETARIA**

3 Spikelets ± spiny-hispid and usually also awned; ligule none ..... .................................................. **ECHINOCHLOA**

3 Spikelets glabrous or pubescent but not coarsely hispid and not awned; ligule present, distinct or nearly absent (of hairs or membranous) .................................................. 4

4 Each spikelet subtended by a small cup-like involucre **ERIOCHLOA**

4 Spikelets lacking a cup-like involucre .................................................. 5

5 Inflorescence composed of 1-sided spikes or spike-like racemes, the rachis of each winged or at least flat on the side opposite the spikelets .................................................. 6

5 Inflorescence an open panicle, not spike-like or distinctly one-sided .................................................. 7

6 Ligule membranous, conspicuous (0.8–2.2 mm long), without a dense row of hairs; fertile lemma rather leathery in texture, acute, about twice as long as wide (or longer), with thin flat translucent margins .................................................. **DIGITARIA**

6 Ligule a short (less than 1 mm, usually ca. 0.5 mm) membranous band behind which is a dense row of much longer hairs; fertile lemma very hard, broadly rounded, less than 1.5 times as long as wide, with thickened, ± inrolled, hardened margins .. **PASPALUM**

7 Ligule a membranous collar 1–1.5 mm high, without hairs; base of leaf blade and very summit of sheath without special zone of short pubescence or long hairs or cilia; first glume minute or vestigial; fertile lemma leathery in texture, with thin flat translucent margins .................................................. ***Digitaria cognata***

7 Ligule usually partly or entirely of short or long hairs; (if ligule membranous, plants not otherwise as above: ligule ca. 0.5 mm long or virtually absent; or summit of sheath or basal margin of blade pubescent or ciliate; and/or first glume more than 0.5 mm long); fertile lemma hard and shiny .................................................. 8

8 Spikelets at least sparsely pubescent towards their margins ...... .................................................. **DICHANTHELIUM**

8 Spikelets glabrous .................................................. 9

9 Terminal panicle 8–40 cm long, (smaller in occasional depauperate individuals of annual species); annuals or perennials, but without clear remnants of overwintering basal rosette leaves; flowering and fruiting summer-fall .................................................. **PANICUM**

9 Terminal panicle 2.5–8 (–12) cm long; tufted perennials with clear remnants of old, dead leaves from the previous year present at the base, these sometimes formed into a clear overwintering rosette; flowering spring and fruiting in late spring–early summer ...... .................................................. **DICHANTHELIUM**

**KEY 5.** Tribe ARISTIDEAE (*Aristida*). Tribe BRACHYELYTREAE (*Brachyelytrum*). Tribe CHASMANTHEAE (*Chasmanthium*). Tribe POEAE. Subtribe Agrostidinae (*Agrostis, Ammophila, Calamagrostis,*

*Polypogon*); Subtribe Miliinae (*Milium*); Subtribe Phleinae (*Phleum*); Subtribe Poinae (*Alopecurus, Apera, Beckmannia, Cinna, Ventenata*). Tribe STIPEAE. Subtribe Stipinae (*Oryzopsis, Piptatherum, Hesperostipa*). Tribe ZOYSIEAE. Subtribe Sporobolinae (*Calamovilfa, Crypsis, Sporobolus*)

1 Lemma with awn (or awns) strictly terminal .................... 2
1 Lemma with awn absent or dorsal or subterminal.............. 8

2 Awns of lemma 3 (lateral ones sometimes very short) . **ARISTIDA**
2 Awn of lemma solitary........................................ 3

3 Body of lemma 8–23 mm long ............................... 4
3 Body of lemma less than 7 mm long .......................... 5

4 Glumes 9.5–45 mm long ...................... **HESPEROSTIPA**
4 Glumes rudimentary or one of them up to 5 mm long ........... ...................................... **BRACHYELYTRUM**

5 Glumes acute to obtuse, more than 1 mm wide, scarcely if at all keeled, the spikelets nearly terete; lemma rounded on the back, ± firm and hardened ......................................... 6
5 Glumes acuminate, not over 1 mm wide, keeled, the spikelets somewhat compressed (or glumes vestigial in *Muhlenbergia schreberi*); lemma ± keeled, membranous or thin................ 7

6 Principal leaf blades basically flat (just the margins involute), basal or nearly so, densely and very finely rough-puberulent, with strong closely spaced veins ............................ **ORYZOPSIS**
6 Principal leaf blades involute or, if flat, all cauline and glabrous .. ........................................ **PIPTATHERUM**

7 Inflorescence various (if spike-like, the plants with scaly rhizomes); ligules up to 2 mm long; glumes gradually tapered into awn or awnless ................................ **MUHLENBERGIA**
7 Inflorescence a dense thick spike-like panicle, the plants annual, without rhizome; ligules at least 3 mm long; glumes ± 2-lobed or rounded at tip, not tapered into the very slender awn (ca. 5–8 mm long) ........................................ **POLYPOGON**

8 Spikelets 10–15 mm long; anthers (4–) 5–8 mm long; panicle crowded and ± spike-like, 10–20 (–28) mm across at the middle ... ........................................... **AMMOPHILA**
8 Spikelets less than 8 mm long (excluding awns); anthers to 4.5 mm long (usually much shorter); panicle various .................. 9

9 Spikelets sessile or nearly so, crowded in a very dense spike-like panicle (branches of panicle suppressed, scarcely if at all visible without dissection of panicle) .............................. 10
9 Spikelets in ± open or contracted (but not densely spike-like) inflorescences, with evident pedicels and/or panicle branches .. 14

10 Glumes awned ............................................... 11
10 Glumes awnless .............................................. 13

11 Plants with scaly rhizomes; glumes gradually tapered into awn; ligule to ca. 1 mm long ...................... **MUHLENBERGIA**

11 Plants without scaly rhizomes; glumes abruptly rounded or truncate, the awn distinct; ligule over 1 mm long .............. 12

12 Awn of glume rather stout and stiff, not over ca. 3 mm long; anthers ca. 1–2 mm long; spikelets articulated above the glumes; glumes prominently pectinate-ciliate on keel basally, otherwise glabrous or variously (but not so prominently) pubescent or ciliate; lemmas awnless .............................................. PHLEUM

12 Awn of glume very slender, 5–8 mm long; anthers less than 1 mm long; spikelets articulated below the glumes; glumes ± evenly hispidulous basally; lemmas often with delicate awn ........... ............................................ POLYPOGON

13 Panicle ± ovoid, ca. 2–3.5 times as long as wide; spikelets articulated above the glumes; ligule a fringe of hairs; lemmas awnless (if ligule membranous, try *Phalaris*) ........................... CRYPSIS

13 Panicle cylindrical (in common species slender and pencil-like), 3–15 times as long as wide; spikelets articulated below the glumes; ligule membranous; lemma with slender awn attached near the middle or base of keel (in common species, often very inconspicuous, shorter than the glumes (if spikelets articulated above the glumes and lemmas stoutly awned, try *Anthoxanthum*) . ........................................... ALOPECURUS

14 Spikelets rounded on back, not keeled (neither glumes nor lemma with a midvein more prominent than other nerves), at least 2.5 mm long; lemma ± shiny, distinctly firmer in texture than the glumes 15

14 Spikelets keeled (glumes and/or lemmas with midvein more prominent than other nerves) or less than 2.5 mm long; lemma no firmer in texture than the glumes ........................... 16

15 Lemmas with appressed pubescence; leaves with blades usually involute; upper ligules not over 3 mm long... ***Piptatherum pungens***

15 Lemmas glabrous; leaves with blades broad and flat; upper ligules mostly 4–6 (–8) mm long ............................ MILIUM

16 Ligule a fringe of short hairs; lemmas awnless ................ 17

16 Ligule membranous (at most minutely ciliate at summit of membrane); lemmas awned or awnless ...................... 18

17 Lemma (4.7–) 5–6.7 mm long, surrounded with a tuft of long hairs (more than half its length) at its base .......... CALAMOVILFA

17 Lemma 1.5–5.5 mm long, without long hairs at its base .......... ........................................... SPOROBOLUS

18 Lemma with long hairs at base (on or near callus) ............. 19

18 Lemma without long hairs at base (at most with hairs on callus less than 0.5 mm long) ........................................ 20

19 Long hairs at least in part arising from lower portion of lemma; glumes (excluding awn-tips if present) shorter than lemma ...... ............................................ MUHLENBERGIA

19 Long hairs restricted to callus at base of floret; glumes slightly exceeding lemma ......................... CALAMAGROSTIS

20 Glumes both distinctly shorter than lemma; lemma awnless ..... ............................................ MUHLENBERGIA

20 Glumes (one or both of them) equaling or exceeding the lemma and/or the lemma awned .................................... 21

21 Floret raised above base of glumes on a minute stalk; spikelet articulated below the glumes; lemma with a small subterminal awn; stamen 1 .............................................. **CINNA**

21 Floret not stalked; spikelets articulated above the glumes; lemma awnless or with long subterminal awn or with dorsal awn; stamens 3 ..................................................... 22

22 Lemma with a long subterminal awn, much exceeding the body; rachilla prolonged (scarcely 0.5 mm) behind the palea . . . **APERA***

22 Lemma awnless or with mid-dorsal awn; rachilla not prolonged . ................................................ **AGROSTIS**

**KEY 6.** Tribe ORYZEAE (*Leersia, Zizania*)

1 Spikelets bisexual; plant variously scabrous-pubescent, at least the nodes retrorsely bearded; stamens 3 .................. **LEERSIA**

1 Spikelets all unisexual, segregated into different parts of the inflorescence, upper panicle branches bearing awned pistillate spikelets, lower branches bearing staminate spikelets; plant glabrous; stamens 6 ................................. **ZIZANIA**

**KEY 7.** Tribe ARUNDINEAE (*Phragmites*). Tribe BROMEAE (*Bromus*). Tribe CHASMANTHEAE (*Chasmanthium*). Tribe CHLORIDEAE. Subtribe Gouiniinae (*Triplasis*); Subtribe Tridentinae (*Tridens*). Tribe DIARRHENEAE (*Diarrhena*). Tribe ERAGROSTIDEAE. Subtribe Eragrostidinae (*Eragrostis*). Tribe MELICEAE (*Glyceria, Melica, Schizachne*). Tribe POEAE. Subtribe Aveninae (*Graphephorum, Trisetum*); Subtribe Coleanthinae (*Puccinellia, Sclerochloa*); Subtribe Cynosurinae (*Cynosurus*); Subtribe Dactylidinae (*Dactylis*); Subtribe Holcinae (*Deschampsia*); Subtribe Loliinae (*Festuca, Schedonorus, Vulpia*); Subtribe Torreyochloinae (*Torreyochloa*)

1 Plants tall and stout (usually over 1.5 m tall) with larger leaf blades 1–3.5 cm wide; spikelets ca. 11–17 mm long; ligule a densely ciliate brown band; rachilla (above the lowest floret) with silky beard about equaling or exceeding the lemmas ............... **PHRAGMITES**

1 Plants generally less than 1.5 m tall with narrow leaves less than 1 cm wide; spikelets and ligule various; rachilla with beard shorter or absent ........................................................ 2

2 Spikelets sessile or at most very short-pediceled, crowded into dense clusters, these either at the ends of elongate panicle branches or in a single congested, rather spike-like inflorescence .............. 3

2 Spikelets short- to long-pediceled in a ± open panicle........... 7

3 Lemma with a prominent, somewhat twisted or spreading dorsal awn; rachilla villous .............................. **TRISETUM**

3 Lemma with awn absent or short and strictly terminal; rachilla not villous ..................................................... 4

4 Clusters of spikelets at the ends of elongate naked branches of the panicle; sheaths closed much of their length; ligule 2–8 mm long ................................................ **DACTYLIS**

4 Clusters of spikelets all crowded into a congested, rather spike-like inflorescence; sheaths open their entire length; ligule 1 mm or less long .......... **5**

5 Spikelets of two kinds in a cluster: normal fertile and special sterile fan-like ones; fertile lemmas mostly short-or long-awned .......... **CYNOSURUS**

5 Spikelets all similar, fertile; lemmas not awned .......... **6**

6 Foliage glabrous; flowering stems to 20 cm, inflorescence to 3.5 cm; prostrate annuals .......... **SCLEROCHLOA***

6 Foliage (at least the lowermost sheaths) pubescent or puberulent; flowering stems 20–100 cm tall, inflorescence (3–) 5–25 cm long; ± erect, perennials .......... *see* **KEY 3**

7 Callus at base of floret with dense beard of straight hairs 0.5 mm or more long .......... **8**

7 Callus glabrous, minutely puberulent, or cobwebby (not bearded with straight hairs) .......... **13**

8 Awn of lemma arising near base .......... **DESCHAMPSIA**

8 Awn of lemma absent, terminal, subterminal, or arising between terminal teeth.......... **9**

9 Lemmas awnless, weakly 5-nerved; sheaths open .......... **GRAPHEPHORUM**

9 Lemmas with short or long awn, either 3-nerved or plants with closed sheaths .......... **10**

10 Sheaths open; lemmas 3-nerved, ± truncate (ragged or lobed) at tip, the nerves hairy .......... **11**

10 Sheaths closed; lemmas 5–7-nerved, tapering to an apparently 2-lobed or sharply bifid tip, glabrous or hairy .......... **12**

11 Panicles terminal and axillary, small, the former often partly and the latter entirely included in the swollen sheaths; palea villous on apical half; nodes bearded; plants annual, with the leaves, sheaths, and culms toward base upwardly scabrous .......... **TRIPLASIS**

11 Panicles terminal, large, exserted; palea not villous on apical half; nodes glabrous; plant a stout perennial smooth toward base .......... **TRIDENS**

12 Callus with distinct beard, the lemma glabrous or nearly so; grain glabrous .......... **SCHIZACHNE**

12 Callus lacking a distinct beard, the pubescence like that of the lemma; grain pubescent at the summit .......... **BROMUS**

13 Glumes (at least one of them) and usually also lemmas strongly keeled (lemmas in a few species rounded on the back); awns absent or not over 2 mm .......... **14**

13 Glumes and lemmas rounded on back, not keeled (or obscurely so toward tip); awns absent or present .......... **19**

14 Larger glumes ca. 4.5–7 mm long .......... **15**

14 Larger glumes not over 4.4 mm long (to 5 mm in *Leptochloa,* with ± 1-sided inflorescence branches) .......... **16**

15 Spikelets nearly sessile, numerous, ascending in a ± crowded panicle; ligules ca. 2–8 mm long; sheaths closed much of their length .......................................... **DACTYLIS**

15 Spikelets on long pedicels, ascending to drooping in a much expanded raceme or panicle; ligules less than 1.5 mm long; sheaths open (if confronted at this point with a plant bearing an open panicle and closed sheaths, try *Bromus*) ..... **CHASMANTHIUM**

16 Larger glumes ± obovate, broadest above the middle ... *see* **KEY 3**

16 Larger glumes broadest at or below the middle ............... **17**

17 Ligule a fringe of hairs; lemmas with 3 prominent nerves, glabrous; spikelets 2–30-flowered .......................... **ERAGROSTIS**

17 Ligule a membranous scale, the cilia, if any, shorter than the scale; lemmas with 3–5 nerves, glabrous, hairy, and/or cobwebby at base; spikelets various (if specimens will not fit here, try couplet 19) .. **18**

18 Glumes very unequal, the smaller (first) usually only slightly more than half the length of the larger; lemmas obscurely 2-toothed, with minute awn between teeth, the nerves silky-pubescent basally; callus glabrous ................................ **LEPTOCHLOA**

18 Glumes slightly unequal; lemmas rounded or pointed, but neither toothed nor awned, the nerves pubescent or glabrous; callus often with a tuft of cobwebby hairs ............................. **POA**

19 Lemmas distinctly 3-nerved, thick and leathery ... **DIARRHENA**

19 Lemmas 5- (or many-) nerved (the nerves sometimes very indistinct), in most species thin ............................ **20**

20 Lemmas at least as broad as long; mature glumes and florets spreading nearly at right angles to rachilla, the spikelets nearly or quite as broad as long ............................ ***Briza media**** 

20 Lemmas longer than broad; glumes and florets not so widely spreading, the spikelets in most species much longer than broad **21**

21 Lemmas usually 2-toothed or minutely 2-lobed at the tip and usually with at least a short awn arising from just below or between the teeth (if teeth apparently united, as in some species of *Bromus,* the awn thus subterminal); sheaths closed nearly to their summit ............................................................ **22**

21 Lemmas not 2-lobed or 2-toothed at tip, awnless or with strictly terminal awn; sheaths open (or closed in *Glyceria,* with prominently nerved and awnless lemmas, and in the youngest shoots of *Festuca rubra*) ........................................................ **23**

22 Spikelets narrowly linear-lanceolate on much shorter, densely hispid pedicels; lemmas awned, minutely strigose or scabrous, at least on the nerves; sheaths retrorsely scabrous; ligules 3–6 mm long; grain glabrous ................................. **MELICA**

22 Spikelets broadly linear to oblong, usually on ± elongate pedicels; lemmas various; sheaths glabrous or pubescent but not scabrous; ligules less than 2.5 (–4) mm long; grain pubescent at the summit ........................................................ **BROMUS**

23 Lemmas acute at the tip, awned ............................ **24**

23 Lemmas acutish or obtuse, awnless .......................... **26**

24 Blades of leaves flat (or merely once-folded), at least the larger ones (2.5–) 3–8 mm broad .......................... **SCHEDONORUS**
24 Blades of leaves ± strongly involute, less (usually much less) than 3 mm broad .................................................. **25**

25 Plants perennial, usually in dense tufts including numerous dry sheaths of previous years; florets open at anthesis, with 3 anthers ................................................ **FESTUCA**
25 Plants annual, usually in small tufts or solitary; florets cleistogamous, with usually one included anther (rarely 3 anthers) ................................................ **VULPIA**

26 Nerves of lemma prominent, straight and becoming parallel at the tip; sheaths closed or open .................................. **27**
26 Nerves of lemma very weak (or if visible, then converging, not parallel, at the tip); sheaths open ............................ **28**

27 Sheaths closed much of their length (but easily splitting); second (larger) glume with one distinct nerve; plants rhizomatous ....... ................................................ **GLYCERIA**
27 Sheaths completely open; second glume with 3 (–5) nerves distinct at its base; plants without rhizomes (though culms may be decumbent or prostrate) .................... **TORREYOCHLOA**

28 Lemmas ca. 2 mm long ......................... **PUCCINELLIA**
28 Lemmas 2.5–8 mm long ........................................ **29**

29 Blades of leaves strongly involute, usually much less than 3 mm wide ................................................ **FESTUCA**
29 Blades of leaves flat (or merely once-folded), at least the larger ones 3–8 mm wide .............................................. **30**

30 Larger lemmas 2.5–4.5 mm long; anthers 0.8–1.4 mm long; spikelets mostly containing 2–4 (–5) florets and borne beyond the middle of the primary panicle branches ............. ***Festuca subverticillata***
30 Larger lemmas 5.5–8 mm long; anthers 2.2–3.5 (–3.8) mm long; spikelets often containing 5 or more florets, borne below as well as above the middle of the primary panicle branches ............... ........................................ **SCHEDONORUS**

**KEY 8.** Tribe POEAE. Subtribe Phalaridinae (*Anthoxanthum, Hierochloe, Phalaris*)

1 Panicle open, pyramidal, the branches spreading or drooping; glumes nearly equal in length, with lateral nerves obscure or prominent only on basal half; lower florets staminate, at least as large as bisexual floret, awnless .................. **HIEROCHLOE**
1 Panicle contracted, the branches ascending or suppressed; glumes equal or not, with lateral nerves (at least on larger glumes) prominent beyond the middle; lower florets sterile, either vestigial or large and awned .......................................... **2**

2 Glumes very unequal; lower lemmas with prominent dorsal awns, concealing the awnless bisexual floret ...... **ANTHOXANTHUM**
2 Glumes nearly or quite equal; lower lemmas awnless, small and inconspicuous, only the awnless bisexual floret evident .......... ................................................ **PHALARIS**

**KEY 9.** Tribe DANTHONIEAE (*Danthonia*). Tribe POEAE. Subtribe Aveninae (*Arrhenatherum, Avena, Graphephorum, Koeleria, Sphenopholis, Trisetum*); Subtribe Holcinae (*Deschampsia, Holcus*)

1 Lemmas all awnless; larger glumes ± obovate (broadest above the middle), generally shorter than the lowest floret................ 2

1 Lemmas with distinct twisted, jointed, or curved awn (sometimes largely hidden by the glumes or absent on some florets of a spikelet); glumes mostly ovate to lanceolate, at least one of them longer than the lowest floret ........................................... 4

2 Rachilla and callus prominently bearded with long straight hairs ......................................... **GRAPHEPHORUM**

2 Rachilla and callus glabrous or at most with short hairs (under 0.5 mm long) ................................................. 3

3 Axis and branches of inflorescence glabrous, at most scabrous; larger glumes not over 3 (–3.2) mm long........ **SPHENOPHOLIS**

3 Axis and branches of inflorescence densely short-pubescent; larger glumes 3–4.2 (–4.7) mm long ...................... **KOELERIA**

4 Larger glumes 6–27 mm long .............................. 5

4 Larger glume less than 6 mm long ........................... 7

5 Ligule a fringe of short hairs with a long tuft at each side; lemma with awn between terminal teeth ................ **DANTHONIA**

5 Ligule membranous, hairless; lemma with awn arising dorsally . 6

6 Spikelets less than 10 mm long (excluding awns), the lower floret staminate with strong awn and the upper floret bisexual with (usually) weak awn ..................... **ARRHENATHERUM**

6 Spikelets ca. 20–27 mm long, the florets all bisexual or the upper rudimentary; awns various ............................ **AVENA**

7 Bisexual (lowermost) floret awnless; awn subterminal on a reduced staminate floret; nodes pubescent .................... **HOLCUS**

7 Bisexual florets awned; awn twisted or spreading; nodes glabrous or (sometimes in *Trisetum*) pubescent ......................... 8

8 Awn arising above middle of lemma; panicle ± crowded and spike-like ............................................ **TRISETUM**

8 Awn arising well below middle of lemma; panicle at maturity very open and diffuse ........................... **DESCHAMPSIA**

## AGROPYRON *Wheatgrass*

***Agropyron cristatum*** (L.) Gaertn. crested wheatgrass Introduced. Disturbed areas; commonly used for land restoration in the western states. *Agropyron desertorum* auct. non (Fisch. ex Link) Schult. Other species formerly included in *Agropyron* are here placed in *Elymus* or *Pascopyrum.*

## AGROSTIS *Bent-Grass; Bent*

Perennial grasses, clumped or spreading by rhizomes or sometimes by stolons. Leaves soft, auricles absent, ligules membranous, sheaths open, usually smooth and glabrous. Head an open panicle. Spikelets small, 1-flowered.

*Key to species*

1 Palea present, about half as long as the lemma or longer; anthers ca. 0.8–1.5 mm long .......... 2

1 Palea absent or vestigial; anthers ca. 0.6 mm long or shorter .... 3

2 Plants rhizomatous but not stoloniferous, i.e., stems arising from underground rhizomes, straight or curved at the very base, otherwise erect and nearly or quite straight; larger leaf blades mostly 3–7 (–10) mm wide; spikelets usually flushed with red or purplish; bases of middle panicle branches mostly meeting the axis of the panicle at an angle of 30–45° (except when very immature) .......... ***A. gigantea***

2 Plants stoloniferous but not rhizomatous, i.e., stems usually decumbent at their bases, the lower nodes often strongly bent and/or rooting, but underground rhizomes absent; larger leaf blades 1.7–3 (–4) mm wide; spikelets pale, greenish; bases of middle panicle branches usually strongly ascending or appressed to axis of panicle, at most diverging about 15° (but panicle branches often spreading distally). .......... ***A. stolonifera***

3 Longest panicle branches less than 6 (–12) cm long and the uppermost leaf blade more than 5 cm long; leaf blades flat, the wider ones 1.5–3.5 mm wide; panicle branches forked about or below the middle, often smooth or only sparingly hispidulous-scabrous; panicle pale, greenish (very rarely red-tinged) .......... ***A. perennans***

3 Longest panicle branches more than 6 cm long or uppermost leaf blade less than 5 cm long (or both conditions); leaf blades usually ± involute, the widest to 1.5 (rarely 3) mm wide; panicle branches often not forked until beyond the middle, copiously hispidulous-scabrous; panicle ± flushed with red .......... 4

4 Lemmas 1.3–1.8 mm long .......... ***A. scabra***

4 Lemmas 0.9–1.2 mm long .......... ***A. hyemalis***

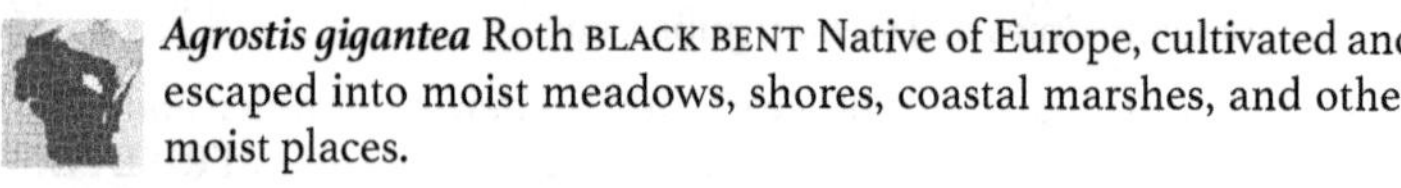

***Agrostis gigantea*** Roth BLACK BENT Native of Europe, cultivated and escaped into moist meadows, shores, coastal marshes, and other moist places.

***Agrostis hyemalis*** (Walt.) B.S.P. WINTER BENT June–Aug. Wet meadows, bogs, ditches, streambanks, shores; more commonly in dry, sandy places.

***Agrostis perennans*** (Walt.) Tuckerman UPLAND BENT Various habitats, usually in dry soil.

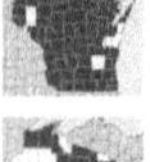

***Agrostis scabra*** Willd. ROUGH BENT Abundant, widely distributed in many habitats, and variable.

***Agrostis stolonifera*** L. [•127] REDTOP; SPREADING BENT Introduced (naturalized). July–Sept. Wet meadows, ditches, streambanks and shores; disturbed areas. *Agrostis alba* var. *palustris* (Huds.) Pers., *Agrostis palustris* Huds.

## ALOPECURUS *Meadow-Foxtail*

Annual or perennial grasses. Stems erect or more or less horizontal at base. Leaves mostly from lower 1/2 of the stems; sheaths open; auricles absent; ligules membranous, entire to lacerate. Heads densely flowered, cylindric, spike-like panicles. Spikelets 1-flowered, flattened. The narrow panicles resemble those of timothy (*Phleum*).

*Key to species*

1 Spikelets (excluding awns) ca. 4–6.5 mm long; awns mostly exserted ca. 3.5–6 mm beyond tips of glumes; anthers ca. 2.4–3.5 mm long . .......................................................... ***A. pratensis***

1 Spikelets not over 3 mm long; awns at most exserted ca. 2–3 mm; anthers less than 2 mm long ................................ **2**

2 Awn exserted at most about 1 mm beyond tips of glumes, usually included, inserted about a third or half the distance from base of lemma ........................................... ***A. aequalis***

2 Awn of most lemmas exserted ca. 2–3 mm, inserted near the base of lemma (on lower 1/5–1/4) ................................ **3**

3 Anthers 0.3–0.7 mm long; spikelets 2–2.5 mm long A. carolinianus

3 Anthers 1.4–1.8 mm long; larger spikelets 2.6–3 mm long ......... ............................................... ***A. geniculatus***

***Alopecurus aequalis*** Sobol. SHORT-AWN FOXTAIL June–Aug. Shallow water or mud of wet meadows, marshes, ditches, springs, open bogs, fens, shores and streambanks; sometimes where calcium-rich.

***Alopecurus carolinianus*** Walt. TUFTED MEADOW-FOXTAIL Introduced. May–July. Mud flats, temporary ponds, wet meadows, marshes, low prairie, fallow fields.

***Alopecurus geniculatus*** L. MARSH-FOXTAIL Introduced. Mud and shallow water.

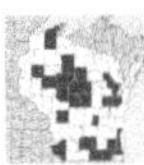

***Alopecurus pratensis*** L. FIELD MEADOW-FOXTAIL Native of Eurasia; naturalized in moist meadows, fields, and waste places.

## AMMOPHILA *Beach-Grass*

***Ammophila breviligulata*** Fern. AMERICAN BEACH-GRASS Coarse perennial grass, from long running rhizomes. Panicle dense; spikelets 1-flowered, strongly flattened, articulated above the glumes. Dunes and dry sandy shores along the Great Lakes; useful as a sand-binder in dune control.

## ANDROPOGON *Bluestem*

***Andropogon gerardii*** Vitman [•127] BIG BLUESTEM, TURKEY-FOOT Perennial grass, forming large bunches or extensive sod. Racemes 2–6, on a long-exserted peduncle, 5–10 cm long. Spikelets of two kinds, in pairs at the joints of the rachis, one sessile and perfect, the other pediceled and staminate, sterile, or abortive. Moist or dry soil of prairies, roadsides, railroads; in dry open woods, old fields, rarely in fens and sedge meadows.

## ANTHOXANTHUM *Sweet Vernal Grass*

***Anthoxanthum odoratum*** L. LARGE SWEET VERNAL GRASS Sweetly scented, tufted perennial grass. Panicle spike-like. Spikelets 1-flowered, articulated above the glumes. Native of Europe; shores, meadows, roadsides, and waste places. *Hierochloe* sometimes placed in this genus.

## ARISTIDA *Three-Awn*

Annual or perennial grasses, often weedy. Stems usually branched from some or all of the nodes. Leaf blades narrow, often involute, sheaths open; auricles lacking; ligules long-ciliate. Flowers in terminal panicles or racemes. Spikelets 1-flowered. Awns elongate, normally 3. Species mostly of dry, sterile or sandy soil.

ADDITIONAL SPECIES ***Aristida desmantha*** Trin. & Rupr. (Curly three-awn), adventive in La Crosse County.

*Key to species*

1 Awns tightly twisted and ± connate, forming a column ca. 5–9 mm long at summit of lemma before diverging into 3 ± equal and much longer free portions .............................. ***A. tuberculosa***

1 Awns not forming a column, separate from their bases ......... 2

2 First glume with 3–5 distinct nerves, 16–24 mm long (plus awn if present); body of lemma 15–20 mm long; awns 3.5–6 cm long .... .................................................. ***A. oligantha***

2 First glume with 1 distinct nerve, 2.5–12.5 mm long; body of lemma 3.5–11 mm long; awns less than 3.5 cm long .................... 3

3 Middle awn on most lemmas loosely spiraled (at least when dry) in 1 or 2 loops toward its base ................................... 4

3 Middle awn on most lemmas bent, slightly twisted, or straight, without spiraled loops at base ................................ 5

4 Body of lemma 7–11 mm long, with middle awn 9–18 mm long and lateral awns 6–12 mm long; glumes clearly unequal, the first usually equaling or shorter than the body of the lemma ..... ***A. basiramea***

4 Body of lemma 5–7 mm long, with middle awn mostly 4–8 mm long and lateral awns less than 2 mm long; glumes mostly subequal, both longer than body of lemma ......................... ***A. dichotoma***

5 Middle awn of lemma 7–12 mm long, strongly divergent or somewhat reflexed; lateral awns 1–4 mm long, ± erect or slightly spreading; first glume 2.5–4 (–6) mm long (excluding awn-tip if present); body of lemma 3.5–5.5 mm long ........... ***A. longespica***

5 Middle awn of lemma 15–33 mm long; lateral awns 9–26 mm long; all awns somewhat spreading or divergent; first glume 8.5–13.5 mm long; body of lemma 5–8 mm long ............... ***A. purpurascens***

***Aristida basiramea*** Engelm. FORKED THREE-AWN Dry sterile or sandy soil.

***Aristida dichotoma*** Michx. CHURCHMOUSE THREE-AWN Dry sandy soil.

***Aristida longespica*** Poir. SLIMSPIKE THREE-AWN Dry sterile or sandy soil.

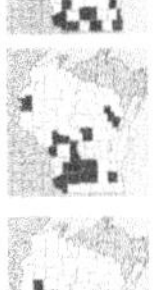

***Aristida oligantha*** Michx. PRAIRIE THREE-AWN Dry open ground.

***Aristida purpurascens*** Poir. [•127] ARROWFEATHER THREE-AWN Dry sandy soil and prairies.

***Aristida tuberculosa*** Nutt. SEASIDE THREE-AWN Dry sterile soil, especially on dunes.

### ARRHENATHERUM *Oatgrass*

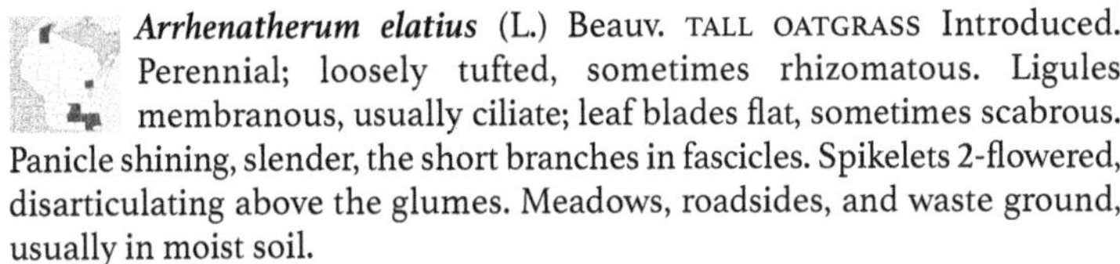

***Arrhenatherum elatius*** (L.) Beauv. TALL OATGRASS Introduced. Perennial; loosely tufted, sometimes rhizomatous. Ligules membranous, usually ciliate; leaf blades flat, sometimes scabrous. Panicle shining, slender, the short branches in fascicles. Spikelets 2-flowered, disarticulating above the glumes. Meadows, roadsides, and waste ground, usually in moist soil.

### AVENA *Oat*

Mostly annual grasses with broad flat blades and ample panicles of large spikelets. Sheaths open, auricles absent, ligules membranous. Spikelets 2–3-flowered.

*Key to species*

1 Lemmas with a stout, strongly twisted awn and often with stiff hairs on the back; florets falling from the spikelet by a distinct oval disarticulation surface .................................. *A. fatua*

1 Lemmas with the awn usually straight, weak, or absent, and the back glabrous; florets falling by fracture of rachilla at base of spikelet ........................................... *A. sativa*

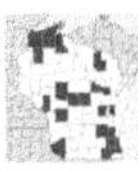

***Avena fatua*** L. WILD OAT European native; disturbed sites such as railroads, roadsides, beaches.

***Avena sativa*** L. OAT Introduced. An important cultivated species apparently derived from *A. fatua* (with which it hybridizes); often adventive along roads and railways, probably not persistent.

### BECKMANNIA *Slough Grass*

***Beckmannia syzigachne*** (Steud.) Fern. AMERICAN SLOUGH GRASS Stout annual grass. Ligule membranous; leaf blades flat, rough-to-touch. Head of many 1-sided spikes in a narrow panicle, the panicle branches erect. June–Sept. Wet meadows, marshes, ditches, shores and streambanks; more common in Great Plains region.

### BOUTELOUA *Grama-Grass*

Perennial, usually tufted grasses, the relatively short spikes solitary or 2–many, racemose on a common axis. Leaves mostly basal; ligules of hairs, membranous, or membranous and ciliate. Spikelets with 1 perfect flower and 1 or more sterile rudiments.

*Key to species*

1 Spikes 10–50, forming a long erect inflorescence .. ***B. curtipendula***
1 Spikes 1–3 .......................................................... 2

2 Rachis of the spike prolonged as a stiff point beyond the summit of the uppermost spikelet ................................ ***B. hirsuta***
2 Rachis of the spike not exceeding the uppermost spikelet ***B. gracilis***

***Bouteloua curtipendula*** (Michx.) Torr. SIDE-OATS GRAMA Dry open places and prairies; often included in prairie seed mixtures.

***Bouteloua gracilis*** (Willd.) Lag. BLUE GRAMA Dry grasslands, much more common westward in the USA where an important component of short-grass prairie.

***Bouteloua hirsuta*** Lag. HAIRY GRAMA Dry prairies and sand hills.

## BRACHYELYTRUM *Shorthusk*

Perennial forest understory grasses from knotty rhizomes. Leaves broad, mostly along the stem; ligules short, membranous; lower leaf blades absent or reduced; upper leaf blades flat. Panicles narrow, few-flowered. Spikelets readily deciduous, 1-flowered.

*Key to species*

1 Lemmas scabrous, the hairs short, to 0.2 mm long in the middle portions of the lemma; awns mostly 17–24 mm long .. ***B. aristosum***
1 Lemmas strongly hispid, the hairs longer, 0.3–0.8 mm long in the middle portions of the lemma; awns mostly 13–17 mm long ..... ................................................ ***B. erectum***

***Brachyelytrum aristosum*** (Michx.) Beauv. ex Branner & Coville BEARDED SHORTHUSK Moist to dry deciduous forests, lowland forests, moist thickets, sandy pine forests, and coniferous swamps.

***Brachyelytrum erectum*** (Schreb.) Beauv. [•127] BEARDED SHORTHUSK Dry or moist deciduous woods; our two species occupying essentially the same habitats.

## BROMUS *Brome; Chess; Cheat-Grass*

Perennial grasses. Leaves generally flat; sheaths closed to near top, usually pubescent; auricles usually absent; ligules membranous, usually erose or lacerate. Head a panicle of drooping spikelets. Spikelets with several to many flowers.

*Key to species*

1 First glume with one distinct nerve; second glume with 3 (–5) nerves ................................................................ 2
1 First glume with 3 (–5) distinct nerves; second glume with 5–7 nerves ........................................................ 7

2 Awns 10–30 mm long, as long as or longer than their lemmas; apex of lemma beyond insertion of awn 1.5–2.7 mm long; annual weed ................................................ ***B. tectorum***
2 Awns absent or up to 7 (–9) mm long, shorter than their lemmas; apex of lemma less than 1.5 mm long; perennials, mostly native (*B.*

*inermis* and *B. erectus* introduced) .............................. 3

3 Plants with elongate rhizomes; lemmas (at least when fresh) usually ± flushed with purplish, especially toward the margins, the awns absent or less than 4 (–5.5) mm long; anthers 3.3–4.7 (–6) mm long ................................................... ***B. inermis***

3 Plants without elongated rhizomes; lemmas (when fresh) green (very rarely flushed with purple), the larger awns 3–7 (–9) mm long; anthers various ........................................... 4

4 Branches of nearly simple panicle erect or strongly ascending; anthers ca. 4.5–6 (–8) mm long; leaf blades involute; clump-forming adventive of open disturbed ground .................. ***B. erectus***

4 Branches of the compound panicle loosely ascending, spreading, or nodding at maturity; anthers ca. 0.9–5 mm long; leaf blades flat; native, in forests, savannas, and thickets ...................... 5

5 Nodes and also number of leaves usually 8–15; leaf sheaths longer than the internodes, thus overlapping and covering all the nodes, the summit of the sheath with a band of dense pubescence and (when intact) with a pair of prominent tooth-like auricles; anthers 1.5–2.2 mm long ................................. ***B. latiglumis***

5 Nodes and leaves usually not more than 6 (–8 in *B. ciliatus*); leaf sheaths shorter than at least the upper internodes, exposing one or more of them, the summit of the sheath glabrous or pubescent but lacking auricles; anthers various ............................ 6

6 Lemmas ± uniformly hairy (very rarely glabrous); anthers 2.5–5 mm long; glumes pubescent at least on keel (sometimes only scabrous) ................................................. ***B. pubescens***

6 Lemmas with long hairs along the margin, especially toward the base, glabrous or only minutely pubescent on the back; anthers 0.8–1.7 mm long; glumes glabrous or at most scabrous ..... ***B. ciliatus***

7 Lemmas pubescent all across the back, at least near tip; glumes pubescent; awns straight ................................. 8

7 Lemmas glabrous or scabrous on the back; glumes glabrous; awns usually divaricate or undulate (or absent)..................... 9

8 Larger awns 2–3.5 mm long; primary branches of inflorescence mostly longer than the spikelets; anthers 1.5–2 mm long; ligule less than 0.7 mm long, glabrous on the back; plant a native perennial . ................................................... ***B. kalmii***

8 Larger awns 3.5–14 mm long; branches of inflorescence mostly much shorter than the spikelets; anthers 0.5–1.2 mm long; ligule 0.5–2 mm long, pubescent on the back (side next to the blade); introduced annuals ............................. ***B. hordeaceus***

9 Lemma equaling or slightly shorter than the tip of the mature palea; sheaths glabrous (or occasionally the lowermost with some short hairs); margins of ripe lemmas strongly inrolled, exposing the rachilla; lemmas ca. 7–8.2 mm long, the awns ± undulate, sometimes as long as lemma but usually much shorter, rudimentary, or occasionally absent ................................ ***B. secalinus***

9 Lemma at least slightly exceeding tip of palea; sheaths of at least middle and lower leaves ± densely (though sometimes finely) hairy; lemmas and awns various .................................. 10

10 Sides of larger lemmas (from middle of back to margin) 3–3.5 mm wide, including a broad hyaline border; tips of at least the upper lemmas in a spikelet exceeding their paleas by more than 2 mm; awns either absent (or less than 1 mm long) or very strongly divaricate (even recurved) at maturity .............. ***B. squarrosus***

10 Sides of larger lemmas not over 2.5 mm wide; tips of upper lemmas exceeding paleas by less than 2 mm (occasionally 2.5 mm) ..... 11

11 Longest awns in a spikelet longer than their lemmas and more than twice as long as awn on lowest lemma of the spikelet; branches of inflorescence lax and flexuous; hairs of sheath very fine and delicate, and tending to be ± crooked or tangled toward their tips (though basically ± retrorse); tip of palea 1–2.5 mm shorter than tip of lemma; anthers 0.5–1.5 mm long .................. ***B. arvensis***

11 Longest awns in a spikelet about as long as their lemmas or shorter, and usually less than twice as long as awn on lowest lemma of the spikelet; branches of inflorescence rather stiff (whether spreading or ascending); hairs of sheath usually fine but stiffish and straight (spreading to retrorse); tip of palea less than 1.5 mm shorter than tip of its lemma (rarely 2.5 mm in *B. racemosus*); anthers 1–2 mm long .......................................................... 12

12 Branches of panicle, at least the lower ones, ± widely spreading or drooping at maturity, forming a broad open inflorescence usually nodding at the summit; larger lemmas 8–10 mm long, attached ca. 1.5–2 mm apart on the rachilla ................... ***B. commutatus***

12 Branches of panicle erect or very strongly ascending, the entire inflorescence narrow and compact; larger lemmas 6.5–7.5 mm long, attached 1–1.5 mm apart on the rachilla ............. ***B. racemosus***

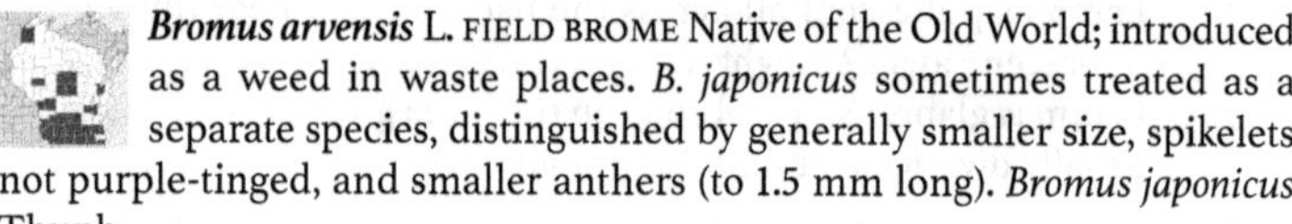

***Bromus arvensis*** L. FIELD BROME Native of the Old World; introduced as a weed in waste places. *B. japonicus* sometimes treated as a separate species, distinguished by generally smaller size, spikelets not purple-tinged, and smaller anthers (to 1.5 mm long). *Bromus japonicus* Thunb.

***Bromus ciliatus*** L. FRINGED BROME July–Sept. Streambanks, shores, thickets, sedge meadows, fens, marshes; also in moist woods.

***Bromus commutatus*** Schrad. HAIRY CHESS Native of Europe, introduced in disturbed places in Wisc. Sometimes considered synonymous with *B. racemosus (no map)*.

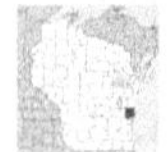

***Bromus erectus*** Huds. UPRIGHT BROME Introduced in Sheboygan County.

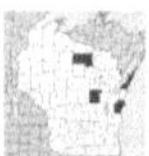

***Bromus hordeaceus*** L. SOFT CHESS Native of Europe; introduced in our range in roadsides, cultivated ground, and waste places. *Bromus mollis* L.

***Bromus inermis*** Leyss. SMOOTH BROME Native of Europe, cultivated for forage and often escaped.

***Bromus kalmii*** Gray Kalm's Brome Dry woods, rocky banks, and sandy or gravelly soil.

***Bromus latiglumis*** (Shear) A.S. Hitchc. [•128] EARLY-LEAF BROME Floodplain forests, thickets and streambanks, sometimes in rocky woods. Similar to *B. pubescens,* a species of mostly mesic woods, but stems with more leaves, the top of sheaths with a ring of dense hairs, and leaf blades with well developed auricles. *Bromus altissimus* Pursh.

***Bromus pubescens*** Muhl. Hairy WOODLAND BROME Rich moist woods. Variable.

***Bromus racemosus*** L. BALD BROME Native of Europe; introduced in fields, roadsides, and waste ground.

***Bromus secalinus*** L. RYE BROME; CHESS Native of Europe; introduced in grainfields, roadsides, and waste places.

***Bromus squarrosus*** L. CORN BROME Native of Europe; occasionally introduced in waste places.

***Bromus tectorum*** L. CHEAT GRASS Native of s Europe, widely established as a weed in waste ground and on roadsides; now a prominent feature of many sagebrush ecosystems in the western states.

## CALAMAGROSTIS *Reed-Grass*

Perennial grasses, spreading by rhizomes. Leaves flat or inrolled, green or waxy blue-green; ligule large, membranous, usually with an irregular, ragged margin. Head a loose to dense panicle. Spikelets 1-flowered.

*Key to species*

1 Callus hairs and glumes at least 1.5 times as long as the lemma; uncommon in Wisc ................................ ***C. epigeios***

1 Callus hairs and glumes barely if at all exceeding lemma; widespread in Wisc ........................................... **2**

2 Leaf blades rather lax, to 10 mm wide; panicle mostly open with rather loosely ascending to spreading branches at flowering time; lemma nearly or quite smooth, membranous and translucent for at least the apical half; awn nearly or quite smooth, at least on basal half; callus hairs about as long as lemma (occasionally shorter), ± uniform in length and distribution; palea not over 2 mm long .... ............................................... ***C. canadensis***

2 Leaf blades stiff, to 4 mm wide; panicle mostly narrow and contracted with strongly ascending branches at flowering time; lemma usually firm and prominently scabrous, colorless and translucent only toward the tip; awn distinctly but minutely antrorsely scabrous its entire length (at 20×); callus hairs generally shorter than lemma, ± unequal in length or distribution (those immediately below the middle of the lemma shorter than those at the side, or absent; do not confuse the hairy prolongation of the rachilla behind the palea); palea often longer than 2 mm ***C. stricta***

***Calamagrostis canadensis*** (Michx.) Beauv. [•128] BLUEJOINT June–Aug. Common grass of wet meadows, shallow marshes, calcareous fens, streambanks, thickets.

***Calamagrostis epigeios*** (L.) Roth FEATHERTOP Native of Eurasia; uncommon in disturbed places.

***Calamagrostis stricta*** (Timm) Koel. SLIM-STEM REED-GRASS June–Sept. Wet meadows, shallow marshes, shores, streambanks; rocky shore of Lake Superior. *Calamagrostis inexpansa* Gray.

## CALAMOVILFA *Sand-Reed*

***Calamovilfa longifolia*** (Hook.) Scribn. SAND-REED Perennial grass, from creeping rhizomes, the rhizomes covered with shiny, scale-like leaves. Ligule a ring of short hairs 1–2 mm long; leaf blades flat at base, involute above, tapering to a fine point. Panicle open, with ascending branches. Spikelets 1-flowered. Dry sandy prairies and dunes. Var. *magna,* primarily of dunes along Lake Michigan, is state threatened, and distinguished from typical var. *longifolia* as follows:

*Key to species*

1 Panicle open and spreading; most spikelets overlapping no more than 1 other spikelet, usually brownish ............... var. *magna*

1 Panicle more tightly contracted; most spikelets overlapping 2-3 other spikelets, usually not brownish tinged ........ var. *longifolia*

## CENCHRUS *Sandbur*

***Cenchrus longispinus*** (Hack.) Fern. COMMON SANDBUR Annual grass. Inflorescences spikelike panicles of highly reduced branches termed fascicles ("burs"); fascicles consisting of 1-2 series of many, stiff, sharp bristles surrounding 1-4 spikelets. Disturbed places, especially where dry and sandy. The spines are painfully sharp if walking barefoot.

## CHASMANTHIUM *Wood-Oats*

***Chasmanthium latifolium*** (Michx.) Yates INDIAN WOOD-OATS Perennial grass, from short rhizomes. Panicle open, drooping. Adventive in Forest County (*no map*).

## CINNA *Wood-Reed*

Tall perennial grasses, rhizomes weak or absent. Leaves wide, flat and lax; auricles absent; ligule brown, membranous, with an irregular, jagged margin. Head a large, closed to open panicle, the branches upright to spreading or drooping. Spikelets small, 1-flowered, laterally compressed.

*Key to species*

1 Panicle more or less crowded and narrow, the branches upright; second glume 4–6 mm long ..................... ***C. arundinacea***

1 Panicle open, the branches spreading to drooping; second glume 2–4 mm long ......................................... ***C. latifolia***

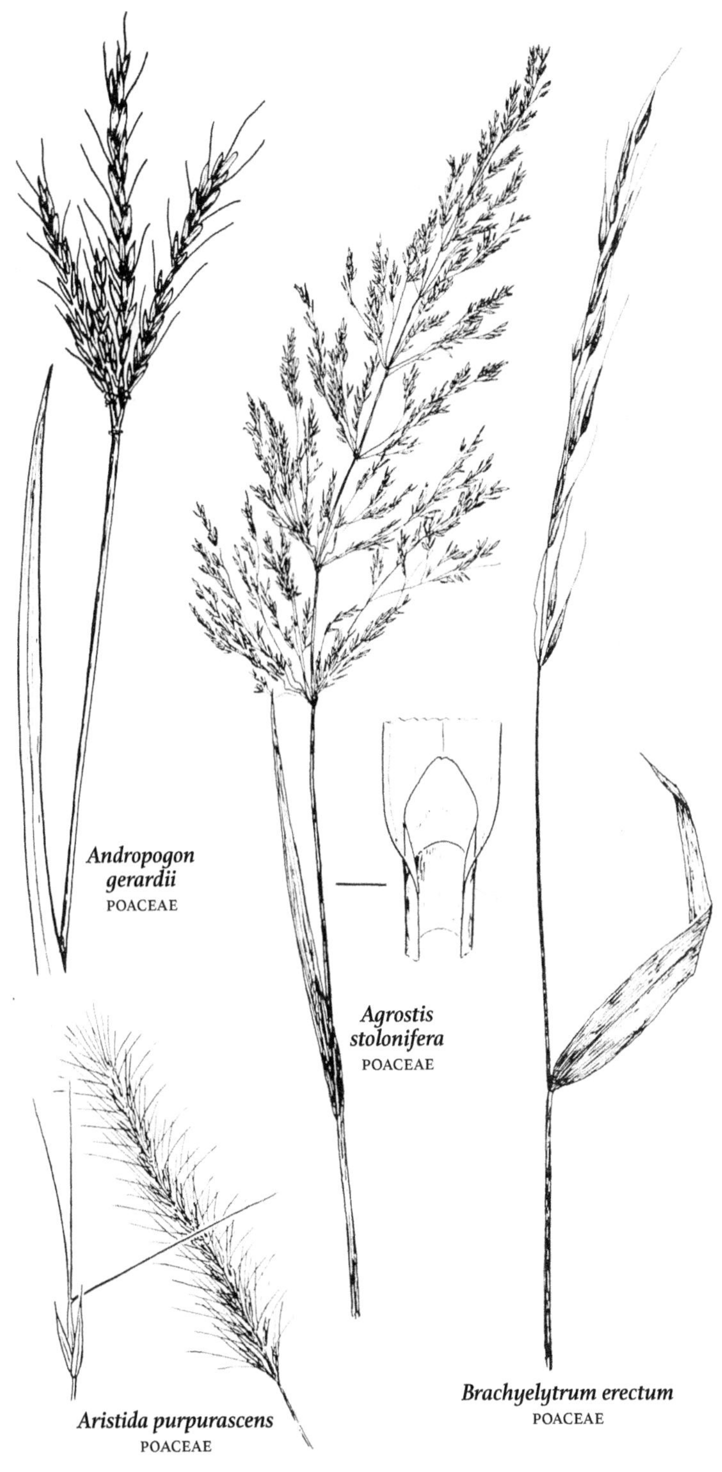
Andropogon gerardii
POACEAE
Agrostis stolonifera
POACEAE
Aristida purpurascens
POACEAE
Brachyelytrum erectum
POACEAE

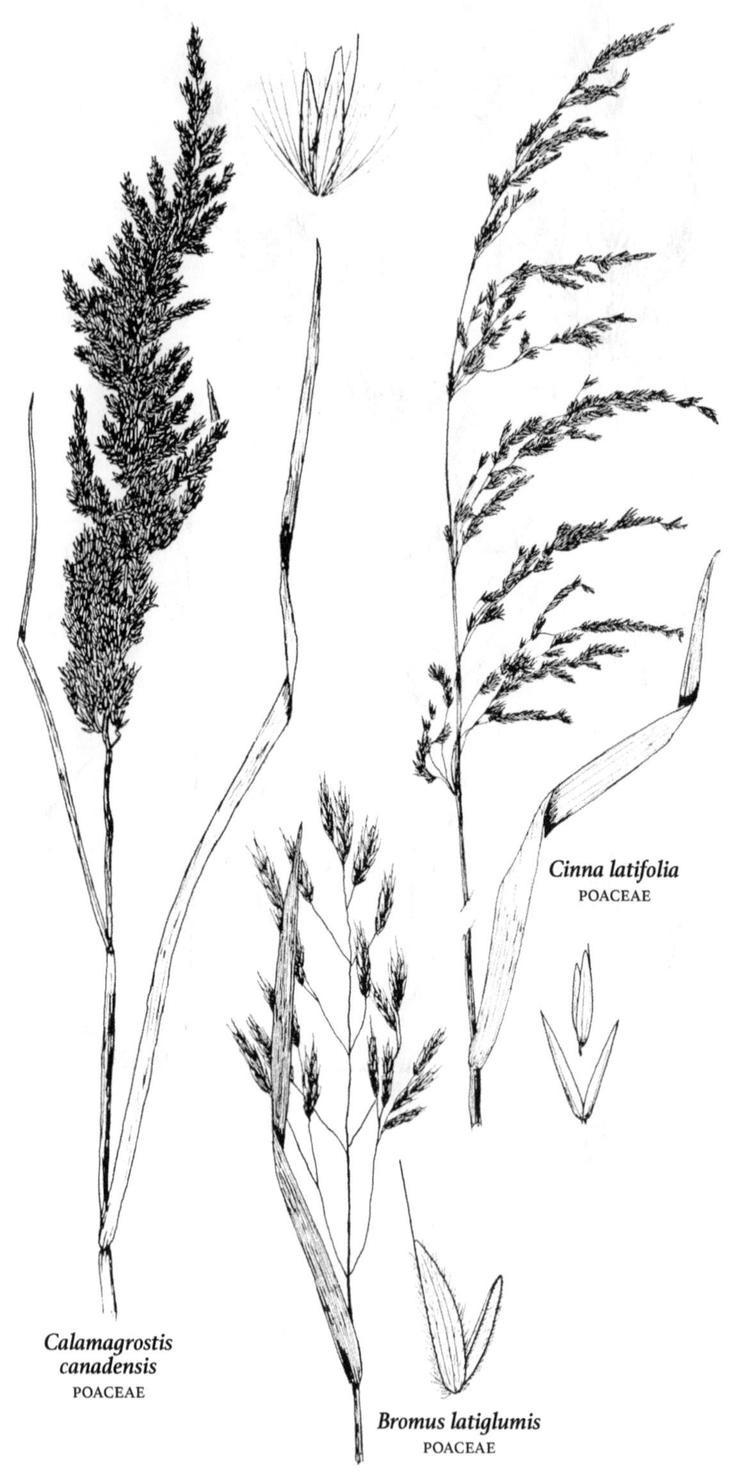

Calamagrostis canadensis
POACEAE
Bromus latiglumis
POACEAE
Cinna latifolia
POACEAE

***Cinna arundinacea*** L. SWEET WOOD-REED Aug–Sept. Swamps, floodplain forests, streambanks, pond margins, moist woods. Distinguished from *C. latifolia* by its 3-veined upper glumes and larger spikelets.

***Cinna latifolia*** (Trev.) Griseb. [•128] DROOPING WOOD-REED July–Aug. Wet woods, swamps, springs.

## CRYPSIS *Prickle Grass*

***Crypsis schoenoides*** (L.) Lam. SWAMP PRICKLE GRASS Native of s Europe; introduced in waste ground. closely resembles *Alopecurus,* but has no awn and differs in the articulation of the spikelet.

## CYNOSURUS *Dog's-Tail Grass*

***Cynosurus cristatus*** L. CRESTED DOG'S-TAIL GRASS Introduced from Europe in fields and waste places; Milwaukee County (*no map*).

## DACTYLIS *Orchard-Grass*

***Dactylis glomerata*** L. ORCHARD-GRASS Introduced from Europe, cultivated for hay or pasture; occasional escape to moist fields, meadows, lawns, and roadsides.

## DANTHONIA *Wild Oatgrass*

Tufted perennial grasses with narrow, often involute leaves, the ligule reduced to a tuft of hairs, auricles absent, and small panicles of large spikelets. Spikelets several-flowered (usually 4–6).

*Key to species*

1 Longest pedicels on lowest panicle branches 1.1–1.8 times as long as the spikelet (including awns); lower panicle branches spreading to reflexed at maturity; ne Wisc ..................... ***D. compressa***

1 Longest pedicels on lowest panicle branches 0.3–0.9 times as long as the spikelet; lower panicle branches ascending to erect at maturity; statewide ................................ ***D. spicata***

***Danthonia compressa*** Austin FLAT OATGRASS Northern hardwood forests, especially on margins and in openings. Forest County. Similar to *D. spicata;* the leaves of *D. compressa* do not curl tightly upon drying like those of *D. spicata* (*no map*).

***Danthonia spicata*** (L.) Beauv. [•129] POVERTY WILD OATGRASS Common in dry woods in sandy or stony soil, especially on jack pine plains, where it may form extensive colonies following disturbance; occasionally found in marshy or boggy places. The basal leaves tend to curl and form distinctive tufts.

## DESCHAMPSIA *Hairgrass*

Tufted perennial grasses. Leaves usually mainly basal, narrow, flat or involute; sheaths open; auricles absent; ligules membranous. Flowers in panicles; spikelets yellowish or purple, 2-flowered.

*Key to species*

1 Leaf blades involute, 1–2 mm wide; ligule 1–2.5 mm long; lemmas minutely scabrous-pubescent, bearing a conspicuously bent awn 1–3 mm longer than the lemmas; palea not bifid at tip . ***D. flexuosa***

1 Leaf blades flat or conduplicate, 1–5 mm wide; ligule usually 3–12 mm long; lemmas glabrous, bearing a ± straight awn shorter than to slightly exceeding the lemmas; palea bifid at the tip ***D. cespitosa***

***Deschampsia caespitosa*** (L.) Beauv. TUFTED HAIRGRASS June–July. Wet meadows, streambanks, shores, calcium-rich seeps, rocky shores of Great Lakes.

***Deschampsia flexuosa*** (L.) Trin. HAIRGRASS Dry woods, fields, and sand hills. *Avenella flexuosa* (L.) Drej.

## DIARRHENA *Beakgrain*

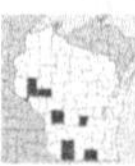

***Diarrhena obovata*** (Gleason) Brandenburg HAIRY BEAKGRAIN *Endangered.* Perennial grass, from a thick scaly rhizome. Stems with a long, slender, few-flowered, drooping panicle. Moist woods, floodplain forests.

## DICHANTHELIUM *Panic-grass*

Perennial grasses, tufted or sometimes rhizomatous, sometimes with hard, corm-like bases. Stems hollow, usually erect or ascending, sometimes decumbent in the fall. Basal rosettes of winter leaves sometimes present. Flowers in terminal panicles (vernal) developing late spring to early summer, and sometimes lateral panicles (autumnal) in late-summer or fall.

ADDITIONAL SPECIES ***Dichanthelium clandestinum*** (L.) Gould (Deer-tongue grass), known from a sandy, grassy clearing in Vilas County and from adjacent Michigan Upper Peninula; more common south and east of Wisc.

*Key to species*

1 Basal leaf blades similar in shape to the lower stem leaves, usually erect to ascending, clustered at the base, sometimes vestigial; stems branching from near the base in the fall, with 2-4 leaves, only the upper 2-4 internodes elongated . . . . . . . . . . 2

1 Basal leaf blades usually well-differentiated from the stem blades, spreading, forming a rosette, or basal blades absent; stems usually branching from the midculm nodes in the fall, with 3-14 leaves, usually all internodes elongated . . . . . . . . . . 4

2 Upper glumes and lower lemmas forming a beak extending 0.2-1 mm beyond the upper florets; spikelets 3.2-4.3 mm long; primary panicles with 7-25 spikelets . . . . . . . . . . ***D. depauperatum***

2 Upper glumes and lower lemmas equaling or exceeding the upper florets by no more than 0.3 mm, not forming a beak; spikelets 2-3.4 mm long; primary panicles with 12-70 spikelets . . . . . . . . . . 3

3 Stem blades 4-8 cm long, all alike . . . . . . . . . . ***D. wilcoxianum***

3 Uppermost stem blades 10-20 cm long, distinctly longer than the lower blades . . . . . . . . . . ***D. linearifolium***

4 Lower glumes thinner and more weakly veined than the upper

glumes, attached about 0.2 mm below the upper glumes, the bases clasping the pedicels; spikelets attenuate basally .. *D. portoricense*

4 Lower glumes similar in texture and vein prominence to the upper glumes, attached immediately below the upper glumes, the bases not clasping the pedicels; spikelets usually not attenuate basally 5

5 Ligules with a membranous base, ciliate distally; stems usually arising from slender rhizomes; lower florets often staminate; stem blades 5-40 mm wide, often with a cordate base ............... 6

5 Ligules of hairs; stems arising from caudices; lower florets sterile; stem blades 1-18 mm wide, bases usually tapered, rounded, or truncate at the base, sometimes cordate ...................... 8

6 Spikelets ellipsoid, not turgid, with pointed apices; stem blades 4-6, cordate at the base; sheaths without papillose-based hairs ..... ............................................ *D. latifolium*

6 Spikelets obovoid, turgid, with rounded apices; stem blades 3-4, tapered, rounded or truncate to cordate at the base; sheaths with papillose-based hairs ...................................... 7

7 Blades and spikelets with papillose-based hairs; panicles usually slightly longer than wide, with spreading to ascending branches . ............................................ *D. leibergii*

7 Blades glabrous; spikelets puberulent to almost glabrous; panicles usually more than 2x longer than wide, with nearly erect branches ........................................ *D. xanthophysum*

8 Spikelets 2.5-4.3 mm long, usually obovoid, turgid; upper glumes usually with an orange or purple spot at the base, the veins prominent ................................. *D. oligosanthes*

8 Spikelets 0.8-3 mm long, ellipsoid or obovoid, not turgid; upper glumes lacking an orange or purple spot at the base and the veins not prominent ........................................... 9

9 Ligules 1-5 mm long, or the stems and sheaths with long hairs and also puberulent; spikelets variously pubescent to subglabrous.. 10

9 Ligules absent or to 1.8 mm long, without adjacent pseudoligules; stems and at least the upper sheaths glabrous or sparsely pubescent with hairs of 1 length only; spikelets glabrous or pubescent .... 11

10 Spikelets 1.1-2.1 mm long; sheaths glabrous or pubescent with hairs no more than 3 mm long ...................... *D. acuminatum*

10 Spikelets 1.8-3 mm long; sheaths with hairs to 4 mm long *D. ovale*

11 Spikelets glabrous or, if pubescent, either the nodes bearded or the stems weak and prostrate; blade of the flag leaf usually spreading ............................................ *D. dichotomum*

11 Spikelets pubescent; nodes glabrous; stems erect or ascending; blade of the flag leaf erect or ascending .............. *D. boreale*

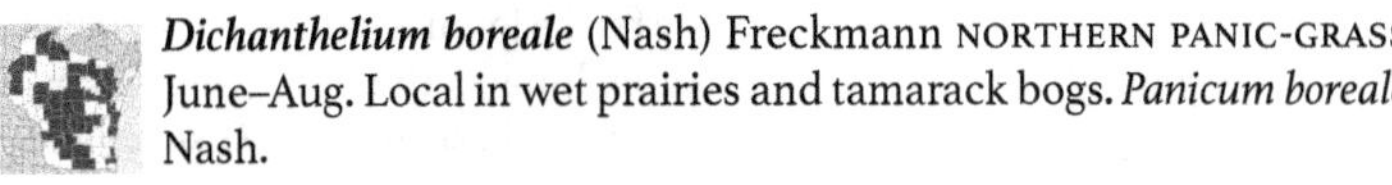

***Dichanthelium acuminatum*** (Sw.) Gould & C. A. Clark HAIRY PANIC-GRASS Moist or dry situations, open woods, dunes, shores, and prairies. *Panicum acuminatum* Sw.

***Dichanthelium boreale*** (Nash) Freckmann NORTHERN PANIC-GRASS June–Aug. Local in wet prairies and tamarack bogs. *Panicum boreale* Nash.

***Dichanthelium depauperatum*** (Muhl.) Gould STARVED PANIC-GRASS Dry or sandy soil, usually in open woods. *Panicum depauperatum* Muhl.

***Dichanthelium dichotomum*** (L.) Gould FORKED PANIC-GRASS Dry or moist woods. *Panicum dichotomum* L.

***Dichanthelium latifolium*** (L.) Gould & C. A. Clarkmap BROADLEAVED PANICGRASS Moist or dry woods and thickets. *Panicum latifolium* L.

***Dichanthelium leibergii*** (Vasey) Freckmann LEIBERG'S PANICGRASS Moist to dry grassy meadows and prairies. *Panicum leibergii* (Vasey) Scribn.

***Dichanthelium linearifolium*** (Scribn.) Gould LINEAR-LEAVED PANIC-GRASS Dry or stony soil, open woods and banks. *Panicum linearifolium* Scribn.

***Dichanthelium oligosanthes*** (J.A. Schultes) Gould [•129] FEW-FLOWERED PANIC-GRASS Dry or moist, often sandy soil, open woods and prairies. *Panicum oligosanthes* J.A. Schultes.

***Dichanthelium ovale*** (Elliott) Gould & C.A. Clark STIFF-LEAVED PANIC-GRASS Dry, sandy openings and meadows.

***Dichanthelium portoricense*** (Desv. ex Ham.) Hansen & Wunderlin BLUNT-GLUMED PANIC-GRASS Moist or dry, especially sandy soil. *Panicum columbianum* Scribn.

***Dichanthelium wilcoxianum*** (Vasey) Freckmann WILCOX'S PANIC-GRASS Dry prairies, often where sandy or gravelly.

***Dichanthelium xanthophysum*** (Gray) Freckmann PALE PANIC-GRASS Dry sandy soil of open woodlands. *Panicum xanthophysum* A. Gray.

## DIGITARIA *Crabgrass*

Annual or perennial grasses. Leaves wide, flat, with the tips ascending; sheaths open, ligules membranous. Flowers in several terminal, digitate, spike-like racemes. Spikelets 1-flowered, single or in clusters of 2 or 3 on one side of an elongate rachis.

*Key to species*

1 Inflorescence an open panicle, neither spike-like nor distinctly one-sided ........ ***D. cognata***

1 Inflorescence composed of 1-sided spikes or spike-like racemes, the rachis of each winged or at least flat on the side opposite the spikelets ........ **2**

2 Spikelets ca. 2–2.3 mm long, the fertile lemma dark brown; second glume nearly or fully as long as the floret; sheaths and blades usually nearly or quite glabrous (except around summit of sheath) ........ ***D. ischaemum***

2 Spikelets ca. 2.5–3 mm long, the fertile lemma light or dark grayish; second glume only about half as long as the floret; sheaths and usually blades ± pilose, at least toward base of plant ***D. sanguinalis***

***Digitaria cognata*** (J.A. Schultes) Pilger WITCH-GRASS Dry, especially sandy soil. *Leptoloma cognatum* (J.A. Schultes) Chase.

***Digitaria ischaemum*** (Schreb.) Schreb. SMOOTH CRABGRASS Eurasian weed of lawns, waste places.

***Digitaria sanguinalis*** (L.) Scop. HAIRY CRABGRASS Eurasian; often a troublesome weed of fields, gardens, lawns, and waste ground.

## ECHINOCHLOA *Barnyard-Grass*

Large, weedy, annual grasses. Leaves flat, wide and smooth; sheaths smooth or hairy; ligules usually absent. Head a dense panicle, the branches crowded with spikelets forming racemes or spikes.

ADDITIONAL SPECIES ***Echinochloa frumentacea*** Link. (Japanese millet); introduced cereal grain; several Wisc locations.

*Key to species*

1 Lower leaf sheaths rough-hairy; spikelets each with 2 awns; s portion of our region ................................ *E. walteri*
1 Leaf sheaths smooth; spikelets with usually 1 awn (from sterile lemma); widespread ........................................ 2

2 Fertile lemma rounded or broadly tapered to a thin, membranous, withered beak ...................................... *E. crus-galli*
2 Fertile lemma tapered to a stiff, persistent beak ....... *E. muricata*

***Echinochloa crus-galli*** (L.) Beauv. LARGE BARNYARD-GRASS Introduced and naturalized throughout most of USA. *E. muricata* is similar in form and habitat, but distinguished by features of the lemma (see key).

***Echinochloa muricata*** (Beauv.) Fern. BARNYARD-GRASS July–Sept. Shores, streambanks and ditches, where sometimes in shallow water.

***Echinochloa walteri*** (Pursh) Heller Saltmarsh COCKSPUR GRASS Aug–Sept. Streambanks, lakeshores, ditches; especially along Fox River.

## ELEUSINE *Goose Grass*

***Eleusine indica*** (L.) Gaertn. INDIAN GOOSE GRASS Native of the Old World; weedy in lawns, gardens, and waste places. Often confused with hairy crabgrass (*Digitaria sanguinalis*).

## ELYMUS *Wild Rye*

Tufted perennial grasses. Leaves flat, sheaths open for most of their length, auricles often present, ligules short. Head a densely flowered spike. Spikelets usually 2 at each node of spike.

ADDITIONAL SPECIES ***Elymus diversiglumis*** Scribn. & Ball (Minnesota wild rye); spikes somewhat nodding; glumes and lemma awns outwardly curved. Suspected to be a hybrid between *E. canadensis* and *E. villosus.* Moist soil in mostly nw Wisc. ***Elymus glaucus*** Buckl. (Blue wild rye); common in western USA, reported from Rusk County. ***Elymus macgregorii*** R.E. Brooks & J.

Campbell, a recently described species formerly included with *E. virginicus;* appears to flower and fruit slightly earlier than *E. virginicus,* and the inflorescence is more exserted from the sheath; mesic forests, especially along streams, south half of Wisc.

*Key to species*

1 Spikelets clearly 1 at each node (or most nodes) of the rachis; glumes not more than 2 .......... 2
1 Spikelets mostly 2–3 at each node of the rachis .......... 5

2 Lemmas densely hairy; leaves with narrow (rarely as much as 4.5 mm wide) often involute blades, the whole plant usually strongly glaucous .......... ***E. lanceolatus***
2 Lemmas glabrous (rarely slightly pubescent), smooth or scabrous; leaves various .......... 3

3 Stems tufted, rhizomes absent; anthers 1–2.2 (–2.4) mm long; rachilla readily disarticulating between the florets when mature (on dry specimens, the florets very easily dislodged and empty glumes often remaining on older plants) .......... ***E. trachycaulus***
3 Stems from elongate rhizomes; anthers 3–6 mm long; rachilla often not readily disarticulating (florets not easily dislodged on dry specimens except over-ripe ones, empty glumes seldom if ever present) .......... 4

4 Spikelets 15-30 mm long, with 6-16 florets; leaf blades stiff, deeply grooved on the upper surface; cartilaginous band of upper nodes of stem shorter than thick .......... ***E. smithii***
4 Spikelets 10-18 mm long, with 3-6 florets; leaf blades lax, not deeply grooved; cartilaginous band of upper nodes of stem as long as thick .......... ***E. repens***

5 Glumes absent or vestigial, or, if present, slenderly awn-like their entire length and at least onemuch shorter than the others at a node; spikelets horizontally spreading at maturity (± ascending when young), well separated, clearly revealing the entire rachis .......... ***E. hystrix***
5 Glumes present, awn-like to lanceolate, of about equal length; spikelets ascending at maturity, usually concealing much of the rachis .......... 6

6 Larger paleas (lowest in each spikelet) 8.6–13 mm long; awns of lemmas usually widely spreading at maturity .......... 7
6 Larger paleas 5.5–8.5 mm long; awns of lemmas mostly straight . 9

7 Body of glume about twice as long as its awn, or longer; awns of lemmas usually straight at maturity; spike curved to erect .......... *(go to couplet 10)*
7 Body of glume about equaling its awn, or shorter; awns of lemmas ± curved at maturity (straight when young); spike curved to strongly nodding .......... 8

8 Leaves 5–8 on a stem, the broadest blades rarely as much as 15 mm wide, glabrous above .......... ***E. canadensis***
8 Leaves 10–12 on a stem, the broadest blades 15–19 mm wide, finely hairy above .......... ***E. wiegandii***

9 Glumes, at least the broadest, 1–2 mm wide, clearly expanded and flattened above the base .................................... 10

9 Glumes less than 1 mm wide, scarcely if at all widened above the base .................................................... 13

10 Spikelets spreading to form a spike 2.5–5 cm wide (measured awn tip to awn tip, pressed); longer glume awns 12–25 mm long ....... .......................................... *E. macgregorii**

10 Spikelets ascending to form a spike 1–2.2 (–2.5) cm broad; longer glume awns less than 12 mm long .......................... 11

11 Base of glumes not conspicuously bowed out, but flattened, hardened for less than 1 mm; glumes not thickened above the base on inner face, with very narrow, thin, translucent margins, often slightly overlapping; stem leaves 5–6 ................. *E. glaucus*

11 Base of glumes ± bowed out, terete and hard for 1 mm or more; glumes also thickened, pale, and hardened on inner face for about the basal half or more, with firm margins, not at all overlapping; stem leaves (6–) 7–10 ......................................... 12

12 Lemmas awnless or with short awns less than 4 mm longE. curvatus

12 Lemma awns 8–20 mm long ...................... *E. virginicus*

13 Palea of lowest floret in spikelet 7–8.5 mm long; leaves 8–10, glabrous ....................................................... *E. riparius*

13 Palea 5.5–7 mm long; leaves 6–7, the sheaths and upper surface of blades finely villous ............................... *E. villosus*

***Elymus canadensis*** L. NODDING WILD RYE Dry or moist soil, often where sandy or gravelly, usually in full sun. Variable.

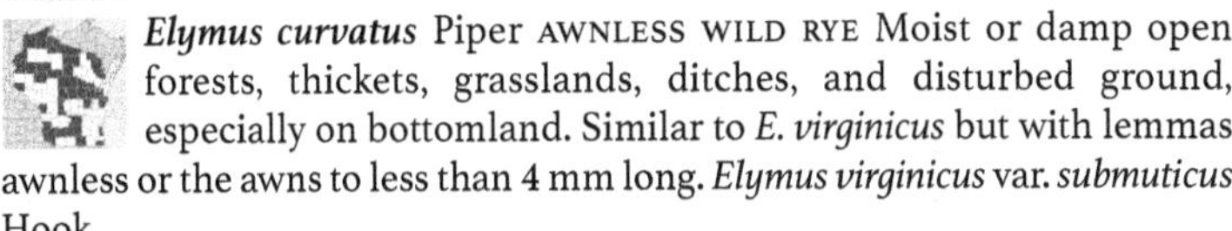

***Elymus curvatus*** Piper AWNLESS WILD RYE Moist or damp open forests, thickets, grasslands, ditches, and disturbed ground, especially on bottomland. Similar to *E. virginicus* but with lemmas awnless or the awns to less than 4 mm long. *Elymus virginicus* var. *submuticus* Hook.

***Elymus hystrix*** L. BOTTLEBRUSH-GRASS Moist deciduous woods, especially in wet or slightly disturbed areas. Spikelets soon horizontally divergent, the lemmas easily detached. *Hystrix patula* Moench.

***Elymus lanceolatus*** (Scribn. & J.G. Sm.) Gould STREAMSIDE WILD RYE Subsp. *psammophilus,* with lemmas densely hairy, the hairs to 1 mm long or longer, is state threatened; it occurs on sandy beaches and dunes along Lake Michigan. *Agropyron dasystachyum* (Hook.) Scribn. & J.G. Sm.

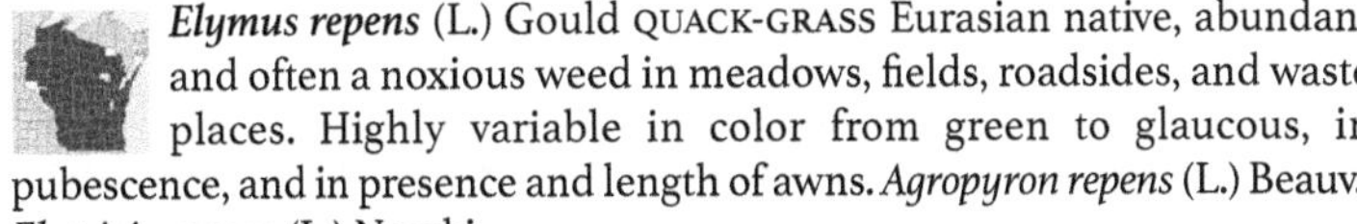

***Elymus repens*** (L.) Gould QUACK-GRASS Eurasian native, abundant and often a noxious weed in meadows, fields, roadsides, and waste places. Highly variable in color from green to glaucous, in pubescence, and in presence and length of awns. *Agropyron repens* (L.) Beauv., *Elytrigia repens* (L.) Nevski.

***Elymus riparius*** Wieg. RIVERBANK WILD RYE Streambanks, floodplain forests. Similar to nodding wild rye (*E. canadensis*), a species of drier, sandy places, but awns straight rather than bent and curved.

***Elymus trachycaulus*** (Link) Gould ex Shinners SLENDER WILD RYE Dry, open, rocky woods, sandy shores and barrens; rarely in fens and tamarack swamps. Variable but distinguished by the short anthers (when young), and by the readily disintegrating spikelets (when mature); the rachilla is also nearly always villous. *Agropyron caninum* (L.) Beauv., *Agropyron trachycaulum* (Link) Steud.

***Elymus villosus*** Muhl. HAIRY WILD RYE Swampy forests and riverbanks; also in drier woods. Glumes and lemmas usually conspicuously hirsute.

***Elymus virginicus*** L. VIRGINIA WILD RYE Floodplain forests, thickets, streambanks.

***Elymus wiegandii*** Fern. WIEGAND'S WILD RYE Moist forests, especially along streams.

## ERAGROSTIS *Lovegrass*

Annual grasses (ours), perfect-flowered or with staminate and pistillate flowers on different plants. Heads usually many, in an open or narrow panicle. Spikelets few- to many-flowered.

ADDITIONAL SPECIES ***Eragrostis mexicana*** (Hornem.) Link (Mexican lovegrass), adventive in Milwaukee County. ***Eragrostis trichodes*** (Nutt.) Wood (Sand lovegrass), a Great Plains species, adventive in Juneau County.

*Key to species*

1 Plants prostrate basally, rooting at lower nodes; nodes of stem bearded (very rarely glabrate) . . . . . . . . . . . . . . . . . . . . . ***E. hypnoides***

1 Plants ± erect or spreading from the base, not rooting at the nodes; nodes of stem glabrous . . . . . . . . . . . . . . . . . . . . . . . . . . . . . . . . . . . . 2

2 Margins (often inrolled) of leaves and also (usually at least sparsely) pedicels and keels of lemmas and glumes ± glandular-warty . . . . 3

2 Margins of leaves, pedicels, and keels of lemmas and glumes not glandular-warty . . . . . . . . . . . . . . . . . . . . . . . . . . . . . . . . . . . . . . . . . 4

3 Well-developed spikelets 2.5–3.5 mm wide; larger glume 1.7–2.5 mm long; sheaths essentially glabrous except at summit . E. cilianensis

3 Well-developed spikelets 1.5–2 mm wide; larger glume 1–1.5 mm long; sheaths sparsely pilose . . . . . . . . . . . . . . . . . . . . . . . . . . ***E. minor***

4 Spikelets reddish to purplish; plants perennial, with hard knotty base; lowest panicle branches usually with a long-pilose white to yellowish or red pubescence in the axil . . . . . . . . . . . . ***E. spectabilis***

4 Spikelets greenish gray to dark lead-colored (occasionally with purplish flush besides); plants annual, with relatively soft base; lowest panicle branches glabrous to sparsely pilose . . . . . . . . . . . 5

5 Larger spikelets mostly 6–11 (–15)-flowered, usually on ± appressed pedicels (though panicle branches may be widely spreading); lowest lemma ca. 1.4–2 mm long; lateral nerves of lemma distinct (in *E. pectinacea*) . . . . . . . . . . . . . . . . . . . . . . . . . . . . . . . . . . . . . . . . . . . . . 6

5 Larger spikelets mostly 2–4 (–6)-flowered, on spreading pedicels; lowest lemma ca. 1.2–1.4 (–1.6 or very rarely 1.9) mm long; lateral nerves of lemma obscure. . . . . . . . . . . . . . . . . . . . . . . . . . . . . . . . . . 7

6 Lateral nerves of lemma distinct, at least on lower half; larger mature spikelets (1.3–) 1.5–2 mm wide; axils of panicle glabrous or rarely the lowermost with a few hairs; panicle branches usually alternate or subopposite at lowest two nodes of inflorescence .... .......................................... *E. pectinacea*

6 Lateral nerves of lemma usually obscure; larger mature spikelets ca. 1–1.4 mm wide; axils of lower primary branches of panicle sparsely pilose; panicle branches whorled or clustered at one of the two lowest nodes of inflorescence ...................... *E. pilosa*

7 Sheaths pilose; grain with a groove the length of one edge; length of stem below lowest branch of terminal panicle less than the height of the panicle .................................... *E. capillaris*

7 Sheaths essentially glabrous except at summit; grain not grooved; length of stem below lowest branch of terminal panicle usually more than the panicle .................................. **8**

8 Axils glabrous; tip of second glume ± opposite tip of lowest lemma .............................................. *E. frankii*

8 Axils of at least the lower primary branches of panicle sparsely long-pilose; tip of second glume usually much shorter than the lowest lemma (across from it) ......................... *E. pilosa*

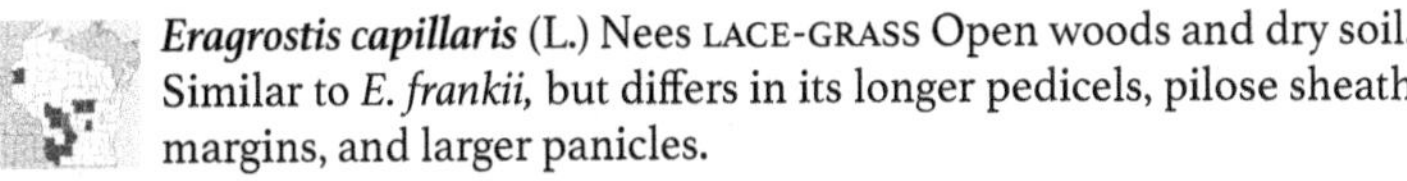

***Eragrostis capillaris*** (L.) Nees LACE-GRASS Open woods and dry soil. Similar to *E. frankii,* but differs in its longer pedicels, pilose sheath margins, and larger panicles.

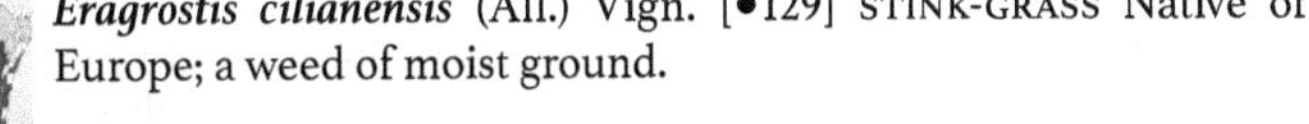

***Eragrostis cilianensis*** (All.) Vign. [•129] STINK-GRASS Native of Europe; a weed of moist ground.

***Eragrostis frankii*** C.A. Mey. SANDBAR LOVEGRASS Aug–Sept. Wet, muddy areas, streambanks, sandbars, roadside ditches, cultivated fields.

***Eragrostis hypnoides*** (Lam.) B.S.P. TEAL LOVEGRASS July–Sept. Wet, sandy or muddy shores and streambanks, sandbars, mud flats.

***Eragrostis minor*** Host LITTLE LOVEGRASS Introduced from Europe in moist soil, waste land, gardens, railways, and roadsides.

***Eragrostis pectinacea*** (Michx.) Nees CAROLINA LOVEGRASS Moist ground, especially as a weed in gardens, roadsides, railways, and waste places.

***Eragrostis pilosa*** (L.) Beauv. INDIAN LOVEGRASS Native of Europe; a weed of moist or dry open places.

***Eragrostis spectabilis*** (Pursh) Steud. PURPLE LOVEGRASS Dry soil, fields and open woods. Whole panicle eventually detached and behaving as a tumbleweed. Available commercially for planting as an ornamental grass.

### ERIOCHLOA *Cupgrass*

***Eriochloa villosa*** (Thunb.) Kunth WOOLLY CUPGRASS An Asian introduction, first reported in the USA in the 1940s; weedy,

especially in corn and soybean fields, and often spread by farm equipment; also along fencerows and roadsides. A small cup-like involucre subtending each spikelet is distinctive.

### FESTUCA *Fescue*

Annual or perennial grasses, often densely tufted. Leaves flat to involute, auricles absent; ligules membranous, usually truncate, usually ciliate. Flowers in open or contracted panicles. Spikelets 3–11-flowered.

ADDITIONAL SPECIES ***Festuca paradoxa*** Desv., historical records only from s Wisc, mostly from moist sedge meadows near the lower Wisc River, more common southward; resembles *F. subverticillata* in general habit, the stems usually somewhat stouter and panicle more freely branched and drooping.

*Key to species*

1 Blades of leaves flat (or merely once-folded), at least the larger ones 3–8 mm wide; lemmas awnless or rarely with awn less than 1 mm long ........ ***F. subverticillata***

1 Blades of leaves strongly involute, usually much less than 3 mm wide; lemmas awned or awnless ........ 2

2 Lemmas awnless, ca. 2.5 (–3) mm long; leaves hair-like, at most 0.2 mm thick, mostly more than half as high as the stem F. filiformis

2 Lemmas awned, the body ca. 3–6.5 mm long; leaves often stiff, up to 0.8 mm thick, and in some species mostly less than half as high as the stem ........ 3

3 Margins of lemmas thin and membranous; tip of ovary bristly-pubescent; awns mostly more than 3 mm long, nearly equaling or longer than the bodies of their lemmas; mature panicle open and lax ........ ***F. occidentalis***

3 Margins of lemmas at most very narrowly membranous-bordered, the lemmas firm and thick throughout; summit of ovary glabrous; awns all less than 3 mm long, shorter than the bodies of their lemmas; mature panicle rather narrow, crowded, and compact, the branches strongly ascending or, if spreading, very short ........ 4

4 Sheaths closed in young leaves, the old ones ± dark reddish brown basally, becoming fibrous by splitting between the prominent pale veins; basal shoots usually arising laterally, the stems thus tending to be strongly curved or bent at the base; anthers mostly 2–3.5 mm long ........ ***F. rubra***

4 Sheaths open most of their length even in young leaves (margins ± overlapping), the old ones mostly pale or drab brown, not becoming fibrous; basal shoots erect, the stems thus nearly or quite straight from the base upwards; anthers various ........ 5

5 Lower panicle branches often spreading; anthers 2–3 mm long ........ ***F. trachyphylla***

5 Lower panicle branches strongly ascending; anthers less than 2 mm long ........ ***F. saximontana***

***Festuca filiformis*** Pourret FINE-LEAF SHEEP FESCUE Introduced from Europe; lawns and waste places.

***Festuca occidentalis*** Hook. WESTERN FESCUE *Threatened.* Cobble beaches and stabilized dunes along Lake Michigan, dry woods.

***Festuca rubra*** L. RED FESCUE Widely distributed in n Europe and North America; Wisc plants considered adventive.

***Festuca saximontana*** Rydb. ROCKY MOUNTAIN FESCUE Dry forests, shores, dunes, and disturbed places; rock crevices near Lake Superior. Similar to *F. trachyphylla* in general appearance. *Festuca brachyphylla* J.A. Schultes.

***Festuca subverticillata*** (Pers.) Alexeev NODDING FESCUE Moist forests of beech-maple or oak-hickory; occasionally in wet conifer woods. Resembles *F. paradoxa,* but its spikelets are less crowded on the branches.

***Festuca trachyphylla*** (Hack.) Krajina HARD FESCUE; SHEEP FESCUE Native of Europe, widely introduced as a turf grass and sometimes weedy. *Festuca ovina* auct. p.p. non L.

## GLYCERIA *Manna Grass*

Perennial grasses, loosely clumped or spreading by rhizomes. Leaves flat or folded; sheaths closed for most of their length; ligules scarious, erose to lacerate. Head an open panicle. Spikelets 3-flowered, ovate to linear, round in section or somewhat flattened. *Glyceria, Puccinellia,* and *Torreyochloa* are often confused because of similarities in form and their occurrence in wetlands; only *Glyceria* has closed leaf sheaths and 1-veined upper glumes (the other two genera have open leaf sheaths and 3-veined upper glumes). Only *Puccinellia* has inconspicuous veins on the lemmas (the other two genera generally have conspicuous veins on the lemmas).

ADDITIONAL SPECIES ***Glyceria maxima*** (Hartman) Holmb. (Reed sweetgrass), native to Eurasia and an aggressive invader of wetlands, was first discovered in the USA in Racine County in 1975. In Wisc, the species is now known from several additional locations. Spreading by rhizomes, the grass may form large patches and grow to 2.5 m tall. It is similar to large plants of *G. grandis,* but differs in its firmer, more prow-tipped lemmas as well as its larger lemmas and usually larger anthers.

*Key to species*

1 Spikelets linear-cylindric, 10 mm long or longer ............... 2
1 Spikelets ovate, 2–7 mm long ............................... 3

2 Leaves less than 5 mm wide; lemmas more or less smooth ....... .................................................. ***G. borealis***
2 Leaves 5 mm or more wide; lemmas finely hairy . ***G. septentrionalis***

3 Spikelets 3–4 mm wide; veins of lemma not raised .. ***G. canadensis***
3 Spikelets 2–2.5 mm wide; veins of lemma raised ............... 4

4 Spikelets 4–7 mm long ............................ ***G. grandis***
4 Spikelets 2–4 mm long ............................ ***G. striata***

***Glyceria borealis*** (Nash) Batchelder NORTHERN MANNA GRASS June–Aug. Marshes, ponds, stream, ditches, often in shallow water or mud.

***Glyceria canadensis*** (Michx.) Trin. [•130] RATTLESNAKE MANNA GRASS Marshes, swamps, thickets, open bogs, fens. Common.

***Glyceria grandis*** S. Wats. [•130] AMERICAN MANNA GRASS June–Sept. Marshes, ditches, streams, lakes and ponds, open bogs, fens; usually in shallow water or mud.

***Glyceria septentrionalis*** A.S. Hitchc. FLOATING MANNA GRASS June–Aug. Swamps, thickets, shallow water of pond margins, wet depressions in forests.

***Glyceria striata*** (Lam.) A.S. Hitchc. [•130] FOWL MANNA GRASS June–Aug. Swamps, thickets, low areas in forests, wet meadows, springs, streambanks.

## GRAPHEPHORUM *False Oat*

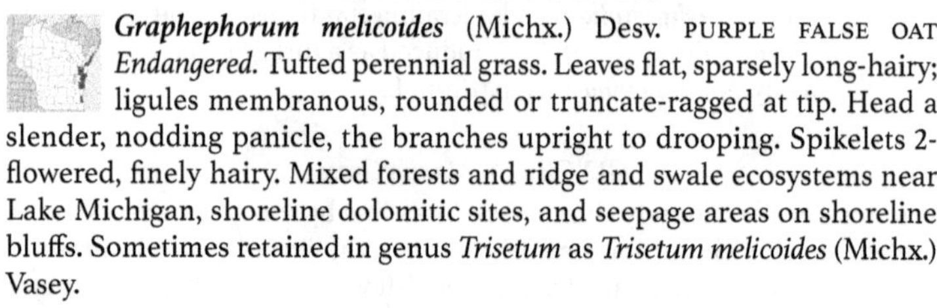

***Graphephorum melicoides*** (Michx.) Desv. PURPLE FALSE OAT *Endangered.* Tufted perennial grass. Leaves flat, sparsely long-hairy; ligules membranous, rounded or truncate-ragged at tip. Head a slender, nodding panicle, the branches upright to drooping. Spikelets 2-flowered, finely hairy. Mixed forests and ridge and swale ecosystems near Lake Michigan, shoreline dolomitic sites, and seepage areas on shoreline bluffs. Sometimes retained in genus *Trisetum* as *Trisetum melicoides* (Michx.) Vasey.

## HESPEROSTIPA *Needlegrass*

Tufted perennial grasses with narrow, elongate, often involute blades and open or contracted panicles of large spikelets; auricles absent; ligules membranous, frequently ciliate. Spikelets 1-flowered.

1 Lemmas usually evenly white-pubescent, sometimes glabrous immediately above the callus; lower ligules often lacerate ***H. comata***

1 Lemmas unevenly pubescent with brown to beige hairs; lower ligules not lacerate .................................... ***H. spartea***

***Hesperostipa comata*** (Trin. & Rupr.) Barkworth NEEDLE-AND-THREAD Native to the western USA where an important grass of prairies and high deserts; considered adventive in Wisc. *Stipa comata* Trin. & Rupr.

***Hesperostipa spartea*** (Trin.) Barkworth PORCUPINE GRASS Sandy, often calcareous places; dune ridges, oak savanna, dry prairies, along railways. *Stipa spartea* Trin.

## HIEROCHLOE *Sweetgrass*

***Hierochloe odorata*** (L.) Beauv. SWEETGRASS Perennial grass, from creeping rhizomes; plants nicely sweet-scented, especially when dried. Ligules membranous; upper surface of leaf blades glabrous and shiny, undersurface pilose. Head a pyramid-shaped panicle, the branches spreading to drooping. Spikelets 3-flowered. May–July. Wet meadows, shores, low prairie; often where sandy. The fragrance emitted when fresh plants are crushed or burned is from coumarin, an anti-coagulant agent. *Hierchloe* sometimes treated as *Anthoxanthum hirtum* (Schrank) Y.Schouten & Veldkamp. *Hierochloe hirta* (Schrank) Borbás.

## HOLCUS *Velvet-Grass*

***Holcus lanatus*** L. COMMON VELVET-GRASS Perennial tufted grass. Ligules membranous, erose-ciliolate; leaf blades pale green, flat, softly pubescent. Panicle narrowly ovoid, dense. Spikelets 2-flowered. Native of Europe; meadows, roadsides, and waste places in moist soil.

## HORDEUM *Barley*

Annual or perennial grasses with flat blades, scarious truncate ligules, and dense bristly spikes which disarticulate at each joint. Spikelets 1-flowered or rarely 2-flowered.

*Key to species*

1 Body of larger lemmas ca. 8–11 mm long; leaves glabrous, with prominent auricles at base of blade; awns of lemmas much stouter than those of glumes; rachis of spike not disintegrating ***H. vulgare***

1 Body of larger lemmas ca. 3.5–6 mm long; leaves (at least lower sheaths) ± pubescent, without auricles; awns of lemmas as slender as those of glumes; rachis of spike readily disintegrating as it matures **2**

2 Awns much longer than 2 cm; glumes all bristle-like (reduced to awns) ........ ***H. jubatum***

2 Awns less than 2 cm long; glumes in part broadened (± lanceolate) ........ ***H. pusillum***

***Hordeum jubatum*** L. FOXTAIL-BARLEY June–Sept. Wet meadows, ditches, shores, shallow marshes, disturbed areas; often where brackish.

***Hordeum pusillum*** Nutt. LITTLE BARLEY Dry or sterile, especially alkaline soil. Native in much of USA; adventive in Dane Co. (*no map*).

***Hordeum vulgare*** L. COMMON BARLEY Introduced. Cultivated grain and an occasional waif along roads and railways.

## KOELERIA *Junegrass*

***Koeleria macrantha*** (Ledeb.) J.A. Schultes PRAIRIE JUNEGRASS Perennial tufted grass. Stems pubescent below the panicle. Leaves mostly basal; ligules membranous, erose. Panicle spike-like, shining, silvery-green. Spikelets normally 2-flowered. Variable, especially in pubescence. Dry soil, prairies, sand hills, open woods. *Koeleria pyramidata* (Lam.) Beauv.

## LEERSIA *Cut-Grass*

Perennial grasses, spreading by long rhizomes. Stems slender, somewhat weak. Leaves flat, smooth to hairy or rough-to-touch; sheaths open; auricles absent; ligules membranous, short. Head an open panicle. Spikelets 1-flowered, laterally compressed.

*Key to species*

1 Spikelets ovate, 3–4 mm wide ........ ***L. lenticularis***

1 Spikelets linear, 1–2 mm wide ........ 2

2 Stems round in section; leaves very rough-to-touch; spikelets 4–6 mm long ........................................ *L. oryzoides*

2 Stems flattened in section; leaves smooth or finely roughened; spikelets to 3.5 mm long ............................ *L. virginica*

***Leersia lenticularis*** Michx. CATCHFLY GRASS River floodplains. Local in sw Wisc, especially along Mississippi and lower Wisconsin Rivers.

***Leersia oryzoides*** (L.) Sw. [•130] RICE CUT-GRASS July–Sept. Muddy or sandy streambanks, shores, swales and marshes; sometimes forming large patches.

***Leersia virginica*** Willd. [•130] WHITE GRASS July–Sept. Swamps, floodplain forests, shaded forest depressions, streambanks.

## LEPTOCHLOA *Sprangletop*

***Leptochloa fusca*** (L.) Kunth SPRANGLETOP Tufted annual grass. Leaves flat to loosely inrolled, finely rough-to-touch; sheaths open, often purplish; ligules membranous, becoming lacerate at maturity. Head a more or less cylindric panicle, the branches upright and bearing spikelets in racemes. Spikelets 6–12-flowered. Native in much of central and western USA; adventive in Wisc. *Diplachne fusca* (L.) P. Beauv. ex Roem. & Schult., *Leptochloa fascicularis* (Lam.) Gray.

## LEYMUS *Lyme Grass*

***Leymus arenarius*** (L.) Hochst. EUROPEAN LYME GRASS Perennial grass from a stout rhizome, strongly glaucous. Auricles usually present; ligules membranous. Spike stout, dense, 1–2 dm long, 1–2 cm thick. Spikelets with 2–5 florets, scabrous to villous or almost tomentose. Native of Europe, locally introduced on sandy shores of Lake Michigan. Sometimes mistaken for *Ammophila breviligulata,* which appears similar and grows in the same shoreline habitat; but spikelets in that species have only a single floret. *Elymus arenarius* L.

## LOLIUM *Rye Grass*

***Lolium perenne*** L. ENGLISH RYEGRASS Perennial tufted grass. Ligules membranous, to 4 mm long; leaf blades flat, glossy. Spike slender, 1–2 dm long; spikelets solitary at each node, placed edgewise to the rachis, the edge fitting into a concavity in the axis. Native of Europe; cultivated in meadows and lawns and often included in commercial seed mixes; and escaped and established on roadsides and in waste places.

## MELICA *Melic Grass*

Perennial, loosely tufted grasses. Leaf blades flat, sheaths closed almost to the top, auricles sometimes present; ligules conspicuous, erose to lacerate. Flowers few in sparsely branched panicles. Spikelets 2–5-flowered. The pubescent pedicels in most of our species are contorted, causing the spikelets to spread away from the axis or to nod.

*Key to species*

1 Southern Wisc; spikelets disarticulating below the glumes; pedicels sharply bent just below the spikelets .................. *M. nitens*

*Dichanthelium oligosanthes*
POACEAE

*Eragrostis cilianensis*
POACEAE

*Danthonia spicata*
POACEAE

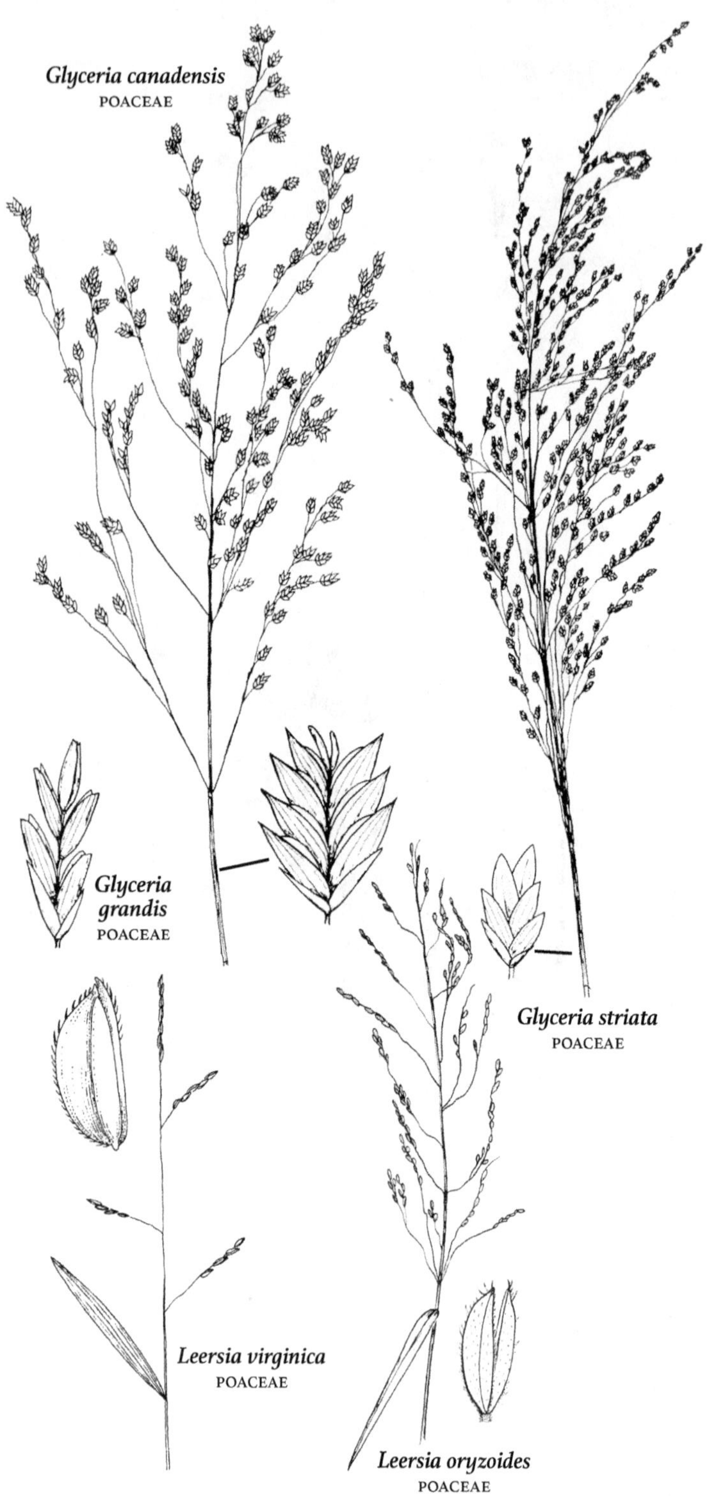
Glyceria canadensis
POACEAE
Glyceria grandis
POACEAE
Glyceria striata
POACEAE
Leersia virginica
POACEAE
Leersia oryzoides
POACEAE

1 Rare in northern Wisc; spikelets disarticulating above the glumes; pedicels more or less straight ........................ *M. smithii*

***Melica nitens*** (Scribn.) Nutt. THREE-FLOWER MELIC GRASS Rocky or dry upland woods, railways.

***Melica smithii*** (Porter) Vasey SMITH'S MELIC GRASS *Endangered.* Moist hardwood forests, wooded dunes.

## MILIUM *Millet Grass*

***Milium effusum*** L. [•131] AMERICAN MILLET GRASS Perennial rhizomatous grass. Ligule membranous, obtuse-erose; leaf blades flat, broad. Panicle ovoid or pyramidal, the branches in fascicles of 2 or 3, widely spreading and bearing drooping spikelets beyond their middle. Spikelets 1-flowered. Rich, moist or dry woods.

## MISCANTHUS *Silver-Grass*

***Miscanthus sacchariflorus*** (Maxim.) Benth. AMUR SILVER-GRASS Plants perennial, tufted, spreading from coarse rhizomes to sometimes form large, dense stands. Stems to 2.5 m tall. Native of wet places in e Asia; escaped from cultivation to roadsides, ditches, and old fields. Plants have a large, plumose panicle with recurving leaves that turn orange in the fall.

## MUHLENBERGIA *Muhly*

Perennial grasses, clumped or with creeping rhizomes. Stems erect or reclining at base, often branching from base. Leaves smooth to hairy, ligules membranous. Head a panicle, usually narrow and spike-like, sometimes open and spreading, at ends of stems and sometimes also from leaf axils. Spikelets 1-flowered.

ADDITIONAL SPECIES ***Muhlenbergia sobolifera*** (Muhl.) Trin. (Rock muhly), stems solitary or few together from scaly rhizomes; leaf blades soft and flat, scaberulous, panicles elongate, very slender; dry woods; local in s and c Wisc, more common southward.

*Key to species*

1 Lemmas not pilose at the base, glabrous (or with minute, even pubescence on back), awnless; stems loosely or densely tufted or matted, rhizomes, if present, thin and wiry, not densely clothed with overlapping scales ........................................ 2

1 Lemmas pilose at base, glabrous or short-pubescent on back, awned or awnless; stems arising from conspicuous, elongate scaly rhizomes (except in *M. schreberi*) .............................. 5

2 Spikelets less than 2 mm long, mostly on pedicels more than twice as long, in an open panicle .................................. 3

2 Spikelets ca. 2.4–3.5 mm long, mostly on pedicels less than twice as long, in a slender contracted panicle .......................... 4

3 Panicles ca. 8–20 cm wide, nearly as wide as long when fully expanded ....................................... *M. asperifolia*

3 Panicles 1–6 cm wide, clearly longer than wide ........ *M. uniflora*

4 Ligules less than 0.5 mm long; lemmas with a little minute pubescence on back ........ *M. cuspidata*

4 Ligules ca. 1.5–2.5 mm long; lemmas usually glabrous across back ........ *M. richardsonis*

5 Glumes minute, the larger one less than 0.5 mm long, the other vestigial or absent; stems often rooting at nodes of decumbent bases, but without elongate scaly rhizomes ........ *M. schreberi*

5 Glumes at least half as long as body of lemma; stems from elongate scaly rhizomes ........ 6

6 Glumes (including prominent awn-tip) 3.5–7.5 mm long, mostly distinctly longer than the body of the lemma; lemma at most short-awned; anthers 0.5–1.3 mm long ........ 7

6 Glumes generally less than 3.6 mm long (rarely, especially on lower spikelets of panicle, up to 4 mm), mostly about equaling or shorter than the body of the lemma; lemma awnless to long-awned; anthers not over 0.5 mm long (except in *M. tenuiflora* with distinctive short, wide glumes)........ 7

7 Internodes minutely puberulent or roughened over much of their surface (rarely nearly glabrous); ligule (excluding cilia) 0.5–0.7 mm long or shorter; anthers 0.8–1.3 mm long ........ *M. glomerata*

7 Internodes of stem smooth and glabrous over most of their surface; ligule 0.7–1 mm long; anthers ca. 0.5–0.8 mm long ... *M. racemosa*

8 Larger glumes 0.6–1 mm wide, less than 4 times as long, hence ovate and usually ± abruptly tapered at the tip; stems puberulent below the nodes........ 9

8 Larger glumes not over 0.6 mm wide, more than 4 times as long, hence narrowly lanceolate and usually ± attenuate at the tip; stems puberulent or glabrous below the nodes ........ 10

9 Anthers not over 0.5 mm long; broader leaf blades 3–5 mm wide; ligules 1–2 mm long; sheaths all glabrous ........ *M. sylvatica*

9 Anthers 1–1.5 mm long; broader leaf blades 6–13 mm wide; ligules 1 (–1.2) mm long or shorter; sheaths (at least some of them) usually ± pubescent ........ *M. tenuiflora*

10 Stem smooth and glabrous throughout, sometimes decumbent at base and rooting at nodes, generally much branched and bushy above; inflorescences (except terminal one) with base often enclosed in upper leaf sheath ........ *M. frondosa*

10 Stem puberulent below the nodes, ± erect, simple or branched; inflorescences all generally exserted ........ 11

11 Ligules ca. 1 mm long or shorter; some spikelets in panicle sessile or subsessile ........ *M. mexicana*

11 Ligules, at least the longest, 1.3–2 mm long; spikelets all on distinct (though sometimes rather short) pedicels ........ *M. sylvatica*

***Muhlenbergia asperifolia*** (Nees & Meyen) Parodi ALKALI MUHLY July–Sept. Mostly weedy in disturbed areas such as roadside ditches and along railroads; native to w USA, adventive in Wisc.

***Muhlenbergia cuspidata*** (Torr.) Rydb. STONY-HILLS MUHLY Prairies and open hillsides, in dry or gravelly soil.

***Muhlenbergia frondosa*** (Poir.) Fern. WIRESTEM MUHLY Aug–Sept. Floodplain forests, streambanks, thickets, shores; also somewhat weedy in disturbed areas such as along railroads.

***Muhlenbergia glomerata*** (Willd.) Trin. MARSH MUHLY Aug–Sept. Swamps, wet meadows, marshes, springs, open bogs, fens, calcareous shores.

***Muhlenbergia mexicana*** (L.) Trin. MEXICAN MUHLY Aug–Sept. Swamps, floodplain forests, thickets, wet meadows, marshes, springs, fens and streambanks.

***Muhlenbergia racemosa*** (Michx.) B.S.P. GREEN MUHLY Moist or wet soil in open places.

***Muhlenbergia richardsonis*** (Trin.) Rydb. MATTED MUHLY Endangered July–Sept. Low prairie, wet meadows, marshes and seeps; often where brackish.

***Muhlenbergia schreberi*** J.F. Gmel. NIMBLEWILL Moist ground, especially where shaded; sometimes weedy in lawns and gardens.

***Muhlenbergia sylvatica*** Torr. WOODLAND MUHLY Aug–Sept. Streambanks, shaded wet areas.

***Muhlenbergia tenuiflora*** (Willd.) B.S.P. SLENDER MUHLY Drier oak and beech-maple woods, forested dunes, riverbanks.

***Muhlenbergia uniflora*** (Muhl.) Fern. BOG MUHLY Wetland margins, exposed sandy shores.

## ORYZOPSIS *Mountain Ricegrass*

***Oryzopsis asperifolia*** Michx. [•132] WHITE-GRAIN MOUNTAIN RICEGRASS Loosely tufted perennial grass. Leaves mostly basal; auricles absent. Raceme slender, the paired branches each with a single spikelet. Spikelets 1-flowered. Moist or dry open woods, forested dunes. New leaves start to develop in mid-summer, the blades at first erect, then bending downward and remaining green through winter. Sheaths below the level of the duff are usually bright purple. Our other species formerly in this genus are now placed in *Piptatherum.*

## PANICUM *Panic-Grass*

Annual or perennial grasses. Heads narrow to open panicles (ours). Spikelets small, with 1 fertile flower.

*Key to species*

1 Spikelets all or mostly 3 mm or more in length, strongly nerved . 2

1 Spikelets less than 3 mm long, strongly nerved or not .......... 4

2 First glume more than half as long as second glume; plants over 5 dm tall, essentially glabrous (except for margin and throat of leaf sheath), from strong scaly rhizome, with panicle terminal and over 15 cm tall ........................................ *P. virgatum*

2 First glume not over half as long as second glume (except in *P. miliaceum*); plants shorter, usually ± pubescent, not rhizomatous, with panicles usually shorter or several ........................ 3

3 Leaves (blades and sheaths) essentially glabrous *P. dichotomiflorum*

3 Leaves pubescent (at least on sheaths) ............... *P. miliaceum*

4 Sheaths sparsely to heavily pilose on back ..................... 5

4 Sheaths of middle and upper leaves glabrous on back (may be ciliate on margins)........................................ 8

5 Panicle 2–3 times as long as wide, with clearly ascending branches ....................................................... *P. flexile*

5 Panicle less than 2 times as long as wide, branches ± spreading at maturity ................................................ 6

6 Peduncles only slightly exserted from sheaths, exserted portion less than half as long as panicle .......................... *P. capillare*

6 Peduncles long-exserted from sheaths, exserted portion half as long as panicle or longer.......................................... 7

7 Spikelets lanceolate in outline, acuminate at tip, the longest 2.4–3.3 mm long ............................................ *P. capillare*

7 Spikelets ovate or narrowly ovate in outline, acute at tip, the longest 1.9–2.3 mm long ............................ *P. philadelphicum*

8 Ligule an erose membrane ca. 0.5–1 mm long; spikelets 1.7–2.2 mm long; pedicel tip of many spikelets with long ascending hairs (these breaking off with age) .............................. *P. rigidulum*

8 Ligule entirely or mostly a fringe of hairs 1.2–2.5 mm long; spikelets 2.3–3.4 mm long; pedicel tip without long ascending hairs, or these sparse and not more than half the length of the spikelets ........ ........................................ *P. dichotomiflorum*

***Panicum capillare*** L. [•131] COMMON PANIC-GRASS Dry or moist soil, often a weed in fields in gardens, widely distributed and variable.

***Panicum dichotomiflorum*** Michx. FALL PANIC-GRASS Moist soil and shores, sometimes in shallow water, often a weed in cultivated land.

***Panicum flexile*** (Gattinger) Scribn. WIRY PANIC-GRASS Aug–Sept. Sandy and gravelly shores, fens, marshes; often where calcium-rich.

***Panicum miliaceum*** L. BROOMCORN MILLET Asian native; occasionally grown for forage or for bird seed; adventive on roadsides and in waste places.

***Panicum philadelphicum*** Bernh. PHILADELPHIA PANIC-GRASS Dry soil and sandy fields.

***Panicum rigidulum*** Bosc ex Nees REDTOP PANICUM July–Aug. Pond margins, streambanks, ditches. Native and common in s and e USA, considered adventive in Wisc.

***Panicum virgatum*** L. [•131] SWITCHGRASS Open woods, prairies, dunes, and shores, and brackish marshes. Variable in length and shape of the glumes, especially the first.

## PASCOPYRUM *Western Wheatgrass*

***Pascopyrum smithii*** (Rydb.) Barkworth & D.R. Dewey [•132] WESTERN WHEATGRASS Perennial grass, usually glaucous. Stems from long rhizomes; auricles 0.2–1 mm long, often purple; ligules membranous, about 0.1 mm long; leaf blades involute when dry. Roadsides, railways; dry or sandy soil. An important grass of grasslands and high deserts in the w USA. *Agropyron smithii* Rydb., *Elytrigia smithii* (Rydb.) Nevski, *Elymus smithii* (Rydb.) Gould.

## PASPALUM *Crown Grass*

***Paspalum setaceum*** Michx. SLENDER CROWN GRASS Perennial tufted grass. Ligules membranous; sheaths and blades hirsute, varying to nearly or quite glabrous. Racemes spike-like, straight or slightly curved. Spikelets in pairs. Dry or moist, especially sandy soil.

## PHALARIS *Canary-Grass*

Annual or perennial grasses. Leaves glabrous, auricles absent; ligule large, membranous. Flowers in dense or spike-like panicles of medium-sized or large spikelets.

*Key to species*

1 Rhizomatous perennial, with elongate lobed panicle (or the lower branches spreading at anthesis); glumes mostly 4–5.7 mm long, the keel not winged ................................ ***P. arundinacea***

1 Annual, with very dense, compact, ovoid panicle; glumes mostly 6–8 mm long, the keel prominently winged ........... ***P. canariensis***

***Phalaris arundinacea*** L. [•132] REED CANARY-GRASS June–July. Wet meadows, shallow marshes, ditches, shores and streambanks. Reed canary-grass is an aggressive, highly competitive wetland species, now widely naturalized, often to the detriment of our native flora. Our populations are likely a mix of native and Eurasian strains, including cultivars developed for forage.

***Phalaris canariensis*** L. COMMON CANARY-GRASS Native of Europe; introduced and adventive; grown for bird seed. Distinguished by the exposed, nearly semi-circular ends of the glumes.

## PHLEUM *Timothy*

***Phleum pratense*** L. COMMON TIMOTHY Tufted perennial grass. Ligules membranous; leaf blades flat, rough-margined. Panicle spike-like and cylindric. Spikelets 1-flowered, strongly flattened. Introduced from Eurasia as a forage grass, commonly cultivated for hay and pasture; escaped to fields, roadsides, and disturbed places.

## PHRAGMITES *Reed*

***Phragmites australis*** (Cav.) Trin. [•131] COMMON REED Aug–Sept. Fresh to brackish marshes, shores, streams, ditches, occasional in tamarack swamps; sometimes in shallow water. *Phragmites communis* Trin. Two subspecies in Wisc, one native (subsp. *americanus,* whose distribution is poorly understood) and one introduced and invasive (subsp. *australis*):

I Plants rarely forming a monoculture; ligules 1–1.7 mm long; lower glumes 3–6.5 mm long; upper glumes 5.5–11 mm long; lemmas 8–13.5 mm long; leaf sheaths deciduous, exposing stems in winter .. .......................................... subsp. *americanus*

I Plants invasive and often forming a monoculture; ligules 0.4–0.9 mm long; lower glumes 2.5–5 mm long; upper glumes 4.5–7.5 mm long; lemmas 7.5–12 mm long; leaf sheaths not deciduous, stems not exposed in winter .............................. subsp. *australis*

## PIPTATHERUM *Ricegrass*

Tufted perennial grasses, sometimes rhizomatous. Leaf blades with flat or involute, auricles absent, ligules membranous to hyaline. Spikelets often large, in contracted or open panicles. Spikelets 1-flowered.

*Key to species*

I Blades flat, mostly 5–18 mm wide; body of lemma 5.5–7 mm long; ligules absent or to 0.5 mm long ................... ***P. racemosum***

I Blades involute, less than 2 mm wide; body of lemma 2.5–4 mm long; ligules of upper leaves 1.5–3 mm long .................... **2**

2 Awn 6–9 mm long, ± twisted; glumes completely smooth ........ ............................................... ***P. canadense***

2 Awn absent or less than 2 (–3) mm long, nearly straight; glumes very minutely scabrous toward tip (20×) ................... ***P. pungens***

***Piptatherum canadense*** (Poir) Dorn CANADIAN MOUNTAIN RICEGRASS Dry, sandy or rocky woods, often with jack pine and white spruce. The persistent, longer awns distinguish *P. canadense* from *P. pungens*. *Oryzopsis canadensis* (Poir.) Torr. ex A. Gray.

***Piptatherum pungens*** (Torr. ex Spreng.) Dorn SHORT-AWN MOUNTAIN RICEGRASS Sandy dry woods, usually with aspen, oak, jack pine, and red pine; dunes and rocky places. The fragile awn is readily broken off and is lacking in most herbarium specimens. *Oryzopsis pungens* (Torr.) Hitchc.

***Piptatherum racemosum*** (Sm.) Barkworth [•132] BLACK-SEED MOUNTAIN RICEGRASS Moist, rich deciduous forests and wooded dunes, sometimes in disturbed places; not often found in dry woods of jack pine or oak. *Oryzopsis racemosa* (Sm.) Ricker ex Hitchc.

## POA *Bluegrass*

Annual or perennial grasses, with or without rhizomes or stolons, densely to loosely tufted or the culms solitary. Leaves mostly near base, flat to folded, midrib 2-grooved, the tip keeled similar to the bow of a boat; sheaths partly closed, auricles absent, ligules membranous. Head an open panicle. Spikelets small, with 2 to several flowers.

ADDITIONAL SPECIES ***Poa arida*** Vasey (Plains bluegrass), a salt- and drought-tolerant perennial grass of the w USA, in Wisc in heavily salted ditches along I-94 in Kenosha and Racine counties; plants flower early like *Poa annua,* but are pale green in color.

*Panicum virgatum*
POACEAE
*Panicum capillare*
POACEAE
*Phragmites australis*
POACEAE
*Milium effusum*
POACEAE

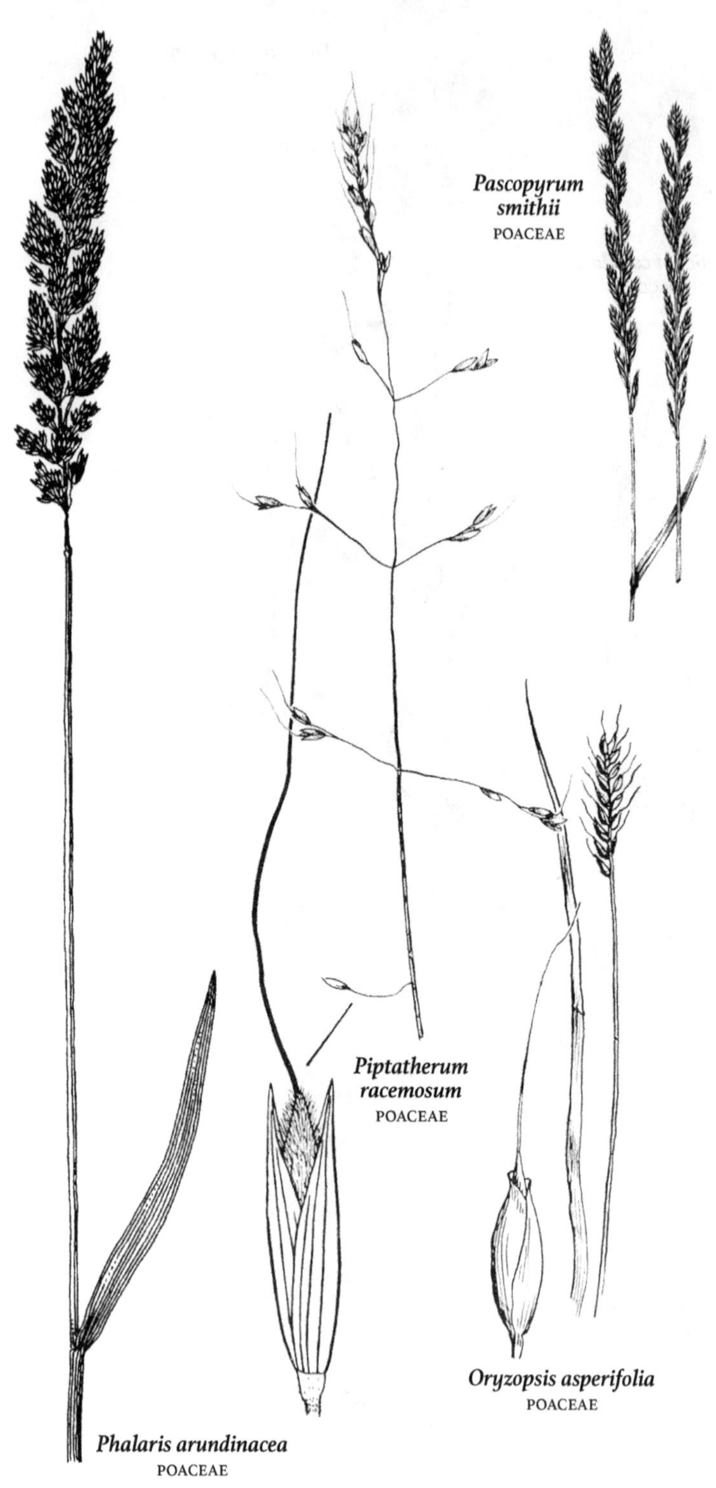
*Pascopyrum smithii*
POACEAE
*Piptatherum racemosum*
POACEAE
*Oryzopsis asperifolia*
POACEAE
*Phalaris arundinacea*
POACEAE

*Poa group key*

Some species appear more than once in the keys; 'flag leaf' refers to the uppermost 'leaf' which is often angled outward from the stem.

1 Stems with bulbous bases; spikelets often bulbiferous .. *P. bulbosa*
1 Stems with non-bulbous bases .............................. 2

2 Plants annual or perennial; anthers 0.1–1 mm long in all florets and well developed, or only the upper 1–2 florets with rudimentary anthers ........................................ SUBKEY 1
2 Plants perennial; some anthers 1.3–4 mm long, or the florets pistillate and all anthers vestigial and to 0.2 mm long, or longer and poorly developed........................................... 3

3 Plants rhizomatous or stoloniferous, rhizomes or stolons usually longer than 5 mm; basal leaves of the erect shoots with well-developed blades; plants densely to loosely tufted or the stems solitary ........................................ SUBKEY 2
3 Plants neither rhizomatous nor stoloniferous; basal leaves of the erect shoots sometimes without blades; plants densely tufted .... ................................................ SUBKEY 3

*Poa subkey 1*

*Plants annual or perennial. Stems not bulbous at base. Basal leaf sheaths not swollen at the base. Spikelets not bulbiferous, florets developing normally. Anthers 0.1–1 mm long.*

1 Plants annual, sometimes surviving for a second season, introduced, weedy species; calluses glabrous; lemmas usually softly puberulent to long-villous on the keel and marginal veins, often also on the lateral veins, glabrous between the veins, non-alpine plants rarely glabrous throughout; palea keels smooth, usually short- to long-villous near the apices, rarely glabrous; panicle branches and glume keels smooth ........................................ *P. annua*
1 Plants perennial, native, sometimes growing in disturbed habitats; calluses webbed or glabrous, if glabrous, the lmma pubescence not as above or the palea keels at least slightly scabrous near the apices; panicle branches and glume keels smooth of scabrous.......... 2

2 Calluses webbed; lemma keels glabrous throughout or, if hairy on the proximal 1/2, the marginal veins glabrous .................. 3
2 Calluses webbed or glabrous, if webbed, the lemmas hairy on the keel and marginal veins ..................................... 4

3 Lemmas hairy only on the keels; branches in whorls of (2) 3–5 (7) ................................................ *P. alsodes*
3 Lemmas usually glabrous, marginal veins rarely sparsely hairy at the base, hairs to 0.15 mm long; branches 1–3 per node *P. saltuensis*

4 Sheaths closed for 1/10–1/5 their length; lower 1–3 leaves along the stems; anthers 0.8–1.2 mm long, sometimes poorly developed ... 5
4 Sheaths closed for 1/5–7/8 their length; stems with or without bladeless leaves; anthers 0.2–1.2 mm long, well developed ...... 6

5 Flag leaf nodes at or above mid-stem length ......... *P. nemoralis*

5 Flag leaf nodes usually in the basal 1/3 of the stem ...... *P. glauca*

6 Panicle branches smooth or sparsely scabrous, usually terete or slightly sulcate; lower glumes subulate to broadly lanceolate; lemmas glabrous between the veins .................. *P. sylvestris*

6 Panicle branches sparsely to densely scabrous, terete or angled; lower glumes subulate or broader; lemmas glabrous or puberulent between the veins ............................................ 7

7 Palea keels puberulent; anthers 0.8–1.2 mm long; lemmas 3–4.7 mm long, lateral veins distinct ............................ *P. wolfii*

7 Palea keels scabrous; anthers 0.2–0.8 mm long; lemmas 2.5–4 mm long, lateral veins faint ........................... *P. paludigena*

*Poa subkey 2*

*Plants with rhizomes or stolons, densely to loosely tufted or the stems solitary.*

1 Stems and nodes strongly compressed; stems usually geniculate; lower stem nodes usually exserted; panicle branches angled, scabrous on the angles; sheaths closed for 1/10–1/5 their length .. ............................................ *P. compressa*

1 Stems terete to somewhat compressed, nodes not or only weakly compressed; stems geniculate or not; lower stem nodes exserted or not; panicle branches angled or terete, smooth or scabrous; sheath closure varied ............................................ 2

2 Lemma keels softly puberulent for 3/5 their length, hairs usually sparse, marginal veins glabrous or puberulent to 1/4 their length, intercostal regions smooth and glabrous; lateral veins prominent; calluses webbed; palea keels smooth, muriculate, tuberculate, or scabridulous; lower glumes 1-veined, usually arched to sickle-shaped; ligules 3–10 mm long, acute to acuminate; panicle branches angled, angles densely scabrous; plants usually weakly stoloniferous ...................................... *P. trivialis*

2 Lemmas glabrous or variously pubescent, if as above, the lateral veins faint or moderately prominent or the calluses glabrous or the palea keels distinctly scabrous or hairy or the lower glumes 3-veined; calluses glabrous or hairy; palea keels scabrous at least near the apices; lower glumes 1–3-veined, not arched, not sickle-shaped; ligules 0.5–18 mm long, truncate to acuminate; panicle branches terete or angled, smooth or scabrous; plants stoloniferous or not 3

3 Southeast Wisc only, along highways where salted; calluses glabrous, diffusely webbed with hairs to 1/2 the lemma length, or with a crown of hairs, or sparsely and dorsally webbed with hairs to 1/4 the lemma length; lemmas glabrous or pubescent .. *P. arida**

3 Statewide in a variety of habitats; calluses dorsally webbed, hairs over 1/2 the length of the lemmas, sometimes with additional webs below the marginal veins; lemma short- to long-villous on the keels and marginal veins ....................................... 4

4 Sheaths closed for 1/10–1/5 their length; spikelets 3–5 mm long; lemmas glabrous between the keels and marginal veins; panicle branches angled, angles densely scabrous; plants sometimes stoloniferous, sometimes branching above the stem bases; florets bisexual ............................................ *P. palustris*

4 Sheaths closed for 1/5–9/10 their length; spikelets 3.5–12 mm long; lemmas glabrous or hairy between the keels and marginal veins; panicle branches terete or angled, smooth or scabrous; plants rarely stoloniferous, usually rhizomatous, never branching above the stem bases; florets bisexual or unisexual .................. *P. pratensis*

*Poa subkey 3*

*Plants perennial, loosely to densely tufted, rhizomes and stolons absent. Stems not bulbous at base. Basal sheaths not swollen. Spikelets not bulbiferous, florets developing normally.*

1 Calluses usually dorsally webbed .................... *P. trivialis*
1 Calluses glabrous or with a crown of hairs .................... 2

2 Lemma lateral veins pronounced, keels pubescent, marginal veins glabrous or softly puberulent at the base, lemmas glabrous elsewhere; lower glumes 1-veined, subulate to narrowly lanceolate, usually arched to sickle-shaped; callus web well-developed ...... ................................................. *P. trivialis*
2 Lemma lateral veins obscure to pronounced, keels glabrous throughout or, if pubescent, the marginal veins distinctly pubescent for more than 1/4 their length, lemma lateral veins and intercostal regions glabrous or pubescent, or, if pubescent as in *P. trivialis*, then the callus web short, scant, poorly developed and the lower glumes 3-veined and lanceolate or broader .......................... 3

3 Panicles open, conical, with whorls of 3–10, spreading to eventually reflexed, scabrous-angled branches at the lower nodes; lemmas hairy on the keel and veins; callus webs well developed P. sylvestris
3 Panicles contracted to open, if open then not conical and without whorls of (2) 3–10, eventually reflexed, scabrous-angled branches at the lower nodes; branches smooth or scabrous-angled; lemmas glabrous or hairy; calluses glabrous, with diffuse hairs, or with a scanty or well-developed web ............................... 4

4 Sheaths closed for 1/3–3/4 their length ............... *P. saltuensis*
4 Sheaths closed for up to 1/4 their length ...................... 5

5 Flag leaf nodes usually in the lower 1/10–1/3 of the stems; flag leaf blades usually distinctly shorter than their sheaths; lemmas sometimes softly puberulent between the veins, lateral veins usually with at least a few minute hairs; ligules 1–4 mm long .... *P. glauca*
5 Flag leaf nodes usually in the upper 2/3 of the stems; flag leaf blades shorter or longer than their sheaths; lemmas glabrous between the veins, lateral veins usually glabrous, rarely with 1 to several minute hairs; ligules 0.2–6 mm long ............................... 6

6 Spikelets lanceolate; glumes subulate to narrowly lanceolate, gradually tapering to narrowly acuminate tips; ligules to 0.5 mm long, truncate; flag leaf nodes at or above the middle of the stems; flag leaf blades usually longer than their sheaths; rachillas usually hairy, hairs to 0.15 mm long; webs usually short, scanty *P. nemoralis*
6 Spikelets and glumes not as above or, if so, the ligules 1.5–6 mm long, truncate to acute, and the rachillas glabrous; flag leaf nodes at or above the lower 1/3 of the stem; flag leaf blades longer or shorter than their sheaths; webs short or long, scanty or not..... 7

7 Panicles 10–30 cm long, branches 4–15 cm long; stems closely spaced to isolated at the base; lower glumes tapering to the apices; lemma keels abruptly inwardly arched beneath the scarious tips; lemma margins distinctly inrolled; rachillas usually muriculate, rarely sparsely hispidulous; web hairs usually longer than 2/3 the length of the lemmas ................................ *P. palustris*

7 Panicles 3–15 cm long, branches 0.4–8 cm long; stems closely spaced at the base; lower glumes abruptly narrowing to the apices, lengths 4.5–6.3 times the widths; lemma keels not abruptly inwardly arched beneath the scarious apices; lemma margins not or slightly inrolled; rachillas usually softly puberulent; web hairs shorter than 1/2 (2/3) the length of the lemmas ............................. *P. interior*

***Poa alsodes*** Gray [•133] GROVE BLUEGRASS May–July. Alder thickets, swamp hummocks, most common in moist deciduous or mixed conifer-deciduous forests.

***Poa annua*** L. ANNUAL BLUEGRASS Native of Eurasia and a widely distributed weedy species of roadsides, lawns, forest trails, clearings, shores, and disturbed places.

***Poa bulbosa*** L. BULBOUS BLUEGRASS Native of Europe; introduced in fields, lawns, and roadsides.

***Poa compressa*** L. CANADA BLUEGRASS Native of Europe; open, usually dry places, especially in acidic soil. Along with *P. pratensis,* a very common grass in Wisc; *P. compressa* differs from *P. pratensis* in its flattened, less tufted stems, lemmas with sparse or even absent web at the base, and a more slender panicle with fewer branches at each node.

***Poa glauca*** Vahl WHITE BLUEGRASS Open, sandy forests; rock crevices and rocky shores, sometimes where underlain by limestone.

***Poa interior*** Rydb. INTERIOR BLUEGRASS Shallow rocky or sandy soil of outcrops and talus slopes. Distinguished from *P. nemoralis* by its longer ligules and wider glumes and lemmas; differs from *P. palustris* in having a densely tufted habit, scantly webbed calluses, and lemmas with wider hyaline margins.

***Poa nemoralis*** L. WOODLAND BLUEGRASS Introduced (naturalized). Dry, sandy or rocky soil, forest borders and clearings, old farmsteads.

***Poa paludigena*** Fern. & Wieg. [•133] BOG BLUEGRASS June–July. Swamps, alder thickets, sedge meadows, open bogs, cold springs; usually in sphagnum moss and often under black ash (*Fraxinus nigra*). Easily mistaken for the common *Poa compressa.*

***Poa palustris*** L. FOWL BLUEGRASS June–Sept. Wet meadows, marshes, shores, streambanks, ditches and low prairie; also moist woods. Native to boreal regions of North America and n Eurasia.

***Poa pratensis*** L. KENTUCKY BLUEGRASS Moist or dry soil, disturbed places, woods, fields, avoiding acidic soils and heavy shade, often cultivated in lawns and meadows. Introduced from Europe and naturalized in much of North America.

***Poa saltuensis*** Fern. & Wieg. OLD-PASTURE BLUEGRASS Dry or rocky deciduous and mixed woods.

***Poa sylvestris*** Gray WOODLAND BLUEGRASS Rich deciduous woods.

***Poa trivialis*** L. ROUGH-STALK BLUEGRASS Native of Europe; meadows, moist woods, roadsides, along shaded trails.

***Poa wolfii*** Scribn. WOLF'S BLUEGRASS Moist woods and streambanks.

## POLYPOGON *Rabbit's-Foot Grass*

***Polypogon monspeliensis*** (L.) Desf. ANNUAL RABBIT'S-FOOT GRASS Annual. Panicles narrowly ellipsoid, dense, sometimes lobed. Native to s Europe and Turkey, now a common weed especially in the w USA, where it grows in damp to wet, often alkaline soils.

ADDITIONAL SPECIES ***Polypogon interruptus*** Kunth, native to w USA, adventive in Sheboygan County; distinguished as follows:

1 Plants annual; ligules 2.5–16 mm long ............ ***P. monspeliensis***
1 Plants perennial; ligules 2–6 mm long .............. ***P. interruptus***

## PUCCINELLIA *Alkali-Grass*

Tufted perennial grasses, usually in brackish habitats. Ligules membranous. Head an open panicle, the branches upright to spreading. Spikelets several-flowered, oval to linear, nearly round in section.

*Key to species*

1 Lower panicle branches horizontal or angled downward when mature; lemmas broad, not tapered to the blunt or rounded tip ........................................................ ***P. distans***
1 Lower panicle branches usually upright; lemmas narrow, tapered to a rounded tip .................................... ***P. nuttalliana***

***Puccinellia distans*** (Jacq.) Parl. EUROPEAN ALKALI-GRASS Introduced (naturalized). May–Aug. Occasional in brackish waste areas and ditches along salted highways.

***Puccinellia nuttalliana*** (J.A. Schultes) A.S. Hitchc. NUTTALL'S ALKALI-GRASS Introduced. June–July. Moist flats, sometimes in shallow water, often where salty. *Puccinellia airoides* S.Watson & J.M.Coult.

## SCHEDONORUS *Tall Fescue*

Previously in *Festuca, Schedonorus* includes the large, broad- and flat-leaved species with awned or at least sharply pointed lemmas. Very closely related to *Lolium,* with which it hybridizes.

*Key to species*

1 Auricles at top of leaf sheath ciliate, having at least 1 or 2 hairs along the margins; panicle branches at the lowest node usually paired, the shorter with 1–13 spikelets, the longer with 3–19 spikelets; lemmas 5.5–7 mm long, usually scabrous at least distally, unawned or with an awn up to 4 mm long ................. ***S. arundinaceus***

1 Auricles glabrous; panicle branches at the lowest node 1 or 2, if paired the shorter with 1–2(3) spikelets, the longer with 2–6(9) spikelets; lemmas 7–8.5 mm long, usually smooth, sometimes slightly scabrous distally, unawned or with a mucro to 0.2 mm long .......................................................... ***S. pratensis***

***Schedonorus arundinaceus*** (Schreb.) Dumort. TALL RYE GRASS Native of Europe; cultivated for forage and as a turfgrass; established in fields and meadows. *Festuca arundinacea* Schreb., *Lolium arundinaceum* (Schreb.) S.J. Darbyshire.

***Schedonorus pratensis*** (Huds.) Beauv. MEADOW RYE GRASS Native of Europe, cultivated for forage and established in fields, meadows, and moist soil. *Festuca elatior* L. p.p., *Festuca pratensis* Huds., *Lolium pratense* (Huds.) S.J. Darbyshire.

## SCHIZACHNE *False Melic Grass*

***Schizachne purpurascens*** (Torr.) Swallen [•133] FALSE MELIC GRASS Loosely tufted perennial grass. Ligule membranous. Panicle with few drooping branches each bearing 1–3 slender spikelets about 2 cm long. Spikelets 3–5-flowered, usually purplish. Drier, sandy or rocky woods and openings; deciduous forests.

## SCHIZACHYRIUM *Little Bluestem*

***Schizachyrium scoparium*** (Michx.) Nash LITTLE BLUESTEM Perennial grass; tufted or with rhizomes, green to purplish, sometimes glaucous. Ligules membranous. Racemes solitary, usually long-exsert, bearing 5–20 pairs of spikelets on a straight or flexuous, white-ciliate rachis. An important prairie species; in Wisc often in drier sandy woods and openings, old fields, sand dunes and shores. *Andropogon scoparius* Michx.

## SECALE *Rye*

***Secale cereale*** L. RYE An important Eurasian cereal grass, also widely used for soil stabilization and, especially in Canada, for whisky. Mostly along roadsides, where planted for erosion control following construction; also on shores, dunes, along railroads, and in old fields; not long-persisting.

## SETARIA *Bristle Grass*

Tufted annual grasses (ours). Ligules membranous and ciliate or of hairs. Spikelets all alike, with 1 perfect flower, turgid or plano-convex, subtended by an involucre of 1 to many slender bristles.

ADDITIONAL SPECIES ***Setaria verticilliformis*** Dumort. (Barbed bristlegrass), European annual adventive grass found at scattered, mostly urban, locations in the USA; in Wisc, reported from Buffalo, Crawford, and Dane counties.

*Key to species*

1 Bristles, summit of stem, and axis of panicle scabrous with retrorse barbs; panicle branches tending to appear whorled, the panicle ± interrupted toward its base ........................ ***S. verticillata***

1 Bristles, summit of stem, and axis of panicle scabrous or pubescent with antrorse barbs or hairs; panicle very compact throughout .. 2

2 Fertile lemmas mostly ca. (2.7–) 3 (–3.4) mm long, rugose with distinctly transverse ridges, the upper half exposed at maturity; bristles 5 or more per spikelet, becoming orange or golden-brown; sheaths glabrous .................................... ***S. pumila***

2 Fertile lemmas less than 3 mm long, evenly and finely rugose or reticulate or smooth (without transverse ridges), the upper half largely or entirely concealed at maturity; bristles fewer than 5 per spikelet, pale greenish or purple (rarely yellow) at maturity; sheaths ciliate with long hairs on the margins ........................ 3

3 Spikelet articulated above the glumes and sterile lemma; fertile lemma distinctly yellow or darker at maturity; panicle very dense, often ± lobed in appearance .......................... ***S. italica***

3 Spikelet articulated below the glumes; fertile lemma pale green or brown; panicle not lobed ................................. 4

4 Panicle strongly nodding, bent below the middle; spikelets mostly over 2.5 mm long, the fertile lemma ± tapering to a distinctly exposed tip; leaf blades ± hairy above .................. ***S. faberi***

4 Panicle straight and erect or rarely slightly nodding; spikelets not over 2.5 mm long, the blunt fertile lemma nearly or quite concealed by the second glume; leaf blades glabrous above ........ ***S. viridis***

***Setaria faberi*** Herrm. JAPANESE BRISTLE GRASS Unintentionally introduced into North America from China in the 1920s, now a serious weed in corn and soybean fields of the midwest USA.

***Setaria italica*** (L.) Beauv. FOXTAIL-MILLET Native of the Old World; sometimes cultivated and escaped to ditches, fields and disturbed places.

***Setaria pumila*** (Poir.) Roem. & Schult. PEARL-MILLET Native of Europe, introduced in lawns, roadsides, railroads, cultivated fields, and disturbed places. *Setaria glauca* (L.) Beauv.

***Setaria verticillata*** (L.) Beauv. ROUGH BRISTLE GRASS European native, weedy in cultivated or waste ground.

***Setaria viridis*** (L.) Beauv. GREEN FOXTAIL-GRASS Native of Eurasia; weedy in gardens, cultivated fields, and disturbed places.

## SORGHASTRUM *Indian Grass*

***Sorghastrum nutans*** (L.) Nash [•134] YELLOW INDIAN GRASS Large perennial grass from short scaly rhizomes. Stems in loose tufts, 1–2 m tall, the nodes densely pubescent; ligules membranous, usually with thick, pointed auricles. Panicle narrow, 10–25 cm long, the ultimate branches of the panicle bearing short racemes of 1–5 spikelets. Moist or dry prairies, open woods, fields, shores, and rarely, in marshes; sometimes spreading in disturbed places as along roadsides and railroads.

## SORGHUM *Sorghum*

***Sorghum halepense*** (L.) Pers. JOHNSON GRASS Tall, stout perennial grass, from creeping rhizomes. Ligules membranous, conspicuously ciliate. Panicles terminal, primary branches compound, terminating in clusters of 1-5 spikelet pairs. Native of the Mediterranean region; escaped from cultivation in fields, roadsides, and waste places, a difficult to eradicate weed, especially in the s USA.

ADDITIONAL SPECIES ***Sorghum bicolor*** (L.) Moench (Sorghum), is the cultivated annual forage crop, rarely adventive in Wisc.

## SPARTINA *Cord-Grass*

***Spartina pectinata*** Bosc [•134] FRESHWATER CORD-GRASS Stout perennial grass, strongly rhizomatous, the rhizomes scaly, purplish-brown or light brown. Stems tough; ligules membranous, ciliate; margins strongly scabrous. Head a spike-like raceme of mostly 10–30, 1-sided spikes, the spikes upright to sometimes appressed. Spikelets 1-flowered, flattened. July–Sept. Shallow marshes, wet meadows.

## SPHENOPHOLIS *Wedgescale*

Perennial grasses. Leaf blades flat, sheaths open, auricles absent; ligules membranous, erose. Panicles slender or spike-like, shining. Spikelets 2-flowered.

*Key to species*

1 Larger (second) glume distinctly swollen or distended, abruptly truncate and usually shallowly 2-lobed at tip .......... ***S. obtusata***
1 Larger glume not swollen or distended, obtuse to acute at tip .... .............................................. ***S. intermedia***

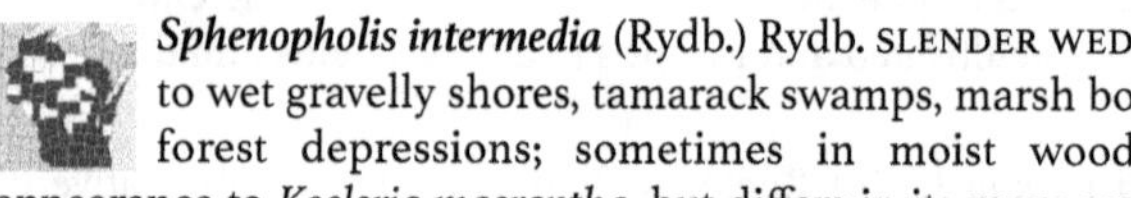

***Sphenopholis intermedia*** (Rydb.) Rydb. SLENDER WEDGESCALE Moist to wet gravelly shores, tamarack swamps, marsh borders, thickets, forest depressions; sometimes in moist woods. Similar in appearance to *Koeleria macrantha,* but differs in its more open panicle, the very narrow first glume, and the essentially glabrous foliage and panicle. *Sphenopholis obtusata* var. *major* (Torr.) K.S. Erdman.

***Sphenopholis obtusata*** (Michx.) Scribn. PRAIRIE WEDGESCALE June–Aug. Dry forests, moist to wet meadows, gravelly shores. The inflorescence is more dense and contracted compared to the more open and lax panicle of *S. intermedia.*

## SPOROBOLUS *Dropseed*

Annual or perennial grasses. Leaf blades narrow, often involute; sheaths open, usually glabrous, often ciliate at the top; ligules of short hairs. Panicles open or contracted. Spikelets 1-flowered.

*Key to species*

1 Plants annuals or short-lived perennials flowering in the first year 2
1 Plants perennial ............................................ 3

2 Lemmas strigose; spikelets 2.3-6 mm long; mature fruits 1.8-2.7 mm long ........................................ ***S. vaginiflorus***
2 Lemmas glabrous; spikelets 1.6-3 mm long; mature fruits 1.2-1.8 mm long ............................................ ***S. neglectus***

3 Plants with rhizomes; rare in sw Wisc ............ ***S. clandestinus***
3 Plants without rhizomes; mostly widespread species ........... 4

4 Spikelets 1-2.5 mm long ........................ *S. cryptandrus*
4 Spikelets 2.5-10 mm long ................................ 5

5 Mature panicles to 30 cm wide, pyramidal; panicle branches appressed or spreading .......................... *S. heterolepis*
5 Mature panicles to 4 cm wide, spikelike; panicle branches appressed ........................................................ 6

6 Lemmas minutely pubescent or scabridulous, chartaceous and opaque; pericarps loose but neither gelatinous nor slipping off the seeds when wet; fruits 2.0-3.5 mm long ........... *S. clandestinus*
6 Lemmas usually glabrous and smooth, membranous to chartaceous and hyaline; pericarps gelatinous, slipping off the seeds when wet; fruits 1-2 mm long ............................... *S. compositus*

***Sporobolus clandestinus*** (Biehler) A.S. Hitchc. ROUGH DROPSEED Dry sandy prairies and barrens.

***Sporobolus compositus*** (Poir.) Merr. TALL DROPSEED Dry or sandy soil of prairies, roadsides and along railways. *Sporobolus asper* (Beauv.) Kunth.

***Sporobolus cryptandrus*** (Torr.) Gray [•134] SAND-DROPSEED Dry, especially sandy soil, cedar glades, barrens, fields and dunes; often in sandy disturbed areas such as roadsides and railways.

***Sporobolus heterolepis*** (Gray) Gray PRAIRIE-DROPSEED Moist to dry prairies, sometimes in fens and in shallow soil on dolomite pavement. Plants have a distinctive musky smell especially noticeable in hot weather.

***Sporobolus neglectus*** Nash SMALL DROPSEED Dry sterile or sandy soil of roadsides and fields; also along shores and on mudflats.

***Sporobolus vaginiflorus*** (Torr.) Wood POVERTY-GRASS Dry sandy or sterile soil as along roadsides (especially where gravelly) and in fields. Very similar to *S. neglectus* and impossible to distinguish without spikelets; *S. vaginiflorus* differs in having strigose lemmas, sheaths that are sparsely hairy towards the base and, usually, longer spikelets. Both differ from our other species in their annual habit, and by having nearly equal glumes.

## TORREYOCHLOA *False Mannagrass*

***Torreyochloa pallida*** (Torr.) Church FALSE MANNAGRASS Perennial rhizomatous grass. Stems slender and flaccid, usually more or less decumbent and creeping at base; ligules membranous. Panicle with relatively few branches. Spikelets narrowly ovate. Cat-tail marshes, bogs, shorelines, wet forest depressions, often in shallow water. *Puccinellia pallida* (Torr.) Clausen.

## TRIDENS *Purpletop*

***Tridens flavus*** (L.) A.S. Hitchc. PURPLETOP Perennial grass, with firm, knotty, shortly rhizomatous bases. Collars densely pubescent; ligules membranous, ciliate. Panicle viscid, with spreading or

drooping branches. Sandy fields, roadsides, and dry open woods.Sandy fields, roadsides, and dry open woods. A common grass south and east of Wisc.

### TRIPLASIS *Sandgrass*

***Triplasis purpurea*** (Walt.) Chapman PURPLE SANDGRASS Tufted annual grass. Ligule of hairs to 1 mm long. Panicle at first included in the upper sheath, eventually exsert with a few spreading branches each bearing several purplish spikelets. Spikelets 2–6-flowered. Sandy shores and openings, dunes, blow-outs, cut-banks.

### TRISETUM *False Oat*

***Trisetum spicatum*** (L.) Richter NARROW FALSE OAT *Threatened.* Tufted perennial grass. Ligules membranous. Panicle from spikelike to open, green, purplish, or tawny, usually silvery-shiny. Spikelets usually 2-flowered. Highly variable. Exposed or partly shaded sandstone ledges and crevices.

### TRITICUM *Wheat*

***Triticum aestivum*** L. COMMON WHEAT Introduced. Commonly cultivated, sometimes appearing on roadsides from spilled grain; probably never persisting in our flora.

### VULPIA *Six-Weeks Fescue*

Our two species tufted annuals, withering by mid-summer.

*Key to species*

1 Lower (shorter) glumes less than half the length of the upper glumes; local in e Wisc . . . . . . . . . . . . . . . . . . . . . . . . . . . . . . . *V. myuros*

1 Lower (shorter) glumes more than half the length of the upper glumes; widespread in Wisc . . . . . . . . . . . . . . . . . . . . . . . . . *V. octoflora*

***Vulpia myuros*** (L.) K.C. Gmel. RAT-TAIL SIX-WEEKS FESCUE Native of Europe; dry sandy fields, roadsides, and waste ground. *Festuca myuros* L., *Vulpia megalura* (Nutt.) Rydb.

***Vulpia octoflora*** (Walt.) Rydb. SIX-WEEKS FESCUE Dry sandy places, often where disturbed. *Festuca octoflora* Walt.

### ZIZANIA *Wild Rice*

Large annual grasses (ours) of marshes and shallow water, with tall stems, wide flat blades, and fleshy yellow roots. Sheaths open, not inflated; ligules membranous or scarious. Spikelets 1-flowered.

*Key to species*

1 Pistillate inflorescence branches usually divaricate at maturity; pistillate lemma thin and membranous and at least sparsely hispid-scabrous between the strong nerves; widest leaves 1.5–4.5 cm wide . . . . . . . . . . . . . . . . . . . . . . . . . . . . . . . . . . . . . . . . . . . . . . . . . . *Z. aquatica*

1 Pistillate inflorescence branches usually appressed at maturity, or with 1 to few, somewhat spreading branches; pistillate lemma firm

Poa paludigena
POACEAE
Poa alsodes
POACEAE
Schizachne purpurascens
POACEAE

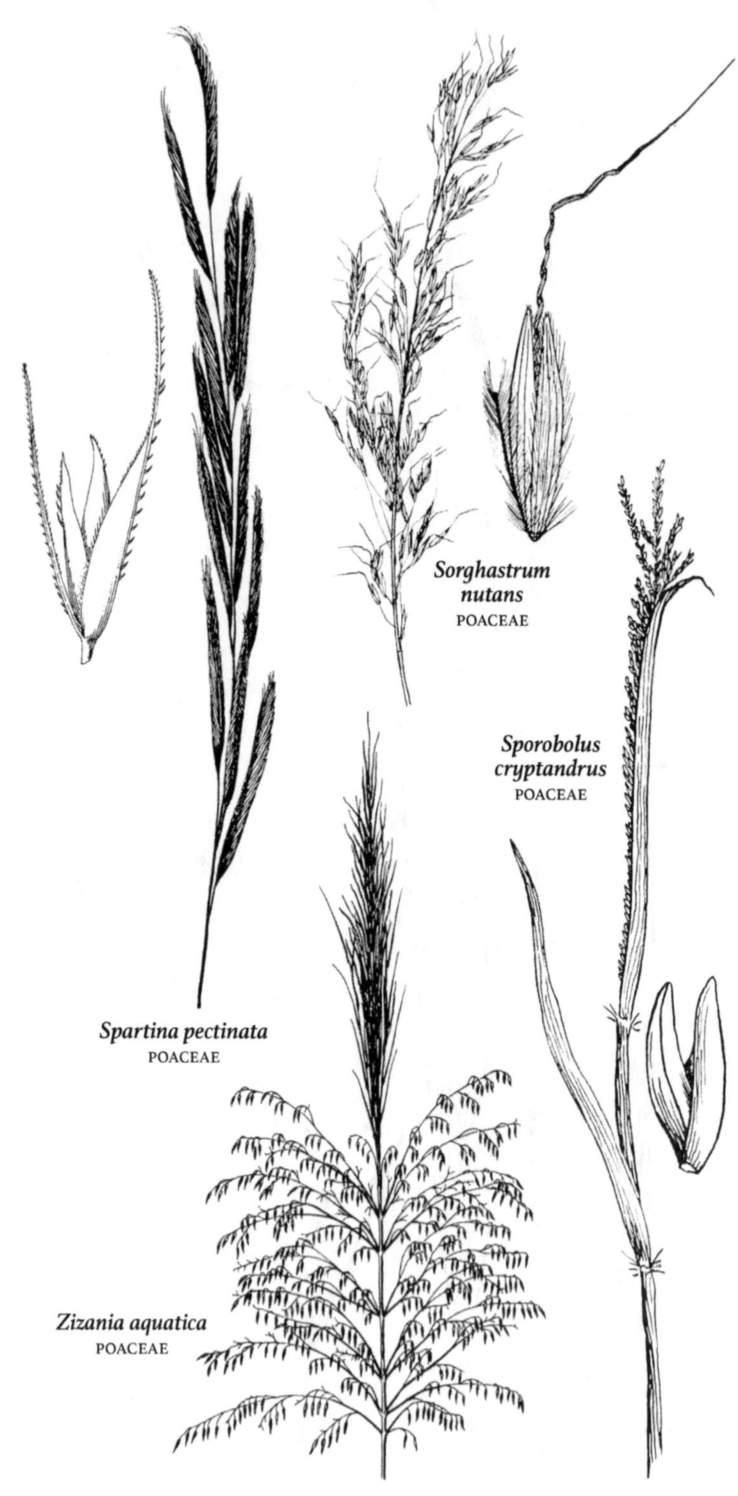
*Sorghastrum nutans*
POACEAE
*Sporobolus cryptandrus*
POACEAE
*Spartina pectinata*
POACEAE
*Zizania aquatica*
POACEAE

and tough, scabrous-hispid only on the nerves and at most at the base and apex; widest leaves 0.5–1.7 cm wide ......... *Z. palustris*

***Zizania aquatica*** L. [•134] WILD RICE July–Sept. Shallow water (up to 1 m deep) or mud of streams, rivers, lakes, ponds; where water is slightly flowing and not stagnant; soils vary from muck to silt, sand, or gravel, with best establishment of plants on a layer of soft silt or muck several cm thick.

***Zizania palustris*** L. NORTHERN WILD RICE Rangewide, *Z. palustris* grows mostly to the north of *Z. aquatica,* but their ranges overlap in the Great Lakes region. Two varieties occur in Wisc:

1 Lower pistillate branches with 9–30 spikelets; pistillate part of the inflorescence 10–40 cm or more wide, the branches ascending to widely divergent; plants 1–3 m tall; blades 10–40 mm wide or more ................................................ var. *interior*

1 Lower pistillate branches with 2–8 spikelets; pistillate part of the inflorescence usually less than 10 cm wide, the branches appressed or ascending, or a few branches somewhat divergent; plants to 2 m tall; blades 3–21 mm wide .......................... var. *palustris*

*Z. palustris* is the source of commercial wild rice (California is the nation's largest producer); in the Great Lakes region, harvesting is most common in Minnesota, especially by Native Americans where large areas of lakes and shallow marshes may be dominated by this plant. The grain is also an excellent food for waterfowl. Many of our populations are intentional introductions.

## PONTEDERIACEAE *Pickerelweed Family*

Mostly perennial, aquatic or emergent herbs. Leaves alternate, stalkless and straplike, or with a petiole and broad blade. Flowers perfect, regular or irregular, single from leaf axils or in spikes or panicles, subtended by leaflike bracts (spathes), light yellow, white or blue-purple.

*Key to genera*

1 Flowers 2-lipped, each lip 3-lobed, the 3 lower lobes spreading; stamens 6, 3 longer than petals, 3 shorter; fruit 1-seeded ......... ............................................ PONTEDERIA

1 Flowers regular, the lobes more or less equal; stamens 3, all longer than petals; fruit a many-seeded capsule ..... HETERANTHERA

### HETERANTHERA *Mud-Plantain*

***Heteranthera dubia*** (Jacq.) MacM. GRASS-LEAF MUD-PLANTAIN Aquatic perennial herb, with lax stems and leaves, or plants sometimes exposed and forming small, leafy rosettes. Leaves alternate, linear, translucent. Flowers 1, opening on water surface, light yellow, enclosed in a spathe from upper leaf axils. Fruit a many-seeded capsule about 1 cm long. July–Sept. Shallow water, muddy shores of ponds, lakes, streams and marshes. Distinguished from the pondweeds (*Potamogeton*) by lack of a leaf midrib. *Zosterella dubia* (Jacq.) Small.

### PONTEDERIA *Pickerelweed*

***Pontederia cordata*** L. [•135] PICKERELWEED Perennial emergent herb, spreading from rhizomes and forming colonies. Stems stout, with 1 leaf. Leaves lance-shaped to ovate, heart-shaped at base; petioles sheathing on stem. Flowers blue-purple (rarely white), many in a spike, subtended by a bractlike spathe. June–Sept. Shallow water (to 1 m deep) of lakes, ponds, rivers and swamps.

---

## POTAMOGETONACEAE *Pondweed Family*

This treatment includes two genera, *Ruppia* and *Zannichellia*, previously included in separate families.

*Key to genera*

1 Submersed leaves opposite or whorled, floating leaves absent .... .......................................... ZANNICHELLIA

1 Submersed leaves alternate, floating leaves (sometimes present) alternate or opposite .................................... 2

2 Flowers 2, at first enclosed in sheathing leaf base, the peduncle elongating and often spiraled or coiled at its base; fruit long-stalked; stipular sheath lacking free ligule at summit (the stipule wholly adnate to the leaf blade and merely rounded at the summit); leaf blade terete .......................................... RUPPIA

2 Flowers several to many in a peduncled head or spike; perianth of 4 tepals; fruit ± sessile; stipular sheath absent (stipules entirely free from leaf) or with a short ligule-like extension if stipules fused to the leaf blade ................................................ 3

3 Stipules adnate to the leaves for 10–30 mm or more (at least on the larger leaves), adnate for ca. 2/3 of the length of the stipule; leaves all submersed, filiform to narrowly linear (up to 2.5 mm wide) ... ............................................... STUCKENIA

3 Stipules free from the leaves or adnate for less than half the length of the stipule (adnate for 5 mm or less except in *P. robbinsii*); leaves submersed or floating, filiform to ovate, oblong, or elliptic ....... .......................................... POTAMOGETON

### POTAMOGETON *Pondweed*

Aquatic perennial herbs, with only underwater leaves, or with both underwater and floating leaves, from rhizomes or tubers, sometimes reproducing and over-wintering by free-floating winter buds. Stems long, wavy, anchored to bottom by roots and rhizomes. Leaves alternate, or becoming opposite upward in some species, simple, with an open or closed sheath at base. Underwater leaves usually linear and threadlike, sometimes broader, margins often wavy, usually stalkless. Floating leaves, if present, oval or ovate, with a waxy upper surface. Flowers perfect, regular, green to red, in stalked spikes at ends of stems or from leaf axils, usually raised above water surface, the spikes with few to many small flowers. The narrow-leaved pondweeds (leads 8–16 in Group 2 key) are often difficult to positively identify in the field, the distinguishing features being somewhat hard to see.

*Key to species*

I Plants with underwater leaves only, these all alike ...... GROUP 1
I Plants with 2 kinds of leaves: broad floating leaves and broad or narrow underwater leaves ........................... GROUP 2

**_Potamogeton group 1_**

*Plants with underwater leaves only, these all alike.*

I Leaves broad, lance-shaped to oval or ovate, never linear ....... 2
I Leaves linear.................................................... 7

2 Leaf margins wavy-crisped, finely toothed .............. *P. crispus*
2 Leaf margins flat or sometimes wavy, entire (or rarely finely toothed at tip) ................................................ 3

3 Base of leaf blade tapered, not clasping stem .................. 4
3 Base of leaf blade clasping stem ............................... 5

4 Plants green; upper leaves stalked; leaf margins finely toothed near tip .......................................... *P. illinoensis*
4 Plants red-tinged; upper leaves more or less stalkless; leaf margins entire ................................................ *P. alpinus*

5 Stems whitish; leaves 10–30 cm long; fruit 4–5 mm long .......... ............................................... *P. praelongus*
5 Stems green; leaves 1–12 cm long; fruit 2–4 mm long ........... 6

6 Leaves ovate, mostly 1–5 cm long, margins flat; stipules small or absent; plants drying olive-green .................. *P. perfoliatus*
6 Leaves lance-shaped, mostly more than 5 cm long; margins wavy-crisped; stipules conspicuous, persisting as shreds; plants drying light green .................................... *P. richardsonii*

7 Stipules joined with lower part of leaf to form a sheath at least 1 cm long ............................................... *P. robbinsii*
7 Stipules free from leaf, or rarely joined to leaf base for only 1–2 mm ..................................................... 8

8 Plants with slender creeping rhizomes ......................... 9
8 Plants with short rhizomes or rhizomes absent (plants often rooting at lower nodes of stem) ................................... 10

9 Flower clusters on stalks at ends of stems, the stalks mostly 5–25 cm long; leaves threadlike, narrower than stems ....... *P. confervoides*
9 Flower clusters on stalks from leaf axils, the stalks less than 3 cm long; leaves linear, wider than stems .................. *P. foliosus*

10 Leavess 9- to many-veined (with 1–2 main veins and many finer ones) ........................................ *P. zosteriformis*
10 Leaves 1–7-veined ............................................ 11

11 Leaves without glands at base ........................ *P. foliosus*
11 At least some of leaves with pair of glands at base ............ 12

12 Leaves with 5–7 nerves ................................. *P. friesii*
12 Leaves with 3 (rarely 1 or 5) nerves........................... 13

13 Leaves gradually tapered to a bristlelike tip .................. 14
13 Leaves rounded at tip or tapered to a point, not bristle-tipped... 15

14 Leaf margins rolled under; widespread ..... *P. strictifolius*
14 Leaf margins flat, not rolled under ..... *P. hillii*

15 Leaves 1–4 mm wide, rounded at tip; body of achene 2.5–4 mm long ..... *P. obtusifolius*
15 Leaves to 2.5 mm wide, usually tapered to a sharp tip; body of achene to 2 mm long ..... *P. pusillus*

### *Potamogeton group 2*

*Plants with 2 kinds of leaves: broad floating leaves and broad or narrow underwater leaves.*

1 Underwater leaves broad, never linear ..... 2
1 Underwater leaves linear ..... 7

2 Floating leaves with 30–55 nerves; underwater leaves with 30–40 nerves ..... *P. amplifolius*
2 Floating leaves with fewer than 30 nerves; underwater leaves with less than 30 nerves..... 3

3 Underwater leaves with more than 7 nerves, all leaves stalked ..... 4
3 Underwater leaves mostly with 7 nerves, at least the lower leaves stalkless ..... 5

4 Base of floating leaves more or less heart-shaped ..... *P. pulcher*
4 Base of floating leaves tapered or rounded, not heart-shaped ..... *P. nodosus*

5 Margins of underwater leaves finely toothed near tip *P. illinoensis*
5 Margins of underwater leaves entire ..... 6

6 Plants red-tinged; underwater leaves 5–20 cm long and at least as wide as floating leaves, mostly on main stem..... *P. alpinus*
6 Plants green; underwater leaves 3–8 cm long and narrower than floating leaves, often numerous on short branches from leaf axils ..... *P. gramineus*

7 Spikes of 1 kind only; fruit not (or only slightly) compressed; stipules not joined with leaf base ..... 8
7 Spikes of 2 kinds: those in axils of lower underwater leaves round, on short stalks; those in axils of upper or floating leaves cylindric, often emersed on long stalks; fruit flattened; stipules of leaves (or at least some of lower leaves) joined with leaf base ..... 12

8 Floating leaves less than 1 cm wide and less than 2 cm long *P. vaseyi*
8 Floating leaves more than 1 cm wide and more than 2 cm long ..... 9

9 Underwater leaves flat and tapelike, 2–10 mm wide ..... *P. epihydrus*
9 Underwater leaves round in cross-section, often reduced to a petiole, mostly less than 1.5 mm wide ..... 10

10 Blade of floating leaves oval, tapered to base; fruit 3-keeled ..... *P. nodosus*
10 Blade of floating leaves ovate to nearly heart-shaped at base; fruit barely keeled ..... 11

11 Floating leaves mostly 3–10 cm long; spikes 3–6 cm long . *P. natans*
11 Floating leaves 2–5 cm long; spikes 1–3 cm long ..... *P. oakesianus*

12 Underwater leaves hair-like, to only about 0.3 mm wide, acute to long-tapering at tip; tips of floating leaves acute.. . . . ***P. bicupulatus***

12 Underwater leaves hair-like but slightly wider (more than about 0.5 mm), leaf tips obtuse to acute; floating leaf tips rounded . . . . . . . 13

13 Underwater leaves blunt-tipped; floating leaves with a small notch at tip . . . . . . . . . . . . . . . . . . . . . . . . . . . . . . . . . . . . . . . . . . . . . . ***P. spirillus***

13 Underwater leaves tapered to a pointed tip; floating leaves not notched at tip . . . . . . . . . . . . . . . . . . . . . . . . . . . . . . . . . . . . ***P. diversifolius***

***Potamogeton alpinus*** Balbis REDDISH PONDWEED July–Sept. Shallow to deep (usually cold) water of lakes and streams.

***Potamogeton amplifolius*** Tuckerman [•135] LARGE-LEAF PONDWEED July–Aug. Shallow to deep water of lakes and rivers.

***Potamogeton bicupulatus*** Fern. SNAIL-SEED PONDWEED Early summer–fall. Soft water lakes (water low in dissolved minerals). P. bicupulatus is similar to *P. diversifolius.* Both species have extremely fine, hair-like underwater leaves, but in *P. diversifolius,* these leaves very slightly wider and the floating leaves somewhat rounded at tip. *Potamogeton diversifolius* var. *trichophyllus* Morong

***Potamogeton confervoides*** Reichenb. TUCKERMAN'S PONDWEED *Threatened.* June–Aug. Shallow water of lakes, kettle hole ponds and peatlands. Unique among our pondweeds in its much-branched stems with linear leaves and the flower spike atop an elongate, leafless stalk.

***Potamogeton crispus*** L. [•135] CURLY PONDWEED Introduced (invasive). April–June. Shallow to deep water of lakes (including Great Lakes) and rivers; pollution-tolerant.

***Potamogeton diversifolius*** Raf. [•135] WATERTHREAD June–Sept. Shallow water of ponds. Similar to *P. spirillus,* which see.

***Potamogeton epihydrus*** Raf. [•136] RIBBON-LEAF PONDWEED July–Sept. Water to 2 m deep in lakes, ponds and rivers.

***Potamogeton foliosus*** Raf. LEAFY PONDWEED June–Aug. Shallow to deep water of lakes, ponds, rivers and streams.

***Potamogeton friesii*** Rupr. FLAT-STALK PONDWEED June–Aug. Shallow to deep water of lakes, ponds, rivers and streams.

***Potamogeton gramineus*** L. GRASSY PONDWEED June–Aug. Shallow to deep water of lakes and ponds.

***Potamogeton hillii*** Morong HILL'S PONDWEED Shallow water of ponds and streams, often where calcium-rich. Douglas and Florence counties.

***Potamogeton illinoensis*** Morong ILLINOIS PONDWEED July–Sept. Shallow to deep water of lakes and rivers.

***Potamogeton natans*** L. FLOATING PONDWEED June–Aug. Usually shallow water (to 2 m deep) of ponds, lakes, rivers and peatlands.

***Potamogeton nodosus*** Poir. LONG-LEAF PONDWEED July–Aug. Shallow water to 2 m deep, mostly in rivers; lakes.

***Potamogeton oakesianus*** J.W. Robbins OAKES' PONDWEED Ponds and streams, peatland pools. Similar to floating pondweed (*P. natans*) but plants smaller and the fruit more or less smooth on sides (vs. depressed in *P. natans*).

***Potamogeton obtusifolius*** Mert. & Koch BLUNT-LEAF PONDWEED Lakes, ponds, streams, peatland pools.

***Potamogeton perfoliatus*** L. REDHEAD-GRASS Lakes and streams. Brown County.

***Potamogeton praelongus*** Wulfen WHITE-STEM PONDWEED June–Aug. Shallow to deep water of lakes (including Great Lakes), streams.

***Potamogeton pulcher*** Tuckerman [•136] SPOTTED PONDWEED *Endangered.* Muddy shores and shallow water of lakes.

***Potamogeton pusillus*** L. [•136] SLENDER PONDWEED June–Aug. Shallow water (to 2 m deep) of lakes and ponds, less often in streams.

***Potamogeton richardsonii*** (Benn.) Rydb. RED-HEAD PONDWEED July–Aug. Shallow to deep water of lakes (including Great Lakes), streams. Similar to *P. perfoliatus* but that species uncommon in Wisc, its leaves narrower and often longer, and the stipules persisting as fibers, vs. soon decayed in *P. perfoliatus.*

***Potamogeton robbinsii*** Oakes FERN PONDWEED July–Aug. Shallow to deep water of lakes, ponds and streams.

***Potamogeton spirillus*** Tuckerman SPIRAL PONDWEED Shallow water of lakes and ponds. Similar to *P. diversifolius,* but the underwater leaves typically blunt-tipped, and the floating leaves with a small notch at tip. In *P. diversifolius,* the underwater leaves are generally tapered to a pointed tip, and floating leaves are not notched at tip.

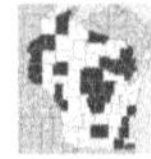

***Potamogeton strictifolius*** Benn. STRAIGHT-LEAF PONDWEED June–Aug. Shallow to deep water of lakes and rivers.

***Potamogeton vaseyi*** J.W. Robbins VASEY'S PONDWEED Shallow to deep water of ponds.

***Potamogeton zosteriformis*** Fern. FLAT-STEM PONDWEED July–Aug. Shallow to deep water of lakes (including Great Lakes) and streams.

### RUPPIA *Ditch-Grass*

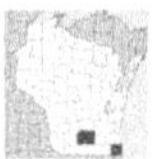

***Ruppia cirrhosa*** (Petag.) Grande SPIRAL DITCH-GRASS Aquatic perennial herb. Stems white-tinged, the internodes often zigzagged.

Leaves simple, alternate or opposite, threadlike, with a sheathing stipule at base. Flowers very small, perfect, in small, 2-flowered spikes from leaf axils, the spikes enclosed by the leaf sheath at flowering time, the flower stalks elongating and usually coiling as fruit mature. July–Aug. Lakes and ponds, often where brackish.

## STUCKENIA *False Pondweed*

Small genus of perennial aquatic herbs, now segregated from *Potamogeton.* In *Stuckenia,* the stipules are joined to the blade for 2/3 to nearly the entire length of the stipule; in *Potamogeton,* the stipules in most species are free, or if adnate, joined for well less than half the length of the stipule. Also, submersed leaves of *Potamogeton* are translucent, flat, and without grooves or channels; in *Stuckenia,* submersed leaves are opaque, channeled, and turgid.

ADDITIONAL SPECIES ***Stuckenia vaginata*** (Turcz.) Holub, historical record from Dane Co., now likely no longer present in Wisc.

*Key to species*

1 Leaves gradually tapered to tip; rhizomes tuber-bearing; stigmas raised on a tiny style ................................ ***S. pectinata***

1 Leaves rounded, blunt-tipped or tipped with a short, sharp point, stigmas inconspicuous, broad and not raised .................. 2

2 Plants short, to 0.5 m long; sheaths tight around stem; spikes with 2–5 whorls of flowers ................................ ***S. filiformis***

2 Plants large and coarse, 2–5 m long; sheaths enlarged to 2–5 times diameter of stem; spikes with 5–12 whorls of flowers; historical record only, Dane County ......................... ***S. vaginata****

***Stuckenia filiformis*** (Pers) Boerner THREADLEAF FALSE PONDWEED July–Aug. Mostly shallow water (to 1 m) in lakes (including Great Lakes) and rivers. *Potamogeton filiformis* Pers.

***Stuckenia pectinata*** (L.) Boerner [•136] SAGO FALSE PONDWEED June–Sept. Shallow to deep water of lakes, ponds and streams; tolerant of brackish water. The large fruit an important waterfowl food. *Potamogeton pectinatus* L.

## ZANNICHELLIA *Horned-Pondweed*

***Zannichellia palustris*** L. HORNED-PONDWEED Perennial aquatic herb, with creeping rhizomes, often forming extensive underwater mats. Stems slender and delicate, wavy. Leaves simple, opposite (or upper leaves appearing whorled), threadlike. Flowers small, produced underwater, either staminate or pistillate, separate on plant but from same leaf axil, with 1 staminate flower and usually 4 (varying from 1–5) pistillate flowers at each node. Fruit a brown to red-brown, crescent-shaped nutlet, gently wavy on margins, tipped by a beak 1–2 mm long. June–Aug. Submerged in fresh or brackish water of streams, reservoirs, muddy lake and pond bottoms, marshes and ditches.

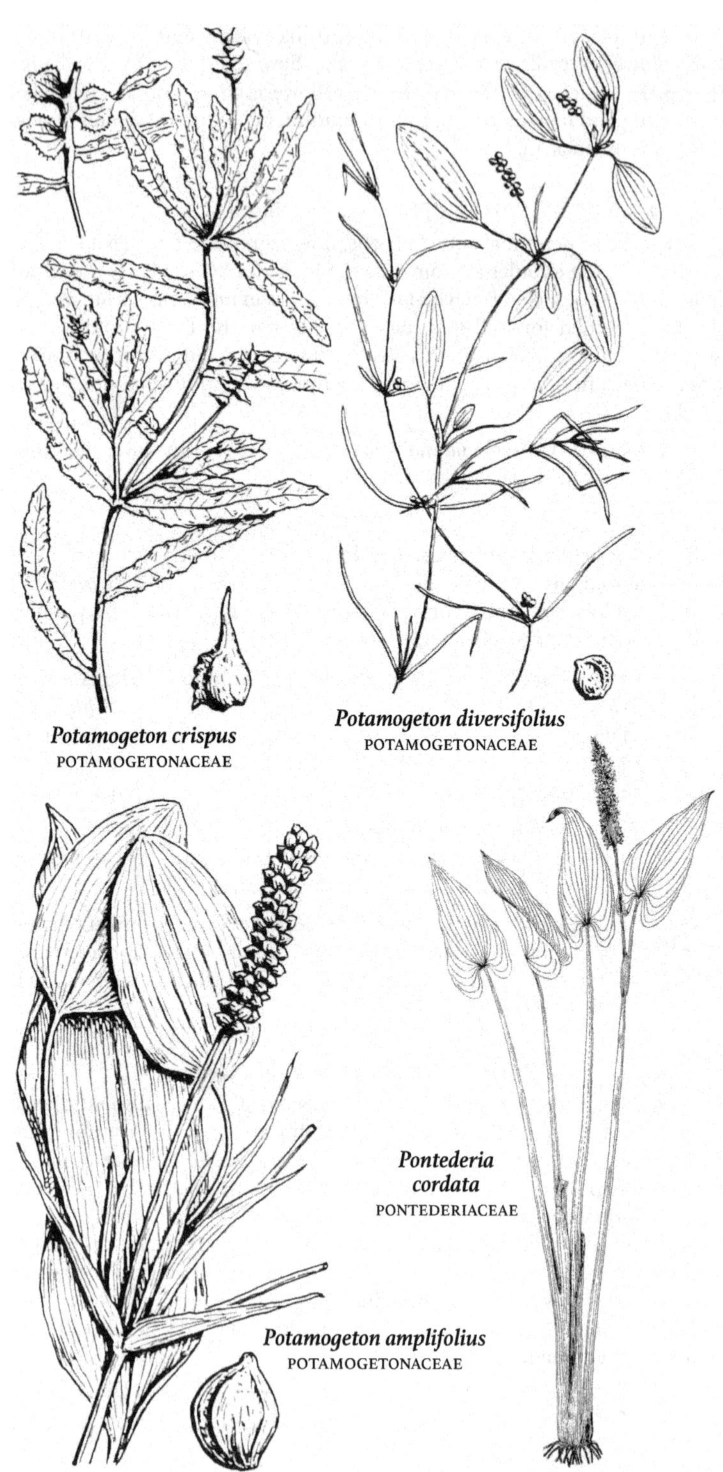

*Potamogeton crispus*
POTAMOGETONACEAE

*Potamogeton diversifolius*
POTAMOGETONACEAE

*Potamogeton amplifolius*
POTAMOGETONACEAE

*Pontederia cordata*
PONTEDERIACEAE

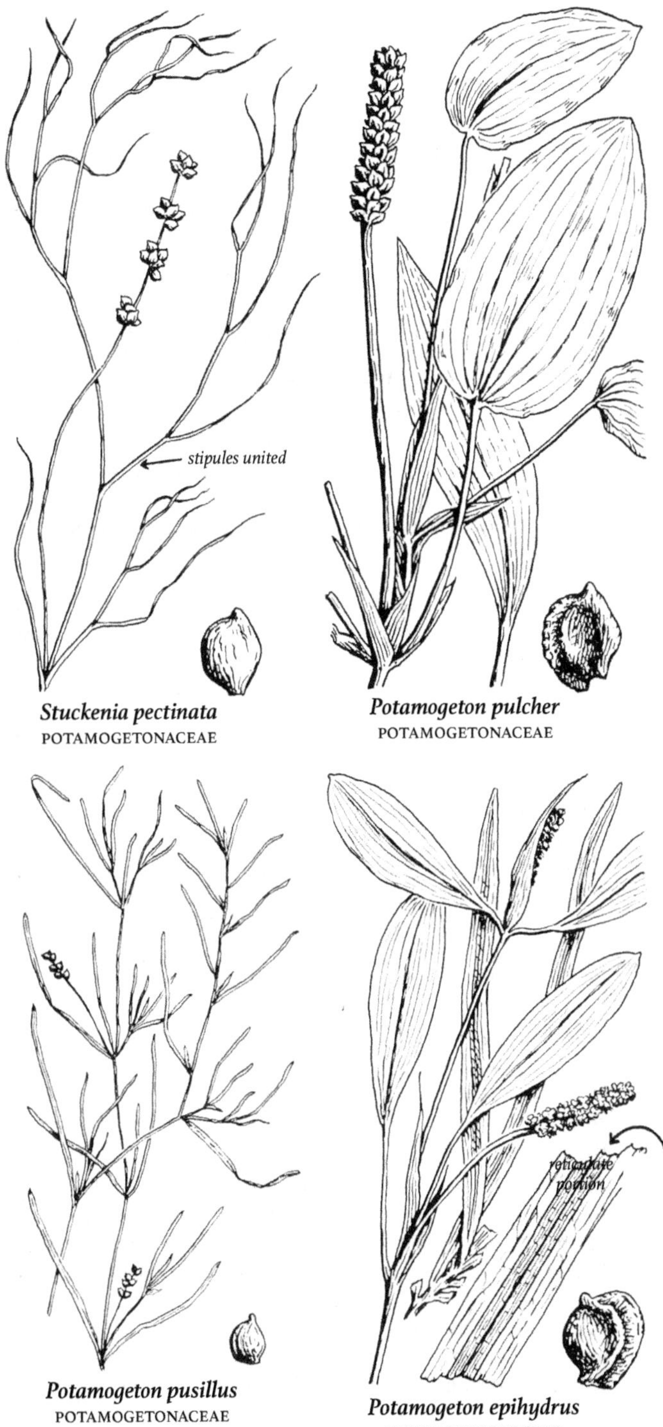


***Stuckenia pectinata***
POTAMOGETONACEAE

***Potamogeton pulcher***
POTAMOGETONACEAE

***Potamogeton pusillus***
POTAMOGETONACEAE

***Potamogeton epihydrus***
POTAMOGETONACEAE

## SCHEUCHZERIACEAE *Scheuchzeria Family*

### SCHEUCHZERIA *Pod-Grass*

***Scheuchzeria palustris*** L. [•137] POD-GRASS Perennial rushlike herb, from creeping rhizomes, remains of old leaves often persistent at base of plant. Leaves alternate, several from base and 1–3 along stem. Flowers perfect, regular, green-white, in a several-flowered raceme. May–June. Wet sphagnum peatlands.

## SMILACACEAE *Greenbrier Family*

### SMILAX *Greenbrier*

Perennial herbs (with annual stems), or vining shrubs, climbing by tendrils terminating the stipules, with wide, longitudinally nerved, net-veined, alternate leaves and axillary peduncled umbels of small yellow or greenish yellow flowers. Flowers dioecious, the staminate often the larger. Fruit a 1–6-seeded berry. Leaves of all species vary greatly in size and shape.

*Key to species*

1 Stems woody and prickly; leaves glabrous beneath (sometimes roughened on main veins) ........................ ***S. tamnoides***

1 Stems herbaceous, never prickly; leaves finely puberulent, at least on the veins beneath .......................................... 2

2 Stem of mature plants more than 1 m long, the main stem or elongate branches climbing (or resting on other objects for support); plant almost always branched, with total of more than 25 leaves; tendrils conspicuously curled, present at most nodes, including those from which peduncles arise; peduncles longer than petioles (sometimes several times as long), all or most arising from axils of foliage leaves; flowers (at least on main stem) more than 25 in an umbel (but not all develop into fruit) ................ ***S. lasioneura***

2 Stem less than 1 m tall, stiffly erect much of its length; plant unbranched, with fewer than 25 leaves (in *S. illinoensis* rarely more); tendrils absent or at most poorly developed and limited to uppermost nodes (never at the lower nodes from which peduncles arise); peduncles longer or, more often, shorter than petioles, at least the lowest ones usually arising from scale-like bracts on the stem below the foliage leaves; flowers more or fewer than 25 in an umbel ...................................................... 3

3 Pistillate (and usually also staminate) flowers fewer than 25 in an umbel; leaves fewer than 20 (usually 7–9) on a plant; stems under 50 cm tall; peduncles usually shorter than the petioles or slightly longer; tendrils completely absent (rarely on upper 2–3 nodes) ... ...................................................... ***S. ecirrata***

3 Pistillate and staminate flowers usually more than 25 in an umbel and plants with one or more other exceptions to the above (i.e., leaves more than 20, stems over 50 cm tall, peduncles more than 2 cm longer than petioles, tendrils present on several upper nodes) ...................................................... ***S. illinoensis***

***Smilax ecirrata*** (Engelm.) S. Wats. UPRIGHT CARRION-FLOWER May–June. Rich deciduous woods, floodplain forests.

***Smilax illinoensis*** Mangaly ILLINOIS GREENBRIER May–June. Woods, thickets.

***Smilax lasioneura*** Hook. [•137] BLUE RIDGE CARRION-FLOWER May–June. Moist soil of open woods, roadsides, and thickets. Variable. *Smilax herbacea* var. *lasioneura* (Hook.) A. DC.

***Smilax tamnoides*** L. CHINAROOT May–June. Moist woods and thickets. *Smilax hispida* Muhl. ex Torr.

## TYPHACEAE *Cat-Tail Family*

Family now includes genus *Sparganium* from former family Sparganiaceae (discontinued under APG III).

*Key to genera*

1 Pistillate flowers in one to several spherical heads; perianth of greenish sepals; leaves strongly keeled (3-angled in cross-section) .......... **SPARGANIUM**

1 Pistillate flowers in an elongate densely flowered spike; perianth of white hairs; leaves flat-elliptic in cross-section .......... **TYPHA**

### SPARGANIUM *Bur-reed*

Perennial sedgelike herbs, floating or emergent in shallow water, from rhizomes and forming colonies. Stems stout, usually erect, unbranched, round in section. Leaves long, broadly linear, sheathing stem at base. Flowers crowded in round heads, the heads with either staminate or pistillate flowers. Fruit a beaked, nutlet-like achene.

*Key to species*

1 Plants large, about 1 m tall; leaves usually erect; stigmas 2; achenes broadly oblong pyramid-shaped.......... ***S. eurycarpum***

1 Plants smaller, leaves erect or floating; stigmas 1; achenes slender . 2

2 Fruiting heads about 1 cm wide; staminate head 1 (often absent by fruiting time); achene beaks less than 1 mm long .......... ***S. natans***

2 Fruiting heads 1.5 cm or more wide; staminate heads 2 or more; achene beaks 2 mm or more long .......... 3

3 Fruiting heads 1.5–2 cm wide; anthers and stigma less than 1 mm long; leaves mostly flat .......... 4

3 Fruiting heads larger mostly 2–3 cm wide; anthers and stigma 1–4 mm long; leaves often keeled .......... 5

4 Staminate heads several, separate from the pistillate heads; achene not shiny .......... ***S. fluctuans***

4 Staminate heads usually 1 (sometimes 2) and near upper pistillate head; achene shiny .......... ***S. glomeratum***

5 Fruiting heads or branches all from leaf axils .......... 6

5 At least some fruiting heads or branches borne above leaf axils . 7

6 Inflorescence unbranched or the branches short with 1–2 staminate heads; achenes dull, with a beak 3–4 mm long ..... ***S. americanum***

6 Inflorescence branched, the branches jointed, with 3 or more staminate heads; achenes shiny, with a beak 5–7 mm long ........................................................ ***S. androcladum***

7 Leaves floating; achene beak 1–3 mm long ....... ***S. angustifolium***

7 Leaves usually stiffly erect and emersed; achene beak 3–5 mm long ........................................................ ***S. emersum***

***Sparganium americanum*** Nutt. AMERICAN BUR-REED Marshes, shallow water, streambanks.

***Sparganium androcladum*** (Engelm.) Morong [•137] BRANCHED BUR-REED Marshes, lakeshores, fens. *Sparganium lucidum* Fern. & Eames.

***Sparganium angustifolium*** Michx. NARROW-LEAF BUR-REED Lakes, ponds and shores. Sparganium acaule (Beeby) Rydb.; *Sparganium chlorocarpum* var. *acaule* (Beeby) Fern.; *Sparganium emersum* var. *angustifolium* (Michx.) Taylor & MacBryde; *Sparganium multipedunculatum* (Morong) Rydb.

***Sparganium emersum*** Rehmann NARROW-LEAF BUR-REED Shallow water or mud of marshes, streams, ditches, open bogs, ponds. *Sparganium chlorocarpum* Rydb.; *Sparganium simplex* Huds.

***Sparganium eurycarpum*** Engelm. [•137] BROAD-FRUIT BUR-REED Usually in shallow water of marshes, streams, ditches, ponds and lakes, often with cat-tails (*Typha*); our most common bur-reed.

***Sparganium fluctuans*** (Morong) B.L. Robins. FLOATING BUR-REED Shallow water of ponds and lakes.

***Sparganium glomeratum*** (Laestad.) L. Neum. CLUSTERED BUR-REED *Threatened.* Shallow water of marshes and bogs.

***Sparganium natans*** L. ARCTIC BUR-REED Shallow water, pond margins. *Sparganium minimum* (Hartman) Wallr.

## TYPHA *Cat-tail*

Large reedlike perennials, from fleshy rhizomes and forming colonies. Stems erect, unbranched, round in section, sheathed for most of length by overlapping leaf sheaths. Leaves mostly near base of plant, alternate in 2 ranks, erect, linear, spongy. Flowers tiny, either staminate or pistillate, separate on same plant. Heads with staminate flowers above pistillate in a single, dense, cylindric spike, the staminate and pistillate portions of the spike unalike, contiguous in broad-leaf cat-tail (*T. latifolia*) or separated in narrow-leaf cat-tail (*T. angustifolia*). A hybrid between *T. angustifolia* and *T. latifolia* is termed ***Typha* × *glauca*** Godr. Usually larger than either parent, staminate and pistillate portions of hybrid plants are usually separated by a space to 4 cm long. The staminate portion of the spike is light brown, 0.5–2 dm long and about 1 cm wide at flowering time; the pistillate portion is dark brown, 10–20 cm long and 1–2 cm wide. Since *Typha* × *glauca* is sterile, reproduction is vegetative by rhizomes. The hybrid can occur wherever populations of *T. angustifolia* and *T. latifolia* overlap.

*Key to species*

1 Staminate and pistillate portions of spike usually separated; leaves to 1 cm wide; stigmas long and slender, pale brown . ***T. angustifolia***

1 Staminate and pistillate portions of spike usually contiguous, not separated; leaves mostly 1–2 mm wide; stigmas broad and flattened, dark brown .......................................... ***T. latifolia***

***Typha angustifolia*** L. NARROW-LEAF CAT-TAIL Introduced (naturalized). June. Marshes, lakeshores, streambanks, roadside ditches, pond margins, usually in shallow water; more tolerant of brackish conditions than *Typha latifolia.*

***Typha latifolia*** L. BROAD-LEAF CAT-TAIL June. Marshes, lakeshores, streambanks, ditches, pond margins, usually in shallow water; less tolerant of brackish conditions than *Typha angustifolia.*

## XYRIDACEAE *Yellow-Eyed-Grass Family*

### XYRIS *Yellow-Eyed-Grass*

Perennial rushlike herbs. Stems erect, leafless, straight or sometimes ridged. Leaves all from base of plant, upright to spreading, linear, often twisted, usually dark green. Flowers small, perfect, yellow, in rounded or cylindric heads at ends of stems.

*Key to species*

1 Plants swollen and hard at base ......................... ***X. torta***

1 Plants flattened and soft at base ..................... ***X. montana***

***Xyris montana*** Ries NORTHERN YELLOW-EYED-GRASS Wet sandy shores, pools in sphagnum peatlands.

***Xyris torta*** Sm. SLENDER YELLOW-EYED-GRASS June–Aug. Wet sandy shores.

## ADDITIONAL SPECIES

*Members of families not treated elsewhere.*

### AIZOACEAE

***Tetragonia tetragonioides*** (Pallas) Kuntze (New Zealand-spinach), occasionally cultivated; reported from Bayfield County.

### BUXACEAE

***Pachysandra terminalis*** Siebold & Zucc. (Japanese mountain-spurge); a commonly cultivated leathery-leaved, evergreen Asian sub-shrubby ground cover, forming dense carpets by creeping rhizomes; spreading into forests, Dane County.

### LAURACEAE

***Sassafras albidum*** (Nutt.) Nees (Sassafras), small tree once known from Grant and Kenosha counties, now assumed to be extirpated from the state (but common immediately south of Wisc and widespread in southern and southeastern USA); easily recognized by the aromatic foliage and roots, the variably lobed leaves, and the yellow-green branchlets.

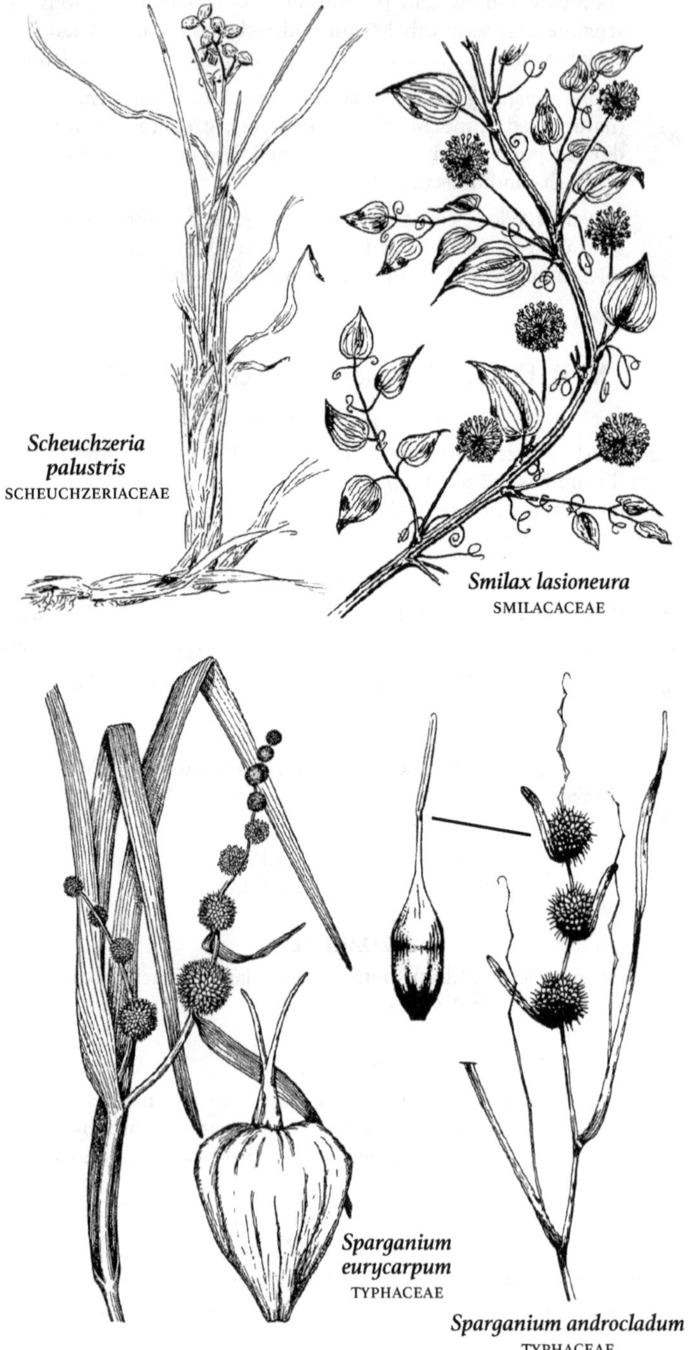

*Scheuchzeria palustris*
SCHEUCHZERIACEAE

*Smilax lasioneura*
SMILACACEAE

*Sparganium eurycarpum*
TYPHACEAE

*Sparganium androcladum*
TYPHACEAE

# GLOSSARY

**abaxial** On the side away from the axis, usually refers to the underside of a leaf (compare with adaxial).

**acaulescent** Without an upright, leafy stem.

**achene** A one-seeded, dry, indehiscent fruit with the seed coat not attached to the mature wall of the ovary.

**acid** Having more hydrogen ions than hydroxyl (OH) ions; a pH less than 7.

**acuminate** Tapering to a narrow point, more tapering than acute, less than attenuate.

**acute** Gradually tapered to a tip.

**adaxial** On the side toward the axis, usually refers to the top side of a leaf (compare with abaxial).

**adnate** Fused with a structure different from itself, as when stamens are adnate to petals (compare with connate).

**adventive** Not native to and not fully established in a new habitat.

**alkaline** Having more hydroxyl ions than hydrogen ions; a pH greater than 7.

**alluvial** Deposits of rivers and streams.

**alternate** Borne singly at each node, as in leaves on a stem.

**ament** Spikelike inflorescence of same-sexed flowers (either male or female); same as catkin.

**androgynous** Spike with both staminate and pistillate flowers, the pistillate located at the base, below the staminate (compare with gynaecandrous).

**angiosperm** A plant producing flowers and bearing seeds in an ovary.

**annual** A plant that completes its life cycle in one growing season, then dies.

**anther** Pollen-bearing part of stamen, usually at the end of a stalk called a filament.

**anthesis** The period during which a flower is fully open and functional.

**anthocyanic** Pigmented with anthocyanins, this usually manifested as a tinging or suffusion of pink, red, or purple.

**aphyllopodic** Having basal sheaths without blades; with new shoots arising laterally from parent shoot (compare with phyllopodic).

**apiculate** Having an apiculus.

**apiculus** An abrupt, very small, projected tip.

**appressed** Lying flat to or parallel to a surface.

**aquatic** Living in water.

**areole** In leaves, the spaces between small veins.

**aril** A specialized appendage on a seed, often brightly colored, derived from the seed coat.

**aristate** Tipped with a slender bristle.

**armed** Bearing a sharp projection such as a prickle, spine, or thorn.

**aromatic** Strongly scented.

**ascending** Angled upward.

**asymmetrical** Not symmetrical.

**attenuate** Tapering gradually to a prolonged tip.

**auricle** An ear-shaped appendage to a leaf or stipule.

**awl-shaped** Tapering gradually from a broad base to a sharp point.

**awn** A bristle-like organ.

**axil** Angle between a stem and the attached leaf.

**barb** Sharp, thorn-like projection.

**basal** From base of plant.

**basic** A pH greater than 7.

**beak** A slender, terminal appendage on a 3-dimensional organ.

**beard** Covering of long or stiff hairs.

**berry** Fruit with the seeds surrounded by fleshy material.

**biennial** A plant that completes its life cycle in two growing season, typically flowering and fruiting in the second year, then dying.

**bifid** Cleft into two more or less equal parts.

**blade** Expanded, usually flat part of a leaf or petiole.

**bloom** A whitish powdery or waxy coating that can be rubbed away.

**bog** A wet, acidic, nutrient-poor peatland characterized by sphagnum and other mosses, shrubs and sedges.

Technically, a type of peatland raised above its surroundings by peat accumulation and receiving nutrients only from precipitation.

**boreal** Far northern latitudes.

**brackish** Salty.

**bract** An accessory structure at the base of some flowers, usually appearing leaflike.

**bractlet** A secondary bract (*Typha*).

**branchlets** A small branch.

**bristle** A stiff hair.

**bud** An undeveloped shoot, inflorescence, or flower, in woody plants often covered by scales and serving as the overwintering stage.

**bulb** A group of modified leaves serving as a food-storage organ, borne on a short, vertical, underground stem (compare with corm).

**bulbil** A bulb-like structure borne in the leaf axils or in place of flowers.

**bulblet** Small bulb borne above ground, as in a leaf axil.

**ca.** About, approximately (Latin *circa*).

**caducous** Falling off early, as stipules that leave behind a scar.

**callosity** A hardened thickening.

**callus** A firm, thickened portion of an organ; the firm base of the lemma in the Poaceae.

**calcareous fen** An uncommon wetland type associated with seepage areas, and which receive groundwater enriched with primarily calcium and magnesium bicarbonates.

**calcium-rich** Refers to wetlands underlain by limestone or receiving water enriched by calcium compounds.

**calyx** All the sepals of a flower.

**campanulate** Bell-shaped.

**capillary** Very fine, hair-like, not-flattened.

**capitate** Abruptly expanded at the apex, thereby forming a knob-like tip.

**capsule** A dry, dehiscent fruit splitting into 3 or more parts.

**carpel** Fertile leaf of an angiosperm, bearing the ovules. A pistil is made up of one or more carpels.

**caruncle** An appendage at or near the hilum of some seeds.

**caryopsis** The dry, indehiscent seed of grasses.

**catkin** Spikelike inflorescence of same-sexed flowers (either male or female); same as ament.

**caudex** Firm, hardened, summit of a root mass that functions as a perennating organ.

**cauline** Of or pertaining to the above-ground portion of the stem.

**cespitose** Growing in a compact cluster with closely spaced stems; tufted, clumped.

**chaff** Thin, dry scales; in the Asteraceae, sometimes found as chaffy bracts on the receptacle.

**cilia** Hairs found at the margin of an organ.

**ciliate** Provided with cilia.

**circumboreal** Refers to a species distribution pattern which circles the earth's boreal regions.

**clasping** Leaves that partially encircle the stem at the base.

**clavate** Widened in the distal portion, like a baseball bat.

**claw** The narrow, basal portion of perianth parts.

**cleistogamous** Type of flower that remains closed and is self-pollinated.

**clumped** Having the stems grouped closely together; tufted.

**colony-forming** A group of plants of the same species, produced either vegetatively or by seed.

**column** The joined style and filaments in the Orchidaceae.

**coma** A tuft of fine hairs, especially at the tip of a seed.

**composite** An inflorescence that is made up of many tiny florets crowded together on a receptacle; members of the Aster Family (Asteraceae).

**compound leaf** A leaf with two or more leaflets.

**concave** Curved inward.

**conduplicate** Folded lengthwise into nearly equal parts.

**cone** The dry fruit of conifers composed of overlapping scales.

**conifer** Cone-bearing woody plants.

**connate** Two like parts that are fused (compare with adnate).

**connivent** Converging and touching but not actually fused, applies to like organs.

**convex** Curved outward.

**convolute** Arranged such that one edge is covered and the other is exposed, usually referring to petals in bud.

**cordate** With a rounded lobe on each side of a central sinus; heart-shaped.

**coriaceous** With a firm, leathery texture.

**corm** A short, vertical, enlarged, underground stem that serves as a food storage organ (compare with bulb).

**corolla** Collectively, all the petals of a flower.

**corymb** An indeterminate inflorescence, somewhat similar to a raceme, that has elongate lower branches that create a more or less flat-topped inflorescence.

**costa** (plural costae) A prominent midvein or midrib of a leaflet.

**crenate** With rounded teeth.

**crenulate** Finely crenate.

**crisped** An irregularly crinkled or curled leaf margin.

**crown** Persistent base of a plant, especially a grasses.

**culm** The stem of a grass or grasslike plant, especially a stem with the inflorescence.

**cuneate** Tapering to the base with relatively straight, non-parallel margins; wedge-shaped.

**cyme** A type of inflorescence in which the central flowers open first.

**deciduous** Not persistent.

**decumbent** A stem that is prostrate at the base and curves upward to have an erect or ascending, apical portion.

**decurrent** Possessing an adnate line or wing that extends down the axis below the node, usually referring to leaves on a stem.

**dehiscent** Splitting open at maturity.

**deltate** Triangle-shaped.

**dentate** Provided with outward oriented teeth.

**depauperate** Poorly developed due to unfavorable conditions.

**dicots** One of two main divisions of the Angiosperms (the other being the Monocots); plants having 2 seed leaves (cotyledons), net-venation, and flower parts in 4s or 5s (or multiples of these numbers).

**dioecious** Bearing only male or female flowers on a single plant.

**dimorphic** Having two forms.

**disarticulation** Spikelets breaking either above or below the glumes when mature, the glumes remaining in the head if disarticulation above the glumes, or the glumes falling with the florets if disarticulation is below the glumes.

**discoid** In composite flowers (Asteraceae), a head with only disk (tubular) flowers, the ray flowers absent.

**disjunct** A population of plants widely separated from its main range.

**disk** In the Asteraceae, the central part of the head, composed of tubular flowers.

**dissected** Leaves divided into many smaller segments.

**disturbed** Natural communities altered by human influences.

**divided** Leaves which are lobed nearly to the midrib.

**dolomite** A type of limestone consisting of calcium magnesium carbonate.

**dorsal** Underside, or back of an organ.

**driftless area** Portions of sw Wisconsin, ne Iowa, and se Minnesota that are not covered by glacial drift.

**drupe** A fleshy fruit with a single large seed such as a cherry.

**echinate** With spines.

**eglandular** Without glands.

**elliptic** Broadest at the middle, gradually tapering to both ends.

**emergent** Growing out of and above the water surface.

**emersed leaf** Growing above the water surface or out of water.

**endangered** A species in danger of extinction throughout all or most of its range if current trends continue.

**endemic** A species restricted to a particular region.

**entire** With a smooth margin.

**erect** Stiffly upright.

**erose** With a ragged edge.

**escape** A cultivated plant which

establishes itself outside of cultivation.

**evergreen** Plant retaining its leaves throughout the year.

**excurrent** With the central rib or axis continuing or projecting beyond the organ.

**exserted** Extending beyond the mouth of a structure such as stamens extending out from the mouth of the corolla.

**falcate** Sickle-shaped

**false indusium** A modified tooth or reflexed margin of a fern leaf that covers the sorus.

**fen** An open wetland usually dominated by herbaceous plants, and fed by in-flowing, often calcium- and/or magnesium-rich water; soils vary from peat to clays and silts.

**fern** Perennial plants with spore-bearing leaves similar to the vegetative leaves and bearing sporangia on their underside, or the spore-bearing leaves much modified.

**fibrous** A cluster of slender roots, all with the same diameter.

**filament** The stalk of a stamen which supports the anther.

**filiform** Thread-like.

**flexuous** An elongate axis that arches or bends in alternating directions in a zig-zag fashion.

**floating mat** A feature of some ponds where plant roots form a carpet over some or all of the water surface.

**floodplain** That part of a river valley that is occasionally covered by flood waters.

**floret** A small flower in a dense cluster of flowers; in grasses the flower with its attached lemma and palea.

**follicle** A dry, dehiscent fruit that splits along one side when mature.

**floricane** the second-year flowering stem of *Rubus* (compare with primocane).

**genus** The first part of the scientific name for a plant or animal (plural genera).

**glabrate** Nearly glabrous or becoming so.

**glabrous** Lacking hairs.

**gland** An appendage or depression which produces a sticky or greasy substance.

**glandular** Bearing glands.

**glaucous** Having a bluish appearance.

**glumes** A pair of small bracts at base of each spikelet the lowermost (or first) glume usually smaller the upper (or second) glume usually longer.

**grain** The fruit of a grass; the swollen seedlike protuberance on the fruit of some *Rumex*.

**gymnosperm** Plants in which the seeds are not produced in an ovary, but usually in a cone.

**gynaecandrous** Having both staminate and pistillate flowers on the same spike, the staminate located at the base, below the pistillate (compare with androgynous).

**gynophore** The central stalk of some flowers, especially in cat-tails (*Typha*).

**halophyte** A plant adapted to growing in a salty substrate.

**hastate** More or less triangular in outline with outward-oriented basal lobes.

**haustorium** A specialized, root-like connection to a host plant that a parasite uses to extract nourishment.

**hardwoods** Loosely used to contrast most deciduous trees from conifers.

**herb** A herbaceous, non-woody plant.

**herbaceous** Like an herb; also, leaflike in appearance.

**hilum** The scar at the point of attachment of a seed.

**hirsute** Pubescent with coarse, somewhat stiff, usually curving hairs, coarser than villous but softer than hispid.

**hispid** Pubescent with coarse, stiff hairs that may be uncomfortable to the touch, coarser than hirsute but softer than bristly.

**hummock** A small, raised mound formed by certain species of sphagnum moss.

**humus** Dark, well-decayed organic matter in soil.

**hybrid** A cross-breed between two species.

**hydric** Wet (compare with mesic, xeric).

**hypanthium** A ring, cup, or tube around the ovary; the sepals, petals and stamens are attached to the rim of the hypanthium.

**imbricate** Overlapping, as shingles on a roof.

**indehiscent** Not splitting open at maturity.

**indusium** In ferns, a membranous covering over the sorus (plural indusia).

**inferior** The position of the ovary when it is below the point of attachment of the sepals and petals.

**inflorescence** A cluster of flowers.

**insectivorous** Refers to the insect trapping and digestion habit of some plants as a nutrition supplement.

**interdunal swale** Low-lying areas between sand dune ridges.

**internode** Portion of a stem between two nodes.

**introduced** A non-native species.

**invasive** Non-native species causing significant ecological or economic problems.

**involucral bract** A single member of the involucre; sometimes called phyllary in composite flowers (Asteraceae).

**involucre** A whorl of bracts, subtending a flower or inflorescence.

**irregular flower** Not radially symmetric; with similar parts unequal.

**joint** A node or section of a stem where the branch and leaf meet.

**keel** A central rib like the keel of a boat.

**lance-shaped** Broadest near the base, gradually tapering to a narrower tip.

**lateral** Borne on the sides of a stem or branch.

**lax** Loose or drooping.

**leaf axil** The point of the angle between a stem and a leaf.

**leaflet** One of the leaflike segments of a compound leaf.

**lemma** In grasses, the lower bract enclosing the flower (the upper, smaller bract is the palea).

**lens-shaped** Biconvex in shape (like a lentil).

**lenticel** Blisterlike openings in the epidermis of woody stems, admitting gases to and from the plant, and often appearing as small oval dots on bark.

**ligulate** Having a ligule; in the Asteraceae, the strap-shaped corolla of a ray floret.

**ligule** In grasses and grasslike plants, the membranous or hairy ring at top of sheath between the blade and stem.

**linear** Narrow and flat with parallel sides.

**lip** Upper or lower part of a 2-lipped corolla; also the lower petal in most orchid flowers.

**lobed** With lobes; in leaves divisions usually not over halfway to the midrib.

**local** Occurring sporadically in an area.

**low prairie** Wet and moist herbaceous plant community, typically dominated by grasses.

**margin** The outer edge of a leaf.

**marl** A calcium-rich clay.

**marsh** Wetland dominated by herbaceous plants, with standing water for part or all the growing season, then often drying at the surface.

**megaspore** Large, female spores.

**mesic** Moist, neither dry nor wet (compare with hydric, xeric).

**microspore** Small, male spores.

**midrib** The prominent vein along the main axis of a leaf.

**mixed forest** A type of forest composed of both deciduous and conifer trees.

**moat** The open water area ringing the outer edge of a peatland or floating mat.

**monecious** Having male and female reproductive parts in separate flowers on the same plant.

**monocots** One of two main divisions of the Angiosperms (the other being the Dicots); plants with a single seed leaf (cotyledon); typically having narrow leaves with parallel veins, and flower parts in 3s or multiples of 3.

**muck** An organic soil where the plant remains are decomposed to the point where the type of plants forming the soil cannot be determined.

**mucro** A sharp point at termination of an organ or other structure.

**naked** Without a covering; a stalk or stem without leaves.

**native** An indigenous species.

**naturalized** An introduced species that is established and persistent in an ecosystem.

**needle** A slender leaf, as in the Pinaceae.

**nerve** A leaf vein.

**neutral** A pH of 7.

**node** The spot on a stem or branch where leaves originate.

**nutlet** A small dry fruit that does not split open along a seam.

**oblanceolate** Reverse lance-shaped; broadest at the apex, gradually tapering to the narrower base.

**oblique** Emerging or joining at an angle other than parallel or perpendicular.

**oblong** Broadest at the middle, and tapering to both ends, but broader than elliptic.

**obovate** Broadly rounded at the apex, becoming narrowed below.

**ocrea** A tube-shaped stipule or pair of stipules around the stem; characteristic of the Smartweed Family (Polygonaceae).

**opposite** Leaves or branches which are paired opposite one another on the stem.

**organic** Soils composed of decaying plant remains.

**oval** Elliptical.

**ovary** The lower part of the pistil that produces the seeds.

**ovate** Broadly rounded at the base, becoming narrowed above; broader than lanceolate.

**palea** The uppermost of the two inner bracts subtending a grass flower (the lower bract is the lemma).

**palmate** Divided in a radial fashion, like the fingers of a hand.

**panicle** An arrangement of flowers consisting of several racemes.

**papilla** (plural: papillae) A short, rounded or cylindrical projections.

**pappus** The modified sepals of a composite flower which persist atop the ovary as bristles, scales or awns.

**parallel-veined** With several veins running from base of leaf to leaf tip, characteristic of most monocots.

**peat** An organic soil formed of partially decomposed plant remains.

**peatland** A wetland whose soil is composed primarily of organic matter (mosses, sedges, etc.); a general term for bogs and fens.

**peltate** More or less circular, with the stalk attached at a point on the underside.

**pepo** A fleshy, many-seeded fruit with a tough rind, as a melon.

**perennial** Living for 3 or more years.

**perfect** A flower having both male (stamens) and female (pistils) parts.

**perianth** Collectively, all the sepals and petals of a flower.

**perigynium** A sac-like structure enclosing the pistil in *Carex* (plural perigynia).

**petal** An individual part of the corolla, often white or colored.

**petiole** The stalk of a leaf.

**phyllary** An involucral bract subtending the flower head in composite flowers (Asteraceae).

**phyllode** An expanded petiole.

**phyllopodic** Having the basal sheaths blade-bearing; with new shoots arising from the center of parent shoot (compare with aphyllopodic).

**pinna** The primary or first division in a fern frond or leaf (plural pinnae).

**pinnate** Divided once along an elongated axis into distinct segments.

**pinnule** The pinnate segment of a pinna.

**pistil** The seed-producing part of the flower, consisting of an ovary and one or more styles and stigmas.

**pith** A spongy central part of stems and branches.

**pollen** The male spores in an anther.

**prairie** An open plant community dominated by herbaceous species, especially grasses.

**primocane** The first-year, vegetative stem in *Rubus* (compare with floricane).

**pro sp.** When a taxon is transferred from the non-hybrid category to the hybrid category, the author citation remains unchanged, but may be followed by an indication in parentheses of the original category.

**prostrate** Lying flat on the ground.

**raceme** A grouping of flowers along an elongated axis where each flower has its own stalk.

**rachilla** A small stem or axis.

**rachis** The central axis or stem of a leaf or inflorescence.

**radiate heads** In composite flowers, heads with both ray and disk flowers (Asteraceae).

**ray flower** A ligulate or strap-shaped flower in the Asteraceae, where often the outermost series of flowers in the head.

**receptacle** In the Asteraceae, the enlarged summit of the flower stalk to which the sepals, petals, stamens, and pistils are usually attached.

**recurved** Curved backward.

**regular** Flowers with all the similar parts of the same form; radially symmetric.

**rhizome** An underground, horizontal stem.

**rib** A pronounced vein or nerve.

**rootstock** Similar to rhizome but referring to any underground part that spreads the plant.

**rosette** A crowded, circular clump of leaves.

**samara** A dry, indehiscent fruit with a well-developed wing.

**saprophyte** A plant that lives off of dead organic matter.

**scale** A tiny, leaflike structure; the structure that subtends each flower in a sedge (Cyperaceae).

**scape** A naked stem (without leaves) bearing the flowers.

**section** Cross-section.

**secund** Flowers mostly on 1 side of a stalk or branch.

**sedge meadow** A community dominated by sedges (Cyperaceae) and occurring on wet, saturated soils.

**seep** A spot where water oozes from the ground.

**sepal** A segment of the calyx; usually green in color.

**sheath** Tube-shaped membrane around a stem, especially for part of the leaf in grasses and sedges.

**shrub** A woody plant with multiple stems.

**silicle** Short fruit of the Mustard Family (Brassicaceae), normally less than 2x longer as wide.

**silique** Dry, dehiscent, 2-chambered fruit of the Mustard Family (Brassicaceae), longer than a silicle.

**simple** An undivided leaf.

**sinus** The depression between two lobes.

**smooth** Without teeth or hairs.

**sorus** Clusters of spore containers (plural sori).

**spadix** A fleshy axis in which flowers are embedded.

**spathe** A large bract subtending or enclosing a cluster of flowers.

**spatula-shaped** Broadest at tip and tapering to the base.

**sphagnum moss** A type of moss common in peatlands and sometimes forming a continuous carpet across the surface; sometimes forming layers several meters thick; also loosely called peat moss.

**spike** A group of unstalked flowers along an unbranched stalk.

**spikelet** A small spike; the flower cluster (inflorescence ) of grasses (Poaceae) and sedges (Cyperaceae).

**sporangium** The spore-producing structure (plural sporangia).

**spore** a one-celled reproductive structure that gives rise to the gamete-bearing plant.

**sporophyll** A modified, spore-bearing leaf.

**spreading** Widely angled outward.

**spring** A place where water flows naturally from the ground.

**spur** A hollow, pointed projection of a flower.

**stamen** The male or pollen-producing organ of a flower.

**staminode** An infertile stamen.

**stem** The main axis of a plant.

**stigma** The terminal part of a pistil which receives pollen.

**stipe** A stalk.

**stipule** A leaflike outgrowth at the base of a leaf stalk.

**stolon** A horizontal stem lying on the soil surface.

**style** The stalklike part of the pistil between the ovary and the stigma.

**subspecies** A subdivision of the species forming a group with shared traits which differ from other members of the species (subsp.).

**subtend** Attached below and extending upward.

**succulent** Thick, fleshy and juicy.

**superior** Referring to the position of the ovary when it is above the point of attachment of sepals, petals, stamens, and pistils.

**swale** A slight depression.

**swamp** Wooded wetland dominated by trees or shrubs; soils are typically wet for much of year or sometimes inundated.

**talus** Fallen rock at the base of a slope or cliff.

**taproot** A main, downward-pointing root.

**tendril** A threadlike appendage from a stem or leaf that coils around other objects for support (as in *Vitis*).

**tepal** Sepals or petals not differentiated from one another.

**terete** Circular in cross-section.

**terminal** Located at the end of a stem or stalk.

**thallus** A small, flattened plant structure, without distinct stem or leaves.

**thicket** A dense growth of woody plants.

**threatened** A species likely to become endangered throughout all or most of its range if current trends continue.

**translucent** Nearly transparent.

**tree** A large, single-stemmed woody plant.

**tuber** An enlarged portion of a root or rhizome.

**truncate** Abruptly cut-off.

**tubercle** Base of style persistent as a swelling atop the achene different in color and texture from achene body.

**tundra** Treeless plain in arctic regions, having permanently frozen subsoil.

**turion** A specialized type of shoot or bud that overwinters and resumes growth the following year.

**umbel** A cluster of flowers in which the flower stalks arise from the same level.

**umbelet** A small, secondary umbel in an umbel, as in the Apiaceae.

**upright** Erect or nearly so.

**urceolate** Constricted at a point just before an opening; urn-shaped.

**utricle** A small, one-seeded fruit with a dry, papery outer covering.

**valve** A segment of a dehiscent fruit; the wing of the fruit in *Rumex*.

**variety** Taxon below subspecies and differing from other varieties within the same subspecies (var.).

**vein** A vascular bundle, as in a leaf.

**velum** The membranous flap that partially covers the sporangium in *Isoetes*.

**venation** The pattern of veins on an organ.

**ventral** Front side.

**ventricose** Inflated or distended.

**verrucose** Covered with small, wartlike projections.

**verticil** One whorled cycle of organs.

**verticillate** Arranged in whorls.

**villous** Pubescent with long, soft, bent hairs, the hairs not crimped or tangled.

**vine** A trailing or climbing plant, dependent on other objects for support.

**viscid** Sticky, glutinous.

**whorl** A group of 3 or more parts from one point on a stem.

**wing** A thin tissue bordering or surrounding an organ.

**woody** Xylem tissue (the vascular tissue which conducts water and nutrients).

**xeric** Dry (compare with hydric, mesic).

## INDEX (*Scientific Names*)

*Synonyms listed in italics*

***Amerorchis rotundifolia***
ORCHIDACEAE, *page 543*

www.ingramcontent.com/pod-product-compliance
Lightning Source LLC
Chambersburg PA
CBHW062107020426
42335CB00013B/884
* 9 7 8 1 9 5 1 6 8 2 6 5 1 *